WA
100
O985
v.1
1991

D1317020

OXFORD MEDICAL PUBLICATIONS

OXFORD TEXTBOOK OF PUBLIC HEALTH

EDITORS

WALTER W. HOLLAND, MD, FRCGP, FRCP Edin, FRCP, PFPHM
Professor of Clinical Epidemiology and Social Medicine,
Department of Public Health Medicine, and
Honorary Director, Social Medicine and
Health Services Research Unit,
United Medical and Dental Schools,
St. Thomas's Campus,
London SE1 7EH, England.

ROGER DETELS, MD, MS
Professor of Epidemiology,
School of Public Health,
University of California, Los Angeles, CA 90024, USA.

GEORGE KNOX, MD, BS, FRCP, FFPHM, FFOM
Professor of Social Medicine, Department of Social Medicine,
Health Services Research Centre,
University of Birmingham Medical School,
Edgbaston, Birmingham, B15 2TJ, England.

BEVERLEY FITZSIMONS, BA
Research Assistant, Department of Public Health Medicine,
United Medical and Dental Schools,
St. Thomas's Campus,
London SE1 7EH, England.

LUCY GARDNER, BA
Research Assistant, Department of Public Health Medicine,
United Medical and Dental Schools,
St. Thomas's Campus,
London SE1 7EH, England.

OXFORD TEXTBOOK OF PUBLIC HEALTH

SECOND EDITION

VOLUME 1

Influences of Public Health

Edited by

WALTER W. HOLLAND,
ROGER DETELS, and GEORGE KNOX

with the assistance of

BEVERLEY FITZSIMONS and LUCY GARDNER

SEP 13 1991

MCW Libraries

Oxford New York Toronto
OXFORD UNIVERSITY PRESS
1991

Oxford University Press, Walton Street, Oxford OX2 6DP
Oxford New York Toronto
Delhi Bombay Calcutta Madras Karachi
Petaling Jaya Singapore Hong Kong Tokyo
Nairobi Dar es Salaam Cape Town
Melbourne Auckland
and associated companies in
Berlin Ibadan

Oxford is a trade mark of Oxford University Press

Published in the United States
by Oxford University Press, New York

© The contributors listed on p. xvii, 1991

All rights reserved. No part of this publication may be reproduced, stored in a retrieval system, or transmitted, in any form or by any means, electronic, mechanical, photocopying, recording, or otherwise, without the prior permission of Oxford University Press

British Library Cataloguing in Publication Data
Oxford textbook of public health.—2nd ed.
1. Public health
I. Holland, Walter W. (Walter Werner) II. Detels, Roger
III. Knox, George 1926–614

ISBN 0 19 261706 0 Vol. 1
ISBN 0 19 261707 9 Vol. 2
ISBN 0 19 261708 7 Vol. 3
ISBN 0 19 261926 8 (3 vol set)

Typeset by Promenade Graphics Ltd., Cheltenham
Printed in Great Britain by
Wm. Clowes Ltd., Beccles, Suffolk

Preface to the first edition

It is not an easy task to follow in the footsteps of such a renowned editor as Bill Hobson. We were however very honoured when, on the retirement of Professor Hobson, the Oxford University Press approached us about taking up the challenge of revising Hobson's *Theory and practice of public health*. Since this work first appeared in February 1961 Professor Hobson was responsible for taking it through no less than five editions. Many eminent public health academics and practitioners have contributed to this book and it has been recognized as a standard textbook on the subject. Sadly, Professor Hobson died after a long illness at the end of November 1982. After an early training in public health starting as a medical officer of health and then as a specialist in hygiene and epidemiology in the army, he went on to be a lecturer in social medicine at Sheffield University, becoming professor in 1949. From 1957 until his retirement, he served in a variety of posts at the WHO, where his major responsibilities were always concerned with education and training. His interest in this and in the international aspects of health were well exemplified by the first edition of *Theory and practice of public health*. One of the major strengths of the book has been its international nature and its link to the WHO.

On accepting the daunting task of revising this major work our first step was to look dispassionately at its role within public health, a field which has evolved and changed greatly over the last 25 years. We decided that although this book is held in great esteem in the western world it was appropriate now to introduce major revisions and thus increase its relevance to the problems facing us as we approach the twenty-first century. A particularly important advance has been the recognition in recent years that the problems in public health facing developing countries are quite different to those facing the developed world. The interests of WHO, quite correctly, have been focused on developing countries. We consider that this book should concentrate on presentıng a comprehensive view of public health as it relates to developed countries. (Perhaps there is a place now for a comparable textbook concerned specifically with developing countries.) This is not to say however, that the content will not prove relevant and of interest to the student of public health from a developing country.

The *Oxford textbook of public health* attempts to portray the philosophy and underlying princples of the practice of public health. The methods used for the investigation and the solution of public health problems are described and examples given of how these methods are applied in practice. It is aimed primarily at postgraduate students and practitioners of public health but most clinicians and others concerned with public health issues will find some chapters relevant to their concerns. It is intended to be a comprehensive textbook present in the library of every institution concerned with the health sciences. The term 'public' is used quite deliberately to portray the field. Public health is concerned with defining the problems facing communities in the prevention of illness and thus studies of disease aetiology and promotion of health. It covers the investigation, promotion, and evaluation of optimal health services to communities and is concerned with the wider aspects of health within the general context of health and the environment. Other terms in common use, such as community medicine, preventive medicine, social medicine, and population medicine have acquired different meanings according to the country or setting. This gives rise to confusion and we have avoided their use since this book is directed to a worldwide audience. Public health, we believe, is more evocative of the basic philosophy which underlies this book.

The first volume aims to lead the reader through the historical determinants of health to the overall scope and strategies of public health. Through knowledge of historical aspects of the subject we may gain an understanding of what it is possible to achieve now and in the future. Only by grasping the underlying strategies of public health can we determine whether specific actions are feasible or not. In outlining the scope of public health we have emphasized that this covers not only the environment but also the social and genetic determinants of disease, which may ultimately enable us to identify those at greatest risk and thus to prevent the development and onset of disease and disability. The scope has now been broadened further by the growing concern with the provision and development of health services. This is clearly a function of public health. The approaches towards public health and the underlying political realities differ from country to country. However there are a few basic concepts common to every situation and these are outlined.

The major determinants of health and disease are dealt with in the second section of this volume, which paints, with a broad brush, the factors concerned with the development of disease, such as the physical environment, infectious agents, the social environment, war and social disorder. It is also important to have an understanding of the methods governments employ to control health hazards as well as their overall policies towards health. This we illustrate by describing the widely different approaches adopted on the two sides of the Atlantic. The strategies both for tackling modern hazards and for modifying behaviour through education are considered. To draw all these themes together into the same perspective, the final three chapters, prepared by three specialists with broad experience of public health in the context of the world as a whole, consider overall public health strategies relevant to the western as well as the developing world.

Volume 2 deals in depth with the processes of health promotion. Historically, public health has been primarily concerned with the identification and prevention of disease as the primary means to safeguard the health of a population. The need for this traditional approach still remains, but within the last decade public health has also come to incorporate a broader perspective as our understanding of the factors which determine the health of the public has increased.

This greater understanding of factors which enhance health has led to a reconceptualization of health as a condition which must be actively maintained. Thus public health has been able to adopt a proactive approach with health, rather than absence of disease, as a goal. This aim is reflected in the World Health Organization motto, 'Health for all by the Year 2000'. The agenda for public health in the 1980s will be to actively promote health and enhance the quality of life through a wide range of activities, both traditional and new.

Commensurate with this proactive approach is the assumption of new responsibilities. For example, public health is now also concerned with the quality of health care, adequate access to medical care by all segments of society, the health effects of the many chemicals and other agents which are released into the environment as a result of the many new technical advances, and promotion of healthy life-styles. These responsibilities call for both new strategies and processes to promote the health of the public and innovative use of traditional strategies.

In this volume we present a discussion of the processes currently being used to promote the health of the public. These range from the scientific and regulatory strategies used to control the physical environment and the spread of infectious diseases to the involvement of national governments and international organizations in promotion of public health.

In the first two chapters of the volume, the control of the physical environment and the control of infectious diseases are discussed. Although these have been important traditional concerns of public health, there have also been new strategies developed as new health problems emerge and former health problems are resolved.

The next several chapters address more recent concerns of public health: modifying the social environment, intervening in population dynamics in order to limit growth, and providing more effective health services to include provisions for specialty care and for care of the mentally and emotionally handicapped.

The third section discusses the organization of public health services, the staffing of these services, and the training of personnel for public health. The volume concludes with a discussion of the co-ordination and development of strategies and policy on the national level and to some extent, on the international level.

In conclusion, this volume should give the reader some appreciation of the dynamic, evolving nature of public health and the diversity of approaches which are being used to assure the highest levels of public health in developed countries.

Volume 3 deals with the investigative methods used in public health.

The final volume of this series attemps to pull together all the threads of the earlier volumes which have considered the theory, policies, strategies and research methodologies that form the basis of public health endeavour. Specific diseases and the needs of specific groups within a community have been touched on in greater or lesser detail in earlier volumes. Here, the major disease groups, systems of the body, and special care groups are treated systematically to review the public health issues they raise, and the extent to which the methodologies described earlier have been applied to their control and prevention.

The volume is divided into three sections, the first of which is concerned with the application of public health methods to specific disease processes including acute infectious episodes, diseases related to nutrition, trauma, developmental defects, degenerative neurological diseases, psychiatric conditions, and neoplasms. Each chapter attempts to review the public health impact of the diseases in question, including their epidemiology and the contribution of public health measures. Of particular emphasis is the potential for prevention that enquiry into the aetiology has revealed, and the role of public health interventions in implementing programmes of prevention.

Leading on from specific disease states, the next section looks at the role of public health in relation to the various systems of the body. The intention is not to provide an exhaustive description of the conditions affecting each system, but to discuss the facets of the system where public health investigation has had (or indeed in some cases has failed to make) a major contribution. It will be evident to the reader that the extent to which the public-health approach has contributed to research and strategies to prevent and treat conditions affecting the various systems of the body varies tremendously. The cardiovascular system and the respiratory system provide classical examples of the application of the public health approach to a health problem. Extensive epidemiological research has clarified the causation of these conditions and public health measures form the basic framework for implementing control measures. Investigations of gastrointestinal disease, metabolic and endocrine disorders, and conditions affecting the genitourinary system have on the other hand, relied more on clinical than public health orientated research. The chapters dealing with these systems however describe the epidemiological work that has been carried out and its relevance to clinical practice. The chapter on the genito-urinary system, for example, illustrates the value of national and international comparative studies in informing policy decision makers of different management practices and outcomes. The problems presented by the investigation of musculoskeletal and dental conditions are clearly demonstrated in this section of Volume 4.

The final section of this volume treats the unique problems and health service needs of special client groups. Again, this is not intended to be comprehensive or exhaustive, the emphasis being placed on demonstrating the application of public health approaches to a variety of different problems. These include investigations to identify the needs of special client groups and the extent to which these needs are satisfied, and the application of public health measures to solving the problems of these groups.

The chapter on acute emergencies is an excellent example of the ways in which epidemiological investigation can be applied to surgical conditions, providing evidence from which conclusions can be drawn about the most appropriate form of treatment. The chapter on adolescence demonstrates the relevance of a multidisciplinary approach to health problems, and the difficulties of tackling some of the problems experienced by adolescents in the changing society around us. The discussion of handicap—both physical and mental—illustrates the application of epidemiological and public health methodology to both the prevention as well as the development of care policies for specific groups of individuals.

Health care policy decision-making is fraught with difficulties as is illustrated by the chapter on maternity care, theoretically the simplest of examples, but one which demonstrates the complexity of the issues involved. The chapter on the special needs of the elderly also shows how complicated the problems facing modern public health are. In outlining the difficulties faced by the unemployed and disadvantaged Illsley demonstrates the way the social scientist approaches a public health problem.

This volume should give the reader some idea of the vast scope of public health, the range of fields and problems for which public health has a major contribution to make, and the multidisciplinary nature of much of the work. It attempts to review how we can approach and develop a methodology for the investigation, prevention, and control of the major public health problems in our society at this time.

The development of public health policy is dependent upon a series of scientific methods, and we do not attempt in this book to cover all the methods and their applications. However it is to be hoped that those examples that have been chosen will illustrate to the reader the way in which particular problems can be approached. Each chapter includes a comprehensive list of further reading which should equip the reader with the means of obtaining a deeper knowledge should he or she wish to pursue any theme further.

This is the first of what we hope will be many editions. As each chapter was submitted to the editors we have attempted to identify gaps and areas of overlap. There is no doubt however that some remain. It is only through feedback from readers that we will be able to adapt, modify, and improve further editions. If the book is successful it will be entirely due to the effort of the contributors who undertook with great patience a tremendous amount of work. They were bombarded with instructions, advice, reminders, and modifications and we would like to express our thanks and extend our apologies to all of them. Our gratitude also goes to our secretaries and assistants who coped so admirably with the enormous task of compiling this work. We hope that it will be widely read by all those concerned with the formulation and execution of public health policy and that it will provide a suitable framework for devising approaches to some of the problems challenging public health today.

W.W.H.
R.D.
G.K.

London
August, 1984

Preface to the second edition

The first edition of the *Oxford textbook of public health* attempted to provide a sense of the history and philosophy of the subject, the underlying forces that condition the subject, as well as the basic methodologies and their application to specific problems or care groups.

It became clear to the editors from reviews and discussion with leaders in the field that the ideas they had about the format, content, and structure of the book did not achieve their original intentions as well as they had hoped. One of the problems was that the relationships between the philosophy of public health and its applications in the solution of individual problems were not clearly expressed in the book. Further, there were problems in effectively conveying an understanding of what the editors were trying to achieve.

In spite of these problems, most of the readers and reviewers were generally satisfied with the content of the first edition, partly because the chapters in themselves were extremely good, thanks to the quality of the contributing authors.

Furthermore, in the intervening time period, between the publication of the first edition and the plans for the second, there have been major changes in the way in which public health is regarded. The appearance of a major new infectious disease, AIDS, has influenced opinion on the importance of public health measures in the structure of services and the ways in which disease prevention is approached.

We recognize that the first edition did not clearly express our intention of considering primarily the problems of the developed world and had few contributors from non-English-speaking countries.

To try to overcome these problems, and to attempt to rectify the deficiences that were apparent to the editors in the first edition, we have undertaken a radical change in both the authorship and structure of the second edition. This in no way implies that the authors in the first edition were inadequate. It merely demonstrates the editors' wish to broaden the contributions and to make the second edition somewhat more coherent. We do not feel able to deal adequately with the problems of the developing world, but in an effort to broaden the scope of this edition, we have expanded Volume 1 to include more examples from non-English-speaking countries.

Two other changes have been introduced into this edition. The first has been to reduce the textbook from four volumes to three. The second, in response to comments from reviewers, has been to try to provide syntheses of different approaches to public health management and application of public health methodologies, thus we have added concluding chapters for the major sections on policy, management, and methodologies which both summarize and evaluate the different approaches currently in use to resolve public health problems.

Volume 1 examines various influences on public health. This volume attempts to set the scene of the subject of public health, describe its determinants and the methods used in its improvement. We have attempted to include most of the contributions from Volumes 1 and 2 of our previous edition within this volume. In addition we have broadened the description of a variety of health systems in the western world by providing more examples from countries from Europe and Asia, as well as the US, UK, and Australia. The volume gives a historical account of the subject's development. Similarly, we have tried to provide a description of some of the major forms of provision of public health services. Finally there is a series of chapters that illustrate how strategies and policies are co-ordinated and the legal and ethical public health issues.

The second volume is almost entirely concerned with the several methodologies basic to public health including information systems, epidemiologic approaches, social science approaches, and field investigations.

The third volume demonstrates the application of public health strategies and methods, presented in the first two volumes, both to broad areas of public health concern (the environment, social issues etc.) and to specific public health problems. These include both diseases of public health importance (for example mental and cerebrovascular illness) and groups requiring special attention from public health leaders. The volume concludes with two broad overviews on the performance of the public health function.

We hope that the discussion of the structure and philosophy of the book will assist the reader to place the individual contributions of the public health leaders contributing to the textbook into the broader context of public health.

W.W.H.
R.D.
G.K.

London
September, 1990

Contents

Contents of volumes 2 and 3

Volume 2

D Field Investigations of Physical, Biological, and Chemical Hazards

Volume 3

A Interventions in Public Health

B Health Service Planning and Evaluation

C Major Public Health Problems

D Needs of Special Client Groups

E The Performance of the Public Health Function

Contributors

ANTHONY C. ALLISON DPhil BM FRCPath
Vice President for Research, Syntex Research, Division of Syntex (USA) Inc., 3401 Hillview Avenue, PO Box 10850, Palo Alto, CA 94304, USA.

ALBERT E. BENJAMIN
Institute for Health and Aging, School of Nursing, University of California, Box 0612, San Francisco, CA 94143, USA.

JAN E. BLANPAIN MD
Professor and Chairman European Health Policy Forum, School of Public Health, Leuven University, PO Box 214, Leuven 3000, Belgium.

WINDELL R. BRADFORD BA MPA
Deputy Director, Center for Prevention Services, Centers for Disease Control, Atlanta, GA 30333, USA.

WILLIAM BRASS CBE MA FBA
Emeritus Professor of Medical Demography, Centre for Population Studies, London School of Hygiene and Tropical Medicine, 99 Gower Street, London WC1E 6AZ, UK.

LESTER BRESLOW MD MPH
Professor of Public Health and Director, Health Promotion Center at Los Angeles, 1100 Glendon Avenue, Suite 711, Los Angeles, CA 90024–3511, USA.

SIMON CHAPMAN BA PhD
Lecturer, Department of Community Medicine, The University of Sydney, Westmead Hospital, Westmead, New South Wales 2145, Australia.

MIA DEFEVER PhD
European Health Policy Forum, School of Public Health, Leuven University, PO Box 214, Leuven 3000, Belgium.

ROGER DETELS MD MS
Professor of Epidemiology, School of Public Health, University of California, Los Angeles, CA 90024–1772, USA.

ZENG DONG-LU MD MPH
Former Health Program Officer, UNICEF Office for China. Technical Specialist, Johns Hopkins Institute for International Programs, The Johns Hopkins University School of Hygiene and Public Health, 103 East Mount Royal Avenue, Baltimore, MD 21202, USA.

ELIZABETH FEE
Department of Health Policy and Management, School of Hygiene and Public Health, The Johns Hopkins University, 624 North Broadway, Baltimore, MD 21205–1901, USA.

JOHN F. FORBES PhD
Lecturer, Department of Community Medicine, University of Edinburgh, Medical School, Teviot Place, Edinburgh EH8 9AG, UK.

JOHN FRY CBE MD FRCS FRCGP
General Practitioner, 138 Croydon Road, Beckenham, Kent BR3 4DG, UK.

JOHN S. GARROW MD PhD FRCP
Professor of Human Nutrition, Medical College of St Bartholomew's Hospital, Charterhouse Square, London EC1M 6BQ.

CLIFFORD GRAHAM LLB Barrister at Law
Priority Care Group, Department of Health, Alexander Fleming House, Elephant and Castle, London SE1 6BY, UK.

LAWRENCE W. GREEN BS MPH DrPH
Vice President, Henry J Kaiser Family Foundation, 2400 Sand Hill Road, Menlo Park, CA 94025, USA.

ROD GRIFFITHS
Professor of Public Health, The Medical School, University of Birmingham, Edgbaston, Birmingham B15 2TT, UK.

ATSUAKI GUNJI
Professor, Department of Health Administration, School of Health Sciences, Faculty of Medicine, The University of Tokyo, 7–3–1 Hongo, Bunkyo-ku, Tokyo, 113 Japan.

NICKY HART BSc PhD
Associate Professor, Department of Sociology, University of California, 405 Hilgard Avenue, Los Angeles, CA 90024, USA.

O. MICHIO HASHIMOTO MD MPH
President, Overseas Environment Co-operation Center (Corporation), 2–701 Dorfaobadai, 1–11 Aobadai, Midori-ku, Yokohama City, Kanagawa Prefecture, 227 Japan.

GENE I. HIGASHI MD CSD
Professor of Epidemiology, School of Public Health, The University of Michigan, 109 Observatory Street, Ann Arbor, MI 48109, USA.

ALAN R. HINMAN MD, MPH
Director, Center for Prevention Services, Centers for Disease Control, Atlanta, GA 30333, USA.

MICHAEL S.T. HOBBS MBBS DPhil FRACP
Associate Professor in Social and Preventive Medicine, Department of Medicine, University of Western Australia,

Stirling Highway, Nedlands, Western Australia 6009, Australia.

J. G. R. HOWIE MD PhD FRCPE FRCGP
Professor, Department of General Practice, University of Edinburgh, Levinson House, 20 West Richmond Street, Edinburgh EH8 9DX, UK.

L. B. HUNT, MA MBBCh DPH DHMSA MFCM
Senior Medical Officer, Department of Health, Alexander Fleming House, Elephant and Castle, London SE1 6BY, UK.

NOAKI IKEGAMI MD MA
Professor of Health and Public Service Management, Faculty of Policy Management, Keio University of Shonan Fujisawa, 5322 Endoh Fujisawa 252, Japan.

JOHN H. JAMES BA DipEconPolSci
Director of Health Authority Finance, Department of Health, Friar House, 157–168 Blackfriars Road, London SE1 8EU, UK.

KONRAD JAMROZIK MBBS BMedSci DPhil
Lecturer in Public Health, Department of Medicine, University of Western Australia, Stirling Highway, Nedlands, Western Australia 6009, Australia.

MILTON KOTELCHUCK PhD MPH
Chairman and Associate Professor, Department of Maternal and Child Health, School of Public Health, University of North Carolina, Chapel Hill, NC 27599–7400, USA.

KENNETH LEE MA BSc
Professor and Director, Centre for Health Planning and Management, University of Keele, Keele, Staffs ST5 5BG, UK.

PHILIP R. LEE MD
Professor of Social Medicine and Director, Institute for Health Policy Studies, School of Medicine, University of California, 1326 Third Avenue, San Francisco, CA 94143, USA.

STEPHEN R. LEEDER BSc (Med) MB BS PhD FRACP
Professor of Public Health and Community Medicine, Director Department of Community Medicine, The University of Sydney, Westmead Hospital, Westmead, New South Wales 2145, Australia.

JANE E. LEWIS PhD
Reader, Department of Social Science and Administration, London School of Economics and Political Science, Houghton Street, Aldwych, London WC2A 2AE, UK.

ADETOKUNBO O. LUCAS MD, FFCM
Professor of International Health, Harvard School of Public Health, 655 Huntington Avenue, Boston, MA 02115, USA.

CARL F. MARRS PhD
Assistant Professor of Epidemiology, School of Public Health, The University of Michigan, 109 Observatory Street, Ann Arbor, MI 48109, USA.

NOBUHIRO MARUCHI MD Dr Med Sci
Professor and Chairperson, Department of Public Health, Shinshu University School of Medicine, Asahi 3–1–1, Matsumoto, Nagano, 390, Japan.

MASAMI MATSUDA Dr Hlth Sci
Assistant Professor, Department of Public Health, Shinshu University School of Medicine, Asahi 3–1–1, Matsumoto, Nagano, 390 Japan.

ANGUS McGREGOR MA MD FRCP FFCM
Regional Medical Officer/Director of Planning, West Midlands Regional Health Authority, Arthur Thomson House, 146–150 Hagley Road, Birmingham B16 9PA, UK.

ARNOLD S. MONTO MD
Professor of Epidemiology and International Health, Department of Epidemiology, School of Public Health, The University of Michigan, 109 Observatory Street, Ann Arbor, MI 48109, USA.

JOHANNES MOSBECH MD
Department of Health Statistics, National Board of Health, Amaliegede 13, Copenhagen 1256K, Denmark.

JOHN R. MOSS MSocSci BEC MB BS AHA
Lecturer, Department of Community Medicine, University of Adelaide, North Terrace, Adelaide, SA 5000, Australia.

T. G. C. MURRELL MB BS MD DTM&H FRACGP
Foundation Professor of Community Medicine, Department of Community Medicine, University of Adelaide, North Terrace, Adelaide, SA 5000, Australia.

NOBUO ONODERA
President, Saitama College of Health, 519 Kamiokubo, Urawa-shi, Saitama-ken, 338 Japan.

ROBERT L. PARKER MD MPH
Senior Program Officer for Health, UNICEF Office for China, Beijing, People's Republic of China.

CALUM PATON MA MPP DPhil
Senior Lecturer, Centre for Health Planning and Management, University of Keele, Keele, Staffs, ST5 5BG UK.

MIKE PORTER
Lecturer, Department of General Practice, University of Edinburgh, Levinson House, 20 West Richmond Street, Edinburgh EH8 9DX, UK.

JULIUS B. RICHMOND MD BS MS AM
Professor Emeritus, Department of Social Medicine, Harvard Medical School, 641 Huntington Avenue/3rd Floor, Boston, MA 02115, USA.

RUTH ROEMER JD
Adjunct Professor of Health Law and Past President, American Public Health Association, School of Public Health, University of California, Los Angeles, CA 90024–1772, USA.

E. ROSEMARY RUE DBE FRCP FFPHM
Regional General Manager/Regional Medical Officer (Retired), Oxford Regional Health Authority, Old Road, Headington, Oxford OX3 7LF, UK.

A. J. P. SCHRIJVERS PhD
Professor, Department for Health Sciences and Epidemiology, Faculty of Medicine, University of Utrecht, Bijlhouwerstraat 6, 3511 ZC, Utrecht, The Netherlands.

DENISE SIMONS-MORTON MD MPH
Assistant Professor, Department of Community Medicine, Baylor College of Medicine, 1 Baylor Plaza, Houston, Texas 77030, USA.

COR SPREEUWENBERG PhD
Professor, Weegbree 2, 3434 ER, Nieuwegein, The Netherlands.

BARBARA STARFIELD MD MPH
Professor and Head, Division of Health Policy, Department of Health Policy and Management, School of Hygiene and Public Health, The Johns Hopkins University, Hampton House, 624 North Broadway, Baltimore, MD 21205–1901, USA.

KOZO TATARA MD
Professor, Department of Public Health, Osaka University Medical School, 3–57, 4 Chome, Nakanoshima, Kita-ku, Osaka, 530 Japan.

CARL E. TAYLOR
Emeritus Professor, Department of International Health, The Johns Hopkins School of Public Health, 615 North Wolfe Street, Baltimore, MD 21210, USA.

ALBERT VAN DER WERFF
Professor, Adviser on International Health Policy, Organization and Information, Vrouwgeestweg 20, 2481 KN Woubrugge, The Netherlands.

Introduction to volume 1
The influences of public health

This volume attempts to set the scene of the subject of public health and describe the determinants of public health and the methods used in its improvement.

The first section consists of a historical account of the development of public health in the US, the UK, and Japan. Contrasts and similarities are demonstrated in these chapters. The clearly ordered development of the subject in Japan is demonstrated. The much less ordered approach to the subject is seen in the historical analysis of the development in the US and the UK. The conflicts between the medical and lay groups are clearly outlined and important questions are raised of the influence of public health medical practitioners in the improvement and fight for public health, particularly at times of shortage of resource. Too many appear to be willing to go with the status quo, too few appear to have had a crusading role.

These chapters lead straight into the description and discussion of the current scope and concerns of public health, including the methods used in the subject. The conflicts between the crusading imperatives of public health and the political opposition to much of these is evident in the assessment of where public health can play a part. These chapters, in addition, attempt to demonstrate briefly the methods applicable and used by public health practitioners at this time.

The next section is largely concerned with the determinants of health and disease. For example, population factors, nutrition, infection, genetic, physical, behavioural, and social factors, and even medical care services. These chapters clearly outline those determinants that play a role in the level of health of a population, demonstrate the importance, particularly in western countries, of behavioural factors in the initiation of disease, and demonstrate that social and economic factors are important in health determination. The role of health services now in the improvement of public health is evident and demonstrable in our society and it is therefore important that we consider these as being one of the determinants and components of the public health rather than being separated off as they have been in the past.

The next series of chapters attempts to describe how Governments attempt to develop policies for improvement of health and how these are constrained by a variety of different activities. The contrasts between the US, Europe, Japan, and China are clearly demonstrated. Both similarities and contrasts are revealed in the individual chapters. The delivery and methods of financing of health care are demonstrated in Section 'D' including systems dependent entirely upon insurance such as in Europe, dependent on national systems of financing as in the UK, the variations in countries with incomplete systems as in the US, and systems changing often, as in Australia.

The provision of public health services in the various countries is described in Section 'E' which demonstrates how these services can be applied under different Governments.

Section 'F' attempts to discuss how public health medical care and social policies are coordinated in different societies. This section shows some overlap with the previous sections, but in every case demonstrates different attitudes to this subject by the different authors. Although in a perfect world these chapters should have been coordinated with the previous chapters and integrated in such a way that a uniform message should have been received by the reader, we have deliberately left them in this form to illustrate the differences in approach and in understanding—evident in the different countries according to who has written the particular chapter.

The final chapter in this volume attempts to explain some of the legal and ethical aspects of public health which should govern all our activities.

The lesson that this volume should give to the reader is that if public health is to make progress in the western world, it will make enemies amongst specific pressure groups and often Governments. The major determinants of the health of the public are behavioural factors in our societies. These include what we eat, what we drink, what we smoke, and where we live. Although infectious agents still play an important role in the aetiology of disease these are still influenced by our behavioural factors, as evidenced through the rise of AIDS at this time. The organization of medical care and public health services can influence the improvement in public health. The tremendous strides made in countries such as Japan and China demonstrate that a willingness to experiment, change, and apply basic principles can influence improvements in health that are measurable.

A

The Development of the Discipline of Public Health

1

The origins and development of public health in the United States

ELIZABETH FEE

In the US, prior to the twentieth century, there were few formal requirements for public health positions, no established career structures, no job security for health officials, and no formalized ways of producing new knowledge. Public health positions were usually part-time appointments at nominal salary; those who devoted much effort to public health typically did so on a voluntary basis. Until the mid-nineteenth century, public health, like other governmental functions, was usually the responsibility of the social élite. The public health officer was expected to be a 'statesman' acting in the public interest, not a 'politician' answering to a class constituency; men of independent property and wealth were believed to be independent of special interests and therefore capable of disinterested judgement.

Charles Rosenberg (1962) has eloquently described an earlier conception of both poverty and disease as consequences of moral failure at the individual and social level. Disease attacked the dirty, the improvident, the intemperate, the ignorant; the clean, the pious, and the virtuous tended to escape. Epidemic diseases were the consequence of a failure to obey the laws of nature and of God: they were the indicators of social and moral dissolution. As cleanliness was linked to godliness, virtue was an essential qualification for managing the state; the conscientious, the respectable, the educated, and the affluent were naturally qualified for public office. Physicians were frequently chosen as public health officers, but lawyers or gentlemen of independent means could also be appointed.

Public health programmes, when organized at all, were organized locally: as Robert Wiebe (1967) has argued, the US in the nineteenth century was a society of 'island communities' with considerable economic and political autonomy. The first public health organizations were those of the rapidly growing port cities of the eastern seaboard in the late eighteenth century. Here, the American republic intersected with the world of international trade. Local authorities tried to protect the population from the threat of potentially catastrophic epidemic diseases, such as the yellow fever epidemic that had crippled Philadelphia in 1793, while they also tried to maintain the conditions for successful economic activity (Powell 1949).

From 1832, the recurring threat of cholera pandemics provided a major impetus to create boards of health in many parts of the eastern US. Public health in this period was largely a police function. Traditionally, port cities had dealt with epidemics by means of quarantine regulations, confining ships suspected of carrying disease in the harbour for up to 40 days. However, quarantine regulations clearly interfered with shipping and were energetically opposed by those whose economic interests were tied to trade (Ackerknecht 1948). Opponents of quarantine argued that disease was internally generated by the filthy conditions of the docks, streets, and alleys that provided an ideal environment for 'putrefactive fermentation'. City health departments attempted to regulate the worst offenders: graveyards, tallow chandleries, tanneries, sugar boilers, skin dressers, dyers, glue boilers, and slaughter houses; they tried to clean the privies and alleys and to remove dead animals and decaying vegetable matter from the streets and public spaces (Howard 1924).

The causes of disease were in dispute throughout much of the nineteenth century. The contradictory evidence available suggested no clear victor between those who believed that diseases were brought in from overseas—and thus should be fought by quarantine regulations—and those who believed that diseases were internally generated and should be fought by cleaning up the cities. Health regulations were written and revised more in response to political pressures than to shifts in medical thinking. Quarantine regulations were alternately relaxed in response to pressure from merchants and strengthened under the immediate threat of epidemics.

Periodically, the dread of yellow fever, plague, and cholera galvanized city authorities into action. The more common endemic diseases with less spectacular mortality rates—typhoid fever, typhus, measles, diphtheria, influenza, tuberculosis, and malaria—were met with a stolid indifference born of familiarity and a sense of helplessness. For these diseases, little was done beyond attempts to maintain general cleanliness, backed up by prayer, fasting, and exhortations to virtue. In Baltimore in 1819, for example, the clergy claimed to have saved the city from an epidemic of yellow fever by turning the wind to a northwesterly direction (Howard 1924, p. 87).

In the southern States, white slave-owners perforce regulated—or failed to regulate—health conditions on the plantations through direct control of working conditions, housing, food, clothing, 'discipline', sanitation, maternal and infant care, and the provision of medical care (Savitt 1988). Communicable diseases were largely seasonal: in winter, respiratory infections; in summer, intestinal infections and fevers. Musculoskeletal deformities due to flogging were a specific public health problem produced by the slave system of production.

The Civil War enforced a national consciousness of epidemic disease: two-thirds of the 360 000 Union soldiers who died were killed by infectious diseases rather than by enemy bullets (Adams 1952). The ravages of dysentery, spread by inadequate or non-existent sanitary facilities, were appalling. The US Sanitary Commission, a voluntary organization inspired by Florence Nightingale's work in the Crimean War, tried to promote the health of the Union army by inspecting army camps, distributing educational materials, and providing nursing care and supplies for the wounded. Joseph Jones, a surgeon in the medical department of the Confederate Army, estimated that three-quarters, or 150 000, of the Confederate soldiers' deaths were due to disease; others believed he may have underestimated these losses (Cunningham 1958). In either case, the majority of losses on both sides were the result of disease rather than of battle (Woodward 1863). According to contemporary accounts, the main causes of death were 'typho-malaria' (perhaps a combination of typhoid fever and malaria), camp diarrhoea, and 'camp measles'; scurvy, acute respiratory diseases, venereal diseases, rheumatism, and epidemic jaundice were also widespread in army encampments (Woodward 1863).

By 1860, public health activities were just beginning to move beyond the confines of local city politics. Between 1857 and 1860, quarantine and sanitary conventions were held in Philadelphia, Baltimore, New York, and Boston (Proceedings 1859). The first state board of health, largely a paper organization, was created in Louisiana in 1855. In the 1870s and 1880s most of the states created their own boards of health. The first working state health board was formed in Massachusetts in 1869, followed by California (1870), the District of Columbia (1871), Virginia and Minnesota (1872), Maryland (1874), and Alabama (1875) (Paterson 1939). The impact of these state boards of health should not, however, be overemphasized; by 1900, only three states (Massachusetts, Rhode Island, and Florida) spent more than two cents per caput for public health services (Abbott 1900).

As state boards of health were organizing, public health reformers also urged the formation of a national health board. In 1879, a disastrous yellow fever epidemic, sweeping up the Mississippi Valley from New Orleans, prompted the US Congress to create a National Board of Health consisting of seven physicians and one representative each from the Army, the Navy, the Marine Hospital Service, and the Department of Justice. Responsible for formulating quarantine regulations concerning trade and travel between the states, the National Board of Health soon became embroiled in fierce battles over states' rights. In 1883, it was disbanded and quarantine powers reverted to the Marine Hospital Service. Gradually, the Marine Hospital Service expanded its public health activities into public health research. In 1887, it set aside a single room as a 'hygienic laboratory' to be gradually expanded into an important centre for the investigation of infectious diseases. From this small beginning would later grow the National Institutes of Health (Harden 1986).

In this period after the Civil War, northern industrialists had begun the process of transforming the country into a single national market. Agricultural and industrial mechanization irrevocably altered the traditional patterns of production and consumption; railroad companies competed to cross the country with railway lines; small companies merged and collapsed into large corporations. Between 1860 and 1894, the value of manufactured goods multiplied by five. The US was moving into first place as the most powerful industrial country in the world, bypassing England, Germany, and France.

The belief that epidemic diseases posed only occasional threats to an otherwise healthy social order was shaken by the industrial transformation of the late nineteenth century. The burgeoning social problems of the industrial cities could not be ignored: the overwhelming influx of immigrants crowded into narrow alleys and tenement housing, the terrifying death and disease rates of working class slums, the total inadequacy of water supplies and sewerage systems for the rapidly growing population, the spread of endemic and epidemic diseases from the slums to the neighbourhoods of the wealthy, and the escalating squalour and violence of the streets. Almost all families lost children to diphtheria, smallpox, or other infectious diseases. Poverty and disease could no longer be treated simply as individual failings.

The early efforts of city health department officials to deal with health problems represented an attempt to mitigate the worst effects of unplanned and unregulated growth: a rearguard action against the filth and congestion created by anarchic economic and urban development (Blake 1959; Duffy 1968, 1974; Galishoff 1975; Leavitt 1982; Rosenkrantz 1972). As cities grew in size, as the flow of immigrants continued, and as public health problems became ever more obvious, pressures mounted for more effective responses to the problems. New York, the country's largest city, and the one with some of the worst health conditions, produced many of the most energetic and progressive public health leaders; Boston and Providence were also noted for their active public health programmes; Baltimore and Philadelphia trailed far behind (Cassedy 1962; Jordan *et al.* 1924; Rosenkrantz 1972; Winslow 1929).

Public health as social reform

Industrialization had brought new sources of affluence as well as of misery. America no longer fit its self-image as a country of independent farmers and craftsmen; like the European countries, it displayed extremes of wealth and privilege, social misery, and deprivation. Labour and social unrest forced awareness of the need for social and health reforms. The great railroad strike of 1877, the assassination of President Garfield in 1881, the Haymarket bombing of 1886, the Homestead strike of 1892, and the Pullman strike of 1894 were just a few of the reminders that all was not well with the Republic. The Noble Order of the Knights of Labor—dedicated to such measures as an income tax, an eight-hour day, social insurance, labour exchanges for the unemployed, the abolition of child labour, workers' compensation, and public ownership of railroads and utilities—grew from a membership of 11 persons to over 700 000 within a few years. Massive strikes for better wages and working conditions revealed deep class divisions and warned of potential social disorder. At the same time, the development of democratic 'machine' politics challenged the dominance of the political and social élite, permitting some labour leaders to establish local bases of influence and power. The perceived social anarchy of the large industrial cities mocked the pretensions to social control of the traditional forces of church and state, and highlighted the need for more activist responses to the multiplicity of problems.

An increasing number of reform groups devoted themselves to social issues and improvements of every variety. Health reformers, physicians, and engineers urged improved sanitary conditions in the industrial cities. Medical men were prominent in reform organizations, but they were not alone (Rosenberg and Rosenberg 1968; Shryock 1937). Barbara Rosenkrantz (1974) has contrasted public health in the late nineteenth century with the internecine battles within general medicine: 'the field of public hygiene exemplified a happy marriage of engineers, physicians, and public-spirited citizens providing a model of complementary comportment under the banner of sanitary science'. The most formally organized and professional body, the American Public Health Association, was founded in 1872; it included scientists, municipal officials, physicians, engineers, and the occasional architect and lawyer (Ravenel 1921; Smith 1921).

Middle and upper class women, seizing an opportunity to escape from the narrow bounds of domestic responsibilities, joined in campaigns for improved housing, for the abolition of child labour, for maternal and child health, and for temperance; they were active in the settlement house movement, trade union organizing, the suffrage movement, and municipal sanitary reform. Sanitary reform activity, when renamed 'municipal housekeeping', was viewed as a natural extension of women's training and experience as 'the housekeepers of the world' (Ryan 1975). By the early years of the twentieth century, many such voluntary health organizations were established around specific issues, thus providing the impulse and energy behind public health reforms. The American Red Cross had been formed in 1882, the National Tuberculosis Association in 1904, the American Social Hygiene Association in 1905, the National Committee for Mental Hygiene in 1909, and the American Society for the Control of Cancer in 1919 (Smillie 1955).

The progressive reform groups in the public health movement advocated immediate change tempered by scientific knowledge and humanitarian concern. Sharing the revolutionaries' perception of the plight of the poor and the injustices of the system, they counselled less radical solutions (Hays 1968, 1980; Rogers 1982; Wiebe 1967). They advocated public health reforms on political, economic, humanitarian, and scientific grounds. Politically, public health reform offered a middle ground between the cut-throat principles of entrepreneurial capitalism and the revolutionary ideas of the socialists, anarchists, and utopian visionaries. As William H. Welch (1920, p. 598) explained to the Charity Organization Society, sanitary improvement offered the best way of improving the lot of the poor, short of the radical restructuring of society.

Economically, the progressive reformers argued that public health should be viewed as a paying investment, giving higher returns than the stock market. In Germany, Max von Pettenkofer (1941) had first calculated the financial returns on public health 'investments' to prove the value of sanitary improvements in reducing deaths from typhoid, and his argument would be repeated many times by American public health leaders. As Welch explained: 'merely from a mercenary and commercial point of view it is for the interest of the community to take care of the health of the poor. Philanthropy assumes a totally different aspect in the eyes of the world when it is able to demonstrate that it pays to keep people healthy' (Welch 1920, p. 596).

Whether progressives stressed the humanitarian need for reform or the business efficiency of improving public health, they emphasized the need for more scientific training of those responsible for public health activities. Charles Chapin (1934*a*), the superintendent of health of Providence, Rhode Island, and a leading light in the progressive movement, argued that public health should be a profession with appropriate training and income: 'We hope that every local unit of government will have its health officer and that the iceman and the undertaker will not be considered suitable candidates, but that every health officer will be trained for his work. We hope that he will receive a reasonable reward for his services, and that the pay for saving a child's life with antitoxin will at least equal that received by a plumber for mending a leaky pipe; and that for managing a yellow fever outbreak a man may receive as much per week as a catcher on a baseball nine'.

Public health leaders argued that the demand for centralized planning and business efficiency required scientific knowledge rather than the undisciplined enthusiasms of voluntary groups (Rotch 1909). Public health decisions should be made by an analysis of costs and benefits 'as an up-to-date manufacturer would count the cost of a new process'. The health officer, like the merchant, should learn 'which line of work yields the most for the sum expended' (Chapin 1934*b*).

Existing health departments were dominated more by patronage and political considerations than by economic or administrative efficiency. Progressives regretted 'the evil of politics' and wanted to increase the pay and minimum qualifications for health officers to attract personnel on the basis of skill rather than influence. Their attempt to insulate boards of health from local political control was part of a broader movement to make all forms of public administration more 'rational' and 'efficient' by reducing the influence of political bosses and by promoting a new group of professional administrators (Schiesl 1980). The goal was for a well-trained professional élite to conduct social reform on scientific lines.

National and international health

Public health was quickly becoming a national and even international issue. Although the US Congress was reluctant to enact federal health legislation, there were mounting pressures for the US to pay attention to public health abroad. As American businessmen were seeking enlarged foreign markets, a vocal group of intellectuals and politicians argued for an assertive foreign policy. The US began to challenge European dominance in the Far East and Latin America, seeking trade and political influence more than territory, but taking territory where it could. National defence goals included broadening control of trade routes, building a canal across Central America, and establishing strategic bases in the Caribbean and western Pacific.

In 1898, the US entered the Spanish–American War, expanded the army from 25 000 to 250 000 men, and sent troops to Cuba. That war demonstrated that the US could not afford military adventures overseas unless more attention was paid to sanitation and public health: 968 men died in battle, but 5438 died from infectious diseases (Cosmas 1971; Sternberg 1912*a*). None the less, the US defeated Spain and installed an army of occupation in Cuba. Cuba had long been regarded as a source of the yellow fever epidemics threatening the US. When yellow fever endangered the US troops in Cuba in 1900, the response was efficient and effective. An army commission under Walter Reed was sent to Cuba to study the disease and, in a dramatic series of human experiments, confirmed the hypothesis of a Cuban physician, Carlos Finlay, that it was spread by mosquitoes, that there was an incubation period in the mosquito, and that the mosquito itself bred in close proximity to human habitats. This was quickly followed by a quasi-military operation, undertaken by Surgeon-Major William Gorgas as the Sanitary Officer, which did indeed eliminate yellow fever from Havana (Kelley 1906).

The US Army in the Philippines faced even greater problems: a climate that sapped the strength of the strongest soldiers while exposing them to disease-bearing insects, parasites, and bacteria-laden water supplies (Gillett 1987). George Sternberg appointed the Tropical Disease Board in 1889 which worked for years on the new diseases found in the Philippines as well as on the more familiar problems of malaria, diarrhoeal diseases, and venereal disease. Success in the war and in the Philippine Insurrection was dependent on gaining some degree of control over malaria, dysentery, dengue fever, and beriberi (Gates 1973).

These experiences illustrated the importance of public health for successful US efforts overseas. The attempt by the French to build a canal across Panama had been abandoned because of enormous mortality rates from disease, and the US, having decided to undertake the effort, was facing the same problems (Sternberg 1912*b*). In 1904, Gorgas, now promoted to General, took control of a campaign against the malaria and yellow fever that threatened canal operations. He was finally able to persuade the Canal Commission to institute an intensive campaign against mosquitoes; in one of the great triumphs of practical public health, yellow fever and malaria were brought under control and the canal was successfully completed in 1914.

US industrialists brought some of the lessons of Cuba and the Panama canal home to the southern US. The South at that time resembled an underdeveloped country within the US, characterized by poor economic and social conditions. Northern industrialists were already investing heavily in southern education as well as in cotton mills and railroads; John D. Rockefeller had created the General Education Board to support 'the general organization of rural communities for economic, social and educational purposes' (Fosdick 1962). Charles Wardell Stiles managed to convince the Secretary of the General Education Board that the real cause of misery and lack of productivity in the South was hookworm, the 'germ of laziness'. In 1909, Rockefeller agreed to provide $1 million to create the Rockefeller Sanitary Commission for the Eradication of Hookworm Disease, with Wickliffe Rose, originally a philosophy professor in Tennessee, as Director (Ettling 1981). This was to be the first instalment in Rockefeller's massive national and international investment in public health.

Rose went beyond the task of attempting to control a single disease and worked to establish an effective and permanent public health organization in the southern states (Rose 1910). At the end of five years of intensive effort, the campaign had failed to eradicate hookworm but had greatly expanded the role of public health agencies. Between 1910 and 1914, county appropriations for public health work in the South increased from a total of $240 to $110 000 (Ettling 1981). In 1914, the organizational experience gained in the southern States enabled the Rockefeller Foundation to extend its hookworm control programme to the Caribbean, Central America, and South America.

Meanwhile, in Washington the Committee of One Hundred on National Health, composed of such notables as Jane Addams, Andrew Carnegie, William H. Welch, and Booker T. Washington, campaigned for the federal regulation of public health (Marcus 1979; Rosen 1972). Its president, the economist Irving Fisher (1909), argued that a public health service would be good policy and good economics and would help conserve 'national vitality'. In 1912, the Federal Government made its first real commitment to public health when it expanded the responsibilities of the Public Health Service, empowering it to investigate the causes and spread of diseases and the pollution and sanitation of navigable streams and lakes (Williams 1951). By 1915, the Public Health Service, the US Army, and the Rockefeller Foundation were the major agencies involved in public health activities, supplemented on a local level by a network of city and State health departments, many of whose members were organized in the American Public Health Association.

The professionalization of public health

These developments all led to an increasing demand for people trained in public health to direct the new programmes being developed on local, state, and national levels. Those attempting to develop such programmes became increasingly critical of the lack of properly trained personnel; part-time public health officers were simply not adequate to staff the ambitious new programmes being planned and implemented. Public health reformers agreed that full-time practitioners, specially trained for the job, would be needed to staff public health positions in the future. In 1913, the New York State legislature passed a law requiring public health officers to have specialized training. Where were such people to be found, and what training should they receive? As William Sedgwick (1924) argued: 'If, as I believe, we are in fact moving irresistibly towards a bureaucracy, while clinging to the ideals of a democracy, we shall do well to pause and inquire what kind of bureaucracy we are building up about ourselves . . . scientists and technicians alike . . . must be employed and paid by the people, to rule over them as well as to guide and to guard them, to constitute a kind of official class, a kind of bureaucracy constituted for themselves by the people themselves . . . what kind of scientists and technicians shall we have in our public service? . . . I honestly believe that upon our ability to solve, and solve wisely, these fundamental problems of our American life will depend in large measure our comfort and success as a people in the 20th century'.

Public health had been defined in terms of its aims and goals—to reduce disease and maintain the health of the population—rather than by any specific body of knowledge. Many different disciplines contributed to effective public health work: physicians diagnosed contagious diseases; sanitary engineers built water and sewerage systems; epidemiologists traced the sources of disease outbreaks and their modes of transmission; vital statisticians provided quantitative measures of births and deaths; lawyers wrote sanitary codes and regulations; public health nurses provided care and advice to the sick in their homes; sanitary inspectors visited factories and markets to enforce compliance with public health ordinances; and administrators tried to organize everyone within the limits of health department budgets. Public health thus involved economics, sociology, psychology, politics, law, statistics, and engineering, as well as the biological and clinical sciences. However, in the period immediately following the brilliant experimental work of Louis Pasteur and Robert Koch, the bacteriological laboratory became the first and primary symbol of a new, scientific, public health.

Bacteriology and alternative views of health and disease

The clarity and simplicity of bacteriological methods and discoveries gave them tremendous cultural and ideological importance: the agents of particular diseases had been made visible under the microscope. The identification of specific bacteria seemed to have cut through the misty miasmas of disease to define the enemy in unmistakable terms. Bacteriology thus became an ideological marker, sharply differentiating the 'old' public health, the province of untrained amateurs, from the 'new' public health, which would belong to scientifically trained professionals.

Young Americans who had studied in Germany brought back the new knowledge of laboratory methods in bacteriology and started to teach others: William Henry Welch and T. Mitchell Prudden in New York, George Sternberg in Washington, and A.C. Abbott in Philadelphia were among the first to introduce the new bacteriology to the US. These young scientists were convinced that physicians should stop squabbling over medical ethics and politics and commit themselves to the purer values of laboratory research.

The laboratory ideal rapidly influenced leading progressives in public health. By the 1880s, Charles Chapin had established a public health laboratory in Providence, Rhode Island; Victor C. Vaughan had created a state hygienic laboratory in Michigan; and William Sedgwick (1901) had used bacteriology to study water supplies and sewage disposal at the Lawrence Experiment Station in Massachusetts. Sedgwick demonstrated the transmission of typhoid fever by polluted water supplies and developed quantitative methods for measuring the presence of bacteria in the air, water, and milk. Describing the impact of bacteriological discoveries, he said: 'Before 1880 we knew nothing; after 1890 we knew it all; it was a glorious ten years' (Jordan *et al.* 1924).

The powerful new methods of identifying diseases through the microscope drew attention away from the larger and more diffuse problems of water supplies, street cleaning, housing reform, and the living conditions of the poor. The approach of locating, identifying, and isolating bacteria and their human hosts was a more elegant and easier way of dealing with disease than environmental reform. The public health laboratory demonstrated the scientific and diagnostic power of the new public health. But by focusing on the diagnosis of infectious diseases, it narrowed the distance between medicine and public health and brought public health into potential conflict with private medical practice. Physicians began to resent the public health officials' claim to diagnose, and often to treat, infectious diseases.

The new epidemiology, like the new bacteriology, was firmly oriented to the control of specific diseases. Charles Chapin, the superintendent of health of Providence, Rhode Island, was one of the leading proponents of the new epidemiology. Chapin (1901) had published a comprehensive text on municipal sanitation but soon concluded that much of the effort devoted to cleaning up the cities was wasted; instead, public health officers should concentrate on controlling specific routes of infection. Chapin (1910) published a new text, *The sources and modes of infection*, which became a gospel of infectious disease control. Hibbert Winslow Hill, director of the division of epidemiology of the Minnesota Board of Health, popularized Chapin's work in a lively series of articles first printed in 1100 newspapers across the US, and later published as a book, *The new public health*. Hill (1916) explained why modern scientific methods in public health were more efficient than old-fashioned social reforms. To control tuberculosis, for example, it was hardly necessary to improve the living conditions of the 100 000 000 people in the US—it was only necessary to supervise the 200 000 active tuberculosis cases '*merely to the extent of confining their infective discharges* . . . Need any more be said to indicate the superiority of the new principles, as practical business propositions, over the old?' (Hill 1916). The vital statistician, said Hill, would be the future scientific manager of public health expenditures: 'Much abused, laughed at, neglected, he is, or will be, like the cost-of-production scientific manager of modern business, 'the most indispensable man on the staff' . . . a man who knows costs in each department in proportion to production, and where to cut cost, increase production, save time, unnecessary work, and waste in general' (Hill 1916).

The dominance of the disease-oriented approach to public health was evident in the first handbook for practising public health officers, *A manual for health officers*, by J. Scott MacNutt (1915), which echoed the views of Chapin and Hill. MacNutt devoted approximately half of his 600-page handbook to the contagious diseases, four pages to industrial hygiene, and gave only passing notice to housing, water supplies, public education, and environmental health.

But while the narrow bacteriological view was dominant, there were several competing models for public health research and practice extant at the time. Public health was characterized by a diversity of views and approaches. Compare, for example, Hill's narrow focus with the broad and expansive gaze of Charles-Edward A. Winslow (1920) of Yale University: 'Public health is the science and art of preventing disease, prolonging life, and promoting physical health and efficiency through organized community efforts for the sanitation of the environment, the control of community infections, the education of the individual in principles of personal hygiene, the organization of medical and nursing service for the early diagnosis and preventive treatment of disease, and the development of the social machinery which will ensure to every individual in the community a standard of living adequate for the maintenance of health'.

Winslow's was not the only alternative view. In the same year that Hill published his book on the new public health, Alice Hamilton in Illinois conducted a survey of industrial lead poisoning and established the fact that thousands of American workers in pottery glazing, bath-tub enamelling, cut glass polishing, cigar wrapping, can sealing, and dozens of other industrial processes were slowly being killed by white lead (Hamilton 1943; Sicherman 1984). Unaided by legislation, Hamilton argued, persuaded, shamed, and flattered individual employers into improving working conditions. Almost single-handedly, she created the foundations of industrial hygiene in America.

Joseph Goldberger's epidemiological studies of pellagra for the Public Health Service offer an example of yet another approach to public health. In 1914, Goldberger announced that pellagra was due to dietary deficiencies and not to some unknown micro-organism; he and his colleagues had cured endemic pellagra in a Mississippi orphanage by feeding the children milk, eggs, beans, and meat. He then teamed up with the economist, Edgar Sydenstricker, to survey the diets of southern wage-workers' families. They showed how the share-cropping system had impoverished tenant farmers, led to dietary deficiencies, and thus produced endemic pellagra (Terris 1964). The economic system of cotton production in the South could thus be seen as directly responsible for the high prevalence of pellagra.

Elmer V. McCollum was a leading exponent of the view that nutrition was of central importance to human health. McCollum had started research on animal feeding at the College of Agriculture of the University of Wisconsin; he had wanted to find out why corn-fed cows were healthier than cows fed only on wheat or oats, a problem of obvious practical interest to farmers. He conducted his experiments on rats since they could easily be kept in large numbers in the laboratory. When the little rats grew fat and healthy on a diet containing butter fat, but sickened on the same diet containing lard or olive oil, McCollum had his first clue leading to the discovery of vitamin A. In 1918, he accepted a professorship at the Johns Hopkins School of Public Health where he continued hundreds of nutritional experiments. He found an organic substance in cod-liver oil, later called vitamin D, that protected rats from rickets. The discovery of vitamin D led to a generation of children brought up on cod-liver oil and the virtual elimination of rickets as a childhood disease. McCollum was convinced that many diseases of unknown origin were caused by nutritional deficiencies. Not only were rickets, scurvy, beriberi, and pellagra clearly related to diet, but he thought other more subtle conditions, including emotional states, could be traced to inorganic elements in the diet. McCollum was deeply involved with popular health education. As a member of Herbert Hoover's Advisory Council on Nutrition, he travelled around the country giving public lectures, urging mothers to serve milk, leafy vegetables, eggs, liver, and wheat bread.

Alice Hamilton, Joseph Goldberger, Edgar Sydenstricker, and Elmer McCollum were minority voices amid the growing majority who focused exclusively on bacteria. As most bacteriologists and epidemiologists concentrated on specific disease-causing organisms and the individuals who harboured them, only a minority continued to relate the problems of ill-health and disease to the larger social environment, and to issues of occupational health, poverty, and nutrition.

The relationship between public health and medicine

While the broader conceptions of public health required an understanding of economics and politics, the dominant model of public health knowledge was based almost exclusively on the biological sciences. This redefinition of public health in bioscientific terms reinforced the medical professions' claim to pre-eminence in the field. Physicians felt that, since they were the experts in infectious diseases, they were uniquely qualified to become the ultimate authorities in the new, scientific, public health.

By the second decade of the twentieth century, non-medical public health officers were beginning to protest the dominance of public health by medical men. By this time, the sanitary engineers were the only professional group strong enough to challenge the physicians' assumption that the future of public health should be theirs. Civil and sanitary engineers had created clean city water supplies and adequate sewerage systems, major factors in the declining death-rates from infant diarrhoea and other infectious diseases. With the benefit of hindsight, we can say that the sanitary engineers, through their work in providing relatively clean water, probably deserve much of the credit for the decline of infectious disease mortality in the late nineteenth century (Duffy 1971; Haines 1979; Hoffman 1907; Meeker 1972).[1]

Like the public health physicians, sanitary engineers debated alternate theories of disease aetiology; they constructed sewerage systems in the light of available knowledge and the financial constraints of municipal employers (Tarr 1979). Thus, the sanitary engineers as well as the public health physicians claimed that their specialized expert knowledge should be influencing the course of public health; both professional groups often fought among themselves and with the city authorities over the most appropriate actions. The professional competition between the sanitary engineers and physicians became intense in the early years of the twentieth century as sanitary engineers vociferously complained about the increasing 'medical monopoly' of public health.

By 1912, 15 states required that all members of their boards of health be physicians, 23 states required at least one physician member, and 10 states had no professional requirement for eligibility (Knowles 1913). The medical profession was well organized and was beginning to exert control over public health practice. Physicians were willing to concede to the sanitary engineers, responsibility for public sanitation and water supplies, but little else. Yet the majority of physicians demonstrated little interest in public health; many were hostile to public health and suspicious lest any expansion of public health activities threaten either their autonomy or their economic interests.

The relationship between the emerging profession of public health and the well-established profession of medicine continued to be problematic and controversial. The increased activity of health departments in the identification and control of infectious diseases tended to bring health officers into conflict with private practitioners; as soon as public health left the confines of sanitary engineering and took on the battle against specific diseases, it challenged the boundaries of medical autonomy. As Duffy (1979) has argued, the medical profession moved from a position of strong support for public health activities to a cautious, and sometimes suspicious, ambivalence.

Public health officers understood that, although their interests and responsibilities might at times be opposed to those of private practitioners, it was nevertheless important to cultivate co-operative relationships with the medical profession. Indeed, Duffy has argued that this attitude had the effect of making public health officers 'cautious to the point of timidity', so reluctant were they to undertake any programmes that might disturb the interests of their medical colleagues (Duffy 1979).

The Flexner reforms in medical education were one symptom of the larger transformation occurring in medical knowledge and practice in the early years of the twentieth century (Brown 1979; Starr 1982). As medical practice became dependent on developing scientific knowledge and technology, it was institutionalized in hospital settings (Rosenberg 1987). Hospitals became dependent on physicians, and physicians, in turn, became dependent on access to hospital facilities. As physicians became ever more interested in the capacities of scientific medicine and the possibilities of specialization, they became correspondingly less interested in community and preventive activities. As the standards of education and criteria for admission to the profession became more controlled and demanding, the numbers of practitioners fell and their incomes correspondingly began to rise. Medical practice was a challenging and increasingly lucrative field. This was not a time to expect large numbers of

[1] It is difficult to be confident about mortality rates in the US before 1900, when the death registration areas began regular reporting. The evidence seems, however, to suggest that mortality rates between 1850 and 1880 remained relatively constant, with wide annual variations depending on the presence of epidemics. In the 1880s the mortality rates began to decline, and continued this decline, with minor fluctuations, throughout the period from 1890 to 1915. The major component of the decline was in infant mortality, especially mortality rates from the infectious diseases and infant diarrhoea. This pattern is consistent with the thesis that the extension of municipal water systems and the filtration of water supplies played a major role in the decline in mortality rates. The pasteurization of milk was probably also an important contributing factor.

physicians to be attracted to public health practice with its low incomes, political controls, and lack of individual autonomy.

When schools of public health were created in the 1910s and 1920s, they had hoped that young physicians would take advanced training in public health after graduating from medical school. But very few young medical graduates entered schools of public health; the prospects and remuneration in public health paled beside the glamorous inducements of curative medicine and surgery. Johns Hopkins, Harvard, Yale, and Columbia universities all started schools or divisions of public health and all experienced the same problems; those seeking to enrol as students were either experienced older men who had worked in public health positions without specialist qualifications, or young scientists interested in bacteriology, epidemiology, and other public health disciplines, but who lacked the medical degree generally regarded as an essential qualification for public health leadership (Curran 1970; Fee 1983, 1987; Viseltear 1982*a*, 1986; Williams 1976). Important positions in public health were thus offered to physicians without specialist training in preference to non-physicians with doctorates in public health; the demand for physicians was such that they rarely needed public health training as a professional job requirement. As a result, the incentives for physicians to take specialized degrees in public health were further reduced, and schools of public health were forced to admit the ever larger numbers of nurses, engineers, statisticians, and biologists who sought public health training. This structural problem in the relationship between medicine and public health, already clear by 1920, was never entirely resolved. Public health in the US would continue to be open to many professional groups and disciplines, while it maintained a special and privileged status for those with medical qualifications. Certainly, those who have made major contributions to public health in the twentieth century have been both physicians and non-physicians, some with specialized public health training, and some without.

Public health organization and practice

The practical importance of public health was well recognized by the early decades of the twentieth century. The incidence of tuberculosis, diphtheria, and other infectious diseases was falling, apparently in response to energetic public health campaigns. School health clinics and maternal and child health centres were established in many cities with active public support. Registration for the draft in the First World War—like draft registration for every war—revealed that a substantial proportion of young men were either physically or mentally unfit for combat; this perception led to increased political support for public health activities. The influenza epidemic that devastated families and communities in 1916–18 underlined the continuing threat of infectious disease.

In the 1920s, state and municipal health departments developed new organizational units and increased their hiring of public health personnel, especially public health nurses. While bacteriological laboratories continued to be important, divisions of tuberculosis, child and maternal health, venereal diseases, public health administration, and health education played a major role in most state and city health departments, along with divisions of sanitation and vital statistics. The Sheppard–Towner Act of 1921, which provided funds for prenatal and child health clinics, represented a new level of involvement by the Federal Government in providing public health services for women and children. After the first flush of enthusiasm for the achievements of bacteriology, many health departments were now paying more attention to community-based health activities and popular health education. Charles-Edward A. Winslow (1923, 1926) went so far as to announce the ending of the bacteriological age and to describe popular health education as the keynote of the 'new public health': 'The dominant motive in the present-day public health campaign is the education of the individual in the practices of personal hygiene. The discovery of popular education as an instrument in preventive medicine, made by the pioneers in the tuberculosis movement, has proved almost as far-reaching as the discovery of the germ theory of disease thirty years before'.

In fact, public health practice varied greatly throughout the states and cities across the country. A report of the Committee on Municipal Health Department Practice of the American Public Health Association (*Public Health Bulletin* 1923) illustrated the wide variation in forms of public health organization across 83 US cities in the period.

The most important federal organization in public health was the US Public Health Service, which had been created in 1912 through a development of the Marine Hospital Service. The Marine Hospital Service, formed in 1798, had initially provided medical care for merchant seamen and had gradually expanded to include the enforcement of quarantines and research on and treatment of epidemic, infectious, and nutritional diseases (Koop and Ginzburg 1989). As the Public Health Service, it aided the development of state health departments by giving grants-in-aid, loaning expert personnel, and providing advice and consultation on specific problems. If, for example, a state was facing an unexplained outbreak of typhoid fever or other epidemic disease, the Public Health Service would send epidemiologists to trace the source of the disease and suggest means of preventing its further spread. Public Health Service assistance was sent only on request by the States and thus was not seen as a threat to often closely guarded States' rights.

Harry S. Mustard, Director of the School of Public Health at Columbia University, noted: 'The United States Public Health Service occupies an enviable and commendable position in its relationships with the health authorities of the several states . . . through high standards of performance and through demonstration and efficiency, the Public Health Service has raised the level of work performed in every county, city, and state health department with which it has had even indirect contact' (Mustard 1945). By 1930 the Public Health Service had established a strong position of leadership and performed its tasks with 'wisdom, fairness, and efficiency' (Mustard 1945).

Other health activities were scattered throughout the federal bureaucracy. About 40 agencies participated to some degree in public health activities, including the Department of Agriculture (home economics, entomology, plant quarantine, etc.), Department of Commerce (Bureau of the Census), Department of the Interior (Office of Indian Affairs in sanitation and medical care, Bureau of Mines in industrial hygiene), Pure Food and Drug Administration, and the Department of Labor (Children's Bureau, Bureau of Labor Statistics).

A major stimulus to the development of public health practice came in response to the Depression, with the New Deal and the Social Security Act of 1935. The Social Security Act expanded financing of the Public Health Service and provided federal grants to states to assist them in developing their public health services. Federal and State expenditures for public health actually doubled in the decade of the Depression. The expansion of local health units was especially fuelled by the Social Security Act. In most parts of the country, provision of public health services to local communities depended on county health organizations, simpler units than the larger state health departments. In 1934, only 541 counties out of the 3070 counties in the US had any form of local public health service, but by June 1942, 1828 counties could boast of health units directed by a full-time public health officer (Kratz 1943). (Much of this gain would be lost during the war; by the end of the war only 1322 counties had an organized health service (Corwin 1949; Mustard 1945).)

A significant part of the Social Security Act and the accompanying changes in the Public Health Act was that for the first time, the Federal Government would provide funds, administered through the states, for public health training. Federal regulations now required states to establish minimum qualifications for the public health personnel employed through new federal grants. Thus, it was no longer sufficient for state programmes to employ any willing physician; some form of professional public health training could be expected. As a result of the growing demand for public health training, several state universities began new schools or divisions of public health, and existing schools of public health expanded their enrolments. By 1936, ten schools offered public health degrees or certificates requiring at least one year of attendance (Leathers *et al.* 1937). By 1938, more than 4000 individuals, including about 1000 doctors, had received some public health training with funds provided by the Federal Government through the states. The economic difficulties of maintaining a private practice during the Depression had pushed some physicians into public health; others were attracted by the new availability of fellowships or by increased social awareness of the plight of the poor and their need for public health services. For the first time, physicians eagerly sought public health positions and competed on civil service examinations. In 1939, the Federal Government allocated over $8 million for maternal and child health programmes, over $9 million for general public health work, and over $4 million for venereal disease control.

Several important trends were stimulated by these federal funds: first was the development of programmes to control specific diseases and for services targeted to specific population groups—the 'categorical' approach to public health. Second was the expansion in the number of local health departments; third, the increased training of personnel; and fourth, the assumption of responsibility for some phases of medical care on the part of health departments (Corwin 1949). The Children's Bureau, led by Martha May Eliot, played an especially important national role in focusing attention on the health of women and children. The categorical approach to public health proved politically popular; Members of Congress were willing to vote funds for specific diseases or for particular groups—health and welfare services for children were especially appealing—but they showed less interest in general public health or administrative expenditures. Although state health officers often felt constrained by targeted programmes, they rarely refused federal grants-in-aid and thus adapted their programmes to the available pattern of funds. Federal grants came for maternal and child health services and crippled children (1935), venereal disease control (1938), tuberculosis (1944), mental health (1947), industrial hygiene (1947), and dental health (1947). The pattern of funding started in the 1930s would thus shape the organization of public health departments through the postwar period. As institutionalized in the National Institutes of Health, it would also shape the future patterns of biomedical research.

Public health and the Second World War

Mobilization for war acted as another major force in the expansion and development of public health in the US. Public health was declared a national priority for the armed forces and for the civilian population engaged in military production. As James Stevens Simmons (1943), Brigadier General and Director of the Preventive Medicine Division of the US Army, announced: 'A civil population that is not healthy cannot be prosperous and will lag behind in the economic competition between nations. This is even more true of a military population, for any army that has its strength sapped by disease is in no condition to withstand the attack of a virile force that has conserved its strength and is enjoying the vigor and exhilaration of health'.

Politicians were now willing to vote appropriations for public health as essential to national defence (Mustard 1941). For the first time, the Federal Government provided maternal and child health services to the families of recruits to the armed services. As the Federal Government began planning and organizing many new health programmes, State and local governments were urged to orient their activities to the war effort (Burgdorf 1943; Swartout 1944). Joseph W. Mountin (1942), Assistant Surgeon-General, underlined the sense of urgency: 'If a machine is idle because the worker who should tend it is sick, that machine is doing a job for Hitler'. Health departments again suffered from a critical shortage of personnel as physicians, nurses, engineers, and other trained and experienced professionals joined the armed services (Mountin 1943).

The US Public Health Service in 1940 expanded its programmes of grants to States and local communities, sending personnel to particularly needy areas (Furman 1973; Williams 1951). The Community Facilities Act provided $300 million in funds for construction of public works, including health and sanitation facilities, in communities with rapidly expanding populations because of military camps and war industries. The Office of Defense Health and Welfare Services, created as part of the Office for Emergency Management, co-ordinated efforts to protect the health of the nation at war, while the Office of Scientific Research and Development organized the national research effort (Sullivan 1942).

The shock of the discovery that many of the young men being called into the Army were physically unfit for military service provided a powerful impetus for increased national attention to public health. Fully 40 per cent of the young men being examined by the Selective Service Boards had been declared physically or mentally unfit for service (Perrott 1944). The Selective Service examinations represented the most massive health survey ever undertaken, with over 16 000 000 young men examined. The leading causes of rejection were defective teeth, vision problems, orthopaedic impairments (e.g. from polio), diseases of the cardiovascular system, nervous and mental diseases, hernia, tuberculosis, and venereal diseases. As G. St. J. Perrott (1946) of the US Public Health Service noted, mortality from tuberculosis and some other infectious diseases had declined since the First World War, but morbidity rates had changed little, if at all: in both cases, war had demonstrated an enormous amount of ill-health in the American population. The war also brought increasing public awareness and discussion of health issues (Underwood 1942). As one health officer remarked: 'The public is more health-minded at the present time than ever before in our generation. Publicity of the results of Selective Service examinations, the national nutrition program, more widespread pre-employment health examinations, mass health surveys, Workmen's Compensation legislation . . . and many others have made health a news item' (Burgdorf 1943). Hubert O. Swartout (1944), Health Officer of Los Angeles County, said the public was more anxious about health issues than ever before: ' . . . there is the public, more upset and more fearful about almost everything now than ever before, and more often and more insistently than ever demanding that the health department "do something about it" '.

Major population shifts had occurred with the mobilization for war, with the movement of both troops and workers to areas with defence industry plants. Peaceful villages near army camps had turned into boom towns and many cities had doubled their population within a couple of years (Maxcy 1942). In many cases, the existing infrastructure of water supplies and sewerage systems were completely inadequate to cope with these population movements (Goudey 1941; Weir 1945). Army training camps had often been placed in areas with warm climates, where the *Anopheles* mosquito bred in profusion and malaria was endemic. In order to control malaria in the South, the Public Health Service established the Center for the Control of Malaria in War Areas. After the war, when substantial funds were made available for malaria eradication efforts, this organization was transformed into the Centers for Disease Control which would come to play the dominant national role in the effort to control both infectious and non-infectious diseases.

Early in the war, the US Army began preparations for a possible emergency such as the influenza pandemic of 1918. In 1941, the division of preventive medicine service in the Surgeon-General's office created the Board for the Investigation and Control of Influenza and other Epidemic Diseases in the Army (Paul 1973). The Board recruited academic epidemiologists, clinicians, and clinical investigators to serve on its various commissions. These commissions made considerable progress during the Second World War in their investigations and efforts to control influenza, meningococcal meningitis, measles, mumps, sandfly fever, hepatitis, louse-borne and scrub typhus, malaria, and filariasis—among other infections.

Post-war reorganization

In the immediate post-war period, considerable optimism and energy were devoted to the possible reorganization of public health and medical care. Many of the discussions of a future national medical care system posited the potential unification of preventive and curative medicine (Viseltear 1973). Some public health leaders were advocating the direct administration of tax-supported medical care by health departments. Others opposed such a development, feeling that if public health and medical care adminstration were combined, preventive and educational efforts would be submerged under the demand for costly therapeutic services (Stern 1946).

While public health officials were debating whether they wanted to take responsibility for medical care services, the American Hospital Association and the American Medical Association had been strongly pushing a bill for the construction of hospitals. The Hospital Survey and Construction Act, more popularly known as the Hill–Burton Act, after Senators Lister Hill of Alabama and Harold H. Burton of Ohio, was passed in 1946 with the support of many public health leaders. Hospital construction, especially in rural areas, promised to bring the benefits of medical science to all the people—without in any way disturbing the freedoms of the medical profession or the patterns of paying for their services. One-third of the costs of building hospitals would be paid by the Federal Government, with $75 million set aside for each of the first five years. The largest building programme ever initiated in the health sector, it was highly popular with the organized health professions.

Although the Act included funds for public health centres, the public health component of the bill was far overshadowed by its emphasis on acute care facilities. Hill–Burton was a symbol of the national demand for access to medical care and in no way challenged the private organization of medical practice nor the relative disregard of public health services. The US could have been completely served by local health departments for a fraction of the cost of Hill–Burton, had there been a strong political constituency for public health able to compete effectively for resources with curative medicine.

In 1942, the American Public Health Association had studied the problem of providing local health services across the nation and had formulated a plan for organizing district health units with appropriate staffing (Emerson 1945). Haven Emerson's report, *Local health units for the nation*, emphasized the fact that only two-thirds of the people of the US were served by local health units and estimated the cost of a modest but adequate basic health service for each of the 1197 additional units proposed. His committee argued that communities of over 50 000 people should be able, at the cost of only one dollar per caput, to employ the number and quality of persons necessary to assure reasonably adequate local services. A superior service could be provided for two dollars per caput (Emerson 1945). A survey of state health departments had meanwhile found that a multitude of agencies, state boards, and commissions were involved in public health activities, as many as 18 different agencies being involved in a single state (Mountin and Flook 1941, 1943). The money spent for public health work also varied widely, ranging from $0.13 per caput in Ohio to $1.68 in Delaware. In most cases, the states spending the largest sums were spending most of these funds on hospital services rather than on prevention.

The epidemiological transition

In the post-war years, the public health community clearly understood that the disease patterns of the country had changed: in 1900, the leading causes of death had been tuberculosis, pneumonia, diarrhoeal diseases, and enteritis; by 1946, the leading causes of death were heart disease, cancer, and accidents. Great progress had been made in controlling infectious diseases and especially the diseases of early childhood, but more attention had now to be paid to the diseases of middle and later life. Recognition of the importance of the chronic diseases had been growing in the 1930s but had been temporarily eclipsed by the more urgent demands of infectious disease control during the war. With the return to peace, however, health departments recognized that they must now come to terms with the prevalence of chronic diseases. Health departments realized that communicable disease control no longer provided a sufficient *raison d'être*; the major infectious disease problems of the early twentieth century—tuberculosis, syphilis, typhoid, and diphtheria—were now effectively controlled. At the same time, health departments had little idea of how to prevent cancer or heart disease. There was little agreement or clarity about the relevance of nutritional, occupational, or the newer environmental health factors. When federal grants were made available for cancer control in 1946 and for heart disease in 1948, the main approach to prevention advocated screening and early diagnosis. Such programmes drew public health departments into providing medical services. Many public health officers were unhappy about becoming involved in the direct provision of medical care, uncertain about the best approach to chronic disease problems, and unsure in dealing with new and uncharted territories. Harry Mustard (1945) warned that if medical care became the responsibility of public health departments, 'Many comfortably established routines would be rudely shaken and it is possible that a hypertrophied medical-care tail would soon wag the none too robust public health dog'.

Some representatives of preventive medicine further claimed that public health departments should have nothing to do with the chronic diseases. In their view, private medical practice was appropriate to the task of prevention on an individual level: 'For example, disturbances of the climacteric, cancer, metabolic disorders, such as diabetes, cardiovascular disease, and the like are utterly beyond the influence of public health measures . . . prevention, early detection, and control of these disorders are and will continue to be individualistic and not *en masse*' (Stieglitz 1945). To most physicians, preventive medicine meant periodical examinations—and they preferred to keep the State out of the doctor's office.

Struggles over the definition of preventive medicine mirrored the broader tension between medicine and public health. Charles-Edward A. Winslow, professor of public health at Yale, thus defined preventive medicine as part of public health, whereas John R. Paul, Yale's professor of preventive medicine, defined it as part of clinical medicine (Viseltear 1982*b*). In the first view, preventive medicine must be allied to epidemiology as part of a broad perspective on community health; in the second, preventive medicine started with an individual sick patient and examined the social or family context in which sickness occurred with the aim of making clinical practice more effective. The first was the view of public health, the second, of clinical medicine.

Social medicine

In the late 1940s and early 1950s, some American public health people welcomed the concept of social medicine as seeming to offer a fresh perspective on the problems of chronic illness. Iago Galdston, Secretary of the New York Academy of Medicine, organized the Institute on Social Medicine in 1947, later publishing the papers as *Social medicine: Its derivations and objectives* (Galdston 1949). In the introduction to that volume, Howard Reid Craig, Director of the New York Academy, described social medicine as the integration of the social and medical sciences, and reassured physicians that it had little to do with socialized medicine (Galdston 1949). John A. Ryle, Professor of Social Medicine at Oxford University, emphasized the distinctions between the new social medicine and the old public health. Public health, he said, was concerned with environmental improvement, while social medicine extended its view to 'the whole of the economic, nutritional, occupational, educational, and psychological opportunity or experience of the individual or of the community' (Ryle 1949). Whereas public health was concerned with communicable diseases, social medicine was concerned with all diseases—ulcers and rheumatism, heart disease and cancer, neuroses and accidents. Ryle stated that social medicine, in close alliance with clinical practice, posed the exciting challenge of the future. Edward J. Stieglitz responded that social medicine was far from a new field; it was simply 'public health maturing . . . The maiden, public health, has married sociology and changed her name' (Stieglitz 1949). Ernest L. Stebbins (1949), Dean of the Johns Hopkins School of Public Health, argued that epidemiology was the essential discipline for dealing with both chronic and infectious diseases. Margaret Merrell and Lowell J. Reed (1949), statisticians from the Hopkins School, made a similar point in a brief paper that would become a classic statement on 'The epidemiology of health'. They suggested a graded scale for measuring degrees of health, and not simply the absence of illness: 'On such a scale people would be classified from those who are in top-notch condition with abundant energy, through the people who are well, to fairly well, down to people who are feeling rather poorly, and finally to the definitely ill' (Merrell and Reed 1949).

The idea that health could be quantitatively measurable, and that health could be advanced in the total absence of disease, helped make connections between the new social medicine and the older public health. Epidemiology, broadening its scope to place more emphasis on the social environment, became newly fashionable as 'medical ecology'. The terms 'ecology of health' and 'epidemiology of health' were also popular (Corwin 1949; Galdston 1953). John E. Gordon (1950, 1953), Professor of Preventive Medicine and Epidemiology at the Harvard School of Public Health and a prominent exponent of the 'newer epidemiology', explained how the triumvirate of 'environment, host, and disease' could be applied to non-communicable organic diseases such as pellagra, cancer, psychosomatic conditions, traumatic injuries, and accidents. The notion of a single cause of disease (the agent) was now firmly rejected in favour of multiple causation (Stebbins 1949).

The small but energetic group of people interested in social medicine was optimistic about the possibilities for new approaches to the chronic diseases, for the integration of preventive and curative medicine, and for the extension of comprehensive health programmes to the whole population (Smillie 1950). But others were less enthusiastic. Eli Ginzberg (1950), for example, introduced a tone of pessimism and caution: 'As an economist, I would like to stress my concern with the difficulties which I see in implementing the many good programs that have been sponsored. Adequate personnel resources are not available'. Ginzberg was sceptical about any statements that preventive medicine and curative medicine were becoming integrated: he noted that in New York State, an excess of $1.5 billion was being spent annually on medical budgets, whereas the combined public health budgets amounted to only $36 million, or 2.4 per cent of the total (Ginzberg 1950). Even in England, he said, the British Health Service was neglecting preventive services in favour of curative medicine. Ginzberg warned optimistic thinkers of an 'antigovernment attitude' in the US and of the prevalent assumption that health depended on medical care, with the ever-increasing provision of doctors and hospital beds. He urged public health professionals to do a more effective job of persuading the public that advances in diet, housing, and public health nursing were more important to health than the construction of hospitals. He noted that while hospitals were being built across the country, local health officer positions stood vacant because communities refused to provide reasonable salaries.

Ginzberg's prognosis proved correct in the political climate of the 1950s. The hopes and aspirations of the advocates of social medicine were not translated into effective health programmes; as acute care facilities and biomedical research expanded dramatically in the post-war period, public health departments struggled to maintain their programmes on inadequate budgets and with little political support. The post-war reconstruction meant massive expenditures for biomedical research and hospitals, the partial payment for medical care by expanding private insurance coverage, the relative neglect of public health services, and a complete failure to implement the more radical ideas of social medicine through focused attention on the social determinants of health and disease.

The decline of public health in the post-war era

There are many reasons why the US moved towards ever more sophisticated biomedical research and high technology medicine in the post-war era. Stephen Strickland (1972), among others, has examined the politics of research funding and shown how the priorities of research were set by the basic sciences and clinical medicine. Perhaps the more idealistic visions of social medicine, of the integration of the social and biomedical sciences in relation to health, and of the integration of preventive and curative services in a reorganized health system, were unrealistic given the balance of political forces both inside and outside medicine. But some of the blame for the neglect of public health must be shared by the public health profession for failing to articulate its goals and programmes, to build the political support it needed, to communicate more effectively to the general public, to document the cost-benefit of public health activities, or to promote a strong and persuasive case for adequate funding of public health services.

The Committee on Medicine and the Changing Order (Report of the New York Academy of Medicine 1947), supported by the Commonwealth Fund, the Milbank Memorial Fund, and the Josiah Macy Jr. Foundation, recommended the extension of public health services but also noted that the quality of public health officers must be improved by better recruitment, training, assured tenure, and adequate salaries: 'At present, most state health officers are subject to appointment by each succeeding governor. The mean salary of state health officers is now only a little over $5000, which is certainly inadequate. Only when tenure and better salaries are provided will it be possible to attract the more competent individuals and to insist upon higher training standards'.

Harry Mustard (1945) argued that the problems of public health were largely political. State health officers were of relatively low-grade rank in the hierarchy of state officials, and were limited in their freedom to introduce new proposals: 'strained relations are inevitable if the state health officer does not hold the furthering of his health program within the bounds of the governor's interests, which is not necessarily great'. Too often, state health officers accepted political constraints and bureaucratic barriers as natural and inevitable. Too seldom were they willing to risk their positions by appealing to a larger constituency.

In retrospect, it seems clear that public health failed to claim sufficient credit for controlling infectious diseases. The major scientific achievements of the war in relation to health—the discovery of penicillin and the use of DDT—were especially relevant to public health. In popular perception, however, scientific medicine took credit for both the specific wartime discoveries and the longer history of combating epidemic disease; in public relations terms, medicine and biomedical research seized the public glory, the political interest, and the financial support given for further anticipated health improvements in the post-war world.

Public health departments needed to claim some share of the credit for declining infectious diseases and to move quickly to develop programmes for the chronic diseases. Most health departments did neither of these things but simply continued running the same programmes and clinics within already established bureaucratic structures. There were some notable exceptions, such as the pioneering efforts of the New York State and California State Health Departments in chronic disease control. The more populous and progressive states were able to respond more effectively to new public health problems.

The political atmosphere of the 1950s was not, however, supportive of public health initiatives, and most health department budgets were stagnant, without the funding needed to develop broad new health programmes. Health departments did implement, or try to implement, one important, new and very cost-effective public health measure; the fluoridation of water supplies to protect children's teeth. But despite virtually unanimous support from scientific authorities and professional organizations, fluoridation was effectively halted in many cities and towns through vocal local opposition. Since such a simple and obviously effective measure could be so energetically opposed, health departments must have perceived the difficulty of instituting more adventurous or expensive interventions.

The one great triumph of the 1950s was the successful development of the polio vaccine and its implementation on a mass scale. The success of the polio campaign was in large part due to private funding and a massive public relations campaign by the Foundation for Infantile Paralysis which raised public awareness and developed public support, interest, and enthusiasm. The appeal for crippled children proved extremely popular and the polio vaccination campaign, despite some major setbacks, was a remarkable success (Paul 1971).

Despite this public triumph, the real expenditures of public health departments in the 1950s failed to keep pace with the increase in population (Sanders 1959). Federal grants-in-aid to the states for public health programmes steadily declined, the total amounts falling from $45 million in 1950 to $33 million in 1959. Given inflation, the decline in purchasing power of these dollars was even more dramatic (Terris 1959). At a time when public health officials were facing a whole series of new and poorly understood health problems, they were also underbudgeted and understaffed. Jesse Aronson (1959), director of local health services in New Jersey, offered an honest if rather devastating account of the state of small public health departments: 'The full-time health officer is frequently, because of inadequate budget and staff, limited in his activities to a series of routine clinical responsibilities in a child health station, a tuberculosis clinic, a venereal disease clinic, an immunization session, and communicable disease diagnosis and treatment. He has little or no time for community health education, the study of health problems and trends, the initiation of newer programs in diabetes control, cancer control, rheumatic fever prophylaxis, nutrition education, and radiation control. In a great many areas the health officer position has been vacant year after year with little real hope of filling it. In these situations, even the pretense of public health leadership is left behind and local medical practitioners provide these services on an hourly basis'.

Many state legislatures were setting up new agencies to build nursing homes, abate water pollution, or promote mental health, and simply bypassed the health department as not active or not interested in these issues. Public health officials were expressing 'frustrations, disappointments, dissatisfactions, and discontentments' said John W. Knutson (1957), in his Presidential Address to the American Public Health Association. Public health officials, he said, must develop more imagination, political skills, and knowledge of human motivation and behaviour. 'Our graduate students', he said, 'need a better understanding of the social and political forces swirling about them as they work professionally. In place of rapidly outdated factual information, I would have them gain from graduate education in public health a fundamental knowledge of four broad fields: cultural anthropology, human ecology, epidemiology, and biostatistics' (Knutson 1957). Yet even with the best possible preparation, the bureaucratic controls of state health departments tended to insure conformity and to discourage young professionals from initiating new programmes or activities.

The annual meeting of the American Public Health Association for 1955 adopted the theme 'Where are We Going in Public Health?' Leonard Woodcock, international vice-president of the United Auto Workers, opened the meeting. In a rousing speech, Woodcock (1956) both praised public health and damned public health departments. State divisions of occupational health were hopelessly under-financed, he said, workplace inspections were rarely carried out, and occupational disease reporting was totally inadequate. Labour leaders had the impression that health departments had no leaders with the passion and commitment for which Alice Hamilton had been so admired.

Labour unions wanted public health officials to assess the quality of medical services, evaluate prepayment plans, and provide comprehensive preventive services. The public health departments, complained Woodcock, were standing aside and letting commercial insurance companies dictate the future of medical care services (Woodcock 1956). Subsequent discussions showed that many of those present at the American Public Health Association meetings admitted the truth of Woodcock's critique (Leavell 1956).

Most commentators agreed that public health departments should become active in the new areas of public health—the chronic diseases, including rehabilitation, mental health, industrial health, accident prevention, environmental issues, and medical care. But many who accepted the new consensus seemed to have few specific ideas of how to begin. Herman E. Hilleboe (1955), the New York State Commissioner of Health, was one of few to report new programmes of research on the chronic diseases. The New York State Health Department had started to examine all male state employees between the ages of 40 and 55 years, to learn how best to detect cardiac disease at an early stage and to develop effective and economical methods of screening for cardiac problems. The State Health Department had also built facilities at the Roswell Park Memorial Institute in Buffalo with research facilities, a hospital, laboratories, and animal farms, to support research on the causes and treatment of cancer. Hilleboe mentioned other new state chronic disease programmes such as multiple screening efforts in Virginia and pilot studies of atherosclerosis and nutrition in Minnesota; thus, while some states seemed bemused by the prospect of having to confront the chronic diseases, a few were already initiating new programmes. The California State Health Department set a positive example through its pioneering efforts in setting up the Bureau for Chronic Disease. Hilleboe (1957) urged more states to undertake research on new and more effective methods of public health practice.

A few years later, Milton Terris (1959) offered a forceful summary statement of the dilemma of public health. The communicable diseases were disappearing; their place had been taken by the non-infectious diseases which the public health profession was ill-prepared to prevent or control. The public understood the fact that research was crucial, and federal expenditures for medical research had multiplied from $28 million in 1947 to $186 million 10 years later. This money, however, was being spent for clinical and laboratory research; there was little understanding of the importance of epidemiological studies in addressing these problems. Schools of public health had been slow to deal with these issues, as had health departments, with a few notable exceptions, as in the case of New York and California. Yet even the small sums spent on epidemiological research had produced dramatic successes: the discovery of the role of fluoride in preventing dental caries, the relation of cigarette smoking to lung cancer, and the suspected relation of serum cholesterol and physical exercise to coronary artery disease (Terris 1959).

In the late 1950s, public health leaders recognized and lamented the failure of their profession to assert a strong political presence or even to perceive the importance of politics to practical public health. The American Public Health Association devoted its annual meeting in 1958 to 'The politics of public health' (American Public Health Association Symposium 1959). The editor of the *American Journal of Public Health*, George Rosen, wrote that the education of public health workers should begin with teaching them to think politically about their work, to see it in relation to the political process: 'An understanding of the political process as it affects the handling of community health problems is as crucial an area of knowledge to health workers as the scientific underpinning of public health practice'. Raymond R. Tucker (1959), Mayor of St. Louis—the city hosting the American Public Health Association convention—insisted that public health officials learn not to confuse the organized vocal opposition of special interest groups with public opinion, for despite the noisy antagonism of a few, the general public solidly supported public health reforms. And at least some sociological studies had found that the majority of the public supported the expansion of public health services—more so than did the power élite in their communities (Prince 1958).

Schools of public health

In the 1950s, schools of public health had become increasingly distant from public health practice. Under the impetus of federal funding for research, mainly through the National Institutes of Health, the schools of public health had greatly expanded their research base through the financing of specific research projects. Federal funding supported basic research in the biomedical sciences, more than in the applied sciences, and schools of public health participated in this growth by strengthening their basic science departments. Money flowed for research rather than for teaching, and for basic research rather than for administration or practice. As a result, many schools of public health were starting to turn into research institutes, oriented toward the priorities of the National Institutes of Health and their associated scientific communities, rather than to the practical problems and needs of health departments. Those who were selected and rewarded for their abilities in research were often relatively uninterested in the immediate problems and activities of local health departments.

The numbers of students being trained at schools of public health fell by half in the period from 1946 to 1956. In 1956, the US Congress recognized that public health departments were suffering from a severe lack of adequately trained personnel. The Senate Committee on Labor and Public Welfare noted a 'startling and shocking drop' in the annual number of public health trainees between 1947 and 1955 (Report of the National Conference on Public Health Training 1958). The Health Amendment Act then authorized the Public Health Service to provide a programme of traineeships in public health for physicians, nurses, sanitary engineers, nutritionists, medical social workers, dentists, health educators, veterinarians, and sanitarians.

In 1958, the official public health agencies still reported well over 2500 vacancies in professional categories due to the lack of trained personnel. In some areas, there were as many as four vacant positions for every trained applicant and many vacancies had thus been filled with inadequately trained people (Ennes 1957). Over 20 000 professional workers employed by governmental and voluntary health agencies had absolutely no specialized training in public health. Few had any idea of how to cope with new health hazards such as radiation: 'The public health agencies are not staffed as they should be now. Much less are they prepared for even the known hazards that lie ahead' (Report of the National Conference on Public Health Training 1958).

The new funds for public health training undoubtedly helped address a critical situation as more than 1000 new traineeships were provided in 1957 and 1958. The Association of Schools of Public Health lobbied energetically and effectively for additional funding and received sympathetic attention from US Congressmen Hill and Rhodes. The Hill–Rhodes Act authorized $1 million per year to the schools of public health, and the First National Conference on Public Health Training held in Washington in 1958 suggested that $15 million was needed to aid construction costs for new teaching facilities and for training grants. Its report (Report of the National Conference on Public Health Training 1958) concluded with a stirring appeal to consider public health training as an important aspect of national defence: 'The great crises of the future may not come from a foreign enemy . . . 'D' day for disease and death is everyday. The battle line is already in our own community. To hold that battle line we must daily depend on specially trained physicians, nurses, biochemists, public health engineers, and other specialists properly organized for the normal protection of the homes, the schools, and the work places of some unidentified city somewhere in America. That city has, today, neither the personnel nor the resources of knowledge necessary to protect it'.

There were then ten accredited schools of public health in the US at Berkeley, Columbia, Harvard, Johns Hopkins, Michigan, Minnesota, North Carolina, Pittsburgh, Tulane, and Yale. None of them had adequate budgets. Faculty salaries were so low that advertised faculty positions went unfilled, and about 10 per cent of all faculty positions remained vacant (Smillie and Luginbuhl 1959). As federal support was limited almost entirely to research grants, the research budgets in most of the schools were increasing while the teaching budgets were static or declining. Some of the schools were tending to limit admissions to those students seeking a doctoral degree—thus emphasizing the shift from teaching to graduate research (Smillie and Luginbuhl 1959). More than half of the doctoral candidates were non-physicians studying for the degrees of Doctor of Science or Doctor of Philosophy in specialized scientific fields.

State-supported schools, which still placed a major emphasis on practical training for public health personnel, were having difficulty attracting physicians to their programmes and were increasingly admitting non-physicians to the Master of Public Health degree. None of the schools were training large numbers of physicians for public health practice. Nationally, any real increases in the student body came from public health nurses, health educators, sanitarians, and other types of non-physician specialists.

The 1960s and the war on poverty

The 1960s saw the collapse of the conservative complacency of the 1950s, the growing power of the civil rights movement, riots in the urban black ghettos, and federal support for the 'war on poverty'. The antipoverty effort and other Great Society programmes soon became deeply involved with medical care. Growing concern over access to medical care and hospitalization, especially by the elderly population, culminated in Medicare and Medicaid legislation in 1965 to cover medical care costs for those on social security and for the poor. Both programmes were built on a 'politics of accommodation' with private providers of medical care, thus increasing the incomes of physicians and hospitals and leading to spiralling costs for medical services (Starr 1982). Other antipoverty programmes, such as the neighbourhood health centres that were intended to encourage community participation in providing comprehensive care to underserved populations, fared less well because they were seen as competing with the interests of private care providers (Davis and Schoen 1978). Debates over the public responsibility to provide medical care for those unable to afford private services often echoed nineteenth century distinctions between the 'deserving' and the 'undeserving' poor (Rosner 1982).

Most of the new health and social programmes of the 1960s bypassed the structure of the public health agencies, and set up new agencies to mediate between the Federal

Government and local communities. Medicare and Medicaid reflected the usual priorities of the medical care system in favouring highly technical interventions and hospital care while failing to provide adequately for preventive services. Neighbourhood health centres and community based mental health services were established without reference to public health agencies. When environmental issues attracted public concern and political attention in the 1960s and 1970s, separate agencies were also created to respond to these concerns. At the federal level, the Environmental Protection Agency was created to deal more aggressively with such issues as solid wastes, pesticides, and radiation. At the State level, environmental agencies were often separate from public health departments and failed to reflect specific health concerns or public health expertise. Mental health agencies were often separate from public health agencies, in part, because they dealt with institutional treatment or custodial care of the mentally ill. The broader functions of public health were again split between numerous different agencies; losing a clear institutional base, public health also lost visibility and clarity of definition. For a field that depends so heavily on public understanding and support, this loss was disastrous.

The schools of public health responded to the new federal initiatives by developing new programmes in medical care administration, mental health, population control, and environmental health. Private foundations, such as Ford and Rockefeller, supported early international health and population control programmes, later followed by the Agency for International Development (AID). In the 1960s, enrolments at schools of public health climbed. Between 1958 and 1973, enrolments in schools of public health quadrupled. The vast majority of students were in masters degree programmes, but the numbers of doctoral candidates also quadrupled between 1960 and 1971. Beginning in the 1960s, there was a greatly expanded interest in population control and medical care administration, made possible by new federal grants in these areas (Hume 1974). The majority of students were now going into health adminstration, hospital adminstration, mental hygiene, health education, maternal and child care, and family planning. Many were attracted to the field of international health. About one-third of the graduates in the 1960s went into public health agencies, one-quarter into medical care facilities, one-quarter into colleges and universities, and the remainder into international health agencies and voluntary organizations (Sheps 1974). By 1960 there were 12 accredited schools of public health in the US, the new ones being California at Los Angeles, Puerto Rico, and Pittsburgh. Seven more schools were accredited between 1965 and 1972 at the universities of Hawaii, Oklahoma, Texas, Washington, Illinois, Massachusetts, and Loma Linda, and five more schools between 1973 and 1989 at the universities of Alabama, Boston, San Diego, South Carolina, and South Florida. Other schools are currently being established and some are in process of applying for accreditation.

By the 1960s, schools of public health had become dependent on the Federal Government and were training officials as much for federal health agencies as for local and State health departments. This orientation to federal programmes meant new growth and development for the schools but also made them vulnerable to changes in federal priorities; when federal funds were cut, the schools would have difficulty maintaining their programmes (Hogness 1974). The federal grant system, which had allowed the schools to expand their teaching and research programmes, also influenced their organization, function, and orientation. Programmes tended to appear and disappear in response to changing federal fashions in research funding (Ramsey 1974). By the early 1970s, more than half the schools' financial support came from the Federal Government, and in some schools, the figure was as high as 85 per cent (Bowers and Purcell 1974).

Public health today

In the 1970s public health departments became providers of last resort for uninsured patients and for Medicaid patients rejected by private practitioners. By 1988, almost three-quarters of all State and local health department expenditures went for personal health services (Institute of Medicine 1988). As Harry Mustard (1945) had predicted some 40 years earlier, direct provision of medical care absorbed much of the limited resources—in personnel, money, energy, time, and attention—of public health departments, leading to a slow starvation of public health and preventive activities. The problems of care for the uninsured and the indigent loomed so large that they eclipsed the need for a basic public health infrastructure in the minds of many legislators and the general public.

The organization of public health has continued to show huge variation among the States: in some States, public health may be combined with mental health, with environmental services, with Medicare and Medicaid agencies, or with social welfare agencies. Where public health activities are combined with personal health care or with social welfare services, public health tends to be swamped by these large and more expensive programmes. There continue to be large numbers of people serving in the public sector who are untrained in public health; salaries have remained uncompetitive so that it is difficult for health departments to attract and hold the most talented and best-trained people when they can readily find more attractive positions elsewhere. Those who stay in the field of public health do so out of a feeling of strong social commitment, often at considerable personal cost.

In the Reagan revolution of the 1980s, federal funding for public health programmes was cut. Through the mechanism of the block grants, power was returned to State health agencies, but in the context of funding cuts this was the unpopular power to cut existing programmes (Omenn 1982). In the middle of general budget cuts, State health departments were often left the task of managing Medicaid programmes or of delivering personal health services to the uninsured and indigent populations. State health departments also had to

MCW Libraries

attempt to deal with the adverse health consequences of reductions in other social programmes, and the problems of a growing poverty population, as evidenced in drug abuse, alcoholism, teenage pregnancy, family violence, and homelessness, as well as the health and social needs of growing populations of illegal immigrants.

As Daniel Fox (1988) has argued, the acquired immune deficiency syndrome (AIDS) epidemic made obvious a national crisis of authority in the health system and revealed the structural contradictions and weaknesses of national and federal health policy—or lack thereof. For State and local health agencies, the AIDS epidemic exacerbated existing problems and also gave a new visibility and urgency to their public health efforts. The public health community agreed that a major national effort was needed in education and prevention. Much of the AIDS funding, when it did come, went into research and medical care; as usual, education and prevention received much less attention. At the same time, the mobilization of public concern provides an opportunity for renewed attention to public health and increased political support. The recent report by the Institute of Medicine (1988), *The future of public health*, notes that: 'In a free society public activities ultimately rest on public understanding and support, not on the technical judgement of experts. Expertise is made effective only when it is combined with sufficient public support, a connection acted upon effectively by the early leaders of public health'. The growth in the technical knowledge of public health in the past 100 years has been extraordinary—and insufficiently addressed in this brief account—but our ability to implement this knowledge in health and social reform has but little advanced. Public health today needs to mobilize public support, to communicate forcefully the broader needs and importance of public health, to build an effective public health infrastructure to respond to endemic health and social problems as well as to new crises, and to demonstrate convincingly the benefits of prevention to its constituency, the public at large.

Acknowledgements

The author would like to thank D. A. Henderson and Myron Wegman for their helpful suggestions on an earlier version of this chapter.

References

Abbott, S.W. (1900). *The past and present conditions of public hygiene and state medicine in the United States*. Wright and Potter, Boston, Massachusetts.

Ackerknecht, E.L. (1948). Anticontagionism between 1821 and 1867. *Bulletin of the History of Medicine* **22**, 562.

Adams, G.W. (1952). *Doctors in blue*. Henry Schuman, New York.

American Public Health Association Symposium 1958 (1959). The politics of public health. *American Journal of Public Health* **49**, 300.

Aronson, J.B. (1959). The politics of public health—reactions and summary. *American Journal of Public Health* **49**, 311.

Blake, J. (1959). *Public health in the town of Boston, 1630–1822*. Harvard University Press, Cambridge, Massachusetts.

Bowers, J.Z. and Purcell, E.F. (1974). Summary. In *Schools of public health: Present and future* (ed. J.Z. Bowers and E.F. Purcell) p. 172. Josiah Macy Jr. Foundation, New York.

Brown, E.R. (1979). *Rockefeller medicine men*. University of California Press, Berkeley, California.

Burgdorf, A.L. (1943). War and the health department. *American Journal of Public Health* **33**, 26.

Cassedy, J.H. (1962). *Charles v. Chapin and the public health movement*. Harvard University Press, Cambridge, Massachusetts.

Chapin, C.V. (1901). *Municipal sanitation in the United States*. Snow and Farnham, Providence, Rhode Island.

Chapin, C.V. (1910). *The sources and modes of infection*. John Wiley and Sons, New York.

Chapin, C.V. (1934*a*). Pleasures and hopes of the health officer. In *Papers of Charles v. Chapin, MD*, p. 11. Compiled by F.P. Gorham and C.L. Scamman, New York.

Chapin, C.V. (1934*b*). How shall we spend the health appropriation? In *Papers of Charles v. Chapin, MD*, p. 28. Compiled by F.P. Gorham and C.L. Scamman, New York.

Corwin, E.H.L. (ed.) (1949). *Ecology of health*. The Commonwealth Fund, New York.

Cosmas, G.A. (1971). *An army for empire: The United States Army in the Spanish–American War*. University of Missouri Press, Columbia, Missouri.

Cunningham, H.H. (1958). *Doctors in gray: The Confederate medical service*. Louisiana State University Press, Baton Rouge, Florida.

Curran, J. (1970). *Founders of the Harvard School of Public Health, with biographical notes, 1909–1946*. Josiah Macy Jr. Foundation, New York.

Davis, K. and Schoen, C. (1978). *Health and the war on poverty*. Brookings Institution, Washington, DC.

Duffy, J. (1968). *A history of public health in New York City, 1625–1866*. Russell Sage Foundation, New York.

Duffy, J. (1971). Social impact of disease in the late nineteenth century. *Bulletin of the New York Academy of Medicine* **47**, 797.

Duffy, J. (1974). *A history of public health in New York City, 1866–1966*. Russell Sage Foundation, New York.

Duffy, J. (1979). The American medical profession and public health: From support to ambivalence. *Bulletin of the History of Medicine* **53**, 1.

Emerson, H. (1945). *Local health units for the nation*. The Commonwealth Fund, New York.

Ennes, H. (1957). Manpower—The Achilles' heel in public health. *American Journal of Public Health* **47**, 1390.

Ettling, J. (1981). *The germ of laziness: Rockefeller philanthropy and public health in the new South*. Harvard University Press, Cambridge, Massachusetts.

Fee, E. (1983). Competition for the first school of hygiene and public health. *Bulletin of the History of Medicine* **57**, 339.

Fee, E. (1987). *Disease and discovery: A history of the Johns Hopkins School of Hygiene and Public Health, 1916–1939*. The Johns Hopkins University Press, Baltimore, Maryland.

Fisher, I. (1909). *A report on national vitality, its wastes and conservation* (bulletin 30). Committee of One Hundred on National Health, Government Printing Office, Washington, DC.

Fosdick, R.B. (1962). *Adventure in giving: The story of the General Education Board*. Harper and Row, New York and Evanston.

Fox, D.M. (1988). AIDS and the American health polity: The history and prospects of a crisis of authority. In *AIDS: The*

Burdens of History (ed. E. Fee and D.M. Fox) p. 316. University of California Press, Berkeley, California.

Furman, B. (1973). *A profile of the public health service, 1798–1948*, p. 418. National Institutes of Health, Bethesda, Maryland.

Galdston, I. (ed.) (1949). *Social medicine: Its derivations and objectives*. The Commonwealth Fund, New York.

Galdston, I. (1953). *Epidemiology of health*. New York Academy of Medicine, Health Education Council, New York.

Galishoff, S. (1975). *Safeguarding the public health: Newark, 1895–1918*. Greenwood Press, Westport, Connecticut.

Gates, J.M. (1973). *Schoolbooks and krags: The United States Army in the Philippines, 1898–1902*. Greenwood Press, Westport, Connecticut.

Gillett, M.C. (1987). Medical care and evacuation during the Philippine Insurrection, 1899–1901. *Journal of the History of Medicine and Allied Sciences* **42**, 169.

Ginzberg, E. (1950). Public health and the public. In *Tomorrow's horizon in public health*, p. 101. Transactions of the 1950 conference of the Public Health Association of New York City. Public Health Association, New York.

Gordon, J.E. (1950). The newer epidemiology. In *Tomorrow's horizon in public health*, p. 18. Transactions of the 1950 conference of the Public Health Association of New York City. Public Health Association, New York.

Gordon, J.E. (1953). The world, the flesh and the devil as environment, host and agent of disease. In *The epidemiology of health* (ed. I. Galdston) p. 60. New York Academy of Medicine, Health Education Council, New York.

Goudey, R.F. (1941). Wartime protection of water supplies. *American Journal of Public Health* **31**, 1174.

Haines, M.R. (1979). The use of model life tables to estimate mortality for the United States in the late nineteenth century. *Demography* **16**, 289.

Hamilton, A. (1943). *Exploring the dangerous trades: The autobiography of Alice Hamilton, M.D.* Little, Brown, Boston, Massachusetts.

Harden, V.A. (1986). *Inventing the NIH: Federal biomedical research policy, 1887–1937*. The Johns Hopkins University Press, Baltimore, Maryland.

Hays, S.P. (1968). *Conservation and the gospel of efficiency: The progressive conservation movement, 1890–1918*. Beacon Press, Boston, Massachusetts.

Hays, S.P. (1980). The politics of reform in municipal government in the progressive era. In *American political history as social analysis*, p. 205. University of Tennessee, Knoxville, Tennessee.

Hill, H.W. (1916). *The new public health*. Macmillan, New York.

Hilleboe, H.E. (1955). Public health in a changing world. *American Journal of Public Health* **45**, 1517.

Hilleboe, H.E. (1957). Editorial: Research in public health practice. *American Journal of Public Health* **47**, 216.

Hoffman, F.L. (1907). The general death rate of large American cities, 1871–1904. *Publications of the American Statistical Association* **10**, 1.

Hogness, J.R. (1974). The future of schools of public health in relation to government. In *Schools of Public Health* (ed. J.Z. Bowers and E.F. Purcell) p. 124. Josiah Macy Jr. Foundation, New York.

Howard, W.T. (1924). *Public health administration and the natural history of disease in Baltimore, Maryland, 1797–1920*. Carnegie Institution, Washington, DC.

Hume, J.C. (1974). The future of schools of public health: The Johns Hopkins School of Hygiene and Public Health. In *Schools of public health: Present and future* (ed. J.Z. Bowers and E.F. Purcell) p. 64. Josiah Macy Jr. Foundation, New York.

Institute of Medicine, Committee for the Study of the Future of Public Health (1988). *The future of public health*. National Academy Press, Washington, DC.

Jordan, E.O., Whipple, G.C., and Winslow, C.-E.A. (1924). *A pioneer of public health: William Thompson Sedgwick*. Yale University Press, New Haven, Connecticut.

Kelley, H.A. (1906). *Walter Reed and yellow fever*. Medical Standard Book Company, Baltimore, Maryland.

Koop, C.E. and Ginzburg, H.M. (1989). The revitalization of the Public Health Service Commissioned Corps. *Public Health Reports* **104**, 105.

Knowles, M. (1913). Public health service not a medical monopoly. *American Journal of Public Health* **3**, 111.

Knutson, J.W. (1957). Ferment in public health. *American Journal of Public Health* **47**, 1489.

Kratz, F.K. (1943). Status of full-time local health organizations at the end of the fiscal year 1941–1942. *Public Health Reports* **58**, 345.

Leathers, W.S., McIver, P., Smillie, W.G., *et al.* (1937). Committee on Professional Education of the American Public Health Association. Public health degrees and certificates granted in 1936. *American Journal of Public Health* **27**, 1267.

Leavell, H.R. (1956). Where are we going in public health? Association Symposium (Part II)—Resolving the basic issues. *American Journal of Public Health* **46**, 408.

Leavitt, J.W. (1982). *The healthiest city: Milwaukee and the politics of health reform*. Princeton University Press, Princeton, New Jersey.

MacNutt, J.S. (1915). *A manual for health officers*, p. 85. John Wiley and Sons, New York.

Marcus, A.I. (1979). Disease prevention in America: From a local to a national outlook, 1880–1910. *Bulletin of the History of Medicine* **53**, 184.

Maxcy, K.F. (1942). Epidemiologic implications of wartime population shifts. *American Journal of Public Health* **32**, 1089.

Meeker, E. (1972). The improving health of the United States, 1850–1915. *Explorations in Economic History* **9**, 353.

Merrell, M. and Reed, L.J. (1949). The epidemiology of health. *Social medicine: Its derivations and objectives*, p. 105. The Commonwealth Fund, New York.

Mountin, J.W. (1942). Evaluation of health services in a national emergency. *American Journal of Public Health* **32**, 1128.

Mountin, J.W. (1943). Responsibility of local health authorities in the war effort. *American Journal of Public Health* **33**, 35.

Mountin, J.W. and Flook, E. (1941). Distribution of health services in the structure of state government: The composite pattern of state health services. *Public Health Reports* **56**, 1676.

Mountin, J.W. and Flook, E. (1943). Distribution of health services in the structure of state government: State health department organization. *Public Health Reports* **58**, 568.

Mustard, H.S. (1941) Editorial: Yesterday's school children are examined for the army. *American Journal of Public Health* **31**, 1207.

Mustard, H.S. (1945). *Government in public health*. The Commonwealth Fund, New York.

Omenn, G.S. (1982). What's behind those block grants in health? *New England Journal of Medicine* **306**, 1057.

Paterson, R.G. (ed.) (1939). *Historical directory of state health departments in the United States of America*. Ohio Public Health Association, Columbus, Ohio.

Paul, J.R. (1971). *A history of poliomyelitis*. Yale University Press, New Haven.

Paul, J.R. (1973). An account of the American Epidemiological Society: A retrospect of some fifty years. *Yale Journal of Biology and Medicine* **46**, 52.

Perrott, G. St. J. (1944). Findings of selective service examinations. *Milbank Memorial Fund Quarterly* **22**, 358.

Perrott, G. St. J. (1946). Selective service rejection statistics and some of their implications. *American Journal of Public Health* **36**, 336.

Pettenkofer, M. von (1941). *The value of health to a city* (translated and introduced by H.E. Sigerist). The Johns Hopkins University Press, Baltimore, Maryland.

Powell, J.H. (1949). *Bring out your dead: The great plague of yellow fever in Philadelphia in 1793*. University of Pennsylvania Press, Philadelphia, Pennsylvania.

Prince, J.S. (1958). A public philosophy in public health. *American Journal of Public Health* **48**, 903.

Proceedings and debates of the third national quarantine and sanitary conference (1859). Edward Jones, New York.

Public Health Bulletin, no. 136 (1923). Report of the Committee on Municipal Health Department Practice of the American Public Health Association, in cooperation with the United States Public Health Service. Government Printing Office, Washington, DC.

Ramsey, F.C. (1974). Observations of a recent graduate of a school of public health. In *Schools of public health* (ed. J.Z. Bowers and E.F. Purcell) p. 130. Josiah Macy Jr. Foundation, New York.

Ravenel, M.P. (1921). The American Public Health Association: Past, present, future. In *A half century of public health*, p. 13. American Public Health Association, New York.

Report of the National Conference on Public Health Training to the Surgeon General of the Public Health Service (1958). US Department of Health, Education, and Welfare, Washington, DC.

Report of the New York Academy of Medicine, Committee on Medicine in the Changing Order (1947). *Medicine in the changing order*. The Commonwealth Fund, New York.

Rogers, D.T. (1982). In search of progressivism. *Reviews in American History* **10**, 115.

Rose, W. (1910). *First annual report of the administrative secretary of the Rockefeller Sanitary Commission*, p. 4. Cited in Fosdick, R.B. (1952) *The story of the Rockefeller Foundation*, p. 33. Harper and Brothers, New York.

Rosen, G. (1959). Editorial: The politics of public health. *American Journal of Public Health* **49**, 364.

Rosen, G. (1972). The Committee of One Hundred on National Health and the campaign for a national health department, 1906–1912. *American Journal of Public Health* **62**, 261.

Rosenberg, C.E. (1962). *The cholera years: The United States in 1832, 1849 and 1866*. University of Chicago Press, Chicago.

Rosenberg, C.E. (1987). *The care of strangers: The rise of America's hospital system*. Basic Books, New York.

Rosenberg, C.E. and Rosenberg, C.S. (1968). Pietism and the origins of the American public health movement. *Journal of the History of Medicine and Allied Sciences* **23**, 16.

Rosenkrantz, B. (1972). *Public health and the state: Changing views in Massachusetts, 1842–1936*. Harvard University Press, Cambridge, Massachusetts.

Rosenkrantz, B. (1974). Cart before horse: Theory, practice and professional image in American public health. *Journal of the History of Medicine and Allied Sciences* **29**, 57.

Rosner, D. (1982). Health care for the 'truly needy': Nineteenth-century origins of the concept. *Milbank Memorial Fund Quarterly* **60**, 355.

Rotch, T.M. (1909). The position and work of the American Pediatric Society toward public questions. *Transactions of the American Pediatric Society* **21**, 12.

Ryan, M.P. (1975). *Womanhood in America: From colonial times to the present*. Franklin Watts, New York.

Ryle, J.A. (1949). Social pathology. In *Social medicine: Its derivations and objectives* (ed. Galdston, I.) p. 64. The Commonwealth Fund, New York.

Sanders, B.S. (1959). Local health departments: growth or illusion? *Public Health Reports* **74**, 13.

Savitt, T.L. (1988). Slave health and southern distinctiveness. In *Disease and distinctiveness in the American South* (ed. T.L. Savitt and J.H. Young) p. 120. University of Tennessee Press, Knoxville.

Schiesl, M.J. (1980). *The politics of efficiency: Municipal administration and reform in America, 1880–1920*. University of California Press, Berkeley, California.

Sedgwick, W.T. (1901). The origin, scope and significance of bacteriology. *Science* **13**, 121.

Sedgwick, W.T. (1924). Scientists and technicians in the public service. Cited in Jordan, E.O., Whipple, G.C., and Winslow, C.-E.A. (1924). *A pioneer of public health: William Thompson Sedgwick*, p. 133. Yale University Press, New Haven, Connecticut.

Sheps, C.G. (1974). Trends in schools of public health in the United States since World War II. In *Schools of Public Health: Present and Future* (ed. J.Z. Bowers and E.F. Purcell) p. 1. Josiah Macy Jr. Foundation, New York.

Shryock, R.H. (1937). The early American public health movement. *American Journal of Public Health* **27**, 965.

Sicherman, B. (1984). *Alice Hamilton: A life in letters*. Harvard University Press, Cambridge, Massachusetts.

Simmons, J.S. (1943). The preventive medicine program of the United States Army. *American Journal of Public Health* **33**, 931.

Smillie, W.G. (1950). The responsibility of the state. In *Tomorrow's horizon in public health*, p. 95. Transactions of the 1950 conference of the Public Health Association of New York City. Public Health Association, New York.

Smillie, W.G. (1955). *Public health: Its promise for the future*. Macmillan, New York.

Smillie, W.G. and Luginbuhl, M. (1959). Training of public health personnel in the United States and Canada: A summary of ten years' advance in schools of public health. *American Journal of Public Health* **49**, 455.

Smith, S. (1921). The history of public health, 1871–1921. In *A half century of public health* (ed. M.P. Ravenel) p. 1. American Public Health Association, New York.

Starr, P. (1982). *The social transformation of American medicine*. Basic Books, New York.

Stebbins, E.L. (1949). Epidemiology and social medicine. In *Social medicine: Its derivations and objectives* (ed. I. Galdston) p. 101. The Commonwealth Fund, New York.

Stern, B.J. (1946). *Medical services by government: Local, state, and federal*. The Commonwealth Fund, New York.

Sternberg, G.M. (1912*a*). Sanitary lessons of the war. In *Sanitary lessons of the war and other papers*, p. 2. Byron S. Adams, Washington, DC.

Sternberg, G.M. (1912*b*). Sanitary problems connected with the construction of the Isthmian Canal. In *Sanitary lessons of the war and other papers*, p. 39. Byron S. Adams, Washington, DC.

Stieglitz, E.J. (1945). *A future for preventive medicine*, p. 32. The Commonwealth Fund, New York.

Stieglitz, E.J. (1949). The integration of clinical and social medicine. In *Social medicine: Its derivations and objectives* (ed. I. Galdston) p. 76. The Commonwealth Fund, New York.

Strickland, S.P. (1972). *Politics, science, and dread disease*. Harvard University Press, Cambridge, Massachusetts.

Sullivan, F. (1942). Public health planning for war needs: Order or chaos? *American Journal of Public Health* **32**, 831.

Swartout, H.O. (1944). Wartime problems of a county health officer. *American Journal of Public Health* **34**, 379.

Tarr, J.A. (1979). The separate vs. combined sewer problem: A case study in urban technology design choice. *Journal of Urban History* **5**, 308.

Terris, M. (1959). The changing face of public health. *American Journal of Public Health* **49**, 1113.

Terris, M. (ed.) (1964). *Goldberger on pellagra*. Louisiana State University Press, Baton Rouge, Florida.

Tucker, R.R. (1959). The politics of public health. *American Journal of Public Health* **49**, 300.

Underwood, F.J. (1942). The role of public health in the national emergency. *American Journal of Public Health* **32**, 530.

Viseltear, A.J. (1973). The emergence of the Medical Care Section of the American Public Health Association, 1926–1948. *American Journal of Public Health* **63**, 986.

Viseltear, A.J. (1982*a*). C.-E.A. Winslow and the early years of public health at Yale, 1915–1925. *Yale Journal of Biology and Medicine* **55**, 137.

Viseltear, A.J. (1982*b*). John R. Paul and the definition of preventive medicine. *Yale Journal of Biology and Medicine* **55**, 167.

Viseltear, A.J. (1986). The Yale plan of medical education: The early years. *Yale Journal of Biology and Medicine* **59**, 627.

Weir, W.H. (1945). Lessons learned from the internal security program of the war department. *American Journal of Public Health* **35**, 353.

Welch, W.H. (1920). Sanitation in relation to the poor. An address to the Sanitation Organization Society of Baltimore, November 1892. In *Papers and addresses by William Henry Welch*, Vol. 3. The Johns Hopkins Press, Baltimore, Maryland.

Wiebe, R.H. (1967). *The search for order, 1877–1920*. Hill and Wang, New York.

Williams, G. (1976). Schools of public health: Their doing and undoing. *Milbank Memorial Fund Quarterly* **54**, 489.

Williams, R.C. (1951). *The United States Public Health Service, 1798–1950*. Government Printing Office, Washington, DC.

Winslow, C.-E.A. (1920). The untilled fields of public health. *Science* **51**, 23.

Winslow, C.-E.A. (1923). *The evolution and significance of the modern public health campaign*. Yale University Press, New Haven, Connecticut.

Winslow, C.-E.A. (1926). Public health at the crossroads. *American Journal of Public Health* **16**, 1075.

Winslow, C.-E.A. (1929). *The life of Hermann M. Biggs: Physician and statesman of the public health*. Lea and Febiger, Philadelphia, Pennsylvania.

Woodcock, L. (1956). Where are we going in public health? *American Journal of Public Health* **46**, 278.

Woodward, J.J. (1863). *Chief camp diseases of the United States armies*. J.B. Lippincott, Philadelphia, Pennsylvania.

2

The origins and development of public health in the UK

JANE LEWIS

The 1988 Government Committee of Inquiry into 'the future development of the public health function' defined public health as 'the science and art of prolonging life and promoting health through the organized efforts of society' (PP 1988, para 1.3). In practice the committee saw the task of public health as twofold: the prevention of disease and the promotion of health on the one hand, and the planning and evaluation of health services on the other. These twin pillars of public health are also identified in the preface to this volume. However, the history of British public health in the nineteenth and twentieth centuries shows clearly, first, that the balance between the two major emphases has shifted over time, and second, the extent to which there has been conflict, rather than compatibility, between them. These points are crucial to an understanding of the changes in the concept of public health reflected in the changes of nomenclature, with the introduction of social medicine during the Second World War and community medicine in the mid 1970s; in 1988, the committee of inquiry advocated a return to 'public health medicine'.

In fact, the public health movement did not become firmly medical until the last decade of the nineteenth century. The first section of this chapter looks at the 'heroic age' of public health and suggests the extent to which nineteenth century public health smacked more of what today might be termed a 'healthy public policy' approach, the most crucial element being the efforts made on the part of the central state (albeit in an often *ad hoc* and piecemeal fashion) to regulate—especially buildings, nuisances and foods—and of local authorities (albeit with substantial regional variation) to implement the sanitary idea. The encouragement given by Thomas McKeown's (1976) analysis to a monocausal explanation of the dramatic decline in death rates due to infectious disease in terms of rising living standards, and in particular standards of nutrition, has been subjected to severe scrutiny in recent work. Undoubtedly more credit is due to the collective efforts of nineteenth century government and administration to prevent disease and promote health.

The nineteenth century public health doctor was by no means absent from these endeavours, but nor was he necessarily the critical variable. As public health was professionalized, so its focus narrowed and, in keeping with medical practice, the focus in the early twentieth century became increasingly the individual and what was termed 'personal prevention'. In practice, as the second part of the chapter demonstrates, this became hard to distinguish from the work of other medical doctors, especially general practitioners, and by the inter-war period public health practitioners found themselves engaged in battles that were more related to the content and methods of delivering medical services than to prevention and promotion. Indeed, public health doctors were hopeful that the new National Health Service (NHS) would be organized around the local authorities and the clinic-based and municipal hospital services developed by public health departments during the 1920s and 1930s. But this was not to be and public health in the 1950s and 1960s found itself searching for an identity within a framework of a rigid and unreformed local government structure. The civic pride of the nineteenth century had all too often given way to municipal decline. Academic leaders in public health began to seek a new role for the specialty, first in social medicine, and then in medical administration. Finally, as the third part of the chapter shows, they promoted the development of community medicine as a specialty of population medicine that was immediately recognizable and significantly different from both general practice and hospital medicine.

When the new Faculty of Community Medicine (formed in 1972) referred to the specialty's origins, it usually mentioned the nineteenth century pioneers rather than the pre-1974 public health departments. There was a sense in which academics in the field believed that public health had 'lost its way', having been diverted into the provision of personal preventive services in the early part of the twentieth century. Yet, in the end, the new specialty was born more of administrative fiat than of professional strength. Its identity was intimately bound up with the new structure of the NHS and was severely undermined by the subsequent reorganizations of the NHS in 1982 and 1984.

The major sources of tension that community physicians continue to experience reflect the long-standing problems of public health in terms of its relationship to the state and to the rest of the medical profession. In 1974, governmental policy-makers were hopeful that the community physician would be primarily a manager, working within the NHS bureaucracy, looking at the need for medical services in a particular community and recommending a more rational allocation of resources at one with the Government's stated aim of giving more support to the 'Cinderella' services. But this took community medicine further than ever from the idea that the concern of public health was the health and welfare of the people, and threatened to turn the community physician into someone concerned above all with the efficient management of services. Much of the debate in the community medicine literature since 1974 has been about how the community physician may best advise on health problems and health needs, and how he or she might become more accountable to the local community. But community medicine faces the problem that the adoption of a broader mandate would inevitably entail political conflict. The kind of collective effort required to promote health in the late twentieth century is very different from that of a century ago, as shown by the controversy over the recommendations of the Black Report on inequalities in health (Townsend and Davidson 1982), which advocated greater public expenditure on matters such as housing, education, and income maintenance. Furthermore, to be successful, community medicine would have to be treated 'not so much as a specialty within medicine as the way in which health services should be considered within a welfare state' (Francis 1978), but this is immediately to invoke the other major spectre of medico-political conflict.

The heroic age? Victorian and Edwardian public health

For many, the term public health still conjures up the names of Chadwick, Farr, and Simon, charismatic figures doing battle with Victorian vested interests for pure food and water and for sewerage systems, and against infectious disease. However, the reasons for the dramatic decline in mortality have been the subject of fierce debate. Most influential has been Thomas McKeown (1976), who analysed the changes in cause-specific mortality and then inferred from the aetiology of the diseases concerned the most likely factors causing their decline. Working with the returns of deaths classified by age and certified causes of death, which are available for England and Wales from 1837 onwards, McKeown calculated that the overall mortality rate fell by some 22 per cent between 1848–54 and 1901. Dividing cause-specific mortality into four categories, McKeown argued that airborne diseases accounted for 44 per cent of the decline in late nineteenth

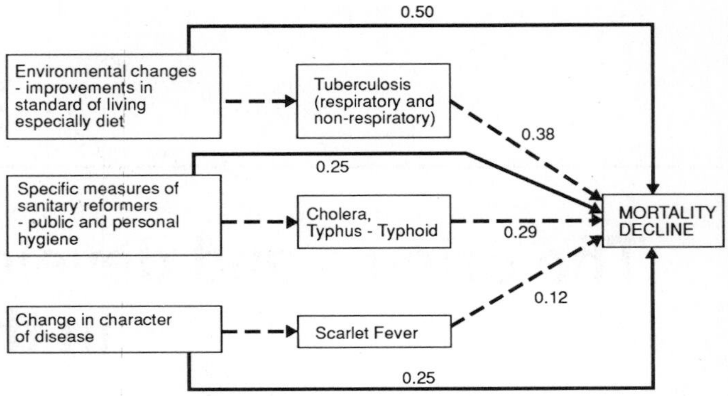

Fig. 2.1. A generalized model of Thomas McKeown's interpretation of the reasons for the decline of mortality in late nineteenth-century England and Wales. (Source: Woods and Woodward 1984, with permission.)

century mortality; water- and foodborne diseases for 33 per cent; other conditions attributable to micro-organisms for 15 per cent; with only 8 per cent due to conditions not attributable to micro-organisms. In regard to the airborne category, McKeown's case hinged on the dramatic decline in deaths due to tuberculosis, and within the water- and foodborne category, on the decline of cholera, typhus, and typhoid mortality.

In searching for an explanation of this pattern of mortality decline, McKeown proceeded deductively. First and most simply, he showed that mortality from all the important diseases other than smallpox pre-dated significant scientific advance in terms of immunization, chemotherapy, and hospital treatments. Second, he considered the possibility of a spontaneous change in the virulence of some of the infective micro-organisms and concluded that this was significant in the nineteenth century only for scarlet fever and influenza. Third, he considered the influence of sanitary and hygienic measures of reform, concluding that these were crucial to the diminution of water- and foodborne disease mortality. But, according to McKeown's data, these diseases accounted for proportionately less of the decline in mortality than airborne disease. Pride of place in McKeown's explanation was therefore reserved for improvements in the standard of living, especially in nutrition, which raised the potential victim's resistance to airborne diseases. Woods and Woodward (1984) have constructed a useful model summarizing the essence of McKeown's argument (Fig. 2.1).

Thus, in McKeown's view, the part played by public health measures in reducing mortality from infectious disease comes a rather poor second, but the criticisms of his argument suggest that he may have fallen victim to the epidemiologist's desire to isolate a single cause, whereas in this instance the historian's instinctive rejection of monocausal explanations may be a sounder guide. As Woods and Woodward (1984) have pointed out, the central problems in any assessment of McKeown's analysis are, first, the need to establish that there was a rise in standards of living and nutritional status during the late nineteenth century, and, second, the need to prove the extent of the quantitative contribution of the various possible factors contributing to the decline in mortality. On the first, historical debate has been prolonged (Taylor (1975) provides a useful summary), but the burden of argument now suggests that while the fall in real wages was reversed at some point during the second decade of the nineteenth century, differentials between the skilled and unskilled workers nevertheless remained virtually unchanged up to the 1880s (Lindert and Williamson 1983). Wrigley and Schofield's (1981) data show virtually no improvement in life expectancy between the 1820s and the 1870s. In Liverpool, a city whose population had trebled in size between 1801 and 1841, life expectancy was a mere 25.7 years in mid-century, compared with 36.7 years for London and 45.1 years in Surrey (Szreter 1986). Although new technologies allowed Britain to escape from the Malthusian trap, industrial workers paid the price of overcrowded and insanitary living conditions in the fast-growing towns and saw a relatively high proportion of their rising wages disappear on rent. McKeown's emphasis on the part played by rising nutritional standards in causing the decline of tuberculosis in particular is additionally problematic because the aetiology of tuberculosis is very complex. In her study of mortality from tuberculosis among women in urban areas, Cronje (1984) has attributed improvements to a combination of factors, including better diet, housing, and changes in work patterns. Bryder's (1988) exhaustive study of the late nineteenth and early twentieth century anti-tuberculosis movement favours the idea of rising living standards as the chief explanation of the decline in mortality, but includes in the term living standards improvements in housing as well as nutrition. Certainly, from 1870 onwards the 'rising living standards' argument is on stronger ground, especially if it is defined to include more than just nutritional status. In his study of the decline of infant mortality during the First World War, Winter (1982, p. 729) concludes that 'the prime agency at work . . . was a rise in family incomes, especially among the poorest sections of the population'.

In the early and mid-nineteenth century, not only is it more difficult to demonstrate the validity of this sort of conclusion, but also there is considerable doubt about the relative quantitative contribution of the factors involved. In this regard Simon Szreter's (1986) contribution appears to be one of the most powerful. He has suggested that the evidence adduced by McKeown to demonstrate the early decline of tuberculosis is far from convincing. In particular he stresses the importance of the strong counter-trend in deaths due to the acute bronchitis group of diseases which registered an increase in mortality rate of over 20 per cent to 1901, due largely to the conspicuous failure of the Victorians to tackle the problem of air pollution. This trend is important because either the trend in bronchitis deaths has to be accepted as genuine, thereby constituting a contradiction of McKeown's argument that a general fall in airborne diseases was the major element during the nineteenth century, or if, as is likely, some change in the practice of certification of death took place in mid-century from tuberculosis to bronchitis, then the fall in mortality due to tuberculosis becomes less spectacular (see Hardy (1988) on the problems of disease classification more generally). Szreter (1986, p. 19) concludes that it may be that water- and foodborne disease mortality account for 'at least half as much again, and perhaps almost twice the absolute quantity of mortality reduction during the nineteenth century', which argument reiterates the importance of according more attention to factors other than improvements in nutritional status when seeking to explain the decline in nineteenth century mortality. In particular, Szreter (1986, p. 5) argues for the importance of 'human agency in the form of politically negotiated expansion of preventive public health provisions and services, rather than the impersonal invisible hand of rising living standards'. His argument is convincing, but it is important, first, that in prioritizing public health measures the argument is not permitted to swing towards another form of monocausal explanation, and second, that the limitations of Victorian public health efforts at both the national and the local level be acknowledged.

The response of Victorian central government to health problems may be interpreted as largely *ad hoc* and permissive. Epidemic disease, particulary of cholera in 1831–2 and 1865–6, acted as the catalyst to reform. It was the report of three doctors in the aftermath of the 1832 cholera epidemic that led the ubiquitous Victorian civil servant, Edwin Chadwick, to investigate 'the sanitary condition of the labouring poor'. His famous report of 1842 emphasized the crucial link between dirt due to insanitary conditions and overcrowding, and disease, and stressed the need for a central administrative structure to oversee health issues. A central board of health was set up in 1848, but was abandoned in 1854. Indeed, the 1848 Public Health Act was nothing if not tentative; no national system of sanitation, sewerage, drainage, and street cleansing or public health departments was set up. Rather, the legislation made action at the local level permissive. Communities without local councils were compelled to set up local boards of health only when the death rate reached 23 per 1000, which encouraged a crisis intervention approach.

While nineteenth century central governmental intervention on health issues cannot be described as either strong or consistent in terms of administration, it was nevertheless informed by a clear body of economic and social ideas, which helps to explain the wide-ranging nature of the issues subjected to central governmental scrutiny. Chadwick's 1842 report stressed both the economic cost of ill-health and the social cost in terms of its effects on morals and habits (Finer 1952; Flinn 1965). As Smith (1979) has remarked, Chadwick was the first man since William Petty to use the economic value of man to justify improved health-related services. Classical economic theorists saw labour as the source of value and abhorred unproductive people and expenditure. Chadwick himself came to the study of health problems from his involvement with reforming the poor law, and the main question that interested him was that of the relationship between poverty and ill-health. In common with most

Victorian and Edwardian reformers, he was convinced that the causal relationship between these two variables operated in only one direction: ill-health was the major cause of pauperism, which necessitated large amounts of unproductive expenditure on poor relief, which in turn served only further to demoralize the poor. Measures to improve the health of the people, whether through better sanitation, the tightening of building regulations, or broadening the scope of poor law medicine may be seen within the context of this human capital approach. William Farr's work at the General Registry office was also intimately linked to the contemporary preoccupation with the productivity of the population and the fear of physical degeneration among town-dwellers (Eyler 1979).

As Sutcliffe (1983) has perceptively argued, the urban variable acted as a spur to Victorian state intervention and government growth because a large number of social questions concerned with poverty and housing as well as health were packed into the fear of urban degeneration and physical deterioration. Public health reform increasingly served as a filter for more general social reform. This was clearly visible in the provisions of the 1872 and 1875 public health legislation, which established a sanitary authority for every district and gave local authorities powers to deal with water, sewerage, nuisances, the quality of food, and regulations concerning lodging houses, bakehouses, and artisans' dwellings (Brand 1965). While the last three decades of the nineteenth century have been generally interpreted by historians (Flinn 1965; Lambert 1963; MacLeod 1968) as ones in which Treasury control and the poor law mentality of the Local Government Board (the central governmental department which controlled public health and poor law matters from 1871 to 1919) stifled new initiatives in the public health field, it is nevertheless possible to see the housing legislation of the 1870s and the increasing local effort to introduce elements of the sanitary idea as proof of the way in which concern about the conditions of the urban population continued to act as a spur to health reform. Wohl (1983) and Szreter (1986) have stressed the extent to which local government spending on sanitary provisions increased. Between 1858 and 1870 only £11 million in public works loans were requested by local authorities, whereas between 1871 and 1897, £84 million was borrowed. Of course some local authorities were considerably more active than others in this respect. It is no coincidence that the investigations of infant mortality during the 1910s by doctors employed by the Local Government Board revealed the incidence of mortality to be highest in those areas where least had been accomplished in providing mains drainage, sewerage, cleansing, and scavenging. Luckin's (1984) study of typhus and typhoid in London has shown that in neither case can the diminution of mortality be correlated successfully with the provision of pure water or of sewerage. Typhoid showed a substantial and continuous decline between 1871 and 1885, but it was only in the 1890s that more than 50 per cent of London's population had access to a constant water supply, while in the case of typhus, extra-urban sewerage was not introduced until the 1860s, and overcrowding persisted through the 1880s. Such findings serve to reinforce the importance of seeking multicausal explanations; indeed Luckin suggests it is necessary to take into account the decrease in migration from urban Ireland after the 1870s.

Nevertheless, in reaching some assessment of public health provisions in the nineteenth century, it is important to acknowledge the breadth of the public health remit. 'Slum' and 'fever den' were terms used interchangeably in the nineteenth century (Wohl 1983). Both they and their inhabitants were feared as agents of infection before it was even understood exactly how this occurred. As Starr (1982) has pointed out, all dirt was considered dangerous. By the end of the century, social investigators were convinced that physical well-being was a necessary prerequisite for further social progress. The urban environment was feared to be producing a race of degenerates, physically stunted and morally inferior. The slippage between social and moral categories, so characteristic of Victorian social science, served only to intensify the fear of contamination. Fear, together with religious zeal and civic pride (albeit often moderated by ratepayer parsimony), combined to effect the sanitary reform associated with the early public health movement. Collective action, most importantly at the local level, was the means of accomplishing reform, and those best served by modern local government after 1835 got the best protection against disease (Flinn 1965).

The role of doctors was muted in the Victorian public health effort. Flinn (1965) has argued though that their real contribution consisted in the promotion of an awareness of the association between dirt and disease and has suggested that there was little in Chadwick's 1842 report that had not already been discovered by doctors. Chadwick's own propensity to see the medical profession as merely 'the first among equals' rather than the sole source of professional expertise on issues of public health is well known. The 1848 Public Health Act permitted local boards of health to appoint a medical officer of health (MOH), but it was not until the 1872 Act that the new local sanitary authorities were required to do so, and not until 1909 that the county councils were also compelled to appoint county MOsH. As Watkins' (1984) revealing analysis of the professionalization of the MOH has shown, the definition of the role of the medical officer caused considerable heart-searching during the last quarter of the nineteenth century. The problem was how to make the MOH an independent medical practitioner, as opposed to a poor law doctor or an engineer. The main duties of the MOH were twofold (Wilkinson 1980), involving first the duties of sanitary inspection and improvement (which were to earn the MOH the derogatory title of 'drains doctor'), and second, disease control, emphasizing primarily isolation and removal to hospital and the tracing of the foci of infection in epidemics. The lonely battles of some late nineteenth century medical officers to raise the awareness of a backward council or to take on negligent landlords were real enough, and before legislation was passed to give MOsH security of tenure (in 1909 to county medical officers and in 1921 to district MOsH) they ran the risk of being dismissed for their pains.

However, it was not until the 1890s that the public health doctors emerged as a clear professional group, and as late as 1893 some were arguing against security of tenure and full-time employment because they felt this would stop them from practising curative medicine as general practitioners. The emergence of a definable group of public health doctors was grounded firmly in the development of medical qualifications and an effort to distinguish preventive from curative medicine. The 1888 Local Government Act, which created the exclusive appointment of only suitably qualified officers to county posts, together with the Medical Act of 1886, which required candidates for the Diploma of Public Health to have a medical licence, combined to produce the basis for a new professionalism. Above all, training in public health stressed the importance of understanding the aetiology of disease, and by the 1900s advances in bacteriology had come to dominate the public health curriculum. Fee (1990) shows that the story is similar in the US. Indeed Starr (1982) has characterized the shift in the changing nature of public health work in the twentieth century as a move towards a 'new concept of dirt'. As a result of germ theory, the twentieth century concept of dirt 'narrowed' so that it proved considerably cheaper to clean up. This analysis is valuable for the way in which it acknowledges the importance of the political imperative to a more limited, less costly, mandate for public health, in addition to that of developments in medical science.

C. E.-A. Winslow, the early twentieth century American authority on public health, identified three phases in the development of public health: the first, from 1840 to 1890 was characterized by environmental sanitation; the second, from 1890 to 1910, by developments in bacteriology, resulting in an emphasis on isolation and disinfection; and the third, beginning around 1910, by an emphasis on education and personal hygiene, often referred to as personal prevention (Starr 1982). This chronology is broadly congruent with developments in Britain. For despite a dramatic growth in the statutory powers of public health departments, twentieth century developments resulted in a narrowing of public health's mandate. Scientific advances in bacteriology redefined the kind of intervention appropriate for public health. Once it was realized that dirt *per se* did not cause infectious disease, the broad mandate of public health to deal with all aspects of environmental sanitation and housing as the means of promoting cleanliness disappeared. Germ theory deflected attention from the primary cause of disease in the environment and from the individual's relationship to that environment, and made a direct appeal from mortality figures to social reform much more difficult (Hart 1985). Increasingly public health authorities and the growing numbers of MOsH in them focused on what the individual should do to ensure personal hygiene. During the Edwardian period, there was much more legislative activity in the field of social policy—Harris (1983) has argued that during these years social policy issues, for example in the form of national insurance and old-age pensions, entered the realm of 'high politics'—but public health reform no longer served as the filter for more general social reform. Health and welfare were firmly separated in Edwardian Britain and a tighter mandate was imposed on both. Thus, in narrowing its focus, public health was arguably responding to the changing framework of state intervention as well as to changes in medical science.

No clearer example of the effects of these developments can be found than in the campaign to reduce the infant mortality rate (Lewis 1980). Epidemiological studies of the problem conducted by public health doctors in the 1910s revealed the death rate to be highest in poor inner-city slums, where insanitary and overcrowded conditions prevailed. Yet government officials and public health doctors tended to view maternal and child welfare in terms of a series of discrete personal health problems, to be solved by the provision of health visitors, infant welfare centres, and better maternity services. The bulk of their attention as well as that of the large voluntary maternal and child welfare movement was focused on health education, encouraging mothers to breast-feed and to strive for higher standards of domestic hygiene. Dwork (1987) has argued that this emphasis was justified by the bacteriological work on infant diarrhoea. Furthermore, public health doctors could only do what was politically feasible. Unable to abolish poverty and secure better environmental conditions, they concentrated on mothers. This is fair enough; public health was no longer the vehicle through which a variety of collective provisions could be justified. But opinions differ over the effectiveness of the new twentieth century focus on the individual, justified by public health doctors as 'applied physiology' and a new kind of preventive medicine. Dwork (1987) and Szreter (1986) are probably overly optimistic in the praise they reserve for the educational efforts of public health workers; the number of mothers reached must have been relatively few, and in this instance Winter's (1982) emphasis on the importance of rising living standards during The First World War, and especially on rising real wages, appears more crucial.

The vision of nineteenth century public health was broad, not least because in the period before other social policy issues such as housing and income maintenance entered the realm of 'high politics', public health legislation provided the only legitimate means of attacking them. Leaders such as Chadwick and Simon were able to exploit this situation, albeit that Simon experienced considerable civil service opposition to his expansionist ideas in the last part of the century. But the existence of such a broad vision does not necessarily mean that proponents of health were bent on pursuing an optimal strategy to secure the health and welfare of the people. It would be more accurate to describe nineteenth century aims as minimalist, designed to secure a functioning working population. Chadwick, after all, was inspired to action on health reform by the idea that disease brought large numbers on to the poor law. Nineteenth century public health measures isolated infectious people and began clearing the slums that were the product of poverty, but they made no attempt to tackle the issue of poverty directly. Public health was defined in such a way that a wide range of social issues fell within its compass, and its effectiveness was due in large part to the degree of state intervention and collective provision that was achieved. But this did not mean that its proponents saw it necessary to take positive action to tackle social issues or to put accountability to the people before accountability to the state. The nineteenth century sanitarians were not afraid to tackle water companies, landlords, and other vested interests, but it was not part of their plan to change radically the social and political fabric. By the early twentieth century, the concerns of public health were chiefly those of personal hygiene. Indeed, the Fabian model for reform of state medical services developed by the Webbs sought to use the public health departments because they felt that public health doctors created 'in the recipient an increased feeling of personal obligation and even a new sense of social responsibility . . . the very aim of the sanitarians is to train the people to better habits of life' (Webb and Webb 1910). While in 1911 a national insurance rather than a public health model of reform was adopted in the restructuring of publicly provided medical services, the Webbs were correct in their perception of the direction of public health practice in the early twentieth century, with its increasing emphasis on the delivery of personal preventive medical services. This was to render public health's impact on the structural variables crucial to the prevention of disease and the promotion of health more limited.

Community watchdog or 'Third Grade Doctors'? Public health 1919–68[1]

By 1918, the chief medical officer to the Local Government Board, Sir George Newman, was arguing that preventive medicine must be given a greater place in the education of every medical student. He insisted that public health was no longer concerned with sewerage, disinfection, the suppression of nuisances, the notification and registration of disease, and the implementation of by-law regulations, but rather was about 'the domestic, social and personal life of the people' (Newman 1928). In his memorandum on the practice of preventive medicine, first issued in 1919, Newman argued for a new 'synthesis and integration' in medicine and, in particular, a closer integration of preventive and curative medicine. But emphasis on the idea that the prevention of disease had become less a matter of removing external environmental 'nuisances' and more a personal concern brought the practice of public health very close to that of the general practitioner.

Public health doctors welcomed a wider recognition of the importance of the ideas they promoted. However, if it could be argued that a 'preventive consciousness' was something that all doctors should have, it became additionally difficult to distinguish the core of public philosophy and practice from the work of other medical practitioners. Throughout the inter-war years, there was considerable antagonism between public health doctors and general practitioners over the role of the public health clinics, with general practitioners accusing medical officers of health of 'encroachment' on their private practices. However, in terms of medical politics, public health departments felt themselves to be in a position of strength during this period. Governments of the inter-war years failed to extend health service provision under the National Health Insurance Act, and instead added piecemeal to services provided by the local authorities via the public health departments. By 1939, local authorities were permitted to provide maternal and child welfare services, including obstetric and gynaecological specialist treatment; a school medical service, including clinics treating minor ailments; dentistry; school meals and milk; tuberculosis schemes, involving sanatorium treatment, clinics, and aftercare services; health centres, the most elaborate being that built by the Finsbury Borough Council in 1938; and local regional cancer schemes. The most important addition to the medical officer of health's responsibility came in 1929 when local authorities were permitted to take over administration of the poor law hospitals and many MOsH found themselves taking on the role of medical superintendent. From a position of growing strength, in terms of the tasks they were being called upon to perform, MOsH spoke with increasing confidence of the importance of the public health service in leading the way in preventive medicine, primarily through the work of educating the public in personal hygiene, and of the importance of educating general practitioners to play their part. But at the end of the day, it may be argued that public health failed adequately to distinguish the content and direction of its work from that of other practitioners, especially general practitioners, which made its position extremely vulnerable when, in the post-war reorganization of health services, government decided not to take the public health service as the model for the new NHS. In addition, public health's preoccupation with, first, the hygiene of the individual and, second, the administration of a growing number of services, resulted in a neglect of the MOH's traditional task of 'community watchdog' in respect to sources of danger to the people's health.

There was considerable discussion as to the meaning of preventive medicine in the public health journals during the inter-war years, but new thinking about health as opposed to sickness and about the determinants of both came not so much from the public health practitioners as from privately funded experiments such as the Peckham Health Centre (which aimed to promote health rather than to prevent disease); pressure groups, such as the Women's Health Inquiry and the Children's Minimum Council (a forerunner of the Child Poverty Action Group); and most importantly, from academics in medical and social science, who began talking in the 1940s about the importance of a concept they called social medicine rather than public health. It is significant that public

1. The issues raised in the next two sections are dealt with more fully in Lewis (1986*a*).

health practitioners were at first puzzled by the discussion of social medicine and by the late 1940s had rejected it.

Throughout the 1920s, public health departments were able to sustain their claim that 'public health work is mainly clinical medicine but clinical medicine of a special kind' by following the principles of Newman. The division of labour between MOsH and general practitioners rested on the separation of health education and advice from treatment. Despite the charges of general practitioners, the evidence suggests that public health departments were, in fact, careful not to offer any treatment other than for the mildest ailments, but the boundary between the two types of provision was obviously hard to draw and was indeed sustained largely by a system of health services in which treatment was not free at the point of access. Nevertheless, increasingly MOsH looked forward confidently to the time when, in the words of the MOH for Willesden, 'the very large provisions and concentrations in respect of public health and medical work made by the Local Government Act of 1929 are likely to lead to a state medical service' (Buchan 1931, p. 9).

By the 1930s, many MOsH were greatly involved in their new administrative responsibilities for the former poor law medical institutions. Fears were expressed that the work of hospital administration was diverting MOsH from their main task of prevention. The editor of the *Medical Officer* (Editorial 1930, p. 21; 1931, p. 1) wondered whether MOsH would be able to return 'from the pursuit of pathology to their proper allegiance to physiology', commenting that 'much recent public health work seems to aim at converting it into a gigantic hospital'. But, by defining preventive medicine as including all measures devised to prevent premature death and to maintain optimum health, at least one MOH produced a spirited defence of the MOH's involvement in hospitals (Ferguson 1938). Most MOsH were content to define public health in terms of the tasks it was prepared to assume.

The history of diphtheria immunization gives point to the criticism that MOsH had become too involved in day-to-day administrative duties, particularly in respect of hospitals. Responsibility for immunization rested very much with individual local authorities and the local MOH, whose task it was to persuade the local public health committee to pursue an active immunization campaign. This often required special persistence in the financially straitened circumstances of the 1930s. Effective immunization agents were available by the early 1920s and reports of successful large-scale trials in Canada and the US were published at the end of the decade. Yet between 1927 and 1930 the medical journals show that large numbers of MOsH were preoccupied with more traditional approaches to the control of the disease and were deeply distrustful of immunization. A significant number seem to have concentrated their efforts on swabbing throats and noses in an effort to identify carriers, and on confining victims in isolation hospitals. Bryder's (1988) research has also shown that public health doctors employed as tuberculosis officers tended to identify with the institutional treatment of the disease. In the case of diphtheria, the result was that while the death rate in Canada fell steadily in the 1920s and the 1930s, in Britain the rate showed no decline until 1941, when a national immunization scheme was eventually implemented (Lewis 1986*b*). Certainly, R.M.F. Picken, the professor of preventive medicine at the Welsh School of Medicine, felt that the case of diphtheria proved that the MOH had taken on too much administrative work to be an effective proponent of preventive medicine (Picken 1939).

Recent research provides considerable evidence that MOsH neglected many aspects of their duties as community watch-dogs during the 1930s, both in regard to more traditional areas such as immunization and to the new-found concerns over the effects of long-term unemployment on nutritional standards and levels of morbidity and mortality. For the most part, MOsH filed optimistic annual reports on the health of their communities. Government opposition to increased public expenditure during the Depression made the Ministry of Health wary of supporting any findings that might have been interpreted as necessitating more spending on welfare benefits, and it consistently refuted evidence provided by pressure groups and social scientists as to the existence of a relationship between high unemployment and deteriorating health standards, reserving particular condemnation for the handful of MOsH who expressed similar opinions. Some fifty authorities, mainly in the depressed areas, sent in returns to the Ministry of Health suggesting that they were experiencing less than half the average incidence of subnormal nutrition. Some MOsH were philosophically opposed to giving nutritional supplements despite the apparently good results in this regard achieved by the National Birthday Trust among pregnant women in South Wales (Lewis 1980). It is hard to avoid Charles Webster's conclusion that in adopting an optimistic view, MOsH were telling the Ministry of Health what it wanted to hear (Webster 1982, 1985).

The lead in raising questions about the health status of the population during the 1930s was taken by political lobby groups such as the Children's Minimum Council, the Committee against Malnutrition, and the National Unemployed Workers Movement, all of which called for a higher level of unemployment benefit to enable families to secure the minimum nutritional requirements set out by the British Medical Association. Groups such as the Women's Health Inquiry surveyed the health status of some 1250 working-class wives and found that only 31.3 per cent could be considered to be in 'good health' (Spring Rice 1981), while a professional social scientist like Richard Titmuss (rather than a public health doctor as was the case in the 1910s) undertook a survey of infant mortality and concluded that the decline in the overall infant mortality rate was not matched by a narrowing of the gap between social classes (Titmuss 1943). Finally, a small number of consultants, particularly obstetricians and gynaecologists, attempted to draw attention to the high rate of maternal mortality and morbidity. Sir James Young estimated about 69 per cent of hospital gynaecology to be a legacy from vitiated childbearing and despaired of 'the apathy of organized medicine towards the positive value of health ideals' and 'the profession's devotion to disease [which] has blinded us to the duties of health' (Young 1933, pp. 119–20).

G.C.M. McGonigle was one of the very few MOsH who attempted to link public health more to ideas of positive health, rejecting the idea of 'personal preventive clinical medicine'. For example, he was virtually alone among public health doctors in insisting that the general decline in infant mortality had begun long before the advent of generalized child welfare work. In his own district of Stockton-on-Tees he undertook an influential study of a group of families who were moved from slum houses to a new housing estate and showed that their health status deteriorated relative to those who stayed behind, largely because of the greater proportion of their income that was absorbed by the higher rents that they had to pay (McGonigle and Kirby 1936). Essentially McGonigle was defining public health's task as a concern with the determinants of health and their promotion at a time when the majority of MOsH were content to expand the range of the services provided by their public health departments, expecting that the balance of medical care provision would soon swing in favour of a state service, thereby increasing the influence of the MOH as the only salaried doctor. However, in view of both the British Medical Association's resolute opposition to such a vision (British Medical Association 1938) and the Ministry's disinclination to opt for a full state-salaried service, such hopes were unrealistic.

In the meantime, public health practitioners failed to support the first major initiative to provide a new direction for public health in the twentieth century: social medicine. The roots of social medicine were to be found in the work of social investigators and pressure groups concerned about health status during the 1930s. John Ryle paid tribute to their work when he became the first professor of social medicine at Oxford in 1942. Lecturing to his medical students in Cambridge in 1940, he stressed the importance of developing a social conscience and of considering the larger social problems (Ryle 1940, p. 657). Richard Titmuss played a major part in developing the concept of social medicine and in 1942 drafted a paper explaining the idea as 'yet another stage in the growing recognition of the social relations of Health. Our vision is broadening; men are being pictured against a man-made environment; the multiple factor in disease and disorder is replacing the single causation concept; the study of life is replacing a morbid concentration of death' (Titmuss 1942). Titmuss was consciously seeking to make public health departments take the lead in promoting renewed collective action against the structural impediments to health.

However, after the establishment of the Oxford chair in 1942, the development of social medicine was conditioned by the fact that its voice was confined to a few university departments, and in the search for academic credibility it moved further away from a concern with health policy and social science. Thus social medicine failed in two crucial respects to fulfil its early promise. In part because of this and in part because of their own narrowness of vision, public health practitioners did not take up the idea of social medicine. This in turn resulted in a damaging rift between the leading teachers of social medicine and practitioners of public health.

Ryle's own later work emphasized not only the links between social medicine and clinical medicine and epidemiology at the expense of social science and health policy, but also the importance of the study of 'social pathology'—the quantity and cause of disease—at the expense of the more radical and difficult aim of promoting health. As an American observer remarked in 1951, the Oxford Institute concerned itself more and more with factors affecting mortality and morbidity, shying away from 'the allegedly sentimental aspects of social medicine . . . often stigmatized as the "unmarried mother" category of social problems' (Weinerman 1951). For example, Ryle criticized J.N. Morris and R. Titmuss's study of the epidemiology of rheumatic heart disease for paying too much attention to 'the poverty factor'. Leff (1953) noted that the practice of social medicine had increasingly come to mean the collection of medical statistics and that one of its main weaknesses lay in the arbitrary selection of problems for study which were often unrelated to the practice of medicine or to the life of the community.

Medical officers of health were offended by the criticisms levied at public health departments by academic leaders of social medicine and registered their impatience with its high academic tone. At the same time social medicine failed to have the kind of impact on the medical schools that Ryle had hoped for. The 1944 Report of the Inter-Departmental Committee on Medical Schools recommended the development of departments of social medicine, seeing them as a means of reorienting the whole medical curriculum (Ministry of Health and Department of Health for Scotland 1944). However, most medical schools responded by slightly modifying their departments of public health but without fundamentally changing their approach to medical education. Indeed, Ryle's chair was not filled when he died in 1950.

In the way that it developed, social medicine was arguably deeply flawed. However, in the early 1940s it offered an opportunity for challenging the whole nature of medical education and for creating an exciting synthesis between social science and medicine. As it was, the influence of social medicine departments remained limited and the schism between academics and practitioners persisted, attracting the attention of the 1968 Commission on Medical Education (PP 1968*a*), which recommended that a new specialty of community medicine be established to bridge the gap between the two groups.

In the new NHS of the post-war period, public health doctors found themselves searching for a new direction, having lost control over the municipal hospitals and facing the inevitable decline in their clinic work because of the universal access to general practitioner services provided under the NHS Act. The immediate inclination of MOsH was to look for new services to administer. Responsibility for running ambulance services, home helps, and old people's homes became particularly time-consuming tasks. As a result, local authorities expressed the view that MOsH were but 'administrators with medical knowledge', and public health doctors spent much of the 1950s and 1960s fighting the insistence of local authorities that they should be paid on a scale comparable to other administrative officers rather than to other doctors.

Increasingly MOsH found themselves squeezed between pressures from within—in the form of the local government hierarchy and the desire on the part of sanitary inspectors, health visitors, and social workers for greater professional freedom—and pressures from without. The latter included general practitioners, with whom they had to share the extra-hospital territory, and from academics and social scientists who expressed increasing impatience in respect of the perceived failure of the public health departments to deliver effective community care. As Walker (1982) has observed, community care meant different things at different times and in relation to different groups in need. In respect of the elderly, where it was originally intended to mean domiciliary care, it was reinterpreted to include local authority residential care. Thus both the Hospital Plan of 1962 and the local authorities' Health and Welfare Plans of 1963 envisaged the expansion of residential provision. In the meantime, the shortage of beds in both sectors resulted in increasing confusion as to the boundaries between the two types of care. With the failure either firmly to distinguish community care from institutional provision or to increase the flow of resources to domiciliary care, the Ministry resorted to exhorting the three parts of the NHS—general practitioners, hospitals, and public health—to co-operate and co-ordinate their work. MOsH were seen as the principal co-ordinators and increasingly found themselves condemned as unimaginative and narrow in their approach. In the view of Titmuss and Morris, for example, the description of 'administrators with medical knowledge' was broadly accepted and MOsH were seen primarily as managers of services, doing little to investigate properly the health status of their populations and to plan services accordingly. The public image of the MOH was unhappily personified in the dreary and obstructionist character of Dr Snoddy in the popular television series, 'Dr Finlay's Casebook'.

Academics in departments of social medicine and of public health and social scientists began as early as the 1950s to urge substantial reform in the training of public health recruits and in public health practice in order to reinvigorate the specialty. In particular a case was made for 'medical administration' as specialized work, not in the sense of institutional administration, as was the case in the 1930s, but rather in the hope that MOsH would become broadly based 'health strategists' (e.g. Irvine 1954; Wofinden 1959). The Department of Social Medicine at Edinburgh University was the first to offer a diploma in medical services administration in 1959. However, in the debates over the possibilities of medical administration, the relationship between the executive (or management) and advisory roles inherent in the work of the medical administrator was never made clear. These issues were to continue to bedevil the conceptualization of community medicine and the role of the community physician.

Specialist advisers or managers? Community medicine 1968–88

In the context of the strong arguments for reform and revitalization being provided by the academics, the recommendations of the Seebohm Report on social services and of the Government's Green Paper on NHS reorganization, both published in 1968 (Ministry of Health 1968; pp. 1968*b*), provided the final push. J.N. Morris was the only medical member of the Seebohm inquiry, and both he and Titmuss were convinced of the weakness of local authorities in general and of public health departments in particular in achieving progress in the field of community care and in developing new approaches to social work. In recommending the setting up of new social service departments, the Seebohm Report threatened to remove the fastest growing services that came under the MOH's control. It was no coincidence that the Green Paper sought to reassure public health doctors that they would find a new expanded (albeit unspecified) role as community physicians within the reorganized NHS.

It was J.N. Morris who first defined the role of the community physician. He believed strongly that public health practice should be grounded more firmly in the principles of modern epidemiology. His textbook on epidemiology identified the major uses of the subject as historical study, community diagnosis, analysis of the workings of health services, analysis of individual risks and chances, the identification of syndromes, and the completion of the clinical picture (Morris 1969*a*). From this he evolved the concept of a community physician responsible for community diagnosis and thus providing the 'intelligence' necessary for the efficient and effective administration of the health services (Morris 1969*b*).

Morris did not agree with the attempt of some American epidemiologists to 'rescue' epidemiology from public health and bring it back to the 'laps of practising physicians' (Paul 1958), but he nevertheless approached prevention through the needs of the individual, believing that a multicausal, epidemiological approach would ensure consideration of socio-economic and environmental variables and eliminate the danger of 'blaming the victim' for illness. Using the example of coronary heart disease, he argued that the barriers between prevention and cure were crumbling and that 'public health needs clinical medicine—clinical medicine needs a community' (Morris 1969*b*). Like Ryle, Morris emphasized the importance of co–operation with clinicians.

His ideas were fed directly into two crucial policy documents of the late 1960s, the Seebohm Committee and (via Richard Titmuss) the Todd Commission on Medical Education, which also reported in 1968. The latter clearly articulated the two main strands of Morris's formulation of community medicine when it defined it as 'the specialty practised by epidemiologists and administrators of medical services' (PP 1968*a*). It also recommended closer links with clinical medicine and between academics and practitioners in the field and, like the Seebohm Committee, envisaged environmental health services and social work services leaving the public health department, and the community physician moving away from clinic work.

MOsH showed considerable awareness of the problems of coming under the control of central government and of forging working relationships with other doctors in the NHS, particularly in the hospital sector, but were nevertheless attracted to the idea of community medicine chiefly because they understood that it meant a substantial rise in status for the specialty. They also interpreted the job description for the community physician as recognition of their past work in administering services; at no time were they able to conceptualize the nature of their management role in the reorganized NHS.

This was particularly crucial because it would seem that policy-makers' understanding of the role of the community physician differed in emphasis from that of academics. The key documents published prior to NHS reorganization in 1974 saw the new community physician as the key to effective integration of the health services, linking lay administrators to clinicians and co-ordinating the work of the NHS with that of the local authorities. The community physician was recognized as a specialist adviser, with particular skills in epidemiology, but a substantial number of community physicians were to be given management responsibilities in the new consensus management teams in order properly to utilize their expertise (Department of Health and Social Security 1972*a*, *b*). It seems clear that while the Faculty of Community Medicine stressed the community physician's specialist/advisory role and stressed the complementarity between community and clinical medicine, policy-makers stressed the importance of the community physician recommending changes in the deployment of resources and of management.

MOsH moved into the role of community physician believing that they were to be the linchpins of the new NHS, coordinating and administering services, but with little idea as to the meaning of their formal role in the new management structure or the place of 'management' in their total package of tasks and concerns. From the beginning, community physicians found the 'community hat' a difficult one to wear. While the Todd Commission on Medical Education had proposed that the term community should embrace the whole population, including those in institutions, increasingly 'community' came to described the non-hospital services. Thus while it was intended that the community physician should provide the necessary 'intelligence' for adjudicating the resource needs of various types of health services including the hospital, in practice the title of community physician often meant that other members of the management teams expected them to speak for the community services outside the hospital. On the other hand, the battle to come to terms with the problems of the hospital services meant that many community physicians who continued to feel considerable commitment to the extra-hospital health services and to the work of prevention and promotion, felt that their work was determined more by the needs of the NHS than by those of the communities they served (Scott Samuel 1979). Thus the position of community physicians was subject to serious conflicts in terms of both their relationship with other members of the medical profession and the nature of their primary responsibility, whether for the management of health services or for the analysis of health problems and health needs.

The image of the community physician as primarily a part of the NHS management structure was reinforced when in 1976 community medicine posts were included in a review of management costs. The idea that community physicians should look at the health needs of the district and the allocation of resources across the whole spectrum of the NHS would have led to difficulties in relations with other members of the medical profession at the best of times, for as Gill (1976) pointed out, their role could easily be 'interpreted as an additional mechanism for increasing the accountability of the profession through internal review and evaluation'. But during the mid-1970s, when severe financial restraints were imposed on the NHS, the position of the community physician became considerably more difficult.

Furthermore, the crucial policy documents of the 1980s have shown little awareness or appreciation of the community physician's role. Increasingly integration of the health service has ceased to be the focus of attention and the concept of management has shifted away from the achievement of consensus towards a more straightforward preoccupation with careful administration and clear lines of accountability. The 1979 Government document which signalled the 1982 reorganization of the NHS made no mention of community medicine and the emphasis was clearly on better management of the hospitals (Department of Health and Social Security and Welsh Office 1979). The Griffiths Report of 1983 (Department of Health and Social Security 1983) also focused firmly on the hospitals and recommended the appointment of a single general manager, readily identifiable at all levels of administration.

Because the role of community physicians was determined in large part by the place they occupied in the 1974 NHS structure, this shift has clear implications for community medicine. While the tensions between the community physician's role as specialist adviser or manager has largely disappeared since 1984—very few community physicians became general managers—so also have the tasks of community medicine become fragmented and the numbers of community physicians been reduced in many districts. Nor is there any shared view among the new general managers as to community medicine's purpose. In many districts it seems that the mandate of the community physician has been further narrowed, to evaluation and audit, or to medical staffing, for example.

These changes in the fortunes of community medicine must be located more widely. The reorganization of the NHS in 1974 was motivated in large part by the Treasury's desire to gain more control over public spending by the Department of Health and Social Security. In this attempt, the community physician was perceived by policy-makers as the linchpin. When costs in the hospital sector continued to rise, further reorganizations followed, and in 1984 professional managers rather than community physicians were seen as the answer to containing costs and achieving a more rational allocation of resources. In all this, it would appear that Government has effectively overlooked the other side of community medicine's task—that of preventing disease and promoting health through the provision of specialist advice. But Government in the 1980s has not been oblivious to the idea of prevention and promotion, although it has not been inclined to link it either to the need for a body of specialist medical practitioners or to the need for collective provision, as the poor response to the Black Report on inequalities in health (Townsend and Davidson 1982) has shown.

One of the reasons that Governments have gained confidence in their dealings with the medical profession is the publicity accorded to studies that suggest that medical services have played very little part in raising the health status of populations (e.g. McKeown 1976). In the light of such evidence, Governments have tried first to switch resources away from the expensive acute sector to preventive medicine, community care, and the 'Cinderella' specialties, and in so doing have harnessed the rhetoric of prevention to the cause of cost-control. When proponents of social medicine talked about prevention in the 1940s, they meant the identification of social and environmental factors inimical to health. However, the concept of prevention in the last decade has concentrated on the individual's responsibility to maintain a healthy lifestyle. Second, Governments have sought to invoke the right to consumers to increased choice and have sought to achieve this negatively, by decreasing the power of providers rather than by empowering consumers (Davies 1987).

Because community medicine was embedded in the structure of the NHS, it can be argued that it requires political will to revitalize the specialty. But current Government thinking is along very different lines from that of 1974 and the role of community medicine does not figure largely on the political agenda, notwithstanding the 1988 Government inquiry into public health or the renewed interest in communicable disease, the traditional preoccupation of the MOH, because of AIDS. Any attempt to broaden the mandate of community medicine to a full-blown consideration of health problems and health status must have both political and medico-political implications. First, the task of taking a holistic view of health services and of assessing the best balance to be achieved between services means that public health doctors run the risk of conflict with other members of the medical profession. The second task of analysing patterns of health and health needs is likely to involve consideration of factors outside the scope of the NHS, such as work, environment, income, and housing. In the last quarter of the twentieth century, unlike in the mid-nineteenth century, these factors are not usually thought of as health problems *per se* and for doctors to talk about them may require the abandonment of scientific neutrality. A return to 'public health medicine' in accordance with the recommendations of the 1988 committee of inquiry (PP 1988) may help to clarify the task of public health doctors, but in and of itself is unlikely to resolve these fundamental problems.

References

Brand, J.L. (1965). *Doctors and the state*. Johns Hopkins University Press, Baltimore, Maryland.

British Medical Association (1938). *A general medical service for the nation*. BMA, London.

Bryder, L. (1988). *Below the magic mountain*. Oxford University Press, Oxford.

Buchan, G. (1931). British public health and its present trend. *Public Health* **45**, 9.

Cronje, G. (1984). Tuberculosis and mortality decline in England and Wales, 1851–1910. In *Urban disease and mortality* (ed. R. Woods and J. Woodward). Batsford Academic, London, p. 79.

Davies, C. (1987). Things to come: the NHS in the next decade. *Sociology of Health and Illness* **9**, 302.

Department of Health and Social Security (1972*a*). *Management arrangements for the reorganized NHS*. HMSO, London.

Department of Health and Social Security (1972*b*). *Report of the working party on medical administrators*. HMSO, London.

Department of Health and Social Security (1983). *Report of the management inquiry*. HMSO, London.

Department of Health and Social Security and Welsh Office (1979). *Patients first: consultative paper on the structure of the NHS in England and Wales*. HMSO, London.

Dwork, D. (1987). *War is good for babies and other young children*. Tavistock, London.

Editorial (1930). Medicine and the state. *Medical Officer* **44**, 21.

Editorial (1931). Preventive medicine in 1930. *Medical Officer* **45**, 1.

Eyler, J. (1979). *Victorian social medicine*. Johns Hopkins University Press, Baltimore, Maryland.

Fee, E. (1990). The values and impace of the Fox and Welch reports: alternative conceptions of public health education in the United States, 1910–1939. In *Public health education in nineteenth and twentieth century America and Britain* (ed. R. Acheson and E. Fee). Oxford University Press, Oxford.

Finer, S.E. (1952). *The life and times of Sir Edwin Chadwick*. Methuen, London.

Flinn, M.W. (1965). Introduction to Edwin Chadwick's *The sanitary condition of the labouring population of Great Britain*. Edinburgh University Press, Edinburgh, pp. 1–73.

Francis, H. (1978). Towards community medicine: the British experience. In *Recent advances in community medicine* (ed. A. E. Bennett). Livingstone, Edinburgh, p. 1.

Gill, D. (1976). The reorganization of the NHS: some sociological aspects with special reference to the role of the community physician. In *The sociology of the NHS* (ed. M. Stacey). University of Keele, Keele, p. 9.

Hardy, A. (1988). Diagnosis, death and diet: the case of London, 1750–1909. *Journal of Interdisciplinary History* **XVIII**, 387.

Harris, J. (1983). The transition to high politics in English social policy, 1880–1914. In *High and low politics in modern Britain* (ed. M. Bentley and J Stevenson). Clarendon Press, Oxford, p. 58.

Hart, N. (1985). *The sociology of health and medicine*. Causeway Books, Ormskirk, Lancs.

Irvine, E.D. (1954). Medical administration. *Public Health* **67**, 172.

Lambert, R. (1963). *Sir John Simon 1816–1906*. MacGibbon and Kee, London.

Leff, S. (1953). *Social medicine*. Routledge and Kegan Paul, London.

Lewis, J. (1980). *The politics of motherhood: child and maternal welfare in England, 1900–1939*. Croom, London.

Lewis, J. (1986*a*). *What price community medicine?* Wheatsheaf, Brighton.

Lewis, J. (1986*b*). The prevention of diphtheria in Canada and Britain, 1914–45. *Journal of Social History* **20**, 163.

Lindert, P.H. and Williamson, J.G. (1983). English workers' living standards during the Industrial Revolution: a new look. *Economic History Review* **36**, 1.

Luckin, B. (1984). Evaluating the sanitary revolution: typhus and typhoid in London, 1851–1900. In *Urban disease and mortality* (ed. R. Woods and J. Woodward). Batsford Academic, London, p. 102.

McGonigle, G.C.M. and Kirby, J. (1936). *Poverty and public health*. Gollancz, London.

McKeown, T. (1976). *The modern rise of population*. Edward Arnold, London.

MacLeod, R.M. (1986). *Treasury control and social administration*. Clarendon Press, Oxford.

Massey, A. and Ferguson, J. (1938). Hospital policy in relation to preventive medicine. *Public Health* **51**, 235.

Ministry of Health (1968). *National Health Service: the administrative structure of the medical and related services in England and Wales*. HMSO, London.

Ministry of Health and Department of Health for Scotland (1944). *Report of the Inter-Departmental Committee on Medical Schools*. HMSO, London.

Morris, J. (1969*a*). *The uses of epidemiology*. Livingstone, Edinburgh. (1st ed. 1957).

Morris, J. (1969*b*) Tomorrow's community physician. *Lancet* **ii**, 811.

Newman, Sir G. (1928). *The foundation of national health*. Ministry of Health, London.

Paul, J.N. (1958). *Clinical epidemiology*. University of Chicago Press, Chicago, Illinois.

Picken, R.M.F. (1939). The changing relations between the medical officer of health and the medical profession. *Public Health* **52,** 261.

PP (1968*a*). *Report of the Royal Commission on Medical Education*. Cmnd. 3569.

PP (1968*b*). *Report of the committee on local authority and allied personal social services*. Cmnd. 3703.

PP (1988). *Public health in England. The report of the committee of inquiry into the future development of the public health function*. Cmnd. 289.

Ryle, J. A. (1940). Today and Tomorrow *British Medical Journal*, **2**, 240.

Scott Samuel, A. (1979). The politics of health. *Community Medicine* **1**, 123.

Smith, F.B. (1979). *The People's Health, 1830–1910*. Croom, London.

Spring Rice, M. (1981). *Working class wives*. Virago, London. (1st edn. 1939).

Starr, P. (1982). *The social transformation of American medicine*. Basic Books, New York.

Sutcliffe, A. (1983). In search of the urban variable. In *The pursuit of urban history* (ed. D. Fraser and A. Sutcliffe). Arnold, London, p. 2341.

Szreter, S. (1986). *The importance of social intervention in Britain's mortality decline c. 1850–1914: a re-interpretation*. Centre for Economic Policy Research discussion paper no. 121, London.

Taylor, A.J. (ed.) (1975). *The standard of living in Britain in the industrial revolution*. Methuen, London.

Titmuss, R.M. (1942). Untitled paper on social medicine. TS 21/12/42. Papers of the late Mrs Kay Titmuss.

Titmuss, R.M. (1943). *Birth, poverty and wealth*. Hamish Hamilton, London.

Townsend, P. and Davidson, N. (1982). *Inequalities in health*. Penguin, Harmondworth.

Walker, A. (ed.) (1982). *Community care*. Blackwells, Oxford.

Watkins, D.E. (1984). The English revolution in social medicine, 1880–1991. Unpublished PhD thesis, University of London.

Webb, S. and Webb, B. (1910). *The state and the doctor*. Longmans, London.

Webster, C. (1982). Healthy or hungry thirties? *History Workshop* **13**, 110.

Webster, C. (1985). 'Health, welfare and unemployment during the depression'. *Past and Present* **109**, 204.

Weinerman, E.R. (1951). *Social medicine in Western Europe*. School of Public Health, University of California.

Wilkinson, A. (1980). The beginning of disease control in London: the work of the medical officers in three parishes, 1856–1900. Unpublished DPhil thesis, University of Oxford.

Winter, J.M. (1982). Aspects of the impact of the First World War on infant mortality in Britain. *Journal of European Economic History* **11**, 713.

Wofinden, R.C. (1959). Medical administration. *Public Health* **73**, 343.

Wohl, A. (1983). *Endangered lives: public health in Victorian Britain*. Harvard University Press, Cambridge, Massachusetts.

Woods, R. and Woodward, J. (ed.) (1984). *Urban disease and mortality in nineteenth century England*. Batsford Academic, London.

Wrigley, E.A. and Schofield, R.S. (1981). *The population history of England 1541–1871*. Arnold, London.

Young, J. (1933). The medical schools and the nation's health. *Lancet* **i**, 61 and 119.

3

The origins and development of public health in Japan

KOZO TATARA

Introduction

Japan is a small country at the eastern edge of the Asian continent, and is one of the countries furthest away from western society.

Buddhism is reported to have been introduced to Japan in AD 538. The first well-established period in the history of Japan as a nation was Nara (710–94), when there was a capital called Heijokyo at the site of the present Nara city. The following periods up until Meiji were Heian (794–1192), Kamakura (1192–1333), Muromachi (1338–1573), Azuchi-Momoyama (1573–1603), and Edo (1603–1868). In these periods, the capital of the country, where the Emperor resided, was in Kyoto.

The first legislative code of the nation, the *Taiho Ritsuryo*, was decreed in 701. There was already a chapter for a medical care system, the *Ishichi Ryo*, in this code. The first formal written history of the nation, the *Kojiki*, was published by the central Government at Nara in AD 712.

With reference to exchange with foreign countries, there exists a golden seal that is evidence that Japan had formal contacts with China in AD 57, and there was also active communication with Korea. However, there was almost no direct exchange with western countries.

On 25 August 1543, a ship came ashore on the southernmost island of the country with about 100 passengers led by two Portuguese. They were the first western people to visit Japan. This was also the year in which Andreas Vesalius (1514–64) published *De Humani Corporis Fabrica*, placing the science of anatomy on a firm basis. After this time, some limited communications between Japan and western countries began. In 1555, a Portuguese surgeon, Luis de Almeida, visited Japan, first opening an orphanage in Oita, and in 1557 building a hospital that is said to be the first western-style hospital in Japan. In 1584, he died at Amakusa in Kyushu.

In 1639, the central government at Edo decided to close Japan's doors to the rest of the world except for one specified place called Deshima in Nagasaki, where Dutch merchants were allowed to continue trading. At about the same time, William Harvey performed an experiment to show that the action of the heart circulates blood through the arteries. This finding had a great impact on medical science, and symbolized the beginning of a new approach to medicine in the western world. However, from this time until 1853, when four American gunboats under the command of Commodore Perry visited Uraga, a small village near Edo, the Japanese people continued their relatively peaceful lives based on traditional ways of thinking and working.

Even during these 220 years, the visiting physicians and surgeons at Deshima had a strong influence on the history of medicine and science of the country. Such visitors had totalled about 150 by 1850. The most influential were Caspar Schamberger [1649–50]*, Willem ten Rhyne [1674–6], Engelbert Kaempfer [1690–2], Carl P. Thunberg [1775–6], Jonkheer Philipp Franz Balthasar von Siebold [1823–9, 1859–62], Otto Gerhard Mohnike [1848–51], Jonkheer Johannes Lydius Catharinus Pompe van Meerdervoort [1857–62], Anthonius Francois Bauduin [1862–70], and Koenraad Wolter Gratama [1866–71]. Von Siebold published *Nippon*, *Fauna Japonica*, and *Flora Japonica*, which described the life of people, animals, and plants in Japan, introducing them to the West.

In 1774, Genpaku Sugita (1733–1817), a Dutch scholar, published the first western-style anatomical textbook in Japan, the *Kaitaishinsho*, a translation of *Tabulae Anatomicae* by Johann Adam Kulumus (1689–1745) with the help of Ryotaku Maeno (1723–1803) and Junan Nakagawa (1739–86).

It is of special interest in the history of public health that Jenner's theory of vaccination for smallpox had already reached this country by 1801 or 1802, which was only a few years after the first successful vaccination by Jenner himself. Mohnike introduced the practice of vaccination with a cowpox vaccine imported from Batavia in Java in 1849. Up to

* Square brackets indicate time spent in Japan.

then, the only inoculation that had been tried used fluid from human pox, the original method of which was introduced by a Chinese in Nagasaki in 1744.

In 1838, a famous school for the teaching of Dutch culture and medicine, the Tekijuku, was founded by Koan Ogata (1810–63) in Osaka. In 1858, an official place for vaccination for smallpox, called the Shutosho, was built at Otamagaike in Kanda, Tokyo, with government approval. This Shutosho was renamed the Seiyo Igakusho, meaning 'the place for learning western medicine', under a chancellor, Shunsai Otsuki (1803–62), in 1861. The Seiyo Igakusho became a medical school with a hospital after annexation to the army hospital in Yokohama in 1869, and was renamed the Daigaku Toko in the same year. It became the Tokyo Igakko, or Tokyo Medical School, the origin of the Faculty of Medicine of the University of Tokyo, in 1874.

In 1868, the Seitokukan in Nagasaki, which had its origin in the Yojosho, the medical institution founded by Pompe in 1861, was renamed the Nagasaki Medical School, and in 1869 the Osaka Medical School was established.

Origins of public health

Administrative system

The Meiji Government (1868–1912) organized for the first time the modern political system in Japan and opened the door to the world. The new Government moved the capital from Kyoto to Edo, which it renamed Tokyo. The Government encouraged open political discussions by the people, and stated the need for co-operation among all persons, irrespective of class or social status, and new national government policies. The Government also expressed its intention to discard obsolete customs and to absorb scientific knowledge from western countries in particular.

The Government founded the Department of Medical Affairs in the Ministry of Education in 1872, and it became a Bureau in 1873. In 1875, the Bureau of Medical Affairs was moved to the Home Office and its name was changed to the Bureau of Hygiene.

A new system of local government under the Meiji administration was begun by the abolition of the feudal system and the plans for a modernized prefectural hierarchy in 1871.

The main landmark that signifies the start of public health in Japan is the decree named the Isei ('medical system') in 1874. This was drafted by Sensai Nagayo (1838–1902), who became the Chief Medical Officer of the Bureau in 1873. Nagayo had stayed in the US in 1871–2 and in Europe in 1872–3 as a member of a mission from the government. At first he had planned to learn about the system of medical education in western countries, but having heard the expressions of Sanitaetswesen or oeffentliche Hygiene in various places, he came to think that it was more important to learn about public health. Japan had a long history of communication with the Netherlands, and Nagayo had studied Dutch medicine at Nagasaki, so he stayed in Amsterdam and learned about the sanitation system that was run by the police department and the local government.

When Nagayo developed the idea of the Isei, it was natural that its contents were strongly influenced by western concepts, because in that period public health movements were being enthusiastically promoted in western countries; for example, in England, the Local Government Board was founded in 1871 and the Public Health Act was passed in 1875.

The Isei consisted of 76 sections and covered a wide range of legislative needs, from the field of public health administration to that of medical education. Provisions for public health administration were outlined in sections 1–12 of the Isei. The contents of the Isei concerned:

1. *Public health administration*. A Chief Medical Officer was to be appointed to the Bureau of Medical Affairs of the Ministry of Education. The country was to be divided into seven regions, and a Department of Hygiene was to be established in each region. Front-line Medical Supervisors, selected from among local doctors, pharmacists, and veterinarians, were to be appointed to each Department of Hygiene.
2. *Medical education*. In each university region, one medical school was to be established together with an affiliated hospital. Medical schools were to teach a 3-year preliminary course and a 5-year regular course, which would include anatomy, physiology, pathology, pharmacology, medicine, surgery, and legal and preventive medicine.
3. *Hospitals*. To open a hospital, it was necessary to receive approval, through the Department of Hygiene, from the Ministry. The superintendent of the hospital was required to have a medical licence.
4. *Doctors, midwives, and acupuncturists*. A system for medical licences was to be established. Dispensing by doctors was not to be approved. A system for midwifery licences was to be established.
5. *Pharmaceutical affairs*. Permission to dispense drugs was to be limited to licensed pharmacists.

Although the Isei was introduced in Tokyo, Kyoto, and Osaka in 1875, it was not a formal Act, and not all of its sections were actually implemented. However, with the ideas set out in the Isei, Japan had an enlightened start in establishing public health administration by following the examples of European countries.

Cholera epidemics

The first large-scale epidemic of cholera recorded in Japan was in 1822, and the next one was in 1858. In the epidemic of 1858, cholera spread from Nagasaki to the entire country, and the number of deaths in Edo is reported to have been 28 421.

The first cholera epidemic in the Meiji period was in 1877. The number of cases was 13 816, and the total number of deaths was 8027. Epidemics were again seen in 1879 and 1886, with 162 637 and 155 923 cases, and 105 786 and 108 405 deaths, respectively.

With the history of numerous great epidemics of cholera, and with the danger of similar outbreaks at any time, the Government was under pressure to carry out effective policies in the field of public health.

In 1877, the Home Office issued a guide for the prevention of cholera, which was a predecessor of the Infectious Diseases Prevention Act. Also, Regulations for the Prevention of Infectious Diseases were issued in 1880. In 1879, the Central Sanitary Council was established at the Home Office to discuss methods to combat cholera with special reference to the quarantine of ships, and a Local Sanitary Council was also founded in each prefecture. This movement began about 30 years after the establishment in England of the General Board of Health and the Local Boards of Health by Edwin Chadwick (1810–90). At the same time, in each prefecture, a Department of Hygiene was established, and in each town and village a Sanitary Supervisor was appointed. In 1883, the Great Japanese Sanitary Association was founded in the private sector to promote the sanitary movement by the general populace.

These measures constitute the backbone of public health in Japan that has continued to this day. However, the Meiji Constitution was decreed in 1889 and the first parliament assembled in 1890. To replace the traditional system, a new cabinet system was introduced in 1885. The new city and town/village systems for local government were introduced in 1888, with a prefectural system in 1890. Japan had become a young-adult country 20 years after the Meiji Restoration. At this point, the central Government began gradually to strengthen its control of local governmental activities.

In 1885, the system of central administration was largely reformed upon the commencement of the new cabinet system. As a result of this reform, the Local Sanitary Councils' and the Sanitary Supervisors' posts were abolished, and in 1893 it was decided that all such activities were to be supervised by the police authorities except in Tokyo; this system continued until 1942.

Population surveys

Sanitary statistics for Japan have two origins: the Household Act (1871) and the Isei (1874).

The Household Act established a system for population surveys by defining districts and appointing district supervisors, called Kucho, who were assigned to report the numbers of households, births, deaths, and families moving into and out of the district. The Isei also required doctors to report cases of infection and the causes of deaths to the Medical Supervisor.

The influential *Annual Report of the Bureau of Hygiene* was first published in 1877. This is the oldest record in Japan of the sanitary conditions of the public and has been published annually since then, except during the Second World War.

In 1886, the Home Office developed a model form of sanitary report, on which modern sanitary statistics in Japan are based. In 1898, a new Household Act was passed, and a population survey was carried out under a central collection system by the gathering, in cities, towns, and villages, of cards detailing births, deaths, stillbirths, marriages, and divorces. Starting in 1920, a national census has been taken every 5 years.

The population of Japan is estimated to have been 30–32 million during the 120–130 years after the early part of the eighteenth century. In 1872, the population was 34.8 million; it gradually increased to about 50 million in 1910, 70 million in 1935, and 100 million in 1970. The birth rate was 32.4 per 1000 population in 1900, peaking at 36.2 in 1920, and decreasing to 18.8 by 1970. The death rate was about 20–21 per 1000 population during the Meiji period, and was 6.9 in 1970. The infant mortality rate was 155 per 1000 live births in 1900, when infant deaths accounted for 20 per cent of the total deaths, and it was 162 in 1910, 90.0 in 1940, 60.1 in 1950, and 13.1 in 1970.

Prevention of infectious diseases

The new Government sent circulars to the prefectures to promote the vaccination of the public in 1870.

In 1876, the Home Office issued regulations for smallpox prevention. It was made compulsory to receive a vaccination within 1 year of birth and then twice more at intervals of 5–7 years. In 1909, the Smallpox Vaccination Act was passed, which ordered that a record of vaccinations was to be written in the household documents kept by the local authorities.

In 1880, as mentioned above, the Regulations for the Prevention of Infectious Diseases were drawn up. This document included 24 sections for the prevention of cholera, typhoid fever, dysentery, diphtheria, epidemic typhus, and smallpox. It is reported that the number of deaths from these six infectious diseases in 1886 was 150 771, which was 16 per cent of the total number of deaths (938 343). Starting in 1890, following the establishment of the city and the town/village systems in 1888, it became the general rule in Japan that the responsibility for the prevention of infectious diseases was that of the city, town, or village. In 1897, the Infectious Diseases Prevention Act was passed to reflect progress in medicine and changes in the system of local government. With this Act, plague and scarlet fever were added to the official list of infectious diseases.

For the prevention of venereal diseases, health examinations of prostitutes were carried out for the first time at Yokohama in 1867. Regulations for the Supervision of Prostitutes were passed in 1900.

The Leprosy Prevention Act was passed in 1907, the Trachoma Prevention Act in 1919, and the Helminthiasis Prevention Act in 1931.

Prevention of tuberculosis

According to one of the Annual Reports from the Bureau of Hygiene, the number of deaths from tuberculosis in 1882 in the Tokyo area was 2355. The Government founded a treatment room for tuberculosis in Tokyo Medical School, and opened the first sanatorium at Suma in Hyogo Prefecture in 1889.

In 1899, the Government carried out a national survey of the number of deaths by tuberculosis. The total number was 66 408, which was 15.3 per 10 000 population and 71.2 per 1000 deaths. In 1908, Robert Koch (1843–1910) visited Japan, and in 1911 the Emperor expressed a desire to promote welfare services for the poor. As a result of these events, general social interest in the prevention of tuberculosis was greatly stimulated.

In 1913, the Japanese Association for Tuberculosis Prevention was established in the private sector to promote health education concerning the prevention of tuberculosis.

In the Taisho period (1912–26), the annual number of deaths from tuberculosis was more than 100 000. The death rate was more than 200 per 100 000 population, and the number of cases was estimated to be more than 500 000. These numbers increased steadily and, in response, the Tuberculosis Prevention Act was passed in 1919.

In the Showa period (1926–1989), the death rate from tuberculosis in 1943 was 235.3 per 100 000 population; in the same year it was 60.9 in England and Wales, 42.6 in the US, and 147.0 in France.

Foundation of the medical system

In the first year of the Meiji period, the new Government proclaimed that western medicine was to be adopted, and that a medical licence system should also be established. The largest institution to adopt western medicine was the army. From the end of the Edo period, Dutch medicine had had a particularly strong influence. However, the new Government first thought of introducing British medicine by inviting William Willis [1861–81] to teach in the future, but decided in 1869 to adopt German medicine instead, inviting Leopold Mueller [1871–5], a teacher of surgery, and Theodor Eduard Hoffmann [1871–5], a teacher of medicine, from Germany to lecture at the Daigaku Toko. Erwin Baelz [1876–1905], in particular, contributed to medical education during its formative years in Japan.

The Tokyo Medical School was reorganized and renamed the Faculty of Medicine of the University of Tokyo in 1877, and the department system was introduced there in 1893 with 23 departments. The professors of these departments were all Japanese. The first Professor of Hygiene was Masanori Ogata (1853–1919), who had studied under Professor Max von Pettenkofer (1818–1901) in the Department of Hygiene of the University of Munich.

Hospitals

The Seyaku-in and Hiden-in, built with the support of the Empress Komyo-kogo at the Kofukuji Temple in Nara in 723, were the most famous and oldest places in Japan to be provided with the staff and facilities to take care of and shelter the sick and poor.

In 1722, the Koishikawa Yojosho was built to offer hospital care to the poor in Edo. However, the first western -style hospital was the Nagasaki Yojosho mentioned above. It had eight wards with fifteen beds each, four wards for isolation and operations, and a preparatory room. In 1874, the total number of hospitals in the country was 52, and by 1882 this had increased to 626. In 1879, a hospital for cholera patients was founded; by 1911, the number of hospitals for infectious diseases in Japan was 1532.

A mental hospital attached to the Kyoto Prefectural Hospital was founded in Kyoto in 1868, and another was founded in Tokyo in 1879.

Doctors

In 1875, an examination system for issuing licences to practise medicine was started in Tokyo, Kyoto, and Osaka, but under this system practitioners who were already practising were allowed to continue without examination. In 1878, the examination system was extended to cover almost the entire country. The establishment of an examination system that adopted western medicine, to the exclusion of traditional Chinese-style medicine, was the basis for the dependence on western medicine of the hierarchy of medical education and medical practice in Japan.

The total number of doctors in 1874 was 28 262, of whom 23 015 were traditional-style doctors and 5247 were western-style.

In 1906, the Medical Act was passed. The regulations for medical licences and practice were organized by this Act. In the process of establishing a medical system in Japan, controversy between traditional-style and western-style doctors caused serious problems. Traditional Chinese-style medicine had a history of 1000 years of achievements, and so it was difficult for non-traditional doctors to import western medicine and to put it into practice. At first, such doctors in Japan tried to separate dispensing from medical practice, as was the custom in European countries, and as was set down in the Isei. However, this difference was not acceptable to the patients, who expected their doctors to dispense medicine like traditional-style doctors. It was suggested that traditional-style doctors should be prevented from dispensing drugs, but this would have amounted to recognition of traditional medicine as a formal science. So there was a sudden change in policy, and western medical practice was made competitive with traditional medical practice by giving western-style doctors permission to dispense drugs. This policy has had lasting repercussions on medical care in Japan.

The number of medical practitioners who had studied modern western medicine was 9.5 per cent of the total of 40 880 practitioners in 1884, 55.3 per cent of the 35 289 practitioners in 1904, and 99.5 per cent of the 64 234 practitioners in 1939.

The Japan Medical Society was founded in 1890, the Japan

Anatomical Society in 1893, the Japan Surgical Society in 1898, the Japanese Society of Internal Medicine in 1903, and the Japanese Society for Hygiene in 1904.

During this period, Japanese workers of international fame in the field of experimental research in medicine were Shibasaburo Kitasato (1852–1931), who succeeded in culturing the tetanus bacillus under the auspices of Robert Koch in 1889, and who developed serum therapy with Emil A. von Behring in 1890; Kiyoshi Shiga (1870–1957), who discovered *Shigella dysenteriae* in Tokyo in 1897; Jokichi Takamine (1854–1922), who purified epinephrine in New York in 1900; Sunao Tawara (1873–1952), who discovered the atrioventricular node; Sahachiro Hata (1873–1938), who prepared arsphenamine in co-operative research with P. Ehrlich in 1910; and Hideyo Noguchi (1876–1928), who succeeded in making a pure culture of *Spirochaeta pallida* in 1911.

Dentists

Dentistry was introduced to Japan in 1860 by the American dentist William Clark Eastlake [1860–9]. Modern dental practice was begun by Einosuke Obata (1850–1909) in 1875. The words 'dental practice' appeared formally for the first time in the Regulations of Examination for Medical Practice in 1879. In 1883, the examination for licences to practise dentistry was completely separated from the one for medical practice, and the Dentists' Act was passed in 1906 at the same time as the Medical Act. The number of dentists was 1125 in 1910 and their number per 100 000 population was 2.3. These figures were 9983 and 17.0, respectively, in 1924, and 23 311 and 32.7 in 1939.

Pharmacists

The Isei in 1874 set out for the first time the required qualifications of pharmacists. The Japan Pharmacopoeia was begun in 1886, and a control system for pharmacists and their dispensaries was established by the Regulations for Drug Trade and Drug Management in 1889.

Beginning in 1890, it was compulsory for those having a pharmacy to take the pharmacists' examination, and the Pharmacists' Act was passed in 1925.

The number of pharmacists was 4643 in 1910 and their number per 100 000 population was 9.4. These figures were 12 267 and 20.8, respectively, in 1924, and 29 833 and 41.8 in 1939.

Nurses

Nursing education in Japan began in Tokyo in 1884, when the Tokyo Voluntary Hospital Nursing School was founded with the help of the American nurse E. Read [1884–7], and in Kyoto in 1886, when the Kyoto Nursing School was founded with the help of the American physician John Cuting Berry [1872–93] and the American nurse Linda Richards [1886–90].

Such facilities gradually appeared throughout the country. These schools all followed the education system of the Nightingale Nursing School in London. In 1915, regulations for nurses were established by which nursing qualifications were made uniform nationally. The number of nurses was 11 574 in 1910 and their number per 100 000 population was 23.5. These figures were 42 367 and 72.0, respectively, in 1924, and 127 466 and 178.6 in 1939.

Environmental health

Removal of nuisances

The epidemics of cholera and typhoid that occurred around 1870 led people to acknowledge the necessity for cleanliness in the community. In 1899, the first patients with plague appeared in Japan; the disease spread to 62 people, and killed 55 of them. The Government promoted the eradication of rats from communities by circulars issued by the Home Office.

In 1900, the Nuisances Removal Act was passed, defining nuisances and assigning the responsibility for their removal to the owner, the user, and the occupier of the land; in municipalities, it made provision for the methods of removal and transportation of nuisances.

Water supply and sewerage

Edo, it is estimated, already had a population of 1 million by the seventeenth century; it was one of the largest cities in the world. The Kanda-josui and the Tamagawa-josui, constructed in the early part of the Edo period to supply clean water to the residents of Edo, were very advanced water supply systems.

When Japan opened her doors to the world, various virulent diseases immediately entered the country. Existing facilities could not prevent these attacks.

In 1887, after several cholera epidemics, the Central Sanitary Council submitted a document to the Government that strongly advocated the construction of systems for water supply and sewerage.

The Government passed the Water Supply Act in 1890 and the Sewerage Act in 1900. The 1890 Act gave responsibility for construction of water supply systems to the cities, towns, and villages.

Sanitation of food

The first sanitary measure carried out by the Government for protection of the quality of food was a regulation published in 1878 about the use of aniline for colouring food and drink.

The Regulations of Food and Drink Act was passed in 1900. This Act had national, comprehensive measures for food sanitation and was the basis for the administration of sanitation of food in Japan thereafter.

Factory legislation

Many factories were built after the Meiji Restoration, and the number of factory workers rapidly increased. However, general conditions in the factories were poor and workers suffered various physical impairments and occupational or infectious diseases. At this time, the regulation of factory sanitation was under the control of the police department, which was part of the local government.

In 1877, the Regulations for Manufacturing Industries were set out in Osaka and, in 1890, the Mining Industries Regulations Act was passed. This Act made provisions for the nation-wide improvement of conditions of miners by regulating the policing of miners, protecting the lives of miners, restricting their hours of work, and, especially, restricting the hours and the kinds of work undertaken by women and young persons. This Act was introduced in 1892, but some parts of its legal procedures were not implemented by the Ministry of Agriculture and Industry; in 1905, the Mining Industry Act was passed.

In 1913, Osamu Ishihara (1885–1947), a consultant physician at the Ministry of Agriculture and Industry, gave an address entitled *Joko to Kekkaku* ('female operatives and tuberculosis') at the Assembly of the Society of State Medicine, reporting that many young factory girls aged 12–13 who worked for 12 hours without breaks on a two-shift basis were dying from tuberculosis.

The Factory Act was passed in 1911 and was put into effect in 1916. This Act protected workers in enterprises employing 15 or more workers as follows:

1. Persons aged under 12 years were not permitted to work.
2. Various restrictions were put on workers aged under 15, women, persons receiving medical treatment, and pregnant women.
3. The administrative authorities were to regulate and advise employers concerning safety, sanitation, discipline, and public morals.

With this Act, responsibility for the health of workers in factories moved to the Ministry of Agriculture and Industry, having previously been that of the police in the prefectures.

About 230 000 workers were affected by the Mining Industry Act, and about 630 000 workers by the Factory Act. These Acts, even though not completely satisfactory, improved factory sanitation throughout Japan.

School health

In 1872, the modern education system in Japan was started. In 1891, an officer for school health was appointed for the first time at the Ministry of Education, and a survey of school health was carried out. In 1897, the Regulations for School Cleanliness and for Periodical Physical Examinations were decided, and, in the following year, the system for providing school doctors in public schools was begun. In 1898, the Regulations for the Prevention of Infectious Diseases and Sanitation in Schools were promulgated. In 1900, the Department of School Health was founded at the Ministry of Education.

Social health insurance

The Japanese had various kinds of mutual aid systems even during the Edo period, but a new social insurance system was developed by the Government for the protection of workers during the rapid growth of industry that followed the wars with China in 1894 and with Russia in 1904.

With the Factory Act of 1911, employers were obliged to help their workers if they became ill or were injured, provided that the disorder arose when the worker was working in the factory.

The social insurance system in Germany that was started in 1883 by Chancellor Bismarck (1815–1898) was introduced and much discussed in Japanese academic societies. The most influential proponent of such a system in Japan was Shinpei Goto (1857–1929), a Chief Medical Officer at the Home Office, who advocated the need for an insurance system for workers' sickness after studying in Germany in 1892, and who submitted to the Prime Minister, Hirofumi Ito (1841–1909) a Workers' Sickness Insurance Bill in 1898.

In 1905, one private spinning company organized a mutual aid system for 12 000 workers to offer financial help at times of illness, injury, and bereavement; all workers contributed 2 per cent of their salary to the system. A similar system was also organized for workers at a public iron works in that year.

Although Japan's economic strength was largely established by the time of the First World War, inflation eroded the quality of the workers' lives and prompted movements for the creation of a welfare system for the labour force. In particular, various reports and recommendations published by the International Labour Organization (founded in 1919) had great impact on the discussion and implementation of the measures advocated by these movements.

Health insurance for workers

The Health Insurance Act was passed in 1922 and came into effect in 1926. The insurance system established by the Health Insurance Act took the following form:

1. It covered sickness, injury, and death, whether from an occupational cause or not, and also the cost of giving birth.
2. Workers who were employed in factories under the Factory Act or the Mining Industry Act and who were paid 1200 yen per year or less had to be insured.
3. The insurer was, as a rule, the Government or a self-management co-operative. Industries that employed more than 300 workers were permitted to organize co-operatives.
4. The benefits were restricted to the insured. They included medical care services, a sickness allowance, funeral costs, the cost of giving birth, and an allowance for giving birth.
5. The contribution was, as a rule, shared by the insurer and the insurant.

These features and the methods of management of the system show that the Act had been influenced by the National Health Insurance System started in 1913 in the UK. Soon after the system began, 1 140 865 workers were insured by the government-managed system, and 800 581 were insured by the co-operative-managed system, which included 316 co-operatives. Of the population, 3.2 per cent was insured. The number of panel doctors was 32 155, which was about 70 per cent of all doctors. Panel practice was a small part of the work of most physicians. The remuneration for panel doctors in the Government-managed insurance was paid to the Japan Medical Association on a capitation basis until 1942. Remuneration in the co-operative-managed system was paid by various methods up to the same year. The capitation system in both insurance systems was then abolished, and each panel doctor was remunerated on an item-for-services basis. Claims were settled 1 635 499 times, and the total payment was 7 081 187 yen in 1926 (about 17 700 million yen in 1988 currency). The number of the insurants was 3 368 902 in 1936, and 5 625 034 in 1941.

The Seamen's Health Insurance Act was passed in 1939, and the number covered by this Act was 102 140 in 1940.

Health insurance for farmers

The world economic crisis in 1929 left rural society in Japan severely damaged. The government had to promote policies to protect the welfare of farmers, with special reference to their state of health. This was the situation when the National Health Insurance Act was passed in 1938. The insurance system for farmers differed from that for other workers in the following ways:

1. The insurer was an Insurance Co-operative established in cities, towns, and villages.
2. The establishment of an insurance system by each local government was not compulsory.
3. Each household in the district could be a member of the Insurance Co-operative; the members and their dependants were insured by the system; and it was not compulsory for persons to join the system.
4. Legally provided benefits were medical care services, and optional benefits were medical care in giving birth, funeral costs, and maternity allowances.
5. Medical care services provided as benefits were governed by contracts between the insurer and the medical profession.

The numbers of insurers and insurants were 168 and 578 759 in 1938, respectively, and 10 158 and 37 291 721 in 1943.

In 1944, 68.5 per cent of the total population of 73 060 000 was covered by some kind of medical insurance.

Foundation of the Ministry of Health and Welfare

In 1916, a Council for the Investigation of Health and Hygiene was started, and 34 members of the Council were appointed under the chairmanship of the Vice-Minister of the Home Office. The subjects for investigation were:

(1) babies, infants, schoolchildren, and juveniles;
(2) tuberculosis;
(3) sexually transmitted diseases;
(4) leprosy;
(5) mental diseases;
(6) food, clothing, and housing;
(7) sanitary conditions in rural areas;
(8) statistics.

The Council was established because of the impact of discussions on the growth of population in western countries, where the death rates of infants, children, and young people had already decreased, while those in Japan had rapidly increased. The recommendations of the Council had strong effects on Government policies thereafter. In particular, the health conditions of the people in local areas were to be reported annually to the Home Office. Thus, the central administration system for public health was gradually strengthened. In 1921, the number of departments in the Bureau of Hygiene was increased to five: health, prevention, quarantine, medicine, and investigation.

In 1937, the Public Health Centre Act was passed. A plan was made to construct 550 public health centres in the following 10 years for the purposes of health education and counselling. The staff of each centre was to be made up of two doctors, one pharmacist, one clerk, three instructors, three public health nurses, and one assistant. In the regulations of the Act, the position of public health nurse was mentioned formally for the first time. The nurses were required to be at least 18 years old. By 1925, two model health centres had already been built in Tokyo and Saitama with the financial support of the Rockefeller Foundation. In 1937, 49 public health centres had been built, and 770 were built by 1944.

In the same year as the birth of the National Health Insurance System, 1938, the Ministry of Health and Welfare was also born, unifying the various kinds of affairs controlled under the Bureaux of Hygiene and Society at the Home Office, the Ministry of Education, and the Ministry of Trade and Industry. The new Ministry had five bureaux: the Physical Activity Bureau, Hygiene Bureau, Prevention Bureau, Society Bureau, and Labour Bureau.

The Relief for the Poor Act was passed in 1929, the Workers' Impairments Assistance Act in 1931, and the Maternity and Child Protection Act in 1937.

Public health in the post-war period

In 1946, a new constitution was passed. In chapter 25 of the constitution, it is written that 'all people shall have the right to maintain a certain standard of healthy and cultured life, and, to achieve this purpose, the state shall try to promote and improve the conditions of social welfare and security, and of public health'.

After the Second World War, the policies for public health in Japan were based on the ideas written in the constitution.

Administration system

In 1946, three bureaux, Public Health, Medical Affairs, and Prevention, were established at the Ministry of Health and Welfare to promote health and provide social services.

At the suggestion of General Headquarters of the Occupying Army, the Council for Medical Education was founded at the Ministry of Education in 1946. The Council recommended the establishment of a Department of Public Health to be independent of the Department of Hygiene in each medical school. The first, at the University of Tokyo, Osaka University, and Niigata University, were established in July 1947. In 1967, 35 medical schools among the total of 46 had Departments of Hygiene and Public Health. The Japanese Society of Public Health was established in 1947.

The main national institutions for public health to be reformed or established in this period were the National Institute of Hygienic Sciences (established in 1874), Institute of Public Health (1938), Institute of Population Problems (1939), National Institute of Nutrition (1947), National Institute of Health (1947), National Institute of Hospital Administration (1949), National Institute for Mental Health (1952), and National Institute of Leprosy Research (1954).

The system of local government was changed from one based on Continental law to one based on Anglo-American law; the change was a move from centralization to decentralization. The Local Government Act of 1947 made it compulsory for the Government to establish prefectural Bureaux of Hygiene and Welfare.

In 1947, a new Public Health Centre Act was passed, and public health policies founded on the needs of local residents were promoted.

In 1960, public health centres were classified as being one of four types: urban, urban–local, local, and rural. By 1970, the number of public health centres was about 830.

Public health centres are generally expected to promote the planning of health services for residents in an area, to collect information on the health of the residents, to implement projects for the prevention of public nuisances, to inspect environmental health, and to conduct laboratory examinations. The major activities of health centres reported for 1986 are shown in Table 3.1.

Table 3.1. Major activities of health centres, 1986

Activity	Absolute number	Number per health centre*
Group health examinations	375 128	442
Health examinations (multiple count)	13 831 795	16 292
Pregnant women and mothers given health guidance (multiple count)	554 355	653
Infants and children provided with health guidance (multiple count)	3 516 257	4 142
Pregnant women and mothers given guidance by visiting public health nurses (multiple count)	297 521	350
Newborn babies cared for by visiting public health nurses (multiple count)	385 041	454
Mothers given guidance on bringing up children	86 400	102
Dental examination or guidance	3 085 547	3 634
Counselling on nutrition	4 742 527	5 586
Counselling on mental health	747 018	880
Consultation with medical social workers	216 473	255
Meetings for health education	299 225	352
Home visits by public health nurses	1 246 425	1 468
Counselling by public health nurses at health centres	17 470 319	20 578
Immunization	6 206 168	7 310
Tests for dysentery, typhoid fever, and paratyphoid fever	5 925 775	6 980
Food samples inspected	304 400	359
Inspections of other materials	37 147 376	43 754

* The number of health centres was 849 in 1985.

Source: Statistics and Information Department, Ministry of Health and Welfare, Japan (1988). Statistics on activities of health centers. In *Health and welfare statistics in Japan*, p. 84. Health and Welfare Statistics Association, Tokyo.

The Public Health Nurses', Midwives', and Nurses' Act was passed in 1948.

Prevention of infectious diseases

The number of cases of epidemic typhus was more than 32 000 in 1946, and cholera also appeared for the first time since the epidemic in 1920. Smallpox was common at this time.

It was necessary for the government to promote inoculation to prevent epidemics of these diseases. The Inoculation Act was passed in 1948, based on the findings of the effects of inoculations used during the epidemics of typhus in 1945. The Act made it compulsory for persons to receive periodical inoculation for smallpox, typhoid fever, paratyphoid fever, diphtheria, pertussis, and tuberculosis; responsibility for carrying this out lay with the heads of the cities, towns, and villages. Inoculation was required for epidemic typhus, cholera, plague, scarlet fever, influenza, and Weil's disease if necessary. In 1951, inoculation for tuberculosis was continued under the control of a new Tuberculosis Prevention Act. In 1958, scarlet fever, and in 1976, typhoid fever, paratyphoid fever, epidemic typhus, and plague were deleted from the list of required inoculations. In 1961, poliomyelitis was added to the list, as were measles, rubella, and Japanese encephalitis in 1976. This Act is still in effect today and has been invaluable for the prevention of infectious diseases in Japan from the Second World War onwards.

The Venereal Diseases Prevention Act was passed in 1948.

In 1985, the first patient in Japan officially confirmed to have acquired immunodeficiency syndrome (AIDS) was reported. By December 1988, 97 patients with AIDS and

1065 persons infected by human immunodeficiency virus (HIV) had been reported. However, AIDS is not a reportable disease in Japan. The Government submitted an AIDS Prevention Bill to Parliament in March 1988.

Tuberculosis

The death rate from tuberculosis decreased after the Second World War. However, it was 146.4 per 100 000 population in 1950 (having been 212.5 in 1940) and was still the largest cause of death until 1950.

In 1951, a new Tuberculosis Prevention Act was passed, the main items of which were:

(1) annual health examinations for persons who work in special services,or who live and work in groups, and for people who live in areas where tuberculosis prevails;
(2) annual inoculation, if needed, of persons under 30 years of age and those who live or work in groups;
(3) systems for the registration of patients and for home visits by public health nurses when necessary;
(4) public payment of the cost of medical care;
(5) improvement of public medical institutions.

The newly established public health centres were very important in the implementation of the policies of the new Act.

Environmental health

Environmental conditions in Japan have rapidly deteriorated as the growth of the economy has speeded up.

In the summer of 1955, a curious disease appeared in the west part of Japan among bottle-fed babies who were found to have been fed on milk processed at one particular factory. The causative substance, arsenic, was identified in the milk. By February 1956, the number of babies poisoned was reported to be 12 159, in a total of 27 prefectures; 131 of the babies died.

From 1953 to 1961, patients with Minamata disease appeared in a fishing area in Kumamoto Prefecture, Kyushu. The cause of the disease was organic methyl mercury discharged from a chemical factory. The number of patients who died was already 51 in the year of the appearance of the disease. Numbers increased greatly and the incident was one of the most serious problems caused by the deterioration in environmental standards in Japan after the Second World War. The same disease appeared in Niigata in 1964, when 26 patients were diagnosed and 5 patients died.

The Government began to promote the oil–chemical industry in 1955, and one site at which such a production system was opened was Yokkaichi. The main cause of respiratory disorders in this city was SO_2 in the air. Patients, whose disorder was called 'Yokkaichi asthma', gradually appeared from 1959, and the number peaked in 1965 when about 8 per cent of residents aged over 40 years and 10 per cent of those aged over 50 years suffered from this asthma.

In 1955, a disease called itai-itai-byo ('very painful disease') was reported at an academic meeting. It was acknowledged by the Ministry that the cause of this disease was chronic poisoning from cadmium released from a mining operation. The number of cases identified by the Special Research Group by 1975 was 129, of which only 3 were men.

Around 1955, a new neurological disease appeared in various places in Japan. The first case was reported at an academic meeting in 1958. In 1964, the disease was named subacute myelo-opticoneuropathy (SMON). The incidence of this disease has increased since around 1967, and the number of officially recognized cases was 4355 at the end of 1968. In September 1970, the Ministry of Health and Welfare decided to suspend the sale of clinoquinol because of the finding that this drug, used to treat digestive disorders, was the cause of the disease. After this measure, no new cases of the disease have appeared.

Patients poisoned by polychlorinated biphenyls (PCBs) appeared in 1968. The direct cause was the intake of oil contaminated with PCBs in a food processing factory. The number of officially recognized cases was 1540 by 31 May 1976.

These were major failings in public health in Japan, and, to remedy the situation, the Basic Project against Public Hazards Act was passed in 1967, and the Prevention of Air Pollution Act and the Prevention of Water Pollution Act were passed in 1968 and 1970, respectively. In July 1970, a Project Office for the Prevention of Public Hazards was organized under the chairmanship of the prime minister. The Ministry for the Environment was established in 1971.

Maternity and child health

The Child Welfare Act was passed in 1947. The regulations for child health stated in the Act were:

1. The governor of each prefecture is to promote policies for health education during pregnancy and about baby care.
2. The governor is to ensure that health examinations for babies and infants are made.
3. Pregnant women are to register their pregnancy and receive a Maternity and Child Health Note.
4. Institutional care for all pregnant women who need it, even if they cannot afford it, is to be made available.

In the promulgation of these regulations. public health centres have been indispensable. Starting in 1958, many Maternity and Child Health Centres were built in places far from hospitals or health centres by cities, towns, and villages.

Based on the excellent results of the Child Welfare Act, in 1965, the Maternity and Child Health Act was passed. This Act called for health examinations of infants aged 3 years old to be carried out by public health nurses in the public health centres, and from 1977 examinations for infants aged one and a half years were begun by public health nurses in cities, towns, and villages. These Acts were the bases of present-day maternity and child health services in Japan.

For schoolchildren, the School Meals Act was passed in 1954 and the School Health Act in 1958.

Japan attained almost the lowest infant death rate in the world in 1975, when it was 10.0 per 1000 live births, although it was 162 in 1910, 90.0 in 1940, and 13.1 in 1970; the rate was 6.2 in 1983.

Mental health

The Mental Health Act was passed in 1950, After this year, the number of beds for mental disorders gradually increased, reaching 44 250 in 1955.

The National Survey of Mental Health in 1963 showed that the estimated number of persons diagnosed as having mental disorders was 1 240 000 and the number per 1000 population was 12.9. Of these, 280 000 needed institutional care and 480 000 needed out-patient care. Based on the findings of the survey, the Mental Health Act of 1950 was amended in 1965. The main amendments were:

1. The prefectures are to construct mental health centres.
2. The public is to pay the cost of out-patient care.
3. The supervisor of the public health centre has responsibility for ensuring that there is counselling and instruction for patients with mental disorders and, to achieve this purpose, mental health counsellors are to work at the centres.
4. A Council for Mental Health is to be established in each prefecture.

Although policies based on these points were implemented in Japan, the Mental Health Act was amended in 1987 to stress the promotion of procedures for the protection of the human rights of persons with mental disorders.

Prevention of diseases of middle age

According to the World Health Statistics Annual of the World Health Organization, Japan's revised death rate in 1978 was reported to be 3.8 per 1000 population, the lowest in the world, compared with 4.9 for the US, 5.0 for the Federal Republic of Germany, and 4.0 for Sweden. This result shows that health conditions of the Japanese improved dramatically after the Second World War. However, the prevalence of diseases found by the National Health Survey has been increasing year by year; for example, in 1955, the crude prevalence of disease per 1000 population was 37.9, in 1965, 63.6, and in 1985, 145.2. This tendency was most conspicuous in the diseases characteristic of middle age, such as diseases of the circulatory system or the digestive system, and respiratory diseases. In 1985, cancer, heart disease, and strokes caused 61.7 per cent of all deaths. The death rate from malignant neoplasm per 100 000 population was 187.4 for men and 125.9 for women in 1985. For heart diseases, these figures were 121.6 for men and 113.2 for women; for stroke, they were 110.6 and 113.9.

In 1961/2 and 1971/2, a basic survey of the diseases of middle age, and in 1980, a fundamental survey of the diseases of the circulatory system were carried out; in 1958, 1960, 1963, and 1979, surveys were made of actual conditions of patients with malignant neoplasms.

Strokes were the main cause of death for Japanese from 1951 to 1980. Various kinds of community activities to prevent strokes were organized in many places, among which the best known are those in Sawauchi village, Ikawa town, Yachiho village, Yao city, Yakumo village, Noichi town, and Asakura town. These activities initially depended on the health maintenance budget and staffing of the national health insurance system in each place. This situation, in which the initiative is taken locally, is not common in other countries. Screening for hypertension and also health education (particularly education about dietary factors in the development of stroke) have been promoted by public health nurses.

The Ministry of Health and Welfare implemented the Special Stroke Prevention Programme designed to halve stroke incidence, for the 3 years starting in 1969. Special government subsidies were provided for programmes for the early detection of hypertension in cities, towns, and villages starting in 1973, and for a Cardiovascular Disease Priority Programme starting in 1977.

In Japan, the death rate from stomach cancer is particularly high. The rate was 52.2 per 100 000 population for men and 31.7 for women in 1955, 55.5 and 34.4 in 1975, and 51.1 and 30.6 in 1985.

To help prevent deaths from stomach cancer, early detection programmes that involve mass screening by X-ray examination of the stomach have been strongly supported in Japan. The effectiveness of these programmes has been generally recognized. In 1975, 2 779 399 people were screened for stomach cancer. Examinations for uterine cancer have also been specially promoted; in 1975, the number of women who received such examinations was 1 524 944.

In 1978, the percentage of the cities, towns, and villages where mass screening for cancer was implemented was 98.3 for stomach cancer, 98.6 for uterine cancer, and 15.9 for lung cancer.

In June 1983, the Government published its 10-Year Plan for General Anti-Cancer Strategies.

Health for workers

In 1947, the Labour Standards Act was passed, and at the same time the Factory Act was abolished. In the new Act, standards were decided for working conditions, such as working hours (8), a minimum age for workers (15 years), compensation for absence from work because of injury (60 per cent of salary), lump-sum compensation for 1200 days or more when the 60 per cent compensation period has passed; a system was drawn up for health examinations and health education; and provisions were made for environmental health for workers.

In this year, the Ministry of Labour was established, and

the Workmen's Accident Compensation Insurance Act was passed.

In 1955, the Special Protection for Silicosis and Traumatic Spinal Impairments Act, which was amended to the Pneumoconiosis Act in 1960, was passed.

In 1956, the Ministry decided on Regulations for Special Health Examinations for Workers in a dangerous work environment. In 1964, the Act for Co-operative Bodies for the Prevention of Work Accidents was passed. With this Act, the Central Association for the Prevention of Occupational Accidents was established.

The administration of the system for workers in Japan is undertaken by the Bureau of Labour Standards in the Ministry, the 47 Offices for Labour Standards in the prefectures, and the 348 Centres for Labour Standards in the different districts.

Reflecting the increased productivity in industries and the rapid increase in occupational diseases among workers, the Labour Safety and Health Act was passed in 1972. The main provisions enjoined on employers made in this Act were:

(1) the creation of a pleasant working environment;
(2) the employment of occupational health doctors in or their assignment to factories where 50 or more workers are employed;
(3) the promotion of health examinations for workers and measurements to monitor environmental conditions;
(4) the registration of dangerous or harmful occupations;
(5) the planning of the promotion of activities to help prevention of occupational accidents in each industry.

Table 3.2. Number of medical care institutions and beds per 100 000 population, 1970–85

Institution	No. of institutions (No. of beds per 100 000 population)			
	1970	**1975***	**1980***	**1985***
Total	106 882	113 973	125 500	134 075
	(1265.5)	(1276.1)	(1373.2)	(1469.9)
Hospitals	7 974	8 294	9 055	9 608
	(1024.4)	(1039.9)	(1127.1)	(1235.5)
General hospitals	6 869	7 235	8 003	8 527
	(580.4)	(644.9)	(765.0)	(892.7)
Mental hospitals	896	929	977	1 026
	(238.4)	(248.5)	(263.6)	(276.5)
Communicable-disease hospitals	35	27	20	12
	(22.3)	(18.8)	(15.6)	(12.1)
Tuberculosis sanatoria	160	87	39	27
	(170.6)	(115.3)	(72.5)	(45.6)
Leprosaria	14	16	16	16
	(12.7)	(12.5)	(10.5)	(8.7)
General clinics	68 997	73 114	77 611	78 927
	(240.7)	(235.9)	(245.9)	(234.2)
Dental clinics	29 911	32 565	38 834	45 540
	(0.4)	(0.3)	(0.2)	(0.2)

* Includes Okinawa Prefecture.
Source: Statistics and Information Department, Ministry of Health and Welfare, Japan (1988). Report on survey of medical care institutions. In *Health and welfare statistics in Japan*, p. 95. Health and Welfare Statistics Association, Tokyo.

Medical services

Medical care system

The Medical Care Act, passed in 1948, is the basis of present-day medical care in Japan. It made the following provisions:

(1) the differentiation between hospitals and clinics on the basis of the number of beds; institutions with 20 beds or more are considered to be hospitals, and those with fewer than 20 beds, clinics, where patients are not allowed to stay for more than 48 hours;
(2) the foundation of general hospitals with 100 beds or more, and at least the following five departments: medicine, surgery, obstetrics, ophthalmology, and otorhinolaryngology;
(3) the foundation of places in which to give birth;
(4) the establishment of an inspection system for medical institutions;
(5) the regulation of the number of personnel and the facilities in hospitals;
(6) the establishment of public medical institutions owned by local governments or persons given permission by the Minister of Health and Welfare;
(7) the foundation of councils for medical institutions, remuneration of doctors, and the management of public medical institutions.

The number of beds in hospitals and the number per 100 000 population were 512 688 and 574.3, respectively, in 1955, and 1 164 098 and 1039.9 in 1975. The number of beds per capita almost doubled during these 20 years.

The number of doctors and their number per 100 000 population were 94 563 and 105.9, respectively, in 1955, and 132 479 and 118.3 in 1975. The number of nurses and their number per 100 000 population were 129 860 and 145.5, respectively, in 1955, and 361 604 and 323.0 in 1975. The number of public health nurses and their number per 100 000 population were 12 369 and 13.9, respectively, in 1955, and 15 962 and 14.3 in 1975.

The numbers of medical institutions and health personnel are given in Tables 3.2 and 3.3, respectively.

Medical insurance

During the Occupation (1945–52), General Headquarters invited an American Mission for the Investigation of Social Services in Japan, which published a report in July 1948. The report advised the establishment of a Council for Social Services to have the role of recommending governmental policies. Following this recommendation, the Government established the Council for Social Services as a consultative body to the prime minister in 1949. The Council published

'Recommendations on the System of Social Services', in which the unification of policies for social services and the compulsory establishment of national health insurance by local authorities were advocated. The recommendations had great impact on the promotion of social services in Japan during this period.

The Council for Social Services published its recommendations on the medical system in 1956, which included the promotion of an insurance system to cover everyone. The new National Health Insurance Act was passed in 1958, and with this Act it became compulsory for all cities, towns, and villages to establish their own insurance system in the fiscal year 1960 with a benefit rate of 50 per cent. All persons were covered by some kind of insurance by April 1961.

Starting in 1968, the benefit rate rose to a minimum of 70 per cent for all insurants, and starting in 1973, all people aged 70 years or more were offered free medical care. Starting in 1975, a high-cost payment system for monthly costs higher than 30 000 yen covered all insurants.

In 1987, the Health Insurance for Workers gave a coverage rate for medical benefits of 90 per cent, but for dependants, 80 per cent for ambulatory care and 70 per cent for hospital care; National Health Insurance gave 70 per cent coverage for either ambulatory or hospital care. Those aged over 70 years have to pay 800 yen for ambulatory care per month, and 400 yen a day for hospital care.

The numbers of insurants classified by type of insurance for every 5-year period since 1955 are given in Table 3.4.

Table 3.3. Number of health personnel and ratio per 100 000 population, 1970–86

Health personnel	Number (Ratio) 1970	1975†	1980†	1982†	1984†	1986†
Physicians*	117 195	131 010	154 578	166 212	179 358	189 531
	(113.0)	(117.0)	(132.1)	(140.0)	(149.2)	(155.8)
Dentists*	36 914	42 577	52 369	57 148	61 911	65 605
	(35.6)	(38.0)	(44.7)	(48.1)	(51.5)	(53.9)
Pharmacists*	65 179	77 084	95 319	102 913	108 806	114 680
	(62.8)	(68.9)	(81.4)	(86.7)	(90.5)	(94.3)
Public health nurses	14 007	15 962	17 957	19 137	20 858	22 050
	(13.5)	(14.3)	(15.3)	(16.1)	(17.3)	(18.1)
Midwives	28 087	26 742	25 867	25 416	24 649	24 056
	(27.1)	(23.9)	(22.1)	(21.4)	(20.5)	(19.8)
Registered nurses and enrolled nurses	273 572	361 604	487 169	540 971	590 177	639 936
	(263.8)	(323.0)	(416.2)	(455.8)	(490.9)	(526.0)

* Numbers of physicians, dentists, and pharmacists do not include those who were engaged in other occupations or who were unemployed.
† Includes Okinawa Prefecture.
Sources: Statistics and Information Department, Ministry of Health and Welfare, Japan (1988). Report on survey of physicians, dentists and pharmacists; Statistics and Information Department, Ministry of Health and Welfare, Japan (1988). Statistics on public health services. In *Health and welfare statistics in Japan*, p. 106. Health and Welfare Statistics Association, Tokyo.

Table 3.4. Number of insurants (thousands) by type of insurance

Year	Population (thousands)	Total number of insurants	Insurance for workers: Total	Insurants	Dependants	National Health Insurance
1955	89 767	59 023*	32 491	12 475	20 016	28 711
1960	94 308	88 876*	43 506	18 415	25 091	46 171
1965	98 710	96 965	53 822	24 255	29 567	43 143
1970	104 320	103 634	60 271	28 145	32 126	43 363
1975	112 461	111 794	67 798	29 796	38 002	43 996
1980	117 415	117 037	72 501	31 754	40 747	44 536
1985	121 315	120 741	75 447	33 629	41 818	45 294

* Some of the insurants are counted twice under 'Insurance for workers' and 'National Health Insurance'.
Source: Ministry of Health and Welfare, Japan (1987). *Kokumin eisei no doko*, p. 207. Kosei Tokei Kyokai, Tokyo.

New departures based on the principle of 'Health For All By The Year 2000'

The World Health Organization published the declaration 'Health for all by the year 2000' at the International Conference on Primary Health Care at Alma-Ata in 1978. This declaration has had a significant impact on public health policy in Japan.

In 1986, the Japanese life expectancy at birth reached 75.23 years for men and 80.93 years for women.

In 1987, there were 80 medical schools in Japan and almost all of these had departments of hygiene and public health.

With an insurance system covering everyone, the network system of 850 health centres with community health activities has continued, and public health is being promoted in Japan today based on the philosophy of Alma-Ata. The main policies carried out at present are:

1. *Health promotion*. Since 1978, the National Health Promotion Movement has had three main objectives: the establishment of a framework of health services incorporating examinations and advice throughout life; the provision of community health centres closely involved with the people; and the promotion of a health education programme among the people.
2. *Implementation of health services for the elderly*. Based on the Health Services for the Elderly Act passed in 1982, plans for health education, health counselling, health examinations, medical care, home nursing, and rehabilitation are decided in each city, town, or village, depending on the age distribution and patterns of illness of the residents. New institutions such as nursing homes are to be built for the care of the elderly in the community.
3. *Measures for maternal and child health*. Health examinations for pregnant women and babies are to be consolidated and promoted, and follow-up care after examinations, in particular, is to be promoted.
4. *Medical plans*. With the Medical Care Act, amended in 1985 to maintain the most appropriate institutional care for the people, a medical plan is to be made and pub-

lished by each prefecture and, in that plan, the numbers and area covered by the districts in the prefecture and the number of beds in each district are to be decided.

5. *Promotion of international co-operation.* The numbers of Japanese travelling abroad and of foreigners visiting Japan were 4 090 000 and 1 790 000, respectively, in 1982. Human interchanges with foreign countries are rapidly increasing. For the health and welfare of the people, it is essential to promote international co-operation in the field of public health, perhaps by deepening mutual understanding of problems in other countries by public health personnel as a first step.

Table 3.5. Milestones in public health in Japan

Before Nara period (–710)	
701	Taiho Ritsuryo
Nara period (710–94)	
723	Seyaku-in and Hiden-in were built at the Kofukuji Temple to take care of and shelter the sick and poor
Muromachi period (1338–1573)	
1555	Luis de Almeida (a Portuguese surgeon) visited Japan
1557	Almeida opened an orphanage in Oita
Edo period (1603–1868)	
1722	Koishikawa Yojosho was built to offer hospital care to the poor in Edo
1774	Genpaku Sugita published the first western-style anatomical textbook
1801/2	Jenner's theory of vaccination for smallpox reached Japan
1838	Tekijuku was founded by Koan Ogata
1849	O. G. Mohnike introduced the practice of vaccination
1858	An official place for vaccination for smallpox, the Shutosho, was built
1861	Shutosho was renamed the Seiyo Igakusho
	Yojosho was founded in Nagasaki by Pompe van Meerdervoort
Meiji period (1868–1912)	
1869	Seiyo Igakusho was renamed Daigaku Toko
1871	Household Act was passed
1872	Department of Medical Affairs was founded in the Ministry of Education
1873	Department of Medical Affairs became Bureau
1874	Daigaku Toko became Tokyo Medical School, the origin of the Faculty of Medicine of the University of Tokyo
	Isei was decreed
1875	Bureau of Medical Affairs was moved to the Home Office and its name was changed to the Bureau of Hygiene
1876	Regulations for smallpox prevention were decided
1877	A guide for the prevention of cholera was issued
1880	Regulations for the prevention of infectious diseases were decided
1883	Great Japanese Sanitary Association was founded
1886	Japan Pharmacopoeia was begun
1888	New city and town/village systems for local government were introduced
1890	New prefectural system was introduced
	Water Supply Act was passed
	Mining Industries Regulations Act was passed
1897	Infectious Diseases Prevention Act was passed
	Regulations for School Cleanliness and for Periodical Physical Examinations were decided
1900	Nuisances Removal Act was passed
	Sewerage Act was passed
	Regulations of Food and Drink Act was passed
1905	Mining Industry Act was passed
1906	Medical Act was passed
	Dentists' Act was passed
1907	Leprosy Prevention Act was passed
1911	Factory Act was passed
Taisho period (1912–26)	
1913	Japanese Association for Tuberculosis Prevention was established
1915	Regulations for nurses were established
1919	Tuberculosis Prevention Act was passed
	Trachoma Prevention Act was passed
1922	Health Insurance Act was passed
Showa period (1926 to the present)	
1931	Helminithiasis Prevention Act was passed
1937	Public Health Centre Act was passed
1938	Ministry of Health and Welfare was founded
	National Health Insurance Act was passed
1939	Seamen's Health Insurance Act was passed
1947	Local Government Act was passed
	Child Welfare Act was passed
	Labour Standards Act was passed
1948	Inoculation Act was passed
	Venereal Diseases Prevention Act was passed
	Medical Care Act was passed
	Public Health Nurses', Midwives' and Nurses' Act was passed
1950	Mental Health Act was passed
1954	School Meals Act was passed
1958	School Health Act was passed
1965	Maternity and Child Health Act was passed
1972	Labour Safety and Health Act was passed
1982	Health Services for the Elderly Act was passed

Select bibliography

Author's translations of titles of papers or books written in Japanese are given in brackets.

Aoshima, O. (1973). *Shika no ayumi* [*History of dental care*]. ABC Kikaku, Tokyo.

Ban, T. (1987). *Tekijuku to Nagayo Sensai* [*Tekijuku and S. Nagayo*]. Sogensha, Osaka.

Doyokai Rekishibukai (1973). *Nihon kindai kango no yoake* [*History of nursing in modern times*]. Igaku-shoin, Tokyo.

Fujikawa, Y. (1972). *Nihon igaku shi* [*Medical history in Japan*]. Keiseisha, Tokyo.

Fujikawa, Y. (1974). *Nihon igaku shi yoko* [*Summary of medical history in Japan*]. Heibonsha, Tokyo.

Fujino, K. (1984). *Nihon saikingaku shi* [*History of bacteriology in Japan*]. Kindai Shuppan, Tokyo.

Fujiwara, M. and Watanabe, G. (1978). *Sogo eisei-koshueisei gaku* [*General hygiene and public health*]. Nankodo, Tokyo.

Furuichi, K. (ed.) (1988). *Health and welfare statistics in Japan*, Health and Welfare Statistics Association, Tokyo.

Harashima, S. (1967). *Rodoeiseigaku josetsu* [*Hygiene for workers*]. Koseikan, Tokyo.

Hashimoto, M. (1973). Japan. In *Health service prospects—an international survey*, p. 211. (ed. I. Douglas-Wilson and G. McLachlan). The Lancet, London.

Hashimoto, M. (1981). National health administration in Japan. *Bulletin of the Institute of Public Health* **30**, 1.

Hashimoto, M. (1984). Health services in Japan. In *Comparative health systems*, p. 335. (ed. M. W. Raffel). Pennsylvania University Press, Pennsylvania.

Hasuda, S. (ed.) (1960). *Kokumin kenko hoken shi* [*History of national health service in Japan*]. Japan Medical Association, Tokyo.

Health Insurance Bureau, Ministry of Health and Welfare (1958). *Kenkohoken 30 nenshi* [*History of insurance for workers for 30 years*]. Zenkoku Shakaihoken Kyokai Rengokai, Tokyo.

Ikematsu, S. (1932). *Iyaku seidoron to bungyo undo shi* [*History of discussions about dispensing doctors in Japan*]. Gankaido, Tokyo.

Kawakami, T. (1965). *Gendai nihon iryo shi* [*History of modern medical care in Japan*]. Keisoshobo, Tokyo.

Kitamura, M. (1967). Gyokairui o kaishita suigin chudoku [Mercury poisoning through the eating of fish and shellfish]. *Journal of the Japan Medical Assocation* **57**, 488.

Koga, J. (1972). *Seiyou ijutsu denrai shi* [*History of introduction of Western medicine into Japan*]. Keiseisha, Tokyo.

Koizumi, A. (1982). Development of public health in Japan. *Asian Medical Journal* **25**, 14.

Komachi, Y. (ed.) (1987). *Junkanki shikkan no henbo* [*Changes in diseases of the circulatory system in Japan*]. Hoken Dojinsha, Tokyo.

Kono, R. (1972). Sokatsu-hokoku [General report]. In *SMON chosa kenkyu kyogikai kenkyu hokokusho* [*Report on survey of SMON*] No. 12, p. 45. Ministry of Health and Welfare, Tokyo.

Kosei Tokei Kyokai (ed.) (1987). *Kokumin eisei no doko* [*Annual report on movements about conditions of public health in Japan*]. Kosei Tokei Kyokai, Tokyo.

Kuratsune, M. (1963). Epidemiologic study on *yusho*. In *PCB poisoning and pollution*, p. 9. (ed. K. Higuchi). Kodansha, Tokyo and Academic Press, London.

Kusumoto, M. (ed.)(1971). *Hokensho 30 nenshi* [*History of 30 years of public health centres*]. Nihon Koshueisei Kyokai, Tokyo.

Medical Affairs Bureau, Ministry of Health and Welfare (1955). *Isei 80 nenshi* [*History of 80 years after the Isei*]. Presswork Bureau Choyokai, Tokyo.

Medical Affairs Bureau, Ministry of Health and Welfare (1976). *Isei 100 nenshi* [*History of 100 years after the Isei*]. Gyosei, Tokyo.

Ministry of Health and Welfare (1974). *Health services in Japan*. Japan Public Health Association, Tokyo.

Ministry of Health and Welfare (1983). *Annual report on health and welfare for 1983—The trend of a new era and social security*. Japan International Corporation of Welfare Services, Tokyo.

Ministry of Health and Welfare (ed.) (1987). *Kosei hakusho* [*White paper on health and welfare in Japan*]. Kosei Tokei Kyokai, Tokyo.

Nakae, K., Yamamoto, S., Shigematsu, 1., and Kono, R. (1973). Relation between subacute myelo-optic neuropathy (S.M.O.N.) and clioquinol: Nation-wide survey. *Lancet* **i**, 171.

Ogawa, T. (ed.) (1967). *Tokyo daigaku igakubu 100 nenshi* [*History of 100 years of the University of Tokyo*]. University of Tokyo Press, Tokyo.

Ohira, M. and Aoyama, H. (1973). Epidemiological studies on the Morinaga powdered milk poisoning incident. *Japanese Journal of Hygiene* **27**, 500.

Otani, F. (1980). *21 seiki kenko eno tenbo* [*Perspectives of health in the 21st century*]. Medical Friends, Tokyo.

Sakai, S. (1982). *Nihon no iryo shi* [*History of medical care in Japan*]. Tokyo Shoseki, Tokyo.

Shigematsu, 1., Ishizaki, A., Kiba, T., *et al.* (1968). Itai-itai-byo no genin ni kansuru kenkyu [Research on the cause of itai-itai -byo]. In *Koseisho kenkyuhan hokokusho* [*Report of Ministry research groups*]. Ministry of Health and Welfare, Tokyo.

Shigematsu, I., Yanagawa, H., Tanemura, M., and Tani, S. (1970). SMON kanja zenkoku jittai chosa seiseki [Actual conditions of SMON patients in the entire country]. In *SMON chosa kenkyu kyogikai kenkyu hokokusho* [*Report of the committee for research on SMON*] No. 1, p. 4. Nihon Koshueisei Kyokai, Tokyo.

Soda, T. (ed.) (1980). Drug-induced sufferings — Medical, pharmaceutical and legal aspects. Excerpta Medica, Oxford.

Stephen, W.J. (1979). *An analysis of primary medical care — an international study*. Cambridge University Press, Cambridge.

Suita City (1986). *Hoken keikaku* [*Health plan*]. Nihon Koshueisei Kyokai, Tokyo.

Takada, K. (ed.) (1960). *Koseisho 20 nenshi* [*History of 20 years of the Ministry of Health and Welfare*]. Kosei Mondai Kenkyukai, Tokyo.

Tatara, K., Shinsho, F., Asakura, S., and Hashimoto, M. (1984). *Shi-cho-son no hoken jigyo* [*Local health services in Japan*]. Nihon Koshueisei Kyokai, Tokyo.

Tsubaki, T., Toyokura, Y., and Tsukagoshi, H. (1964). Subacute myelo-optico-neuropathy following abdominal symptons. *Journal of the Japanese Society of Internal Medicine* **53**, 779.

Wakatsuki, S. (1971). *Noson igaku* [*Medicine in rural areas*]. Keiso Shobo, Tokyo.

World Health Organization (1978). *Alma-Ata 1978: primary health care*. WHO, Geneva.

Yoshida, K., Oshima, H., and Imai, M. (1966). Air pollution and asthma in Yokkaichi. *Archives of Environmental Health* **13**, 763.

4

Current scope and concerns in public health

R. DETELS and L. BRESLOW

Introduction

Public health is the process of mobilizing local, state, national, and international resources to solve the major health problems affecting communities. In the nineteenth and early twentieth centuries these health problems were primarily associated with the undernutrition, crowding, exhaustion, and faecal contamination of water supplies that accompanied early industrialization. These conditions resulted in a high prevalence of tuberculosis, enteric infections, infant mortality, and acute respiratory diseases. In response, communities, provinces, and nations developed successful ways of dealing with these important problems through public health action. Public health, from the outset, embraced both social action and scientific knowledge. This partnership meant linking the anti-poverty (reform) movement with the findings from epidemiologic and bacteriologic investigations, for example, to combat such diseases as tuberculosis and typhoid fever.

Now at the end of the twentieth century many of the major health problems facing Britain, Japan, the US, and other highly industrialized nations stem from an overly rich diet, cigarette use, excessive alcohol consumption, and other conditions which often accompany living in these countries. In this chapter we will present broadly the current scope and concerns of public health as well as issues which confront public health organizations in industrialized society. Subsequent chapters will present specific topics in greater detail.

The first part of this chapter will outline the major health problems facing the developed countries, including infectious diseases, chronic diseases, trauma, and mental health. Determinants of health such as nutritional problems, environmental hazards, and disorders resulting from life-style choices constitute the next part. The third part will deal with the scientific responses which public health uses to cope with the problems, including strategies basic to public health such as epidemiology, and those which are borrowed and modified from other disciplines such as the social, biological, and physical sciences. The programmatic scope of public health will then be outlined: preventing disease and promoting health, improving medical care, promoting health-related behaviour, and controlling the environment—these are four public health ways of influencing health. The fifth section discusses the strategies for applying these scientific approaches to public health problems, and the final section will deal with the interaction of the various governmental and voluntary actions aimed at improving the health of the community.

Public health is of course only one of the major influences on a community's health. The basic economic and social conditions of existence directly impact people's level and mode of living, and thus constitute the foundation of health. These conditions limit and, to a considerable extent, direct the resources which can be devoted specifically to health promotion and disease intervention. Prevailing economic and social conditions also affect health in ways beyond the level of living and the concomitant ability of people to obtain the necessities of healthy life. Strong economic forces expressed in agriculture, manufacturing, commerce, and politics, for example, may sway people to use tobacco and thus injure their health.

The magnitude and success of public health efforts will vary both in time and place in different areas of the world. Nevertheless, the principles of public health remain the same. The actions which should be taken are determined by the nature and magnitude of the problems affecting the health of the community. What can be done will be determined by the scientific knowledge and resources available. What is done will be determined by the social and political commitment existing at the particular time and place.

Health problems

Infectious disease

Before 1982 it appeared that pandemics of infectious disease other than influenza had been eliminated as a major problem in developed countries. The decline in the incidence of the traditional infectious diseases had been controlled largely through provision of safe drinking water, better handling of sewage, effective vaccine campaigns, improved personal hygiene, and improved nutrition, especially among children. The recognition of the world-wide epidemic of acquired

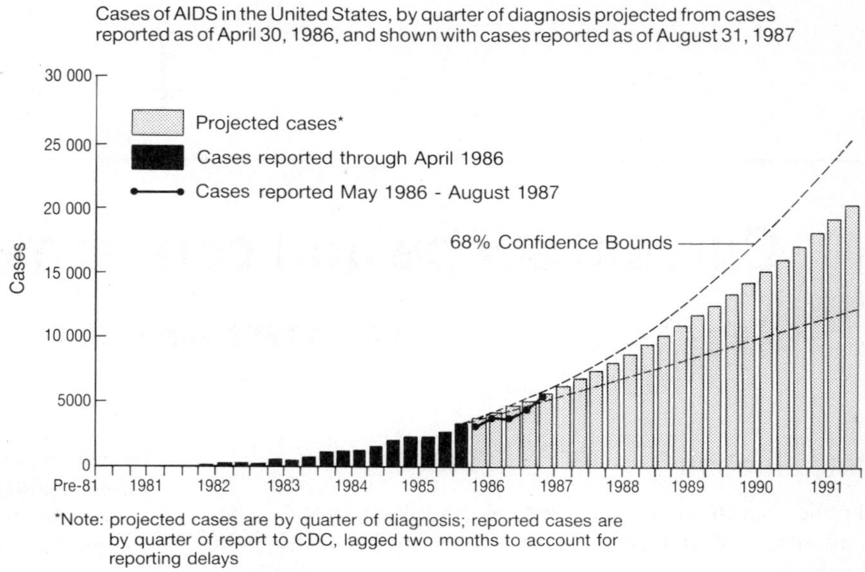

Fig. 4.1. Cumulative incidence of AIDS cases—USA (Source: Meade, W.M. and Curran, J.W. (1986).)

immune deficiency syndrome (AIDS), however, dramatically underscored that infectious diseases are likely to remain an important problem even in developed countries for many years to come (Gottlieb *et al.* 1981; Mann *et al.* 1988; Piot *et al.* 1988).

The major infectious disease problems in developed countries today are related to changing life-styles, technical advances which create groups with increased susceptibility to infectious disease agents, changes in the age distribution of susceptible populations resulting from incomplete immunization programmes, the emergence of new agents such as the human immunodeficiency virus type 1 (HIV–1), newly recognized manifestations of known organisms such as the venereal and respiratory manifestations of infection by chlamydia, and continued poverty resulting in crowded, unhealthy living conditions, and difficult access to medical care and prevention programmes.

Changing life-styles, especially increased sexual freedom and use of drugs over the past several decades, are responsible for the continuing high incidence of diseases such as syphilis, genital tract chlamydial infections, and hepatitis, as well as for the emergence of AIDS as a world-wide epidemic (Curran *et al.* 1985). Although there are two major patterns of transmission of AIDS currently recognized (heterosexual in Africa and homosexual/intravenous drug use in Europe and North America), both are related to life-style (Heyward and Curran 1988; Piot *et al.* 1988). Traditionally, diseases related to life-style have been the most difficult to control. The emergence of antibiotic resistant strains of syphilis and other organisms further complicates the control of many of these diseases.

The development of new technologies and drugs such as steroids, to keep alive individuals who would otherwise have died, has created groups of people who are particularly susceptible to infectious diseases, especially to those agents which do not usually cause disease in uncompromised individuals. The persistence of pneumonia and influenza as the sixth leading cause of death in the United States may in part be due to infections occurring in these susceptible groups as well as in the elderly.

Although immunization campaigns in most developed countries have included coverage of children, not all have been immunized successfully. Because the reduction of susceptibles in the population has lowered the incidence of these diseases, many of the unvaccinated children will remain susceptible until adulthood when the clinical manifestations of diseases such as mumps, measles, rubella, and chicken pox are more severe. Thus, the overall incidence of these diseases may have been reduced, but the proportion of infected individuals expressing unusual clinical manifestations may increase. Because of differences in the infectivity and other characteristics of infection with many of these agents, their eradication will be more difficult than was the eradication of smallpox (Fenner *et al.* 1988).

With the recognition of the AIDS epidemic in the 1980s has come the realization that the public health profession must constantly be alert to the emergence of new infectious disease agents. From the description of the first cases in 1982, the number of cases reported in the United States as of the fall 1988 has risen to over 70 000 cases, and world-wide to more than 250 000 cases (Centers for Disease Control 1988; Mann *et al.* 1988). The United States Centers for Disease Control has projected that the cumulative number of cases in the United States will reach 270 000–350 000 by 1991 (Fig. 4.1) (Morgan and Curran 1986; Heyward and Curran 1988) and WHO has projected 500 000 cases by 1992 in Africa alone (Mann *et al.* 1988).

In addition to the emergence of new agents, new clinical

manifestations are being recognized which are caused by agents previously associated with other diseases. The chlamydia group of agents, for example, has long been recognized to cause the ocular disease trachoma, but it is now recognized as the leading genital tract infection in the United States, and recently it has been reported as a cause of severe respiratory disease, TWAR (Grayston and Wang 1975). Some agents are now known to cause diseases previously thought to have a non-infectious aetiology. Infection in early life with hepatitis B virus, for example, has been demonstrated epidemiologically to be a cause of liver cancer, and schistosomiasis infection has been associated with squamous cell cancer of the bladder (Cheever 1976; Chevlen *et al* 1979; Elsebai 1977).

Poverty remains an important co-factor for infectious diseases in most developed countries. Crowding and poor nutrition which are common among the poor increase both exposure and susceptibility to infectious disease agents. Furthermore, poor children are less likely to have been immunized and less likely to have been brought to medical attention early in the course of infectious disease when intervention is most likely to be successful. For these reasons the incidence of infectious diseases is appreciably higher in the poor.

Public health advances have resulted in dramatic reductions in the incidence of infectious diseases in developed countries. However, the public health professional needs to be alert to the emergence of new agents, new manifestations of previously known agents, and the presence of groups in the population which are particularly susceptible to disease through poverty as a result of advances in medical technology or due to other factors. The control of infectious diseases in the future will call for the implementation of new strategies such as the intensive use of mass media for health education used to lessen the AIDS epidemic, the development of new technology for identification of infection, and the development of new drugs and vaccines for cure and prevention.

Chronic disease

Beginning in the nineteenth and continuing through the twentieth century, industrialization has vastly changed the way people live and, correspondingly, the nature of their health problems. This change becomes apparent in comparing the leading causes of death in the US in 1900 and 1985 (McGinnis 1982, Table 4.1). While heart disease, cancer, stroke, and accidents now cause more than two-thirds of all deaths, a different pattern appears when illness rather than death is used as the measure of health.

Table 4.2 shows the chronic conditions with greatest prevalence. After the age of 45 years chronic conditions become a more important cause for restriction of activity than acute conditions (Department of Health, Education and Welfare 1980). Beyond age 65 years the average person suffers restriction of activity from one or several chronic conditions for almost one month a year, about one-third of that time confined to bed. Heart disease is the major chronic condition resulting in limitation of activity (Department of Health, Education and Welfare 1977). Arthritis and rheumatism follow, and impairments of the back, spine, hips, and lower extremities or other musculoskeletal disorders all result in substantial limitation of activity.

Table 4.1. Leading causes of death in the United States, 1900 and 1985

1900		1985	
Cause	**Rate***	**Cause**	**Rate***
Influenza and pneumonia	210	Diseases of heart	181
Tuberculosis	199	Cancer	134
Heart disease	167	Accidents	35
Stroke	134	Cerebrovascular disease	32
Diarrhoea and related diseases	113	Chronic obstructive lung diseases	19
Cancer	81	Pneumonia and influenza	13
Accidents	76	Suicide	12
Diabetes	13	Cirrhosis	10
Suicide	11	Diabetes	10
Homicide	1	Homicide	8
All other causes	775	All other causes	92
All causes	1779	All causes	546

* Age adjusted, per 100 000.
Source: McGinnis, J.M. (1988).

Table 4.2. Most prevalent conditions in the United States, 1983–1985

Chronic condition	Rank order
Chronic sinusitis	1
Arthritis	2
High blood pressure	3
Orthopedic impairments	4
Deafness	5
Hay fever	6
Heart disease	7
Chronic bronchitis	8
Hemorrhoids	9
Dermatitis	10

Source: Collins, J.G. (1988*b*).

One success story in the prevention of chronic disease has been achieved through public health dentistry. The introduction of fluoride into municipal water supplies and dentifrices has dramatically reduced the incidence of dental caries over the last several decades. The major dental public health problem in developed countries is now periodontal disease. Reduction of this disease will require motivating the public to practice better oral hygiene and to improve nutritional habits. In this respect the strategies which must be used by public health dentistry are similar to those needed to reduce the incidence of the other major chronic diseases.

Other important chronic disease trends reflecting public health need and effort include the rise and fall of ischaemic heart disease mortality in various countries, reflecting changes in living conditions since the Second World War (Table 4.3) (Pisa and Uemara 1981).

Health problems among the elderly are an increasing concern of public health, both because more people are living

Table 4.3. Mortality rate for coronary heart disease among men age 35–74, for several countries in 1969 and 1977

Country	1969 rate	1977 rate	Per cent difference
		Decrease	
US	865	670	−23
Australia	844	683	−19
Japan	126	103	−19
Canada	703	624	−11
		Increase	
Poland	187	308	+65
Bulgaria	299	424	+42
Romania	171	237	+39
Yugoslavia	185	228	+23

Source: Report of the working group on arteriosclerosis of the US National Heart, Lung and Blood Institute (1981). Vol. 1, p. 4.

longer and because of the greater frequency of disabling illness among them. Until recently the main element extending the average duration of life in industrialized nations was the reduction in fatal diseases in infancy and the early years of life; now, decrease in mortality during the later years has become the major factor. The average duration of life in the US after age 65 years, for example, remained fairly steady until about 1960. During the recent 25 years, 1960–1985, the life expectancy of individuals at age 65 years increased by 2.4 years, the same as the amount gained during the 60-year period, 1900–1960. The major interest in preventing and controlling chronic disease is, of course, not just extending the duration of life; maintaining the quality of that extended life is increasingly important.

The major measure of health that has been used traditionally has been mortality. Using that criterion, heart disease and cancer are the major concerns of public health. But both these diseases tend to occur and kill in the last decades of life when relatively few years of productive life remain. Recently, a newer system of evaluating the impact of disease and other factors affecting survival has been introduced; that is, years of potential life lost (YPLL). Years of potential life lost indicates the total number of years of life lost before age 65 years due to each cause of death. Thus, causes of death which tend to occur in early life are weighted more heavily than those that occur later in life. Further, years of potential life lost is also dependent on the age distribution of the population; populations with higher proportions of young people have greater years of potential life lost than populations with smaller proportions of young people even if the age-specific rates by cause of death are the same for the two populations. None the less, this is a useful measure for estimating the public health impact of the various causes of death and it can be used by decision makers to plan prevention strategies.

Trauma

Unintentional injuries are the leading cause of years of potential life lost and account for 19.6 per cent of years of potential life lost (Centers for Disease Control 1986*a*). The leading causes of unintentional injuries are listed in Table 4.4. Motor vehicle traffic accidents head the list accounting for well over half of the years of potential life lost due to unintentional injuries (Table 4.5). The Carter Center estimates that motor vehicle related deaths and injury could be reduced by 75 per cent, and injuries due to accidents occurring at home could be reduced by 50 per cent, by applying a broad based mixed strategy for prevention (Smith 1985).

Suicide and homicide are the fourth leading cause and account for an additional 10.6 per cent of years of potential life lost (Table 4.4, Centers for Disease Control 1986*b*). The rates of suicide and homicide, however, are not evenly distributed within a population, tending to occur more among the young males and at different rates in the various racial groups (Centers for Diseases Control 1985). In 1983 suicide alone was the fifth leading cause of years of potential life lost. Suicide among white males alone accounted for 70.6 per cent of the total years of potential life lost due to suicide in the US. Suicide is a major cause of premature death in most developed countries, especially in northern Europe and Scandinavia. The causes for suicide are not well understood

Table 4.4. Estimated years of potential life lost before age 65 and cause-specific mortality, by cause of death—United States, 1984

Cause of mortality (ninth revision ICD)	Years of potential life lost by persons dying in 1984*	Cause-specific mortality† (rate/100 000)
All causes (Total)	11 761 000	866.7
Unintentional injuries§ (E800–E949)	2 308 000	40.1
Malignant neoplasms (140–208)	1 803 000	191.6
Diseases of the heart (390–398, 402, 404–429)	1 563 000	324.4
Suicide, homicide (E850–E978)	1 247 000	20.6
Congenital anomalies (740–759)	684 000	5.6
Prematurity¶ (765, 769)	470 000	3.5
Sudden infant death syndrome (798)	314 000	2.4
Cerebrovascular diseases (430–438)	266 000	65.6
Chronic liver diseases and cirrhosis (571)	233 000	11.3
Pneumonia and influenza (480–487)	163 000	25.0
Chronic obstructive pulmonary diseases (490–496)	123 000	29.8
Diabetes mellitus (250)	119 000	15.6

* For details of calculation, see footnotes for Table V. *MMWR* (1986) **35**; 27.
† Cause-specific mortality rates as reported in the MVSR are compiled from a 10% sample of all deaths.
§ Equivalent to accidents and adverse effects.
¶ Category derived from disorders relating to short gestation and respiratory distress syndrome.
Source: Centers for Disease Control (1986*b*) p. 365.

Table 4.5. Years of potential life lost (YPLL) rates per 100 000 population, YPLL rate ratios, and average YPLL per death, by the 10 leading causes of unintentional injury YPLL and by sex—United States, 1983

Cause of death	YPLL rate	(YPLL rate ratio*)	Average YPLL per death
Motor vehicle traffic crashes			
Male	953.1		35.1
Female	334.3	(2.9)	34.6
Drownings†			
Male	156.7		39.9
Female	33.7	(4.7)	43.9
Fire and flames			
Male	73.9		33.7
Female	47.8	(1.5)	38.4
Poisonings			
Male	81.7		30.4
Female	33.4	(2.4)	29.1
Falls			
Male	59.5		23.6
Female	16.1	(3.7)	21.6
Firearms			
Male	49.0		36.8
Female	7.8	(6.3)	38.2
Choking on food or object			
Male	26.8		27.7
Female	14.5	(1.8)	29.5
Air transport			
Male	29.3		28.0
Female	5.9	(5.0)	31.5
Water transport			
Male	32.7		30.6
Female	3.8	(8.6)	34.5
Motor vehicle nontraffic crashes			
Male	26.6		39.1
Female	8.3	(3.2)	45.8

* For males compared with females within each cause-specific category.
† Includes those not related to water transport.
Source: Centers for Disease Control (1986*b*) p. 355.

nor have satisfactory predictors of who will commit suicide been developed, making the development of preventive strategies particularly difficult. Often, suicides among young people tend to occur in clusters, prompting the US Centers for Disease Control to issue recommendations for dealing with suicide in a community (O'Carroll *et al.* 1988).

In the US, homicide is currently the leading cause of death among young black males aged 15–24 years: it is six times more common among black males than in white males in the same age group (Centers for Disease Control 1988). The differences between white and black males is reduced but not eliminated when adjusted for socio-economic status.

The public health agenda for the 1990s must address the issue of premature mortality due to trauma including suicide, homicide, war, and natural disasters such as the devastating 1988 earthquake in Armenia. With the advances in public health over the last decades, injuries have become a major public health problem which must be resolved in developed countries. They are particularly important because they often affect the young disproportionately. Improvements in highways, seat belt laws, and improved car design have reduced injuries and death due to motor vehicle traffic accidents, and improvements in such things as design of ladders have reduced accidents in the home. Providing an effective forum and process for resolution of international differences has prevented some conflicts but more needs to be done. Better planning may not prevent disasters, but it has and can further reduce the toll from them. More vigorous application of such intervention strategies should further reduce deaths and injuries from these types of trauma, but the nature of factors causing death and injury from many types of trauma are still not known. Further success in reducing injuries will therefore require multidisciplinary research to find the causes of these various types of injury. Such research should identify strategies which can be used for effective intervention programmes.

Mental disorders

The concept of mental disorders has been extending to include not only psychoses, neuroses, and mental retardation, but also alcoholism, dependency disorders, child abuse, and learning disabilities. Progress in understanding these disorders, especially the biological basis for many of them, and what can be done about them is accelerating. The early discovery of syphilis as the cause of general paresis, and nutritional deficiency as the cause of pellagra, were early, notable achievements. One can add the more recent identification of a specific genetic abnormality (trisomy 21) and lead intoxication as key factors in mental retardation, and inappropriate use of pharmaceuticals as the cause of a vast amount of mental disorder.

Despite the considerable progress in elucidating the etiology and therapy of mental disorders, however, the public health response has not kept pace with the huge problem it represents. The lag in implementing what can be done about mental disorders seems largely to reflect the lingering, hostile attitude toward 'strange' people. A more enlightened approach, not simply to place them in custodial institutions and thereby exclude them from society, has developed in recent decades as more effective therapy has become available. For example, general paresis can be avoided through treatment of syphilis, the birth of children with Down's syndrome may be avoided through acting on amniocentesis results, and many cases of psychoses can be avoided through the use of appropriate pharmaceuticals.

The problem as a whole, however, is not diminishing. Increased longevity among persons with congenital forms of mental retardation means more life-years with that disorder, and the extension of longevity among older persons is enlarging the problem of dementia, including Alzheimer's disease and arteriosclerosis of the brain, toward the end of life.

Significant developments in the past few decades that are likely to result in improved methods of prevention and treatment of mental disorders are:

(1) the increased numbers of mental health professionals and the increased range of services they can provide;

(2) understanding that personal and community support systems can help individuals with mental health problems;

(3) awareness that early intervention may avoid chronicity;

(4) the advances made in understanding the biological mechanisms of mental disorders.

The means to provide the mentally ill with the rights taken for granted by the vast majority of people in industrialized nations are now becoming available. Assuring these rights to those suffering from mental illness must become part of the public health agenda.

Determinants of health

Nutrition

Fifty years ago the major nutritional problems throughout the world were the lack of adequate, safe, affordable supplies of food, inadequate knowledge of the dietary needs of humans, and wide-scale ignorance among the public concerning the relationship of nutrition to health and disease. Large segments of the population in industrialized nations were not aware of the need for a balanced diet which included components from all the major food groups. Even if they had been aware of nutritional needs, essential foods were often not available due to poverty, or because of transportation and distribution problems affecting a wide range of foods. Fresh fruits and other major sources of vitamin C for example, were often almost unobtainable in northern climates for major portions of the year. Finally, provision of safe foods—uncontaminated by parasites, bacteria, and viruses—was difficult in the absence of refrigeration and pesticides.

The high prevalence of infective and parasitic disease conditions led to undernutrition and, in turn, made individuals more susceptible to infectious disease agents because of their compromised nutritional status. Individuals with ascaris infection, for example, could be undernourished in the face of what appeared to be an adequate diet and would thus be more susceptible to infections with other disease agents.

Most of the nutritional problems of fifty years ago have been resolved in the developed countries. On the other hand, these problems persist in the developing countries and even among certain groups of people in developed countries; for example, among the poor and immigrants whose knowledge of fundamental nutrition may be limited.

The major nutritional problem in the developed countries now, however, appears to be overnutrition, particularly an excessive consumption of fats, refined carbohydrates, and salt, and a decreased consumption of fibres and cereals; this situation promotes diseases such as coronary heart disease, diabetes, hypertension, and dental caries. In 1985, 25 per cent of Americans weighed more than 120 per cent of the ideal weight for their height (National Center for Health Statistics 1988). The importance of nutritional factors has been emphasized by various scientific bodies, such as the US National Academy of Sciences, and health organizations such as the American Heart Association (National Research Council 1982). On the other extreme are the food faddists who, out of a desire to lose weight or for other reasons, subject themselves to nutritionally unsound diets which adversely affect their health.

The resolution of several current health problems requires intensive, additional research into the relationships between nutritional factors, health, and disease. Diet varies in populations with high and low risks of cancer, for example. Several studies of different populations have linked fat intake with colon cancer, whereas a high fibre diet appears to protect against it. Studies of these differences may lead to insight into how nutritional factors promote or inhibit cancer, for example.

Nutritional surveys and surveillance such as those carried out by the US National Center for Health Statistics can identify problems of malnutrition using questionnaires and anthropometric measurements such as height–weight index, ponderal index, and skin-fold thickness. Education about properly balanced diets and diseases induced by poor nutrition should be expanded, especially among expectant mothers, children, and the elderly, in whom nutritional problems are more likely to occur. Finally, opportunities for nutritional intervention through fortification and supplementation of common foods, such as the addition of vitamin D to milk, growth hormones and antibiotics to beef, and vitamins to bread, need continuing examination.

Environmental and occupational hazards

In the early part of the twentieth century the public health professional was largely concerned with ensuring the provision of biologically safe water and food to the public, and the safe removal of sewage and garbage. The hazards of exposure to such substances as asbestos, lead, dust, and radiation were unrecognized.

Systems are now in place in most developed countries to assure the supply of biologically safe water and foods, and the proper removal of sewage and garbage. The environmental problems of importance in the developed countries in the 1980s now largely result from the explosion of technology over the last several decades. The production of chemicals in the US, for example has grown from 1.9 billion pounds in 1940 to over 375 billion pounds in 1980. This increase in the production of chemicals has also been accompanied by a tremendous expansion of the spectrum of chemicals to which the public is exposed. It has been estimated that 8 million workers in the US alone are exposed to neurotoxic chemicals (Department of Health and Human Services 1987).

The major current environmental threats arise from chemical pollution of the air, water, and land (see Volume 2, Chapters 25, 26, and 27, and Volume 3, Chapter 2). The major threats to workers arise from exposure to substances not yet identified as hazardous and the inadequate use of protective devices. Unfortunately, health effects from such pollution may occur years after exposure and they are thus difficult to document. These health effects may be the result of cumulative burdens of chemicals in tissues, or the deposition of chemicals in parts of the body which are not readily accessible for evaluation; for example, the brain. More research is needed to identify the potential health effects of acute and chronic exposure to a wide range of substances both in the general environment and in the workplace.

Safeguarding the health of the public requires regulation of pollutants. Establishing and maintaining acceptable levels of pollutants is complicated by two factors. The first is the difficulty of establishing dose–response relationships, particularly for diseases with a long induction period. The second is the economic burden which may result from regulatory actions. Whereas intervention in infectious diseases has met with enthusiastic support from the public, control of toxic substances in the environment has, until very recently, not been vigorously supported by the public. Inadequate understanding of the relationship between chemical pollutants and resultant disease, and the expense related to surveillance, regulation, and control of toxic substances has deterred action. The application of stringent controls may also cause increased costs as well as unemployment. Thus, regulation often rests more with the courts and legal procedures than on scientific expertise and judgement.

Resolution of important issues confronting society may themselves introduce health hazards. The need to develop energy efficient housing, for example, has increased the levels of pollutants such as radon and formaldehyde in homes which, in turn, increases the potential for the occurrence of related disease. In addition, certain personal habits may promote the action of specific toxic substances. The likelihood of developing lung cancer from exposure to asbestos, for example, is increased about sevenfold in the smoker and, as well, in the non-smoking individual exposed to his smoke.

Protecting the health of the public against environmental and occupational hazards in the future will depend upon:

(1) research into the acute and long-term health effects of the thousands of substances being released into the environment;

(2) surveillance for the occurrence of these hazards in the environment and the workplace;

(3) development and implementation of techniques for eliminating or neutralizing these hazards;

(4) laws to mandate the implementation of control strategies for the reduction of hazardous exposures.

Such efforts will be both expensive and unpopular so that careful considerations should be given to implementing those which will have a high probability of yielding a positive, recognizable effect on the health of the community.

Growing recognition of life-style

Although environmental measures and medical care have prevented much disease and improved health, it is becoming increasingly apparent that individuals themselves play a major role in determining their own health. They can do so through decisions about nutrition and other aspects of life-style. The spectacular achievements of microbiology and other biomedical sciences have evoked widespread and deserved admiration, but unfortunately these achievements have tended to obscure the important influence of life-style: the actions people adopt for the circumstances of their lives.

Far too many people still live in extremely restricted circumstances that limit their life-styles. Most people in industrialized society, however, have access to possibilities of consumption that, especially if followed to the extreme, can generate serious health problems. These include exposure to cigarettes, alcohol, excessive calories, and reduced physical demand. Choices in these and similar matters exert a profound influence on whether an individual in the latter part of the twentieth century will suffer and die prematurely from lung cancer, coronary heart disease, cirrhosis, chronic lung disease, trauma, and other current major causes of illness and death. Analysis of Canadian experience, for example, indicates that 18 per cent of all deaths and 18 per cent of all premature years of life lost from 1 to 70 years of age are, conservatively, attributable just to cigarette smoking and excessive consumption of alcohol (Ouellet *et al.* 1977). Cigarette smoking alone is considered responsible for more than 300 000 deaths annually in the US.

A 1965 general population survey in Alameda County, California, identified seven personal habits:—exercising at least moderately; drinking alcohol moderately, if at all; maintaining optimum weight; eating regularly; eating breakfast; not smoking cigarettes; sleeping 7–8 hours regularly—which are strongly associated with health (Belloc and Breslow 1972). At every age, from 20–70 years, persons who followed all seven of these habits had better physical health status than those who followed six, six better than five, five better than four, four better than three, and three better than two or fewer (Table 4.6). The same relationship held with respect to mortality rates expressed in longevity (Table 4.7).

Health-related habits, of course, do not develop in a vacuum. The extent to which a person acquires them depends on circumstances such as, for example, the advertising, price, and peer support of alcohol consumption. Social policy affecting these matters hence becomes an important issue for public health. In this connection, of course, public health practitioners must avoid compromising their credibility by advocating changes in life-style which are not supported by scientifically sound studies.

A second aspect of life-style significantly associated with health embraces people's relationships to their social networks. Considerable evidence now links health to marital status, degree of closeness to friends and relatives, and social group involvement. Table 4.7 displays the extent to which

Table 4.6. Comparison of average remaining lifetime* for groups of men and women stratified by number of good habits practised, and for all Californians

	Age (years)	Number of health habits practised			
		0–3	4–5	6–7	California 1959–61
Men					
	45	22	28	33	28
	65	11	14	17	13
Women					
	45	29	34	36	33
	65	12	17	20	17

* Based on life table death rates.
Source: Belloc, N.B. and Breslow, L. (1972). pp. 409–21.

Table 4.7. Age-adjusted mortality rates* for men and women aged 30–69, stratified by social network index and number of good health habits practiced, Alameda County, 1965–74.

Social network index	Health practices			
	0–4	5	6–7	Total
Men				
I (fewest connections)	21.5	9.9	10.5	15.6
II	14.6	11.7	9.9	12.2
III	10.3	9.9	6.6	8.6
IV (most connections)	7.8	9.5	4.2	6.2
Total	12.3	10.2	6.7	9.5
Women				
I (fewest connections)	15.2	8.5	10.0	12.1
II	10.4	7.5	4.0	7.2
III	5.8	4.7	4.3	4.9
IV (most connections)	6.5	2.4	4.4	4.3
Total	9.3	5.6	4.7	6.4

* Per 100 for all causes.
Source: Berkman, L.F. and Syme, S.L. (1979).

one study showed that social connections are associated with mortality (Berkman and Syme 1979). This association is largely independent of physical health status, health practices, use of health services, socio-economic status, age, sex, and race.

Population

Success in public health initiatives, particularly in controlling infectious diseases, has reduced death rates world-wide with a resultant increase in longevity resulting in population pressures. In developed countries the decline in mortality took place gradually with a commensurate decline in birth rates as survival of infants increased. In developing countries the drop in mortality has occurred over a shorter time period and without a commensurate drop in birth rates. This has resulted in rapidly expanding populations often in those countries where food and other vital resources are most limited.

Currently, the less developed countries are expanding their populations four times as rapidly as the developed countries. These countries also have a much higher proportion of younger persons which means that the higher birth rates are likely to continue for at least the next several decades. The recent epidemic of AIDS which affects primarily adults of child-bearing age may, however, have some impact in areas of high endemicity such as western and central Africa where as much as 20–30 per cent of women of child-bearing age may be infected. None the less, by the year 2000, it is estimated that nearly 80 per cent of the world's population may be concentrated in developing countries (Mann *et al.* 1988).

It is clear that unless efforts to control population growth are successful, the population of the earth will outstrip its ability to sustain itself. Therefore a major public health effort must be directed at controlling population growth. These efforts must include the continued development and implementation of more effective, safe contraceptive methods, as well as education efforts, both in the need for them and in their correct use.

In the more developed countries where population growth has declined to replacement levels the proportion of elderly has increased as the proportion of those in the productive years has decreased, introducing new problems. In the US, for example, the social security system was designed on the assumption that the working age population would be large enough relative to the elderly population to provide support for the elderly. This assumption may not hold in the future. These shifts in age distributions must be anticipated and their potential adverse effects prevented, especially as population growth is decreased even in the developing countries.

Scientific approaches

Effective public health actions are based on scientifically-derived information about factors influencing health and disease. The basic sciences of public health are epidemiology and biostatistics, but the effective use of these depends in turn on the knowledge and strategies derived from the biologic and physical sciences, the social sciences, and the demographic sciences, including vital statistics.

Epidemiologic strategies

Epidemiology is the core science of public health and preventive medicine. It is the scientific method used to describe the distribution, dynamics, and determinants of disease and health in human populations. Although there are many definitions, the Greek root of the word *epidemiology* delineates well the scope of the discipline: 'The study of that which is upon the people'. The epidemiologist seeks to identify those characteristics of people, the agents of disease, and the environment which determine the occurrence of disease and health. In order to accomplish that objective, the epidemiologist describes:

(1) disease occurrence (time characteristics);
(2) the population affected (person characteristics);
(3) the nature of the environment in which the disease is

occurring (place characteristics) which contribute to knowledge about the natural history of the disease and ways to control it. For example, epidemiologists have observed that coronary heart disease occurs primarily among men middle-aged and older in developed countries, who overeat, have high blood pressure, smoke cigarettes, do little exercise, and have a family history of heart disease.

For studying these matters, epidemiologists must have good information on the occurrence of disease, on the relevant characteristics of the population, and of the environment in which the disease is occurring. The need for this information has stimulated the development of health information systems for co-ordinating existing sources of data and guiding the development of necessary new sources of information relevant to the health of the community.

Epidemiologists depend to a considerable extent upon comparing disease frequencies in different populations. To make these comparisons it is necessary to estimate rates of disease occurrence. Information about populations is usually obtained from periodic censuses, or sometimes by a survey of a probability sample of the population. Rates which depend on census data are likely to be increasingly inaccurate with the number of years which have elapsed since the data were collected. In addition, detailed information on populations derived from a periodic census is often not available for one or more years following the actual collection of the data. Information on population characteristics may be obtained at more frequent intervals by examining an appropriately selected probability sample of a population. This information can be particularly useful at times distant from the date of collection of census data.

The potential of epidemiology for documenting disease occurrence and developing and testing hypotheses has expanded rapidly over the last decade. This advance is a result not only of the rapid development of computer technology but also because of the entry into the field of individuals whose major discipline is epidemiology. Although non-physicians in the UK made several significant contributions to epidemiological methodology (for example, the great biostatistician, Bradford Hill) prior to the last decade, the majority of epidemiologists in the US were physicians who took an additional year or more of training in epidemiology to supplement their biomedical training. Both physician and non-physician epidemiologists are essential to the field.

Recently, there has been a rapid increase in the number of techniques that distinguish factors which are truly related to disease from those factors which are related only indirectly to disease and, thus, may confound a true causal association. For example, using recently developed methods, the role of oestrogen in promoting endometrial cancer was shown not to be due to detection bias (Hutchison and Rothman 1978) and electronic fetal monitoring has been shown to lead to unnecessary Ceasarean sections (Neutra *et al*. 1980).

The capacity of the epidemiologist for ascertaining the distribution, dynamics, and determinants of disease depends upon the availability of several other scientific disciplines. Laboratory procedures derived from chemistry, biochemistry, microbiology, and immunology, for example, can be used for obtaining information about the environment, the agents of disease, and the changes which occur in man. Epidemiology looks to statistics, on the other hand, for mathematical methodologies which describe the strength of correlations between the multiple factors which may promote the occurrence of disease, and to the social sciences for techniques for obtaining accurate information from respondents.

In summary, rapid advances in the field of epidemiology have resulted from new epidemiological methods as well as from the availability of new techniques in biostatistics, the laboratory sciences—microbiology, chemistry, and engineering—and the social sciences. Epidemiology will continue to draw upon these disciplines to help provide the basic information needed for the development and application of effective public health strategies for disease control and health promotion.

Biostatistics

Biostatistics is the science used to quantify relationships observed in public health and medicine. Through the correct application of biostatistical techniques, public health professionals can test and quantify the magnitude of a factor or factors to the health of the community.

Advances in epidemiological methodology have been accompanied by rapid progress in biostatistics, particularly the development of computer technology. Through its application, biostatisticians have developed multivariate techniques to determine the relationship of multiple factors to disease occurrence while simultaneously observing the relationship of these variables to each other. Sophisticated techniques for the analysis of events in relation to time are enhancing the value of the cohort study design. Because the computer can process massive amounts of information rapidly, it has been possible to develop and test mathematical models that describe hypothetical relationships and disease outcomes based on a variety of assumptions. The degree to which the actual occurrence of disease matches the models confirms or refutes these relationships.

These new statistical techniques, many using newly developed computer technology, have been used to identify factors which have a causal relationship to disease as well as to determine the efficacy of preventive strategies such as vaccines and health education, and the efficacy of drugs through clinical trials. The potential of statistics to contribute to public health even more, through further development of biostatistical strategies and innovative computer methodology, is great.

Biological and physical sciences

The laboratory sciences have long played an essential role in public health. Many of the new advances leading to the control of infectious diseases depend upon microbiology to provide new techniques to isolate disease agents and identify markers of prior infection or exposure. The rapid expansion of vaccines to prevent viral diseases reflects new procedures for isolating viruses using cell cultivation techniques that were developed in the late 1930s. These cell culture techniques facilitated the manufacture of live vaccines using attenuated viruses; that is, viruses which have lost their virulence characteristics for man but not their capacity for stimulating immunity. Recently, microbiologists have fragmented disease agents into specific components and selected those which are responsible for the immune response. Vaccines are now also being developed which utilize genetic recombination and synthetic peptide chains (Plotkin and Mortimer 1988).

Startling as these recent developments in microbiology and immunology have been, equally important contributions to public health are coming from the laboratory sciences of chemistry, biochemistry, and engineering. The studies of the chemical interactions of primary pollutants in the atmosphere which lead to development of photochemical oxidants, for example, can suggest steps in these processes at which intervention strategies can be introduced.

Advances in the laboratory sciences can be rapidly translated into new techniques for identifying infection and disease, as well as environmental and occupational hazards. Further, these new advances often lead to new techniques and strategies for intervention and control of threats to the public health.

Social sciences

In addition to their influence on choice of exercise, levels of personal hygiene, eating patterns, and alcohol consumption, behavioural factors also determine the response to illness, particularly to subtle manifestations of disease. Thus they significantly affect the ability of the individual to live in a healthy way and to respond to disease. The role of the behavioural sciences (including psychology, sociology, and anthropology) in public health is therefore increasing. Experience with AIDS over the past few years dramatically illustrates the need for influencing behaviour that leads to disease occurrence, and then with behavioural aspects of the illness that ensues (Fineberg 1988; Turner *et al.* 1988).

Each of the behavioural sciences approaches behaviour from a different vantage point. Psychologists emphasize individual differences; sociologists draw inferences from analyses of how groups of people, whole communities, or nations or sub-sets of them, behave *en masse*, and anthropologists stress the cultural patterns that influence behaviour from generation to generation.

Behavioural science techniques have proved valuable in understanding influences on health. Social survey methodology has greatly enhanced our capacity to analyse relationships of behaviour to disease and to discern trends that are highly important to public health. Psychological investigations of people's knowledge and attitudes yields insight into the habitual and life-style practices that are related to health, and often suggest ways of promoting health. Sociological investigation of the group processes that determine a community's norms and values, and adherence to them, likewise leads to an understanding of how people behave and thus how they can be influenced to follow a healthy life-style. Anthropology elucidates the cultural traditions that affect what people do in everyday life and suggests approaches to health promotion specific for various cultural groups.

Within the field of public health, health education draws upon these disciplines to develop effective techniques for cultivating health-maintenance behaviour. As emphasized earlier, the social milieu in large measure determines the choices that people make. Economic and other social conditions of life profoundly impact on what people do about health-related actions. Life-style does not consist of behaviour elements selected by an individual in a void, but depends upon their circumstances of life. Hence public health must be concerned with the social conditions in which people live and direct substantial effort toward their improvement on behalf of health.

Demography and vital statistics

John Graunt is commonly considered the father of vital statistics because of his early studies of the Bills of Mortality in London and a parish town of Hampshire. He collected and examined the birth and death records maintained by parish clerks from 1603 to 1662 (Graunt 1939). From that epochal work he drew important inferences about the population and its health. He analysed mortality, including infant mortality, seasonal variation of deaths, and longevity, as well as fertility and the excess of male births. His studies laid the groundwork for what has become vital statistics.

Over the past three centuries demography, the study of human populations has been closely intertwined with public health. Vital statistics include delineation of

(1) births and the rates of their occurrence in various segments of the population;

(2) fertility, that is, the ratio of births to women aged 15–49 years;

(3) mortality, including deaths among infants and in subsequent ages as well as trends, specific causes, and determinants of deaths;

(4) migration patterns.

All of these are important to public health.

Information about the occurrence of disease may be obtained through aggregated data from death certificates, birth certificates, hospitals and clinics, surveys, and registries. Computer technology facilitates analysis of mortality in relation to the characteristics which are coded on each person's death certificate. In addition, information from birth certificates and other sources can be linked to the occurrence of death, thus providing additional information about the characteristics of person or events that may cause death.

Data concerning non-lethal diseases are more difficult to obtain than the birth and death information which must be recorded by law. In the US, the Centers for Disease Control (CDC) publishes a morbidity and mortality weekly report which contains information on certain diseases, obtained through reporting from local health departments. Hospital discharge abstracts and summaries provide further information. Special surveys for specific diseases or factors affecting health may be carried out, in addition to ongoing national health surveys administered to a probability sample of the population. Disease-specific registries also provide information about changing trends in disease occurrence, mortality, and duration of survival. Most well-known among these are the cancer registries, but comparable data bases are also being developed for diabetes, coronary heart disease, and other chronic diseases.

Demography focuses on population trends such as growth, that is, the excess of births over deaths, and in-migration over out-migration. Public health statistics are concerned with information about the health of populations. Both fields are devoted to satisfying social concerns about people. Mutual interest in factors such as those determining fertility illustrates the continuing interrelationships of public health and demography.

Programmatic scope of public health

Goal setting

Success in achieving the World Health Organization's objective of eradicating smallpox throughout the world has inspired other efforts to set and popularize explicit goals in public health (Breslow 1987; Fenner *et al.* 1988). The US Public Health Service, for example, established in 1980 specific objectives for 1990 as a guide to public health efforts during the 1980s. The latter included three action categories: personal preventive services, such as immunizations and high blood pressure control; environmental health protection, such as toxic agent control and fluoridation of water supplies; and health promotion, such as smoking cessation and exercise fitness (Department of Health, Education, and Welfare 1980). Not only were specific targets set for various aspects of mortality and morbidity, but objectives were also established regarding prevention of disease, environmental exposures, medical services, and life-style risk factors. A data collection and publication process were undertaken to track progress in the enterprise (Table 4.8).

Table 4.8. Progress toward 1990 health promotion goals in the United States, 1977–1985

	1977	1980	1985	1990 Goal
Deaths per 1000 live births	14.1	12.6	10.6	9
Deaths per 100 000 ages 1–14 years	42.3	38.5	33.8	34
Deaths per 100 000 ages 15–24 years	114.8	115.4	95.9	92
Deaths per 100 000 ages 25–64 years	532.9	498.0	438.7	400
Bed disability days per person, ages 65+*	14.5	14.0	13.7	12

* Data for years after 1981 may not be comparable to those of previous years because of a change in procedures.

Source: National Center for Health Statistics (1987) p. 78.

Prevention of disease and promotion of health

The ultimate goal of public health has always been and remains the prevention of disease and the promotion of health. Although great strides have been made in achieving these goals there are still many diseases which cause premature death and deterioration in the quality of life. Further, there are many people in both the developed and the developing world who have not yet benefitted from the public health achievements of the last century. The major goal of public health in the developed countries in the future, therefore, will remain the prevention of those diseases such as cancer, heart disease, trauma, and AIDS which are currently the leading causes of premature mortality and the decreased quality of life, and the implementation of the achievements of public health for those groups still suffering from diseases which are preventable or for which morbidity and mortality can be reduced using current knowledge and technology. These include the poor and those not yet adequately integrated into the mainstream of society.

The goals for prevention can be achieved through emphasizing preventive aspects of medical care such as immunizations, health education and behavioural modification, control of the environment, and recruitment of the political will of the community for public health initiatives.

Medical care

Beginning with Bismarck, the western nations have generally provided medical care of varying kinds and degree as a social benefit to industrialized workers. In most countries some care has also been extended to others, particularly to families of workers, the elderly, and the poor. The British National Health Service, for example, covers the whole population. On the other hand, the US relies almost entirely on private arrangements for employed persons; large-scale governmental assistance for health care services goes only to the elderly and the poor.

Medical care can be examined from several perspectives: medical and economic, for example, as well as from the standpoint of public health. The medical profession, reflecting both the centuries-old tradition of healing and recent advances in medical science, looks upon medical care as the main means to relieve suffering and restore health in the individual. Economists view medical care in terms of its cost and therefore examine the increasingly large expenditures for it with concern. Public health considers medical care to be one means of protecting and improving the health of people, but also it is concerned with cost, especially in so far as it constitutes a barrier to health care for some groups. The public health focus on medical care emphasizes its potential for enhancing a community's health.

Provision of medical services is usually determined in a specific country by cultural and traditional patterns. Thus in Britain the individual general practitioner and his or her panel of patients are the predominant care module, with referral to the hospital consultant as necessary. In the US, however, the former life-long, physician–patient relationship is now often replaced by clinics where the patient sees any available physician, by specialists such as pediatricians, general internists, and obstetricians, who see patients only during certain limited periods of their lives, and to a growing extent by free-standing emergency medical services. Different medical service patterns are continuously evolving. In Britain, for example, the trend is for general practitioners to work within a group practice pre-payment plan; these pre-payment plans are also expanding in the US.

Public health uses medical care primarily to achieve prevention. Thus public health agencies have organized immunization activities, and, especially in the US, maternal and child health services. Over the past few decades these efforts have contributed to spectacular achievements in control of communicable disease and infant mortality.

In the past, curative services have generally received priority over preventive services on grounds of urgency. Most physicians and medical care agencies adhere to what may be termed the complaint–response system of medicine; patients are taught to recognize and bring their health complaints to the doctor whose response is to diagnose and treat any illness which may be present. Prevention, if advocated at all, is usually a minor consideration.

A new system of medical care, which gives priority to promoting health and preventing disease, has been slowly emerging. Individuals' health is monitored through periodic appraisal geared to age and other factors which determine both current and future prospects of health. Thus, infant care concentrates on growth, appearance of defects, immunization status, and any necessary corrective action to assure the healthiest possible development. When a person has reached 50 years of age, the focus has shifted to blood pressure, weight/height ratio, blood-sugar, blood-cholesterol, cancer detection, cigarette and alcohol consumption, and other physical and behavioural characteristics.

Although health maintenance through periodic health evaluation and health counselling has yet to be fully evaluated, sufficient grounds now exist for making it available that industrial leaders in the US have started to provide such services for their employees. Arrangements for these services at or near the work-site are perhaps the fastest growing aspect of medical care in the US.

In addition to the relative emphasis that should be accorded preventive versus curative efforts, several other issues currently affect the public health approach to medical care. In past years many procedures and drugs have come into widespread use without sufficient consideration of their effectiveness. Initiatives are now underway, for example, by the Medical Research Council and the Departments of Health and Social Security in Britain, and the Congressional Office of Technology Assessment in the US, to establish better systems for evaluating the effectiveness of medical technologies, including drug regimens.

These initiatives reflect concern about the rapidly rising cost of medical care in the western nations and about items of questionable, or even negative, value to health. Another rising issue is the efficiency of medical service; that is, how can the best possible quality of medical care be provided within a given amount of resources? New facilities for 'ambulatory surgery' make it possible to carry out many procedures without the extra expense entailed in admission to a hospital. Organizing medical personnel into groups, as well as providing incentives to personnel, offer possibilities for increasing the productivity of medical services. Still another cost-related problem is the extent to which medical resources should be used for highly expensive procedures and devices such as kidney or heart replacement which benefit only a few at great expense. Expanding technology and limited resources together are forcing consideration of the ethical as well as the health and economic consequences of medical care. Resistance is growing to the technological prolongation of life when the quality of life has deteriorated beyond a point worth saving.

A major programmatic thrust of public health in the immediate future thus will be renewed emphasis on medical care as a means of improving health. However, the balance seems to be shifting from curative or complaint-oriented services to health promotion/disease prevention.

Influencing behaviour

As noted earlier in this chapter, the way people live determines their health to a considerable extent. Thus a prime responsibility for public health is to develop effective strategies to promote healthy life-styles. Several approaches to this are being tested. One is to convert the national and even international milieu to favour healthy behaviour; for example, the various national and WHO campaigns against cigarette smoking. Such activities often result in confrontation with powerful, entrenched economic interests. Tactics in the struggle to turn public policy explicitly towards the side of health, therefore, must be high on the public health agenda.

A second approach is the so-called medical model; that is, using the doctor–patient relationship, or analogues of it, to influence health-related behaviour. The tactic is a one-to-one, or sometimes a small group, effort in a health-oriented environment to guide individuals toward health behaviour. It offers promise particularly when people have, or can be induced to have, concern about particular health problems such as cancer or heart disease, and then are willing to undertake the indicated change of their habits. Physicians have often been reluctant to devote the effort needed, partly because of discouragement with the results. Effective change, however, can be achieved with adequate protocols (Gritz 1988).

Environmental control

Attempts to control pollution of the environment including the occupational setting are complicated by the problem of identifying those pollutants which pose health hazards to humans. Pollutants such as radiation which have no apparent immediate effect on humans can cause disease after many years of chronic and persistent exposure. Some pollutants may cause disease by accumulating in the body. However, the accumulation of these substances may occur in parts of the body such as the bone in which it is difficult to measure levels.

A highly desirable approach to control is to obtain the voluntary co-operation of industry as has been partially achieved in the control of air pollution. However, control mechanisms which are often expensive have seldom been voluntarily adopted by industry.

In the middle decades of the twentieth century the western nations became increasingly aware of threats to health occurring through contamination of the environment. This concern led in the US, for example, to implementation of the National Environmental Act in 1969 which directed the federal government to plan policies in the light of the effect that these policies would have on the environment. The National Environmental Act was followed in the next decade by a series of legislative actions creating regulatory agencies directed at protection of the environment, and especially reduction of pollutants. These are discussed further in Volume 3, Chapter 2.

During the first part of the 1980s the governments in the developed nations have faced increased challenges to their regulatory activities. The deteriorating fiscal situation of federal and local governments has caused legislatures to question the advisability of handicapping industrial operations at a time when these are deemed necessary for economic recovery. In the 1980s there have been serious challenges to maintaining and improving the quality of the environment. Innovative approaches, such as combining the control of wastes and the development of new energy sources and techniques for recycling waste products, will be needed. An example of this is the use of recharged waste water to augment existing limited water supplies.

In summary, the programmatic scope of public health embraces preventive, medical, behavioural, and environmental measures designed to improve health at the community level. Although particular agencies and personnel address specific aspects of community health, public health embraces the whole range of these activities.

Public health strategies

Surveillance

Effective intervention into factors affecting community health depends upon reliable knowledge concerning the occurrence and distribution of these factors in the community. Thus the backbone of public health strategy is the development and maintenance of an accurate, reliable health information system upon which actions can be based. Such surveillance systems should include information on the occurrence of infectious and chronic diseases, environmental information, including occupational exposures, behavioural characteristics, and medical services. Information must be collected on a regular basis and reported rapidly, particularly for the infectious diseases and hazardous environmental exposures. Tardy reporting of information about disease outbreaks or sudden radiation hazards, for example, will prevent the implementation of effective intervention procedures early when they are most likely to be effective. Although most developed countries have a system of ongoing surveillance for infectious diseases and some environmental hazards, surveillance systems used by public health agencies, in general, do not fully meet the above criteria for reporting.

The most extensive experience in surveillance work has concerned the communicable diseases. There are fewer mechanisms for reporting chronic diseases other than mortality.

Currently, surveillance for environmental hazards and occupational exposures is even less satisfactory than surveillance for either infectious or chronic diseases. Most urban areas have systems for monitoring quality of air and quality of water for human consumption although provisions for monitoring new contaminants such as cadmium and magnesium are inadequate. Considerable attention also needs to be directed to surveillance of recreational waters, toxic dump sites, and radiation exposures. In addition, workers continue to be exposed to unsafe working conditions, particularly workers in small industries which are more difficult to monitor and regulate. Details of the current surveillance systems are discussed in Volume 2, Chapter 11.

Until permanent information systems can be implemented which provide accurate, reliable, rapid reporting on all principal factors affecting community health, effective programmes cannot be fully realized.

Intervention

Effective intervention is the heart of public health. Public health efforts to protect communities from health hazards include reducing the number of individuals susceptible to infectious and chronic diseases, treating people early in the course of disease, modifying the environment, and promoting healthy behaviour of both communities and individuals.

Technologic advances play a key role in developing effective intervention programmes, but often implementation of these programmes depends on the use of innovative epidemiologic strategies, behavioural modification of individual life-styles, and changing of the political will of the community. For example, a satisfactory vaccine for smallpox existed for centuries before eradication was made possible by changing from an untargeted mass vaccination approach to an active surveillance and containment strategy. Epidemiologic research has identified many risk factors for cardiovascular disease, but implementation of intervention strategies to reduce these risk factors depends on convincing people to alter their basic life-styles on such factors as diet and exercise. Methods for prevention of most venereal diseases, including AIDS, are well known, and treatment of many of them has been available for decades. None the less, efforts to reduce the incidence of these diseases have been largely unsuccessful because of the difficulty of modifying this most intimate aspect of life-style. The source of many of the pollutants plaguing the major cities of both the developed and the developing world are known, but techniques to reduce these pollutants involve major expenses by both the public and industry and often cause inconvenience for the public. Implementation through legislation has met resistance. Until the political will has been influenced, it is unlikely that major reductions in pollutants can be accomplished.

In summary, successful public health intervention is the result of the development of technical advances coupled with the use of innovative epidemiologic strategies, implementation of effective behavioural modification techniques, and induction of the political will. Public health intervention thus requires the joining of many different disciplines to achieve promotion of the health of the public.

Evaluation

An essential component of public health strategies is evaluation. The effectiveness of surveillance and intervention programmes changes over time due to changes in the incidence of disease, the development of new health hazards, and the development of new technologies for measurement and control. Thus evaluation should be an ongoing, integral part of all public health surveillance and intervention programmes. For many years vaccination of all persons in the US against smallpox persisted even though the need for it had diminished with the elimination of cases from the United States. Ultimately, the world-wide eradication of this disease has for most individuals eliminated the need for vaccination. Since any immunization is associated with some adverse reactions, continued use must provide more benefit than risk. Since this ratio changes in relation to many factors, the relationship must continually be re-evaluated.

The effectiveness of different strategies of community intervention for promotion of healthy life-styles can also be evaluated. The Stanford Three City Study, for example, has evaluated two different strategies of community intervention for reducing behavioural risk factors for disease in three different cities (Farquhar *et al.* 1977). A similar evaluation has been under way for a strategy to reduce heart disease in Finland (Puska *et al.* 1985).

Evaluation of environmental intervention has progressed less rapidly, in part because of the need for appropriate technology to identify and measure levels of pollutants in air, water, land, and the workplace. Although some success has been achieved in reducing levels of pollutants in air (notably in London) and reducing levels of pollutants in water bodies (for example, Lake Washington in Seattle), strategies for control of toxic dump sites have not been successful to date.

In summary, surveillance, intervention, and evaluation are the backbone of public health strategies to prevent disease, eliminate health hazards and, promote health in the community.

Organization of public health

Government structure

Organization of health services, both public and private, tends to be conditioned by the cultural, political, and organizational patterns of the countries in which they are located. Thus in Britain and many European countries there is a national health service covering preventive, community, and clinical health. On the other hand, in the US the tendency has been toward state and local governmental autonomy in community health services, but individual, clinical services have been left largely in the private sector with governmental payment for certain segments of the population.

United States

The basic concept of government in the US has been for the states to relinquish only those rights of jurisdiction which are essential to maintain the union. Public health has followed this pattern with strong state and local jurisdiction. Most public health programmes have been conducted at the local level, under state regulation, with only broad directions or incentives provided by the federal government. Thus the local jurisdictions—the county, city, or township—through authority delegated from the states, typically assume primary responsibility for communicable disease surveillance and control, maternal and child health services, environmental surveillance and control, and other public health activities.

The role of the federal government in public health has evolved for the most part on a piecemeal basis. Usually it has assumed responsibility for meeting needs which were not otherwise met by private, local, or state agencies. These initiatives have generally been categorical in nature, directed primarily at specific disease problems or segments of the population, such as the poor. Exceptions to this approach have been the creation of the National Institutes of Health, the major research funding source in the US, certain regulatory agencies such as the Environmental Protection Agency and the Food and Drug Administration, and the agencies and programmes stemming from the 1935 Social Security Act, the basic social security legislation for the nation.

The US Department of Health, Education and Welfare has published *Healthy People* (Surgeon General 1979) and *Promoting health and preventing disease: Objectives for the nation* (Centers for Disease Control 1980), and other documents indicating a broad approach to prevention of disease and promotion of health in the US. Such documents, however, do not have the force of legislation and have served mainly as guidelines and encouragement for local health agencies. The role of the federal government remains largely to suggest and encourage (sometimes with specific subsidies) actions which are either implemented (or ignored) at the local level. These are discussed further in Volume 1, Chapter 29, and Volume 3, Chapter 8.

Europe

In the European nations the philosophy of central control of public health has predominated, perhaps because of their smaller size and more homogenous populations. The majority of the European nations have a national health scheme of one type or another which is administered federally. Thus to a larger degree than in the US, public health activities can be implemented centrally through an organized system. In some countries such as those in Scandinavia, federal registration of individuals often facilitates public health actions.

The presence of a national health scheme, however, has not guaranteed a more effective public health programme. Often the agencies within these federal governments do not command the respect and resources accorded the clinical components, and therefore are not as effective as they could be (Evans 1982). For example, many of the European countries lack schools of public health or their equivalents which prepare professionals to lead public health programmes. None the less, access to medical care and equity have often been greater in these systems than in the US.

Whatever the government structure for public health, the need for good management is increasingly recognized. The responsibility for handling budgets that are often substantial, complex organizations involving many different categories of people, and maintaining effective relationships with a wide array of health agencies as well as other bodies, requires great managerial skill. In fact, the inadequate preparation in management skills of many health professionals who have occupied public health administrative posts in the past has induced some governing authorities to call upon 'managers' rather than public health experts for the key positions in public health. Too often, then, the task of public health administration has been reduced to budget control or complying with already adopted laws and regulations. As a result little attention is given to health problems or their solution. The ideal, of course, is to combine the talent for leadership in public health with managerial skill. In conclusion, the organization of public health appears to be determined at every level—local, state and national—largely by cultural and historic factors resulting in a wide array of often complex organizational arrangements.

Non-governmental public health agencies

Voluntary health agencies have flourished in the US and to a somewhat lesser extent in Europe. They tend to be organized around specific entities; for example, the American Cancer Society, the American Heart Association, and the American Lung Association. Their success has encouraged the development of many more such groups, devoted to practically all the major diseases and several lesser ones.

The groups are typically organized at the national level with state divisions and local chapters. These voluntary health agencies bring together physicians who are leaders in their particular fields and interested members of the public. They involve millions of people in fund-raising for, and operation of, disease treatment and control activities. In this way they have contributed much to the level of enlightenment and activity concerning health, particularly in the US. Their programmes usually include support of health research, professional education, public education, and demonstration services devoted to the particular category of disease with which the organization is concerned.

These voluntary health agencies have become a considerable force in American public health. They are able to operate with fewer constraints than governmental departments and thus have often broken new ground in the field. The American Heart Association and the American Cancer Society, for example, have been particularly active in bringing the concepts of risk factors and healthy life-styles before the American public.

Another force in developing public health policy has been the private foundations such as the Robert Wood Johnson Foundation, the Pew Memorial Fund, the Kellogg Foundation, and the Rockefeller Foundation. These foundations have stimulated and supported studies of various alternative systems of health care, medical education, and training of public health experts, which have had important policy implications for the way in which medical care and public health are organized and delivered. The Rockefeller Foundation, for example, has fostered an international network of physicians in developing countries in clinical epidemiology through sponsorship of selected training programmes in medical schools. This has had a major impact both in promoting epidemiology and in increasing its profile within medical schools in the participating countries. By supporting programmes and studies with particular social and medical implications, these foundations will probably continue to play an important role in influencing public health policy.

Summary

The scope of public health in the last part of the twentieth century has expanded greatly. Not only have the number of recognized health hazards to the public increased, the strategies available to solve them have grown commensurately. Public health has borrowed and adapted knowledge from the physiological, biological, medical, physical, behavioural, and mathematical sciences, and has been quick to recognize the potential of new fields such as the computer sciences for improving, safeguarding, maintaining, and promoting the health of the community.

As the major communicable diseases have been brought under control through public health measures, more effort has been directed at chronic disease control, mental health, a safe environment, reduction of accidents, violence and homicide, and promotion of healthy life-styles. Although the biological sciences remain an important underpinning of public health, the contribution of the physical, mathematical, and behavioural sciences is recognized increasingly. As in the past, improvements in the health of the public in the future will be achieved by inducing public awareness and concern which results in the introduction and passage of effective legislation and regulations which are implemented by professionals committed to the principles of public health.

The effectiveness of such efforts in the past and the realization of the cost-effectiveness of preventive strategies for promoting and maintaining health have brought renewed attention to public health and have set the stage for a new public health revolution.

References

Belloc, N.B. and Breslow, L. (1972). Relationship of physical health status and health practices. *Preventive Medicine*, **1**, 141.

Berkman, L.F. and Syme, S.L. (1979). Social networks, host resistance and mortality; a nine-year follow-up study of Alameda County residents. *American Journal of Epidemiology*, **109**, 196.

Breslow, L. (1987). Setting objectives in public health. *Annual Review of Public Health*, **8**, 289.

Centers for Disease Control (1980). *Promoting health and preventing disease: Objectives for the nation*. United States Government Printing Office, Washington, DC.

Centers for Disease Control (1985). Homicide among young black males—United States, 1970–1982. *Morbidity and Mortality Weekly Report*, **34**, 629.

Centers for Disease Control (1986*a*). Premature mortality due to unintentional injuries—United States, 1983. *Morbidity and Mortality Weekly Report*, **35**, 353.

Centers for Disease Control (1986*b*). Premature mortality due to suicide and homicide—United States, 1983. *Morbidity and Mortality Weekly Report*, **35**, 365.

Centers for Disease Control (1988). *AIDS Weekly Surveillance Report—United States*. AIDS Program, Centre for Infectious Diseases, Centers for Disease Control, Atlanta, Georgia, October 24.

Cheever, A.W. (1976). Animal model of human disease: Carcinoma of the urinary bladder in schistosoma haematobium infection. *American Journal of Pathology*, **84**, 673.

Chevlen, E.M., Awwad, H.K., Ziegler, J.L., *et al.* (1979). Cancer of the bilharzial bladder. *International Journal of Radiation, Oncology, Biology, Physics*, **5**, 921.

Collins, J.G. (1988). *Prevalence of selected chronic conditions, United States, 1983–85*. Advance Data from DHHS Publication Department of Health and Human Services Number (PHS) 88–1250.

Collins, J.G. (1988*b*). *United States, 1983–85*. Advance Data from the National Center for Health Statistics, US Department of Health and Human Services.

Curran, J.W., Morgan, M., Hardy, A.M., Jaffe, H.W., Darrow, W.W., and Dowdle, W.R. (1985). The epidemiology of AIDS: Current status and future prospects. *Science*, **229**, 1352.

Department of Health, Education and Welfare (1977). *Health, United States, 1967–1977*. Department of Health, Education and Welfare, Number (HRA) 77–1232. Washington, DC.

Department of Health, Education and Welfare (1980). *Health, United States—1979*. Department of Health, Education and Welfare, Number (PHS) 80–1239. Washington, DC.

Department of Health and Human Services (1987). *Prevention 1986/87, Federal programs and progress*. US. Department of Health and Human Services, Washington, DC.

Elsebai, I. (1977). Parasites in the etiology of cancer—*bilharziasis and bladder cancer*. *Cancer*, **27**, 100.

Evans, J.R. (1982). Measurement and management in medicine and health services: Training needs and opportunities. In *Population-based medicine* (ed. M. Lipkin and W.A. Lybrand), p. 3. Praeger Scientific Publishers, New York.

Farquhar, J.W., Macoby, N., Wood, P.D., *et al.* (1977). Community education for cardiovascular health. *Lancet*, **i**, 1192.

Fenner, F., Henderson, D.A., Arita, I., Jezek, Z., and Ladnyi, I.D. (1988). *Smallpox and its eradication*. World Health Organization, Geneva.

Fineberg, H.V. (1988). The social dimension of AIDS. *Scientific American*, **259**, 128.

Gottlieb, M.S., Schroff, R., Schanker, H.M. *et al.* (1981). Pneumocystis carinii pneumonia and mucosal candidiasis in previously healthy homosexual men. *New England Journal of Medicine*, **305**, 1425.

Graunt, J. (1939). *National and political observations mentioned in a following index, and made upon the bills of mortality*. Printed by Tho. Roycroft for John Martin, James Allestry, and Tho. Dicas, London, 1662.

Grayston, J.T. and Wang, S.P. (1975). New knowledge of chlamydiae and the diseases they cause. *Journal of Infectious Disease*, **132**, 87.

Gritz, E.R. (1988). Cigarette smoking: the need for action by health professionals. CA-A cancer journal for clinicians. *American Cancer Society*, **38**, 194.

Heyward, W.L. and Curran, J.W. (1988). The epidemiology of AIDS in the United States. *Scientific American*, **259**, 72.

Hutchison, G.B. and Rothman, K.J. (1978). Correcting a bias? *New England Journal of Medicine*, **299**, 1129.

Mann, J.M., Chin, J., Piot, P., and Quinn, T. (1988). The international epidemiology of AIDS. *Scientific American*, **259**, 82.

McGinnis, J.M. (1982). Targeting progress in health. *Public Health Reports*, **97**, 295.

McGinnis, J.M. (1988). *Targeting progress in health*. National Center for Health Statistics, US Department of Health and Human Services. Department of Health and Human Services Publication, Number (PHS) 88–1232. Washington, DC.

Meade, W.M. and Curran, J.W. (1986). Acquired immunodeficiency syndrome: current and future trends. *Public Health Reports*, **101**, 459–65.

Morgan, W.M. and Curran, J.W. (1986). Acquired immunodeficiency syndrome: current and future trends. *Public Health Reports*, **101**, 459–65.

Multiple Risk Factor Intervention Trial (MRFIT) Research Group (1982). Multiple risk factor intervention trial—risk factor changes and mortality results. *Journal of the American Medical Association*, **248**, 1465.

National Center for Health Statistics (1988). *Health Promotion and Disease Prevention*, National Center for Health Statistics, Series 10, Number 163. Washington, DC.

National Research Council, Committee on Diet, Nutrition and Cancer. Assembly of Life Sciences (1982). *Diet, nutrition and cancer*. National Academy Press, Washington, DC.

Neutra, R.R., Greenland, S., and Friedman, E.A. (1980). The effect of fetal monitoring on Caesarian section rates. *Obstetrics and Gynecology*, **55**, 175.

O'Carroll, P.W., Mercy, J.A., and Steward, J.A. (1988). CDC Recommendations for a community plan for the prevention and containment of suicide clusters. *Morbidity and Mortality Weekly Report*, **37**, Supplement S–6, 1.

Ouellet, B.L., Romeder, J.M., and Lance, J.M. (1977). *Premature mortality attributable to smoking and hazardous drinking in Canada*, **Vol**. 1. Department of Health and Welfare, Canada.

Piot, P.A., Plummer, F.A., Mhalu, F.S., Lamboray, J.L., Chin, J., and Mann, J.M. (1988). AIDS: an international perspective. *Science*, **239**, 573.

Pisa, Z. and Uemara, K. (1981). *Mortality from IHD and other CVD in 26 countries. Trends in 1969–1977.* WHO CVD/CHD/EC 81.12 WHO, Geneva.

Plotkin, S.A. and Mortimer, E.A., Jr. (ed.) (1988). *Vaccines*. W.B. Saunders Company, Philadelphia, Pennsylvania.

Puska, P., Nissinen J.T., Tuomilehto, J., *et al.* (1985). The community-based strategy to prevent coronary heart disease: Conclusions from the ten years of the North Karelia Project. In *Annual review of public health*, (ed. L. Breslow), Vol. 6, p. 147. Annual Reviews, Inc., Palo Alto California.

Report of the Working Group on Arteriosclerosis of the US National Heart, Lung and Blood Institute (1981). *Arteriosclerosis*. US Dept. of Health and Human Services National Institute of Health Publication, Number 81–2034, Washington, DC.

Smith, G.S. (1985). Measuring the gap for unintentional injuries: The Carter Center health policy project. *Public Health Reports*, **100**, 595.

Surgeon General (1979). *Healthy people: The Surgeon General's report on health promotion and disease prevention*. Department of Health, Education and welfare Publication, Number (PHS) 79:55071. Washington, DC.

Turner, C.F., Miller, H.G., and Moses, L.E. (ed.) (1988). AIDS, sexual behaviour and IV drug use. Committee on AIDS Research and the Behavioural, Social and Statistical Sciences. National Research Council. National Academy Press, Washington, DC.

B

Determinants of Health and Disease

5

Population dynamics

W. BRASS

Introduction

The number and characteristics of people in an area determine the level and nature of the need for that area's health services. They also define the population at risk for particular medical conditions. In any assessment of the importance of these conditions it is necessary to gauge the numbers affected against the total potential, that is, those at risk. The structure of a population (that is, the distributions of persons with significant characteristics), is constantly changing. In order to understand the nature of these alterations, and indeed to anticipate them, it is necessary to relate the changes to the factors which affect them.

Population growth

Populations grow because new members are added through births, and these are usually more than the numbers who disappear through deaths. For comparison of populations which may be of very different sizes, the number of births and deaths in a year are divided by the population size and expressed as rates per thousand. These are called the *crude birth-* and *death-rates* (CBR and CDR). The CBR−CDR is the *natural increase* (NI). Populations also increase from persons entering (immigration) and decrease from those leaving (emigration), but with few exceptions, the effects (net migration) on the population size for whole countries is small relative to the natural increase. However, net migration is important, and often overwhelmingly so for population structure and change in local areas. The increase in the population over a year, divided by the total population and expressed as a rate per thousand is the *growth-rate*.

Crude death-rates are now of the order of ten per thousand in most parts of the world. Only in a few very underdeveloped areas, such as the Sahel countries of Africa, are the rates substantially greater (as much as three to four times). It should be noted that many Third World countries have comparatively modest crude death-rates, not because mortality is low but because the age distribution is young, with few old persons. The natural increase and growth is then dominated by the crude birth-rates which, in turn, depend on the level of fertility. In general, developed countries with strong fertility control and hence small families, have crude birth-rates also of the order of ten per thousand and a natural increase not far from zero (Table 5.1). Fifteen to twenty years ago there was little family limitation in the less-developed countries and crude birth-rates were 40 per thousand or even higher. There are still many of them in this state, particularly in Africa and the Middle East, although there are significant transitions to lower fertility, notably in Latin America and parts of South-East Asia. Selected measures for less-developed countries are shown in Table 5.2. Kenya has often been cited as the country with the highest natural increase—with good justification, since it reached some 40 per thousand (4 per cent per year) around 1980 (World Bank 1986).

The statistical history of world population growth is sketched out in Table 5.3. Before the eighteenth century, overall natural increase was slow, with high death-rates offsetting the uncontrolled fertility, although in some places and times there were spurts of growth. With the industrial revolution and rises in the standard of living, mortality began to fall steadily in the richer countries. The crude birth-rate came down later in the nineteenth century and the combination gave a natural increase which fluctuated slightly around ten per thousand for about 150 years (1800–1950). It was not

Table 5.1. Crude vital rates per thousand in selected developed countries, 1987

Country	CBR	CDR	NI
United Kingdom	13.6	11.3	2.3
France	13.8	9.5	4.3
Germany (Fed. Rep.)	10.5	11.2	−0.7
Italy	9.6	9.3	0.3
Spain*	11.2	7.9	3.3
Sweden	12.5	11.1	1.4
Australia	15.0	7.2	7.8
Japan*	11.4	6.2	5.2
USA	15.7	8.7	7.0

* 1986.

Source: Office of Population Censuses and Surveys (1989*a*).

Table 5.2. Estimated crude vital rates per thousand in selected less-developed countries, 1985–90

Country	CBR	CDR	NI
Brazil	28	8	20
China	21	7	14
Egypt	33	9	24
India	31	11	20
Indonesia	31	11	20
Kenya	53	12	41
Pakistan	44	14	30
Senegal	46	18	28

until near the end of that period that the control of mortality in poorer countries became effective and there was an upsurge of growth, 'the population explosion', reaching a rate of nearly 25 per thousand ($2\frac{1}{2}$ per cent per year), around 1970. Although the pace of increase has slowed, it is still at a level which would have been regarded with astonishment up to the middle of this century. A growth-rate of ten per thousand implies a doubling of the population every 70 years; 25 per thousand gives doubling in 28 years. The economic, social, and environmental impacts of such population pressures in the long run are a complex and controversial issue. However, the severe burden of providing health services to cope with such rapidity of increase in demand, quite apart from raising standards at the same time, is not in doubt.

Distribution by sex and age

The most useful classifications of the population composition for the present purposes are by sex and age, since health needs are so intimately related to these characteristics. Their distributions are determined by vital events and migration over periods approaching one hundred years. For example, the number of the very old (above 85 years, say), which is an important indicator of particular needs, depends on the birth-rates of nearly a century ago, as well as deaths and population movements since then. The *proportion* of the very old is, for many purposes, more significant than the absolute number, since it incorporates the relationship with supporters and other dependents. This measure, therefore, is also affected by the series of births over the century. The process is illustrated by Fig. 5.1, which compares the age distribution of the England and Wales population in 1981 with the births from which it was generated. In a *closed* population, where there is effectively no net migration, the numbers of persons in an age group (conventionally one year to five years in length) tend to be more than in the next older group because of deaths and increases in births over time consequent on growing populations and fairly constant birth-rates. This might be called a *regular* age distribution. The conventional method of representation on a graph then justifies the name *population pyramid*. The horizontal bars are of lengths equal to the numbers in the age groups on the *x*-axis scale. The vertical *y*-axis shows the age groups from the youngest upwards, with the bars as it were stacked on top of each other, pyramid or, more properly, ziggurat arranged. The graph for males is on one side of the *y*-axis and for females on the other.

The nature and utility of the form of representation is illustrated by Fig. 5.2. The Syrian (1977) age distribution is a typical *regular* pyramid for a high growth-rate population where the numbers of births have gone up substantially each year. In Korea (1975), the age distribution is *regular* above age 15 years, but a rapid fall in fertility over some twenty years has reduced births and undercut the pyramid. The England and Wales profile of 1981 reflects the fluctuating births from the beginning of the century, with falling fertility offsetting the reduced survivorship with age until some 65 years and upwards. When migration is an important component (for example, in subpopulations of some ethnic groups or urban locations) the pyramid is likely to have a distorted profile. The shapes vary according to the type and duration of the migration streams. Figure 5.2 also shows the pyramid for New Commonwealth and Pakistani immigrants in England and Wales, 1971. There is a relative excess of young adults (20–34 years of age) and of the children under 10 born to them. The lowered numbers are at 10–19 years and increasingly at ages above 50 years. Although the pattern is quite a common one and also its reversal (for example, in the Caribbean countries which have exported people), wide variations occur.

The sex division of births deviates little, from 105 males to 100 females in different populations. Claims for substantial divergences are based on weak evidence, although there may be distortion from hidden infanticide as in China (more than 110 male births per 100 female reported in the second quarter of this century). In the great majority of populations male death-rates are higher than for females at all ages. As a consequence, sex ratios in age groups tend to be close to 100 (males per 100 females) over the reproductive period. Biologically this implies near equality of numbers of mating partners, but not necessarily socially. In many African

Table 5.3. Size and growth of the world population

Division	Number in millions at date 1750	1800	1850	1900	1950	1975	1990
Total	791	978	1262	1650	2515	4080	5292
Less-developed	590	730	915	1077	1683	2984	4087
More-developed	201	248	347	573	832	1096	1205
	Annual rate of increase per thousand in period 0–1750	1750–1800	1800–1850	1850–1900	1900–1950	1950–1975	1975–1990
Total	0.5	4	5	5	8	19	17
Less-developed	"	4	5	3	7	23	21
More-developed	"	4	7	10	9	11	6

Source: modified from Brass (1970) and updated from United Nations (1989).

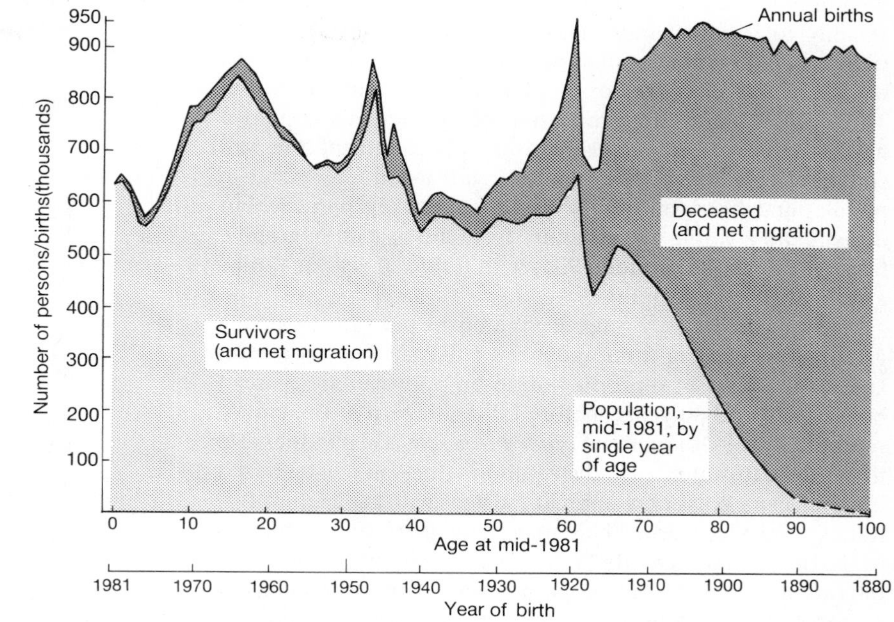

Fig. 5.1. Mid-1981 age distribution, compared with past births: England and Wales.

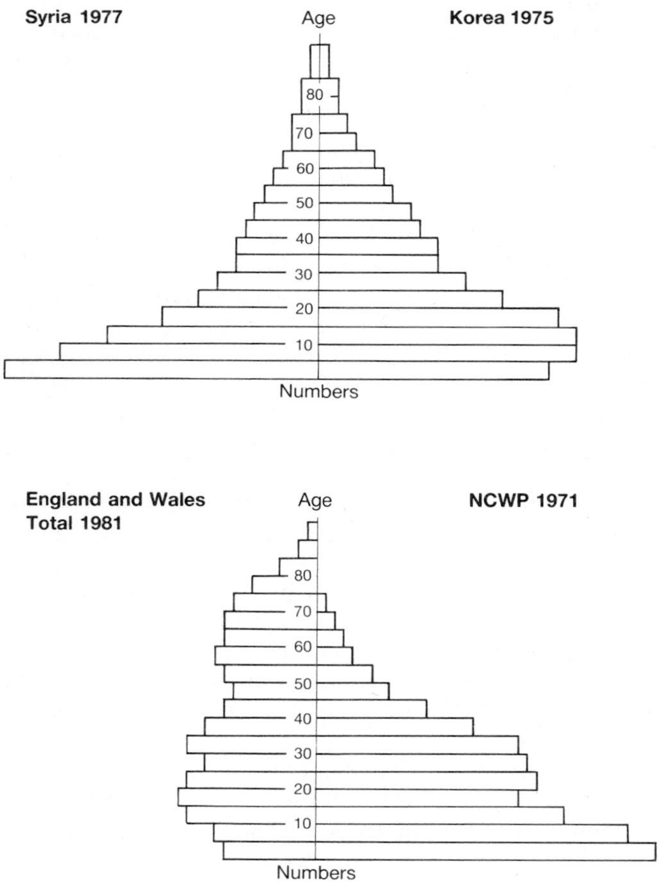

Fig. 5.2. Female age pyramids.

societies women marry at a much earlier age than men and the age pyramids are steep. Female celibacy would be structurally necessary but for the practice of polygamy. At later ages there are generally far more women than men, because of high mortality differentials. Thus, in the United Kingdom in the 1980s there were twice as many women as men over 75 years of age. The sex division for the population as a whole depends strongly on the age distribution. 'Young' populations tend to have a higher proportion of males. For example, in the Chinese census of 1982 there were 105.5 males per 100 females despite a large excess of females at later ages. In England and Wales in 1981, there were only 94.8 males per 100 females because of the high proportions at older ages. There are some populations in less-developed countries where female mortality has been as high or higher than male, with sex ratios over 100 up to late ages and overall. The most notable region where this occurred was the subcontinent of India, presumably because of better care for males, and heavy maternal mortality. However, the trend is now towards a female advantage, as in most of the world. High migration may upset these general configurations but only in extreme cases to a substantial extent.

The measurement of mortality and fertility

The *crude death-* and *birth-rates* (and the *natural increase*) are strongly affected by the age distribution; they are thus unsatisfactory measurements of mortality and fertility respectively, particularly the former. The age-specific death rates (that is, the deaths per year in an age group per thousand persons at risk) are high at the start of life but fall rapidly to a minimum in the range 5 to 15 years. Thereafter they increase steadily to levels at later ages which far exceed childhood rates except in a few residual populations with primitive ways of life. Model examples of death-rates by age are shown in Table 5.4. Thus a low proportion of old persons implies a modest size of *crude death-rate* even where mortality is still comparatively heavy. The smallest *crude death-rates* occur in populations where mortality has been sharply reduced but there is still a young age structure from the past demographic history. Sri Lanka around 1985, with a rate of six per thousand, is a typical example.

Age-specific fertility rates are the births per year to an age group of women per thousand women at risk. They rise from zero at the start of reproduction to an approximate plateau between 20 and 30 years, before falling to zero at the end of reproduction. Although the *crude birth-rate* is less vulnerable than the death-rate to age distribution effects, it will be raised if there is an excess of women at the ages with the highest fertilities, and vice versa. This may arise from sharp changes in fertility some twenty years previously or because of migration. Immigrant subpopulations are a significant example. The persons of New Commonwealth and Pakistani origin in England and Wales have a young age structure, with excess numbers in the middle of the reproductive period. The configuration raises the *crude birth-rate* and lowers the *crude death-rate*, thus adding about seven per thousand per year to the growth-rate compared to what it would have been with a regular age distribution (Brass 1982).

The age-specific death- and fertility-rates are free of the distorting influences of the age distribution and hence give truer depictions of mortality and fertility levels. It is often convenient, however, to summarize the sets of rates by age. For mortality this is commonly done by the construction of a life-table which tabulates significant measures for a group of births, moving through life from age 0 and experiencing the age-specific death schedule (see Table 5.5). The most important derived measures are l_x, the proportion surviving to exact age x, the probability of dying between exact ages x and $x + n$ ($_nq_x$), and e_x, the average years lived after age x by persons alive at age x. e_x is usually called the expectation of life at age x, and e_0, the expectation at birth is the most used single summary index of mortality. Life-tables may be calculated from the age-specific death-rates of a period or for a cohort moving through time. The latter approach provides a description which is closer to realized experience. In particular, it allows for the possible influence of earlier on later mortality, which is important for causes of death where the age pattern of risk is changing (Frost 1939). However, the birth cohorts can only be followed up to their current age, and the simplicity and up-to-dateness of the period life-table measures are lost. The latter are the most frequently applied, but the conceptual nature of the assumption that the age-specific death-rates in a period can be applied over lives in time must be remembered. Expectations of life may be consistently as low as 25 years (some Sahel populations) and much less in periods of heavy epidemics (Iceland in the eighteenth century has reliable data). Currently, in developed countries female expectations of life are as high as 80 years (Japan, The Netherlands, Iceland).

Similar modes of thought lead to summary, age-distribution-free, indices of fertility. If the specific rates for women in each year of age are added over the childbearing period the outcome is the average number of births per woman with this reproductive schedule. It is called the *total fertility*. There are also time-period and cohort versions of the

Table 5.4. Model age-specific death-rates per thousand from the Coale–Demeny system: West region

Age group (years)	M	F	M	F	M	F
Under 1	218.40	182.55	92.34	74.64	22.78	16.11
1–4	28.14	27.99	9.61	9.20	0.92	0.65
5–9	5.92	6.19	2.42	2.30	0.47	0.27
10–14	4.28	4.82	1.80	1.78	0.40	0.23
15–19	6.01	6.43	2.76	2.57	0.78	0.38
20–24	8.56	8.16	3.92	3.38	1.08	0.54
25–29	9.49	9.20	4.20	3.89	1.06	0.68
30–34	10.95	10.44	4.79	4.45	1.20	0.88
35–39	12.97	11.60	5.81	5.13	1.57	1.22
40–44	15.96	12.74	7.51	6.05	2.43	1.85
45–49	19.28	14.31	10.06	7.56	4.15	3.03
50–54	25.29	18.97	14.26	10.50	6.96	4.74
55–59	32.63	24.99	20.46	14.73	11.84	7.51
60–64	46.02	37.31	30.43	22.57	19.16	12.26
65–69	64.12	52.73	45.20	34.77	31.14	21.59
70–74	92.33	80.06	68.79	56.19	50.88	38.33
75–79	136.10	119.79	105.71	93.84	82.49	71.49
80 and over	257.46	216.74	218.76	174.95	178.43	137.32
Life expectancy at birth (years)	39.67	42.50	56.43	60.00	70.84	75.00

Source: Coale and Demeny (1966).

Table 5.5. Selected measures of the life-table: Coale–Demeny Regional model system, West female*

Age, x	l_x	d_x†	q_x	e_x (years)
0	100 000	1593	0.01593	75.000
1	98 407	257	0.00261	75.209
5	98 150	135	0.00136	71.401
10	98 016	110	0.00113	66.495
15	97 906	184	0.00188	61.567
20	97 722	263	0.00269	56.679
25	97 459	332	0.00341	51.825
30	97 127	524	0.00437	46.993
35	96 703	590	0.00610	42.188
40	96 113	886	0.00922	37.432
45	95 227	1434	0.01506	32.757
50	93 793	2199	0.02345	28.220
55	91 594	3376	0.03686	23.837
60	88 218	5248	0.05949	19.654
65	82 970	8498	0.10243	15.739
70	74 472	13 026	0.17491	12.249
75	61 446	18 633	0.30324	9.316
80	42 813	42 813	1.00000	7.282

* Calculated from the age-specific death-rates in the last column of Table 5.4.

† Life-table deaths between x and the next specified age.

Source: Coale and Demeny (1966).

total fertility, the latter close to the average completed family size (children ever born) at the end of reproduction. Slight differences occur because of deaths and migration of women in the childbearing period. Period total fertilities are most commonly quoted but the cohort measures, which are less liable to distortion from movements in the timing of births, are increasingly being used. If only female births are included in the calculation of the summary index the resulting measure is the *gross reproduction rate*, which for practical purposes is the total fertility multiplied by 100/205 to allow, with sufficient accuracy, for the sex division of births. Total fertilities may be as high as eight (Kenya, The Yemen) but in Europe and populations whose main genesis was European colonization, most are now under two. The corresponding gross reproduction rates range from four to something under one.

The measurement of reproduction

It is obvious that if women on average are bearing less than one daughter in their lifetime, the population is not replacing itself and will, at some time, decrease in size. Deaths of children will further reduce the level of replacement. The measure which takes account of both fertility and mortality is the *Net Reproduction Rate* (NRR). It is constructed by calculating the numbers of girls surviving at the current mortality from a thousand births, to the ages of the reproductive period. At each of these ages the corresponding specific fertility rates are applied to the surviving women to give the resulting births. Summation over all ages gives the total births to the original thousand. These are translated into daughters, though multiplication by the proportion of births which are female (100/205) is usually adequate. The *Net Reproduction Rate* is expressed in terms of the initial thousand girls born, or more commonly as per female in the earlier generation. The measure is sometimes called a replacement rate, since it provides a consistent index of how many daughters are in the next generation per woman in the previous one. An NRR of 1 means that the combination of fertility and mortality is such that the generations are just being replaced; an NRR of 2 implies a doubling of girls in the next generation.

The Net Reproduction Rate was invented as an intuitive gauge of reproduction and natural increase before its theoretical justification was established by A.J. Lotka in the 1920s. Lotka showed that a population subject to constant age-specific fertility- and death-rates would ultimately have a fixed proportional sex-age distribution and a constant growth-rate, r. A conceptual population of this kind is called *stable*. It grows by a factor equal to the NRR per generation of length g. A constant growth, r, gives an exponential function of size, and thus NRR $=$ exp rg or $r = (\log_e \text{NRR})/g$. The precise definition and calculation of g, the length of generation, is quite complicated but it is close in value to the mean age of women at the birth of their children (M, say). In most populations M is 26 to 28 years. From the NRR calculated from observations for a year or period, the corresponding r can be found; for most practical purposes g can be taken as 27 in the calculation. The resulting r is the *intrinsic rate of natural increase*, a conceptual measure although based on observations. In general, actual populations do not have the stable age-distributions corresponding to the current fertility and mortality rates; they differ because of past events, including migrations, epidemics, and wars. The NRR and *intrinsic rate of natural increase* are measures of potential which are unlikely to be achieved because of trends and fluctuations in fertility and mortality. Nevertheless, they mark out directions, free from the transient effects of past history, which can strongly influence births, deaths, and growth.

The formal definitions of fertility and reproduction rates may appear to be rather abstract but they can be described approximately in terms which are easily grasped. The *total fertility* can be thought of as the completed family size, children born per woman at the end of reproduction; the *gross reproduction rate* is the daughters per woman, just under half the *total fertility*. The *net reproduction rate* is the *gross reproduction rate* multiplied by l_{27}, the proportion of females surviving from birth to the age at which they replace their mothers on average. In low mortality populations l_{27} is about 0.98; that is, only some 2 per cent of the female births die before the age 27 years. In these conditions the *net reproduction rate* and hence the *intrinsic rate of natural increase* is dominated by fertility. Although the deaths of females by age 27 are much more than 2 per cent in many populations, improvements in mortality are so widespread that 20 per cent is not often exceeded. The scope for the impact of further falls in mortality on population growth-rates is quite small.

To take relevant illustrations, the mean completed family size in Kenya around 1980 was 8, with a *gross reproduction rate* of 3.9 and a *net rate* of about $3.9 \times 0.8 = 3.1$ with l_{27} taken as 0.8. Then $(\log_e 3.1)/27$ is 0.042, giving an intrinsic rate of natural increase of 4.2 per cent per year. A halving of mortality, changing l_{27} to 0.9, with a reduction of total fertility by one birth to 7, gives a *net reproduction rate* which is almost unchanged. The latest studies suggest that these measures may hold for Kenya in the early 1990s. Between 1955 and 1985 the *total fertility* for England and Wales fluctuated greatly. It was a maximum of 2.94 in 1964 and a minimum of 1.66 in 1977. The fall in mortality from 1964 to 1977 has a negligible impact on the calculation of the net reproduction rate at 1.38 in 1964 and 0.79 in 1977, corresponding with intrinsic rates of natural increase of 1.2 and -0.8 per cent. The fertility variations were so massive in terms of potential growth that in fifteen years views about future population sizes were transformed. This happened not only in England and Wales but over the whole of western Europe and in other developed countries such as Japan and Singapore.

The relation of fertility and mortality to age structure

As noted above, the *intrinsic rate of natural increase* does not measure how a population is growing because of the balance of births and deaths, but the implications of the fertility and mortality for the future. Leaving migration out of account, the difference between the observed growth-rate and the intrinsic rate of natural increase depends on the age distribution. Before quantitative examination it is necessary to explain how the age distribution is determined by the level of fertility and mortality. The elucidation of this question was the major advance in population dynamics of the 1950s, (Coale 1957). The key was the increasing power of computers for the performance of extensive repetitive calculations, although mathematical models of demographic processes played a part. Largely on the basis of calculations for a range of conditions, it was concluded that the broad age distribution was a consequence of the level of fertility, and the influence of the level of mortality was small. This was a surprising finding, since it seems intuitively obvious that the lower the mortality the higher the number of persons surviving to older ages. But reduced death-rates also lead to more survivors at younger ages and it is the balance between these two effects which controls the proportion living in the stages of life. The ratio of births in a year (that is, the number of persons in the population around the age 0), to the average number per year of age in the reproductive years around 20–34 is fixed by the level of fertility, since the former are largely born to the latter. Broadly then, the level of fertility controls the lower part of the age distribution, although there may be variations in the regularity of the pyramids from age 0 to the later part of the reproductive period because of birth-rate and mortality trends. The number at ages beyond reproduction do depend on the pattern of mortality, and it is an empirical, not a theoretical, finding that for average patterns the proportions vary little with the *level* of the death-rates. Patterns of mortality which are far from average can have an effect on the age distributions, however. Thus, if death rates in middle to late ages, say from forty years upwards, are considerably higher than average, relative to the mortality in the earlier part of life, the proportion of older persons in the population is reduced. The opposite happens with excess mortality at younger ages. However, since the numbers at older ages are a comparatively small part of a population (very small where fertility is high) these mortality pattern effects have only a modest impact on the broad age distribution.

Thus, the age distribution is largely a consequence of the level of fertility, or more generally of the series of past births. This is well-illustrated by the England and Wales statistics in Figure 5.1. A further important deduction is that a population in which fertility has remained fairly constant and mortality has fallen but not dramatically, has effectively the same age distribution as a *stable population* with a long-term constant fertility and mortality. In this so-called *quasi-stable population* the actual *rate of natural increase* is the same as the *intrinsic rate* calculated from the *net reproduction rate*. Until comparatively recently most populations of the Third World could be regarded as quasi-stable (erratic fluctuations in fertility did not seriously disturb the conclusion). Fertility levels could be deduced from age distributions (usually known from censuses and surveys) and the net reproduction rate could be reasonably estimated, given some information on early childhood mortality (the dominant component in survivorship to age 27 years). A fair assessment of the population dynamics could be obtained from a very small amount of data. Currently, falls in fertility have destroyed this simplicity of structure, although in many populations the age distribution is quasi-stable beyond childhood. Much information can still be recovered from limited data but analysis methods are more complicated (United Nations 1983).

The general consequences of movements from quasi-stability through falls in fertility can be discerned quite easily, although a full delineation requires extensive computations. As the fertility, and hence the births per year, falls the age pyramid is undercut—as can be seen from Korea, 1975, in Figure 5.2. The birth-rate is lowered, but not as fast as the fertility, since the size of the population is also reduced relative to what it would have been without the trend. The death-rate, on the other hand, tends to fall, because the high mortality in the early period of life is operating on progressively fewer births. Thus, the actual rate of natural increase remains above the intrinsic rate. It is not so easy to see what happens when the birth reduction begins to effect the numbers in the reproductive period; in fact, the observed remains above the intrinsic rate of increase until a new level of constant fertility has been in existence for many decades. If fertility increases, the features are reversed, with the actual increase below that implied by the intrinsic measure. The feature whereby a fall in fertility is not reflected fully in a reduced growth-rate for some time is often called population momentum. Its greatest significance is for countries with rapid growth-rates which have policies for reducing these by lowering fertility through family planning and later marriage. Even if the policy is successful the desired response occurs more sluggishly than might be imagined.

Falls in fertility take some time to achieve ultimate effects on growth because they are operating at only one age-point, namely the births. But changes in death-rates normally operate at all ages. The impact on the crude death-rate is immediate. Since the age distribution is little affected, the change in the overall rate is proportional and is reflected fully in the population growth. However, a change in death-rates which has an excess impact at particular ages has the feature of a modification in fertility. Thus, if there is a particularly sharp reduction in mortality of males at middle to late ages (say) there will be a gradual increase in the proportion of older men in the population.

Model life-tables

The variations among populations of sex-age-specific death-rates, organized as life-tables, is constrained by the similarities of biological and social determinants. There is a long history of study of the nature of the variations. Since the 1950s many systems of model life-tables have been constructed. In these the aim is to produce sets of life-tables which can match observation; that is, there is a member of the set which has the same age specific death-rates as any reliably calculated mortality schedule for an actual population. In the pioneering United Nations Model Life Tables of 1955 the set was constructed with only one degree of freedom; that is, for each level of mortality there was a fixed age pattern of specific death-rates. It turned out that this representation did not fit experience very well. Two populations could have the same overall mortality level, as measured by the expectation of life at birth (say), but distinctly different rates by age. Theoretical analysis showed that four types of variation, or degrees of freedom, were required to provide good matches with observations. Model life-tables with four degrees of freedom have been constructed, but they are of limited practical utility because the choice of the appropriate member of the set and the calculation of the required measures is cumbersome (Zaba 1979). The most convenient systems are sets, each with one degree of freedom, but divided into several 'regional' types, exhibiting distinct age patterns of death-rates at each fixed level of mortality. The longest established is the Coale–Demeny model tables in four sets, North, South, East, and West with age schedules of mortality derived largely from geographic areas of Europe (1966). More recently, the new United Nations models are in eight regional sets (1982). The implication that the sets are most appropriate for the geographical regions of the titles has rightly been severely criticized, but the models do provide a useful range of age-specific death-rate patterns. A popular alternative is the Logit system of model life-tables based on a mathematical relation. This has two degrees of freedom and is both flexible and convenient for computer calculations (Brass 1971). However, the range of published tabulations of life-tables and corresponding stable populations is smaller. In these systems the 'best' choice of a model to represent the actual mortality schedule operating may still diverge to some extent from the reality. How much this matters depends on the application, but for many purposes greater simplicity and convenience easily offsets slightly less accurate approximation.

Many of the most important applications of model life-tables are to less-developed countries where there is only sketchy and often indirect information on death-rates. A life-table cannot be calculated from the data, but the data can be used to select a model. However, the value in developed populations should not be underestimated. Knowledge of the future is always uncertain and sketchy; the population dynamics of the future is a major concern in applied demography. Again, although complete and accurate data on mortality, fertility, and populations at risk may be available for a country and its major geographical divisions, this is seldom the case for community and social subgroups which are the focus of interest.

Stable population models

A *stationary* or *life-table* population has a zero rate of increase, the births each year exactly compensating for the deaths; the fertility and mortality are assumed to have continued long enough for stability to be reached, as defined by Lotka. In such a population the number of persons aged x is simply Bl_x where l_x is the life-table survivorship, and B the current births per year. More generally, a *stable population* has a fixed age distribution and a constant rate of natural increase r. The number of persons aged x is now $Bl_x/\exp(rx)$ or $B\exp(-rx).l_x$. The division by $\exp(rx)$ allows for the exponential growth of births at constant rate r from the time at which those aged x were born to the present. Thus, from any set of model life-tables defining the l_x corresponding stable populations can be calculated by applying the $\exp(-rx)$ for any value of r. As well as the age distribution and rate of increase, the conditions fix the birth-rate and also the death-rate. The *net reproduction rate* is $\exp(rg)$, the growth of the population in a generation which requires an extra piece of information the length of g. However, as pointed out, it is usually satisfactory in practical applications take g as 27 years. The *net reproduction rate* effectively determines the *gross reproduction rate* and the *total fertility*. Stable populations can be constructed for persons, but more commonly they are tabulated separately for males and females, with different mortality schedules. The populations for the two sexes are linked through the division of the births in the ratio of 105 males to 100 females.

Table 5.6 illustrates the tabulated measures for the stable populations of the Coale–Demeny system. These provide a variety of ways in which observed data can be used to decide upon an appropriate model to represent the demographic conditions. Populations with a history of large changes in fertility or extensive migration flows do not meet the theoretical constraints for approximate stability of form, but it is often possible to find models which mimic the structures sufficiently well to be useful in applications. An examination of the differences between the observed characteristics of a population and a stable model which fits its major features, is often helpful in the assessment of determinants and implications.

As an illustration, the population of England and Wales in 1981 had been subject over the previous century to large and fluctuating changes in fertility and substantial migration streams both in and out. Table 5.7 compares selected characteristics with a model stable population chosen from the Coale–Demeny set. The agreement is broadly rather good. It could, in fact, be improved by a more refined fitting, interpolating between the published schedules, but this is not required for the present purpose. The model birth, death, and national increase rates are near to the observed values, and the age distributions are very similar for wide groupings. This is perhaps particularly notable at older ages, where the model percentages follow the observed ones closely. Of course, there are deviations, at 15–24 years of age; for

Table 5.6. Proportions of persons and deaths per year by age, with selected indices for two female stable populations

Age group (years)	Natural increases			
	10 per thousand		20 per thousand	
	Persons %	Deaths %	Persons %	Deaths %
Under 1	2.21	12.36	2.93	19.66
1–4	8.26	5.70	10.70	8.84
5–9	9.64	1.66	11.94	2.47
10–14	9.07	1.22	10.70	1.71
15–19	8.54	1.65	9.58	2.21
20–24	8.00	2.03	8.54	2.59
25–29	7.47	2.18	7.58	2.65
30–34	6.96	2.32	6.72	2.68
35–39	6.47	2.49	5.94	2.74
40–44	5.98	2.72	5.23	2.84
45–49	5.50	3.12	4.57	3.10
50–54	5.00	3.94	3.95	3.73
55–59	4.47	4.94	3.36	4.44
60–64	3.88	6.57	2.77	5.62
65–69	3.20	8.35	2.18	6.80
70–74	2.44	10.28	1.58	7.96
75–79	1.61	11.37	0.99	8.37
80 and over	1.30	17.09	0.74	11.59
Birth-rate per 1000	23.33		31.17	
Death-rate per 1000	13.33		11.17	
Average age of popn	31.22		26.81	
Average age of deaths	51.75		41.84	
Gross reproduction rate*	1.55		2.03	

* With *g*, the length of generation, 27 years.
The model death-rates are those shown for females in Table 5.4, with the expectation of life at birth as 60.00 years.

example, where the baby-boom of the 1960s raised the numbers compared with the 'regular' rates of the model. These are obviously important for specific studies, such as of changing entry to the labour force, but not for the many issues where supply and demand are less sharply focused in terms of age.

Population forecasts

Probably the most frequent use of demographic skills in planning is in the forecasting of future populations. The basic characteristics are the numbers by sex and age, since so many other features are related to these—including, of course, births and deaths. In fact, the methods applied for the populations of an area can often be adapted for subpopulations, such as residents in institutions, children in care, by lengths of stay, and so on, with in-movement and out-movements replacing births and deaths. It must be emphasized that all forecasts are speculative and there is no way of removing the fundamental uncertainty which becomes greater as time extends from the known present. The most than can be demanded is that the transitions (that is, elements of change in the forecasts) should be a sensible continuation of the past or that they should be constrained by relevant social and economic developments which are expected to take place. Thus, if a major new source of employment is being planned for an area, there is likely to be substantial net immigration. The clearance of old housing will be associated with loss of population, often with distinctive age-distribution effects. The idea that better forecasts can be made by taking into account local socio-economic conditions is an attractive one but extremely difficult to translate into practice. The potential factors are many and liable to change for forecasts more than a few years ahead.

Most forecasts depend on extrapolations of what has been happening in the past on the argument that similar determinants will continue into the future. Allowance is sometimes made for the attenuation of extreme features towards a more central level. Thus, a mortality pattern with excess adult compared with childhood death-rates may be forecast to approach a more balanced state; the higher fertility of recent immigrants compared with the level for an indigenous population may be taken to fall to meet the latter. In view of the uncertainties of what has yet to come there is a strong argument that forecasting methods should be kept simple and exhibit clearly the links between the future population sizes and structures and the elements of change (births, deaths, and migration). Extra refinements whose effects are comparatively small and lost in the possible errors, should be dispensed with. Unfortunately, many forecasts are more elaborate than can be justified by their precision.

In most countries the Central Statistical Office, or a similar national agency, produces forecasts of the future populations (size, growth, sex, and age distribution). Where there are extensive data and technical resources the estimates are provided for local areas and for other characteristics such as marital status, household number and composition, educational attainment, etc. A population *projection* is a calculation of the future profile from detailed specifications of the compo-

Table 5.7. Percentages of females by age in England and Wales, 1981, and a fitted model stable population

Age group (years)	England and Wales 1981	Model stable population
Under 5	5.75	6.56
5–14	13.53	13.07
15–24	15.16	13.03
25–34	13.81	12.94
35–44	11.67	12.81
45–54	10.87	12.48
55–64	11.25	11.70
65–74	10.20	9.78
75 and over	7.77	7.64
Under 15	19.28	19.63
15–44	40.64	38.78
45–64	22.12	24.18
65 and over	17.97	17.42
Birth-rate per thousand	12.80	13.33
Death-rate per thousand	11.60	13.33
Natural increase per thousand	1.20	0.00

The fitted model death-rates are those shown for females in Table 5.4, with the expectation of life at birth as 75.00 years.

nents of change over time (fertility, mortality, migration). It becomes a forecast if the specifications are regarded as the most likely to occur. Sometimes alternative projections are provided, e.g. low, middle, high, and the judgment of the user can play a part in arriving at a forecast.

The method of calculation most commonly adopted, 'component projection', is straightforward in principle but involves a number of technical devices in practice. The process is of moving forward one unit of time at each step (one year or five years). The age-group length is taken to be the same as the time unit. Then, in a step, the numbers in an age group become the numbers in the next higher age group when allowance is made for deaths. The proportion dying from one age group to the next is found from the life-table which is specified to apply at that particular time. In this way, the numbers in all age groups except the first are generated. The calculations for females provide the numbers at the beginning and end of the time interval and hence the average in between for each age of the reproductive period. The chosen age-specific fertility rates are then multiplied by these average numbers of women at risk to give the births which would have occurred in the interval. Division into females and males by the sex ratio at birth and reduction by the proportions dying before the end of the interval gives the numbers in the first age groups after the step forward in time. Migration is incorporated by adding or subtracting a net change for each sex–age group. Methods for achieving consistency in internal movements of persons from area to area may be complicated. The first step provides the sex–age distribution which is the base for the next, and so on. The main criticism of the component method is that the population *size* appears as the outcome of the separate fertility, mortality, and migration effects. There is no mechanism for incorporating external constraints related to size, although for longer-term forecasts, particularly in subdivisions, these would appear to be necessary.

Projection, particularly with modern computers, is a routine process. The choice of the component measures to enter in the calculations to give sensible forecasts is a mixture of science and art which depends on a deep analysis of tendencies, both demographic and socio-economic. Nevertheless, common sense plus an element of knowledge of the basic rules of population change can provide guidance for the assessment and construction of forecasts for communities and sub-national groups. For relatively short periods ahead, say up to fifteen or twenty years, the effects of changes in mortality will be small; the proportions surviving alter rather slowly over time. There is no great error then in the assumption that the current mortality schedule will continue. This is sufficient to provide numbers at all ages above the length of the projected period, since the persons, apart from migration, have already been born. However, the migration component in district or community forecasts may be large; for example, in urban areas it can overwhelm other changes. The only way to arrive at sensible numbers for future migration is from the experience of the past combined with judgement on whether the determinants will continue. Industrial development, employment opportunities, house building and clearances, and so on, have obvious implications for migration streams. It would be naïve to believe that such assessments have any great precision, but they can protect against the worst errors of extrapolating a trend twenty years into the future when the local conditions controlling it have already changed. Central government forecasts for districts have often failed to take account of alterations in conditions which are well-recognized in the individual communities.

Different considerations govern the evaluation of forecasts of the persons yet to be born. Trends and fluctuations in fertility can have a massive impact on their numbers. At the same time, there can also be particular local features of the age distributions due to migration. For example, young, single men and/or women may move into an area for employment, marry, and produce births but leave for more suitable family housing when the children are still young. Many inner-city districts have such characteristics. The aim in the forecasts is to retain the special configurations but allow for the likely trends in fertility. A simple but useful way of doing this is from the ratios of numbers in the younger age groups, say 0–4, 5–9, and 10–14 years, to the women who bore most of them, aged 20–39, 25–44, and 30–49 years respectively. These ratios are calculated for the most recent age-distribution data which form the base of the projection. Since the number of women aged 20 and over have, it is assumed, already been forecast as explained previously from survival and immigration only, application of the ratios will give the corresponding estimates of children on assumptions of constancy. However, it will usually be possible to make the same calculations from official projections of the whole country or for a large region which contains the local areas of interest. The changes in these child/woman ratios over time are primarily due to the fertility trends. It will generally be reasonable to assume that the local trends are similar to those in the wider population. The numbers of children can then be adjusted accordingly. Thus, if fifteen years ahead in the national projection the ratio of children under 5 to women aged 20–39 years has been reduced by 20 per cent from its base level, the number of children aged 0–4 years in the local projection with constant ratios can be reduced by 20 per cent also.

If two reasonably good counts of the population by sex and age are available from the recent past, the procedure described can be systematized. For simplicity, the counts will be taken to be an interval of five years, but the approach can be adapted to other conditions. The ratio of the numbers in a five-year age group at the second count to the number in the next lower age group at the first count, will provide a factor of change which allows for both mortality and migration. The set of factors obtained can be applied to project forward a further five years, and so on. The younger age-group numbers are filled in as explained above, on the assumption that the ratios of children to the corresponding women at risk of childbearing remain constant. The resulting forecasts are based on the principle that the rates in the future (mortality, migration, fertility) are most likely to be similar to those of the recent past. Modifications can then be introduced if judgement suggests this continuation is unlikely. The numbers of children in relation to women at risk of bearing them can be adjusted to be consistent with the changes in the national projections as suggested previously. Alternatively, the child/woman ratios can be calculated for both the counts (and indeed for earlier sets of data if these are available). If there are steady trends over time the values can be extrapolated to provide estimated ratios at five, ten, and fifteen years ahead to replace the constant measures. The lack of refinement in these procedures is obvious, but there is a cogent argument that the essential uncertainty is so great that no higher precision can be obtained from increased sophistication.

The assessment of fertility trends

The trend in fertility is the most critical factor in the determination of future population growth and age structure. It is also a difficult feature to evaluate and forecast. In the past, changes in fertility (usually reductions) occurred relatively slowly, and disturbances to the age pattern were not large. But with the dominance of control through family planning in all developed and some less-developed societies, the pace and nature of change have been altered. When women, on average, have only about two births in the childbearing period, the scope for movements in timing is large. For example, much of the steep fall in fertility in the 1970s in England and Wales and many other western European countries was a direct result of the reduction of risk exposure through a rapid rise in the ages at marriage. But it was not possible to say to what extent the births were being postponed to later ages as opposed to being permanently foregone. The period total fertilities reflect the immediate trends but they contain a component, which may be substantial, due to the shifts in the ages of women at the births of their children. Over the past fifty years in low fertility populations, the timing of births has been very unstable. Both experience and the nature of the process suggest that the timing variation is bounded; that is, movements upwards or downwards in the ages at births will cease or be reversed within modest spans of time, years rather than decades. Thus, the extrapolation of sharp trends in period total fertilities beyond a few years is liable to give poor forecasts because of the characteristics of the measurements, quite apart from the possible changes in the behaviour governing family sizes (Brass 1989).

Cohort total fertilities, based on the age-specific rates over the reproductive period of women born in the same year or group of years, have no distortion from a timing component, and reflect changes in family sizes directly. Their trends give a better guide to the longer-term tendencies. The common-sense principle in forecasting is to merge the immediate levels and trends in fertility and births of the time period measures with longer-term trends from the cohort ones. There is no established theory on how this can best be done but, in practice, the detailed choice of procedure probably matters little.

Childbearing at different stages of life

When the fertility of a population is high, with most women bearing children over the larger part of their reproductive life, the *total fertility* is a measure which provides nearly all the information required for assessment of future prospects. In low fertility populations the situation is very different. Characteristics of the composition of fertility can then vary considerably among populations in which the levels are similar. The implications of the varied compositions may be very different. Traditionally, interest was in the distribution of fertility by age of woman and its relation with constraints through nuptiality and the use of contraception. Thus, in China in 1964 when the period completed family size was above six births per woman, 30.1 per cent of the fertility occurred before age 25 years and 24.7 per cent after age 35. By 1980, the corresponding proportions were 34.2 and 9.1 per cent in a total fertility a little over 2. In England and Wales, fertility before age 25 years was 38.5 per cent of the total in 1964 but fell to 34.8 per cent in 1986. The older women over 35 years provided 10.9 per cent of the total fertility in 1964 and 8.3 per cent in 1986. These two examples illustrate the feature that in high contracepting, low fertility, populations a small percentage of the births are to women at the later stages of childbearing. In western Europe there has been a steady fall in the contribution to the total fertility of women over 35 years.

The fertility behaviour at the early childbearing ages is less predictable. In some parts of the world, risk exposure in the teens and early 20s has decreased, e.g. in most Asian populations, but the changes in others, widely in Latin America, have been slight. Since induced abortions and illegitimacy, with their severe health and social consequences, are heavily concentrated in the women pregnant at early ages, these fertility trends are of immediate significance. As can be seen from the example of England and Wales, the contribution of young women to the total fertility can rise to a high level in developed populations. In the middle of the baby-boom in 1964 the share of the under-25s was nearing 40 per cent compared with the 30 per cent of the late 1940s.

Family size distribution

Of greater demographic consequence for the assessment of fertility in developed countries is the distribution of women by parity; that is, the total children born to them, or more colloquially, the family size. Formally, this is defined as the distribution at the end of reproduction, but because of the small number of births after age 35 years, as noted above, the measures apply effectively over a considerable part of the life-cycle. They are of live births, but in these populations where child deaths are very small relative to survivors the characteristics of distributions of living children per woman

Table 5.8. Percentage distribution of women by children ever born: England and Wales

Children ever born	Year of birth of woman						
	1920	1925	1930	1935	1940	1945	1950*
0	21.1	17.3	13.8	11.5	11.1	10.3	14.8
1	21.3	21.7	18.0	15.1	12.5	13.2	12.9
2	27.6	28.9	30.7	32.7	35.9	43.0	42.1
3	14.7	15.7	17.6	20.7	23.0	21.2	20.0
4 and over	15.3	16.4	19.9	20.0	17.4	12.3	10.2
Mean completed family size	2.00	2.12	2.35	2.42	2.36	2.20	2.06
Parity progression ratios							
0 to 1	0.789	0.827	0.862	0.885	0.889	0.897	0.852
1 to 2	0.730	0.738	0.791	0.829	0.859	0.853	0.849
2 to 3	0.521	0.526	0.550	0.554	0.530	0.438	0.418
3 to 4	0.510	0.511	0.531	0.491	0.430	0.367	0.338

* Slight extrapolation to end of childbearing.
Source: Office of Population Censuses and Surveys (1989*b*).

are similar. The parity distributions then provide information which is relevant for the study of present and potential child support in both directions, parents to offspring and later, offspring to parents. A linked measure which has now become central to the analysis of fertility trends is the *Parity Progression Ratio*, the proportion of women who having achieved an *n*th birth go on to at least one more. The Parity Progression Ratios can be calculated from the family size distribution and vice versa.

Table 5.8 illustrates the measures for cohorts of women in England and Wales. As can easily be seen, the trends in total fertility are compounded of parity components which reveal variations in behaviour over time. The proportion of women going on to more births after three fell steadily for those born after 1930, although the trend was flattening out as the birth cohorts of the 1950s were approached. There was a rather similar pattern for the movement from the second to third birth. On the other hand, there was a considerable increase in the proportion of women who had a second birth following their first and a substantial reduction in childlessness up to the 1945 cohort, with then a reversal. In summary, there was a greater concentration on family sizes of two at the expense of larger and smaller numbers of children. The strong movement away from larger families is a long-term trend common to all developed countries, but the changes in the popularity of modest-sized families (two or three births) compared with only children and childlessness has varied over time and among populations. The movement of the total fertility above and below replacement levels is a fine balance of these tendencies.

The proximate determinants of fertility

The level of fertility in a population is determined by a combination of biological and social factors. In some less-developed communities there is low to moderate fertility, largely because of widespread pathological sterility from infections with gonorrhoea. This was common in areas of Africa (Zaïre, inland Tanzania, coastal Kenya, lower Sudan, and elsewhere) and the Pacific Islands up to the 1950s, but since then the advent of penicillin has made a major contribution to the reduction of sterility. Now there are only a few residual populations of small size seriously affected. Infertility remains a significant problem for individual couples everywhere but the impact on overall fertility levels is small. In developed countries the low total fertilities are mainly a result of choice and the effective use of family planning and induced abortion to achieve that choice. There is an increasing correspondence between desired and actual family sizes.

There are several systems for relating the level of fertility in a population to the major proximate determinants; that is, the biological and social characteristics with direct effects. The most widely used is the Bongaarts' framework because of its simplicity and convenience (Bongaarts and Potter 1983). The *total fertility rate* (TFR) is expressed as a maximum value (*TF*) reduced by four multipliers which measure the effects of the main inhibiting factors. The formula is TFR $= C_m \times C_c \times C_a \times C_i \times TF$. C_m, the index of marriage or mating, is equal to 1 if all the women of reproductive age are exposed to the risk of conception; it measures the extent to which non-exposure lowers proportionately the aggregate risk. C_c, the index of contraception, ranges from 1 when there is no contraception to 0 when all fecund women use 100 per cent effective contraception. C_a, the index of induced abortion, is the proportion of a birth averted per abortion, taking into account the subsequent interval to conception. C_i, the index of post-partum infecundability measures the extent to which fertility is reduced because women are not immediately exposed to risk following a birth: it would be 1 if such was the case, but the postponement of the resumption of fecundity may be as long as eighteen months on average in some populations, depending mainly on the nature and duration of breast-feeding. *TF* is what the total fertility, the average completed family size, would be if there was no inhibition from

Table 5.9. Estimates of the indices ofthe proximate determinants of total fertility

Country, year	Marriage C_m	Contraception C_c	Abortion C_a	Postpartum infecundability C_i	Model total fertility
Bangladesh, 1975	0.853	0.929	(1.0)	0.539	6.54
Hong Kong, 1978	0.496	0.331	(1.0)	0.930	2.34
Hungary, 1966	0.617	0.327	0.564	(0.93)	1.62
Indonesia, 1976	0.706	0.756	(1.0)	0.577	4.71
Kenya, 1976	0.768	0.976	(1.0)	0.673	7.72
Mexico, 1976	0.610	0.731	(1.0)	0.841	5.73
Poland, 1972	0.437	0.410	0.884	(0.93)	2.26
Sri Lanka, 1975	0.513	0.710	(1.0)	0.608	3.39
United Kingdom, 1967	0.609	0.261	0.989	(0.93)	2.24

Source: Bongaarts and Potter (1983).

the proximate determinants and all the multiplying factors were 1. In principle, the value of the *TF* takes into account the variations in natural fecundability, infertility, and spontaneous abortion among populations, but little is known about these except in extreme cases. For most applications it is taken as 15.3 births, a value derived by Bongaarts empirically. The *C* factors can be estimated for particular populations in a variety of ways, depending on the information available; for example, on types of contraceptive used, failure rates, and post-partum amenorrhoea. In practice, it is necessary to introduce values based on theory and comparative research. There is thus a degree of approximation which can lead to difficulties, particularly in the interpretation of changes over time in the same population. However, the framework is very useful for the broad quantitative evaluation of the impact of the different proximate factors.

Table 5.9 gives the estimates of the *C* factors and the TFR obtained from the formula with *TF* equal to 15.3 for selected populations. The TFR are in reasonable agreement with observed measures. It will be noted that in several cases C_a is taken as 1 because of limited information on induced abortion. In developed countries, post-partum amenorrhoea is short, but there are few direct studies of its duration, and C_i is taken as an approximate 0.93. The populations in the table have been chosen to illustrate the large variations in the *C* factors. The United Kingdom in 1967 had a heavy prevalence of contraception and a strongly inhibiting C_c, but the marriage C_m was moderately high. Hong Kong in 1978 had a similar TFR, but the C_m was considerably lower, and contraceptive C_c raised, compared with the United Kingdom. Hungary in 1966 achieved a very low TFR through a combination of substantial but not dominating contraceptive use and the huge effect of induced abortion.

Illegitimate births

It is common in the European tradition to distinguish births within wedlock (legitimate) from those outside marriage (illegitimate). Generally, the social and health problems presented by the latter are much greater. This division was always a crude surrogate for the conditions of inadequate care which are the cause of the problems, and this has become even more apparent in recent times. In many parts of the world, marriage customs are different from those in Europe and the definition of an illegitimate birth is difficult. It may vary according to the purpose of the classification—legal, social or, indeed, demographic. In most countries of Europe, the proportion of births outside marriage has been rising rapidly. As an illustration, in England and Wales illegitimate births were about 4 to 5 per cent of the total, from the beginning of the century up to the 1960s, apart from slightly raised percentages during the two World Wars. But from the 1960s there has been a transformation. By 1988 more than 25 per cent of live births were technically illegitimate. For mothers under 20 years of age the proportion was three-quarters. In many cases, the parents will marry subsequent to the birth of the children or will remain in stable unions without legal bonds. Accurate, regular statistics on these issues are not collected, although a fair amount of information is available from surveys. The data on illegitimacy now give little guidance for focusing on the sections of the population in which social and health problems are more prevalent. A number of European countries, most notably Sweden, followed the same path of change, earlier and more extensively than Britain.

Differentials in mortality and fertility among subpopulations

The demographic characteristics of the populations of countries are relevant for the assessment of broad social and economic trends and in the framing of policies; but most uses are concerned with the subpopulation—particularly communities or selected groups. The dynamics of these will differ to a greater or lesser extent from those of the total population and data are usually more restricted. Although there has been much study of differentials in both the developed and the less-developed world, attention has been concentrated on only a few factors on which statistical information can easily be obtained; the most prominent of these is area of residence or community. In general, cities tend to have lower fertility and mortality, excess in migration, and higher growth-rates than rural areas, with smaller towns in between. However, it is clear that much of the variation is due to the economic and social composition of the subpopulations rather than the effects of residence as such. In developed countries the area differentials in the components of natural increase are comparatively small and the main cause of divergence is migration. In less-developed countries, regional variations are often associated with ethnic, religious, and cultural diversity. The population dynamics can be strikingly different from one region to another.

In countries of European origin there is a long history of investigation of demographic differentials, particularly in mortality and fertility, by socio-economic class, measured in various ways. Strong gradients in mortality, with rates increasing as income and status fall, have persisted up to the present, but the trends in fertility have been more complex. In the demographic transition of mortality and fertility from high to low levels the fertility decline began among the more educated and prosperous groups. Substantial differentials with the highest birth- and growth-rates in the lowest socio-economic classes emerged. There were many claims that this would lead to a fall in intelligence and biological fitness, although the arguments were indirect; survey evidence was either inconclusive or contradictory. As fertility continued to come down, however, the social-class differentials narrowed and became more complicated. In some populations a 'U-shaped' configuration appeared, with fertility lowest where the husband was a non-manual worker on a modest salary. In the 1970s and 80s it was not uncommon for fertility to be highest in the professional and upper management classes; because of the continued mortality gradient, natural increase tended to be greater, or the decrease smaller, among the economically advantaged. These trends have been of significance in the assessment of the demand for services which vary with the social classification of the family. However, the changing economic and occupational structure of populations, with development as well as social mobility, have altered the proportional division into socio-economic groups much more than the natural increase differentials.

In the Third World, data which can be used to study socio-economic class in the traditional way are hard to find. However, there has been widespread use of educational level of parents, particularly of mothers, as a characteristic to which fertility and child mortality can be related. The differentials are very similar in kind to those for socio-economic class in the developed countries, with strong gradients in childhood mortality and the same tendencies, but less consistent and regular in fertility. In many surveys the sizes of the child-mortality differentials have been striking, with a level some five times higher for the births to illiterate women compared with those to mothers with completed secondary education. The extent to which the classification by educational level demonstrates direct effects, as contrasted with the many other conditions associated with this characteristic, remains unclear. Equally, in developed countries there is no agreement on the precise aspects of socio-ecomonic class which lead to fertility and mortality differentials. In the assessment of the relation between population dynamics and community health needs, currently and in prospect, the heterogeneity of the constituent subgroups must not be ignored.

References

Bongaarts, J. and Potter, R.G. (1983) *Fertility, biology and behaviour: An analysis of the proximate determinants*. Academic Press, New York.

Brass, W. (1970). The growth of world population. In *Population control* (ed. A. Allison) pp. 131–51. Penguin Books, Harmondsworth, Middlesex.

Brass, W. (1971). On the scale of mortality. In *Biological aspects of demography* (ed. W. Brass) pp. 69–110. Taylor & Francis, London.

Brass, W. (1982). The future population of new commonwealth immigrant descent: numbers and demographic implications. In *Demography of immigrants and minority groups in the United Kingdom* (ed. D. A. Coleman) pp. 105–18. Academic Press, London.

Brass, W. (1989). Is Britain facing the twilight of parenthood? In *The changing population of Britain* (ed. H. Joshi) pp. 12–26. Blackwell, Oxford.

Coale, A.J. (1957). The effects of changes in mortality and fertility on age composition. *Milbank Memorial Fund Quarterly* **34**(1), 79–114.

Coale, A.J. and Demeny, P. (1966). *Regional model life tables and stable populations*. Princeton University Press, New Jersey.

Frost, W.H. (1939). The age selection of mortality from tuberculois in successive decades. *American Journal of Hygiene* **30**, 91–6.

Office of Population Censuses and Surveys (1989*a*). *Population Trends*, **57**.

Office of Population Censuses and Surveys (1989*b*). *Birth statistics 1987*, Table 10.5. FMI, No. 16. HMSO, London.

United Nations (1955). *Age and sex patterns of mortality: model life tables for underdeveloped countries*. United Nations, New York.

United Nations (1982). *Stable populations corresponding to the new United Nations model life table for developing countries*. Dept of International Economic and Social Affairs (ST/ESA/SER.R/44), New York.

United Nations (1983). *Indirect techniques for demographic estimation*. Manual *x*, Population Studies, No. 81. Dept of International Economic and Social Affairs (ST/ESA/SER.R/A/81), New York.

United Nations (1989). *Global estimates and projections of population by sex and age: the 1988 revision*. Department of International Economic and Social Affairs (ST/BSA/SER.R.93/93), New York.

World Bank (1986). *Population growth and policy in sub-Saharan Africa*. World Bank, Washington, D.C.

Zaba, B. (1979). The. four parameter Logit life table system. *Population Studies*, **33**(1), 79–100.

6

Nutrition

J. S. GARROW

Introduction

Every living organism requires a substrate which provides a source of energy and material for growth. The interaction of the organism with this substrate is the subject-matter of nutrition. The effect of restricted nutrition on the growth of plants which otherwise have a favourable environment is shown strikingly by the Japanese art of Bonsai: miniature apple trees which are healthy and long-lived if properly cared for show the genetically-determined characteristics of apple trees, but are perhaps one-hundredth of the normal size because they have been maintained on very restricted quantities of nutrients. In animals nutritional stunting of this severity is not compatible with survival, but the growth of piglets can be retarded to about one tenth of normal by under nutrition (see Fig. 6.1). In the Third World protein energy malnutrition is common among children on marginal food intakes. A marasmic child may survive to the age of 9 months without any increase in weight since birth, but will still show some increase in height. However, this text is mainly concerned with developed countries, and in developed countries children do not develop frank marasmus or kwashiorkor as a result of insufficient food supply. This chapter will therefore address the public health problems relating to the balance and adequacy of nutrients in the diet, rather than the problems arising from famine conditions.

Nutrient requirements

Chemical analysis of the cadavers of new-born babies and of adult men have yielded the results shown in Table 6.1 (Widdowson 1965). It is evident that babies increase in weight twenty-fold during normal developments, and they also change in body composition. An adult has proportionately less water, sodium, and chloride (in other words, less extracellular water), but more protein and other electrolytes than a new-born baby. The material for this increase in weight must be supplied from the diet. Of course nutrients requirements are much greater than that required to account for the change in the magnitude of body stores shown in Table 6.1. Energy is required to drive the metabolic processes, and the body stores of nutrients are maintained in dynamic equilibrium with the diet. Body protein is catabolized to amino acids and resynthesized to protein at the rate of about 300 g protein per day in the normal adult, but this re-cycling involves losses which require an input of at least 30 g/day of dietary protein to maintain the protein stores in the body. Similarly there are losses of minerals in urine and faeces, even in the fasting patient, so these losses need to be made good from dietary sources. Finally, vitamins must be supplied in the diet to act as co-factors for many metabolic processes.

Table 6.2 lists the energy, protein, and other major nutrients which are considered to be necessary for the health of normal adult males. For children and pregnant or lactating women other standards apply. It is quite unlikely that there are essential food factors which are at present undiscovered, because patients have been maintained in good nutritional health by parenteral feeding with mixtures containing only

Fig. 6.1. Pigs from the same litter, about one year old. The large animal was provided with unlimited amounts of a good diet from birth, the smallest had small amounts of the same diet, and the third pig had the same diet as the small one, but also unlimited access to sugar. (Source: from McCance 1968.)

Table 6.1. Chemical composition of the body of a full-term baby and of an adult man

	Full-term baby	Adult man
Body weight (kg)	3.5	70
Water (g/kg body weight)	690	600
Fat (g/kg body weight)	160	160
Protein (g/kg body weight)	115	170
Composition of fat-free mass (FFM)		
Water (g/kg FFM)	820	720
Protein (g/kg FFM)	188	212
Na (meq/kg FFM)	82	80
K (meq/kg FFM)	53	69
Cl (meq/kg FFM)	55	44
Ca (g/kg FFM)	9.6	22.4
Mg (g/kg FFM)	0.26	0.50
P (g/kg FFM)	5.6	12.0
Fe (mg/kg FFM)	94	74
Cu (mg/kg FFM)	5	2
Zn (mg/kg FFM)	20	30

Source: After Widdowson (1965).

these nutrients: indeed it was the study of parenterally-fed patients which lead to our present recognition of the essentiality of several trace elements.

Having identified the essential nutrients it is necessary to estimate the daily quantities required for health by the great majority of the population. This task cannot be achieved with any accuracy for three reasons. First, there is some uncertainty about the level of a nutrient which is just sufficient to prevent signs of deficiency: ideally this information is obtained by feeding volunteers with diets depleted in the nutrient under test. These experiments are difficult to carry out because it is necessary to house the volunteers in a metabolic ward for many weeks, and the requirement of the nutrient under test can be influenced by the composition of the remainder of the experimental diet. For example, thiamin requirements are increased by increasing the carbohydrate content of the diet, and copper requirements are increased if the diet has a high concentration of zinc. Second, the food containing the nutrient under test may affect the bioavailability of the nutrient: it is well known that pellagra occurs in maize-eaters, due to niacin deficiency, but maize contains more niacin than rice, which does not cause pellagra. The explanation is that the niacin in maize is bound in a form not readily absorbed in the human gut, and the maize protein does not provide material from which niacin can be formed. The third problem is to assess the margin of safety above minimum requirements which will cover variations in individual requirements. Very little is known about individual variability in requirements for most nutrients, so this estimate of the appropriate margin of safety is largely guesswork.

Each country has an expert committee of nutritionists who report from time to time their assessment of Recommended Daily Allowances (RDAs) for nutrients. Each committee reviews the sparse experimental data on minimum nutrient requirements, and the reports of all the other committees. The level of consensus can be judged from the last three columns of Table 6.2, which is based on a review by Truswell *et al.* (1983) of the RDAs of 40 countries. While all the countries offer RDAs for energy, protein, vitamin A, thiamin, niacin, vitamin C, calcium, and iron, some have no recommendation for the other nutrients. This is not because any country doubts that, for example, folate is an essential nutrient, but some committees (including that in the UK) are unable to make a judgement on the available evidence about the appropriate RDA for folate.

Dietary surveys to assess nutritional status

It appears obvious that, if nutrient requirements are known (within the limitations mentioned above), the most direct way to assess the adequacy of the diet of a population is to survey the diet of a sample of that population, and to compare their intake with the recommended allowances. In fact this approach is very unrewarding because it is so difficult to obtain a reliable estimate of the diet of an individual, or even of a group of people.

The largest-scale nutrition survey in the world is the National Health and Nutrition Examination Survey (NHANES) in the United States of America. In this survey data on diet are collected by a dietitian who, in the course of a 20-minute structured interview, tries to help the subject to recall his or her food and drink intake over the past 24 hours. Assuming that the recalled diet is accurate, it is still of little value to know what a person ate on a particular day, because there are large variations in the diet of an individual from day to day. Table 6.3 shows the estimated daily variation in intakes of various nutrients within an individual, and hence the number of days which must be sampled to obtain an estimate of mean intake in that individual which is within 10 per cent of the true mean intake. Unfortunately we have little information about true inter-individual variation in daily intake of specific nutrients, so it is not possible to calculate the probable error of estimates of mean intakes of groups. Among groups of people the daily variation in intake must be still larger because there is variation between individuals as well as within individuals, so the required number of days sampling is even greater than that shown in Table 6.3.

So far we have considered the limitations of 24-hour dietary recall data on the assumption that most people can, and will, recall accurately what they ate the previous day. However, even this assumption is poorly supported by evidence. Acheson *et al.* (1980) were in the favourable position of investigating the diet of scientists on the British Antarctic Expedition: the situation was favourable because their diet consisted solely of items from the camp stores, since there was no other source of food, and the subjects were scientists accustomed to the accurate recording of data. For 1085 man-days the 12 members of the expedition kept weighed records of everything they ate or drank, and on 86 occasions they

Table 6.2. Recommended daily intake (RDA) for a young adult male: comparison of 40 national committees

Nutrient	Units	RDA mean	Highest	Lowest	Number of countries making recommendation
Energy	kcal	2840	3200	2530	40
Protein	g	66.1	135	39	40
Vitamin A μg	retinol	910	1500	600	40
Vitamin D	μg	4.4	10	2.5	24
Vitamin E	mg	12.0	20	9	12
Thiamin	mg	1.26	1.8	1.0	40
Niacin	mg	18.3	24	15	40
Vitamin B_6	μg	2.0	2.2	1.8	15
Folate	μg	194	400	100	26
Vitamin B_{12}	μg	2.52	5	1	25
Vitamin C	mg	47.4	95	30	40
Calcium	g	0.61	1.1	0.4	40
Phosphorus	g	0.96	1.6	0.6	10
Magnesium	mg	345	500	250	13
Iron	mg	20.0	24	5	40
Iodine	μg	145.7	200	120	15
Zinc	mg	13.2	15	8	10

Source: Truswell (1983).

Table 6.3. Daily variation in intake of nutrients within individuals, and days of sampling required to obtain a value within ±10 per cent of the true mean intake of the individual

Nutrient	Within-person coefficient of daily variation	Days sampled for mean estimate within 10 per cent of true value
Energy	23	5
Carbohydrate	25	6
Protein	27	7
Fat	31	10
Dietary fibre	31	10
Calcium	32	10
Iron	35	12
Thiamin	39	15
Riboflavin	44	19
Cholesterol	52	27
Vitamin C	60	36

Source: Thomas (1988).

were asked to recall what they had eaten during the previous 24 hours. Even in this situation the energy value of the recalled intake ranged from 33–132 per cent of the true intake, with a 33.6 per cent mean underestimate by recall. Thus the error of dietary recall is very large and shows a systematic bias to under-recall. It is doubtful if 24-hour recall data are ever worth the trouble to collect, even for the crudest type of dietary stratification.

Since dietary recall is so unsatisfactory the alternative is some form of prospective diet record. This involves much more work on the part of the investigators, and much more co-operation from the subjects, but it yields much better data on the foods which the subject consumed over the study period. However, prospective recording introduces a new error since the process of weighing and recording food intake tends to inhibit normal eating behaviour. This is easily shown by the fact that subjects who are required to record their food intake usually lose weight over the period of the study. This effect is more marked in obese people than lean ones. It is not simply that the obese people take a period of dietary study as an opportunity to lose weight: if the energy value of the food which they record as having eaten is compared with their energy requirements, measured by calorimetry, it is evident that there has been systematic under-recording of food eaten (Prentice *et al.* 1986).

When an estimate of the food eaten has been obtained, either by recall or by prospective recording, it is necessary to calculate the nutrient content of the diet by means of food tables (Paul and Southgate 1978). These provide typical values for the nutrients in 100 g portions of most foods, but there is variability in the composition of some foods with the season of the year, or country of origin, and the nutrient content of foods can be considerably altered by cooking. These are factors which cannot be adequately controlled by the use of food tables.

Finally, it is difficult to draw conclusions about the nutritional status of a group of subjects if the estimated intake of the group is compared with the appropriate RDA. Figure 6.2 shows the intake of retinol by a nationally representative sample of children aged 10/11 years. For this group the RDA for retinol is 575 μg/day, so if the mean intake for the group is above this it is reasonable to suppose that retinol deficiency is not a serious problem. The mean intake is above the RDA: it is 854 μg/day, but the distribution is very skewed, so the median intake is 565 μg/day, which is below the RDA. The mean intake is above the RDA because some boys had a very high intake, but it cannot reasonably be suggested that the presence in the group of some boys with very high intakes protects those at the bottom end of the distribution from retinol deficiency. I do not know if a significant proportion of children aged 10/11 years in the UK are retinol deficient or

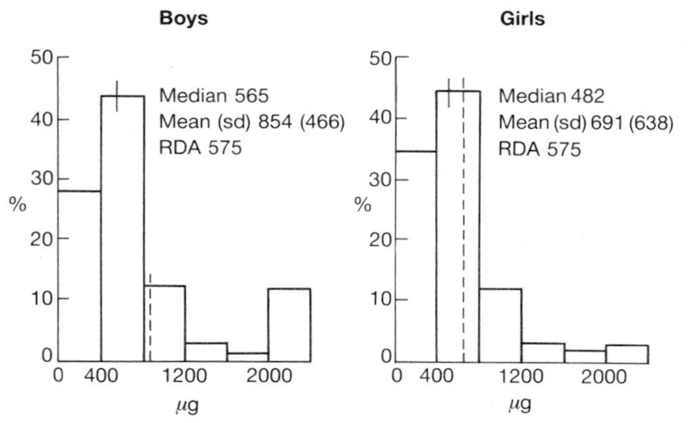

Fig. 6.2. Daily intake of retinol equivalent (μg) by children in the UK aged 10/11 years.

not: this example has been cited to show that the question often cannot be answered simply by measuring the intake of a sample of the population and comparing the mean intake of nutrients with the RDA.

Effects of undernutrition in adults

The effects of undernutrition on normal young adult males has been investigated in two classical studies: 'Human vitality and efficiency under prolonged restricted diet' by Benedict *et al.* (1919), and the 'Minnesota study' by Keys *et al.* (1950). Figure 6.3 shows data from the earlier study: normal volunteers were fed a restricted diet for 126 days so as to achieve and maintain a reduction in body weight of about 10 per cent. An unplanned but very informative feature of the experiment was that the volunteers were allowed to go home at Thanksgiving and Christmas, when evidently the restricted diet was not strictly observed. The results show beautifully that with undernutrition basal heat production (or as we would now say, resting metabolic rate, or RMR) decreases, and a new steady state is achieved with a lower body weight and lower energy requirements. However, when the dietary restriction is temporarily lifted the RMR rapidly increases, but then as rapidly falls back to the adapted state when the diet is reimposed. Energy requirements in man are therefore responsive to energy intake in two ways: there is a gradual adaptation to chronic underfeeding (achieved mainly by loss of body tissue), and also an acute adaptation to energy imbalance, which is achieved by mechanisms not fully understood, but probably involving the hormones regulating sympathetic activity in the body.

These adaptive mechanisms are of value for survival under famine conditions, but they complicate the task of deciding if the energy supply in the diet is adequate: mere maintenance of body weight is not a reliable guide to the adequacy of energy intake, since body weight can be maintained at different planes of nutrition by means of the adaptive mechanisms described above. We need to know at what stage in the adaptation to reduced intake there is a cost in the form of functional impairment: when that point is reached we pass from the zone of adaptation to that of undernutrition. There can be no doubt that during the hungry winter of 1944–45 the population of north-west Holland was undernourished on an energy intake which fell below 1000 kcal/day, because virtually all women of child-bearing age ceased to menstruate (Stein *et al.* 1975). It is one of the adaptive responses to undernutrition in human females to cease reproductive activity, and in experimentally semi-starved men also there is decreased potency and libido (Keys *et al.* 1950). The other physiological changes produced by experimental semi-starvation were postural hypotension, a decrease in the voltage of the electrocardiogram, polyuria, some decrease in auditory acuity, a decrease in muscular strength and endurance, impaired co-ordination, and a marked decrease in capacity to perform severe physical work. These effects are not diagnostic of undernutrition, however, since they may be produced by any debilitating disease.

Undernutrition in children

To maintain health the diet of adults must supply enough energy to fuel necessary metabolic work, and enough of the

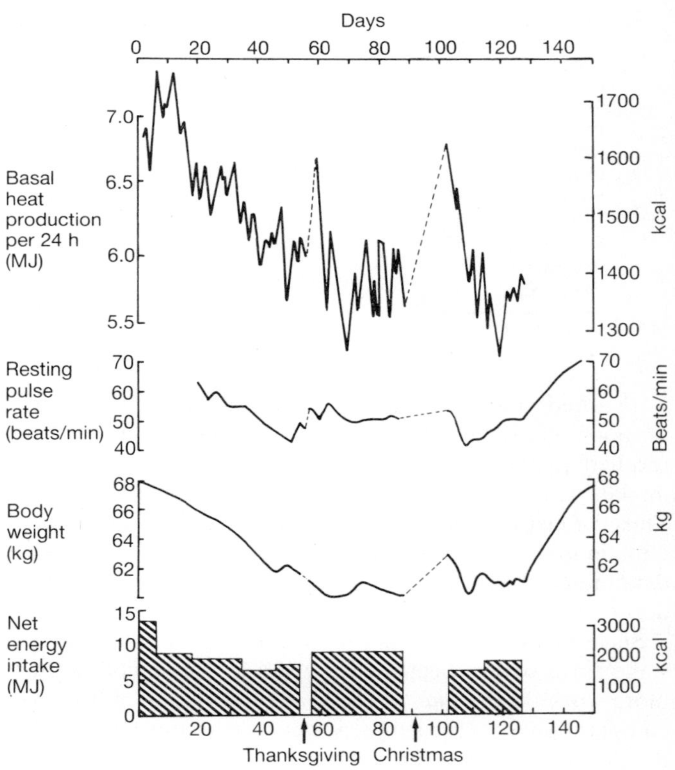

Fig. 6.3. The effect of a restricted diet on the basal metabolic rate, resting pulse rate, and body weight of normal volunteers. (Source: data of Benedict *et al.* 1919.)

Table 6.4. Recommended daily amounts of energy and protein in male and female children and adults

Age range (years)	Energy (MJ)		Protein (g)	
	Male	Female	Male	Female
1	5.0	4.5	30	27
2	5.75	5.5	35	32
3–4	6.5	6.25	39	37
5–6	7.25	7.0	43	42
7–8	8.25	8.0	49	47
9–11	9.5	8.5	57	51
12–14	11.0	9.0	66	53
15–17	12.0	9.0	72	53
18–34 (sedentary)	10.5	9.0	63	54
(very active)	14.0	10.5	84	62

Source: Department of Health and Social Security (1979).

other nutrients to make good the daily losses. However, children have the additional task of maintaining growth, and achieving the change in body weight and composition indicated in Table 6.1. The deposition of new protein involves quite high energy costs, so the recommended daily intakes of children are proportionately much greater than for adults. This is illustrated in Table 6.4: the protein and energy requirements of a child aged 1 year are about half those of a sedentary adult, although the adult will weigh six or seven times as much as the infant. If a growing child is not given enough to eat it ceases to grow at a normal rate: this effect in piglets is strikingly illustrated in Figure 6.1. The velocity of height growth in children is an extremely sensitive index of nutritional status, with the usual proviso that other factors (genetic, infective) may also affect height growth.

Figure 6.4 shows reference centiles for height and weight among boys and girls up to 5 years of age, and Figure 6.5 from 5–18 years (World Health Organization 1983). There is an important difference between these normal standards and centiles for other variables, such as blood pressure or serum cholesterol, concerning which an individual may change in time from a high to a low level and back again. Height (and to a lesser extent weight) at a given time is the cumulative result of previous increments: it is almost impossible for a child to decrease in height. In general a child should grow parallel to the centile lines, and a child who is on a high or low centile for height may be expected to be similarly placed on the weight centile chart. These reference standards provide a means of assessing the relative plane of nutrition of subsections within a surveyed population: unless there are large ethnic differences between the groups the group which is taller and heavier will almost certainly have received a diet supplying more energy and protein, and probably more of all the other nutrients also. In populations in which marginal malnutrition is common, comparisons of height and weight growth in children is an effective way to detect the groups most at risk from malnutrition. However, the more affluent the society, and the higher the general plane of nutrition, the less useful this form of surveillance becomes. We have no grounds to believe that an ever-increasing intake of nutrients,

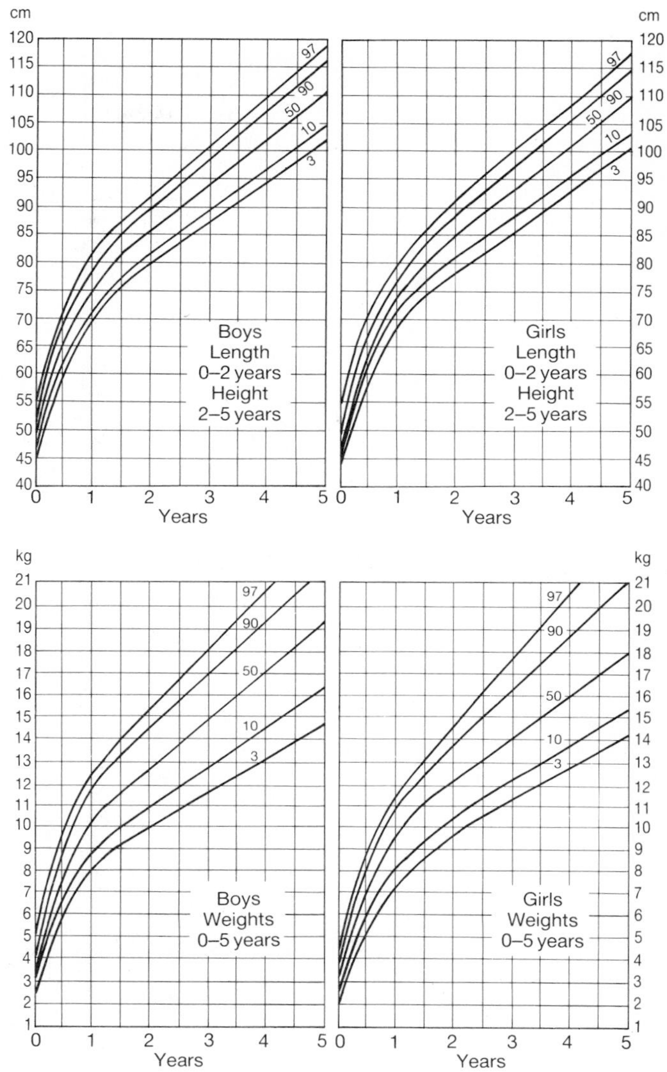

Fig. 6.4. Standard heights and weights of children 0–5 years old.

with a consequently ever-increasing height and weight at a given age among children, is necessarily an index of better and better nutrition.

Effects of overnutrition

If the nutrients listed in Table 6.2 are taken in excess of requirements this excess is excreted (in the case of water-soluble vitamins and electrolytes) or not absorbed (in the case of nutrients which have special transport systems, such as iron or calcium). Excess protein intake is absorbed and deaminated to form urea, which is then excreted in urine. However there are two classes of nutrients for which excess intake inevitably leads to excess storage: these are energy, and the fat-soluble vitamins. The effects of excessive intake of fat-soluble vitamins will be considered later: the point to

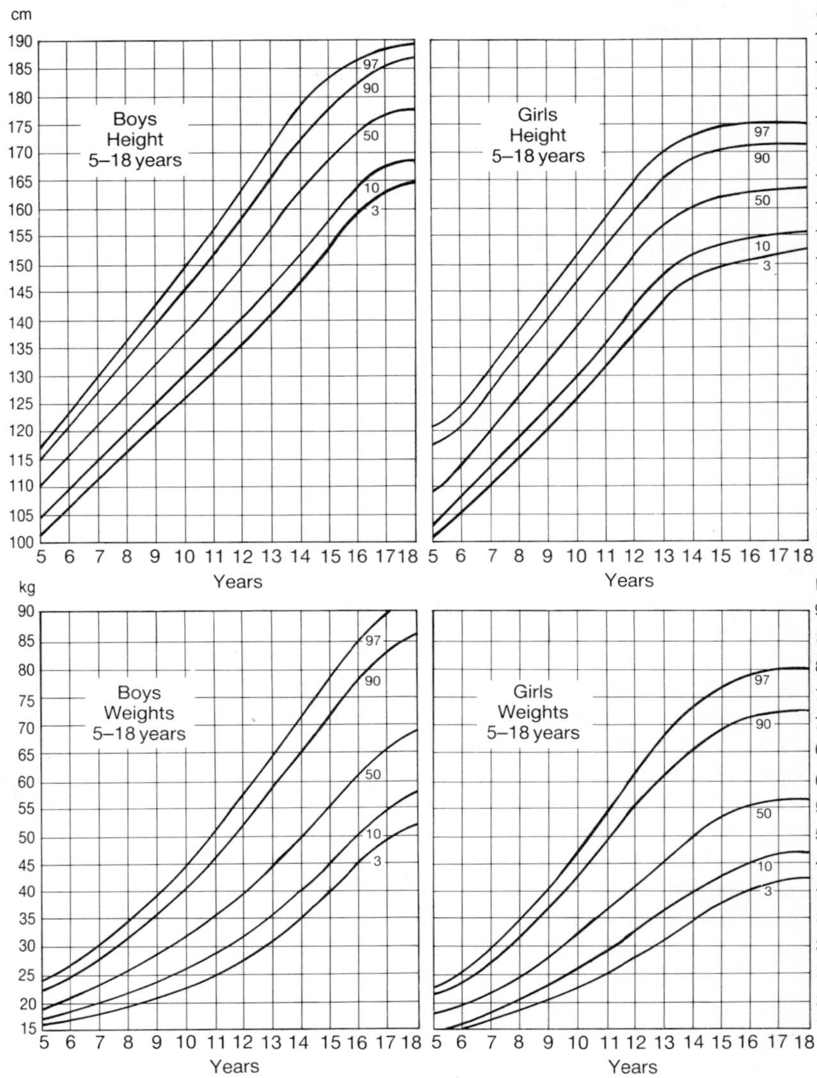

Fig. 6.5. Standard heights and weights of children 5–18 years old.

be made here is that if a normal diet is consumed in excessive quantities the result is that there is an increase in the body energy stores, which in effect means body fat.

A normal adult contains about 12 kg fat, which is an energy store of 108 000 kcal, and about 12 kg of protein, which is equivalent to 48 000 kcal, but of course only a small part (perhaps 25 per cent) of body protein can be sacrificed without severe disability, so body protein is a negligible store of energy compared with fat. Since the daily energy turnover of an adult is about 2000 kcal/day the fat stores would provide this energy during 50 days of starvation even if there was no adaptive response, and as there is an economy of energy metabolism during undernutrition (see Fig. 6.3) a normal subject will usually survive 70–80 days of starvation, provided adequate water is available. The ability to store energy as a reserve in time of famine was no doubt an important factor in man's successful evolution, but in developed countries famines do not occur and obesity, the consequence of excessive energy storage, is the commonest form of malnutrition in regions where food is cheap and plentiful.

Ever since the first actuarial investigations of mortality in 1903 it has been noted that people of above a certain range of weight-for-height tended to die young. The life insurance companies have therefore published 'desirable' ranges of weight-for-height, which agree with large surveys of non-insured populations (Lew 1985) in defining a range of weight which is associated with minimum mortality.

The Belgian astronomer Quetelet (1869) pointed out that adults of normal build, but different heights, tended to have weights in proportion to the square of height: in other words W/H^2 (where W is body weight in kg, and H is height in m) is constant. Quetelet's Index (QI), which was renamed Body Mass Index by Keys *et al.* (1972), is more convenient to use than tables of weight and height, and it happens that the

Table 6.5. Prevalence rates (per cent) of obesity among adults

Country	Age (year)	Grade I		Grade II		Grade III		Author
		M	F	M	F	M	F	
Netherlands	20–34	20	10	2	2	?	?	Van Sonsbeek 1985
	35–49	37	21	4	5	?	?	
	50–64	46	36	5	10	?	?	
Norway	20–24	17	11	1	2	–	0.1	Waaler 1984
	40–44	41	30	5	5	–	0.5	
	60–64	44	43	8	24	0.1	1.1	
UK	16–64	34	24	6	8	0.1	0.3	Rosenbaum *et al.* 1985
Australia	25–64	34	24	7	7	?	?	Bray 1985*a*
Canada	20–29	34	14	4	2	?	?	Canadian Fitness Survey 1981
	30–39	46	22	7	6	?	?	
	40–49	57	30	10	11	?	?	
	50–59	62	42	9	9	?	?	
	60–69	63	44	13	14	?	?	
USA	20–74	31	24	12	12	?	?	Abraham *et al.* 1983

? Grades II and III not separated.
– prevalence <0.05 per cent.

lower boundary of the Metropolitan desirable weight range is similar to $W/H^2 = 20$, and the upper boundary to $W/H^2 = 25$. There is now general agreement that QI 20–24.9 is the desirable range of weight, QI 25–29.9 is Grade I obesity, QI 30–40 is Grade II obesity, and QI >40 is Grade III (Garrow 1988).

The use of different weights at a given height to define obesity implies that these differences in weight are caused by differences in fat content rather than, for example, muscle mass. Indeed weight–height indices may classify as moderately obese very heavily muscled athletes, but the error is not usually large, and it applies to only a few people. In general, measurements of body fat, by estimations of body density, or water, or potassium, or all three of these, have shown that more than 90 per cent of the variation in weight-for-height is explained by variation in body fat (Garrow and Webster 1985). This does not mean that all weight gained is fat, but that the extra weight in an obese person has a fairly constant composition of 75 per cent fat and 25 per cent non-fat tissue.

Prevalence of obesity

The prevalence of obesity in several developed countries is shown in Table 6.5. In all countries for which data are available the prevalence of obesity increases with age. In the UK, among women, there is a steadily increasing proportion of the population in the ranges QI 30–40 between age 20 and 60, but among men there is no increase in those in the range QI 35–40 after age 35 years.

Among men and women in the UK the mean QI increases up to the age of 70 years, but thereafter decreases. Before the age of 60 years women have a lower mean QI than men, but after 65 years the mean value for women is greater than that of men (Burr and Phillips 1984).

Diseases related to obesity

A Consensus Conference in 1985 in the USA (National Institutes of Health Consensus Development Panel 1985) concluded:

> The evidence is now overwhelming that obesity, defined as excessive storage of energy in the form of fat, has adverse effects on health and longevity. Obesity is clearly associated with hypertension, hypercholesterolaemia, non-insulin-dependent diabetes mellitus, and excess of certain cancers and other medical problems. . . . Thirty-four million adult Americans have a body mass index greater than 27.8 (men) or 27.3 (women); at this level of obesity, which is very close to a weight increase of 20 per cent above desirable, treatment is strongly advised. When diabetes, hypertension, or a family history for these diseases is present, treatment will lead to benefits even when lesser degrees of obesity are present.

Several factors have tended to conceal the medical importance of obesity. The effects of overweight on mortality take years to become evident, but they are clear enough to the insurance companies, which necessarily take a long view. Moreover, after the onset of a disease such as diabetes the patient usually loses weight (Knowler *et al.* 1981), so the contribution of obesity to the illness is not so obvious. Another confounding factor is cigarette smoking: smokers tend to be lighter than non-smokers and to die younger, so when studied in a population mixed with non-smokers they distort the true relation of weight to mortality. Finally, there is the logical problem of interpreting multiple regression analysis, which was well shown in the 7-country study (Keys *et al.* 1984). This survey showed that overweight men were more likely to have heart attacks, but that when age, smoking, blood pressure, and serum cholesterol were taken into account, relative weight did not significantly predict coronary

Table 6.6. Changes in fasting blood concentrations, and insulin responsiveness among 19 male volunteers who over-ate a normal diet and increased body weight by 21±1.0 per cent of which fat was 73±4.4 per cent

	Change	$p<$
Fasting blood		
cholesterol	increased	0.1
triglycerides	increased	0.01
aminoacids	increased	
glucose	increased	0.05
insulin	increased	0.005
oral glucose tolerance	decreased	0.005
intravenous glucose tolerance	decreased	0.05
Insulin response		
to oral glucose	increased	ns
to intravenous glucose	increased	0.1
to intravenous arginine	increased	0.01

Source: Data of Sims *et al.* (1973).

heart disease among men aged 40–59 years. This finding may be interpreted to mean that weight is irrelevant to coronary heart disease, or alternatively that it increases the risk of death from coronary heart disease by contributing to hypertension and hypercholesterolaemia. The critical question is: does the obesity cause the hypertension, hypercholesterolaemia, and diabetes, or is it just a non-causal association?

The only way to obtain a definitive answer to this important question is to take normal volunteers without a family history of diabetes or obesity, and overfeed them so they become obese, and see if any or all of them become diabetic. The practical difficulties of such an experiment are obvious, but it was attempted by Sims *et al.* (1973) in Vermont, using volunteers from the State prison. The changes produced by overfeeding all elements of the diet in 19 volunteers in Table 6.6. The study certainly proves that normal-weight people can be made obese by experimental overfeeding, and that they then show biochemical changes in the direction of those found in spontaneous obesity and in non-insulin-dependent diabetes. However, the changes are small and do not take the values outside the normal range, so this degree of experimental obesity does not, on its own, cause diabetes. That would be too much to expect, since spontaneous obesity does not always cause diabetes either. At least we know that obesity (either experimental or spontaneous) causes some degree of insulin resistance, which is the cardinal feature of NIDDM. It is therefore reasonable to conclude, on the available evidence, that obesity plays some part in causing NIDDM in individuals who have a genetic disposition to this disease.

Does obesity cause hypertension? The Vermont study described above was not focused on blood pressure changes, so the evidence to answer this question depends on the observed associations between the two conditions. The prevalence of hypertension among white and black men and women of different weight categories is shown in Table 6.7. The black population has higher blood pressure levels than the white at similar relative weight, but in all groups increasing obesity is associated with increasing prevalence of hypertension (Stamler *et al.* 1978). Obese normotensive subjects (children, adolescent, or adult) when followed prospectively are more likely to become hypertensive than normal-weight normotensive subjects (Berchtold *et al.* 1981).

There are several mechanisms by which obesity might cause hypertension, but the evidence is inconclusive (Berchtold and Sims 1981; Dunstan 1985). It may be that the hyperinsulinaemia associated with obesity causes sodium retention by the kidney (DeFronzo 1981), or the increased sympathetic tone associated with overfeeding may be involved. Whatever the mechanism, obese subjects have an increased risk of hypertension, and this is another important cardiovascular risk factor in obesity. Hyperinsulinaemia may itself be an independent risk factor for atherosclerosis (Stout 1982). Cardiovascular disease is particularly associated with fat deposition in the abdominal region (Bjorntorp 1985).

The idea that obesity is benign unless associated with risk factors such as hypertension and hypercholesterolaemia is challenged by recent results from the Framingham study (Hubert 1984) and other long-term prospective studies. It now appears that obesity is a significant and independent predictor of disease, especially among women. Indeed obesity in women is the next best predictor of cardiovascular disease after age and blood pressure. The Framingham data also show a gradient of cardiovascular risk with adiposity in the subset of men and women who had no major risk factors for cardiovascular disease (Hubert 1984).

Obese people also show an increased mortality and morbidity from disease of the gallbladder (Bray 1985*b*). Increased secretion of cholesterol into the bile is a characteristic of obese patients: in the study of Reuben *et al.* (1985) those with and without gallstones had secretion rates of 107 and 81 μmol/h, which was significantly greater ($p<0.01$) than non-obese patients with and without gallstones (51 and 42 μmol/h). Obese subjects also secreted more cholesterol in relation to bile acid, thus making the bile more liable to form stones. Gallbladder disease was found at operation in about 90 per cent of morbidly obese patients (Amaral and Thompson 1985). Also, a significantly increased prevalence of abnormal liver function tests has been reported among moderately obese people (Nomura *et al.* 1986).

Table 6.7. Frequency of hypertension (diastolic >95 mmHg) per 1000 among white and black men and women aged 20–39 years and 40–64 years

Age	Weight	White men	White women	Black men	Black women
20–39	underweight	38.7	38.6	88.3	88.7
	normal weight	72.4	41.3	122.5	101.9
	overweight	175.9	110.9	259.9	149.1
40–64	underweight	182.4	172.5	300.4	377.2
	normal weight	244.2	221.7	374.7	411.8
	overweight	361.2	352.7	519.0	539.2

Source: Data of Stamler *et al.* (1978).

Obesity also has adverse effects on reproductive function (Friedman and Kim 1985). Very thin and very fat women often have problems with menstrual irregularity and ovulatory failure. The syndrome of polycystic disease of the ovaries is associated with obesity, androgen excess, and failure of ovulation. Ovarian function may improve dramatically with substantial weight loss. Overweight is also associated with an excess mortality from cancers of the colon, rectum, and prostate in men, and the gallbladder, endometrium, cervix, ovary, and breast in women (Garfinkel 1985).

Apart from the metabolic penalties associated with overweight, the extra fat also constitutes a physical burden. Diminished exercise tolerance is an objective consequence of quite modest degrees of obesity, and with severe obesity in older patients osteoarthritis in weight-bearing joints is often a major cause of disability. Finally, obesity is associated with social and psychological disadvantages, since our society tends to have a low valuation of obese individuals and, in the case of only slightly overweight people, low self-esteem may be the most crippling penalty of obesity (Wadden and Stunkard 1985).

Table 6.8. Main recommendations for dietary change for most people

Fats and cholesterol Reduce consumption of fat (especially saturated fat) and cholesterol. Choose foods such as vegetables, fruit, whole-grain cereals, fish, poultry, lean meats, and low-fat dairy produce. Use food preparation methods that add little or no fat.
Energy and weight control Achieve and maintain a desirable body weight. To reduce energy intake, limit consumption of foods relatively high in calories, fats, sugars, and minimize alcohol consumption. Increase energy expenditure through regular and sustained physical activity.
Complex carbohydrates and fibre Increase consumption of whole grain foods and cereal products, vegetables (including dried beans and peas), and fruits.
Sodium Reduce intake of sodium by choosing foods relatively low in sodium and limiting the amount of salt added in food preparation and at the table.
Alcohol To reduce the risk of chronic disease, take alcohol only in moderation (no more than two drinks per day), if at all. Avoid drinking alcohol while pregnant.

Source: Koop (1988).

Excess and deficits of specific nutrients

The ratio of other nutrients to energy in the average mixture of plant and animal material which was eaten by primitive man was probably such that, if energy requirements were satisfied, the requirements for other nutrients would also be satisfied. It was by fulfilling this criterion that material became established as human food. In primitive conditions the supply of food controlled the population density, so there was a balance set up between the availability of food and the number of people to be fed. In different regions of the world the limiting nutrient varies: for example, in India and the Middle East malnourished children very often suffer from xerophthalmia because vitamin A concentration in the staple food is low, whereas in the Caribbean equally severely undernourished children hardly ever show signs of vitamin A deficiency.

In developed countries with extensive food processing it cannot be assumed that the ratio of nutrients to energy in the diet will be that which is physiologically required. Two products of food technology, sugar (sucrose) and alcohol, provide energy with virtually no other nutrient. For the food technologist it is often preferable to use saturated fat in a formulation rather than unsaturated fat, which is more liquid and more susceptible to go rancid. Salt (sodium chloride) occurs in very low concentrations in natural foods, but it is extensively used to enhance flavour and keeping qualities in processed foods. Finally, many food processes involve the removal of the undigestable parts of plant materials, which we have recently come to value as 'dietary fibre'.

These five components of the diet—sugar, alcohol, saturated fat, salt, and fibre—are the subject of much advice to the public concerning a healthy diet. The risk and benefits involved are often overstated by nutritionists with an evangelistic disposition, or understated by food manufacturers whose livelihood is in jeopardy. A profusely referenced report on this difficult topic has recently been published by the US Surgeon General (Koop 1988): the main recommendations are summarized in Table 6.8. Similar conclusions have been reached by other national and international expert committees.

Other contributions to this textbook will indicate in greater detail the connection between the dietary advice given in Table 6.8 and mortality and morbidity in the general population, but the basis for the advice can be summarized briefly here. The intake of saturated fat (and to a lesser extent of cholesterol) clearly influences the level of serum cholesterol, which is itself closely related to mortality and morbidity from coronary heart disease. Intervention trials suggest (but have not proved to the satisfaction of all critics) that reduction in serum cholesterol, and especially LDL-cholesterol, reduces the liability to coronary heart disease. The association between obesity and various diseases has already been reviewed in this chapter.

There is no consensus about the effects of an increased intake of dietary fibre, largely because the term 'fibre' covers many forms of indigestible polysaccaride from different plant sources. In general, insoluble fibre from cereals has the greatest effect on faecal bulk, while soluble fibre has most effect on the rate of absorption of simple sugars from the gut, and hence may be most beneficial to diabetics. There is little evidence that dietary fibre increases satiety, and hence decreases energy intake, unless it is taken in large quantities. However, large fibre intakes, especially of cereal bran, may cause diarrhoea and flatulence, and may inhibit the absorption of minerals such as iron, calcium, and zinc.

There is no doubt that current intakes of sodium chloride are far in excess of physiological requirements, but there is dispute about the justification for severe sodium restriction in the general population to reduce the prevalence of hypertension. Most trials show that sodium restriction causes a small but significant reduction in blood pressure in hypertensive subjects, but has little effect in normotensive subjects. Advice to reduce sodium intake is justified, therefore, on balance.

Alcohol is of importance in the public health aspects of nutrition for several reasons: it is toxic in large quantities and impairs the function of the liver and the normal maturation of red blood cells; it contributes to the aetiology of some cancers, notably of the oral cavity; and it predisposes to malnutrition by its action in affecting social behaviour and providing energy without associated nutrients. The rate at which a normal person can metabolize ethyl alcohol is 100 mg/kg body weight/hour, which yields about 1100 kcal/day for a 70 kg adult. This is the equivalent of 2 litres of wine per day. At higher rates of ingestion the excess goes to increase the blood alcohol concentration. Disregarding any possible toxic effects of alcohol ingestion, it is clearly difficult to maintain an adequate intake of protein, vitamins, and minerals if about half the energy requirements are taken in a form which contributes nothing to the supply of these nutrients.

There is striking evidence that the addition of fluoride to drinking water, in areas where it is not naturally present, greatly reduces tooth decay. This public health measure is opposed by some on the grounds that the individual should not be involuntarily medicated, but that argument might equally be applied to the addition of chlorine to the water supply, which is generally accepted.

Advice in the Surgeon General's report to reduce the frequency and amount of sugar consumption is specifically related to dental caries. However, the observations above on alcohol, about the difficulty of supplying adequate nutrition if a large proportion of energy is taken without other nutrients, applies also to children with a high intake of sugar, and sugary foods and drinks.

Governmental reports advocating changes in diet prompt the question: how did people manage before there were nutrition experts? In man there is a sensory-specific satiety mechanism (Rolls *et al.* 1984) so at a meal we may tire of eating food of a particular taste and texture but still have appetite for food of different sensory properties. This mechanism obviously promotes the intake of a varied diet, and hence tends to protect against specific nutrient deficiencies which might occur if only one staple food was taken. However, this safeguard depends upon foods which differ in taste also differing in nutrients properties. With modern food technology this is not necessarily so. For example, potato crisps may be flavoured to taste like bacon, onion, cheese, and so on, but their nutritional properties are those of potatoes and the fat they are fried in. In principle material can be produced which looks and tastes like virtually any traditional food, but which may have none of the nutritional properties of the food which it resembles.

There are two ways in which the public can be protected from the potentially adverse effects of food technology. One is to ensure that food which appears to be meat, for example, should provide the nutrients normally found in meat, even though the material is originally of fungal origin. However, this alone is not enough, since there are some food products which do not resemble any natural food, or even which are made to resemble a natural material but to have very different nutritional qualities. Examples of this class are the non-nutritive fat substitutes, which are intended to provide the sensory qualities of fat in food, but they resist digestion so they do not provide the energy normally derived from dietary fat. When there is deliberate contradiction between the sensory and nutritive properties of a food it is obvious that the only way to inform the consumer of what is being eaten is by a system of nutritional labelling. There is general agreement that nutritional labelling is desirable (but with some reluctance from manufacturers of foods which are perceived by the public to be 'unhealthy') but it is in practice very difficult to design a system of nutritional labelling for all products which is accurate, helpful, and easily understood.

It was noted above that excessive intake of most nutrients is compensated by increased excretion or diminished absorption, but that fat-soluble vitamins were an exception. It is now common practice in many developed countries to take dietary supplements, and toxic effects have been noted with prolonged intake of Vitamin D at five times the recommended daily allowance. Indeed, poisoning with fat-soluble vitamins was a hazard which faced species from which man was evolved: skeletons of *Homo erectus* have been found with evidence of overdosage of vitamin A, probably due to the excessive consumption of the liver of carnivore animals (Walker *et al.* 1982).

Nutritional surveillance

The most comprehensive system of national nutrition surveillance is NHANES (National Nutrition and Health Examination Survey) in the USA. The goals of this programme are as follows:

(1) to estimate the national prevalence of selected diseases and risk factors;

(2) to provide national population reference distributions of selected health parameters;

(3) to document and investigate reasons for secular trends in selected diseases and risk factors;

(4) to contribute to an understanding of disease aetiology;

(5) to investigate the natural history of selected diseases. The last two goals listed above apply for the first time to the survey cycles starting in 1988 and due to be completed in 1994: previous NHANES surveys had only the first three goals.

It should be noted that these are very wide objectives, encompassing diseases which are not necessarily attributable to dietary factors: however, for many diseases such as cancer and heart disease, nutrition is one of several aetiological factors, and it is not easy (or logical or economical) to distinguish between surveillance for nutrition and for wider issues of public health. To achieve the objectives set out above the NHANES survey involves a physical examination by a physician, an estimate of 24-h food intake, measurements of body size and composition by ultrasound and bioelectrical impedance as well as the usual anthropometry, a dental examination, electrocardiogram, spirometry, an estimate of bone density, audiometry, and a large range of analyses on blood and urine to assess nutritional and immune status, and the presence of various environmental toxins. It is obvious that such a wide net is very expensive. If, to obtain better value for money, we attempt to select a subset of tests, this requires a knowledge of the disorders present in the surveyed population, but it is in order to obtain that knowledge that the survey was required.

In affluent countries the diseases which are common and plausibly linked with diet are coronary heart disease, hypertension, stroke, cancer, diabetes, osteoporosis, gallstones, diverticulitis, constipation, and allergies to foods or food additives. Even in affluent countries there are usually sections of the population, probably in poor inner-city areas, who are at risk from classical undernutrition, so a programme of so-called 'nutritional surveillance' should be designed to detect this form of malnutrition. Since, for reasons given above, it is so difficult to obtain reliable information about the nutrient intake of a population, proxy measurements are made. The foundation of all nutrition surveys has been the measurement of weight and height, to which may be added the measurement of triceps skinfold thickness (T) and mid-upper-arm circumference (MAC). From these data mid-arm muscle circumference (MAMC) can be calculated by:

$$\text{MAMC} = \text{MAC} - (3.14\,\text{T})$$

Anthropometry, related to normal ranges for adults, or weight–height growth charts for children, will provide a good indicator of general over- or under-nutrition in the population, but will not indicate anything more about the nature of the diet.

Although dietary questionnaires are unreliable, it is possible to obtain objective evidence about intake of some nutrients by analysis of blood or urine samples. The 24-hour urine output of Na or K gives a useful measure of the daily intake of these elements, and daily urea output indicates protein intake. The level of intake of fat-soluble vitamins, ascorbic acid, and folic acid can be assessed by blood analyses, but there is still dispute among the experts concerning the interpretation of such data. For example, a person on a diet with $<5\ \mu g$ folate daily will show low serum folate levels in 3 weeks, and thereafter an increasing number of haematological abnormalities, but will not become anaemic until about 20 weeks on this diet (Herbert 1987).

References

Acheson, K.J., Campbell, I.T., Edholm, O.G., Miller, D.S., and Stock M.J. (1980). Measurement of food and energy intake in man–an evaluation of some techniques. *American Journal of Clinical Nutrition*, **33**, 1147.

Amaral, J.F. and Thompson, W.R. (1985). Gallbladder disease in the morbidly obese. *American Journal of Surgery*, **149**, 551.

Benedict, F.G., Miles, W.R., Roth, P., and Smith, M. (1919). *Human vitality and efficiency under prolonged restricted diet*. Carnegie Institution Publication 280, p. 83. Washington, DC.

Berchtold, P., Jorgens, V., Finke, C., and Berger, M. (1981). Epidemiology of obesity and hypertension. *International Journal of Obesity*, **5**, 1.

Berchtold, P. and Sims, E.A.H. (1981). Obesity and hypertension: conclusions and recommendations. *International Journal of Obesity*, **5**, 183.

Bjorntorp, P. (1985). Regional patterns of fat distribution. *Annals of Internal Medicine*, **103**, 994.

Bray, G.A. (1985*a*). Obesity: definition, diagnosis and disadvantages. *Medical Journal of Australia*, **142**, S2.

Bray, G.A. (1985*b*). Complications of obesity. *Annals of Internal Medicine*, **103**, 1052.

Burr, M.L. and Phillips, K.M. (1984). Anthropometric norms in the elderly. *British Journal of Nutrition*, **51**, 165.

Canadian Fitness Survey (1981). Fitness Canada, Ottawa.

De Fronzo, R.A. (1981). Insulin and renal sodium handling: clinical implications. *International Journal of Obesity*, **5**, 93.

Department of Health and Social Security (1979). *Recommended daily amounts of food energy and nutrients for groups of people in the United Kingdom*. Report on Health and Social Security Number 15, p. 27. HMSO, London.

Dustan, H.P. (1985). Obesity and hypertension. *Annals of Internal Medicine*, **103**, 1047.

Friedman, C.I. and Kim, M.H. (1985). Obesity and its effect on reproductive function. *Clinical Obstetrics and Gynecology*, **28**, 645.

Garfinkel, L. (1985). Overweight and cancer. *Annals of Internal Medicine*, **103**, 1034.

Garrow, J.S. (1988). Obesity and related diseases, p. 329. Churchill Livingstone, London.

Garrow, J.S. and Webster, J. (1985). Quetelet's index (W/H^2) as a measure of fatness. *International Journal of Obesity*, **9**, 147.

Herbert, V. (1987). Recommended dietary intakes (RDI) of folate for humans. *American Journal of Clinical Nutrition*, **45**, 661.

Hubert, H.B. (1984). The nature of the relationship between obesity and cardiovascular disease. *International Journal of Cardiology*, **6**, 268.

Keys, A., Brozek, J., Hanschel, A., Mickelson, O., and Taylor, H.L. (1950). *The biology of human starvation*, p. 1385. University of Minnesota Press, Minneapolis.

Keys, A., Fidanza, F., Karvonen, M.J., Kimura, N., and Taylor, H.L. (1972). Indices of relative weight and obesity. *Journal of Chronic Diseases*, **25**, 329.

Keys, A., Menotti, A., Aravanis, C., *et al*. (1984). The seven countries study: 2289 deaths in 15 years. *Preventive Medicine*, **13**, 141.

Knowler, W.C., Pettit, D.J., Savage, P.J., and Bennett, P.H. (1981). Diabetes incidence in Pima Indians: contributions of obesity and parental diabetes. *American Journal of Epidemiology*, **113**, 144.

Koop, C.E. (1988). *The Surgeon General's Report on nutrition and health*. Department of Health and Social Security Publication 88–50210. Washington, DC.

Lew, E.A. (1985). Mortality and weight: insured lives and the American Cancer Society study. *Annals of Internal Medicine*, **103**, 1024.

McCance, R.A. (1968). The effect of calorie deficiencies and protein deficiencies on final weight and stature. In *Calorie deficiencies and protein deficiencies* (ed. R.A. McCance and E.M. Widdowson) p. 319. Churchill, London.

National Institutes of Health Consensus Development Panel (1985). Health Implications of Obesity. *Annals of Internal Medicine*, **103**, 1073.

Nomura, F., Ohnishi, K., Satomura, Y., *et al.* (1986). Liver function in moderate obesity—study in 534 moderately obese subjects among 4613 male company employees. *International Journal of Obesity*, **10**, 349.

Paul, A.A. and Southgate, D.A.T. (1978). McCance and Widdowson's *The composition of foods*, (4th edn), (ed. R.A. McCance and E.M. Widdowson), p. 418. Ministry of Agriculture, Fisheries and Food and Medical Research Council, HMSO, London.

Prentice, A.M., Black, A.E., Coward, W.A., *et al.* (1986). High levels of energy expenditure in obese women. *British Medical Journal*, **292**, 983.

Quetelet, L.A.J. (1869). *Physique sociale*, Vol. 2, p. 92. C. Muquardt, Brussels.

Reuben, A., Maton, P.N., Murphy, G.M., and Dowling, R.H. (1985). Bile lipid secretion in obese and non-obese individuals with and without gallstones. *Clinical Science*, **69**, 71.

Rolls, B.J., Van Duijvenvoorde, P.M., and Rolls, E.T. (1984). Pleasantness changes and food intake in a varied four-course meal. *Appetite*, **5**, 337.

Sims, E.A.H., Danforth, E.Jr, Horton, E.S., Bray, G.A., Glennon, J.A., and Salans, L.B. (1973). Endocrine and metabolic effects of experimental obesity in man. *Recent Progress in Hormone Research*, **29**, 457.

Stamler, R., Stamler, J., Reidlinger, W.R., Algera, G., and Roberts, R.H. (1978). Weight and blood pressure. Findings in hypertension screening of 1 million Americans. *Journal of the American Medical Association*, **240**, 1607.

Stein, Z., Susser, M., Saenger, G., and Marolla, F. (1975). *Famine and human development: the Dutch winter hunger of 1944–1945*. Oxford University Press, New York.

Stout, R.W. (1982). Hyperinsulinaemia as an independent risk factor for atherosclerosis. *International Journal of Obesity*, **6**, 111.

Thomas, B. (1988). *Manual of dietetic practice*, p. 638. Blackwell, Oxford.

Truswell, A.S. (1983). Recommended dietary intakes around the world. *Nutrition Abstracts and Reviews*, **53**, 939 and 1075.

Wadden, T.A. and Stunkard, A.J. (1985). Social and psychological consequences of obesity. *Annals of Internal Medicine*, **103**, 1062.

Walker, A., Zimmermann, M.R., and Leakey, R.E.F. (1982). A possible case of hypervitaminosis A in *Homo erectus*. *Nature*, **296**.

World Health Organization (1983). *Measuring change in nutritional status*. WHO, Geneva.

Widdowson, E.M. (1965). Chemical analysis of the body. In *Human body composition*, (ed. J. Brozek), p. 31. Pergamon Press, Oxford.

7

Infectious agents

ARNOLD S. MONTO, GENE I. HIGASHI, and CARL F. MARRS

Introduction

The concept that transmissible agents might be responsible for the production of at least certain diseases surfaced periodically during the early history of medicine, mainly based on empirical observations. The concept of contagion itself first appeared in the work, *de Contagione*, of Hieronymus Fracastorius in the sixteenth century. Progress was not made until much later, in the nineteenth century, with the work of notable epidemiologists such as Snow and Budd, and microbiologists such as Pasteur and Koch. Only in the twentieth century did rapid advances in microbiology and parasitology with identification and description of many new agents allow precise characterization of infectious diseases and mechanisms of transmission. Certain general principles were developed, such as the idea that asymptomatic infection was frequent, and that the proportion of infections resulting in symptomatic illness or pathogenicity varied with the agent and status of the host. It also became clear that the ability of an agent to survive in the environment determined its transmissibility, which in turn was usually related to its physicochemical characteristics.

In this chapter, the dynamics of the infectious process will first be reviewed. This has classically been described in terms of interaction between the agent of disease and the host, with the outcome affected by the environment. This review will be selective focusing largely on aspects of epidemiological consequence. Thereafter, the features of the agents of infectious diseases of public health importance will be discussed. This will be divided into discussion of viruses, bacteria, fungi and parasites, including protozoa and helminths. For more detailed description of these agents, the reader is directed to standard texts on the particular subject.

Dynamics of the infectious process

The cycle of infection and commonly used measurements

Several aspects of the infectious process are unique and as a consequence produce patterns of disease seen only with replicating agents. It is such replication which, in most cases, must take place for disease to occur; an exception is observed with those agents whose action involves preformed toxins. Following replication, the agent is released into the environment and then is transmitted to a new host. This new host is usually another human or may belong to another genus of animals; in the latter situation, the term 'intermediate host' is frequently used. The cycle of infection therefore involves replication, portal of exit, transmission, and portal of entry. Only rarely, as in the case of rabies in humans, do we encounter 'dead-end hosts', in which no further transmission takes place.

For infection to occur, a sufficient infectious dose must be present to make initiation of the process likely; this dose may vary depending on the portal of entry through which it is presented. For some agents, infection can take place through various portals on entry, but for many others it is limited to one. The dose required is ideally calculated as the infectious dose 50, (ID_{50}), which is the amount that will initiate infection in one-half of those exposed. The ID_{50} can be determined easily in the laboratory using experimental infection in animals; with many infectious diseases it it possible to simulate such infections under conditions similar to those in humans. However, with others no comparable animal model exists and experimental infection in human volunteers has been carried out when appropriate, as in respiratory and enteric infections (Wyatt *et al.* 1974). The ID_{50} is determined to identify which infections require a large dose for initiation, since the control measure taken may depend on the dose required. An illustration of this phenomenon is the difference in transmission patterns of enteric disease observed with *Shigella*, which has a low ID_{50} and can transmit from person to person, and that organism that causes cholera, which requires a large inoculum to initiate infection and generally does not transmit in this manner (DuPont *et al.* 1970).

There are many agents for which the ID_{50} cannot be determined, typically because there is no relevant animal model and experimental infection of humans is not possible. Often, the required dose can only be inferentially categorized as either large or small. However, for those infections that are transmitted from person to person, it is possible to determine infectivity indirectly by calculating the secondary attack rate. This rate is calculated by identifying that proportion of exposed individuals who become infected or diseased. The value, generally expressed as a percentage, if high suggests that the dose required to initiate infection may be small, and if low, the reverse. However, other factors play a role in determining the secondary attack rate, including relative resistance of the hosts and possible inactivation of the agent during transmission. Still, the value gives a valuable index to the relative infectiousness of an agent. It was, in part, recognition that the smallpox virus was not as infectious as previously supposed which led to implementation of the eradication strategy.

When the dose is sufficient to initiate infection, that infection may result in disease of varying severity or in no disease at all. Reasons for this are complex and are frequently referred to under the heading 'host factors'. The concept that infection is often to a large extent inapparent (sometimes termed 'subclinical') has classically been termed the iceberg of infection; infection without disease lies below the waterline while apparent infection rests above. In population groups, agents produce a characteristic proportion of symptomatic infections; they can thus be classified as being of greater or lesser pathogenicity, the term used to describe this proportion. For example, measles infections are nearly always accompanied by symptoms, so that the virus causing measles would be considered highly pathogenic; polio, on the other hand, nearly always produces asymptomatic infection. In the latter case, if disease does indeed occur, it may be life-threatening, indicating that dissociation of pathogenicity and severity is possible (Nathanson and Martin 1979).

Severity of diseases for which death is a potential outcome, is measured by the case fatality rate or ratio (CFR). This calculation identifies that proportion of symptomatic illnesses with a fatal outcome and is a measure of the virulence of the organism. In those diseases for which a fatal outcome is uncommon, virulence is more difficult to quantify since a specific definition of severity is required. Still, it is possible to list various diseases in order of their severity, from rabies and now acquired immune deficiency syndrome (AIDS), in which fatal outcomes appear uniform to mild self-limited illnesses such as the common cold.

Transmission mechanisms

Transmission represents that portion of the cycle of infection in which the infectious agent passes from one host to the next. As such it is often the time when the agent is most vulnerable to inactivation by physical or other environmental means. Mechanisms of transmission differ according to many characteristics of infection, especially the portals of exit and entry. Classification is based principally on the rapidity and directness of an agent's movement from one host to the next (Fox *et al.* 1970). Table 7.1 lists a standard method of classification. The major division between direct and indirect mechanisms is made on the basis of the length of time taken for transmission. Direct transmission is essentially immediate whereas indirect transmission may take place over a long period of time, with the agent in contact with the environment for much of this time. Direct transmission is subdivided into direct contact and, transmission by large droplet. In the former, the agent is passed to the new host without truly coming in contact with the environment. Examples include touching, kissing, and sexual intercourse. Agents which are highly labile can thus be transmitted by this mechanism. Also included in the direct category is droplet spread because droplets settle out rapidly, generally within 1m of the portal of exit. Large droplets thus represent an extension of the direct mechanism since they are really another form of person-to-person transmission.

Table 7.1. Classification of mechanisms of transmission

Direct transmission
Direct contact
Large droplet spread
Indirect transmission
Vehicle spread
Freshly contaminated fomites (indirect contact)
Water, milk, food, other biological products
Vector spread
Mechanical
Biological
Airborne spread
Small particle aerosols
Dusts

Vehicle transmission includes those situations in which the agent is transported to the new host on inanimate objects, or in water, or in biological products such as milk, urine, animal tissue, and various foods. As in all types of indirect transmission, the time between hosts can be long. A special situation exists with respect to fomites or inanimate personal objects. If the fomites are freshly contaminated, transmission patterns will be similar to those seen in direct transmission. Therefore this mechanism is often viewed as another form of contact transmission, in this case termed 'indirect contact'. The term 'contagious' is used collectively for infections which spread both by direct and indirect contact.

Vector transmission refers to spread through animate objects; generally the animate objects are restricted to arthropods, although some would include any animal besides humans. The arthropods typically involved are those that are haematophagous, such as biting flies, ticks, lice, fleas, and mites. Other arthropods such as flies and cockroaches may also be involved, but only in mechanical transmission. This term refers to the passive movement of the agent, on or in the vector, without any multiplication taking place. In contrast, biological transmission requires multiplication of the agent in the vector. With some pathogens, such as the viral encephalitides, no change takes place in the virus except its replication. This is very different from the situation with animal parasites, such as those causing malaria, in which an essential part of the life-cycle takes place in the mosquito. With both, however, amplification of the agent within the vector occurs, so that more widespread infection is possible than would take place through mechanical transmission.

Airborne transmission is considered a form of indirect transmission because, unlike the situation with large droplets which stay suspended for only a short time, the true microbiological aerosols can stay suspended in the air for prolonged periods because of their smaller size. These droplet nuclei generally result when large droplets evaporate to a size of less than 5 μm in diameter. Certain respiratory agents that can be transmitted by large droplets may also be transmitted by droplet nuclei if they can survive desiccation and exposure to the environment; others are transmitted only by large droplet. An example of the former is the influenza virus, which as a result of airborne transmission causes large-scale epidemics, and of the latter, rhinoviruses, which only move slowly through population groups. In addition to aerosols which come mainly from previously infected hosts, airborne transmissions can take place through dust and soil containing infectious material. Fungal infections such as coccidioidomycosis or histoplasmosis can be spread over long distances by this mechanism, as can bacterial pathogens such as *Coxiella burnetii*, the agent of Q fever.

Reservoirs—persistence of agents in the environment

While the cycle of infection for many familiar pathogens involves only humans, the cycle for others may involve an intermediate animal host; alternatively, humans may be involved only incidentally, with the main site of infection occurring in a sylvatic focus. The animal hosts in either case represent potential reservoirs where agents may reside when not infecting humans. Active infection of the host animals is required for the reservoir to operate, but production of clinical disease is not. In fact, in the absence of disease, the reservoir is most dangerous to humans, since infection would be impossible to detect without doing laboratory studies. An example of this phenomenon is Q fever infection of sheep and cattle. Control efforts must take into account the existence of an animal reservoir since, if disease transmission in humans alone is prevented, the infection will eventually re-emerge from the reservoir. Smallpox eradication was possible because the pox virus of non-human primates is distinct, and no animal reservoir for smallpox exists.

Intermediate animal hosts are required for completion of the life-cycle of many animal parasites, and are described below. With these agents, development into a new morphological form occurs in the course of movement from one human host to the next via the reservoir. For control, infection of both humans and the reservoir needs to be considered. Other agents do not change while infecting animals. The extent to which sylvatic hosts are involved determines whether or not the disease is considered a zoonosis. The greater the involvement of animals, the more they have to be taken into account when considering control. Rabies in North America is a true zoonosis with a variety of wild animals enzootically infected by the rabies virus. Nearly complete control of human infection is achieved by vaccinating domestic animals to create a barrier. In contrast, several species of *Salmonella* commonly infect humans, domestic animals, and rodents in contact with humans. Infection of humans is not required to perpetuate the infectious cycles, and control is difficult because of the number of species involved.

The host and its influence of the infectious process

A group of individuals exposed to the same dose of an infectious agent will not uniformly become infected or diseased. A variety of elements, collectively termed host factors, is responsible for this phenomenon. Many are interrelated; age is always an important factor but as age increases so does immunity from serial exposure to infectious agents. As a generality, the young are most likely to be infected. They sometimes also experience more severe disease, as with respiratory and viral enteric infections. However, polio and hepatitis A viruses are less pathogenic and less virulent in young children even though the young are highly likely to become infected if exposed. The elderly, while generally less likely to be infected with enteric and respiratory pathogens, sometimes experience more severe illness. Thus, excess influenza mortality is nearly always confined to the elderly and those with chronic conditions that affect the host's defence mechanisms (Clifford *et al.* 1977). Illness rates also differ typically by sex. Certain occupational pathogens often result in higher rates of infection in males than females. Young boys are known to have more severe respiratory illness with respiratory syncytial virus than females, even though infection rates are similar.

Immunity that is associated with resistance to infection is a discipline in itself and only certain aspects are discussed here. Of particular importance in understanding distribution of infections in populations is the role of humoral antibody in prevention. When a child is born, its mother's immunoglobulin G (IgG) antibodies pass to it transplacentally, and other antibodies as well as immune cells are ingested through mother's milk and colostrum. Because maternal antibodies are passively acquired, they are gradually destroyed so that by approximately 6 months the quantity remaining is insufficient to prevent infection. As a result, many childhood infections, such as measles, are rare in the first 6 months of life. Also, measles vaccine, since it is live attentuated, cannot be given during this period. Thus, placental transmission of antibody can be viewed as a natural form of passive immunization.

Viruses

Mechanisms of viral infection

Viruses are infectious agents that contain only one type of nucleic acid (RNA or DNA) enclosed in a protein coat of

Table 7.2. Major steps in viral replication

Early	Attachment of virus to cell Penetration and uncoating
Middle	Messenger RNA synthesis Protein synthesis Genomic nucleic acid synthesis
Late	Maturation and assembly of virion Release from cell

varied type. They are obligate intracellular parasites and cannot replicate without cells or survive for long outside the host body (Dulbecco and Ginsberg 1980). Thus, transmission is the most hazardous phase of the infection cycle. Within the host, the virus replicates by making use of the metabolic processes of the infected cell. The replication process follows a series of sequential steps common to most animal viruses, and antiviral agents can usually be identified as operating on one or more of these steps (Bachrach 1978; Koch-Weser *et al.* 1980). A summary of these steps is given in Table 7.2.

The virus first attaches to specific receptor sites on the cell membrane; lack of such receptors is an explanation for the insensitivity of certain cells to a particular virus. Penetration occurs by mechanisms which differ among viruses. Thereafter, in the uncoating process, the viral nucleic acid is freed from the coat. Again, the site of this event varies, as does its completeness. The replication itself begins with the transcription process. For DNA-containing viruses, the genome is transcribed to messenger RNA (mRNA) either by using enzymes brought into the cells, as with the poxviruses, or by using cellular RNA polymerases. The nucleic acid in certain single-stranded RNA viruses is already infectious (so-called positive stranded) and no new mRNA is needed to initiate the process; with the negative-stranded viruses mRNA is transcribed using a viral transcriptase (Raghow and Kingsbury 1976). The retroviruses represent a special case, possessing a reverse transcriptase for the production of DNA; included are oncogenic viruses as well as the virus responsible for AIDS. Transcription is followed by translation of the mRNA into viral proteins. Some of these proteins aid in producing new viral genomic RNA or DNA. Often the viral products are produced in excess of what is required for final assembly and release of the completed new virus. Assembly takes place in the cytoplasm and/or in the nucleus, and, with more complex viruses layers are laid down in a stepwise fashion. Release marks the end of the latent period of viral replication, and in the process of leaving the cells viruses may acquire a glycoprotein envelope as they bud through the cell membrane.

As mentioned above, the lack of appropriate receptors is a common cause of resistance. However, resistance can also occur as a result of intrinsic inability of the cell to support the growth of a particular virus. If infection does take place, it can have a number of effects on the cells of the host. Most important in terms of production of acute disease is cytocidal infection. Cell destruction results when the viral proteins produced early in infection shut down cellular RNA and protein synthesis. Transformation may result if susceptible cells are infected with oncogenic viruses (Vogt and Dulbecco 1960). Certain viruses which do not produce either of these effects cause a steady-state infection, a situation in which virus is produced without destruction of the infected cells (Choppin 1964). Latency, another type of persistent infection, can result following acute infection by a number of processes, including integration of viral RNA in the host genome. When an acute infection does not resolve, and virus continues to be produced by cells, the situation is described as a chronic infection.

Cell damage is the most important element in the pathogenesis of viral infection. It can be produced by the action of the virus itself, or by immunological response to the products of viral infection. In the intact host, specific, predictable organ systems demonstrate the effects of viral replication pathologically. The poxviruses are recognized by their growth in the skin but other internal organs can also be involved, since virus is initially disseminated via the bloodstream. Vireamia is not present in other localized skin infections, while infections of the respiratory or enteric tract generally involve only the surface and do not disseminate. Infections of the nervous system may be produced by direct extension, as in rabies, or they may follow more generalized infection and viraemia as in poliomyelitis.

Exact mechanisms of recovery from viral infection are largely unknown although many of the elements are recognized. At one time, it was thought that depletion of the susceptible cells was responsible, but this is now known not to be the case. Circulating antibody is produced in response to viral infection and, for diseases that last for several weeks, appears at a time when it could temporally be involved. However, most viral infections last for too short a period and have largely resolved before antibody is detected. Interferon has been proposed as a likely factor in recovery since it is produced by cells early in response to infection. Cell-mediated immune responses may also be involved; thus with the herpesvirus family, which typically persist as latent infections, depression of cell-mediated immunity, either naturally or as a result of therapy, often results in clinical recurrence. Indeed, the immunodeficient patient in whom viral infection can sometimes not be contained gives important clues as to the role of various types of immunity with specific viruses (Koprowski and Koprowski 1975).

The same factors involved in recovery are involved in prevention, and here the situation is much better defined. Humoral antibody is clearly involved in prevention of a number of infections. Passive administration of antibodies has long been known to prevent measles and hepatitis A, and more recently hepatitis B (Maynard 1978); with influenza, levels of circulating antibody measured in individuals after vaccination or natural infection correlate well with protection (Bell *et al.* 1957). In the surface infections of the respiratory, and presumably of the enteric, tract secretory antibody is

Table 7.3. Isolation techniques for viruses of public health importance

Substrate	Method
Embryonated eggs	Inoculation in amniotic, allantoic cavity, or yolk sac
Suckling mice	Intracerebral or intramuscular inoculation
Cell culture	Detection of viral growth by cytopathic effect, haemadsorption, etc.
Organ culture	Detection by measuring or visualizing virus produced

likely to be involved in the process (Rossen *et al.* 1971). The precise role of cell-mediated immunity in protection is not as well defined, but is undoubtedly of importance sometimes independently and sometimes in association with humoral antibody.

Recognition of viral infection

Recognition is based on detection of the virus by isolation or by identification of its antigens, and/or on detection of a significant antibody response. In Table 7.3 are listed common ways in which viruses are isolated (Lennette and Schmidt 1979). Embryonated or fertile hens' eggs have been employed for many years to isolate influenza viruses by amniotic or allantoic inoculation. This method is still used for vaccine production. The arthropod-borne viruses and other agents have been cultivated in the past by yolk-sac inoculation of suckling mice, although this has largely been supplanted by use of cell culture. Inoculation of suckling mice has been used for isolation of certain coxsackie viruses.

Cell cultures are the principal means for identifying viruses today in laboratories world-wide and especially in more developed countries. They may be of several varieties. Primary cell cultures result when organs or tissues taken from animals are treated to release individual cells and are placed into culture; these cells are capable of only limited further growth, and thus additional subcultivation is rarely practised. Examples are monkey and human embryo kidney or chick embryo cultures. These cultures are generally composed of a number of cell types, as expected from their origin, and they may contain adventitious agents—persistent viruses that had infected the animal from which they were derived. In contrast to these cells, which cannot be propagated well, are the continuous cell lines. These lines are immortal, that is, they can be propagated indefinitely and are mainly derived from tumours (Rafferty 1975). They generally are of a single cell type, and grow readily under proper conditions. Because of their origin, their use for production of vaccines by standard techniques is unacceptable. Finally, there are the diploid or semicontinuous cell lines, derived from normal cells, often embryonic, which maintain the normal chromosome number. They can be propagated in the laboratory but have a finite number of divisions through which they can be carried. They are of one cell type, do not contain adventitious agents, and are acceptable substrates for vaccine production. An example is the human embryo lung line, WI-38 (Hayflick 1965).

Growth of cytocidal viruses in cell culture is mainly recognized by its cytopathic effect. Many viruses produce a characteristic cytopathic effect, specific to that type, while others simply destroy cells without producing a typical microscopic effect. In still other situations, little visible damage to cells occurs, and other means for detecting the presence of virus must be employed. In the haemadsorption technique, red cells are added, which adsorb to certain viruses that bud from the viral membrane. These include influenza and para-influenza viruses as well as the coronaviruses. Other methods used to identify the presence of virus include identification of viral components with antibodies tagged with fluorescent, radioactive, or similar compounds, and visualization of virus by electron microscopy. The interference method has occasionally been used, that is inoculation of a cytocidal virus into a tube already possibly infected with an unknown specimen; the known virus will produce a cytopathogenic effect if the cell culture is not truly infected. Organ cultures, such as explants of human embryo trachea, are used ordinarily for those agents that do not replicate in cell culture (Hoorn and Tyrrell 1969). Virus is identified by visualizing it by electron microscopy or by detecting the antigen by immunological procedures.

When it is difficult to cultivate viruses, or when rapid diagnosis is desired, techniques are employed for direct detection of the agent using antibody that reacts with the viral antigen. This antibody is either itself tagged with a recognizable compound, or a tagged antibody directed against the first antibody is used (Lennette and Schmidt 1979). The tag may be a fluorescent substance, a radioisotope or a compound which, when a substrate is added, produces a colour in the enzyme-linked immunosorbent assay (ELISA) (Herrman *et al.* 1979). A limiting factor in these assays is the quantity of virus contained in a particular specimen. Small amounts of virus may be detected by cell culture, since replication takes place, but in the methods just described the virus can be recognized only if more than a specific amount of virus is contained in the original specimen.

Serological techniques that may be used to identify viral infection by detecting rises in antibody are shown in Table 7.4 (Lennette and Schmidt 1979). In neutralization tests, infectious virus is exposed to antibody and then the mixture is inoculated into cell culture or other susceptible systems. Reduction in infectivity is measured in that system by several methods (Fujita *et al.* 1975). The test is sensitive and specific but is time consuming. For viruses that haemagglutinate, the haemagglutination inhibition test can be performed. Here, the haemagglutinating potential of the virus is blocked by antibody. The test is relatively simple, and is specific and sensitive, although somewhat less than neutralization; non-specific inhibitors may be present in serum, which must be removed for accurate interpretation of the test. Complement fixation is a well-known generally available procedure suitable for most viruses. Complement-fixing antibody tends to

Table 7.4. Methods used to detect rise in antibody titre

Method	Comment
Neutralization text	Neutralization of infectivity—specific and sensitive
Haemagglutination inhibition	Suitable for only certain viruses—must remove non-specific serum inhibitors
Complement fixation	Antibody short-lived—may detect cross-reactive, antibody
Immunofluorescence	Moderately sensitive—requires appropriate microscope
Radio-immunoassay	Sensitive—requires use of radio-isotopes
Enzyme-linked immunosorbent assay (ELISA)	Sensitive—specific equipment not always necessary
Antibody diffustion in agar	Variable sensitivity—requires little specialized equipment

be short lived, and is often directed against whole groups of viruses rather than a specific type. This may make the test more valuable as a screening procedure. However, complement fixation, also tends to be less sensitive, especially in infants and young children.

Fluorescent antibody, radio-immunossay, and ELISA may all be considered together. They are carried out in a manner similar to that described above for detection of antigen, except that known virus is used as antigen for detecting antibody. Sensitivity and specificity of fluorescent antibody is often similar to that seen with complement fixation. However, radio-immunoassay and ELISA sensitivity is high and, depending on the type of antigen used, the test can be made more specific (Murphy *et al.* 1981). The ELISA test is of potential importance in public health, since no radioactive reagents are needed. All of these tests can be specific as to isotype, so that IgG, IgM, and IgA antibodies can be separately identified. Such an ability is of value, for example, in the serodiagnosis of rubella or hepatitis when only a blood specimen is available after infection (Osterholm *et al.* 1980) and the presence of IgM antibody would indicate recent infection.

Diffusion techniques in agar gels are used to detect antibody to a number of viruses. This may involve precipitation based on single radial or double diffusion, or specialized techniques involving incorporation of red cells and complement into the agar, in which the principles of complement fixation apply. Such tests are of moderate sensitivity, and can be made more specific if subunit rather than whole virus antigens are employed.

Specific viruses involved in human infections

The viruses of concern in public health are numerous, and are reviewed briefly in terms of their characteristics in the laboratory. In Table 7.5 are listed families of DNA-containing viruses of public health importance, with information on their structure. Terminology of viruses is in constant flux, and the nomenclature given may change over time (Matthews 1982).

Parvoviruses are the smallest viruses of vertebrates and are not of established significance in human disease. The adeno-associated viruses are satellite viruses which can multiply only in the presence of helper adenoviruses. There is at present no evidence that they play a role in human disease independent of the adenoviruses with which they must grow (Yates *et al.* 1981). Norwalk and similar viruses of gastro-enteritis have been called parvoviruses on the basis of their appearance, but there is little evidence that they truly belong to this family (Greenberg *et al.* 1979). The papovaviridae family includes the papilloma virus of benign human warts and the polyomaviruses. These, in turn, include the polyomavirus itself, which when inoculated into mice and other laboratory animals produces a solid tumour, and the simian virus type 40. The latter agent was present as a contaminant in early batches of inactivated poliovaccines produced in monkey kidney. Because it is a polyomavirus with oncogenic potential, the frequency of tumours in recipients has been evaluated but no clear effect has been found.

The adenoviruses are a large group of viruses mainly involved in respiratory infections of humans, as well as other infections such as keratoconjunctivitis, cystitis, and most recently gastro-enteritis (Jacobsson *et al.* 1979). The behaviour of the virus in humans is dependent on its type. Type specificity is related to the 'fibre protein' of the virus while the cross-reactive antigen is associated with the base protein. The complement fixation test, as well as others, such as ELISA, detects the cross-reactive antigen while haemagglutination inhibition or neutralization is more type specific. Growth of the viruses in permissive cell culture is slow with production of typical cytopathogenic effect. In addition, the penton protein of the virus is toxic, and causes detachment of cells from the surface on which they are growing. Human adenoviruses produce tumours when injected into laboratory animals, and this potential is also dependent on their type; there is no clear evidence that this oncogenicity occurs in natural infection, although it has obvious implications in terms of vaccine development. As is true of certain other DNA viruses, adenoviruses can produce latent infections, and prolonged shedding is characteristic.

The viruses thus far described have been non-enveloped and, since the viral envelope usually contains lipid, the non-enveloped viruses are usually resistant to lipid solvents such as ether or chloroform. Members of the herpesviridae are sensitive to these solvents. These agents in general are similar in that they all cause latent or recurrent infections. They are quite different in many other characteristics; the herpesviruses themselves are related to each other antigenically but not to other members of the family (Nahmias and Roisman 1973). They are readily grown in cell cultures of many types and produce a typical cytopathic effect. Serodiagnosis is only rarely useful because of the recurrent nature of the infection. The varicellazoster virus, in contrast, is much more difficult to grow in cell culture; it is involved in chickenpox and in

Table 7.5. DNA viruses of public health importance

Family	Size (nm)	Structure	DNA	Envelope	Members
Parvoviridae	18–26	Icosahedral symmetry	Single stranded	No	Adeno-associated viruses, ?Norwalk
Papovaviridae	45–55	Icosahedral symmetry	Double stranded	No	Papilloma (wart) virus, polyoma, simian virus type 40
Adenoviridae	70–90	Icosahedral symmetry	Double stranded	No	Adenoviruses of most animals
Herpesviridae	100–200	Envelope around icosahedral capsid	Double stranded	Yes	Herpes, varicella zoster, Epstein-Barr, cytomegalovirus
Poxviridae	230–300	Brick shaped, complex	Double stranded	Yes	Vaccinia, smallpox, cowpox, monkeypox

Table 7.6. RNA viruses of public health importance

Family	Size (nm)	Structure	RNA	Envelope	Members
Picornaviridae	24–30	Icosahedral symmetry	Single stranded	No	See Table 7.7
Flaviviridae	40–50	Spherical with fine projections	Single stranded	Yes	Formerly portion of group B arboviruses
Togaviridae	60–65	Icosahedral capsid—within envelope	Single stranded	Yes	Formerly portion of arboviruses, plus rubella with hog cholera
Bunyaviridae	90–100	Spherical—envelope with spikes	Single stranded Segmented	Yes	Formerly portion of arboviruses—California group
Reoviridae	60–80	Icosahedral double capsid	Double stranded Segmented	No	Reovirus, rotavirus, orbivirus
Coronaviradae	80–130	Spherical—club-shaped projections	Single stranded	Yes	Coronaviruses of humans—animal gastro-enteritis, mouse hepatitis
Orthomyxoviridae	80–120	Spherical or filamentous spikes on envelope	Single stranded Segmented	Yes	Influenza types A, B, and C
Paramyxoviridae	150–300	Spherical—envelope with spikes	Single stranded	Yes	See Table 7.8
Arenaviridae	85–120	Surface projections on envelope contain granules	Single stranded Segmented	Yes	Lymphocytic choriomeningitis, Lassa fever; Tacaribe group
Rhabdoviridae	175 × 70	Bullet shaped	Single stranded	Yes	Rabies, vesicular stomatitis virus
Retroviridae	100	Various	Single stranded	Yes	Oncogenic viruses, human immunodeficiency virus

herpes zoster, its recurrent form. In addition to culture techniques, the agent may be detected around vesicles or in the vesicular fluid by identifying the typical giant cells or by visualizing the virus by electron microscopy. Antibody determinations may be of use in identifying susceptibility, especially in immunocompromised children. Cytomegalovirus infections are widespread and, again, latency is characteristic. While there are cytomegaloviruses of various species, infection with a particular virus seems to be species specific. Congenital infection is the principal reason for public health importance of the agent, which can be identified serologically in the newborn by flourescent antibody or other techniques that distinguish the mother's IgG from the child's IgM antibody. The virus replicates relatively slowly in cell cultures; thus isolation is not as simple a procedure as with many other viruses. It is also possible to detect inclusions in cells on the urine, which have characteristic appearance. Epstein Barr virus was first identified by Epstein and Barr in Burkitt lymphoma cells cultured *in vivo* (Epstein *et al*. 1965). This virus has proved difficult to cultivate, and its identification by direct isolation is limited to the specialized laboratory. A number of serological procedures are available to confirm its aetiological role in cases of suspected infectious mononucleosis.

Vaccinia is the best studied of the poxviruses, while smallpox virus is the one that previously had the greatest impact on the human population. Although they do contain lipid, poxviruses are relatively insensitive to inactivation by lipid solvents, by disinfectants, or by drying, and thus the smallpox virus was well known to survive in crusts from infected individuals. The poxviruses can easily be isolated on the chorioallantoic membrane of chick embryos on which it produces typical pox-like lesions. The isolates of variola virus can be distinguished from those of vaccinia or poxviruses of animals by antigenic characteristics. Virus can also be identified as to family in vesicular fluids by electron microscopy.

The families of RNA viruses are similarly listed in Table 7.6. The picornaviruses are a large family of relatively stable, non-lipid-containing viruses causing many different diseases in humans. The term picornavirus itself indicates that they are small RNA-containing viruses. Table 7.7 lists prominent members of the family. The enterovirus and rhinovirus genera are of greatest importance in human infection and are distinguished in the laboratory by the fact that the former are

Table 7.7. Members of family Picornaviridae

Genus	Common member
Enterovirus	Poliovirus, coxsackie virus A and B, echoviruses, other enteroviruses
Rhinovirus	Rhinoviruses of humans and cattle
Cardiovirus	Encephalomyocarditis
Aphthovirus	Foot and mouth disease

acid stable and the latter are acid labile (Hughes *et al.* 1973). Thus, enteroviruses are able to pass through the stomach and infect the lower enteric tract. The polioviruses, three in type, can be cultivated in cell culture easily and quickly. Infection can also be identified serologically. At present, a recurring public health problem is distinguishing between revertent live attenuated vaccine strains causing paralysis and wild strains. This is accomplished in the laboratory by sensitive neutralization tests. Application of monoclonal antibody techniques has increased the ability to differentiate between different wild strains, so that subtypes can be identified and separated for epidemiological purposes.

Most of the type A and B coxsackie viruses can now be isolated in cell culture. However, use of suckling mice is still necessary to identify some of the 23 types. The group A viruses produce myositis and necrosis in voluntary muscles, while group B produce focal areas of degeneration in the brain plus other more distant changes. The list of diseases caused by these viruses is much longer than the classical herpangina caused by group A and pleurodynia caused by group B. As in suckling mice, coxsackie viruses generally involve the muscles or central nervous system and are characteristically related to specific types.

The name ECHOvirus is an acronym for enteric cytopathic human orphan virus, orphan because these viruses were isolated from the stools of individuals without disease. However, most ECHOviruses have been associated with illnesses, such as aseptic meningitis, febrile illness with rash, and, occasionally in institutional epidemics, enteritis. Cytopathic effects are observed on culture in primary monkey kidney or in other commonly available cell systems. Specific serological diagnosis is difficult with the ECHOviruses, as with the coxsackie viruses, since cross-reactions are common in the tests generally available.

It was decided, after 23 coxsackie and 32 ECHOviruses had been identified, that any future enteroviruses that were not polioviruses would simply be termed enteroviruses and numbered from 68 onward. This was done because of the overlapping characteristics of the genera. Enterovirus 70 is an example of these types, and is the cause of acute haemorrhagic conjunctivitis (Mirkovic *et al.* 1973). Rhinoviruses, of which there are at least 115 types, are the most frequent cause of the common cold. They can be easily isolated in a limited number of cell systems, such as WI-38 (Monto and Cavallaro 1972). Diagnosis is usually carried out by virus isolation since the antibody response is mainly type specific, and without knowing the type of the infecting strain, selection of the correct virus to use is impossible.

The arboviruses were formerly grouped together because they are transmitted biologically by an arthropod vector and exhibited sensitivity to lipid solvents. This grouping did not otherwise consider viral structure. Subsequently, the arboviruses have been subdivided at various times based on differences in their replication strategy and mode of morphogenesis and structure. Currently, there are three families which are made up almost exclusively of agents formerly considered arboviruses. The flaviviruses are mainly composed of the former group B arboviruses (yellow fever, dengue, Japanese B, and St. Louis encephalitis, etc.) and have recently been deemed a separate family. The togaviruses contain former group A arboviruses (eastern equine, Venezuelan and western equine encephalitis, Sindbis virus, and others) as well as genus rubivirus (rubella) and genus pestivirus (hog cholera) (Westaway *et al.* 1985). The third family is the bunyaviruses, which are larger structurally than the others. With all these agents viraemia typically occurs and the arthropod to become infected when it feeds. The agents multiply in the arthropod, and in some instances exhibit transovariantransmission. Isolation was previously carried out by intracerebral inoculation of suckling mice, which has largely been replaced by cell culture in many laboratories. Domestic fowl are also susceptible, and have been used as sentinel animals. The rubella virus shares structural characteristics with the togaviruses and thus has been placed in this family although arthropods are never involved in transmission. The virus does possess a haemagglutinin in common with other togaviruses.

Of the reoviruses, only the rotaviruses are common pathogens for humans. The term derives from respiratory enteric orphan virus, and orphans the reoviruses themselves have remained, since no diseases have been associated with them. The double-stranded, segmental genome is the most characteristic feature of the family. The genera are somewhat different under electron microscopy and do not cross-react immunologically. The human rotaviruses are involved in enteric illnesses, especially of young children. It is difficult to isolate these viruses in cell culture and detection by ELISA is the method of choice, replacing direct visualization by electron microscopy (Yolken *et al.* 1977).

Coronaviruses were named because of their appearance; projections give the virion the appearance of a solar corona. The viruses cause common respiratory illnesses in humans and are difficult to isolate. Organ culture of human embryo trachea has been required for isolation. Much of the information on their behaviour in humans has come from serological studies using complement fixation, haemagglutination inhibition and, more recently, ELISA methods (Monto and Lim 1974). The viruses in animals cause a variety of illnesses, including gastro-enteritis. In humans, the latter association has yet to be clearly documented.

The influenza viruses, classical agents of respiratory infection in humans, are the principal members of the orthomyxovirus family. The types are distinguished from each other on

Table 7.8. Members of family Paramyxoviridae

Genus	Members
Paramyxovirus	Parainfluenza types 1–4, mumps, Newcastle disease virus
Morbillivirus	Measles virus, canine distemper, rinderpest
Pneumovirus	Respiratory syncytial virus

the basis of a common nucleoprotein antigen. The subtypes of type A and B are divided on the basis of their haemagglutinin and neuraminidase located separately on the spikes projecting from the virion. The shifts and drifts characteristic of these glycoproteins are related to the segmented genome of the virus, which has also allowed creation of live vaccine candidates by reassortment–recombination (Odagiri *et al.* 1982). The viruses are easily isolated in embryonated eggs and cell culture, and haemagglutinination inhibition is the most commonly serological procedure.

The paramyxoviruses are larger than the orthomyxoviruses, and their genome is not segmented. As shown in Table 7.8, these viruses are varied in their make-up. The parainfluenzaviruses are of four types, cause respiratory infection, and together with mumps and Newcastle disease virus possess haemagglutinin and neuraminidase. The haemagglutinin and neuraminidase are located together on single spikes surrounding the virion, and the F protein responsible for cell fusion and haemolysis is located on other spikes. Parainfluenzaviruses can be grown readily in cell culture, and identification is facilitated by the haemadsorption technique (Shelokov *et al.* 1958). Complement fixation, haemagglutination inhibition, and neutralization tests can be used for serodiagnosis; specificity is greatest for the last technique and least for the first. The measles virus haemagglutinates but does not have a demonstrable neuraminidase. Like the parainfluenzaviruses, it haemolyses susceptible erythrocytes and on growth of cell cultures can be detected by haemadsorption. Viraemia is associated with infection, and antibody can be detected by a number of commonly available techniques. Respiratory syncytial virus is so named because of the typical cytopathic effect produced in susceptible cell culture. It is quite unstable and cultures containing it must not be subjected to freeze–thaw cycles before inoculation. No haemagglutination has been demonstrated for this agent, which is the most important cause of bronchiolitis in children, but which also produces common respiratory infections in adults. Secretory antibody appears to be of importance in preventing reinfection or modifying symptoms.

The arenaviruses have dense granules in the centre of the virion (arenasand) and many of the agents infect rodent hosts, who shed virus for long periods. The principal member of the rhabdovirus family is rabies virus, which has a unique morphology and is enveloped, rendering it susceptible to lipid solvents. Glycoprotein projections cause haemagglutination of susceptible red blood cells. The virus can be cultured in laboratory animals and in embryonated eggs, but for public health purposes the fluorescent antibody test is relied on for detection of viral antigen in animal brain or human cells. There are several related viruses, but, since they do not infect humans, the test can be considered specific. The retroviridae are a diverse group of agents classically recognized as producing leukaemia and solid tumours in mammals and birds. Their unique feature is possession of a reverse transcriptase; with this enzyme, DNA is produced which is then involved in malignant transformation of the host cell. Retroviruses have been catapulted into great prominence with the discovery of the agents of AIDS (HIV1 (human immunodeficiency virus) and HIV2) and of human T cell lymphotrophic viruses (HTLV). Data on these agents, their interrelationship, and disease expression are accumulating at a rapid rate.

Other viruses which have recently been classified include the virus of hepatitis A, which is now considered to be an enterovirus. Hepatitis B virus, on the other hand, contains double-stranded DNA. The Dane particle, which represents the complete virus of hepatitis B, is 43 nm in diameter and is enveloped (Purcell *et al.* 1973). This virus, as well as the related animal hepatitis virus, is now considered a hepadnavirus. Of additional interest is the unusual virus of delta hepatitis which cannot replicate in the absence of co-infection with hepatitis B. The virus of Ebola has a unique snake-like structure, shared by Marburg fever. The virion contains RNA and may represent a new family, termed filoviridae.

With the identification of new and exotic diseases in various parts of the world, it is probable that the list of agents involved in human disease will continue to expand and will not be restricted to unusual diseases. For example, current work with non-cultivable viruses involved in enteric illnesses continues to identify new agents. It is also probable that additional chronic diseases will be found to have an infectious aetiology, either resulting from an unusual manifestation of a common infection, such as illustrated by subacute sclerosing panencephalitis and measles, or from a new virus specific to the process.

Vaccines will continue to be a principal means of prevention of most infectious diseases. New technologies, such as the ability to clone genes coding for specific portions of a virus, will allow production of purified antigens of high titre, which will then expand the diseases for which such means of prophylaxis are possible. Live attenuated vaccines are also to be introduced and will have greatest potential against surface infections, in which local immunity is critical. Rapid viral diagnosis, often using monoclonal antibodies, will enable early identification of circulating agents. This in turn should stimulate the search for additional antiviral drugs, or other means of aborting outbreaks already in progress. Thus a number of developments at the molecular level are likely to be employed practically for the control of infectious disease transmission.

Bacteria

Normal, healthy individuals have extremely large numbers of bacteria existing on or in them. Adult humans have about

Table 7.9. Gram-positive pathogenic bacteria

Organism	Morphology	Diseases*
Bacillus anthracis	Spore-forming rod	Anthrax (cutaneous or pulmonary)
Bacillus cereus	Spore-forming rod	Food poisoning
Clostridium botulinum	Anaerobic spore-forming rod	Botulism
Clostridium difficile	Anaerobic spore-forming rod	Antibiotic-associated pseudomembranous enterocolitis
Clostridium perfringens	Anaerobic spore-forming rod	Gas gangrene, bacteraemia, food poisoning
Clostridium tetani	Anaerobic spore-forming rod	Tetanus
Corynebacterium diphtheriae	Rod	Diphtheria
Listeria monocytogenes	Micro-aerophilic rod	Listeriosis
Staphylococcus aureus	Cocci	Cutaneous abscess, endocarditis, pneumonia, enterocolitis, food poisoning, skin exfoliation
Streptococcus mutans	Cocci	Dental caries
Streptococcus pneumoniae	Cocci	Pneumonia, sinusitis, otitis media
Streptococcus pyogenes	Cocci	Pharyngitis, scarlet fever, otitis media, meningitis, peritonitis, pneumonia, acute glomerulonephritis, rheumatic fever
Streptococcus viridans	Cocci	Endocarditis

* Does not include all clinical manifestations.

10^{13} cells, and have about 10^{14} bacteria inhabiting their body. Microbial species that commonly inhabit areas of the body are called normal flora and help to prevent colonization by pathogenic organisms. Most other regions of the body, including the blood and lymphatic systems, are considered to be normally sterile with no normal flora. Compared with the very large number of different bacterial species that make up the normal flora, relatively few bacteria are capable of causing disease. Tables 7.9–7.11 list some of the more important pathogenic bacteria of public health interest and their salient features.

The most successful microbial parasites tend to get what they need from the infected host with minimal damage. If an infection is crippling or lethal, there will be a reduction in the numbers of the host species and thus in the numbers of the micro-organism. Thus, although a few species cause disease in a majority of those infected, most are comparatively harmless, causing either no disease or disease in only a small proportion of those infected. Well-established infectious agents have generally reached a state of 'balanced pathogenicity' in the host and cause the smallest amount of damage compatible with the need to enter, multiply, and be discharged from the body.

Those attributes that allow bacteria to overcome the host defences and cause disease are called virulence factors. Overall bacterial virulence is the result of many different virulence factors (Sparling 1983). The expression of virulence factors is often strictly regulated by the bacteria. Thus, pathogenic bacteria grown outside the host often display very different phenotypes than when *in vivo*. Although normal host defences can control many organisms of relatively low virulence, immunocompromised patients are frequently infected by relatively non-pathogenic organisms. An organism that causes infection after a breakdown in normal host defence mechanisms is called an opportunistic pathogen.

In the following sections, we discuss the role of normal flora in disease, and then examine various properties of pathogenic bacteria which may allow them to resist the host defence mechanisms and cause disease.

Role of normal flora in disease

Sometimes, members of the normal flora can be pathogenic. In the presence of the high-sugar diet of modern societies, *Streptococcus mutans* has to be considered normal flora of the teeth. However, it and the complex bacterial colonization called dental plaque that its presence allows make dental caries the most prevalent infectious disease of developed societies. The normal flora of the intestines, when introduced by abdominal injury to usually sterile regions of the body, can act as opportunistic pathogens and cause disease. This kind of transformation from normal flora to pathogen can also occur if there is a breakdown in the normal host defence systems.

Commensals (the residents of normal flora) can play an important role in helping to prevent infection by pathogenic bacteria. The skin is one of the most important general defences against infection, acting as a natural barrier to micro-organisms. Micro-organisms other than commensals are soon inactivated when in contact with the skin. The primary cause of this inactivation is the sebum, a fatty lubricant material secreted by sebaceous glands of the skin. The fatty acids in the sebum (which cause the skin pH to be about 5.5) and by-products of the sebum produced by the commensal species are potent antimicrobial agents. The vagina has no particular cleansing mechanism; however, from puberty to menopause, the vaginal epithelium contains glycogen which is metabolized by commensal lactobacilli to produce lactic acid. The lactic acid gives a vaginal pH of 5.0, and together with other products of metabolism inhibits colonization by all but the lactobacilli and a select number of other bacterial species. The large intestine contains nearly 10^{14} bacteria consisting of several hundred (mostly anaerobic) bacterial species and strains, which coexist without one or a few becoming dominant and displacing the others. As with other complex ecological systems, the initial microbial types inhabiting a baby or a previously 'germ-free' animal change the environment and a succession of different microbial communities progresses until a stable climax stage is reached.

The indigenous microflora of a given body site, being an ecosystem in climax stage, has a stable composition. Stability implies that invading micro-organisms which enter the site from the environment are prevented from colonizing it. If the invader happens to be a pathogen, this effect constitutes a potent host defence mechanism, and has been called 'bacterial antagonism', 'bacterial interference', and 'colonization resistance'. When the normal flora is eliminated, as through the prolonged use of broad spectrum antibiotics, this colonization resistance is lost (van der Waaij 1984). Pseudomembranous colitis can occur after the administration of clindamycin or other antibiotics, often caused by growth of *Clostridium difficile*. In another example, of 14 volunteers given 1000 streptomycin-resistant *Salmonella typhi* by mouth, none developed disease. However, when the antibiotic streptomycin was given at the same time, 25 per cent developed disease. Streptomycin probably promoted infection by its bacteriostatic action on commensal intestinal microorgansisms (Mims 1982).

Extracellular products

Exotoxins are substances released from growing bacteria (generally proteins) which have direct toxic effects on eucaryotic organisms. Enterotoxins are one class of exotoxins which affect the intestinal cells, causing secretion which leads to diarrhoea. Endotoxins are portions of the bacterial cell walls or membranes which can have a toxic effect on the host when present in large amounts. Exotoxins are discussed in this section, while endotoxins and sepsis will be discussed later.

For some bacterial diseases, exotoxin production is the primary cause of disease. In the most extreme cases, toxins can cause all disease symptoms in the absence of live

Table 7.10. Gram-negative pathogenic bacteria

Organism	Morphology	Diseases*
Acinetobacter calcoaceticus	Short rod	Nosocomial infections: pulmonary, urinary tract infections, septicaemia
Bacteroides fragilis	Anaerobic rod	Abscess, peritonitis, endocarditis
Bartonella bacilliformis	Coccobacillus	Bartonellosis
Branhamella catarrhalis	Cocci	Otitis media, pneumonia
Bordetella pertussis	Small rod	Whooping cough
Brucella species	Rod	Brucellosis
Campylobacter fetus ssp *jejuni*	Micro-aerophilic curved rod	Gastro-enteritis
Escherichia coli	Rod	Urinary tract infections, diarrhoea, septicaemia, meningitis, nosocomial infections
Francisella tularensis	Rod	Tularaemia
Haemophilus influenzae	Small rod	Meningitis, septicaemia
Kingella kingae	Short rod	Osteomyelitis, endocarditis
Klebsiella pneumoniae	Rod	Pneumonia, diarrhoea, nosocomial infections
Legionella pneumophila	Coccobacillus	Legionnaire's disease
Neisseria gonorrhoeae	Cocci	Gonorrhoea
Neisseria meningitidis	Cocci	Septicaemia, meningitis
Proteus species	Rod	Urinary tract infections
Pseudomonas aeruginosa	Rod	Urinary tract infections, burn sites, septicaemia
Salmonella cholerae-suis	Rod	Septicaemia
Salmonella enteritidis	Rod	Gastro-enteritis
Salmonella paratyphi	Rod	Paratyphoid fever
Salmonella typhi	Rod	Typhoid fever
Salmonella typhimurium	Rod	Gastro-enteritis
Serratia marcescens	Rod	Nosocomial pneumonia, urinary tract infections
Shigella species	Rod	Dysentery
Streptobacillus moniliformis	Anaerobic rod	Rat-bite fever
Vibrio cholerae	Curved rod	Cholera
Vibrio parahaemolyticus	Curved rod	Food poisoning
Yersinia enterocolitica	Rod	Diarrhoea
Yersinia pestis	Rod	Bubonic and pneumonic plague

* Does not include all clinical manifestations.

Table 7.11. Other pathogenic bacteria

Organism	Morphology	Diseases*
Borrelia burgdorferi	Spirochaete	Lyme disease
Borrelia recurrentis	Spirochaete	Relapsing fever
Chlamydia trachomatis	Very small, varies during development cycle	Trachoma, conjunctivitis, urethritis, pelvic inflammation (female), lymphogranuloma venereum
Coxiella burnetii	Small coccobacillus	Q fever
Mycobacterium leprae	Small rod	Leprosy
Mycobacterium tuberculosis	Small rod	Tuberculosis
Mycoplasma pneumoniae	Very small, spherical to filamentous	Pneumonia
Rickettsia akari	Small coccobacillus	Rickettsialpox
Rickettsia conorii	Small coccobacillus	Boutonneuse fever
Rickettsia prowazekii	Small coccobacillus	Louse-borne typhus, Brill–Zinsser disease
Rickettsia rickettsii	Small coccobacillus	Rocky Mountain spotted fever
Rickettsia tsutsugamushi	Small coccobacillus	Scrub typhus
Treponema pallidum	Anaerobic spirochaete	Syphilis
Treponema pertenue	Anaerobic spirochaete	Yaws

* Does not include all clinical manifestations.

bacteria. More often, bacteria must be growing in the host to produce the toxin(s) which may then act either locally or systemically. Many bacterial exotoxins have a two-component construction: the A chains possess the biological activity, whereas, the B chains mediate receptor binding and entry of the A chain into the cell (Middlebrook and Dorland 1984). The receptors utilized by bacterial exotoxins are often the same ones used in cell uptake of hormones. For example, cholera toxin binds to the GM_1 ganglioside which may also be involved in the receptor pathway for the hormone thyrotropin (Eidels and Hranitzky 1985).

The modes of toxic action are diverse. Botulinum neurotoxin produced by *Clostridium botulinum* is one of the most toxic substances known, 1g being enough to kill 10^{10} mice (Gill 1982). It is an example of a toxin alone causing all the disease symptoms in the absence of live bacteria. The toxin causes paralysis by acting on the peripheral nervous system and interfering with the release of acetylcholine at cholinergic synapses and the neuromuscular junction (Habermann and Dreyer 1986). Tetanus toxin is another neurotoxin of extreme potency (Habermann and Dreyer 1986), so much so that patients recovering from tetanus must still be immunized. If they had received an amount of toxin sufficient to become immunized against it, the toxin would have killed them. Another class of toxins interferes with normal regulation of adenylate cyclase, an enzyme involved in production of cyclic AMP, (Saelinger 1985). Cholera toxin causes an elevation of intracellular cyclic AMP levels, which in turn results in loss of water and electrolytes through the intact epithelial cells and into the small intestine. A profuse watery diarrhoea results with death ensuing in severe cases unless oral rehydration therapy is performed to replace lost water. Another class of toxins inhibits protein synthesis (Saelinger 1985). Elongation factors 1 (EF1) and 2 (EF2) are both involved in the elongation of polypeptide chains (Moldave 1985). Diphtheria toxin is so efficient at stopping protein synthesis by enzymically inactivating EF2 that a single molecule of subunit A in the cytoplasm is sufficient to kill a cell (Burns *et al.* 1986). The heart and peripheral nerves are particularly susceptible, leading to myocarditis and neuritis. Shiga toxin inhibits protein synthesis, but, instead of EF2, its target is EF1 (O'Brien and Holmes 1987). Shiga toxin has been shown to be involved in the fluid accumulation which occurs in the watery diarrhoea sometimes seen in shigellosis. One theory is that this is caused by the toxin primarily binding, being internalized, and killing (by preventing protein synthesis) the mature absorptive epithelial cells of the intestines. Thus, unlike cholera toxin which causes increased fluid secretion, shiga toxin causes diarrhoea by blocking normal fluid reabsorption.

Many genes that encode toxins, as well as other virulence genes, are found on extrachromosomal elements. Some are carried on bacterial viruses which confer the ability to make the toxin on those strains in which the bacterial virus becomes a stable part of the bacterial genome (a process called lysogenization). Examples include production of diphtheria toxin by *Corynebaterium diphtheriae* lysogenized with the bacteriophage beta (Groman 1984), and the shiga-like toxins (SLT I and SLT II) carried on bacteriophage lysogenic for *Escherichia coli* (O'Brien and Holmes 1987). Small, generally circular, independently replicating pieces of DNA called plasmids have also been found to be the sources of toxin production. *E. coli* LT toxin genes are present on large, transmissible plasmids called Ent, which range in size from 90 to 135 kilobases (So *et al.* 1978). The anthrax tripartite toxin of *Bacillus anthracis* is carried on large plasmids ranging in size from 90 to 200 kilobases (Mikesell *et al.* 1983). Toxin genes have also been found on transposons (Lee *et al.* 1985), which are small pieces of DNA capable of moving from one location on a plasmid or chromosome to different locations on the same replicon or on to different plasmids, bacteriophages, or chromosomes.

In addition to exotoxins, other secreted bacterial products have been implicated as virulence factors. One group is the cytolysins, which are agents capable of causing lysis of mammalian cells. The most commonly studied subgroup of cytolysins is the haemolysins. There are three known mechanisms by which cytolysins act against the host cell membranes. Class 1 contains molecules that act as detergent-like agents; class 2 is made up from enzymes that hydrolyse the phospholipids of the cell membrane; and class 3 is proteins that form pores in the membrane (McCartney and Arbuthnott 1978).

Extracellular proteases are also thought to be involved in virulence. Some act directly against the molecules of the immune system, including a variety of IgA_1 proteases that specifically cleave human IgA_1 (Mulks 1985). Two different proteases of *Pseudomonas aeruginosa*, an elastase and an alkaline protease, have been shown to be involved in virulence (Morihara and Homma 1985).

Adherence

Bacterial adhesins bind specific receptors present on some host cell types. Specific adherence plays two important roles: (a) it allows the bacteria to resist the flushing and cleansing mechanisms that protect many epithelial surfaces; and (b) it determines the site of infection by facilitating specific surface-to-surface interaction between the bacteria and the host epithelium. Adhesins are often associated with pili (also called fimbriae), which are long filamentous extracellular appendages composed of up to 10 000 polymerized protein subunits called pilins. Due to their small diameters (2–8 nm), pili can only be seen using an electron microscope. For type 1 and P-pili of *E. coli*, it has been shown that the adhesin proteins are separate from the pilin subunit proteins. The adhesins of P-pili, PapG, and two other minor pilins, PapE and PapF, are localized at the tip of the pilus (Lindberg *et al.* 1987). Afimbrial adhesins help bacteria attach to eucaryotic cells in the absence of pili (Labigne-Roussel and Falkow 1988). In some cases the receptors on the eucaryotic cells to which bacterial adhesins attach have been determined. The binding of a bacterium to two or more erthyrocytes causes visible aggregation of the red blood cells, a process called haemagglutination. *E coli* type 1 pili were found to cause haemagglutination which was inhibited by the presence of the sugar mannose, and thus was called mannose-sensitive haemagglutination. This sensitivity implied that type 1 pili bind to mannose-containing receptors. Some pyelonephritic *E. coli* strains were shown to have pili which had the P blood group antigen as receptors. The P blood group glycolipids contain a digalactoside moiety, α-D-GalP(1–4) β-D-Galp, and the presence of a synthetically made form of the sugar inhibits the binding of P-specific *E. coli* to uro-epithelial cells (Mooi and deGraaf 1985).

Antibodies that recognize and bind to adhesins and/or pili will block the disease process, either by directly preventing binding of the bacteria to the host cells, or by increasing the ability of complement or phagocytes to destroy the bacteria. Some bacteria avoid this host response by antigenically varying their pili. A single strain of *Neisseria gonorrhoeae* can express a huge array of antigenically distinct pilus types. It does so by a complex set of recombinational events between the expressed complete pilin gene and a large number of partial pilin gene sequences scattered throughout the bacterial chromosome (Meyer 1987). The presence of pili on the bacteria can result in increased phagocytosis even in the absence of antibodies. This means that, while the presence of pili can be an important virulence factor in some host environments, their absence can be equally important in other host environments. In *Proteus mirabilis* infections, only piliated bacteria adhere well to uro-epithelial cells and readily produce ascending urinary tract infection in an animal model. But when the route of inoculation is via the bloodstream, the piliated bacteria are ingested by phagocytes, and thus, only non-piliated bacteria are able to produce haematogenous pyelonephritis (Silverblatt 1974). *E. coli* type 1 pili undergo phase variation in which a small region of DNA containing the promoter for the pilin gene is switched on and off by means of a DNA inversion event (Abraham *et al.* 1985).

Invasion and intracellular growth

While some pathogenic bacteria cause disease without ever progressing past attachment to the eucaryotic cell surfaces, many enter the epithelial cells, either to live and grow there, or to pass through into the blood and lymphatic systems. While entry into macrophages and other 'professional' phagocytes can occur by the normal pathway of phagocytosis, 'non-professional' phagocytes, such as epithelial cells, do not normally ingest bacteria. Therefore, bacteria have developed special methods of inducing uptake by non-professional phagocytes. *Yersinia pseudotuberculosis* (a close relative to *Y. pestis*, the agent of plague) has a large protein called invasin present on the bacterial cell surface which allows it to enter non-professional phagocytic cells. When the gene for invasin (*inv*) was cloned into *E. coli* K12, the *E. coli* bacteria gained the capability of invading epithelial cells in culture (Isberg *et al.* 1987). *Legionella pneumophila* enters eucaryotic cells by a special coiling phagocytosis mechanism (Horowitz 1984). This occurs even when the *L. pneumophila* have been killed. It is mediated by specific surface structures on the bacteria, since other bacteria simultaneously ingested still enter phagocytes by normal phagocytosis. In coiling phagocytosis, long phagocyte pseudopods coil around the bacterium as it is internalized. *Chlamydia* enter either professional or non-professional phagocytes by a unique mechanism that is not inhibited by cytochalasin B treatment of the phagocytic cell. Entry of the infectious elementary bodies into host cells is not prevented by inactivating the bodies with ultraviolet light or by inhibiting their synthesis of macromolecules with antibiotics. It has been suggested by some investigators that elementary bodies can enter phagocytes, such as macrophages, by two different endocytic pathways. Infectious elementary bodies enter via a parasite-specific pathway involving receptor-mediated endocytosis, whereas heat-inactivated elementary bodies are taken in by a host-specific route.

The 'professional' phagocytes involved in defending the body from pathogenic micro-organisms include polymorphonuclear leukocytes (mostly neutrophils, and some eosinophils), monocytes, and macrophages. Most unencapsulated bacteria readily attach to neutrophils and macrophages. Most bacterial pili increase attachment frequency. Attachment is greatly stimulated by the binding of complement and/or antibodies to the bacteria. This is mediated by receptors present on the phagocytic cells, which recognize the C3b component of complement, or the Fc portion of the antibody molecule. Attachment of bacteria to the phagocyte surface does not necessarily result in ingestion. For example, encapsulated *E. coli* attached to the macrophages or polymorphonuclear leukocytes by concanavalin A are not ingested unless antibacterial antibody is added. Bacteria coated with IgG antibody or with antibody and complement are said to be opsonized. C3, the third component of complement, is the major phagocytosis-promoting molecule. Ingestion of opsonized particles occurs by a 'zipper mechanism' of phagocytosis. All phagocytic cells contain membrane-bound organelles located in the cytoplasm which contain microbicidal and hydrolytic digestive enzymes and other antimicrobial proteins. The bacterium surrounded by plasma membrane which has been ingested into the phagocytic cell is said to be in a 'phagosome'. Lysosomes fuse with the phagosome, delivering their contents into it, a process termed degranulation. This produces the phagolysosome, within which the pH rapidly drops, reaching a low point of between 3.0 and 6.0 depending on the species. Within the phagolysosome both oxygen-dependent and oxygen-independent killing mechanisms exist for eliminating bacteria.

Once pathogenic bacteria enter a eucaryotic cell they must avoid the above killing mechanisms. Three different pathways of intracellular bacterial survival have been discovered. Some bacteria, including *Shigella flexneri, Rickettsia*, and *Mycobacterium leprae*, escape from the phagosome into the cytoplasm before the fusion with the lysosomes. *Shigella* uses a haemolysin to accomplish its escape (Sansonetti *et al.* 1986), and it has been proposed that *Listeria monocytogenes* also escapes due to the activity of the haemolysin listeriolysin (Portnoy *et al.* 1988). Other bacteria, including *Chlamydia, Legionella pneumophila*, and *Mycobacterium tuberculosis*, inhabit phagosomes that fail to fuse to lysosomes (Moulder *et al.* 1985). A final class of bacteria, including *Salmonella typhimurium, Coxiella burnetii, Yersinia pestis,* and *Mycobacterium tuberculosis* is capable of surviving within the phagolysosome. *C. burnetii* resists the actions of lysosomal enzymes by unknown means and actually grows much better at low pH (5.0 and below) than at pH 7.0.

Dissemination

Bacteria that enter the bloodstream or lymphatic system face a variety of both immune and non-immune host defence systems. The major non-immune mechanism for controlling bacteria is probably by regulation of the amount of free iron available for bacterial use. The iron content of human plasma is high—about 2×10^{-5} M—but the amount of free iron is substantially less than even the low amounts present normally in aerobic environments, being of the order of 10^{-18} M. This extremely low concentration of free iron is insufficient for the growth of bacteria, so only bacteria capable of obtaining host-sequestered iron can grow. Plasma contains complexes of iron as haptoglobin-haem, ceruloplasmin, ferritin, lactoferrin, and transferrin. To acquire the necessary iron, most pathogenic bacteria make ferric-ion-specific ligands known as siderophores (Greek, 'iron bearers'), which bind iron with very high affinities (Finkelstein *et al.* 1983). Several hundred different siderophores have been identified from micro-organisms. These molecules fall into two main groups: those based on phenol catechols, such as enterobactin made by *E. coli*; and those based on hydroxamic acid, such as ferrichrome. The general feature of all siderophores is that they are capable of forming a 'molecular cage', at the centre of which ferric iron is held as part of a six-co-ordinate, octahedral complex. Some bacteria lack siderophores, but can obtain iron directly from host molecules. For example, *Neisseria meningitidis* has specific protein receptors that obtain iron directly from lactoferrin and transferrin (Schryvers and Morris 1988).

An important virulence trait of bacteria is serum resistance. This refers to the ability of the bacteria to avoid being killed by complement (either with or without the help of antibodies). In Gram-negative bacteria, the most important serum resistance factor is the lipopolysaccharide which is constructed from a lipid component, a core carbohydrate, and a carbohydrate antigen called O antigen. The membrane attack complex, which contains complement components C5b,6,7,8, and 9, normally kills bacteria when it inserts into the outer membrane. This complex does form on the surface of the O antigen-containing bacteria, but it fails to insert into the outer membrane and is instead released from the cell. This probably occurs because the O antigens cause the membrane attack complex to form at the end of the polysaccharide chains, a significant distance external to the outer membrane, so that when the complex forms with its increased hydrophobic nature, it fails to insert in the now too distant outer membrane and instead is released from the polar O antigen (Joiner *et al.* 1982). Lipopolysaccharide has another important role in disease in that its lipid component, called lipid A, is the active component of endotoxin. When endotoxin in large quantities is released by killed bacteria into the bloodstream it induces a septic reaction. Estimates of Gram-negative rod bacteraemia in the US range from 70 000 to 300 000 cases each year. Shock complicates about 40 per cent of these cases and has a case fatality rate ranging from 40 to 90 per cent. Therefore, Gram-negative rod bacteraemic shock results in about 20 000–60 000 deaths each year (Parker and Parrillo 1983).

For all pathogenic bacteria unable to survive after phagocytosis by macrophages or other professional phagocytes, avoidance of phagocytosis is essential for survival. The most widespread and well studied antiphagocytic factors are polysaccharide capsules (Timmis *et al.* 1985). While both encapsulated and unencapsulated forms of virulent bacteria are found on the mucous membranes or other body surfaces of healthy individuals, the organisms isolated from the blood and/or tissues are almost all encapsulated. Only some types of capsular polysaccharides within a given species are associated with invasive diseases. For example, of the six known capsular polysaccharides of *Haemophilus influenzae*, only type b is responsible for the majority of invasive human disease (mostly meningitis) caused by this bacterial species. Of more than 100 capsule types of *E. coli*, only a few (e.g. K1, K5, K12) are associated with pathogenic strains. The fact that capsular polysaccharides of similar or identical chemical structure are often found in unrelated pathogens is consistent with the view that only a few capsular types are important in bacterial virulence. Unencapsulated bacteria are susceptible to the bactericidal and opsonizing activity of complement alone. Their encapsulated counterparts, however, are resistant to the direct effects of complement and require an intact immunoglubulin to initiate these protective activities. Inhibition of phagocytosis by the capsule of pathogenic bacteria is probably due to the prevention of opsonization. Specific anticapsular antibodies confer disease protection. Capsules most strongly correlated with virulence tend to be least immunogenic. The *E. coli* K1 polysaccharide, which is composed of polymers of α-(2–8) residues of N-acetyl neuraminic acid, is very poorly immunogenic, perhaps because it is not recognized as foreign due to the fact that these residues are found in gangliosides and in most mammalian serum and cell membrane glycoproteins.

Another antiphagocytic factor is the streptococcal M protein. This is a fibrillar molecule on the surface of group A streptococci. It is composed of two predominantly α-helical protein chains assembled in a coiled coil and extends nearly 60 nm from the cell surface. Streptococci lacking M protein are efficiently opsonized by the alternative complement pathway and in consequence are rapidly ingested and killed by professional phagocytes. Streptococci possessing surface M protein are neither opsonized nor ingested unless antibody to the M protein is present. The M protein-mediated resistance to phagocytosis may be due to prevention of opsonization of the streptococcal cells. M proteins bind fibrinogen present in the blood and this binding prevents opsonization of the M protein-containing bacteria (Whitnack and Beachey 1985). Recently it was reported that M protein-containing bacteria bind to control protein factor H from the alternative complement pathway. It has been suggested that this binding may inhibit effective opsonization (Horstmann *et al.* 1988). M proteins play another role in human disease. Acute rheumatic fever appears to occur due to an auto-immune phenomenon induced by the similarity or identity of certain streptococcal M protein antigens with human tissue antigens, especially those of the heart (Dale and Beachey 1985).

Resistance to antibiotics

The prevalence of transferable plasmids with a wide host range and carrying multiple drug-resistance genes, many of which are also part of transposons, has greatly increased the spread of drug resistance among important pathogenic bacteria (Davies 1981). Reservoirs of resistance plasmids are selected for and maintained in many hospital environments due to the general prevalence of antibiotics. New patients become colonized with the resident hospital resistant strains and these strains may either cause infections or their antibiotic resistance can be transferred to other strains that cause infections. In some hospitals, frequent use of gentamicin has led to the development of resistance to gentamicin in over 80 per cent of *Serratia* strains. These resistances are often due to multiple drug-resistant plasmids; on occasion, strains of *Serratia, Klebsiella*, or *Pseudomonas* resistant to all available antibiotics have been isolated, thus making the treatment of infections caused by these strains extremely difficult. The common use of low doses of antibiotics in animal feed to improve animal growth and weight gain has now been shown to result in the selection of drug-resistant bacteria which can then infect humans. The most clear-cut case of this is the selection for drug-resistant *Salmonella* which has led to epidemics of salmonellosis from drug-resistant strains (Holmberg *et al.* 1984).

Diagnosis of bacterial disease

The diagnosis of bacterial diseases has, for the most part, been dependent upon the cultivation and identification of the pathogen based on growth requirements, Gram-stain morphology using microscopy, and ability to carry out different biochemical reactions. Direct detection and identification of bacteria in clinical specimens is starting to become more practical. Two different approaches have been taken, one based on antibodies, the other on DNA hybridization. Rapid tests utilizing antibodies now allow direct detection of streptococcal group A infections from throat swabs. While there are still sensitivity problems in using DNA probes directly on clinical samples, there is hope that a new DNA amplification technique called polymerase chain reaction may prove to be useful in enhancing the signals from samples which may have a relatively small number of the bacteria of interest (Marx 1988).

Fungi

Fungi are divided into moulds and yeasts. Table 7.12 summarizes the important species responsible for human disease. Traditionally, *Actinomyces* and *Nocardia* have been grouped with the mycotic agents although they are now generally accepted as Gram-positive bacteria.

Infection in all cases is acquired by exposure to the agents by direct contact with soil containing fungi or by inhalation. It is thus understandable that mycoses are primarily cutaneous, mucocutaneous, and pulmonary with respect to sites of infection. In addition, the immunocompromised host is at major risk of severe, systemic mycotic infections as a result of genetic immune deficiencies, severe malignancies, various acquired immune defects, and chemotherapy. In most cases, fungi are not transmitted from patient to patient, the major exception being those that cause dermatophytosis. Although the mycoses are far less frequent than viral, bacterial, protozoal, or helminthic infections, they remain a significant cause of morbidity and mortality worldwide. Some of the agents are restricted in distribution to certain foci in the world, but the majority are found world-wide.

Precise diagnosis remains difficult since many smaller laboratories and many less developed countries do not have the facilities or trained staff to isolate and identify fungi. Thus the diagnosis depends largely upon the clinical picture and identification of characteristic forms of the fungi in properly stained histological sections of biopsy tissues and smears of exudates from lesions. The interested reader should consult specific texts of medical mycology and infectious diseases (Al-Doory 1975; Binford and Connor 1976; Emmons *et al.* 1985; Mandel *et al.* 1985; Rippon 1982) for details on each agent and the disease it causes.

Protozoa

Protozoa are single-celled eucaryotic organisms found widely distributed throughout the world in both parasitic and free-living forms. The parasitic forms are agents of diseases, many of which, e.g. malaria, have afflicted humans since ancient times. Their distributions are shown in Table 7.13.

The protozoa that cause significant morbidity in developed

Table 7.12. Major mycoses of public health importance

Agent	Disease	Primary sites involved	Diagnosis
*Actinomyces israelii**	Actinomycosis	Cervicofacial, thoracic, abdominal tissues	Gram stain of 'sulphur' granule in biopsy of lesion
*Nocardia asteroides**	Nocardiosis	Lungs, brain, meninges	Gram stain of biopsy of lesion
*Nocardia brasiliensis** *Streptomyces somaliensis** *Streptomyces madurae**	Actinomycetoma	Foot, hand	Gram stain of granule in biopsy of lesion
Madurella mycetomi *Acremonium* *Phialophora*	Eumycetoma	Foot, hand	Fungus stain of biopsy of lesion
Aspergillus fumigatus (moulds)	Aspergillosis	Lungs, other organs	Characteristic hyphae and conidiophores on fungus stain of tissue section
Rhizopus *Mucor* *Absidia* (moulds)	Mucormycosis	Various organs	Characteristic hyphae in tissue sections. Culture required for specific identification
Blastomyces dermatitidis	North American blastomycosis	Various organs	Characteristic yeast forms in tissue sections and exudates. Culture confirms identification
Paracoccidioides brasiliensis	South American blastomycosis	Lungs	Characteristic yeast forms in tissue sections. Culture confirms identification
Candida albicans Other *Candida* species	Moniliasis, thrush, candidiasis	Cutaneous, mucocutaneous, gastro-intestinal infection, other organs	Characteristic yeast forms, hyphae, and pseudohyphae in tissue sections. Culture required for specific identification
Coccidioides immitis	Valley fever, coccidioidomycosis	Lungs primarily, but other organs may be involved	Characteristic sporangium in tissue sections. Culture of tissue aspirates
Histoplasma capsulatum	Histoplasmosis	Lungs	Characteristic yeast forms intracellular in mononuclear phagocytes in tissue sections
Histoplasma duboisii	African histoplasmosis	Skin, bones	Characteristic intracellular yeast forms. Culture aids in specific identification
Sporothrix schenckii	Sporotrichosis	Skin, subcutaneous tissue, lymph nodes	Characteristic yeast forms and asteroid bodies in tissue sections, but frequently not found. Culture of tissue or purulent material is usually positive
Cryptococcus neoformans	Cryptococcosis, torulosis, European blastomycosis	Central nervous system, lungs	Smear and culture of cerebrospinal fluid
Phialophor species *Cladosporium* species	Chromomycosis	Skin, subcutaneous tissue	Culture of tissue or purulent material required for species identification
Trichophyton species *Microsporum* species *Epidermophyton* species	Dermatophytosis, tinea	Skin, nails, scalp hair	Microscopic examination of skin scrapings or hair reveals organisms. Culture provides specific identification

* Bacterial species. These agents and their diseases have traditionally been included with the mycoses.

Table 7.13. Geographical distribution and prevalence of major parasitic protozoa

Parasite	Distribution	No. infected (thousands)*
Plasmodium falciparum *P. vivax* *P. malariae* *P. ovale*	Africa, Asia, Latin America	800 000
Trypanosoma brucie gambiense *T. b. rhodesiene*	Tropical Africa	1 000
T. cruzi	Latin America	14 000
Leishmania donovani *L. major* *L. tropica* *L. mexicana* *L. brasiliensis*	Asia, Africa, Latin America	12 000
Toxoplasma gondii	World-wide	500 000
Pneumocystis carinii	World-wide	(?)
Trichomonas vaginalis	World-wide	(20–25% of all adult females)
Giardia lamblia	World-wide	200 000
Entamoeba histolytica	World-wide	400 000

* Taken from references cited in the text and from Walsh and Warren (1979).

countries will be discussed. Several other agents are additionally described because of their obvious importance worldwide. Tables 7.14 and 7.15 summarize the salient features of most parasites. Complete descriptions of these parasites and the diseases may be found in textbooks of parasitology (Beaver *et al.* 1984; Strickland 1984).

Parasites have evolved elegant strategies to avoid destruction by the host immune response, enabling them to persist in humans for years. Thus, the prevalence of infection for most species is high although rates of morbidity are variable. Some of the most exciting current research indicates that, while strain differences with respect to virulence or pathogenicity do exist, the immunological perturbations occurring during infection (MacInnis 1987) and host genetic factors (Wakelin and Blackwell 1988), are important determinants of morbidity.

Transmission of protozoan parasites is effected both by faecally contaminated food and water and by arthropod vectors. This latter form of indirect transmission is obligatory for *Plasmodium, Babesia, Trypanosoma*, and *Leishmania*, providing an extra-human system of parasite amplification and dessemination which contributes to the complexity of the dynamics of transmission and difficulty in control.

Parasitic protozoa are either obligate intracellular or extracellular parasites. Their pathogenic effects are best appreciated by their usual habitat in the host: vascular, extravascular, or intestinal. Table 7.14 lists the organisms of public health importance by their habitat and indicates the most appropriate diagnostic test(s). The balance of this discussion focuses on the most important parasitic agents and the major determinants of disease and resistance.

On the global scale, malaria remains the most important parasite. Over 2000 million persons live in malarious areas with high rates of mortality in children (Gramiccia and Hempel 1972). *Plasmodium falciparum* is by far the most important of the four species infecting humans as it causes the most morbidity and almost all of the mortality.

The infective forms (sporozoites) inoculated from *Anopheles* mosquitoes invade liver cells to multiply. Then they burst out and invade erythrocytes, progressing through the multiplicative stage (schizogony) forming progeny (merozoites) that burst out of the cell to invade others. Rapid multiplication and the ability to invade erythrocytes of all ages are factors that contribute to the malignant characterization of falciparum malaria. The high density of the gametocytes (sexual forms) that occurs in blood is a prominent factor in the density and distribution of infective mosquitoes.

'Natural resistance' has been recognized for a long time in protozoan (especially malarial) infections although until recently its mechanism(s) eluded clarification. Erythrocytes containing the sickle haemoglobin (Friedman 1978), haemoglobin (Freidman *et al.* 1979), fetal haemoglobin (Friedman 1979), haemoglobin E (Nagel *et al.* 1981), or deficiency in glucose-6-phosphate dehydrogenase (Freidman 1979) are inimical environments for *P. falciparum* development. It appears that *P. falciparum* merozoites must utilize the MN blood group erythrocytic membrane glycophorin to invade (Pasvol *et al.* 1982). The near absence of *P. vivax* from West Africans and the descendants of West Africans elsewhere in the world is now thought to result from the lack of erythrocytic Duffy factor which serves as a receptor allowing Duffy-positive cells to be invaded (Miller *et al.* 1976). Ovalocytic erythrocytes appear to be another genetic abnormality that prevents invasion of *P. falciparum*, and possibly of *P. vivax* and *P. malariae* (Kidson *et al.* 1981).

Specific development of immunity to reinfection is indicated in all epidemiological studies of endemic malaria. The experimental evidence points to the major role of serum IgG antibodies, although specific cellular immunity complements the overall picture. Nevertheless, the ability of plasmodial species to subvert the immune process to allow parasite survival provides a dynamic balance between positive and negative forces that is responsible for the outcome of the infection.

Of all the parasitic protozoa that are in humans, infection by *Toxoplasma gondii* is perhaps the most prevelant. Toxoplasmosis is distributed world-wide with a significant proportion of the adult population having experienced exposure as determined by serological tests in immuno-epidemiological surveys (Frenkel 1973). *T. gondii* is intracellular in mononuclear phagocytes, and in immune hosts forms slow-growing cysts in muscle and the central nervous system.

Of greatest significance is that this parasite is intestinal in domestic cats but in all other species of mammal, particularly sheep, *Toxoplasma* exists as cysts in muscle. Depending upon the ecological conditions, human exposure is primarily via environmental contamination by infective cyst-laden cat faeces or by inadequately cooked meat products (Frenkel

Table 7.14. Parasitic protozoa of public health importance

Parasite	Vector/transmission	Diagnosis
Plasmodium falciparum *P. vivax* *P. malariae* *P. ovale*	*Anopheles*	Peripheral blood smear for parasites
Babesia microti	*Ixodes dammini*	Peripheral blood smear for parasites. Serology (ELISA)
Trypanosoma brucei gambiense *P. b. rhodesiense*	*Glossina* (tse-tse)	Peripheral blood smear, lymph node aspiration, cerebrospinal fluid for parasites. Serology of serum and cerebrospinal fluid (IgM, complement fixation test)
T. cruzi	*Triatoma, Rhodnius* (reduviid bugs)	Peripheral blood smear for parasites. Xenodiagnosis ('clean' bugs feed on patient). Serology (complement fixation test)
Leishmania donovani *L. major* *L. tropica* *L. mexicana* *L. brasiliensis*	*Phlebotomus, Lutzomyia, Sergentomyia* (sandflies)	Biopsy of spleen or bone marrow for *L. donovani* only. Smear of lesion exudate
Toxoplasma gondii	Ingestion of meat containing cysts, or oocysts from cat faeces	Biopsy of lymph node. Serology (methylene blue dye test, immunofluorescence, ELISA)
Pneumocystis carinii	Airborne	Smear of tracheobronchial aspirates for parasites. Lung biopsy. Serology (complement fixation test, ELISA)
Trichomonas vaginalis	Sexual contact	Culture or smears of vaginal fluid, urethral discharge
Giardia lamblia	Ingestion of cyst-contaminated water	Smear of duodenal aspirate, stool smear for trophozoites and cysts
Entamoeba histolytica	Ingestion of cyst-contaminated water (and perhaps food)	Stool smear for trophozoites and cysts. Serology (immunodiffusion, counter immunoelectrophoresis, indirect haemagglutination, ELISA)
Cryptosporidum species	Ingestion of oocysts	Colonic biopsy, stool smear for oocysts

Table 7.15. Distribution and prevalence of parasitic helminths

Species	Distribution	No. infected (thousands)*
Diphyllobothrium latum	North America, South America, Europe, Central Africa, foci in Asia	(?)
Taenai saginata	World-wide	(?)
T. solium	North America, South America, Europe	(?)
Schistosoma mansoni	Caribbean, South America, Africa	250 000
S. haematobium	Middle East, Africa, Asia	
S. japonicum	Asia	
Paragonimus westermani	Asia	2 000
Paragonimus species	Africa, Latin America	(?)
Clonorchis (Opisthorchis) sinensis	Asia	20 000 (estimate)
Fasciolopsis buski	Asia	15 000
Heterophyes heterophyes	Africa, Middle East, Asia	(?)
Ascaris lumbricoides	World-wide	800 000
Necator americanus	World-wide	700 000
Strongyloides stercoralis	World-wide	40 000
Enterobius vermicularis	World-wide	500 000
Trichuris trichiura	World-wide	500 000
Trichinella spiralis	World-wide (especially North America, Europe)	600
Onchocerca volvulus	Latin America, Tropical Africa	40 000
Wuchereria bancrofti	South America, Africa, Asia, Pacific Islands	250 000
Brugia malayi	Asia	
Loa loa	Tropical Africa	13 000
Dracunculus medinensis	Africa, Middle East, Indian subcontinent	50 000

* Taken from references cited in the text and from Walsh and Warren (1979).

and Ruiz 1981). Clinically significant morbidity constitutes a very small proportion of the total numbers infected although rates of infection can approach 100 per cent in a few years in certain closed communities (Kean 1972). Although resistance to reinfection develops, the initial infection may persist for years if not throughout a lifetime. The parasites are in cysts in very low density, suggesting a possible source of organisms from which disease may result in the immunocompromised host.

Pneumocystosis due to *Pneumocystis carinii* is world-wide. The protozoa inhabits pulmonary alveoli in a latent form. It is found in all mammals as trophozoites and characteristic cysts. The primary mode of transmission is airborne based on experimental studies. This organism is now of great importance since it is a cause of extensive pulmonary morbidity in the immunocompromised host, particularly in AIDS.

Trichomonas vaginalis is a sexually transmitted agent that chronically infects the urogenital tract of men and women. It is estimated to infect 20–25 per cent of the adult female population and 30–35 per cent of women with abnormal vaginal discharge (Jirovec and Petru 1968).

Amoebiasis due to *Entomoeba histolytica* is distributed world-wide but is most prevelant in tropical climates. The amoeba invades colonic mucosa causing extensive dysentery and ulcers, and in many the amoebae disseminate to extra-intestinal sites, especially the liver, causing abscesses. Transmission via the oral route is effected by faeces containing infective cysts which contaminate water and food supplies. Historically, it was thought that asymptomatic infections predominated, but a major proportion of these were due to a morphologically identical but non-pathogenic amoeba, *E. hartmanni*, which is differentiated on the basis of size (less than 10 μm). Advances made in the identification of pathogenic and non-pathogenic strains are based on iso-enzyme pattern differences (Sargeaunt and Williams 1978 and on the identification of a cytotoxin–enterotoxin (Lusbaugh *et al.* 1979).

Giardia lamblia inhabits the mucosa of the duodenum adhering to villous surfaces by means of a pair of suctorial discs. Transmission is by the oral route from cysts that have been discharged with faeces contaminating water supplies. The distribution of giardiasis has been changing in recent years, especially in the US. Previously, infection was thought to be primarily limited to children but now adults are commonly afflicted. Cyst-contaminated water supplies in small communities result from inadequate water processing procedures and from discharge of cysts into water supplies from infected beavers (Juranek 1979). In the US, *G. lamblia* is a major cause of most waterborne gastro-intestinal outbreaks.

Helminths

Helminthic parasites in humans are broadly divided into two groups: flatworms which include tapeworms, (cestodes) and flukes (trematodes), and roundworms (nematodes). They are widely distributed in the world with most species generally being more prevalent in tropical climates. Table 7.15 lists the important helminths and their general geographical distributions. The most characteristic feature of these parasites is that they may reside in humans for years, in many instances for decades, actively producing eggs or larvae. Almost all species cannot complete their life history in the human host and thus the worm burden is directly dependant upon the number of infective eggs or larvae in the inoculum from each exposure.

Modes of transmission are varied and for many species are complex, utilizing one or two obligatory intermediate vertebrate or invertebrate hosts. This information is summarized in Table 7.16. The importance of particular parasites in certain regions of the world is reflected in how the infections are transmitted to humans and the types of vectors or intermediate hosts required.

In the human host, almost all helminths induce abnormally high levels of immunoglobulin E (IgE) and eosinophils in the peripheral blood. Both have been regarded as indicative of the induction of parasite-related allergic states, although now the eosinophil increase is thought to be important in the development of immunity to reinfection (Butterworth 1980). Those helminths that migrate through the viscera, either via vascular systems or directly through parenchymatous organs, appear to induce the highest levels of eosinophils. Parasites limited to an intra-intestinal existence while in the human elicit minimal host responses. In various ways, parasitic helminths have evolved multiple mechanisms by which they can evade the host defence mechanisms. Interestingly, many helminths in the process of establishing themselves in the human induce protective immunity against further infectious exposures, yet the existing adult worms survive normally. The phenomenon is termed concomitant immunity, and the specific strategies used by each parasite species may differ.

Cestode infections

The three important tapeworm species that inhabit the human intestinal tract are *Diphyllobothrium latum* (broad fish tapeworm), *Taenia saginata* (beef tapeworm), and *T. solium* (pork tapeworm). Of the multitude of intestinal complaints attributed to these organisms, direct competition for vitamin B_{12} by *D. latum* leading to pernicious anaemia is the major pathogenic activity. The other major problem associated with *T. solium* is larval infection, cysticercosis. The hermaphroditic adult tapeworms reside in the small intestine and absorb all of their nutrients from the intestinal contents through their entire body surface, since all species lack alimentary tracts, and live for years attaining lengths of 5–12m.

Trematode infections

Of all the parasitic organisms, schistosomes rank next to the malarial parasites in world-wide importance. Three species

Table 7.16. Major parasitic helminths of public health importance

Species	Life-span (adult)(years)	Habitat	Mode of transmission	Diagnosis
Diphyllobothrium latum	up to 20	Small intestine	Ingestion of larvae (plerocercoids) in flesh of freshwater fish (first intermediate host is microcrustacean)	Eggs in faeces
Taenia saginata	up to 25	Small intestine	Ingestion of larvae (cysticerci) in beef	Proglottids, eggs in faeces
T. solium			Ingestion of cysticerci in pork	
Schistosoma mansoni	4–20	Mesenteric venules		Eggs in faeces, rectal biopsy
S. haematobium	4–20	Vesical plexus	Penetration of skin by larvae (cercariae) in water (intermediate host is snail)	Eggs in urine and faeces, rectal biopsy
S. japonicum	4–20	Mesenteric venules		Eggs in faeces, rectal biopsy
Paragonimus westermani	6	Lungs	Ingestion of larvae (metacercariae) in flesh of freshwater crabs (first intermediate host is snail)	Eggs in sputum, faeces
Clonorchis sinensis	15–20	Bile ducts	Ingestion of metacercariae in flesh of cyprinoid fish (first intermediate host is snail)	Eggs in faeces
Fasciolopsis buski	? several years	Small intestine	Ingestion of metacercariae on water plants, e.g. caltrop, hyacinth, chestnut (intermediate host is snail)	Eggs in faeces
Ascaris lumbricoides	1–2	Small intestine	Ingestion of eggs	Eggs in faeces
Necator americanus	1–2	Small intestine	Penetration of skin by larvae in soil	Eggs in faeces
Ancylostoma duodenale	1–2	Small intestine	Penetration of skin by larvae in soil	Eggs in faeces
Strongyloides stercoralis	30	Small intestine	Penetration of skin by larvae in soil	Larvae in faeces
Enterobius vermicularis	11–35 days	Caecum	Ingestion of eggs	Eggs on perianal skin
Trichuris trichiura	3	Caecum, large intestine	Ingestion of eggs	Eggs in faeces
Trichinella spiralis	30 days	Small intestine (larvae in muscle)	Ingestion of larvae in pork, bear meat, etc.	Muscle biopsy. Serology (bentonite flocculation)
Onchocerca volvulus	7–20	Subcutaneous	Infective larvae from blackflies (*Simulium*)	Microfilariae in skin biopsy, on eye examination
Wuchereria bancrofti	5	Lymphatics	Infective larvae from mosquitoes (*Culex, Aedes, Anopheles, Mansonia*)	Microfilariae in peripheral blood smear
Brugia malayi	5	Lymphatics	Infective larvae from *Mansonia, Anopheles, Aedes*	Microfilariae in peripheral blood smear
Loa loa	4–17	Subcutaneous	Infective larvae from deerflies (*Chrysops*)	Microfilariae in peripheral blood smear
Dracunculus medinensis	1	Subcutaneous	Ingestion of larvae in microcrustaceans (*Cyclops*)	Adult female worm protruding from cutaneous ulcer

are responsible for almost all of the morbidity in schistosomiasis: *Schistosoma mansoni*, *S. japonicum*, and *S. haematobium*. Others infecting humans are *S. mekongi* in South-East Asia and *S. intercalatum* in central Africa. The male and female blood flukes live in constant copulation in the mesenteric venous system (*S. mansoni, S. japonicum, S. mekongi*, and *S. intercalatum*) or in the vesical (bladder) venous plexus (*S. haematobium*). Since specific freshwater snails are obligatory intermediate hosts and the skin-penetrating infective larvae (cercariae) are subsequently released, human exposure is dependent upon water contact for acquiring the infection. Infection of the snails if effected by the discharge of egg-containing faeces and urine.

The adult worms cause minimal pathology but the eggs released by the females are trapped in the liver, intestinal wall, and genito-urinary tract where they incite a specific cellular immune inflammatory response which is the mechanism of the observed lesions. Severe hepatic fibrosis, colonic bleeding, colonic inflammatory polyposis, anaemia, and chronic obstructive urinary tract disease are commonly encountered, depending upon the species and the intensity and duration of infection. *S. haematobium* is further associated with urinary tract bacterial infections and squamous cell carcinoma of the bladder, especially in Egypt (Cheever 1978).

The adult worms become impervious to host defences by the unique strategy of coating their outer surface with host materials. Nevertheless, the human host is able to resist invasion by infective larvae from subsequent exposures. This concomitant immunity enables the host to avoid accumulating large worm burdens and yet allows the resident flukes to survive for years. The infection also modulates the extent of the tissue pathology against the eggs which presumably allows for the greater survival of the host.

Nemotode infections

In prevalence, infection due to *Ascaris lumbricoides* is one of the most frequent world-wide. Male and female roundworms inhabit the small intestine, arriving there after a sojourn by the infective larvae (hatching from ingested eggs) migrating out of the intestine, through the circulation of the lungs, into the air spaces, up the trachea and down the oesophaghus back into the intestine. They incite disease as they migrate through the lungs, causing eosinophilic pneumonitis. Adults in the intestine can cause intestinal obstruction and can aberrantly migrate into the biliary and pancreatic ducts. Evidence is accumulating implicating acariasis as a primary determinant of nutritional deficiency leading to poor growth in children (Stephenson 1980).

Hookworm infection is caused by *Necator americanus* and *Ancylostoma duodenale* which live in the small intestine. Adult males and females attach to the mucosa and imbibe blood and tissue fluid for their nutrients. Each worm ingests from 0.03 to 0.15ml of blood per day leading to chronic blood loss and anaemia, the severity of which is directly attributable to the worm burden, which may number in the hundreds.

Humans acquire the infection by exposure to skin-penetrating larvae in the soil deposited earlier by faeces containing eggs which rapidly embryonate and hatch. The invading larvae migrate as in *Ascaris* infection, finally to reach the small intestine. Oral infection by *A. duodenale* is possible. *A. duodenale* has also been shown to undergo a hypobiotic state during development in the intestine, remaining immature for months. Resumption of maturation and egg laying appears to be correlated with optimal environmental conditions for larval survival in the soil (Banwell and Schad 1978).

Infection by *Strongyloides stercoralis* occurs by skin-penetrating larvae in soil originally deposited by faeces containing larvae. This parasite can live entirely as a free-living nematode in soil as well as being parasitic in humans. The invading larvae migrate to the intestinal tract as in *Ascaris* and hookworms infections. The primary clinical problems are gastro-intestinal with, occasionally, pulmonary complications.

The importance of this agent is shown by frequent infection in the immunocompromised host leading to extensive morbidity and mortality (Scowden *et al.* 1978). Worms can live asymptomatically for over 30 years with occasional reactivation of disease especially due to larvae migrating in the skin, a condition called *larva currens*.

Pinworm infection is due to *Enterobius vermicularis* which lives in the caecum and large intestine with the female migrating out of the anus to deposit eggs at night on the perianal skin. These incite pruritus, the major physical problem. The parasite is found world-wide and infection is easily acquired by ingestion of airborne eggs leading to very high prevalence rates especially in children in institutions, hospitals, and families. Infection is self-limited but the high exposure rate leads to frequent reinfection.

Trichuris trichiura ('whipworm') inhabits the mucosa of the caecum, and infections can consist of hundreds of worms. Although these parasites can cause some blood loss, their primary importance is in children with massive infections leading to diarrhoea, dysentery, rectal prolapse, and weight loss.

Trichinosis has been long recognized in Europe and North America and is being increasingly reported from tropical countries. The causative agent is *Trichinella spiralis*. Infection is acquired by ingestion of poorly cooked meat, especially pork, but also of exotic foods such as bear and walrus meat (Juranek and Schultz 1978). Meat contains encysted larvae which emerge and take residence in the small intestine. Females are 3–4mm in length and deposit live larvae in the mucosa within 3 days after infection. Larvae enter the circulation and disseminate to the skeletal musculature. The clinical picture is protean but is primarily due to larval invasion and to the intensity of infection. Since infection is acquired

Table 7.17. Larval helminths of public health importance

Disease	Aetiology	Mode of infection	Diagnosis
Hydatid	*Echinococcus granulosus* *E. multilocularis*	Ingestion of eggs	Serology (indirect haemagglutination, immunofluorescent assay, immunoelectrophoresis, ELISA), radio-isotope imaging methods and radiographs
Cysticercosis	*Taenia solium*	Ingestion of eggs	Serology (indirect haemagglutination, ELISA), radio-isotope imaging methods and radiographs
Swimmer's itch	Members of the blood fluke family *Schistosomatidae*	Penetration of skin by larvae (cercariae) in water	Occurrence in known endemic foci combined with clinical picture
Anisakiasis (herringworm disease)	*Anisakis simplex*	Ingestion of larvae in flesh of marine fish	Larvae in surgical specimen (stomach)
Visceral larva migrans	*Toxocara canis*	Ingestion of eggs	Serology (ELISA)
Cutaneous larva migrans	*Ancylostoma braziliense* *Uncinaria stenocephala* *Strongyloides*	Penetration of skin by larvae in soil	Clinical picture, history of exposure
Eosinophilic meningitis	*Angiostrongylus cantonensis*	Ingestion of larvae in flesh of terrestrial snails	Larvae in cerebrospinal fluid
Dirofilariasis	*Dirofilaria immitis*	Infective larvae from mosquitoes	Young worm in surgical specimens (lung)

only by ingestion of meat, the human infection cannot be transmitted to others. Thus trichinosis is properly a zoonosis.

Onchocerca volvulus, the sole agent of onchocerciasis, is transmitted to humans by blackflies (*Simulium* sp.) inoculating larvae which mature in the subcutaneous tissues where they live frequently in fibrotic nodules. The females lay embryos (microfilariae) which migrate throughout the skin from which they must be ingested by feeding blackflies for larval development to proceed. The mere presence of microfilariae is not harmful but the host antibody response is the determinant of disease. Immune reaction against microfilariae leads to severe chronic inflammatory reactions in skin, producing dermatitis, and most importantly in all segments of the eyes, resulting in visual impairment leading to blindness. Since the adult worms live for 20 years, constantly discharging embryos, and the worm burden can be extensive, the severity of disease is correlated with both intensity and duration of infection (World Health Organization 1976).

All animal species harbour their own array of parasites, far outnumbering those commonly found in humans. Frequently, human exposure to their infective larvae occurs, resulting in disease. In general, the larvae invade the human body, migrate, and continue to grow as larvae inciting severe inflammation. The most important species are usually those that parasitize pets or animals with whom close human contact occurs. The most important larval helminthic disorders, the causal organisms, mode of infection, and diagnoses are listed in Table 7.17.

References

Abraham, J.M., Freitag, C.S., Clements, J.R., and Eisenstein, B.I. (1985). An invertible element of DNA controls phase variation of type 1 fimbriae of *Escherichia coli*. Proceedings of the National Academy of Sciences, *U.S.A.* **82**, 5724.

Al-Doory, Y. (1975). *The epidemiology of human mycotic diseases*. Thomas, Springfield, Illinois.

Bachrach, H.L. (1978). Comparative strategies of animal virus replication. *Advances in Virus Research* **22**, 163.

Banwell, J.G. and Schad, G.A. (1978). Hookworm. *Clinical gastroenterology* **7**, 129.

Beaver, P.C., Jung, R.C., and Cupp, E.W. (1984). *Clinical parasitology*. Lea and Febiger, Philadelphia.

Bell, J.A., Ward, T.G., Kapikian, A.Z., Shelokov, A., Reichelderfer, T.E., and Huebner, R.J. (1957). Artificially induced Asian influenza in vaccinated and unvaccinated volunteers. *Journal of the American Medical Association* **165**, 366.

Binford, C.H. and Connor, D.H. (1976). *Pathology of tropical and extraordinary diseases*, Vol. 2. Armed Forces Institute of Pathology, Washington, DC.

Burns, G., Abraham, A.K., and Vedeler, A. (1986). Nucleotide binding to elongation factor 2 inactivated by diphtheria toxin. *FEBS Letters* **208**, 217.

Butterworth, A.E. (1980). Eosinophils and immunity to parasites. *Transactions of the Royal Society of Tropical Medicine and Hygiene* **74**,(Supplement), 38.

Cheever, A.W. (1978). Schistosomiasis and neoplasia. *Journal of the National Cancer Institute* **61**, 13.

Choppin, P.W. (1964). Multiplication of a myxovirus (SV5) with minimal cytopathic effects and without interference. *Virology*, **23**, 224.

Clifford, R.E., Smith, J.W.G., Tillet, H.E., and Wherry, P.J. (1977). Excess mortality associated with influenza in England and Wales. *International Journal of Epidemiology* **6**, 115.

Dale, J.B. and Beachey, E.H. (1985). Multiple, heart cross-reactive epitopes of streptococcal M proteins. *Journal of Experimental Medicine* **161**, 113.

Davies, J.E. (1981). Antibiotic resistance—a survey. In *Molecular biology, pathogenicity, and ecology of bacterial plasmids* (ed. S.B. Levy, R.C. Clowes, and E.L. Loenig). p. 145. Plenum Press, New York.

Dulbecco, R. and Ginsberg, H.S. (1980). *Virology*. Harper and Row, Hagerstown.

Dupont, H.L., Gangarosa, E.J., Reller, L.B., Woodward, W.E., Armstrong, R.W., and Hammond, J. (1970). Shigellosis in custodial institutions. *American Journal of Epidemiology* **92**, 172.

Eidels, L. and Hranitzky, K.W. (1985). Overview of bacterial toxin receptors. In *Microbiology—1985* (ed. L. Leive) p. 96. ASM, Washington, DC.

Emmons, C.W., Binford, C.H., Utx, J.P., and Kwon-chung, K.J. (1985). *Medical mycology*. Lea and Febiger, Philadelphia.

Epstein, M., Boav, Y., and Achong, B. (1965). *Studies with Burkitt's lymphoma*. Wistar Institute Symposium Monograph, Philadelphia.

Finkelstein, R.A., Sciortina, C.V., and McIntosh, M.A. (1983). Role of iron in microbo-host interactions. *Reviews of Infectious Diseases* **5** (Supplement 4), S759.

Fox, J.P., Hall, C.E., and Elveback, L.R. (1970). *Epidemiology: Man and disease*. Macmillan, New York.

Frenkel, J.K. (1973). Toxoplasma in and around us. *Bioscience* **23**, 343.

Frenkel, J.K. and Ruiz, A. (1981). Endemicity of toxoplasmosis in Costa Rica: Transmission between cats, soil, intermediate hosts and humans. *American Journal of Epidemiology* **113**, 254.

Friedman, M.J. (1978). Erythrocytic mechanisms of sickle cell resistance to malaria. *Proceedings of the National Academy of Sciences U.S.A.* **75**, 1994.

Friedman, M.J. (1979). Oxidant damage mediates variant red cell resistance to malaria. *Nature (London)* **280**, 245.

Friedman, M.J., Roth, E.F., Nagel, R.L., and Trager, W. (1979). The role of hemoglobin C, S, and N_{balt} in the inhibition of malaria parasite development *in vitro*. *American Journal of Tropical Medicine and Hygiene* **28**, 777.

Fujita, N., Tamura, M., and Hotta, S. (1975). Dengue virus plaque formation on microplate cultures and its application to virus neutralization. *Proceedings of the Society of Experimental Biology and Medicine* **148**, 472.

Gill, D.M. (1982). Bacterial toxins: A table of lethal amounts. *Microbiology Review* **46**, 86.

Gramiccia, G. and Hempel, J. (1972). Mortality and morbidity from malaria in countries where malaria eradication is not making satisfactory progress. *Journal of Tropical Medicine and Hygiene* **75**, 187.

Greenberg, H.B., Valdesuso, J., Yolken, R.H., Gangarosa, E., Gary, W., and Wyatt, R.B. (1979). Role of Norwalk virus in outbreaks of nonbacterial gastroenteritis. *Journal of Infectious Diseases* **139**, 564.

Groman, N.B. (1984). Conversion by corynephages and its role in the natural history of diphtheria. *Journal of Hygiene (Cambridge)* **93**, 405.

Habermann, E. and Dreyer, F. (1986). Clostridial neurotoxins: handling and action at the cellular and molecular level. *Current Topics in Microbiology and Immunology* **129**, 93.

Hayflick, L. (1965). The limited *in vitro* lifetime of human diploid cell strains. *Experimental Cell Research* **37**, 614.

Herrman, J.E., Hendry, R.M., and Collins, M.F. (1979). Factors involved in enzyme-linked immunoassay of viruses and evaluation of the method for identification of enteroviruses. *Journal of Clinical Microbiology* **10**, 210.

Holmberg, S.D., Osterholm, M.T., Senger, K.A., and Cohen, M.L. (1984). Drug-resistant *Salmonella* from animals fed antimicrobials. *New England Journal of Medicine* **311**, 617.

Hoorn, B. and Tyrrell, D.A.J. (1969). Organ cultures in virology. *Progress in Medical Virology* **11**, 408.

Horowitz, M.A. (1984). Phagocytosis of the Legionnaires' disease bacterium (*Legionella pneumophilia*) occurs by a novel mechanism: Engulfment within pseudopod coil. *Cell* **36**, 27.

Horstmann, R.D., Sievertsen, H.J., Knobloch, J., and Fischetti, V.A. (1988). Antiphagocytic activity of streptococcal M protein: Selective binding of complement control protein factor H. *Proceedings of the National Academy of Sciences U.S.A.* **85**, 1657.

Hughes, H.J., Thomas, D.C., and Hamparian, V.V. (1973). Acid lability of rhinovirus type 14: Effect of pH, time and temperature. *Proceedings of the Society of Experimental Biology and Medicine* **144**, 555.

Isberg, R.R., Voorhis, D.L., and Falkow, S. (1987). Identification of invasin: A protein that allows enteric bacteria to penetrate cultured mammalian cells. *Cell* **50**, 769.

Jacobsson, P.A., Johannson, M.E., and Wadell, G. (1979). Identification of an enteric adenovirus by immunoeleltroosmophoresis (IEOP) technique. *Journal of Medical Virology* **3**, 307.

Jirovec, D. and Petru, M. (1968). *Trichomonas vaginalis* and trichomoniasis. *Advances in Parasitology* **6**, 117.

Joiner, K.A., Hammer, C.H., Brown, E.J., Cole, R.J., and Frank, M.M. (1982). Studies on the mechanism of bacterial resistance to complement-mediated killing. *Journal of Experimental Medicine* **155**, 797.

Juranek, D. (1979). Waterborne giardiasis (summary of recent epidemiologic investigations and assessment of methodology). In *Waterborne transmission of giardiasis* (ed. W. Jakubowski and J.C. Hoff) p. 150. U.S. Environmental Protection Agency, Cincinnati.

Juranek, D.D. and Schultz, M.G. (1978). Trichinellosis in humans in the United States: Epidemiologic trends in trichinellosis. In *Proceedings of the Fourth International Conference on Trichinellosis, 26–28 August Poznan, Poland* (ed. D.W. Kim and Z.S. Pawlowski) p. 523. University Press of New England, Hanover.

Kean, B.H. (1972). Clinical toxoplasmosis—50 years. *Tropical Medicine and Hygiene* **66**, 549.

Kidson , C., Lamont, G., Saul, A., and Nurse, G.T., (1981). Ovalocytic erythrocytes from Melanesians are resistant to invasion by malaria parasites in culture. *Proceedings of the National Academy of Sciences U.S.A.* **78**, 5824.

Koch-Weser, J., Hirsch, M.S., and Swartz. M.N. (1980). Medical intelligence: Drug therapy. *New England Journal of Medicine* **302**, 983.

Koprowski, C. and Koprowski, H. (ed.) (1975). *Toward understanding viruses and immunity: Viral immunology and immunopathology* Academic Press, New York.

Labigne-Roussel, A. and Falkow, S. (1988). Distribution and degree of heterogeneity of the afimbrial-adhesin-encoding operon (*afa*) among uropathogenic *Escherichia coli* isolates. *Infection and Immunity* **56**, 640.

Lee, C.H., Hu, S.T., Swiatek, P.J., Moseley, S.L., Allen, S.D., and So, M. (1985). Isolation of a novel transposon which carried the *Escherichia coli* enterotoxin STII gene. *Journal of Bacteriology* **162**, 615.

Lennette, W.W. and Schmidt, N.J. (ed.) (1979). *Diagnostic procedures for viral, rickettsial and chlamydial infections*. American Public Health Association. Washington, DC.

Lindberg, F., Lund, B., Johansson, L., and Normark, S. (1987). Localization of the receptor-binding protein adhesin at the tip of the bacterial pilus. *Nature (London)* **328**, 84.

Lusbaugh, W.G., Kairalla, A.B., Cantey, J.R., Hofbauer, A.F., and Pittman, F.E. (1979). Isolation of cytotoxin–enterotoxin from *Entomoeba histolytica*. *Journal of Infectious Diseases* **139**, 9.

McCartney, A.C. and Arbuthnott, J.P. (1978). Mode of action of membrane-damaging toxins produced by staphylococci. In *Bacterial toxins and cell membranes* (ed. J. Jeljaszewicz and T. Wadstrom) p. 89. Academic Press, New York.

MacInnis, A.J. (1987). *Molecular Paradigms for Eradicating Helminthic Parasites*. Alan R. Liss, New York.

Mandel, G.L., Douglas, R.G., Jr., and Bennett, J.E. (1985). *Principles and practice of infectious diseases*, Vol. 2, Part III, Section F. Wiley, New York.

Marx, J.L. (1988). Multiplying genes by leaps and bounds. *Science* **240**, 1408.

Matthews, R.E.F. (1982). Classification and nomenclature of viruses. Fourth Report of the International Committee on Taxonomy of Viruses. *Intervirology* **17**, 1.

Maynard, J.E. (1978). Passive immunization against hepatitis B: A review of recent studies and comment on current aspects of control. *American Journal of Epidemiology* **107**, 77.

Meyer, T.F. (1987). Molecular basis of surface antigen variation in *Neisseria*. *Trends in Genetics* **3**, 319.

Middlebrook, J.L. and Dorland, R.B. (1984). Bacterial toxins: Cellular mechanisms of action. *Microbiology Reviews* **48**, 199.

Mikesell, P., Ivins, B.E., Ristroph, J.D., and Dreier, T.M. (1983). Evidence for plasmid-mediated toxin production in *Bacillus anthracis*. *Infection and Immunity* **39**, 371.

Miller, L.H., Mason, S.J., Clyde, D.F., and McGinnis, M.H. (1976). The resistance factor in *Plasmodium vivax* in blacks: The Duffy-blood-group genotype, FyFy. *New England Journal of Medicine* **295**, 302.

Mims, C.A. (1982). *The pathogenesis of infectious diseases*. Academic Press, New York.

Mirkovic, R.R., Kono, R., Yin-Murphy, M., Sohier, R., Schmidt, N.J., and Melnick, J.L. (1973). Enterovirus 70: The etiologic agent of pandemic acute hemorrhagic conjunctivitis. *Bulletin of the World Health Organization*. **49**, 341.

Moldave, K. (1985). Eukaryotic protein synthesis. *Annual Reviews of Biochemistry* **54**, 1109.

Monto, A.S. and Cavallaro, J.J. (1972). The Tecumseh study of respiratory illness. IV. Prevalence of rhinovirus serotypes, 1966–1969. *American Journal of Epidemiology* **96**, 352.

Monto, A.S. and Lim, S.K. (1974). The Tecumseh study of respiratory illness. VI. Frequency and relationship between outbreaks of coronavirus infection. *Journal of Infectious Diseases* **129**, 271.

Mooi, F.R. and deGraaf, F.K. (1985). Molecular biology of fimbriae of enterotoxigenic *Escherichia coli*. *Current Topics of Microbiology and Immunity* **118**, 119.

Morihara, K. and Homma, J.Y. (1985). *Pseudomonas* proteases. In *Bacterial enzymes and virulence* (ed. I.A. Holder) p. 42. CRC Press, Boca Raton, Florida.

Moulder, J.W. (1985). Comparative biology of intracellular paristism. *Microbiology Review* **19**, 298.

Mulks, M.H. (1985). Microbial IgA proteases. In *Bacterial enzymes and virulence* (ed. I.A. Holder) p. 81. CRC Press, Boca Raton, Florida.

Murphy, B.R., Phelan, M.A., Nelson, D.L. *et al.* (1981). Haemagglutinin-specific enzyme-linked immunosorbent assay for antibody to influenza A and B viruses. *Journal of Clinical Microbiology* **13**, 554.

Nagel, R.L., Raventos-Suarez, C., Fabry, M.E., Tanowitz, H., Sicard, D., and Labie, D. (1981). Impairment of the growth of *Plasmodium falciparum* in HbEE erythrocytes. *Journal of Clinical Investigation* **64**, 303.

Nahmias, A. and Roisman, B. (1973). Infections with herpes-simplex virus 1 and 2. *New England Journal of Medicine* **289**, 667.

Nathanson, N. and Martin, J.R. (1979). The epidemiology of poliomyelitis; enigmas surrounding its appearance, epidemicity and disappearance. *American Journal of Epidemiology* **109**, 103.

O'Brien, A.D. and Holmes R.K. (1987). Shiga and shiga-like toxins. *Microbiological Reviews* **51**, 206.

Odagiri, T., DeBorde, D.C., and Maassab, H.F. (1982). Cold-adapted recombinants of influenza A virus in MDCK cells: 1. Development and characterization of A/Ann Arbor/6/60 × A/Alaska/6/77. *Virology* **119**, 82.

Osterholm, M.T., Kantor, R.J., Bradley, D.W. *et al.* (1980). Immunoglobulin M-specific testing in an outbreak of foodborne viral hepatitis, type A. *American Journal of Epidemiology* **12**, 8.

Parker, M.M. and J. E. Parrillo. (1983). Septic shock: Hemodynamics and opathogenesis. *Journal of the American Medical Association* **250**, 3324.

Pasvol, G., Jungery, M., Weatherall, O.J., Parsons, S.F., Anstee, D.J., and Tanner, M.J.A. (1982). Glycophorin as a possible receptor for *Plasmodium falciparum*. *Lancet* **ii**, 947.

Portnoy, D.A., Jacks, P.S., and Hinrichs, D.J. (1988). Role of hemolysin for the intracellular growth of *Listeria monocytogenes*. *Journal of Experimental Medicine* **167**, 1459.

Purcell, R.H., Gerin, J.L., Almeida, J.B., and Holland, P.V. (1973). Radioimmunoassay for the detection of the core of the Dane particle and antibody to it. *Intervirology* **2**, 231.

Rafferty, K.A., Jr. (1975). Epithelial cells: Growth in culture of normal and neoplastic forms. *Advances in Cancer Research* **112**, 249.

Raghow, R. and Kingsbury, D.W. (1976). Endogenous viral enzymes involved in messenger RNA production. *Annual Review of Microbiology* **32**, 31.

Rippon, J.W. (1982). *Medical mycology: The pathogenic fungi and the pathogenic actinomycetes*. Saunders, Philadelphia.

Rossen, R.D., Kasel, J.A., and Couch, R.B. (1971). The secretory immune system: Its relation to respiratory viral infection. *Progress in Medical Virology* **13**, 194.

Saelinger, C.B. (1985). ADP-ribosylating enzymes as virulence factors. In *Bacterial enzymes and virulence factors* (ed. I.A. Holder) p. 17. CRC Press, Boca Raton, Florida.

Sansonetti, P.J., Ryter, A., Clerc, P., Maurelli, A.T., and Mounier, P. (1986). Multiplication of *Shigella flexneri* within HeLa cells: Lysis of the phagocytic vacuole and plasmid-mediated contact hemolysis. *Infection and Immunity* **51**, 461.

Sargeaunt, P.G. and Williams, J.E. (1978). Electrophoretic isoenzyme patterns of *Entamoeba histolytica* and *Entamoeba coli*. *Transactions of the Royal Society of Tropical Medicine and Hygiene* **72**, 164.

Schryvers, A.B. and Morris, L.J. (1988). Identification and characterization of the human lactoferrin-binding protein from *Neisseria meningitidis*. *Infection and Immunity* **56**, 1144.

Scowden, E.B., Schaffner, W., and Stone, W.J. (1978). Overwhelming strongyloidiasis, an unappreciated opportunistic infection. *Medicine* **57**, 527.

Shelokov, A., Vogel, J., and Chi, L. (1958). Hemadsorption (absorption–hemagglutination) test for viral agents in tissue culture with special reference to influenza. *Proceedings of the Society of Experimental Biology and Medicine* **97**, 802.

Silverblatt, F.J. (1974). Host–parasite interactions in the rat renal pelvis: A possible role of pili in the pathogenesis of pyelonephritis. *Journal of Expermental Medicine* **140**, 1696.

So, M., Dallas, W.S., and Falkow, S. (1978). Characterization of an *Escherichia coli* plasmid encoding for synthesis of heat-labile toxin: Molecular cloning of the toxin determinant. *Infection and Immunity* **21**, 405.

Sparling, P.F. (1983). Bacterial virulence and pathogenesis: An overview. *Reviews of Infectious Diseases* **5** (Supplement 4), S637.

Stephenson, L.S. (1980). The contribution of *Ascaris lumbricoides* to malnutrition in children. *Parasitology* **81**, 221.

Strickland, G.T. (1984). *Hunter's tropical medicine* (6th edn.) Saunders, Philadelphia.

Timmis, K.N., Boulnois, G.J., Bitter-Suermann, D., and Cabelin, F.C. (1985). Surface components of *Escherichia coli* that mediate resistance to the bactericidal activities of serum and phagocytosis. *Current Topics in Microbiology and Immunity* **118**, 197.

van der Waaij, D. (1984). Effects of antibiotics on colonization resistance. In *Medical microbiology* (ed. C.S.F. Easmon and J. Jeljaszewicz). Vol. 4 Academic Press, New York.

Vogt, M. and Dulbecco, R. (1960). Virus–cell interaction with a tumor-producing virus. *Proceedings of the National Academy of Sciences U.S.A.* **46**, 365.

Wakelin, D.M. and Blackwell. J.M. (ed.) (1988). Genetics and Resistance to Bacterial and Parasitic Infection. Taylor and Francis, London.

Walsh, J.A. and Warren, K.S. (1979). Selective primary health care. *New England Journal of Medicine* **301**, 967.

Westaway, E.G., Brinton, M.A., Gaidamovich, S.Ya., *et al.* (1985). Flaviviridae. *Intervirology* **24**, 183.

Whitnack, E. and Beachey, E.H. (1985). Biochemical and biological properties of the binding of human fibrinogen to M protein in group A streptococci. *Journal of Bacteriology*. **104**, 350.

World Health Organization (1976). *Epidemiology of onchocerciasis: Report of a WHO Expert Committee*. WHO Technical Report Series, No. 597, Geneva.

Wyatt, R.G., Dolin, R., Blacklow, N.R., *et al.* (1974). Comparison of three agents of acute infectious nonbacterial gastroenteritis by cross-challenge in volunteers. *Journal of Infectious Diseases* **129**, 709.

Yates, V.J., Dawson, G.J., and Pronovost, A.D. (1981). Serologic evidence of avian adeno-associated virus infection in an unselected human population and among poultry workers. *American Journal of Epidemiology* **113**, 542.

Yolken, R.H., Kim, H.W., Clem, T., *et al.* (1977). Enzyme-linked immunosorbent assay (ELISA) for detection of human reovirus-like agents of infantile gastroenteritis. *Lancet* **ii**, 263.

8

Genetic evolution and adaptation

ANTHONY C. ALLISON

Introduction

An important determinant of health and disease is inherited predisposition. Sometimes the pattern of inheritance of a disease conforms with expectations for dominant genes with the relatively high penetrance, recessive genes, or sex-linked recessive genes (McCusick 1978). Although most of these genes are uncommon, there are so many of them that their cumulative effects are substantial from the public health point of view. Often the primary gene products are known: usually they are enzymes, but they may be other proteins such as haemoglobins, immunoglobulins, or receptors for lipoproteins or other ligands. This molecular biological information clarifies the pathogenesis of the diseases in question, and the straightforward pattern of inheritance simplifies population genetic analysis. Selection of varying intensity operates against the abnormal phenotype, in the sense that survival to and through reproductive age is reduced or there is a reproductive disadvantage for other reasons. Hence there is in principle an equilibrium between the rate of elimination of these genes by selection and the rate of their replacement by recurrent mutation. However, improved medical care has relaxed selection against some of these genes, as discussed below, so that such an equilibrium will not be attained for a long time.

Other genes affecting susceptibility to disease are polymorphic, in other words the gene frequency reached in the past, when selection was intense, exceeds that attributable to recurrent mutation (see Glossary on p. 131). Such genes are of special interest from the point of view of genetic evolution and adaptation. They show that diseases can themselves act as selective agents, thereby influencing the genetic composition of populations. Three examples will be considered in some detail in this chapter because they illustrate different general points: abnormal haemoglobins, glucose 6-phosphate dehydrogenase deficiency, and genes of the major histocompatibility complex (MHC). Linkage of MHC genes with a number of diseases, or in some cases with subsets of patients having particular diseases, has been firmly established and possible explanations for such associations will be discussed.

For a long time confusion has been generated by attempts to fit complex and difficult human pedigree analyses into simple Mendelian ratios, on the assumption that diabetes mellitus, for example, is a single disease. However, analyses of linkage with MHC genes make it clear that there are several types of diabetes mellitus with different genetic associations and different environmental factors facilitating their expression. The same is true of hepatitis and many other common diseases. A few examples are discussed to illustrate theoretical points such as selective interactions of genes at different loci and practical points such as difficulties of determining the prevalence of diseases and identifying genetically predisposed individuals.

A final point considered in the chapter is the changing intensity and nature of selection. Formerly, infectious diseases killed many persons before reproductive age was complete; in some cases at least, such effects acted differentially on different genotypes, thereby producing selection. During the past century there has been a major demographic transition in the more industrialized countries from a period of high and death rates. The opportunity for selection has been correspondingly reduced, although not eliminated altogether. There has been a much smaller reduction in selection through infectious diseases in the developing world.

As a result of improved medical care, selection against diseases with well-defined genetic causation, such as galactosaemia and phenylketonuria, and those with clear but less well-defined inherited predispositions, such as diabetes mellitus, has been relaxed. This has led to concern about the accumulation of genes for inherited disorders in advanced societies. The maximum rates at which such genes can accumulate in populations can be calculated, and this has some bearing on related ethical problems.

Abnormal haemoglobins

The prototype abnormal human haemoglobin is sickle-cell haemoglobin (HbS). Several human populations are polymorphic for this variant. Each molecule of normal adult human haemoglobin (HbA) is a tetramer composed of two α-chains and two β-chains. In HbS the sixth amino-acid residue from the N-terminus of the β-chain is replaced by a point mutation, presumably resulting from a single nucleic acid base substitution (GAA to GUA). In HbA the glutamic acid residue in this site provides a negative surface charge to the whole tetramer, whereas in HbS the neutral amino acid, valine, in this position provides a combining site allowing polymerization of deoxygenated Hb molecules, which form rigid helical aggregates able to distort the red cell from a discoid to a sickle shape. Moreover, HbS more readily associates with the red cell membrane than does HbA, with consequences discussed below.

Individuals heterozygous for the sickle-cell gene have both HbA and HbS, usually 25–40 per cent of the latter. Each red cell contains a mixture of the haemoglobins, and the transformation to the sickle shape takes place only when the partial pressure of O_2 is lower than usually encountered in the circulation. The heterozygotes are therefore healthy except under extreme anoxic conditions, such as high-altitude flights in unpressurized aircraft. In sickle-cell homozygotes no HbA is formed; the majority of the Hb is HbS, although a small amount of fetal haemoglobin (HbF) if also present. These cells assume the sickle shape at partial pressures of oxygen encountered *in vivo*, so the SS homozygotes are prone to develop sickle-cell disease, a haemolytic and painful vaso-occlusive condition aggravated by infections which produce sickle-cell crises.

Another polymorphic variant is HbC, in which the same glutamic acid residue in the normal molecule, residue 6 of the β-chain, is replaced by lysine. Persons homozygous for HbC have a haemolytic disease, and CS heterozygotes suffer from a variety of sickle-cell disease usually milder than that manifested by SS homozygotes.

A third abnormality is β-thalassaemia, in which the β-polypeptide chain is deficient owing to a structural or regulatory mutation leading to a relative deficit in the formation of β-chain messenger RNA. As a result there is excessive production of α-chains, which are unstable in the absence of β-chains, precipitate, and bind to the erythrocyte membrane. Individuals homozygous for the β-thalassaemia gene (an allele of HbA, HbS, and HbC) have a severe haemolytic disorder, thalassaemia major, and seldom survive to adulthood and to reproduce. Heterozygotes having both HbS and β-thalassaemia suffer from another variant of sickle-cell disease, often of moderate severity.

Although with good medical care some individuals with sickle-cell disease can survive to adulthood and even reproduce (though pregnant women often experience difficulty), there is no doubt that under the conditions prevailing in Africa for centuries very few sickle-cell homozygotes could have reproduced. Hence, the fitness of SS homozygotes has been close to zero and the loss of S genes from populations through death of homozygotes has been much greater than could be replaced by recurrent mutation.

Despite the loss of S genes in this way, the mutation became polymorphic in several populations. Over a large part of Africa south of Sahara and north of Zambia, the frequency of HbS heterozygotes in many tribes is 20 per cent higher, rising to 40 per cent in some populations. Allison (1954*a*) suggested that the sickle-cell polymorphism might be maintained because HbS heterozygotes are relatively resistant to falciparum malaria and have a greater chance of surviving through the dangerous years of first exposure to the disease before immunity is acquired.

In support of this hypothesis, high frequencies of HbS were found in East Africa populations living in regions where falciparum malaria was endemic but not in populations living in non-malarious areas (Allison 1954*a*). This was true of populations with different linguistic affinities and blood groups, suggesting a strong selective effect of a local environmental factor such as malaria. In young Ugandan children, four months to four years of age, who had not yet acquired strong immunity to falciparum malaria, lower parasite counts were found in those with HbAS than in those with HbA (Allison 1954*b*). These observations were confirmed by several groups investigating African children, and potentially lethal infections such as cerebral malaria were observed much less frequently in children with HbAS than in those with HbA (reviewed by Allison 1964).

An important advance was the development by Trager and Jensen (1976) of a method for continuous propagation of *Plasmodium falciparum* in cultures of human erythrocytes. It was then possible to compare under controlled conditions the multiplication of the parasite in normal erythrocytes and in those with abnormal haemoglobins In two laboratories *P. falciparum* was found to grow less well in erythrocytes bearing HbS than those bearing HbA under low partial pressures of O_2 (Friedman 1978; Pasvol *et al.* 1978). Such relatively anoxic conditions would be expected where *P. falciparum* completes its asexual bloodstream cycle of replication attached to endothelial cells of post-capillary venules. Hence the findings with cultured cells nicely complement the observed protection provided by HbS *in vivo*. *Plasmodium falciparum* also grows poorly in erythrocytes bearing HbC and HbE (Friedman *et al.* 1979). The growth of the parasite in thalassaemic cells is strongly inhibited under conditions of oxidant stress (Friedman 1979). Oxidant stress could be produced by effector cells of the immune system bound to parasitized erythrocytes (Allison 1983). Thus acquired cell-mediated immunity, elicited by frequent attacks of falciparum malaria, would be expected to have synergistic effects with abnormal haemoglobins in protecting hosts against infections with this parasite. Sensitivity of erythrocytes to oxidant stress is increased in β-thalassaemia because the excess α-chains oxidise membrane lipids and thereby consume antioxidants such as reduced glutathione (Allison 1983).

The implications for population genetics of the abnormal haemoglobins were pointed out by Allison (1954*c*). Natural selection is expressed by the concept of fitness, which is the contribution of a genotype to the next generation and hence a measure of its capacity to survive and reproduce. For an autosomal locus with two alleles, when the heterozygote has an increased fitness over both homozygotes a stable equilibrium with both alleles can exist (Fisher 1930). It is termed a balanced polymorphism. Among the adult Baamba of Uganda, who have a frequency of about 40 per cent S heterozygotes, no S homozygotes have been observed, confirming the lethality of that condition in the African bush. If these alleles are close to equilibrium, the heterozygotes must have a fitness of about 1.25, which is the highest known fitness for any human genotype. The results suggest a mortality through direct or indirect effects of falciparum malaria of at least 25 per cent of children, in other words a powerful selective effect of the disease. Among the Baamba, 4 per cent of children born are S homozygotes liable to sickle-cell disease. Over vast areas of central Africa the frequency of S heterozygotes is about 20 per cent, implying that 1 per cent of the millions of children born are S homozygotes, creating a substantial public health problem. The S gene is also polymorphic in some non-African populations, including people living in formerly malarious parts of Greece, Turkey, southern Arabia, and India (see Livingstone 1967).

In the absence of selection in favour of the heterozygote (for example when Africans were removed to the US), the frequency of the S gene would be expected to fall exponentially (Allison 1954*c*), representing a transient polymorphism. The HbS frequency in different populations of US Blacks now varies from about 5 to 17 per cent (Livingstone 1967). Taking into consideration genetic variants, the incidence of sickle-cell disease in this population is about 1 to 300 births, an unfortunate genetic legacy of past selection.

The existence of natural selection can be demonstrated by comparing the observed phenotype and expected genotype frequencies in representive samples of young children, before selection has taken place, and samples of adults from the same community (Allison 1956). Where the frequency of S is high and thalassaemia is rare, as in certain East African populations, haemoglobin phenotypes can be equated with genotypes. From the frequency of the S gene in the adult population (p=in effect 0.5×the heterozygote frequency) and the Hardy-Weinberg equilibrium, the expected frequencies of S homozygotes, heterozygotes, and normal homozygotes are p^2, $2p(1-p)$, and $(1-p)^2$, respectively. The observed distribution in young children showed insignificant departure from this expectation whereas in the adult population there was a significant excess of S heterozygotes and deficiency of S homozygotes (Allison 1956).

Similar considerations apply to β-thalassaemia, which is polymorphic in some Mediterranean and other countries. In Sardinia, Greece, New Guinea, and several other countries β-thalassaemia is common in lowland areas, which were formerly malarious, and rare in montane regions which remained free of that disease. This distribution, along with the poor growth of *P. falciparum* in cultured thalassaemic erythrocytes, provides strong evidence that malaria was the major selective factor accounting for the polymorphism of this gene. The consequence is a high frequency of heterozygotes (up to 20 per cent) and a correspondingly high incidence of homozygotes with thalassaemia major (up to 1 per cent) in parts of Italy, Greece, Turkey, and some other countries, as well as migrants from these countries to North and South America and north-western Europe.

In some populations the co-existence of two or more abnormal haemoglobins complicates the population genetics. For example, over a large part of West Africa HbS and HbC co-exist, and in parts of Greece and Turkey HbS and β-thalassaemia co-exist. Because S-β-thalassaemia and SC heterozygotes have decreased fitness, Allison (1956) postulated that these genes must be mutually exclusive in populations; in other words, where the frequency of C or β-thalassaemia is high that of S must be low. Subsequent observations have confirmed that prediction. It is not known whether there is an equilibrium involving three alleles (which is theoretically possible) or whether in West Africa the C gene is being replaced by the S gene, as suggested by Livingstone (1967).

The relationship between the abnormal haemoglobins and malaria provided the first, and still the most direct and convincing, evidence that disease can act as an agent of natural selection, producing a change in gene frequencies, in other words evolution, in human populations.

Malaria and glucose-6-phosphate dehydrogenase (G6PD) deficiency

The existence of this sex-linked genetic trait in Americans of African origin was revealed by the correlation of G6PD deficiency with their sensitivity to the antimalarial drug primaquine (Carson *et al.* 1956). Later it became apparent that there are two polymorphic varieties of G6PD deficiency (the African A- and the Mediterranean B- forms-, a nondeficient form polymorphic in Africans (B+) as well as several rare deficient variants. Furthermore, G6PD individuals show haemolysis when exposed to several different drugs or dietary constituents, such a fava beans, which exert oxidant stress on erythrocytes.

As soon as an assay that could be used under field conditions became available, the distribution of G6PD deficiency was studied in East and West Africa (Allison 1960; Allison and Clyde 1961). The essential findings, which have been repeatedly confirmed, were that G6PD deficiency is common in malarious areas of East Africa (15–20 per cent of males affected) and in West Africa from Nigeria to the Gambia. It is absent from non-malarious parts of East Africa. Many studies (reviewed by Livingstone 1967) have supported the view that high frequencies of both the African and Mediterranean varieties of G6PD deficiency occur only where malaria has been endemic.

Allison and Clyde (1961) found significantly lower *P falciparum* rates and counts in the G6PD$^-$ children. Beinzle *et al.* (1972) reported that Nigerian female children heterozygous for the deficiency (Gd^{A-}/Gd^{A}) showed significantly lower *p. falciparum* counts than other children, and other observations are discussed by Allison (1983).

Roth *et al.* (1983) have shown that *P. falciparum* grows poorly in erythrocytes from persons hemizygous or heterozygous for the Mediterranean variety of G6PD deficiency in the presence of 17 per cent O_2. Presumably the protection would be even greater under conditions of oxidant stress (Friedman 1979), so that synergistic effects with cell-mediated immunity are likely. These observations, together with the distribution of G6PD deficiency and the observed protection against malaria, leave little doubt that malaria has been the major selective factor accounting for this polymorphism. Provided that female heterozygotes have a selective advantage, equilibrium can be maintained. Females heterozygous for G6PD deficiency show mosaicism of their erythrocytes (some having a normal content of G6PD, others full enzyme deficiency), which may be sufficient to provide some protection against malaria while rarely increasingly susceptibility to haemolysis from dietary or other factors, which are haemolytic in enzyme-deficient male hemizygotes.

The best known manifestation of this susceptibility to haemolysis is favism following ingestion by enzyme-deficient persons of fava beans (*Vicia fava*), which is common in Mediterranean and Middle Eastern countries. The susceptibility of G6PD-persons to haemolysis following exposure to drugs can pose public health problems. For example, in the region of Adana in Turkey *Plasmodium vivax* infection still occurs. The exoerythrocytic cycle of infection with this parasite, which allows recurrence, can only be terminated by the use of primaquine, but some 20 per cent of persons in the area are G6PD deficient and liable to severe haemolysis following treatment with the drug.

Since the G6PDd deficiency and abnormal haemoglobins protect against malaria, and individuals with both traits are not a disadvantage, a positive correlation of their frequency in populations would be expected, and is observed (Allison 1964).

The major histocompatibility complex

The HLA and H-2 complexes of man and the mouse respectively are the prototype MHC systems which have been analysed in detail, although comparable systems exist in all vertebrates so far examined. First studied with the intention of finding genetic markers for transplantation, the MHC emerged as a remarkable cluster of linked genes controlling cell surface determinants, which produce a great deal of genetic diversity in natural populations. The MHC determinants influence a variety of cellular interactions, including immune responses.

There are three classes of HLA factors: class 1 comprising HLA-ABC antigens; class 2 comprising D/DR and related antigens; and class 3 comprising certain complement factors (for example; C2, C4a, C4b, Bf). The ABC antigens are controlled by genes at three closely linked loci (Fig. 8.1). The D antigens are detectable by mixed leucocyte culture using homozygous typing cells, and the DR (D-related) antigens by serological methods using B-lymphocytes. Typing for D-anti-

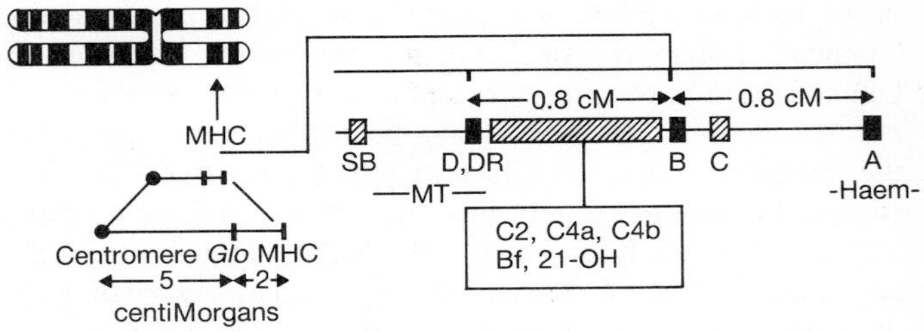

Fig. 8.1. Map showing the location of the MHC complex. The MHC complex is located on the short arm of human chromosome 6, approximately 7 centiMorgans from the centromere and 2 centiMorgans distal to the *Glo* locus which codes for the enzyme glyoxylase. Genes for SB are on the centromeric side of the D/DR region and for MT are in the DR region. The precise order of loci for complement components and 21-OH between DR and B is unknown. The recessive gene *haem* producing haemochromatosis is closely linked to A but its precise location is unknown.

gens with homozygous typing cells probably reveals grouped combinations of lymphocyte-activating determinants rather than single determinants. The D/DR region refers to a chromosome region controlling all antigens resembling DR. Both the α- and β-polypeptide chains of these class 2 molecules are encoded by genes within the D/DR region, which probably contains a number of different α- and β-genes (Shackleford *et al.* 1982). In general the β-chains seem to be more polymorphic than the α-chains. The MT, MB, DC, and LB antigens are thought to be present on β-chains different from those of the DR molecules, encoded by genes at closely linked loci. The primed lymphocyte typing technique has proved useful in studying D/DR antigens, including a new series of lymphocyte-stimulating antigens, termed SB, encoded by genes located on the centromeric side of the DR locus (Shaw *et al.* 1981).

Linked genes code for three classes or families of molecules denoted I, II, and III (Hood *et al.* 1982). The association of these genes over the 500 million years of vertebrate divergence suggests that selection constraints have maintained linkage among these gene families. The most remarkable feature of the MHC in the mouse and man is the extensive genetic polymorphism presented by certain class I and class II genes, but not all of them, suggesting functional requirements for diversity in some of the gene products and uniformity in others.

Class I molecules comprise a 45 000 daltons class I polypeptide, an integral membrane protein, noncovalently associated with β_2-microglobulin, a 12 000 dalton polypeptide encoded by a gene on another chromosome. These are expressed in most cell types. Class II antigens are encoded by genes of the I region in mice and DC, DR, and SB regions in man. Each class II molecule has an α-chain (30 000–33 000 daltons) and a β-chain (27 000–29 000 daltons). Class II molecules are expressed selectively on certain cell types, such as macrophages and B-lymphocytes, and it is known that antigens have to be presented to lymphocytes in association with class II molecules to elicit immune responses. The capacity of mice to respond well to defined antigens depends on their I-region (immune-response-associated) genes, which code for class II molecules. By analogy human DC, DR, and SB genes are likely to control capacity to respond to particular antigenic determinants, and therefore function as immune response genes. In this way they could influence susceptibility to a variety of diseases.

HLA haplotypes and linkage disequilibrium

The term haplotype is used to describe a chromosomal segment inherited *en bloc* from one parent. In a family such a segment is manifested by a particular combination of HLA-A, -B, and -DR phenotypes, from which probable genotypes can be derived. Within a family it can be assumed that the entire segment is identical unless there is evidence of recombination. When different families in a population are compared, combinations of phenotypes vary, but not in a random fashion. There is a tendency for certain genes at different loci to occur together on the same chromosome or haplotype more often than expected by chance. The existence of population associations between products of linked genes, such as antigens of the HLA loci, is termed linkage disequilibrium, which is important when considering associations of HLA with disease.

A well-known example is the A1, B8 haplotype. In north Europeans the frequency of allele A1 is 0.17 and that of B8 is 0.11, so that if they were not associated the expected frequency of the A1, B8 haplotype would be $0.17 \times 0.11 = 0.019$, which is less than one-fourth of the observed frequency, 0.088 (Bodmer and Bodmer 1978). The extent of linkage disequilibrium (Δ or D) can be defined as the difference between the observed and expected haplotype frequencies:

$$\Delta = 0.088 - 0.019 = 0.069.$$

In this case the contribution of linkage disequilibrium to the total haplotype frequency is 0.069/0.088, or nearly 80 per cent. Less than 10 of the nearly 300 possible pairwise combinations of HLA-A and-B locus alleles show significant linkage disequilibrium in western European and North American populations.

Theoretical analysis (see Bodmer and Bodmer 1978) shows that in a random mating population in the absence of selection, δ should be 0 at equilibrium. If δ is at any time not 0 it should approach 0 at a rate $(1\text{-}r)$ per generation, where r is the recombination fraction between the two loci. The more tight the linkage between two genes, the longer it will take linkage disequilibrium to reach zero. Hence one explanation for observed linkage disequilibrium is that there has not yet been time for populations to reach equilibrium with respect to the two relevant alleles. For HLA-A and -B alleles, given a recombination fraction of 0.8 per cent, linkage disequilibrium would in the absence of selection be expected to decrease by a factor of 5 in 200 generations, or some 5000 years. Thus failure to reach equilibrium is unlikely to explain persistent HLA-A and -B combinations found throughout Caucasoid populations which were separated more than 5000 years ago. Likewise, non-random mating, migration, and population admixture are unlikely explanations for observed linkage disequilibrium (Bodmer and Bodmer 1978). It is therefore probable that the persistent linkage disequilibrium observed for some pairs of alleles at the HLA loci is due to the effects of natural selection. This does not necessarily imply interactive selection between the alleles, since linkage disequilibrium can persist for considerable lengths of time as a result of selection affecting alleles at one or more loci closely linked with those under consideration. Nevertheless, as Fisher (1930) showed, natural selection favouring pairwise combinations of alleles at different loci can produce persistent linkage disequilibrium. Such a model is consistent with recent observations on the associations of HLA phenotypes with disease and with the molecule biology of the antigens, as discussed below.

Table 8.1. Estimates of the strength of an association between a genetic marker and a disease

(1) The 2 × 2 Table:

	Percentage of individuals	
	Genetic marker present	Genetic marker absent
Patients	a	b
Controls	c	d

Frequency of marker in patients: $h_p = \frac{a}{a+b}$

(2) Relative risk (RR) $= \frac{a \times d}{b \times c}$

(3) Attributable fraction (AF) $= \left(\frac{RR - 1}{RR}\right)\left(\frac{a}{a+b}\right) = \left(\frac{RR - 1}{RR}\right) h_p$

(Positive associations: RR >1)

(4) Preventive fraction (PF) $= \frac{(1 - RR)\, h_p}{RR(1 - h_p) + h_p}$

(Negative associations: RR <1)

Adapted from Svejgaard *et al.* (1983).

Relative risk

The strength of an association between a genetic marker and a disease is defined by the relative risk, a term originally suggested by Woolf (1955). The relative risk indicates how many times more frequently a disease develops in individuals carrying an antigen or other genetic marker than in individuals lacking it. Traditionally, relatively risk is calculated from the 2×2 table shown in Table 8.1 and is the cross-product ratio or odds ratio for alleles at the loci being considered.

Although this is a useful term, its limitations should be recognized. For example, the ascertainment of patients is a limiting factor, particularly when a disease is heterogeneous and/or is expressed only in ageing subjects; the prevalence of the disease in the population in question must be known if a true relative risk is to be calculated (see Svejgaard *et al.* 1983). Especially when different races are compared, control frequencies of genetic markers and the prevalence of the disease in question may vary greatly and can make a major contribution to cross-product ratios (see Dawkins *et al.* 1983). Thus the whole 2×2 table and derivations from it should be presented and not just cross-product ratios. The 2×2 table should be adjusted so that the terms $(a+b)$ and $(c+d)$ are equal (for example, by using percentages). If it is intended to estimate the likelihood of development of a disease in a random member of the population, it is essential that $(a+b)/(a+b+c+d)$ is approximately equal to the known prevalence of the disease.

Definition of diseases can be difficult for many reasons. Taking the examples discussed more fully below, diabetes mellitus is the end result of several different disease processes, each of which has a distinct set of genetic associations. Ankylosing spondylitis has a component of sacroiliitis but the latter may not be accompanied by ankylosis of vertebrae (Dawkins *et al.* 1983). The disease group should have unequivocal disease, using if possible internationally standardized criteria; the control group should have no sign of disease. If different races are being compared, it should be established that the diseases under study are truly comparable, no easy task.

Another estimate of the strength of an association between HLA and disease is the so-called attributable fraction (AF) calculated as shown in Table 8.1. The attributable fraction provides an estimate of the extent to which a disease is due to a disease-associated factor, and is well known from epidemiology (Miettinen 1976). When the relative risk exceeds unity, the attributable fraction can under certain conditions provide information about the degree of linkage disequilibrium between an HLA gene and a hypothetical disease susceptibility gene (Bengtsson and Thomson 1981). When the relative risk is less than unity, the so-called preventive fraction (PF) can be calculated.

The difference between relative risk and attributable fraction can be illustrated by the increased frequencies of the complement factor BfF1 and of DR3 in insulin-dependent diabetes mellitus (Svejgaard *et al.* 1983). BfF1 confers a higher relative risk for the disease (15.0) than does DR3 (2.9). However, BfF1 has a much lower frequency in the healthy population than DR3, so the attributable fraction for BfF1 is only 0.21 whereas that of DR3 is 0.35. The presence of BfF1 in an individual is a better indicator of insulin-dependent diabetes mellitus than is the presence of DR3. At the population level BfF1 contributes less to the development of the disease than does DR3.

Another approach designed to overcome problems of varying age of onset and incomplete penetrance is sib-pair analysis (Day and Simmons 1976; Thomson and Bodmer 1977). Analysis of HLA haplotype sharing in affected sib pairs eliminates both of the above problems. If HLA and/or genes closely linked to HLA have no influence on the development of a disease, affected sib-pairs will share haplotypes as expected: 25 per cent will share both, 50 per cent will share one and 25 per cent will share no haplotypes. By investigating whether the observed distribution of haplotype sharing differs from that expected, the null hypothesis can be tested. Parents of affected sibs should be HLA typed to establish the expected haplotype sharing.

Association of immune response and immune suppression genes with the major histocompatibility complex

McDevitt and Benacerraf (1969) showed that some strains of mice and guinea pigs are able to respond well, while others respond poorly, to particular synthetic peptide antigens, which present a limited number of antigenic determinants. The immune response (*Ir*) genes controlling such differences were found to map in the I-region of the mouse MHC. Immune suppression (*Is*) genes map in the same region (Benacerraf and Germain 1978; Dorf 1981). Defined products of the I region in the mouse are I-A (immune-response-associated) or class 2 antigens, the composition of which has been described above.

Three major findings have helped to define the role of MHC products in immune responses. The first is the phenomenon of genetic restriction. Rosenthal and Shevach (1973) found that guinea pig T-lymphocytes respond well to antigens only when they are on the surface of histocompatible antigen-presenting cells. Zinkernagel and Doherty (1974) found that sensitized T-lymphocytes kill virus-infected target cells only when the latter are histocompatible. Later analyses of these phenomena showed that the helper subset of T-lymphocytes requires exposure to both foreign antigens and self class 2 MHC products to respond with the production of interleukin-2, a growth factor for other subsets of lymphocytes. The cytolytic subset of T-lymphocytes, which is generated by exposure to antigen in the presence of interleukin-2, releases interferon-γ and other mediators, and kills target cells bearing virus or other antigens on their surfaces, only when these antigens are associated with class 1 MHC products. In other words, the class 1 and class 2 MHC products are recognition units required for cellular co-operation in the elicitation of immune responses and the effector phase of immune responses, respectively. Evidence that class 2 antigens of the MHC complex restrict *Ir*-gene controlled proliferation of T-lymphocytes had been reviewed by Nagy *et al.* (1981), who conclude that certain genotypes lack the capacity to respond to particular antigenic determinants because of gaps in the MHC product repertoire. Such failures to respond might on the one hand increase susceptibility to infectious diseases and on the other hand prevent autoimmune diseases.

For many years it was believed, following Burnet (1949), that clones of antibody-producing B-lymphocytes reacting with autoantigens are eliminated or inactived during ontogeny. However, normal persons were found to have B-lymphocytes with receptors for autoantigens (Bankhurst *et al.* 1973), so that self-tolerance is not due to deletion of auto-antibody-forming clones but to immunoregulation: the absence of T-lymphocyte help for autoantigens and the presence of a subset of T-lymphocytes suppressing autoimmune responses (Allison *et al.* 1971). Abundant experimental evidence in support of this interpretation has accumulated (reviewed by Allison 1977). Hence definition of *Is*-genes linked to the human MHC, and analysis of their role in controlling alloimmune and autoimmune responses, is of special interest. These two processes are not unrelated: growth and differentiation factors produced as a result of immune responses to alloantigens can stimulate proliferation and differentiation of clones of lymphocytes with receptors for autoantigens, with resulting autoantibody formation. Thus the requirements for antigen-presenting cells with class 2 structures reacting functionally with autoantigens, and of helper T-lymphocytes with receptors for autoantigens, can be by-passed (Allison 1977). An example is the formation of autoantibodies during graft-versus-host reactions (Failkow *et al.* 1973; Grebe and Streilein 1976).

Attempts have been made to define *Ir*- and *Is*-genes linked to human MHC complex markers. Hsu *et al.* (1981) studied HLA-associated genetic control of human lymphocyte proliferative responses to several synthetic polypeptides. On the basis of stimulation indices, individuals could be classified as high, intermediate, or low responders. Family studies suggested that high responses to these antigens are inherited as HLA-linked dominant traits. A strong association of Dw2 with IgE and IgG antibody responses to the ragweed allergen Ra5 has been reported (Marsh *et al.* 1982). Analyses of killing of virus-infected and allogenic cells by human T-lymphocytes (McMichael *et al.* 1977; Dickmeiss *et al.* 1977) indicate that some MHC products are superior to others in the presentation of foreign antigens to immunocompetent cells. Although all these findings are preliminary, they are consistent with the presence in humans of *Ir*-genes linked to defined MHC antigens.

Evidence for genetic control of T-lymphocyte-mediated suppression in humans has been published by Sasazuki *et al.* (1983). Proliferative responses of peripheral blood cells to streptococcal cell wall antigen were found to be inherited in a manner suggesting that a high response is recessive and a low response is dominant; suppression in culture was mediated by the generation of antigen-specific Leu-2a$^+$, 3a$^-$lymphocytes, not monocytes. Sasazuki and his colleagues postulate the existence of an *Is*-gene in linkage disequilibrium with HLA-MT1. The same authors described an analogous antigen-specific suppressor cell under the control of an *Is*-gen in linkage disequilibrium with HLA-Bw52, Dw12, in the presence of which the response of lymphocytes to schistosomal antigen is low. Absence of the *Is*-gene in homozygous individuals, who show a strong immune response in the antigen, predisposes them to post-schistosomal cirrhosis.

Sasazuki *et al.* (1983) also analysed 104 families with allergy to clear pollen allergen, and obtained evidence for another HLA-linked *Is*-gene suppressing IgE responses to the allergen through antigen-specific Leu-2A$^+$, 3a$^-$lymphocytes. Absence of this *Is*-gene in the homozygous state predisposed individuals to cedar pollinosis because of their high IgE responses to the allergen. This *Is*-gene was found to be in linkage disequilibrium with HLA-Dw44, DEn-MT1, 2-MB3. Sasazuki and his colleagues suggest that the statistical associations of many diseases with MHC products might be explained by the presence of *Is*-genes in strong linkage disequilibrium with particular alleles at loci in the HLA-D region such as D, DR, MT, and MB.

This is an attractive hypothesis which requires rigorous testing through international co-operative studies. The tests will not be easy, since correlations of *in vitro* suppression tests using human peripheral blood lymphocytes with *in vivi* immune responses, including autoimmune responses, are not straightforward. Nevertheless, let us suppose that several dominant *Is*-genes linked with DR genes exist, and that there are two recessive alleles, one linked with DR3, and concomitantly with B8 through linkage disequilibrium, and the other linked with DR4. When the recessive *Is* alleles are both expressed in homozygous form, in some DR3/4 heterozygotes, major suppressor mechanisms will not be operative: immune responses to both alloantigens and autoantigens will be high and prolonged. Let us postulate, further, that one of the dominant *Is* alleles is linked with DR2 and concomitantly with B7; in the presence of this allele immune responses to alloantigens and autoantigens will be suppressed.

This example can be compared with what is actually found. Eddlestone and Williams (1978) pointed out that several diseases in which immune responses to alloantigens and autoantigens are believed to play a pathogenic role are associated with B8: gluten-sensitive enteropathy, Graves' disease, Addison's disease, chronic active hepatitis, and insulin-dependent diabetes mellitus. Subjects with B8 showed high levels not only of autoantibodies but also of antibodies against exogenous bacterial and virus antigens. Diabetes with B8 likewise had higher titres of antibodies against Coxsackie viruses than those with other genotype. In insulin-dependent diabetes mellitus B8 is associated not so much with the presence of autoantigens reacting with pancreatic islet cells as with their persistence for years after the onset of the disease (see below). B8 is associated with a high relapse rate in thyrotoxic patients after a full course of medical treatment. Insulin-dependent diabetes mellitus, Graves' disease, and Addison's disease are now known to be more strongly associated with DR3 than the B8 (Svejgaard *et al.* 1983). Two other organ-specific autoimmune diseases (the Sicca syndrome and myasthenia gravis) and a generalized autoimmune disease, lupus erythematosus, have also been associated with DR3 (Table 8.2).

Two rare but instructive types of isoimmunization in pregnancy are associated with DR3. One is immunization against the platelet-specific antigen, Zw^a, in mothers lacking this antigen who give birth to children with isoimmune neonatal thrombocytopenia; this condition is strongly associated with B8 and more strongly with DR3 (Svejgaard *et al.* 1983). The second is herpes gestationis, a vesiculo-bullous eruption of the skin of pregnant women associated with autoantibody and complement deposition along the dermo-epidermal junction. Stasny *et al.* (1983) have found herpes gestationis to be

Table 8.2. Associations between HLA and some diseases

Condition	HLA	Frequency (%)		Relative risk	Attributable fraction
		Patients	Controls		
Hodgkin's disease	A1	40	32.0	1.4	0.12
Idiopathic haemochromatosis	A3	76	28.2	8.2	0.67
Behcet's disease	B5	41	10.1	6.3	0.34
Congenital adrenal hyperplasia	B47	9	0.6	15.4	0.08
Ankylosing spondylitis	B27	90	9.4	87.4	0.89
Reiter's disease	B27	79	9.4	37.0	0.77
Acute anterior uveitis	B27	52	9.4	10.4	0.47
Subacute thyroiditis	B35	70	14.6	13.7	0.65
Psoriasis vulgaris	Cw6	87	33.1	13.3	0.81
Dermatitis herpetiformis	D/DR3	85	26.3	15.4	0.80
Coeliac disease	D/DR3	79	26.3	10.8	0.72
	D/DR7 also increased				
Sicca syndrome	D/DR3	78	26.3	9.7	0.70
Idiopathic Addison's disease	D/DR3	69	26.3	6.3	0.58
Graves' disease	D/DR3	56	26.3	3.7	0.42
Insulin-dependent diabetes	D/DR3	56	28.2	3.3	0.39
	D/DR4	75	32.2	6.4	0.63
	D/DR2	10	30.5	0.2	—
Myasthenia gravis	D/DR3	50	28.2	2.5	0.30
	B8	47	24.6	2.7	0.30
SLE	D/DR3	70	28.2	5.8	0.58
Idiopathic membraneous nephropathy	D/DR3	75	20.0	12.0	0.69
IgA deficiency in blood donors	B8	49	24.3	3.0	0.33
	DR3	81	24.6	13.0	0.75
Zw^{a}-immunized mothers	B8	73	24.6	8.4	0.64
Multiple sclerosis	D/DR2	59	25.8	4.1	0.45
Optic neuritis	D/DR2	46	25.8	2.4	0.27
C2 deficiency	D/DR2				
	B18				
Goodpasture's syndrome	D/DR2	88	32.0	15.9	0.82
Rheumatic arthritis	D/DR4	50	19.4	4.2	0.38
Pemphigus (Jews)	D/DR4	87	32.1	14.4	0.81
IgA nephropathy	D/DR4	49	19.5	4.0	0.37
Hydralazine-induced SLE	D/DR4	73	32.7	5.6	0.60
Hashimoto's thyroiditis	D/DR5	19	6.9	3.2	0.13
Pernicious anaemia	D/DR5	25	5.8	5.4	0.20
Juvenile rheumatoid arthritis:					
pauciarticular	D/DR5	50	16.2	5.2	0.40
All cases	D/DRw8	23	7.5	3.6	0.17

The above information has been extracted from the HLA and Disease Registry (Svejgaard *et al.* 1983). Relative risks and attributable fractions are explained in Table 3.1.

strongly associated with DR3, and even more strongly with DR3/4 heterozygotes (relative risk 23.5). In women with herpes gestationis, Stasny and his colleagues found a high frequency of antibodies against foreign (often paternal) HLA antigens, suggesting that maternal immune responses against alloantigens against dermal antigens, perhaps by the mechanism discussed above for graft-verus-host disease.

DR3/4 heterozygotes also show increased susceptibility to insulin-dependent diabetes mellitus (see below). In contrast frequencies of DR2, and concomitantly B7, are decreased in coeliac disease and juvenile diabetes mellitus (Scholz and Albert 1983). All these findings are consistent with the model presented: that in some DR3/4 heterozygotes "switch off" signals of immune responses are deficient. The particular disorder manifested would depend on which powerful "switch on" signals for immune responses operate in that individual, determined by exposure to antigens in the presence of a complex of *Ir*-genes.

These observations do not, of course, confirm the model, and alternative explanations should be considered. For example, the increased susceptibility of DR3/4 heterozygotes to several diseases could result from the presence on antigen-presenting cells of unique class 2 molecules which allow them to interact efficiently with a range of alloantigens and autoantigens. Interactions of two *Ir*-genes based on hybrid I-region antigens have been reported (Fathman 1980). In the example quoted, heterozygotes could have unique antigens produced by combination of α- and β-chains of DR3 and DR4, namely $\alpha_3\beta_4$ and $\alpha_4\beta_3$; if more than one locus is involved, the heterogeneity of products would be greater. A clone of human T-lymphocytes has been obtained selectively recognizing restriction elements encoded by interacting D regions of two different haplotypes (Hansen *et al.* 1982).

The DR3, B8 combination appears again in a remarkable association with IgA deficiency (Oen *et al.* 1982; Hammarstrom and Smith 1982). IgA deficiency may be accompanied by increased susceptibility to pulmonary and/or gastrointestinal infections, but is also observed in about 1 in 700 apparently healthy blood donors. In groups of the latter strong associations with B8 and DR3 were found. Thus failure to secrete an entire class of immunoglulins is HLA related. This again appears to be a disorder of regulation rather than an absence of the structural genes of IgA, since the subjects have B-lymphocytes with surface membrane IgA.

Associations of HLA and disease

An international HLA and Disease Registry has been established in Copenhagen. This registry provides a comprehensive compilation of HLA and disease associations, based on published reports and information made available to the registry. The *Third Report of the HLA and Disease Registry* (Ryder *et al.* 1979) complied information up to 1979, and the list has been updated to 1982 by Svejgaard *et al.* (1983), see Table 8.2. In this table the principal known associations of HLA-A, -B, -C, and DR are summarized. Ankylosing spondylitis is more strongly associated with B27 than DR, suggesting that the former association is primary while the latter follows secondarily from linkage disequilibrium. Likewise the association of psoriasis with Cw6 appears to be primary. However, in many cases the association with D/DR is closer than with HLA-A, -B, or -C,. Since *Ir*- and *Is*-genes are presumed to be located in the D/DR region, it is not surprising that most of the D/DR-associated diseases are believed to have an immune component with their pathogenesis.

However, linkage with HLA or DR does not necessarily have that implication. An example is congenital adrenal hyperplasia due to 21-OH hydroxylase deficiency, which is an autosomal recessive trait. Studies of families with intra-HLA recombinants show that the locus concerned maps between HLA-B and D/DR genes (Dupont *et al.* 1980). Thus the locus controlling congenital adrenal hyperplasia, which may well code for the 21-OH hydroxylase enzyme, is probably associated with HLA-B47 through linkage disequilibrium rather than because of B47 antigen plays any role in the pathogenesis of congenital adrenal hyperplasia. The complete penetrance of 21-OH hydrozylase deficiency and its expression at birth greatly facilitates genetic analysis.

Other HLA-associated disorders have varying ages of onset, and presumably, penetrance of the genes in question. In the case of idiopathic haemochromatosis these difficulties have been to some extent overcome by affected sib-pair analysis (Simon *et al.* 1977) and by linkage analysis of HLA with serum transferrin saturation as marker (Kravitz *et al.* 1979). Both groups of investigators concluded that idiopathic haemochromatasis is an autosomal recessive disorder closely linked to HLA. The locus concerned and the biochemical mechanism controlled by it have not yet been defined, but may not involve HLA antigens directly. Nevertheless, the linkage with HLA allows the probable identification of siblings with haemochromatosis, and iron depletion can be commenced before organ damage develops.

In this chapter there is space to discuss only a few associations of HLA with diseases, to illustrate general points about the analysis of such relationships. One of the principal points to emerge is that many diseases are heterogeneous in their aetiology and genetic associations, for example diabetes mellitus. The heterogeneity of genetic and environmental interactions giving rise to disease, and of their responses to therapy, is a major lesson for social medicine. Two other examples will be mentioned to underline the point. Stasny *et al.* (1983) report that classical rheumatoid arthritis, with rheumatoid factor (RF) in adults, is associated with DR4 whereas in patients with clinical arthritis indistinguishable from rheumatoid arthritis, but who do not produce RF, the frequency of DR4 is the same as in controls. Only the minority of children with polyarthritis, who produce RF, show an increase in DR4. DR3 is more strongly associated with subacute cutaneous lupus erythematosus than with systemic lupus erythematosus.

Although the association of DR3/4 heterozygotes with several diseases is striking, it is not the only combination predisposing to disease. Coeliac disease is more prevalent in DR3/7 heterozygotes, perhaps because an *Ir*-gene for a sensitizing response to gluten is linked with DR7. Drug-induced diseases have genetic associations different from those of the naturally-occurring diseases which they mimic: hydralazine-induced systemic lupus erythematosus is not associated with DR3 (Svejgaard *et al.* 1983) and penicillamine-induced myasthenia gravis is associated with B235 and concomitantly DR1, in contrast to B8 and concomitantly DR3 in the naturally-occurring disease (Dawkins *et al.* 1983).

Ankylosing spondylitis

The finding by Brewerton *et al.* (1973) and Schlosstein *et al.* (1973) that antigen B27 is present in 85–95 per cent of patients withh ankylosing spondylitis, as compared with about 9 per cent of controls, provided convincing evidence for a genetic basis of susceptibility to the disease. With a relative risk of 87 in Caucasoids and 192 in Japanese, it remains the strongest known association of an HLA or D/DR antigen with a disease. An increase in frequency of B27 is found in ankylosing spondylitis associated with two other chronic rheumatic disorders, acute uveitis and Reiter's disease (see Sachs and Brewerton 1978). Even in the absence of sacroilitis, 15 of 38 patients with acute uveitis and 29 of 84 patients with Rieter's disease were also B27-positive. However, the distribution of B27 in non-specific urethritis was similar to that in healthy controls, suggesting that B27 is related to joint involvement rather than urethritis itself. Patients with reactive arthritides following other acute infections (such as those caused by *Salmonella* spp, *Yersinia enterocolitica*, and *Klebsiella pneumoniae*) also have a highly significant increase in B27 in the absence of features of ankylosing spondylitis. Patients with Crohn's disease and ulcerative colitis show an increased frequency of B27 only when they also have spondylitis. Hence the frequency of B27 is increased not only in patients with ankylosing spondylitis, which may be accompanied by Crohn's disease or ulcerative colitis, but also in patients with Reiter's disease, acute uveitis, and reactive arthritides in the absence of ankylosing spondylitis. Although the association of this group of chronic rheumatic disorders with B27 has been recognized for a decade, the underlying aetiological factors are still unknown and illustrate the difficulty of providing a precise genetical and biochemical explanation of HLA associations with diseases.

Genetic analyses of population and family data suggest that susceptibility to ankylosing spondylitis is due to a dominant gene with incomplete penetrance. Kidd *et al.* (1977) estimate that the frequency of the ankylosing spondylitis susceptibility gene in Caucasoids is 0.022 and that it is nearly always (93 per cent of the times) together in B27; the penetrance was estimated to be 38 per cent.

Dawkins *et al.* (1983) reported that in a Caucasoid population of Western Australia, sacroiliitis was essentially restricted to individuals with B27. The sacroiliitis was usually unaccompanied by ankylosing spondylitis and was found in about 18 per cent of B27 subjects of either sex above 50 years of age. Some patients with psoriasis, ulcerative colitis, and other diseases have sacroiliitis in the absence of B27, but even in these diseases, which are not themselves associated with B27, the presence of B27 is associated with sacroiliitis. More than 20 per cent of patients with these diseases develop sacroiliitis. They conclude that B27 confers susceptibility to sacroiliitis but is not sufficient. It seems likely that a gene present in 10–20 per cent of the population interacts with B27 to produce sacroiliitis. In diseases such as ulcerative colitis and psoriasis the second gene is common. A third gene is required for expression of ankylosing spondylitis, which is commoner in males than females, suggesting sex linkage or sex limitation.

Studies of various populations indicate the need for caution before B27 itself is identified as the principal predisposing factor to sacroiliitis and ankylosing spondylitis (Sachs and Brewerton 1978) In Japanese, a mixed Israeli population, and people indigenous to north India, the frequency of B27 in ankylosing spondylitis was as high as in Caucasoids. In Haida Indians of North America, with a 50 per cent frequency of B27, sacroiliitis was found in 10 per cent of males and B27 in all 27 ankylosing spondylitis cases tested. However, 38 per cent of 193 Pima Indians had evidence of sacroiliitis and of these only 26 per cent were B27 positive, which is similar to the frequency observed in the control population. The frequency of both ankylosing spondylitis and B27 is low in American Blacks and only 49 per cent of these patients have B27. In summary, the close association of ankylosing spondylitis with B27 in Caucasoid and Japanese populations suggests a disease susceptibility gene closely linked to B27; the lesser association in Blacks and Pima Indians may have resulted from recombination within the HLA region.

The genetic data therefore cast doubt on the identity of the B27 gene and the gene for susceptibility to sacroiliitis and ankylosing spondylitis. Nevertheless, the striking correlation between this HLA specificity and seronegative arthropathies has encouraged attempts to provide an immunochemical explanation for such associations (see Geczy *et al.* 1983). A serological relationship between the B27 antigen and certain membrane antigens and *Klebsiella pneumoniae* has been found. Antigens in the outer membrane of the organism can modify B27, or a closely related cell surface structure. The authors suggest that modified B27 elicits immune responses able to injure specific target tissues bearing altered self-determinants (for example, synovium and cartilage). The destruction of appropriate target cells may trigger a complex chain of events culmininating in the development of ankylosing spondylitis. In addition to the modification of B27 to provide the target antigen, an immune response gene (perhaps located in the D/DR region) may control the magnitude of the immune response which is elicited. Thus, despite the strong association of B27 with AS and seronegative arthropathies, which has long been known, the chemical and immunological basis of the association remains speculative.

Diabetes mellitus

Diabetes mellitus is a clinical syndrome characterized by an increase in blood glucose and an absolute or relative deficiency in insulin. It is a complex of diseases rather than a single entity, with several distinct genetic and environmental factors interacting to produce the end point of diabetes. It is usual to classify diabetics according to whether their disease can be controlled by oral hypoglycaemic drugs of requires insulin therapy, and whether the disease is of juvenile or adult onset. Some limitations of this classification will be mentioned below. More recentlv the role of viruses in pancreatic islet cell damage (Notkins 1977; Yoon *et al.* 1979), the existence of autoantibodies against islet cells (Doniach and Botazzo 1977; Irvine 1980), and the findings that some types of diabetes are associated with HLA antigens have provided further insight into the pathogenesis of the disease.

In insulin-dependent diabetes mellitus (IDDM) DR4 is increased in all ethnic groups, DR3 is increased in Caucasoids and DR8 in Japanese (Svejgaard *et al.* 1983). Analyses in Caucasoids showed the increases of B8 and B15 to be secondary to those of DR3 and DR4, respectively, with which they are in linkage disequilibrium. About 58.6 per cent of affected sib-pairs share both, 36.9 per cent share one and only 4.6 per cent share no parental haplotypes; this difference from the distribution expected on the assumption of no linkage is highly significant. The relative risk of developing IDDM is much greater in DR3/DR4 heterozygotes than in persons with other genotypes (Contu *et al.* 1982; Svejgaard *et al.* 1983). Juvenile-onset diabetes (JOD) is also highly significantly associated with the heterozygote type DR3/DR4 in all populations tested, including Basques who are classified as non-Caucasoids (Scholz and Albert 1983). The frequency of the DR2 antigen is decreased in JOD in all populations, suggesting linkage of DR2 with a gene protecting against this disease (Scholaz and Albert 1983).

Autoantibodies reacting with the cytoplasm of human pancreatic islet cells (ICAb) are much commoner in patients with IDDM than in those with diabetes that can be controlled by oral hypoglycemic agents (Doniach and Botazzo 1977; Irvine 1980). The prevalence of ICAb is some 60–80 per cent at the time of diagnosis of IDDM but only 20 per cent 2–5 years after diagnosis. ICAb are rare in diabetics whose disease can be controlled by diet and in persons without diabetes. Persistent ICAb are often associated with evidence of other signs of organ-specific autoimmunity, together with an increase in thyroid and gastric parietal cell antibodies in the first-degree relatives of patients. The increased frequency of DR3 again appears in patients with idiopathic Addison's disease (most of whom have adrenocortical autoantibodies, see Irvine 1980) and in Graves' disease (who have thyroid-stimulating autoantibodies), suggesting relaxed suppression of autoantibody formation in this group as a whole, as discussed above.

All these observations provide strong evidence for inherited predisposition to diabetes mellitus, although attempts to analyse the human data in terms of simple genetic models have failed (Svejgaard *et al.* 1983). Experimental animal models provide clues to the complexity of the situation. There has been interest in the role of viruses and toxic chemicals in the pathogenesis of diabetes mellitus. A variant of Coxsackie B4 virus isolated from the pancreatic islet cells of a boy dying acutely of diabetes was found to be more diabetogenic in experimental animals than most strains of Coxsackie B4 virus (Yoon *et al.* 1979). The pancreatic islet cells of different strains of mice vary in susceptibility to infection with certain viruses (Notkins 1977). Viruses not only damage infected cells; they can elicit autoimmune responses (Allison 1977). The inherited predisposition of certain strains of mice, rats, and chickens to autoimmune diseases is well known (see Allison 1977). Toniolo *et al.* (1980) reported that subdiabeotgenic doses of streptozoticin can condition mice so that they are more prone to develop diabetes following infection with Coxsackie virus. Moreover, strains of mice resistant to encephalomyocarditis virus-induced diabetes can be rendered susceptible by pretreatment with sub-diabetongenic doses of streptozoticin (Toniolo *et al.* 1980). The association of diabetes with obesity is well known. Thus, in human diabetes mellitus, inherited susceptibility factors are likely to interact with infectious agents, toxic chemicals, immune, metabolic, and other factors in a complex manner. Nevertheless, attempts should be made to identify major pathogenic factors operating in groups of patients.

The limitations of existing classifications of diabetes mellitus have been emphasized by Irvine (1980). In some centres patients are treated with insulin before systematic attempts are made to control the disease by oral hypoglycaemic agents; such patients may be classified as insulin-dependent when they are not. ICAb are found in IDDM of both juvenile and adult onset, and their presence may be a more reliable guide to pathogenesis than age of onset or insulin requirement, since either could follow from a non-immunological cause such as a virus infection or particular sensitivity to a chemical agent. Most studies of a relationship between ICAb and HLA have concerned the association with B8 (Irvine 1980), whereas in IDDM and JOD linkage with DR3 and DR4 is tighter and deserves further analysis in relation to ICAb. Identification of individuals with ICAb who are likely to develop insulitis and consequent diabetes is an important practical consideration. Rats genetically predisposed to diabetes treated with the immunosuppressive drug cyclosporin A do not develop the disease (Laupacis *et al.* 1983), and preliminary evidence from the same group of investigators suggests that cyclosporin A treatment may also be beneficial in some human patients with diabetes mellitus.

Coeliac disease

Coeliac disease is an enteropathy associated with sensitivity to ingested gluten. Dissecting microscopical and routine histological examination show a flat or nearly flat jejunal mucosa, and gluten withdrawal is followed by clinical, haematological, and morphological improvement. An immune response conferring sensitivity to gluten is believed to be involved in the pathogenesis of the disease, although this is difficult to establish unambiguously. The observed linkage with HLA antigens is consistent with an immunological interpretation but does not prove it.

The association of B8 with coeliac disease was reported in the US by Falchuk *et al.* (1972) and in England by Stokes *et al.* (1972), and has been repeatedly confirmed in children and adults. In populations where the frequency of B8 is high, such as that of the west of Ireland, where it reaches 87 per cent, the prevalence of coeliac disease is also high (Mackintosh and Asquith 1978). Later, DR3 and DR7 were found to be significantly increased in groups of patients with coeliac disease (Scholz and Albert 1983). The primary association with DR3 leads to secondary associations with B8 and A1 through linkage disequilibrium. There is a decrease in the frequency of DR2, and concomitantly that of B7, in coeliac disease, suggesting that DR2 and B7 are associated with a protective gene. A remarkably constant feature in different populations is the high frequency of DR3/DR7 heterozygotes in the different patient populations (Scholz and Albert 1983). The number of DR3/7 heterozygotes in a group of patients exceeded the sum of the two types of homozygotes, DR3/3 and DR7/7. This suggests the simultaneous presence of two susceptibility genes, one associated with DR3 and the other with DR7, rather than a recessive susceptibility gene associated with both DR3 and DR7.

Analysis of affected sib-pairs showed 56 per cent sharing two parental HLA haplotypes, 40 per cent one haplotype, and only 4 per cent sharing no haplotypes, confirming the linkage between HLA and coeliac disease. Combining the population and sib-pair data, Scholz and Albert (1983) conclude that at least three different HLA-linked genes influence susceptibility to coeliac disease: one gene associated with DR3 and another with DR7 increase susceptibility while a third associated with DR2 provides resistance.

From the practical point of view, typing could be useful for identifying patients, but only when the clinical picture and/or the results of other screening tests make the diagnosis of coeliac disease possible (Mackintosh and Asquith 1978). Even susceptible DR3/DR7 heterozygotes should not have jejunal biopsies unless there is a strong indication. In coeliac families typing could be useful to identify individuals at risk before weaning on to gluten-containing foods.

Changing opportunities for selection

In any population natural selection occurs because different genotypes produce different numbers of offspring and because different proportions of their offspring survive to the age of reproduction. Differential reproduction and survival provide the opportunity for selection. The demographic transition from a period of high birth and death rates to the lower birth and death rates in more advanced countries has obviously reduced the opportunity for selection, but has not eliminated it altogether. In these countries both birth and death rates are largely controlled. In 1800 the average American women passing through reproductive life had seven children; today she has three or less (Kirk 1968).

Nevertheless about 1 in 10 married women remains childless, with infertility resulting from infectious disease of the reproductive tract or other causes, so that some opportunity for selection still occurs. Mortality is to a large extent postponed beyond the age of reproduction, which implies that the opportunity for selection in this way is greatly reduced. In 1840 about 50 per cent of American women died before the end of the reproductive years whereas the figure is now only about 4 per cent. For males the present figure is a little higher, about 6 per cent.

Crow (1966) has proposed an *index of opportunity for selection* that tells how much potential genetic selection is inherent in the pattern of births and deaths. Of course these differences may not be heritable—hence the term opportunity for selection. If a trait is completely heritable and perfectly correlated with fitness, the index tells its rate of increase. Otherwise, and this is the situation in practice, it provides only an upper limit.

The index, I, is defined as V/x^2, where V is the variance and x the mean number of progeny per parent. This can be separated into components due to the mortality and to fertility differences. Table 8.3 shows the trends in the fertility component of the index in the US. The pattern has changed from uniform high fertility (x small, I_f small) through low fertility with considerable variability from family to family (x small, I_f large) to the most recent situation where the family size is more uniform and I_f is again small. Meanwhile, the index due to mortality has dropped from about 1 for those born a century ago to less than 0.1 for current death rates.

Table 8.3. Changes in mean number of children and the fertility component of the index of opportunity for selection, I_f, in the US

Date of mother's birth	Mean no. children	I_f
1839	5.5	0.23
1861–65	3.9	0.78
1901–05	2.3	1.14
1911–15	2.2	0.97
1921	2.5	0.63

Note: This includes both white and non-white and both married and unmarried. Data from US Census (Crow 1966).

These values quantify what has already been pointed out, that the opportunity for selection from differential postnatal mortality is greatly reduced. Prenatal mortality has changed much less, and probably contributes at least 0.3 to the total index. The opportunity for selection from fertility differences first increased and then decreased.

If all the differential viability and fertility were genetic, the total index would still be large enough to produce a considerable amount of selection. But the more important question of how much genetically effective selection is still occurring cannot be answered in general terms: specific examples must be considered.

Some trends are clear: infectious diseases have been drastically reduced in many parts of the world. A few decades ago a gene producing a decreased susceptibility to smallpox would have had a selective advantage; now that smallpox is eliminated the advantage has disappeared. Genes for resistance to tuberculosis and other bacterial infections would have had great selective advantages a century ago, whereas in Europe and the US the advantage would now be slight. Nevertheless, malaria and bacterial diseases still exert considerable selection in developing countries.

The greater mobility of contemporary populations will also have genetic consequences; there is certain to be less inbreeding as persons tend to find mates away from their home environs. This should decrease the incidence of rare recessive diseases and cause some increase in general health and vigour—although the latter may not be measurable directly. A second consequence of mobility may be an enhanced degree of assortative marriage. The greater participation of higher education, the stratification of students by aptitude, and the growth of communities with similar interests and attainments all can lead to increased correlations between husband and wife. Added to this is the greater range of choice created by affluence and mobility so that any inherent preferences for assortative marriage are more easily realized. The effect of assortative marriage is to increase the population variability. There is already a high correlation in IQ between husband and wife, and this may well increase. To the extent that this trait is heritable there will be greater variability in the next generation than would otherwise be the case: more geniuses as well as more at the other end of the scale.

The population always carried a number of deleterious genes. Many of them are recessive, or nearly so, and owe their incidence to mutation and opposing selective forces. They may or may not be at equilibrium; probably many are not, for the conditions determining the equilibrium may change faster than the time necessary for equilibrium to be reached. In any case, an environmental change that makes a gene less likely to cause death or sterility will cause that gene to be more frequent in later generations than it would otherwise be. How great is this effect?

Consider first a rare recessive gene causing a disease that can now be treated. Let p be the relative frequency of this gene. Then, with random mating, the proportion of persons homozygous for this gene will be p^2. Let the probability of death or infertility in untreated cases (relative to the normal population) be s and in treated cases t. Then, as a result of the treatment, a fraction $(s-t)p^2$ recessive genes will be transmitted to the next generation that, in the absence of the treatment, would be eliminated by death or failure to reproduce. If p' is the proportion of recessive genes that would otherwise be present in the next generation, the proportion of harmful genes in the next generation is $p'+(s-t)p^2$. The incidence of the disease is the square of this.

Ordinarily p and p' are very similar, for the gene frequency is likely to be near equilibrium. Letting $p'=p$, the incidence in the next generation is $p^2\ (1+ip)^2$, where $i=s-t$ is the improvement produced by the treatment, measured in terms of survival and reproduction.

The proportion by which the incidence is increased is

$$\frac{p^2\ (1+ip)^2}{p^2} = 2ip,$$

approximately, where p is small.

A familiar example is phenylketonuria, which can now be treated with considerable success by a low phenylalanine diet. If untreated persons hardly ever reproduce, s is nearly 1. If the treatment is fully successful, $t=0$; so $i=1$. The gene frequency, p, in this case is about 0.01; thus the proportion of increase is $2ip=0.02$. An increased incidence of 2 per cent per generation would mean about 40 generations for the incidence to double. Even allowing for a geometric, not an arithmetic increase, it would take close to 1000 years for this to occur.

The situation is quite different with a rare dominant gene. In this case, if p is the frequency of the gene in this generation, and p' the frequency in the next generation in the absence of treatment, the fraction of deleterious genes added as a consequence of the treatment is ip. Thus, the proportion in the next generation is $p'+ip$, or if p and p' are approximately equal, the proportion by which the incidence is increased is simply i.

If the mutant has a constant but reduced penetrance, P, the proportion of increase is approximately iP. In summary, the proportion increase in incidence in the next generation for simple inheritance is approximately:

Recessive	$2pi$
Dominant	i
Dominant with penetrance P	Pi

In practice, the relatively rapid increase in frequency of dominant genes due to improved medical care may not be important because of genetic counselling. If the disease is severe or painful, the person with it will not wish to inflict it on his or her children. If the disease is mild or treatment is effective, the parents may not hesitate to expose their children to a risk of up to 50 per cent of developing the disease.

In more complex cases involving interactions of several genes and environmental factors, such as diabetes mellitus, only broad generalizations can be offered. If the condition depends on recessive genes (for example, recessive alleles of *Is*-genes in the example discussed above), the rate of increase will still be slow.

Whether conditions have been so stable in the past that most genes were near equilibrium frequencies, it is evident that environmental factors relevant to the selective value of many genes are changing rapidly—far more rapidly than the gene frequencies can change. Although improved medical care is certainly increasing the frequency of disease-producing genes in human populations, the rate of increase does not justify the alarmist predictions that some distinguished geneticists have made. The benefits clearly outweigh the disadvantages, although the latter must be recognized and dealt with if possible, for example by genetic counselling and prenatal detection of abnormalities.

Glossary

Attributable fraction: The extent to which a disease is due to a genetic or other disease-associated factor (see pp. 123–4 and Table 8.1).

Fitness: The contribution of a genotype to the next generation. It is the product of the probability of survival (to and through reproductive age) and fertility, and is expressed by comparison with other genotypes.

Haplotype: This term (from haploid genotype) describes a combination of genetic determinants that leads to a set of antigenic specificities (or other gene products) which is controlled by one chromosome and so inherited together from one or other parent.

Linkage disequilibrium: This is the existence in populations of associations between products of linked genes such as HLA antigens, for example A1, B8, and DR3 (see p. 123).

Opportunity for selection: The amount of potential genetic selection that is inherent in the pattern of births and deaths in a population. This can be estimated from the index described on p. 130.

Polymorphism: The presence of two or more allelic genes in a population in frequencies higher than expected from recurrent mutation. Polymorphism can be transient or balanced. When it is balanced, opposing selection forces tend to stabilize gene frequencies, for example, when a heterozygote is at a selective advantage over either homozygote (Fisher 1930). In the case of sex-linked genes, if female heterozygotes have an advantage over other genotypes, balanced polymorphism can be attained (Cavalli-Sforza and Bodmer 1978).

Relative risk: The prevalence of a disease in persons carrying a genetic marker as compared with its prevalence in persons from the same population lacking the marker (the latter expressed as unity) (see pp. 123–4 and Table 8.1).

Selection: Unequal survival and reproduction of several genotypes resulting from differences in fitness.

References

Allison, A.C. (1954*a*). The distribution of the sickle-cell trait in East Africa and elsewhere, and its apparent relationship to the incidence of subtertian malaria. *Transactions of the Royal Society of Tropical Medicine and Hygiene* **48**, 312.

Allison, A.C. (1954*b*). Protection by the sickle-cell trait against subtertian malarial infection. *British Medical Journal* **i**, 290.

Allison, A.C. (1954*c*). Notes on sickle-cell polymorphism. *Annals of Human Genetics* **13**, 39.

Allison, A.C. (1956). The sickle-cell and haemoglobin C-genes in some African populations. *Annals of Human Genetics* **21**, 67.

Allison, A.C. (1960). Glucose-6-phosphate dehydrogenase deficiency in red blood cells of East Africans. *Nature* **185**, 531.

Allison, A.C. (1964). Polymorphism and natural selection in human populations. *Cold Spring Harbor Symposium on Quantitative Biology* **29**, 137.

Allison, A.C. (1977). Autoimmune diseases: concepts of pathogenesis and control. In *Autoimmunity, genetic, immunologic, virologic and clinical aspects* (ed. N. Talal) p. 91. Academic Press, New York.

Allison, A.C. (1983). Cellular immunity to malaria and babesia parasites: a personal viewpoint. In *Progress in immunobiology* (ed. J. Marchalonis). Raven Press, New York.

Allison, A.C. and Clyde D.F. (1961). Malaria in African children with deficient erythrocyte glucose-6-phosphate dehydrogenase. *British Medical Journal* **i**, 1345.

Allison, A.C., Denman, A.M., and Barnes, R.D. (1971). Co-operating and controlling functions and thymus-derived lymphocytes in relation to autoimmunity. *Lancet* **ii**, 135.

Bankhurst, A.D., Torrigiani, G., and Allison, A.C. (1973). Lymphocytes binding human thyroglobulin in healthy people and its relevance to tolerance for autoantigens. *Lancet* **i**, 115.

Benacerraf, B. and Germain, R.N. (1978). The immune response genes of the major histocompatibility complex. *Immunology Review* **38**, 70.

Bengtsson, B.O. and Thompson, G. (1981). Measuring the strength of associations between HLA antigens and diseases. *Tissue Antigens* **17**, 356.

Bienzle, V., Ayent, O., Lucas, A.O., and Luzzatto, L. (1972). Glucose-6-phosophate dehydrogenase and malaria. Greater resistance of females heterozygous for enzyme deficiency and of males with a nondeficient variant. *Lancet* **i**, 107.

Bodmer, W.F. and Bodmer, J.G. (1978). Evolution and function of the HLA system. *British Medical Bulletin* **34**, 309.

Brewerton, D.A., Caffrey, M., Hart, F.D., James, D.C.O., Nicholls, A., and Sturrock, R.D. (1973). Ankylosing spondylitis and HLA-27. *Lancet* **i**, 904.

Burnet, F.M. (1949). *The clonal theory of acquired immunity*. Cambridge University Press, London.

Carson, P.E., Flanagan, C.L., Ickes, C.E., and Alving, A.S. (1956). Enzymatic deficiency in primoquine-sensitive erythrocytes. *Science* **124**, 484.

Cavalli-Sforza, L.L. and Bodmer, W.F. (1978). *The genetics of human populations*. W.H. Freeman, San Francisco, California.

Crow, J. (1966). The quality of people: human evolutionary changes. *Bioscience* **16**, 863.

Contu, L., Deschamps, I., Estradet, H. *et al.* (1982). HLA haplotype study of 53 juvenile insulin-dependent diabetic (I.D.D.) families. *Tissue Antigens* **20**, 123.

Dawkins, R.L., Christiansen, F.T., Kay, P.H. *et al.* (1983). Disease associations with complotypes, supratypes and haplotypes. *Immunology Review* **70**, 5.

Day, N.E. and Simons, M.J. (1976). Disease susceptibility genes. Their identification by multiple case family studies. *Tissue Antigens* **8**, 109.

Dickmeiss, E., Soberg, B., and Svejgaard, A. (1977). Human cell-mediated cytotoxicity against modified target cells is restricted by HLA. *Nature* **270**, 526.

Doniach, D. and Bottazzo, G.F. (1977). Autoimmunity and the endocrine pancreas. *Pathobiological Annual* **7**, 327.

Dorf, M.E. (1981). Genetic control of immune responsiveness. In *The role of the major histocompatibility complex in immunobiology* (ed. M. E. Dorf) p. 221. Garland, New York.

Dupont, B., Pollack, M.S., Levine, L.S., O'Neill, G.J., Hawkins, B.R., and New, M.I. (1980). Congenital adrenal hyperplasia. *Joint report in histocompatibility testing 1980* (ed. P.I. Teraski) p. 693. UCLA Tissue Typing Laboratory, Los Angeles, California.

Eddlestone, A.L.W.F. and Williams, R. (1978). HLA and liver disease. *British Medical Journal* **34**, 295.

Falchuk, Z.M., Rogentine, G.N., and Strober, W. (1972). Predominance of histocompatibility antigen HL-A8 in patients with gluten-sensitivity enteropathy. *Journal of Clinical Investigation* **51**, 1602.

Fathman, C.G. (1980). Hybrid I region antigens. *Transplantation* **30**, 1.

Fialkow, P.J., Gilchrist, C., and Allison, A.C. (1973). Autoimmunity in chronic graft-versus-host disease. *Clinical and Experimental Immunology* **13**, 479.

Fisher, R.A. (1930). *The genetical theory of natural selection*. Oxford University Press.

Friedman, M.J. (1978). Erythrocytic mechanism of sickle-cell resistance to malaria. *Proceedings of the National Academy of Science, USA* **75**, 1994.

Friedman, M.J. (1979). Oxidant damage mediates variant red-cell resistance to malaria. *Nature* **280**, 245.

Freidman, M.J., Roth, E.F., Nagel, R.L., and Trager, W. (1979). The role of hemoglobins C,S, and N_{BALT} in the inhibition of malaria parasite development *in vitro*. *American Journal of Tropical Medicine and Hygiene* **28**, 777.

Geczy, A.F., Alexander, K., Bashir, H., Edmonds, J.P., Upfold, L., and Sullivan, J. (1983). HLA-B27, *Klebsiella* and ankylosing spondylitis: biological and chemical studies. *Immunology Review* **70**, 23.

Grebe, S.C. and Streilein, J.W. (1976). Graft-versus-host reactions: a review. *Advances in Immunology* **23**, 120.

Hammarstrom, L. and Smith, C.I.E. (1982). T cell clones restricted to 'hybrid' HLA-D antigens? *Journal of Immunology* **128**, 2497.

Hansen. G., Sönderstrup. G., Svejgaard, A., and Claesson, M.H. (1982). T cell clones restricted 'to hybrid' HLA-D antigens? *Journal of Immunology* **128**, 2497.

Hood, L., Steinmetz, M., and Malissen, B. (1983). Genes of the major histocompatibility complex of the mouse. *Annunal Review of Immunology* **1**, 529.

Hsu, S.H., Chan, M.M., and Bias, W.B. (1981). Genetic control of major histocompatability complex-linked responses to synthetic polypeptides in man. *Proceedings of the National Academy of Sciences, USA* **78**, 440.

Irvine, J. (1980). Immunological aspects of diabetes mellitus: a review (including the salient points of the NDDG report on the classification of diabetes). In *Immunology of diabetes* (ed. J. Irvine). Teviot Scientific Publications Edinburgh.

Kidd, K.K., Bernocco, D., Carbonara, A.O., Daneo, V., Steiger, U., and Ceppellini, R. (1977). Genetic analysis of HLA association diseases: the 'illness susceptible' gene frequency and sex ratios in ankylosing spondylitis. In *HLA and disease* (ed. J. Dausset and A. Svejgaard) p. 72. Munksgaard, Copenhagen.

Kirk, D. (1968). Patterns of survival and reproduction in the United States: implications for selection. *Proceedings of the National Academy of Sciences, USA* **59**, 662.

Kravitz, K., Skolnick, M., Cannings, C. *et al.* (1979). Genetic-linkage between hereditary haemochromatosis and HLA. *American Journal of Human Genetics* **31**, 601.

Laupacis, A., Gardell, C., Dupre, J., Stiller, C.R., Keown, P., and Wallace, A.C. (1983). Cyclosporin prevents diabetes in BB Wistar rats. *Lancet* **i**, 10.

Livingstone, F.B. (1967). *Abnormal hemoglobins in human populations*, p. 470. Aldine Press, Chicago.

McDevitt, H.O. and Benacerraf, B. (1969). Genetic control of specific immune responses. *Advances in Immunology* **11**, 31.

Mackintosh, P. and Asquith, P. (1978). HLA and coeliac disease. *British Medical Bulletin* **34**, 291.

McCusick, V.A. (ed.) (1978). *Mendelian inheritance in man*, p. 533. Johns Hopkins University Press, Baltimore.

McMichael, A.J., Ting, A., Zweerink, H.J., and Askonas, B.A. (1977). HLA restriction of cell-mediated lysis of influenza virus-infected human cells. *Nature* **270**, 520.

Marsh, D.G., Hsu, S.H., Roebber, M. *et al.* (1982). HLA-Dw2: a genetic marker for human immune response to short ragweed pollen allergen Ra5. I. Response resulting primarily from natural antigenic exposure. *Journal of Experimental Medicine* **155**, 1439.

Miettinen, O.S. (1976). Estimability and estimation in case-referent studies. *American Journal of Epidemiology* **103**, 226.

Nagy, Z.A., Baxevanis, C.N., Ishii, N., and Klein, J. (1981). Ia antigens as restriction molecules in Ir-gene controlled T-cell proliferations. *Immunology Review* **60**, 59.

Notkins, A.L. (1977). Virus-induced diabetes mellitus: brief review. *Archives of Virology* **54**, 1.

Oen, K., Petty, R.E., and Schroeder, M.L. (1982). Immunoglobulin A deficiency: genetic studies. *Tissue Antigens* **19**, 183.

Pasvol, G., Weatherall, D.J., and Wilson, R.J.M. (1978). Cellular mechanism for the protective effect of haemoglobin S against *P. falciparum* malaria. *Nature* **274**, 702.

Rosenthal, A.S. and Shevach, E.H. (1973). The function of macrophages in antigen recognition by guinea pig T lymphocytes. *Journal of Experimental Medicine* **138**, 1194.

Roth, E.F., Raventos-Suarez, C., Rinaldi, A., and Nagel, R.L. (1983). Glucose-6-phosphate dehydrogenase deficiency inhibits *in vitro* growth in *Plasmodium falciparum. Proceedings of the National Academy of Sciences, USA* **80**, 298.

Ryder, I.P., Anderson, E., and Svejgaard, A. (eds.) (1979). *HLA and disease registry. Third report*. Munksgaard, Copenhagen.

Sachs, J.A. and Brewerton, D.A. (1978). HLA, ankylosing spondylitis and rheumatoid arthritis. *British Medical Bulletin* **34**, 275.

Sasazuki, T., Nishimura, Y., Muto, M., and Ohta, N. (1983). HLA-linked genes controlling immune response and disease susceptibility. *Immunology Review* **70**, 51.

Schackelford, D.A., Kaufman, J.F., Korman, A.J., and Strominger, J.L. (1982). HLA-DR antigens: structure, separation of subpopulations, gene cloning and function. *Immunology Review* **66**, 134.

Schlosstein, L., Terasaki, P.I., Bluestone, R., and Pearson, C.M. (1973). High association of an HLA antigen, W27, with ankylosing spondylitis. *New England Journal of Medicine* **288**, 704.

Scholz, S. and Albert, E. (1983). HLA and diseases: involvement of more than one HLA-linked determinant of disease susceptibility. *Immunology Review* **70**, 77.

Shaw, S., Kavathas, P., Pollack, M.S., Charmot, D., and Mawas, C. (1981). Family studies define a new histocompatibility locus, SB, between HLA-DR and GLO. *Nature* **293**, 745.

Simon, M., Bourel, M., Genetet, B., and Fauchet, R. (1977). Idiopathic hemochromatosis. Demonstration of recessive transmission and early detection by family HLA typing. *New England Journal of Medicine* **297**, 1017.

Stasny, P., Ball, E.J., Dry, P.J., and Nunez, G. (1983). The human immune response region (HLA-D) and disease susceptibility. *Immunology Review* **70**, 113.

Stokes, P.L., Asquith, P., Holmes, G.K.T., Mackintosh, P., and Brooke, W.T. (1972). Histocompatibility antigens associated with adult coeliac disease. *Lancet* **ii**, 162.

Svejgaard, A., Platz, P., and Ryder, L.P. (1983). HLA and disease 1982—a survey. *Immunology Review* **70**, 193.

Thomson, G. and Bodmer, W.F. (1977). The genetic analysis of HLA and disease associations. In *HLA and disease* (ed. J. Dausset and A. Svejgaard) p. 84. Munksgaard, Copenhagen.

Toniolo, A., Onodera, T., Yoon, J.-W., and Notkins, A.L. (1980). Induction of diabetes by cumulative environmental insults from viruses and chemicals. *Nature* **288**, 383.

Trager, W. and Jensen, J.B. (1976). Human malaria parasites in continuous culture. *Science* **193**, 673.

Woolf, B. (1955). On estimating the relation between blood groups and disease. *Annals of Human Genetics* **19**, 251.

Yoon, J.-W., Austin, M., Onodera, R., and Notkins, A.L. (1979). Virus-induced diabetes mellitus. Isolation of a virus from the pancreas of a child with diabetic ketoacidosis. *New England Journal of Medicine* **300**, 1173.

Zinkernagel, R.M. and Doherty, P.C. (1974). Restriction of *in vitro* T-cell-mediated cytotoxicity in lymphocytic choriomeningitis within a syngeneic or semi-allogeneic system. *Nature* **248**, 701.

9

Physical environment

MICHIO HASHIMOTO

Introduction

The physical environment which surrounds us is composed of both natural and man-made features. Man himself as a component of the biosphere also represents an element of this physical environment.

Traditional medicine had emphasized the micro dimension when viewing physical, chemical, and biological agents as the causative factors in the natural processes which determine both health and disease. But as a consequence of social, economic, and technical development both on a world scale and within individual nations, the overall dimensions of our physical environment have been widened and become more complex at all levels during recent decades. No country can escape the influence of changes in background conditions of the geography, climate, and history of growth and development of a nation.

Geographical background

The maps used by geographers show natural topography and also indicate the boundaries of administrative jurisdiction pertaining to various levels of government. The terrestrial globe defines the location of a country by means of longitude and latitude, providing a global orientation to the local communities in which we live.

Climate and vegetation

The classification of the earth's climate and vegetation zones provides an appropriate orientation of our physical environment and serves to define the basic natural conditions which are the determinants of health and disease. Macro-climates are classified by Köppen as in Table 9.1.

Vegetation is the most sensitive indicator of the climatic environment. Changes in vegetation are associated with high/low temperatures and dry/humid climates. The classification of vegetation zones is shown in Table 9.2.

Temperature, precipitation, humidity, vegetation, and ground condition are the essential environmental factors which determine the growth of flora and fauna. These are closely associated with the ecology of wild animals, insects, and rodents, as well as micro-organisms which act as vectors and/or pathogenic biological agents. This is well illustrated by the aetiology and epidemiology of diseases. Meanwhile, the production of foods and vegetables is significantly affected also by climatic conditions. The nutritional status of a population living under harsh natural conditions is often adversely affected. A severe drought, for example, can result in fatalities.

Adjustment and adaptation

The human body has the capacity to adjust and/or adapt to the external conditions of its physical environment. Physiological response to hot or cold temperatures is an example of this. The number of active sweat glands on the human body is

Table 9.1. Classification of macro-climatic zones

Macro-climate zone	Sign	Local climate
A Tropical climate	Af	Tropical rain forest
	Aw	Savannah
	Am	Tropical monsoon
B Arid climate	BS	Steppe
	BW	Desert
C Temperate climate	CF Cfa	Temperate rainy
	Cfb	West coast ocean
	Cs	Mediterranean
	Cw	Temperate with dry winter
D Sub-polar climate	Df	Sub-polar rainy
	Dw	Sub-polar with summer, Taiga
E Polar climate	ET	Tundra
	EF	Nival

Range of temperatures of macro-climatic zones:
A Average temperature of the coldest month is 18 °C and above
C Average temperature of the coldest month is less than 18 °C but above 3 °C
D Average temperature of the coldest month is 3 °C and below
Average temperature of the warmest month is 10 °C and above
E Average temperature of the warmest month is 10 °C and below

Source: Köppen (1918).

Table 9.2. Classification of vegetation zones

<table>
<tr><td>Low temperature ↑</td><td colspan="4">Zone with ice and snow</td></tr>
<tr><td></td><td colspan="4">Tundra</td></tr>
<tr><td></td><td colspan="4">Coniferous forest zone</td></tr>
<tr><td></td><td rowspan="4">Desert</td><td rowspan="4">Steppe</td><td rowspan="2">Prairie</td><td>Deciduous forest zone</td></tr>
<tr><td></td><td>Laurel-leaf forest zone</td></tr>
<tr><td>↓</td><td rowspan="2">Savannah</td><td>Subtropical rain forest zone</td></tr>
<tr><td>High temperature</td><td>Tropical rain forest zone</td></tr>
<tr><td></td><td colspan="4">Dry, low precipitation ←→ Humid, high precipitation</td></tr>
</table>

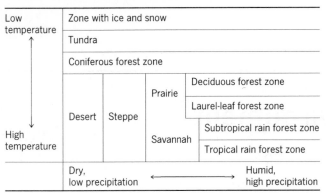

Source: Köppen (1918), (original); Kawamura, T. (1988), *Environmental Science*, **I**, 30.

generally found to be greater among races dwelling in tropical zones than among races dwelling in cold (sub-Polar) climatic zones. An adaptive response to the lower density of oxygen in the atmosphere in high mountains is that the number of erythrocytes measured in a mountain climber at an altitude of 5500 m will be larger than the number measured at an altitude of 215 m. This phenomenon has been observed among climbers tackling Mount Everest. An epidemiological report has shown that the mortality rate for malignant melanoma in the United States is inversely proportional to latitude. A similar observation has been made in Australia. The integrated effective daily UV-B radiation recorded at different latitudes and elevations during the seasonal solar radiation maximum (in J/m^2) reveals a difference between locations at low latitudes with a high elevation and locations at high latitudes with a low elevation. These are just a few examples of the health implications of differing altitudes and latitudes.

The physical environment as an integrated product

The physical environment is an integrated product of both natural and artificial environmental conditions. As is stated in the UNEP Nairobi Declaration of 1982, environmental problems arise as a consequence of the interrelation of population, resources, development processes, and the natural environment. Human settlement and socio-economic development within a finite physical environment subject to specific geographical conditions and undergoing a historical course of development will be examined.

Human settlement and basic sanitation

Location

Human settlement and population concentration are closely related to the geographical merits of a particular location; for example, the presence of a good water supply, its convenience as a base for economic and trading activity, and its general security and access to transportation. Geopolitical factors also have to be taken into consideration in the case of the development and expansion of a capital city responsible for overseeing local and national government. Regional development concerns also influence the overall settlement pattern, with functional differentiation apparent among the areas of human concentration.

The growth of local autonomy also gives rise to features peculiar to individual local communities possessing their own traditional cultural heritage. Conservation efforts and/or the development of local physical environments are sometimes enhanced by initiatives arising from local autonomy.

Land use and siting

Land use and siting are basic concerns in the development of a physical environment, and both are strongly influenced by natural geographical conditions. The physical planning of a region's development provides the framework for the city, town, and county plans of the local autonomic bodies concerned. Legal factors concerning private property rights and political factors involved in the implementation of planning, zoning, siting, and building, via government machinery, play an essential role in the development of a physical environment. In the United Kingdom, during the past century, revision and amendment of the Public Health Act, the Town and Country Planning Act, and the Housing Act have played an important role in the development and management of the nation's physical environment.

Housing

We spend nearly two-thirds of a 24-hour day in our homes. The house is our most basic physical environment. The criteria for healthy housing are given below.

Hygiene

(1) adequate sunshine, lighting, ventilation, and air conditioning (regarding temperature, humidity, and indoor pollution);

(2) noise-proof structure.

Sanitation

(1) safe water supply;

(2) adequate drainage of domestic effluent;

(3) toilet connected to sewerage system or septic tank;

(4) bath;

(5) waste and refuse disposal equipment;

(6) insect and rodent control;

(7) daily cleaning.

Safety and security

(1) design and structure resistant to fire, home accidents (carbon monoxide poisoning by gas, electricity failures), and traffic accidents;

(2) protection against disasters (earthquake, landslide, flood, ground subsidence);
(3) crime-proofing.

Comfortable home life

(1) bedroom for married couple;
(2) rooms for children, aged family members and adults of different sex;
(3) provision for personal privacy;
(4) living room for family;
(5) amenable environment.

Convenience and efficiency

(1) location and siting convenient for daily activities (shopping, transport, school, welfare institutions);
(2) reliable access to emergency services;
(3) energy-saving devices;
(4) telecommunication links;
(5) efficient kitchen.

Good community relations (integrity of local community life)

The physical environments of collective housing projects differ from those of independent dwellings. Green spaces and playgrounds for children, waste disposal/storage facilities, local pedestrian and traffic safety, parking arrangements, and so on, are handled communally by the residents.

The poor physical environments of slum areas and/or squatters' settlements present a difficult social problem, particularly in the rapidly expanding urban areas of developing countries. It is said that approximately one third of all urban populations in the big cities of developing countries live in slums or squatters' quarters. Urban redevelopment, slum clearance and squatter resettlement are the most prominent issues in urban environment management. The UK has employed legislative measures to facilitate slum clearance since early in the nineteenth century. The Housing Act of 1957 features legal requirements relating to the maintenance, repair, and sanitation of housing, slum clearance programmes, the mitigation of overcrowded living conditions, and basic housing facilities. These requirements evolved from the Poor Law Amendment Act of 1834.

Basic infrastructure development

The development and management of public water works, urban sewerage, and urban incinerators, operating in conjunction with efficient water collection services, are the essential requirements of a healthy urban environment. In recent decades, special designs and structures to benefit handicapped or aged people have been integrated into the physical planning and development of roads, stations, public gathering places, and parks. Various sectoral administrations have introduced programmes of community sanitation, health, welfare, and culture to improve the physical environment.

Urban environmental indicators

In 1974, the OECD Environment Committee produced a list of concerns relating to man's urban environment and formulated recommended urban environmental indicators. These urban indicators relate to the following items.

Dwellings

(1) indoor dwelling space;
(2) outdoor dwelling space;
(3) amenities and sanitation;
(4) stability of property rights as fixed assets;
(5) cost and availability of housing.

Services and employment

(1) quality and accessibility of commercial services;
(2) quality and accessibility of health services;
(3) quality and accessibility of education services;
(4) quality and accessibility of recreational services;
(5) quality and accessibility of transport and traffic services;
(6) quality and accessibility of emergency (fire, ambulance, police) services.

Environmental pollution

(1) air quality;
(2) water quality;
(3) noise;
(4) waste disposal;
(5) protection against natural disasters;
(6) climatic conditions;
(7) location and urban structure.

Social and cultural factors

(1) social integration;
(2) community organization;
(3) levels of crime and violence;
(4) quality of cultural activity.

Environmental change and its impact

Development, industrialization, and urbanization

The improvement of social and economic conditions is a common concern of nations around the world. Social and economic development is essential to achieve this goal. The 1960s was a development decade, led by the United Nations. The world economy grew at a rate of 5 per cent per year during the period, but environmental pollution and degradation were the unexpected adverse side-effects. They were the consequences of complex interactions between populations, resources, development, and the environment.

The famous Minamata disease (alkylmercury poisoning) incident in Japan is an extreme example of the adverse effects of industrial activity. The 'green revolution' led by the Food and Agriculture Organization (FAO) achieved a marked increase in food production in developing countries, but on the negative side, natural ecological systems have been degraded or destroyed. Many previously common birds and insects have largely disappeared in various parts of the world. Large-scale population resettlements were implemented, as in the case of the huge Kariba dam construction project in Africa. Urbanization and industrialization prompted the migration of rural populations into urban areas. The rapid growth of Mexico City is an often-cited example of the urban explosion of the early 1970s. Structural changes in production and consumption during the development process led to conflicts and disputes among populations, and between agricultural and manufacturing industries. Damage to human health and/or property damage arising from environmental changes were attributed to these developmental structural changes.

Impact on the life supporting environment system

Air, water, soil, and the biosphere are the basic elements of our natural, life-supporting environment. While the development of artificial environments accompanied by social and economic infrastructures increases, so communities appear to depend less on the life-supporting elements of the natural environment. However, the importance of natural forms of life support become suddenly evident, even in well-developed countries, at the time of natural disasters or accidents, especially those which result in serious environmental pollution.

Developing countries

In the developing countries indigenous populations are very much dependent on the natural environment in respect to cultivation, fishing, hunting, and wood-fuel collection. Their food, water, energy, and shelter are provided almost entirely by the natural environment in which they pursue their traditional life-styles. Desertification endangers their food production, collection of wood fuels, and even their water supply. Deforestation threatens their hunting, wood-fuel collection, sources of building materials, and traditional water sources. Degradation of the quality and quantity of water resources, related to development projects, endangers the traditional forms of water use of indigenous peoples, such as for fishing and bathing. As a consequence, they suffer food and water shortages and a diminished community life. These are the essential issues as concerns basic human needs. The United Nations Environmental Panel (UNEP) has stressed the need for the environmentally sound management of inland waters since 1985.

Developed countries

The recent increase in the number of skyscrapers incorporated in new urban developments has created the special problem of local urban climates. These feature topographic air turbulence and strong winds, down-draughts of smoke from neighbouring buildings, and the obstruction of natural sunlight radiation previously enjoyed by neighbouring residential properties. This is a new form of urban nuisance, sometime involving a decrease in the levels of ultraviolet radiation in domestic environments.

Ground subsidence has sometimes occurred in areas where underground water has been pumped excessively for use in industry, building, and public water works. Properties in those areas can gradually be damaged may become vulnerable to earthquakes. Serious disasters involving thousands of deaths have occurred as the result of typhoons and high tidal floods in Japan in the past.

Transmigration of disease vectors

In Egypt the traditional basic system of irrigation depended upon the natural seasonal rise and fall of the river. The resulting environment was not conducive to the growth of snails, and the incidence of schistosomiasis was low. Following construction of the Aswan high dam, most Egyptian irrigation systems have switched to the perennial type. In this system the water is released from canals as it is needed and drainage ditches carry away the seepage water; the canals never become dry and snails find the habitat favourable. The incidence of schistosomiasis in such areas was found to be as high as 60 per cent of the population, in contrast to a low incidence of only 10 per cent in the areas where a traditional type of irrigation is still practised. After this conversion from the traditional irrigation system to the perennial system as a direct outcome of the Aswan high dam project, the snail hosts of both *S. heamatobium* and *S. mansoni* have extended southwards (Manson-Bahr 1982).

The disease trypanosomiasis is commonly called African sleeping sickness. Trypanosomes cause the death of several thousands of people every year. Various combative measures have been taken, including the use of insecticides, large-scale deforestation (to destroy the tsetse's habitat), and destruction of the large game on which tsetse and trypanosomes depend. Even after this campaign sufficient small animals survived to support the tsetse. Futhermore, as herdsmen moved their cattle into the areas that had been cleared, the tsetse began feeding on both cattle and herdsmen. Thus the transmission of animal and human trypanosomes not only continued but intensified (Morgan *et al.* 1980).

These are typical examples of the transmigration of disease vectors through development projects.

Environmental consequences of socio-economic activity

According to the World Development Report 1988 by the World Bank a below-zero population growth rate is expected to occur in only five countries, among the one hundred and twenty nine member countries between 1986 and 2000. There are also wide difference in the per-capita GNP of the nations of the world. For example, per capita GNP in the 1986 was US $120 for Ethiopia and US $17 680 for Switzerland. This illustrates the current gap in financial terms between the poorest and the richest nations of the world. Socio-economic development is essential for the improvement of welfare throughout the world.

It is important to recognize the potential for environmental effects generated by socio-economic activity, and by a population itself. Populations are increasing. Resources are limited to either renewable or non-renewable assets. Development is continuing, aimed at the improvement of social and economic levels, in both the developing and the developed countries. These present a burden on the environment at local, regional, national, and international levels and even, recently, on a global scale.

The World Commission on Environment and Development (WCED) advocated a future policy direction of sustainable development at the Tokyo Conference in 1987. This is of course a desirable and essential aim for the future, but it is a difficult and complex task to establish effective ways and means to achieve this development goal. This the challenge facing society as the world approaches the twenty-first century.

Potential of environmental effects

The state and composition of the environment are altered by socio-economic activity and human activities such as resource extraction, cultivation, harvesting, hunting, fishing, transportation, storage, manufacturing, energy conversion, construction, trade, consumption, wáste-processing, and the daily biological activities of a population.

Mining

Mining, including the extraction of energy resources, has traditionally affected the environment, bringing landscape deformation and destruction, pollution, and a decline in or the depletion of resources. Heavy air and water pollution caused by smelters are typical examples, damaging the environment both in surrounding areas and in areas downstream or down-wind. Solid mining wastes contain heavy metals, for example, zinc, copper and gold, components of mineral ores. Oil spill accidents at marine oil drilling sites are also typical of pollution incidents. At the same time it is important to note that heavy metals, like the geochemical elements of a mining area, form part of the natural background of the water and soil of the region.

In Japan, the famous Ashio (copper mine) mining pollution case produced a grave local dispute between the smelters and the farmers operating down-stream. The farmers agricultural activities were damaged by copper and arsenic contained in mining effluent which was carried downstream by the Watarase River, beginning in the late nineteenth century. In spite of strong protests in Congress, and the introduction of related legislative measures, no effective controls were achieved. To make matters worse, the farmers were relocated by force in the early 1900s. Anti-pollution measures were finally made enforceable in 1969, at which time the mines were closed because of the depletion of ore. The flow of pollutants has been completely stopped, but residual pollution still persists in the agricultural soil downstream that was irrigated in the past.

In the early days of heavy pollution from smelting, the green forests of the surrounding mountains were devastated, but these are now recovering. There are reports, however, of a high incidence of respiratory disease in neighbouring communities resulting from severe air pollution.

Manufacturing industry

The potential for environmental pollution of manufacturing industries is attributable to the emissions, effluents, and wastes generated through their production processes, and their consumption of fuels, raw materials, and water as part of a technological mechanism. Manufacturing industries which consume extensive energy resources have a higher potential for the generation of environmental pollution, and therefore require stricter controls. The appropriate control of land-use, siting, layout, and design, in advance of construction, play an important role in preventing industrial pollution. Quality control related to fuels and raw materials is also important in ensuring effective pollution control. The improvement of production technologies is a key consideration in minimizing pollution, reducing energy consumption, and rationalizing production costs. The enhancement of green spaces, choice of colourful plants, and harmonious design of background scenery are also effective in improving the physical environment. The development of good community relations is, however, an essential element of industrial enterprise. Industrial wastes, particularly toxic and hazardous wastes, represent the most potentially damaging and critical problem at present. Maximum resource recovery from waste materials is now an integral part of industrial technology.

The nature of industrial pollution has been changing recently. Traditional industries which consume extensive energy resources had, by the late-1970s, been almost satisfactorily brought under control. Visible forms of air and water pollution have been greatly reduced. Modern advanced technological industries, such as integrated circuit production and bioengineering have expanded rapidly, Superficially they appear very clean, but it is important to understand the different nature of the environmental pollution they cause as compared to the older traditional heavy industries. The problem of underground water pollution in Europe, USA, and Japan is a typical example of this new type of pollution.

Catastrophic damage arose from the chemical plant accident at Bhopal in India in 1984. The industrial accident at Seveso in Italy in 1976 was also an important case of an industrial accident devastating a surrounding community. Industrial accidents are usually sudden and unexpected, but predictable all the same.

Stockpiles of toxic industrial wastes have the potential to

pollute underground water and soil, sometimes through the evaporation of toxic and hazardous gases. According to a 1978 report by the Environmental Protection Agency of the USA (EPA), there are about 32 000 old disposal sites of potentially dangerous materials throughout the USA. This is one of the USA's most important environmental problems at present.

Energy

Energy production and consumption is a key potential source of environmental pollution. Similarly, energy control is one of the most effective ways of reducing air pollution. The famous London smog of 1952 resulted from the high incidence of domestic-source coal combustion combined with industrial emissions. The air pollution problem in London was largely solved by the Clean Air Act 1956, which first legislated the mandatory use of 'clean' smokeless coal, then promoted the use of petroleum fuels and electricity. The heavy air pollution which once affected the Tokyo and Osaka bay areas has been greatly reduced, not only by stringent air pollution control measures and enforcement actions, but also by shifting about 30 per cent of electric power generation from fossil fuels to nuclear energy. The expanded use of district heating plants, some of them using waste heat from urban incineration plants, have also made a substantial contribution to the reduction of urban air pollution in Japan during recent decades.

Construction work

The physical environment is being drastically changed by large construction projects such as roads, railways, airports, harbours, bridges, and dams. Traditional communities are often physically divided by road construction projects.

Highways present a lineal source of air pollution caused by automobiles. Roadside residents suffer from traffic noise if adequate silencing measures are not implemented. Appropriate environmental planning of the roadside zones of highways is one of the fundamental measures required to reduce traffic noise and vibration. The architectural structure of highways also have significant implications in traffic noise control measures.

In Japan aircraft and 'bullet train' (Super express train at 220km/h) noise present the most pressing traffic noise problems in the areas adjacent to these transportation facilities. Traffic noise reduction work is the second largest item on Japan's environmental budget at present. Patterns of transport flow are prone to change through the introduction of new ports and major new bridges. These changes in transport flow produce secondary consequences which sometimes intensify traffic noise, vibration, and air pollution in roadside zones.

Drastic changes in physical environments are caused by the construction of large dams. These affect both upstream and downstream locations. The harnessing of water in upstream reservoirs leads to a marked decrease in water flow downstream. This can change the physical environment on a regional scale. In many cases, population relocation is unavoidable in advance of these dam projects. Ecological change is an unavoidable result of any decrease in water flow downstream of a dam. Because emerency water discharges from dam reservoirs are sometimes necessary, for example after heavy rain, the provision of a system flood warnings for populations downstream is an indispensable requirement.

Automobiles

The rapid increase in automobile use places a heavy burden on urban and roadside environments. Automobiles have enhanced the convenience, comfort, and privacy of transportation in modern society. But, on the other hand, air pollution, noise, and vibration from road traffic have been rapidly getting worse. This is a problem not only in developing countries, but also among urban populations in developed nations. Mobile pollution sources present a different form of environmental pollution when compared to stationary sources.

Increased numbers of large-sized diesel trucks and buses, and also passenger cars, are presenting difficult traffic environmental pollution problems in many parts of the world. In spite of stringent emission controls on auto exhausts and on the noise levels of individual vehicles, a reduction in the environmental pollution by automobiles is still very difficult to achieve. Increasing volumes of commercial and passenger traffic negate the incremental improvements expected as a result of lowered emission levels. Furthermore, a decreased renewal rate of existing diesel trucks strengthens the pollution potential.

Agriculture, forestry, and fishery

Agriculture, forestry, and fishery depend on the biosphere, soil, water, and air of the natural environment. Forestry and agriculture play an important role in preserving greenery in our physical environment. Agriculture and fishery supply foods and foodstuffs for use in our human society. These foods and foodstuffs contain not only natural elements, but also geochemical matter, pollutants, and the residues of chemicals used in farming, cultivation, and breeding processes. These become the pathways for a cross-media exposure to chemicals of the population. Biological intake, particularly biological condensation by acquatic organisms at the stage of biological growth, plays the key role in containing these substances within foods and foodstuffs.

Agriculture

Agricultural production has recently developed by means of technological innovation, involving the use of agricultural chemicals such as pesticides, herbicides, and fertilizers, and mechanical and engineering technologies and bio-engineering techniques.

Meanwhile, the potential for environmental impact related to agricultural production has also increased, compared to the days of traditional agricultural practice. Agricultural water systems providing irrigation and effluent control as a feature of land reform have brought increases in production, but have also generated side-effects such as high salinity and/or nutrient discharge into recipient water zones. Increasing populations of both man and cattle have raised the environmental burden on available soil, biosphere, and water.

Soil degradation and over-grazing have also worsened the environmental conditions in the catchment areas of inland fresh water supplies. Eutrophication (excess nutrients in water), sedimentation, and silting accelerate the degradation of inland water resources. Over-grazing in semi-arid areas with dry climates also accelerates the process of desertification.

The transmission of intestinal parasites, such as hookworms, through soil is also a significant infection pathway in the agricultural areas of developing countries.

Forestry

Forestry management plays an important role in water resource conservation in water catchment areas by promoting the retention of water in the soil by vegetation. Water run-off is naturally moderated by forestry within a water catchment area. This is important in flood prevention and in conserving water for multi-purpose use in lower areas. The decline of forestry, particularly complete deforestation, impairs the processes of water resource conservation. It also accelerates soil erosion and siltation, and the danger of flood disasters grows. Sedimentation and siltation upstream of a dam gradually undermine the storage capacity of the water reservoir, thus shortening the expected useful life of the dam. The expected life span of a dam is usually 100 years, but this sometimes drops to 25–30 years as a result of deforestation-caused sedimentation and siltation within the upstream watershed. At this point, water resource development becomes unsustainable.

The worst recent examples of flood disasters attributed to deforestation have been seen in the mountains of Nepal and Bangladesh, affecting the downstream areas of rivers. The numbers of flood disaster victims in Nepal have been increasing in recent years. Meanwhile, forests represent the life-support system of the indigenous peoples of forested areas. Large-scale deforestation arising from logging operations jeopardizes their traditional ways of life.

Forests are also the prime source of energy supplies, in the form of fuel-wood, for the native villagers. Fuel-wood collection has become the essential daily occupation of housewives in certain regions of the sub-Saharan countries because of local deforestation. Fuel-wood is indispensable for the domestic cooking of foods in every household. A scarcity of fuel-wood brings a decline in the quality of domestic cooked foods which in turn leads to poor digestion and malnutrition among the villagers.

Fishery

Fish is an important element of our basic diet. Meanwhile, fish and shellfish exhibit the function and capacity biologically to condense substances in the water and in bottom sediments. In Minamata, the concentration of mercury in the sea water of its bay was below the detection level of the analytical methods available in the late 1950s, but it was detectable in bottom sediment, and in fish and shellfish. The biological condensation is several thousand times that pertaining to sea water. Unfortunately, the fish with a high alkylmercury content in Minamata Bay appeared normal: no one could imagine at the time that fish and shellfish contained high concentrations of the poisonous alkylmercury. In contrast to the Japanese average fish intake of 58 g/day in the 1950s, the local fishermen ate 750 g/day on average and a maximum of 2.5 kg/day. Thus there was an ecological aspect to the Minamata disease incident.

Fish cultivation in inland waters and in sea water zones has been developed and expanded since the 1960s. This has helped to compensate for declining fishery yields. On the other hand, some types of fish cause excessive levels of nutrients to infiltrate the water zones concerned, giving rise to eutrophication. Antibiotics have sometimes been used to treat affected fish populations. This poses another important public health problem in the form of antibiotic residues in food sources.

Tourism and resorts

Tourism is becoming an increasingly important feature of our modern society. Those areas having excellent natural landscapes or a historic and cultural heritage are popular for tourism development and resort construction. There is usually a clear seasonal variation in the number of visitors to these places. The growth of hotels, resort houses, and other facilities in these areas presents another potential source of pollution. The associated generation of wastes is prone to seasonal fluctuation. The conservation of the local environment is essential not only to ensure a high quality of water in bathing places, but also to protect the coastlines and surrounding landscapes. The environment itself represents an essential amenity in these areas.

Environmental pollution

The environment becomes polluted when it is altered in composition or condition, directly or indirectly, as a result of the activities of man, and becomes less suitable for any or all of the functions and purposes for which it would be suitable in a natural state. Since environmental pollution primarily occurs as a local problem, court actions against offenders are also generally conducted at a local level. The UK was the pioneer in initiating a legislative approach to environmental pollution, dealing with the problem as a statutory nuisance under the government's Public Health Act.

Air pollution

Historical experience

The UK's Alkali and Works Regulation Act represented the first legal action taken against the rising inorganic chemical industry of the late nineteenth century. Black smoke resulting from coal burning was a common urban nuisance in the early days of industrialization. As many as 4000 fatalities were recorded in the famous London smog incident of 1952. Maximum concentrations of smoke and sulphur dioxide in the Lambeth area of London between 5 and 10 December 1952 reached a daily average of 4.46 mg/m^3 and 1.34 ppm. The Clean Air Act was enacted in 1956.

The massive petrochemical complexes and high traffic levels of Los Angeles, US, produced photochemical oxidants as the secondary products of photochemical reactions between nitrogen oxides and hydrocarbons in the atmosphere. Eye irritation, damage to plants, and the cracking of rubber were observed as the side-effects of photochemical air pollution in the Los Angeles atmospheric basin. Auto exhaust controls were initiated by the Los Angeles Air Pollution Control District, which was founded in 1955. Air pollution levels at Yokkaichi in Japan, where the nation's first petrochemical complexes were constructed in 1959, reached a record high of 2.3 ppm of sulphur dioxide (as an hourly value) in the Isozu area. These readings were attributed to the down-wash from the relatively low chimneys of the power stations in 1962. Studies of air pollution and its health effects have been continuously and systematically carried out by professor Yoshida of Mie University.

An interdisciplinary expert team assembled under Dr Kurokawa's study committee have conducted special investigations and recommended countermeasures concerning air pollution generated by petrochemical processes and the combustion of petroleum with a high sulphur content. This initiative started the drive towards de-sulphurization in Japan. A district court upheld a group health claim against a petrochemical complex in 1972. Passage of the Pollution Related Health Damage Compensation Law followed in 1973.

Long-range transport

'Acid rain' related to long-range transportation (transboundary long-range transmission by air trajectory currents) was reported in the special Air Sector Meeting of the OECD in Stockholm in 1968. The problem of acid rain has grown as an international political issue in Europe and North America since the 1980s.

Dust storms arriving from the Chinese continent in early spring have been a regular seasonal occurrence in Japan for many years. This also represents a case of trans-frontier pollution.

The radioactive fall-out from the US's hydrogen bomb test at Bikini in 1954 caused the death of a fisherman, and also a wide range of systematic pollution of the air, water, and biosphere. Provisional maximum allowable levels of radioactive pollutant in water and foods were subsequently established by the government. A continuous systematic surveillance of radioactive pollutants in air, water, and the biosphere, including in human bodies, has since been introduced. An unusually high incidence of malignant neoplasma (cancer) among populations affected by radioactive fall-out from the atomic bomb tests in Nevada in the US has been judged to be a result of radioactive pollution. The Special Compensation Law was passed in the US Congress in connection with this issue.

In 1986 the accident at the nuclear power station in Chernobyl, a small town in the western USSR, revealed the huge potential scale of trans-frontier pollution by the radioactive substances which are the product of fission in nuclear reactors. About 1–3 million curies of Ce^{137} (a fission product with a 30-year radioactive half-life) were released. Thirty one people, mostly fire fighters, died, and several hundred were hospitalized. The long-term health effects in the USSR and in Europe remain unclear.

Specific industrial pollutants

The incidence of traditional forms of air pollution has decreased in recent years. Meanwhile it is reported by the Environment Agency of Japan that the concentration of dioxin compounds in the residential areas near industrial zones in 1986 reached 0.0342 ng/m^3 (the average of 24 samples), with a maximum of 0.0685 ng/m^3 and a minimum of 0.0012 ng/m^3 as TCDDs (tetrachlorodibenzo-*p*-dioxin compounds). Rather than traditional pollutants, the micro-pollutants of hazardous chemicals have grown as the object of concern. The ambient levels of asbestos reported by the Environment Agency of Japan, based on 700 samples surveyed over three years, show that average concentrations, classified by area characteristics, are as follows:

Inland mountainous areas	0.43 fibre/l
Residential areas	1.04 fibre/l
Commercial areas	1.41 fibre/l
Building demolition sites	3.24 fibre/l
Waste disposal sites	3.16 fibre/l

It is assumed that asbestos levels in area environments are 10^{-2}–10^{-4} of the Occupational Hygiene Standard of Japan.

Indoor pollution

In the last decade the problem of indoor pollution has been stressed in air pollution health effect studies, particularly in those which relate to the health effects of nitrogen dioxide. Personal exposure levels have been evaluated under classifications which take account of housing structure, heating appliances used, and distance from a roadside. In relation to 'No smoking' campaigns, the concept of 'environmental tobacco smoking' has been adopted as an alternative to the notion of 'passive smoking'. The non-smoker's exposure to environmental tobacco smoking is defined as involuntary smoking.

Efforts toward control

Traditional forms of air pollution attributed to the combustion of fossil fuels have been greatly reduced in many developed countries as the result of efforts in air pollution control. The Clean Air Act of 1956 took the lead in these global efforts, particularly in the field of suspended particulates.

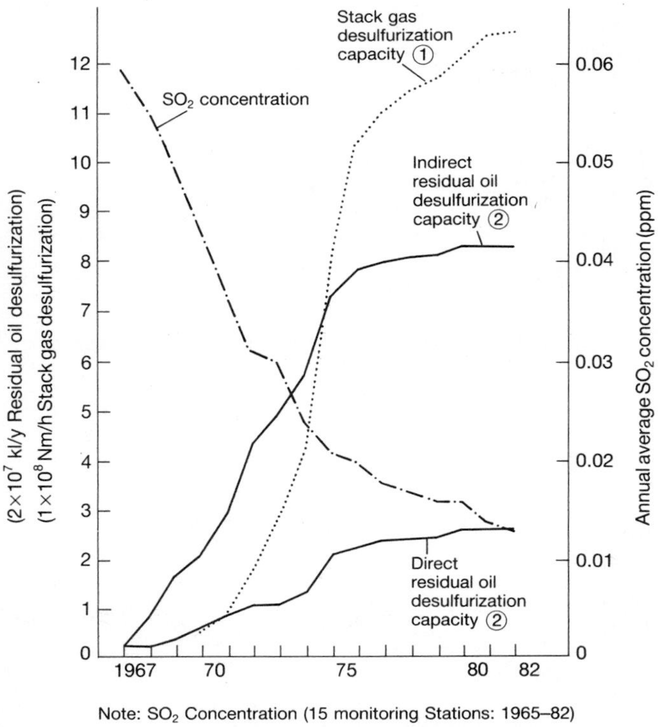

Fig. 9.1. Trends in desulphurization capacity and ambient concentration of SO_2 in Japan (Source: *Environment agency, annual report of environmental quality 1983 FY*, Japanese Government Report to 101st Congress, p. 58).

Japanese efforts towards sulphur dioxide control, based on the recommendations of Kurokawa's 1964 study committee, have resulted in the de-sulphurization programme involving the planned supply of low sulphur fuels, which was established by the ministry of International Trade and Industry in 1969 (Fig. 9.1).

Controlling auto exhausts, particularly those of large-sized diesel trucks, and the numerous low-emission pollution sources, remains a difficult problem in nitrogen oxides control efforts. The expansion of nuclear facilities for electric power generation has been a significant factor in the reduction of traditional forms of air pollution. The increase in LNG use for electric power generation in Japan has also made a significant contribution. Trends in air quality in Tokyo are illustrated in Figure 9.2.

Water pollution

It is common throughout the world for people to select as their site of human settlement a location which has ready access to water. Fresh water is indispensable for household uses, such as for drinking, cooking, bathing, and everyday household needs. Epidemics of water-borne intestinal communicable diseases provided the original impetus for the introduction of the sanitary engineering of public water supplies. Contamination by human excreta and wastes was the problem addressed by the sanitary engineers. Comprehensive sewage treatment systems were developed, and chlorination was introduced for the disinfection of water supplies.

Historical experience

Urbanization and industrialization provoked serious conflicts and disputes related to differing forms of water use. Initially, inter-industrial disputes between agricultural, mining, manufacturing, and fishery industries were prominent.

The main traditional indicators of water pollution are BOD (biological oxygen demand), COD (chemical oxygen demand), and pH measurements. Mining, textile, paper and pulp industries, and so on, were the major polluting industries in the past. Due to the pollution of irrigation water in the lower reaches of the rivers, agriculture and fishery was seriously affected. Meanwhile, water pollution caused by toxic and/or hazardous pollutants, such as hexavalent chromium and cyanide compounds, often led to the sudden mass deaths of fish populations. In addition to the problems faced by fisheries and agriculture, the pollution also posed a serious threat to public water supplies.

Detergents and petroleum

In the early 1960s, the extensive use of synthetic detergents, such as ABS (alkylbenzensulfonate), caused the pollution of domestic and urban effluents, and this pollution was visible in the form of floating foam. Innovations in pesticides and herbicides further worsened the water pollution. The natural ecosystem was disrupted, and once-common insects and fish, and their parasite organisms, rapidly disappeared from the natural environment. Consequently, non-degradable synthetic detergents were replaced by degradable ones.

The expansion of petrochemical industries resulted in cases of petroleum pollution of the marine environment. Marine oil pollution soon became a matter of international concern. The International Treaty for Ocean Oil Pollution was subsequently adopted. Fish-eating populations have expressed dissatisfaction over the oily taste and smell of some fish and shellfish. Bathing places on coasts and rivers have been hit by oil, jeopardizing the amenities of traditional summer resorts.

Sewerage

Sewerage system development requires a huge level of public investment over many years. Investment priorities differ among various countries according to their stage of economic development and the political ambitions of the nation. The installation level of sewerage systems in Japan is only 35 per cent at present, far behind that of Europe and the US. However, night soil collection from toilet pits in individual households, by means of vacuum pump vehicles, and disposal at biological decomposition installations has been rapidly developed as a cheap and efficient alternative measure during the transitional stages in the expansion of sewerage systems. Meanwhile, domestic septic tank installations have increased rapidly since the 1960s.

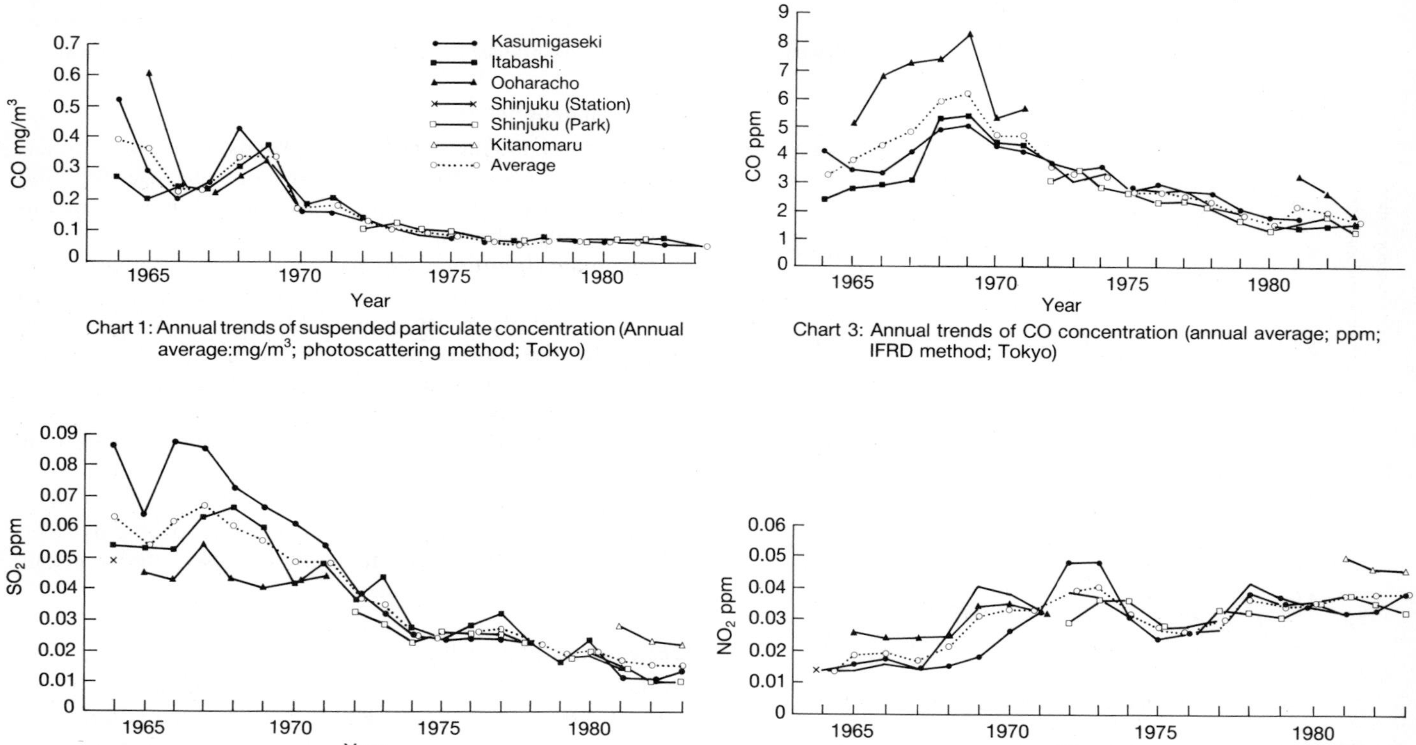

Chart 1: Annual trends of suspended particulate concentration (Annual average:mg/m^3; photoscattering method; Tokyo)

Chart 2: Annual trends of SO_2 concentration (annual average; ppm; electroconductivity method; Tokyo)

Chart 3: Annual trends of CO concentration (annual average; ppm; IFRD method; Tokyo)

Chart 4: Annual trends of NO_2 concentration (annual average; ppm; Saltzman method; Tokyo)

Fig. 9.2. Status and trend of air pollution in Tokyo (1964–1983), from national roadside monitoring stations (Source: *Environment agency, air conservation bureau, division automobile poll pollution, 1983* report. National Hygiene Institute, Ministry of Health and Welfare, unpublished).

Trace chemicals

Japan's desire not to repeat the mistakes which led to the Minamata tragedy (alkylmercury poisoning) focused attention on the urgent need to control the trace (micro) pollutants of environmental toxic chemicals. The problem of biological condensation was examined closely. Accordingly, changes in production processes and the adoption of a closed system for the handling of toxic persistent water pollutants were introduced in the 1970s. Assessment of the environmental toxicity of pesticides and herbicides, prior to the introduction of the Agricultural Pesticides Regulation Law, was covered by legislation introduced in 1970. The OECD Environment Committee also worked out various guidelines related to new environmentally toxic chemicals.

Surveillance of water, bottom sediments, fish, and shellfish has been extensively introduced throughout Japan. Biological monitoring systems for the evaluation of environmentally toxic chemicals were also established in 1974. Improved water treatment methods have been developed to control the formation of trihalomethane during the chlorination of public water supplies. Water pipes containing asbestos are also sometimes identified as the source of asbestos contamination in tap water.

Closed systems and recycling

Because of the scarcity of fresah water resources, the pressing need for rationalization to reduce production costs and the stringent control standards for industrial waste effluent, the recycling of industrial water has become increasingly common, combined with the use of closed systems to economize on water costs. In Japan at present, nearly 75 per cent of industrial water is recycled to reduce fresh water costs.

Eutrophication

Excess nutrients in water cause the eutrophication of the recipient water zone, particularly in closed water zones such as lakes and dam reservoirs. The OECD has carried out research and development into the eutrophication of water since 1965 through the Science Policy Committee. Nutrients are not necessarily man-made in origin, and natural sources play an important role.

The secondary treatment of urban sewage and industrial effluent is not adequate to reduce the nutrient chemicals in effluent such as nitrate and phosphate. Agriculture and fisheries have been regarded as the victims of water pollution in the past. With the advent of the eutrophication problem, farm and grazing lands are potential sources of the nutrients

Table 9.3. The result of survey of underground water pollution 1986

Substance	Number of municipalities surveyed	Number of wells			Number of public wells used for drinking water		
		Number of wells sampled	Number exceeding standard	Margin of excess (%)	Number of wells sampled	Number exceeding standard	Rate of exceeding (%)
Trichloroethylene		2794	146	6.2	1602	69	4.3
Tetrachlorethylene	303	2777	109	3.9	1692	60	3.8
1,1,2-trichloroethane		2763	3	0.1	1691	0	0

Source: *Annual Report of Environmental Pollution* (1987 FY). Environment Agency.
Note: 'Standard' refers to the Provisional Standard designated by the WHO.

which constitute the non-point source of pollution in the catchment areas of recipient water zones. Feeding in fish cultivation also constitutes a non-point source, introducing nutrients into the water zones concerned. Domestic effluent is also a significant source of nutrient supply, and synthetic detergents generally contain nitrate and phosphate. Nitrate in the form of a secondary air pollutant from nitrogen dioxides even falls from the air as precipitation.

Public water supply sources affected by eutrophication often produce drinking water with an offensive taste and smell. More sophisticated forms of water treatment technology, such as those using activated carbon, are required. A wide range of counter-measures are necessary to control eutrophication. In addition to the use of better water treatment technologies to reduce the concentrations of nitrate and phosphate in effluents, consumer product control is required to reduce phosphate contents. There is also a need for the primary disposal of domestic effluent at household level, the improvement of agricultural and fishery practices, agricultural water system modification, and land-use and siting controls. In Scandinavian countries the dredging of bottom sediment in water zones is carried out as a means to reduce one of the natural sources of accumulated nutrients. These countries also apply lime to neutralize the acidification which results from acid rainfall.

Underground water

The problem of the pollution of underground water by organochloride compounds, which originated in the wide use of organochloride synthetic detergents, has presented a difficult and complex problem over the past several years. It has been detected at Silicon Valley, California, US, and also in Europe and Japan. The turnover rate of underground water is extremely slow, which means that it takes many years for underground water to be renewed in natural conditions. Underground water is considered as a stock water resource. According to a report of the Environment Agency of Japan underground water had been polluted by the organochloride synthetic detergents used by the IC (integrated circuit) and metal industries, in dry cleaning, and so on (Table 9.3). Waste synthetic detergents were disposed of by seepage in the past. Disposal is now controlled under provisional enforcement standards for waste effluent, derived from the provisional standards for drinking water of the WHO.

Soil pollution

Stock pollution

Irrigation using river water polluted by mining effluent and wastes containing heavy metals and other hazardous metals, such as copper, cadmium, and arsenic is hazardous, not only to human health directly but also to agricultural plant growth; impurities are absorbed into the crops through biological mechanisms. These metals become stock pollutants of the soil of agricultural land. In these areas, the total human intake of heavy metals can be measured by means of sampling the food, water, and air which is consumed daily by the local residents. Residues of organochloride pesticides represent a typical example of the soil pollution recorded in the past.

History of toxic stock

The poorly-managed disposal of industrial toxic wastes by means of burial is also a significant cause of soil pollution. The pollution incident at the Love Canal near Niagara Falls in New York State, which was first made public in 1979, represented an especially severe case of toxic stock pollution. The New York State Health Department declared a condition of 'grave imminent peril' in August 1978. Eighty two different chemicals were found to be present at the site, and air, water, and soil pollution was confirmed. The playground of the local primary school was affected, and cases of spontaneous abortion, birth defects, cancers, and other adverse health effects believed to be related to the toxic stock pollution were also reported.

President Carter designated the Love Canal zone a disaster area, to be managed under the Federal Disaster Administration. Approximately 500 lbs of carcinogenic chemicals, such as trichlorophenol which contains dioxin, were detected. Thirty seven families were evacuated from the zone and the school was closed. (10th Annual Report of Environmental Quality, USA. 1979).

Noise and vibration

Noise is one of the major nuisances of modern life. Related occupational health standards used to be based on the criteria

Table 9.4. Population exposed to aircraft and road traffic noise; selected countries or regions, mid-1970s (per cent)

Aircraft noise Number of national population exposed to given levels[a,b]				Noise level in Leq (dBA) Outdoor measures		Road traffic noise Percentage of national population exposed to given noise levels[b]											
US[c]	Canada[d]	Japan[c]	Europe[e]			US[a]	Japan[f]	Belgium[f]	Denmark[f]	France[f]	Germany[f]	Netherlands[f]	Norway[a]	Spain[f]	Sweden[a]	Switzerland[f]	UK[g]
13	2	3	3	⩾ 55	Sleep can be disturbed if windows are open	40	80	68	50	47	72	. .	22	74	38	66	50
5	1	1	1	⩾ 60	Sleep and conversation can be disturbed if windows are open	18	58	39	. .	32	40	30	12	50	24	28	27
2	1	0.5	0.2	⩾ 65	Sleep and conversation can be disturbed even if windows are closed	6.4	31	12	20	14	18	7.4	5	23	11	12	11
0.6	0.3	0.2	0.05	⩾ 70	Sleep and conversation disturbance; possible complaints	1.8	10	1	. .	4	4	1.6	2	7	4	1	4
0.2	0.1	0.1	0.01	⩾ 75	Possible long-term danger of hearing impairment	—	1	—	. .	0.5	—	0.1	—	1	1	—	1

[a] Expressed in Leq over 24 h
[b] Data refers to various years in the early seventies for different countries. Since many measurements and surveys do not give results in Leq, equations relating Leq and other indices have been used. The margin of error related to national estimates, differing years and to data transposition is probably very significant, especially for the lower noise levels (± 10%)
[c] For all airports
[d] For 5 major airports (Edmonton, Montreal, Ottawa, Toronto, Vancouver)
[e] For 34 airports. Broad assumptions were made concerning densities around some airports
[f] Expressed in Leq over the period 6–22 h
[g] Expressed in Leq over the period 6–24 h (England only)
Source: Table taken from the OECD Report *The State of the Environment*, Paris, 1979.

of hearing loss. In dealing with the problem of noise nuisance, however, the level of disruption of sleep, conversation, rest, and desk work in daily home life must become the essential criteria in ambient noise control. Construction noise represents a temporary form of nuisance, while traffic noise is produced by a number of mobile sources. Different responses are to be noted in public complaints concerning automobile traffic, railway, and aircraft noise. The perceptions of and responses to noise nuisance vary according to the location, the occasion, and the socio-economic relation between the noise source and the affected individual or group. Responses are not merely physiological, but also involve psycho-somatic reactions of annoyance. The first Report of the Royal Commission of Environmental Pollution, UK, published in 1971, stated that the principal form of noise pollution was that which related to transportation. It concluded that 20–45 per cent of the urban population were subject to traffic noise levels likely to be undesirable in a residential area.

The 1979 report entitled National Exposure to Highway Noise through the year 2000, produced by the Wyle Laboratory of the US EPA Noise Abatement Office, stated that

> so far as road traffic noise in the United States is concerned, its impact in the year 2000 will be 23 per cent greater than it was in 1970 if the present regulations (concerning heavy lorries) remain unchanged—and without regulation it will be twice as great in 2000 as compared with 1970.

The OECD secretariat has produced an international comparison of the status of noise pollution attributed to aircraft and automobiles (Table 9.4).

The EPA (Environment Protection Agency, US) has identified maximum permissible noise levels to safeguard health. They include the following:

(1) to protect against hearing loss: Leq (24h) = 70 dB or less (equivalent to Leq, 8 hours = 75 dB);

(2) to ensure noise does not annoy others or interfere with their activities: Ldn = 55 dB or less, outdoors; Ldn = 45 dB or less, indoors;

In 1981, Japan's Supreme Court upheld a claim by local residents relating to disturbance from aircraft noise in the vicinity of Osaka International Airport. It ordered the national government to pay compensation to the plaintiffs, but rejected an injunction request.

Vibration

Vibration is a nuisance which often accompanies excessive noise. The major criteria for vibration effects are based on the classification of the degree of annoyance caused, particularly as concerns sleep disturbance. Vibration is frequently caused by industrial operations, construction work, road traffic, and railway transportation (notably, in Japan, the 'bullet trains'). Geological conditions are also an important factor in vibration effects.

Low frequency noise, within a range of 1–20 Hz and accompanied by high acoustic pressure, it not audible as noise, but often produces a rattling sound caused by the co-vibration of a building's components. Individual differences in the perception of and response to low frequency noise sometime make it difficult to deal effectively with noise complaints. However, the explosive sound of a sonic boom, resulting from supersonic flight, generally provokes spontaneous expressions of annoyance among the general public.

Urban and industrial wastes

Urban waste

The disposal of excreta, refuse, and waste materials is the basic task of sanitation activities, and represents the first step in urban hygiene. Wastes are generated by production and consumption conducted within the course of daily human activity, and also by the biological metabolism of man and animals.

If adequate provision is not made for waste disposal, the physical environment soon deteriorates, particularly in urban areas. Increases in population, and in the level of consumption, strengthen the potential for urban waste generation in terms of both quantity and quality. According to the 1987 Environmental Data Report of UNEP (United Nations Environment Programme), per capita municipal waste disposal in 1980 amounted to 103 kg per year in the USA, 377 kg in Japan, and 333 kg in the UK. In Japan, waste disposal methods consists of landfill (28 per cent) and incineration (69 per cent). In the UK, 69 per cent goes for landfill and 10 per cent is incinerated. Inefficient urban waste disposal can give rise to offensive odours, breeding places for insects and rodents, with an associated risk of epidemic and a loss of amenities in the urban environment.

Industrial waste

Industrial waste disposal is an increasingly serious environmental issue not only at the local and national levels, but also as a matter of worldwide concern. Industrial waste comes not only from production processes, but also from the operation of environmental pollution control installations.

Industrial waste generation has expanded both in quantity and quality, particularly in the form of toxic and hazardous wastes. These wastes present a serious danger to public health. As in the case of soil pollution (see p. 145), stock toxic and hazardous industrial wastes now present a serious social and political dilemma in many countries. It is necessary to classify these toxic wastes, particularly industrial wastes. Toxic and hazardous substances must be designated as such, and criteria must be introduced which indicate the potential dangers of specific toxic and hazardous materials. The International Treaty on Ocean Pollution Control provides a classification of harmful substances and the requirements for their disposal. These disposal requirements are based on a classification of toxicity potential.

Trans-boundary migration of toxic wastes

Recent reports on the trans-boundary migration of toxic and hazardous wastes, even radioactive wastes, from developed countries to developing nations, should serve as a serious warning, and have provoked public anger on an international scale. The expansion and development requires a large volume of capital investment and manpower, and planning must be undertaken on a long-term basis. International obligations and the social responsibilities of developed countries, particularly of the industrial enterprises concerned, should be strictly adhered to.

Radioactive waste

The disposal of the radioactive waste produced by nuclear power stations is the most crucial problem related to health and public safety in the management of nuclear energy development. In the US, radioactive waste pollution is a serious public health problem which has arisen in connection with the operation of military-related nuclear plants. The accident at the nuclear reprocessing plant at Windscale in the UK in the 1950s should also stand as a salutary warning for the world at large.

Environmentally toxic chemicals

As a result of the Minamata tragedy in Japan, international awareness of the environmental implications of the use of toxic chemicals grew rapidly in the early 1970s among scientists and administrators. In 1970 the Japanese government revised the Agricultural Pesticides Regulation Law, adding new criteria for the assessment of environmental toxicity. These evaluate the potential dangers of pesticides in terms of residual chemicals in crops, and in the soil, and of hazard to fisheries, even where the correct dosages are used in agricultural operations; the OECD Environment Committee published a report in 1970 on the unintended occurrence of pesticides, and at the same time formed a specialist chemical sector group. Environmental criteria covering bioactiveness (biological condensation potential) and adverse health effects were recommended. Ecotoxicity was also taken into account in relation to some countries.

The International Programme of Chemical Safety was established jointly between UNEP (United Nations Environment Programme), WHO (World Health Organization), FAO (Food and Agriculture Organization) and ILO (International Labour Organization). The Law for the Assessment of New Industrial Chemicals was enacted in Japan in 1973, and a biological monitoring programme which employs the systematic sampling of air, water, bottom sediments, and biological species was initiated in 1974. Eight chemicals, including PCB polychlorobiphenyl, DDT p.p.′- dichlorodiphenyltrichlorethane, BHC benenhexadichloride, and so on, have already been designated as harmful products, and manufacture of these designated chemicals has been suspended. The disposal of existing stocks of these chemicals is now subject to strict control.

Global environmental and climatic change

The estimated pre-industrial global CO_2 level was between 260 ppm and 280 ppm. The current level of CO_2 is just over 340 ppm. A steady increase in CO_2 concentrations in the atmosphere has been observed during the era of industrialization. CO_2 is emitted during the combustion of fossil fuels in domestic heating, electricity generation, gas production, and automobile use. These constitute the main artificial sources of CO_2 emission.

The release of CO_2 through deforestation ranks second only in scale to fossil fuel combustion. It is the largest source of carbon release from biota and the soil. The farming practice known as slash-and-burn, traditional in the tropical forests of many developing countries, and also the burning of wood fuels for domestic purposes, are closely linked to deforestation, alongside conventional logging activities for lumber production. Furthermore, The CO_2 assimilation capacity of forests also decreases as a result of deforestation.

CO_2 and other trace gases in the atmosphere are transparent to incoming solar radiation, but opaque to the longer wavelength radiation which is reflected from the earth's surface.This is known as the 'greenhouse effect'. Meanwhile, the production and release of CFC (chlorofluorocarbon) compounds has been rapidly increasing in past decades. These CFC compounds gradually ascend to the stratosphere and become trapped there within the thin layer of ozone which surrounds the earth. CFC compounds gradually destroy the ozone layer, disrupting its function as a barrier to UV solar radiation. Ultraviolet radiation contains an adverse element known as UVb. UVb features a radio-minetic action which has been strongly associated with relatively high incidences of skin cancer and also genetic influences within the ecosystem.

The frequent incidence of abnormal weather on a global scale during recent years is attributed by some to the greenhouse effect and the trend towards ozone depletion in the stratosphere. But a causal relationship has not yet been proved. The scientific consensus on climate change reached as Villach, in October 1985, on the occasion of the Joint Conference of UNEP (United Nations Environment Programme), WHO (World Health Organization), and ICSU (International Council of Sciences Union), was as follows:

> The role of greenhouse effects separate from those related to CO_2 is already almost equal in importance, in terms of climatic change, to that of CO_2 itself. If present trends continue the combined concentration of all the greenhouse-effect gases will be equivalent to a doubling of the CO_2 concentrations of the pre-industrial era. A global warming of 1.5–4.5° C, would lead to a rise is in sea-levels of 20–140 cm by around the year 2030.

Climatic change on a global scale is now predicted for the coming century. The time-scale of these environmental problems is century-long. The global extent of the problem, the diverse basic policy dimensions, and the complex involvement of vested interests on a local, regional and national level world-wide, pose an environmental challenge of enormous magnitude. Solutions are being sought as the end of the twentieth century approaches, to benefit our global family of mankind, resident upon earth.

References

Committee on Causes and Effects of Changes in Stratospheric Ozone, Environmental Studies Board, Commission on Physical Sciences, Mathematics, and Resources and National Research Council (1984). *Causes and effects of changes in stratospheric ozone—update 1983*. National Academy Press, Washington, DC.

Council of Environmental Quality of the US (1979). *The tenth annual report of environmental quality*. United States Government Printing Office, Washington, DC.

Fujiwara, F., Watanabe, G., and Takakuwa, E. (ed.) (1986). *Comprehensive hygiene and public health* (2nd edn). Nanzando, Tokyo. (In Japanese).

Gresser, J., Fujikura, K., and Morishima, A. (1981). *Environmental law in Japan*. MIT Press, Cambridge, Massachusetts.

Hashimoto, M. (1985). *National air management policy of Japan*. World Commission on Environment and Development, Geneva.

Hashimoto, M. (1987) *Development of environmental policy and its institutional mechanisms of administration and finance: Environmental management for local and regional development (the Japanese experience)*, p.57. United Nations Centre for Regional Development and United Nations Environmental Programme, Nagoya.

International Institute for Environment and Development (IIED) (1988). *New Population and health, human settlement, fresh water, atmosphere—climate and managing waste*. New York and London.

Japanese Government (1988). *Annual report of environmental pollution—1987 fiscal year. The report to congress on asbestos, TCDDs, and underground water pollution*. Environment Agency, Japan. (In Japanese).

Kawamura, T. and Iwaki, H. (1988). *Environmental science I*, p.30. Asakura Book Store, Tokyo. (In Japanese).

Koizumi, A., Suzuki, T., Osada, Y., *et al* (1984). *Environment and man*. New Medical Science, Series 11. A. Nakayama Book Store, Tokyo. (In Japanese).

Köppen, W. (1918). Klassification der Klimate vorzugweise nah ihren Besirhungen zur Planzenwelt. *Geographische Zeitschrift* **6**, 593, 657.

Manson-Bahr, P.E.C. and Apted, F.I.C. (1982). *Manson's tropical diseases*. (18th edn). Balliere Tindal, London.

Morgan, J.M., Morgan, M.D., and Wiersma, J.H. (1980). *Introduction to environmental science*. University of Wisconsin, Green Bay. W.H. Freeman and Company, San Francisco.

National Institute for Research Advancement (1985). *Proceedings of the 1984 Shiga conference on conservation and management of world lake environment*. LECS 1984 Conference Proceedings. National Institute for Research Advancement. Shiga, Otsu, Japan.

Organisation for Economic Co-operation and Development (OECD) (1975). *Urban environmental indicator*. OECD Observer, Paris, November/December 1975.

Organisation for Economic Co-operation and Development (OECD) (1980). *Proceedings of the conference on noise abatement policies*. Paris.

Organisation for Economic Co-operation and Development (OECD) (1985). *OECD Environmental Data 1985*. OECD, Paris.

United Nations Environmental Programme (UNEP) (1987). *Environmental data report*. Basil Blackwell, Oxford.

Villach Technical Workshop (1987). *Developing policies for responding to climatic changes*. Policy Issues Workshop Report, Villach, Austria.

World Bank (1988). *World development report 1988*. Oxford University Press, New York and London.

World Commission on Environment and Development (WCED) (1987). *Our common future*. Oxford University Press.

10

The social and economic environment and human health

N. HART

The idea of the environment as a social and economic structure

Tokyo, among the world's largest, most polluted, and congested cities, records the lowest infant mortality and the longest life expectancy of any place on earth. This paradox is testimony to the power of the social and economic environment as a determinant of health. The human habitat in Japan is socially engineered: a synthetic environment made by and for human beings. It is also far more salubrious than any habitat spontaneously produced in nature. Though the capital of Japan symbolizes a highly industrialized, overcrowded, busy, and stressful environment, its inhabitants outlive their rural compatriots and even exceed the longevity of people living in the world's most glamourous cities, Paris, Rome, or San Francisco. Salubrious Tokyo explodes the myth that life in a state of nature is more wholesome and healthy than life in an industrial society.

How does the modern urban environment prove to be healthier than life in a state of nature? The imaginary existence of the 'noble savage' free of exploitation by others, free of repressive social moralities, free from the stresses and strains of clocks and timetables, romanticizes the life of our earliest human ancestors. Clues from paleolithic demography reveal the body of industrial man to be twice as durable as that of his stone age ancestor (Roosevelt 1984). Whatever the measure, stature, or longevity (Fig. 10.1), life in modern society far exceeds the levels of welfare achieved before the dawn of civilization, or at any other time during the last 5000 years of recorded history.

Taking the length of life as a measure of vitality,[1] twentieth century demography shows the modern urban environment to be the most propitious human habitat in recorded history. Likewise, the manner in which people organize their collective life (the social and economic environment) makes all the difference to their physical and mental welfare.

Environment evokes the image of physical surroundings, climatic conditions, and natural resources. Though these are real and important dimensions of the human habitat, they are not the focus of this chapter. The emphasis here is on the social arrangements, which make stable and harmonious co-existence possible for a large number of people sharing the same territory and co-operating together in producing their

[1] Among indicators used to measure the distribution of health in a population, only one, mortality, can be used with any degree of precision to study trends over time between and within societies. Alternative indicators, morbidity or stature, are more useful for analysing intra-national variation and even then they can only be used with caution. Morbidity is the most methodologically problematic of all possible health status indicators because as a complex medico-sociological variable itself, its meaning is influenced by the same economic and normative factors that help shape health outcomes. Thus, for example, the higher morbidity recorded by women is probably more a reflection of gender norms dictating economic roles than any real difference in the physical risk of disease. Men of working age appear to suffer lower rates of morbidity but higher rates of mortality, suggesting that the direct risk of disease itself is not perfectly correlated with the same factors that predispose people to see themselves as unhealthy or in need of medical advice and treatment (cf. Hart 1982). Morbidity, whether self or medically reported, has an important motivational aspect which limits its utility as an objective (rather than subjective) health status indicator. Stature has recently emerged as a possible 'objective' measure of health status. Researchers have shown that it is closely correlated with longevity and of course it has the additional advantage of being a measure of the living not the dead (Flood 1984; Illsley 1986; Rose and Marmott 1981). It has yet to become a routine health status indicator in official statistics, and though in time its use is likely to increase, so far at least few societies produce systematic series of height differentials in the population. However, average height is not exclusively determined by socio-economic factors, nor is it strongly correlated with longevity in an international context. Though, for example, average height in Japan has been increasing rapidly in the last few decades, it remains below that of many other advanced industrial societies where infant mortality and age specific death risk is higher.

The rationale for using mortality as a means of measuring the distribution of vitality and health in the population is, first and foremost,the availability of data, both contemporary and historical. Even if this were not the case there is still good reason for using mortality as the best indicator. Age-specific mortality is a measure of human durability, and phrased thus we must assume that it will reflect human vitality, immune status, and nutritional welfare. More research is needed on the links between longevity and lifetime morbidity but in the meantime, mortality will continue to be employed with good reason as the principal indicator of a population's health status.

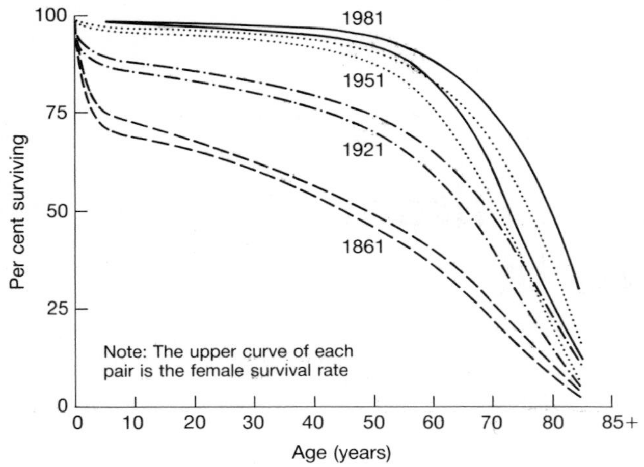

Fig. 10.1. Male and female survival curves for England and Wales. (Source: Hart 1990.)

livelihood. Society is literally the set of arrangements that make stable social and economic life possible and the purpose of this chapter is to estimate its influence on human health.

The social and economic environment is an enduring structure external to and enveloping the individual, pre-dating birth and persisting after death. It comprises *economic institutions* to produce and distribute the material livelihood of the people, *ideological beliefs* (religion, morality, political culture) to uphold shared values, *linguistic codes* to facilitate communication, *social institutions* (marriage, family, education, law, and so on) regulating relationships and protecting the rights of every citizen. Demographic and epidemiological data is a rich source of evidence of health as the product of this social organization. In the pages that follow, the evidence that systematic features of social and economic life exert a pre-eminent influence over the process of human growth, development, and ageing will be laid out.

The evidence of the effect of the environment on vitality

Historical demography

Demographic history leaves no doubt that the length of life, and the probability of resisting and surviving from otherwise fatal disease, is historically specific. We know, moreover, that the last 150 years (4 per cent of recorded history) have revolutionized human vitality. As McKeown (1976, p. 1) dramatically states:

> It took hundreds of thousands of years for the human population to expand to the first thousand million, the second was added in 100 years, the third in 30 and the fourth in 15.

The modern rise of population is the result of a dramatic decline in the risk of 'premature' death. Though medicine has catalogued every form of human degeneration (International classification of disease, ICD), in modern society the majority of people die of natural causes at or near the end of a normal life-span (Fries and Crapo 1981). As Fig. 10.1 illustrates, the probability of survival to the biblical three score plus ten has virtually doubled during the short era of industrial urban civilization.

In 1850, fewer than 25 per cent survived to 70 years. Today more than 60 per cent of men and 75 per cent of women live to this age. The causal ingredients of improved human vitality remain a fascinating academic controversy; what is not in doubt is that people live longer primarily because they are healthier. They are better fed, of larger stature, physically more robust, better educated, and socially and politically more secure. The systematic prolongation of life among whole populations, together with the increase of average stature over the last century (Floud 1984), are incontrovertible signs that human health is the product of social and economic forces external to individuals and their personal habits.

The distribution of vitality between societies

More evidence of the improved durability of human beings lies in the international league table of life expectation. Again, Japan illustrates the synthetic socio-economic character of human health. Among nations in Table 10.1, Japan is ranked bottom in 1950 but ascends to the top by 1980. Between 1950 and 1980 the life expectancy of the Japanese increases by almost 14 and 16 years for males and females respectively, 9 and 11 years of this accruing from improvements after the age of 15 for men and women respectively. The share of the gain at age 65 is 4 and 5 years.

Improved vitality coincides with Japan's so-called economic miracle. As the economy grew from the ashes of the war to become the second largest in the world, the longevity of Japanese people was simultaneously extended. Though Japan is now a global leader in the prevention of infant mortality, life expectancy gains have been achieved mostly by reduced mortality among adults over the age of 15. Even those growing up in the thirties and forties, when mobilization for war caused food scarcities and undernutrition, have shared equally in the gains.

Contrast Japan and Hungary. In 1950 both countries were on a par; a decade later Japan had moved ahead, but Hungary registered some progress, with male life expectancy increasing to the level of the US or Finland. By 1980 Hungary was the focus of the demographic reverse taking place in some parts of Eastern Europe because of its exemplary record in maintaining a free flow of official statistics of mortality. Among both sexes, life expectation at 15 was below the level of the previous decade, indeed among men it was below the level of 1950. Overall, male life expectation fell at all ages while women recorded a tiny increase at birth and at age 65 (Compton 1985).

How do we explain the opposing demographic trends of Japan and Hungary? There is no simple answer. The correct

Table 10.1. Expectation of life 1950–84 in selected countries

Nation	Male			Female		
	Birth	Age 15	Age 65	Birth	Age 15	Age 65
UK						
1950	66.9	54.7	11.8	71.7	59.3	14.4
1960	68.1	55.3	11.9	74.1	61.0	15.4
1970	69.1	55.9	12.2	75.4	61.9	16.2
1980	71.3	57.5	13.2	77.3	63.3	17.2
USA						
1950	66.0	53.9	13.0	71.9	59.3	15.4
1960	66.8	54.4	13.0	73.5	60.3	16.1
1970	67.6	54.5	13.2	75.3	61.9	17.3
1980	70.6	57.0	14.5	78.1	64.3	18.9
Japan						
1950	60.5	51.5	11.4	64.0	54.8	13.5
1960	66.6	54.6	12.0	71.7	59.2	14.6
1970	70.6	57.1	13.3	75.9	62.2	16.2
1980	74.3	60.2	15.3	80.0	65.7	18.7
Finland						
1950	62.9	51.5	11.2	69.3	57.3	13.2
1960	65.5	52.6	11.5	72.5	59.2	13.8
1970	66.5	52.9	11.8	75.0	61.2	15.1
1980	69.9	55.7	13.1	78.5	64.2	17.4
Hungry						
1950	61.4	53.3	11.9	65.7	56.6	13.2
1960	66.4	55.5	12.5	70.9	59.5	14.2
1970	66.5	54.6	12.1	72.4	60.1	14.7
1980	65.3	52.2	11.6	73.1	59.7	14.8
Sweden						
1950	70.3	57.7	13.8	73.1	60.0	14.6
1960	71.5	58.3	13.9	75.5	61.9	15.7
1970	72.2	58.4	14.2	77.7	63.7	17.3
1980	73.4	59.2	14.6	79.6	65.3	18.5

Source: WHO 1986, Table 10, pp. 82–7.

one surely involves a complicated mix of specific factors. In Japan, spectacular economic growth, albeit in a context of high social solidarity, lead to continuous demographic progress. In Hungary the picture is confused. Though economic progress has been far below the Japanese post-war experience, there is no question that living standards today are well above the immediate post-war level. Indeed among Eastern European nations, Hungary has a relatively high standard of living, certainly above its neighbour Czechoslovakia where mortality rates are lower. Regress in Hungarian longevity cannot be reduced to economic forces alone. Almost certainly non-material forces of social life, ideology, morality, religion, and perhaps political culture have made their contribution to this unusual demographic trend.[2]

Table 10.2. Accidents, suicide and homicide in various societies, rates per 1000 population

	Accidents			Suicide			Homicide		
	all	m	f	all	m	f	all	m	f
Hungary 1985	77.2	94.0	61.4	44.4	67.0	23.2	2.7	3.2	2.1
Japan 1985	24.6	36.1	13.6	19.4	26.0	13.1	0.8	1.0	0.7
France 1984	64.3	72.9	56.2	22.0	32.2	12.4	1.3	1.7	1.0
England & Wales 1984	25.3	29.8	21.1	8.7	11.8	5.7	0.7	0.8	0.6
Sweden 1984	41.2	54.2	38.5	19.5	27.4	11.8	1.1	1.4	0.8
US 1984	39.5	56.2	23.8	12.1	19.2	8.4	8.5	13.4	3.9
Czecho-slovakia 1984	55.2	65.4	45.5	18.5	28.1	9.4	1.1	1.3	0.9

Source: WHO 1986, adapted from Tables 11 and 12, pp. 92–630.

The same conclusion may be drawn comparing Japan with other market societies. The yen's value as an international currency gives a misleading picture of average living standards. Japanese families have enjoyed a continuous rise in material circumstances, but their standard of their livelihood is not above the average for Western Europe. Indeed by many consumption and leisure indicators the Japanese worker is relatively deprived in international comparison, reinforcing the impression that vitality cannot be reduced to economic forces alone.[3]

Mortality from accidents, suicide, and homicide provide interesting insights on the relative 'safety' and preservative quality of the social and economic environment. They reveal that an important contribution to recent Hungarian demographic trends is a rising suicide rate. In Table 10.2, Hungary leads for both accidents and suicide.[4] Though the Swedes evoke an image of a suicide prone people, their rate of self-destruction is less than half that of Hungary, and lower than France. Hungarians commit suicide 5 times more often than the British, 3.5 times more often than Americans, and twice as often as the French. For homicide, Hungary slips to second place behind the US, where the risk of being murdered ranks 12th among all causes (see Table 10.2).

What do suicide and homicide reveal about the social and economic environment? Durkheim (1896) demonstrated a

[2] The systematic comparison of health status in different varieties of industrial socialist and capitalist nations is an under-researched area. Though some authors have pointed to political economy as the root source of health inequality (Hart 1986*c*), by restricting the analysis to industrial capitalism, their research passes-by the opportunity to search for insights through a systematic comparison of health status in the existing non-capitalist industrial nations of Europe (Doyal and Pennell 1979; Hart 1985). Though opportunities for research are diminished by the absence of accurate statistics from Eastern Europe, the clear evidence of demographic reverse offers an opportunity to identify non-economic dimensions of the social environment which may impact on human well-being and physical survival.

[3] The healthiness of the Japanese diet is often singled out as the cause of Japan's demographic miracle, but what is often overlooked is the high consumption of tobacco and alcohol (Hirayama 1987). Though the sex mortality ratio in Japan is the lowest in Table 10.1, sex differences in smoking and drinking are well in excess of anything recorded in comparable European industrial nations.

[4] Many researchers have questioned the validity of suicide statistics on the grounds that the frequency of suicide verdicts is more an index of public morality and religious taboo than of acts of violence against the self. For this reason accidents and suicide are often grouped together, though it is clear that the combination involves a disparate assortment of causes (Adelstein and Mardon 1975).

correlation between religion and the rate of self-inflicted death. Drawing on the low suicide rates in predominantly Catholic regions of nineteenth century Europe, he argued that the incidence of suicide reflected the level of social integration, the extent to which motivation and behaviour are governed by shared norms and values. Catholicism emerged as the strongest source of normative integration because of its constraining influence over individualism.

The high suicide rate of Hungary is not new. Hungarians have always ranked high though today their rates are double their pre-war level (Gergely 1987). Moreover, trends exceed that of neighbouring societies by an increasing degree. In 1962, the Czechoslovakian rate was 21.3 per 100 population, compared to Hungary's 26.8. Today it is 18.5 compared to 44.4. National differences are not merely a matter of historical precedent or national character. Nor can they be reduced simply to religion. Though Catholicism may explain low Polish rates (11.3 in 1974) compared to Czechoslovakia (22.4), it cannot explain differences between Hungary and Czechoslovakia.

Durkheim predicted the replacement of the unifying force of religion with secular social ideologies. Opposing free market theorists who saw mutual self interest as the stuff of social co-operation, he insisted that social solidarity depended on shared morality, that stable social life could not be maintained by economic interests alone. The prevalence of violent causes in Hungary suggests that its own Eastern European socialism does not operate as an effective secular substitute for religion. Though promising to unite individuals under a value system based on the spirit of economic justice and equality, state socialism on the Eastern European model does not achieve the degree of social solidarity of Catholicism, nor does it appear to uphold a high degree of commitment to collective or even personal life.

Another sign of culture and economy interwoven in mortality risk is found in the distribution of infant mortality and its relationship to the socio-politics of fertility control. The causes of the rapid decline of infant mortality in Japan are not to be found in medical expertise. Taking maternal mortality as an indicator of bio-medical skill, Japan ranks relatively low in an international context. The descent of infant mortality rates reflects the high degree of rationality attached to childbearing. Average childbearing age is high, most mothers bear their first child in the mid-twenties, and the rate of illegitimacy and teenage motherhood (the strongest correlate of infant mortality) is extremely low. This is facilitated by sanctions against pre-marital intercourse and pragmatic public attitudes to abortion. Abortion in Japan is a major form of family planning and, unlike the West, strong religious taboos do not induce uncertainty, guilt, and remorse in the woman who lacks the resources or the motivation to bear a child. Indeed Japanese Buddhism offers a ritual equivalent to abortion counselling. The result is that the arrival of almost every child is consciously planned by parents prepared for the financial costs of childbearing. Economic welfare combined with a highly rational childbearing culture account for the exemplary public health record of Japan in the field of infant mortality.

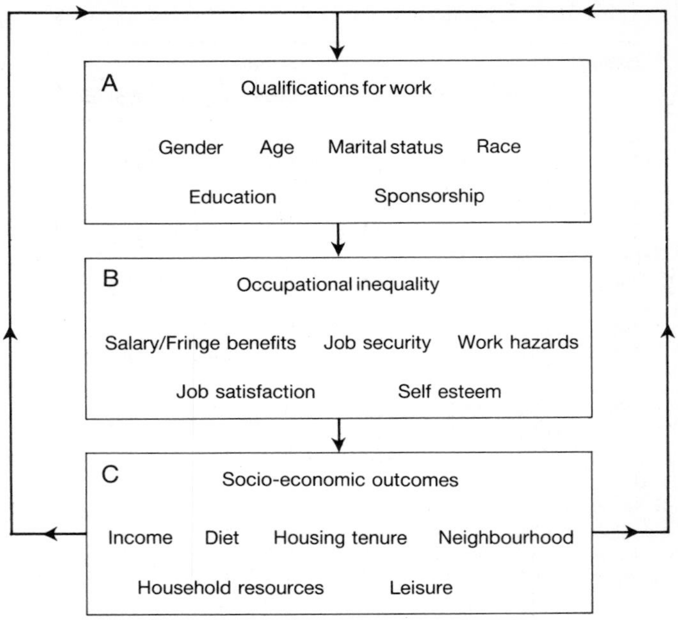

Fig. 10.2. Occupation and livelihood: A system of transmitted privilege and deprivation.

Cross-national analyses of health status raise important methodological problems but offer important insights on the sources of vitality in social and economic life. The evidence reviewed here raises a number of questions but no definitive answers. What is established is the insufficiency of material resources on their own as determinants of trends in human vitality.

The distribution of vitality within society

The distribution of longevity within national populations is the third source of insight on the health giving/denying power of social and economic life. The most significant socio-economic variables stratifying survival chances are sex or gender, race, employment status, occupational class, education, tenure, and marital status. They are all located in Figure 10.2, a representation of inequality as a built-in feature of the way the material livelihood of the population is produced and distributed in industrial society.

In advanced industrial societies, economic production depends upon a vast network of co-operation, the 'division of labour' in social science parlance. The social division of labour is simultaneously a source of solidarity and division. Though millions are brought together for the collective performance of work, inequalities in the terms of participation and the distribution of rewards are a profound source of disunity.

The occupational structure of an industrial capitalist society is organized according to market principles. Productive work is divided into a large number of specialized tasks, and the human labour to perform them is bought and sold under very unequal conditions. The most expensive human labour is highly trained and professionally credentialled. It attracts high prices (wages and salaries) and highly privileged work conditions. The cheapest labour is nominally unskilled and unqualified, poorly rewarded, generally insecure, and often performed under unpleasant if not dangerous work conditions. Between these two extremes, occupations are ranked according to skill and economic reward. Inequalities of wealth, income, and status are not unique to capitalism. Though communism eliminates the market for measuring personal economic worth, occupational class remains a major determinant of the distribution of vitality (see p. 160). Figure 10.2 identifies the links between different axes of social stratification in the division of labour.

Qualifications and access to the labour market

What is the probability of an individual obtaining a privileged niche in the occupational structure? The most desirable jobs are well paid, secure, interesting, and free of risk. Getting one depends on a range of factors, some determined at conception like gender and race, some acquired afterwards like education, marital status, and networks of influential sponsors.[5] Wealth and family connections have always exerted a powerful influence on children's occupation; this continues today even though inheritance tax, nepotism, and anti-discrimination laws seek to limit the transmission of social and economic office and advantage between the generations.

Among 'qualifications' fixed at birth, gender strikingly illustrates the social constructed character of occupational credentials. In the late nineteenth century, when women were supposed to stay at home, their exclusion from higher education and professional training was justified in contemporary scientific journals by reference to smaller brain size and reduced intellectual capacity (Sayers 1982). Racial inequalities were also given the same spurious legitimacy by the science of craniology. Sex and race discrimination in the allocation of educational or occupational opportunity are now illegal in most societies, and though being 'coloured' or female remains something of a disqualification for privileged work, it is no longer justified by moral or 'scientific' arguments. Even so, informal practices still operate to exclude people on the grounds of their sex or skin colour, and to reduce choice and autonomy in job search.

The age of starting and retiring from work is regulated in all advanced societies though this does not mean that the reward system of the occupational structure begins with the first job or ceases with retirement. Human life has a developmental quality. In early life there is little scope for individuals to be the architects of their own physical and intellectual development. Small infants cannot influence their own lives and observed health differentials between them cannot be reduced to voluntary forces within individual control. Childhood is a formative phase in the human lifetime and its outcome is determined by parental class. The mother and father—who they are, where they live, how they generate their livelihood, what they possess—these are the ingredients which launch a child onto a more or less successful life career.

Children experience the occupational structure principally through the occupation of the father, through any specific risks his job may introduce into their home environment,but more importantly through the wage which fixes living standards and material aspirations. Fatherless children, a substantial and growing minority in most advanced societies, tend to grow up in the poorest of economic circumstances because they may have no relationship to the reward system of the occupational structure. For the same reason this group tends to be absent in research on socio-economic indicators on health, although there is no question that healthy development is put at serious risk by economic disadvantage of their statistically deviant familial status. The mortality of 'illegitimate' infants in the UK is almost double that of the 'legitimate'.[6]

In old age, the reward system of the occupational structure continues to shape material welfare. On retirement income falls precipitously even though this is hardly a time when people need to spend less on the maintenance of their living standards. Freedom from financial anxiety in old age impacts directly and indirectly on vitality. People who benefited disproportionately during working life also retire with inflation-proof pensions, other forms of superannuation, in addition to state benefits. People who received the smallest rewards from the division of labour face poverty after retirement because they never earned enough to accumulate personal savings and their jobs offered no retirement pension scheme. The population living below the poverty line in the UK contains a disproportionate number of elderly people who receive nothing but the minimum state pension. This is scarcely enough to pay for weekly food and utilities with nothing spare for other expenses, including even increased heating costs during cold weather.[7]

Relationship to the occupational structure and its reward system, experienced directly through personal participation or indirectly through a spouse, parent, or pension is a lifetime determinant of material welfare. This is why inequalities in health and vitality persist from the cradle to the grave in British society (DHSS 1980).

[5] Though educational qualifications appear to be meritocratic achievements, among the wealthiest section of the community they too are determined at or before birth. This is self evident in England where parents who intend to send their children to the most influential private schools expect to 'put their names down' before the first birthday. The strong associations between the 'old school tie' and the most privileged jobs in the labour market leaves little doubt that education is strongly determined at birth.

[6] In 1975/6 ratios of illegitimate to legitimate infant mortality in England and Wales were as follows: stillbirth, 1.35/1.00; perinatal, 1.49/1.00; neonatal, 1.61/1.00; post-neonatal, 1.64/1.00; OPCS 1978*a*).

[7] Rates of mortality among the elderly are systematically correlated with weekly temperature changes during the winter months in the UK. These deaths from 'hypothermia', in laymen's terms being frozen, are the direct result of low income among the elderly and a quite clear example of poverty directly determining survival in an advanced industrial society.

Education identifies the knowledge, literate skills, and credentials that individuals acquire as they pass through school, college, and university. In a thoroughly meritocratic society, education might represent something close to personal ability, aptitude, skill, and intellect. In most societies it measures these attributes in combination with the volume of resources that parents were able to invest in their children's development. Private schools and additional tuition are the most obvious avenue of extra investment, no less important are the quantity and quality of resources routinely available at home, and the experiences children are exposed to as they grow up. The French sociologist Bourdieu refers to this as 'cultural capital' (Bourdieu and Passeron 1977). In 1980, the cost of raising a child in the US was calculated at around $120 000, an average which varies enormously with the material resources of parents (Zelizer 1985). The idea of investment in children illustrates forcefully the scope for inequality in physical and intellectual development. Some literally have hundreds of thousands of dollars/pounds expended on their development as against tens of thousands or tens of hundreds for others. These inequalities are mediated in some degree by state investment in the name of equality of opportunity. Even so, in all advanced industrial (capitalist and socialist) societies, a curriculum vitae is the best indicator of material advantage and disadvantage in the natal home. It is also the ticket to desirable or undesirable occupation.

Sponsorship refers to the informal mechanisms which well-connected and influential parents use to 'help' their children on in their careers. In some cases sponsorship operates as an adjunct to a privileged and expensive education, the classic 'old boy network', opening doors to occupational opportunities unconnected to formal qualifications. Though such avenues to occupational progress are supposed to be closed in modern 'diploma' oriented societies, there is no doubt of the continued importance of kinship and other networks in furthering individual occupational careers.

Occupational inequality

Occupational inequality takes many forms.[8] Even in the same workplace, wages and salaries may vary by a factor of 10 or more. On top of this, further fringe benefits and pension arrangements make the distribution of economic rewards highly unequal. Job security, the risk of unemployment, is an additional source of inequality. Better paid jobs are almost invariably non-manual (white collar) and typically characterized by security of tenure. Since low wages and insecure work go together, the risk of unemployment is disproportionately borne by workers who are least likely to have financial reserves to tide them over a period of joblessness.

Despite this the lowest paid work is often the most dangerous. Fatal accidents are three or four times more prevalent in manual occupations, and yet people exposed to hazardous work environments are unlikely to receive any public recognition or financial compensation for the risks they have to take to generate their livelihood. This is because non-manual work, besides being safer, is also more highly regarded. Quite literally income and status distinctions between manual and non-manual work 'add insult to injury': the safest jobs offer more opportunity for job satisfaction and for self-esteem.

The inequalities enumerated above reinforce one another in the workplace and they feed directly into the process of consumption and leisure outside it.

Socio-economic outcomes

The week-old British infant discharged from a National Health Service Hospital is about to experience the social and economic environment that will play a formidable role shaping the quantity and quality of future life. When the infant exchanges the warm, modern hospital ward for a damp overcrowded bedroom in an apartment house on a run-down inner city estate, the opportunities for healthy vigorous growth and development will be bleak. The Black Report (DHSS 1980) estimated an average length of life at least 7 years shorter for the child discharged from hospital to unskilled manual compared to a professional home. The causes of this differential are perhaps too obvious to elaborate. They include relative deprivations of every kind; income, diet, warmth, space, domestic and neighbourhood hazards, household facilities, holidays, school, and so on. Moreover, deprivation does not begin as the new-born is carried over the threshold of its parent's dwelling. It has already influenced birth date and maternal age, dictating the degree of conscious planning and provision surrounding the birth, and whether it provokes disruption and strain to its parent(s).

The standard of comfort, warmth, and welfare in a child's natal home and neighbourhood are a matter of consumer power. Consumption is often identified as the key to understanding health inequality. Tobacco, alcohol, and unhealthy food appear as causes of disease and reduced life expectancy. These products of consumer choice need to be carefully distinguished from the larger idea of consumer power which determines much more than what goes into a shopping basket. Consumer power is dictated by income and property, it is the means by which individuals and families literally purchase their socio-economic address, their immediate environment in the society at large.

Consumer power is a product of occupational inequality though it also lies within the scope of governmental policy. Pay, fringe benefits, job security, and work hazards are all within the potential control of government. Public policy can compensate relatively deprived sections of the population—families with dependent children, the elderly, the low paid—for the harsh effects of the market on their lives by redistributing income through tax credits, transfers, allowances, and pensions. Welfare benefits, educational expenditure, housing, and neighbourhood investment are some of the possible

[8] In the British census, the number of titles in the occupational classification has generally stood at around 550, though in the 1961/71 censuses, an attempt at rationalization reduced the number to about 200.

means of redressing inequalities resulting from the processes depicted in Fig. 10.2. Though these are formidable tools for managing the social and economic environment, governments vary in their willingness to alter the effects of market forces on distribution of welfare. Sweden, an example of global excellence in the field of public health, is renowned for its commitment to redistributive economic welfare, and for the degree of economic equality between citizens regardless of age, household status, or occupational class. In the aftermath of the Second World War, many other European governments used redistributive policies to rebuild more egalitarian social structures, so-called lands 'fit for heroes'. More recently, many of the same governments have adopted *laissez faire* attitudes to their public responsibilities, letting poverty, homelessness, and relative inequality increase, even justifying the process as a necessary stimulus to the performance of the economy.

The three sequences in Fig. 10.2 operate as a cycle of relative socio-economic privilege and deprivation, transmitting inequalities in life chances from one generation to the next. At the core of the process is the labour market and the occupational class structure. People with the most privileged jobs secure not only their own but also their children's welfare. At the apex of the income distribution, they have the means of providing the next generation with a good home in a good neighbourhood, good schools, additional tuition, and cultural capital, apart from complete freedom of want in the matter of diet, clothing, warmth, and space. This is how economic privilege is transmitted from one generation to another.

The terms 'cycle of disadvantage' or 'transmitted deprivation' is often employed to describe inter-generational transmission of poverty and deprivation. By focusing on the poorest section of the community, this terminology distracts attention from the other side of the coin, from 'cycles of advantage' or 'transmitted privilege' which ensure that the economic benefits and privileges enjoyed by one generation are reproduced in the next. These processes bring us close to the core of what social class purports to represent, socio-economic inequality as a systematic property of the social order, over which ordinary, and especially young people, can exert a little influence. Children growing up in a cycle of transmitted privilege have the best chance of reproducing their parent's material success, because they have ready access to occupational futures that their age peers at the other end of the socio-economic scale never even dream of. There is a mass of descriptive evidence in every society linking poverty with illness and poor achievement among children. A narrow focus on economic welfare and health defines the problem as a residual lower class, a matter of deprivation among a small minority of very poor people. A wider focus illuminating the entire distribution of chances for physical and intellectual development in children reveals it as a step-like hierarchy systematically linked to the occupational status of the father, and to an increasing extent to that of the mother also. In Britain children of social class I (professional) parents have the highest probability of an eight decade lifetime, social class II (managerial) record the next best chance, and so on down to social class V (unskilled manual) where the infant who lives beyond seven decades defies statistical probabilities (DHSS 1980).

The concept of social class

The concept of class identifies systematic inequality within the social and economic environment.[9] In classical sociological theory, class is specifically an economic relationship, defined by production for Marx, and by the market for Weber. It has also been widely employed as a simple measure for the distribution of material welfare. In this sense, it can be thought of as a tool for measuring the *total impact* on life chances of most of the variables in Fig. 10.2; a summary of social and economic stratification, the distribution of individuals, households, families, even whole neighbourhoods into a recognizable economic hierarchy.

In the study of health, class is used to differentiate points on a range of probability for healthy human growth, development, and longevity. There is no perfect measure. Occupation, education, or tenure are all possible candidates, each producing a slightly different division of the range. The best measure is the one which creates a series of economically homogeneous social groups whose members possess broadly similar levels of command over economic resources. The basis of this homogeneity must not be ephemeral. A series of simple income groups would not be a class because current income is strongly influenced by age, and people earning the same wage or salary may be very heterogeneous in their qualifications, ownership of property, promotion prospects, and so on. *Lifetime earning capacity* is a better basis than current income for the measurement of class (Joshi 1987). The most convenient proxy for lifetime earning capacity is occupation because it is linked to many other possible indicators of class, income, property, education, qualifications, and status/prestige. As a result, it has been widely used to identify socio-economic status in routine government statistics such as the registration of birth, marriage, and death. However, occupation is only one possible means of measuring social class and there are a number of limitations attached to its use, particularly in the analysis of trends.

In recent years, a new proxy for lifelong economic security has emerged. Housing tenure, a simple and largely dichoto-

[9] The words 'life-style' or 'way of life' are sometimes employed loosely as a verbal proxy for class (Fox and Adelstein 1978). This terminology can be misleading since it tends to mix up the basic independent ingredients of class, such as income and wealth, with other more dependent manifestations such as patterns of consumption and leisure. Though these may serve as important mechanisms linking economic circumstances to health and vitality, they are not in themselves part of the institutional fabric of society which generates economic welfare. In any case the term 'life-style' assumes a degree of free choice which usually goes unspecified. To be of any value as a device for explaining the distribution of health, the voluntaristic/non-voluntaristic dimensions of conceptual terms must be specified if analytical precision is to be achieved.

Table 10.3. Chadwick's sanitary map: average age at death by occupational class, England 1842

District	Gentry and professionals	Tradesmen	Labourers
Derby	49	38	21
Bolton	34	23	18
Leeds	44	27	19
Bethnal Green	45	26	16
Whitechapel	45	27	22
Strand Union	43	33	24
Kensington	44	29	26

Source: Lewis 1952, p. 44.

Table 10.4. Mortality by social class 1921–81: England and Wales

Period	Registrar general's social class				
	I	II	III	IV	V
1921	82	94	95	101	125
1931	90	94	95	102	111
1951	98	86	101	94	118
1961	76	81	100	103	143
1971	77	81	104	114	137
1981	66	76	103	116	165

Source: OPCS 1978*a*, Table 8.1, updated with 1981 figures, OPCS 1986.

mous variable, proves to be a formidable indicator of mortality risk. Easy to collect and less prone to error, it must be reckoned as an important addition to the statistical tool box for the measurement of the social and economic environment. We will start our review of inequalities in health with occupation- and tenure-related measures of social class.

Occupation

Edwin Chadwick's 'sanitary' map was an early attempt to measure mortality risk by a crude occupational index. It reveals an enormous gap in life expectancy between the gentry (landed and propertied classes), tradesmen (merchants and shopkeepers, and so on), and labourers. The length of life of the richest occupational group is double that of the poorest; note also the importance of geography as a variable distributing mortality risk (Table 10.3).

One and a half centuries later, inequalities of life expectation are much smaller, though there remains a substantial gap separating the life chances of men and women in different 'walks of life' (see p. 159). Chadwick's sanitary map was published in 1842. By 1860 his colleague William Farr was developing more statistically sophisticated means of identifying the relationship between economic welfare and human vitality. A pioneer in both social statistics and social medicine, Farr's 'Healthy and unhealthy districts' offers an almost graphic image of the environment as a social and economic phenomenon, and identifies the significance of the occupational scale as a means of differentiating life chances in the early stages of industrialization (Farr 1885).

The relationship between occupation, life expectancy, and disease-specific mortality has been the subject of a special enquiry following every census in England and Wales since 1911. Taking the latest census distribution of occupations as the denominator and the occupational titles on death certificates as the numerator, standardized mortality ratios for individual and group occupations have been estimated every decade since the beginning of the century.

The trend of occupational mortality contained in successive decennial supplements (1911 to 1981) reveals a steady improvement in life expectancy in every social class. It also shows the persistence of a substantial gap in survival chances between rich and poor in Britain, despite improved living standards, the growth of the welfare state, and since the war the development of a comprehensive system of free medical treatment. The long-standing public commitment to the study of occupational mortality makes the UK the obvious context for the detailed analysis of trends in longevity by occupational class (Koskinen 1985; Pamuk 1985).

Table 10.4 is the time series of class mortality gradients among males aged 15–64 from 1921–81. Five social classes based on grouped occupations are represented.[10]

Each decade the pattern of stratified mortality risk among British men reappears with the lowest SMR (standardized mortality ratio) in social class I (professionals) and the highest in social class V (unskilled manual workers). In 1981 the rate of premature death (before age 65) among unskilled manual workers was more than two and a half times the rate among professional workers, while the middle-ranked skilled class recorded an SMR 49 per cent in excess of social class I and 62 per cent below social class V. Among married women classified by their husband's occupation, the same pattern of inequality is reproduced though the extent of the differential

[10] The UK Registrar General's occupational class is constructed by grouping similar occupations into 5 ranks, with the middle rank of skilled workers further subdivided into manual and non-manual. Nominally at least, occupational skill appears to be the defining characteristic of the ranking, though the stated rationale for grouping occupations is their 'similar standing in the community'. The five occupational classes are constituted as follows:

Social class I: professionals; for example, doctors, lawyers
Social class II: intermediate; for example, teachers, managers
Social class IIIN: Skilled non-manual; for example clerks
Social class IIIM: Skilled manual; for example, underground coalminers, technicians, ambulance drivers
Social class IV: Partly skilled manual; for example, bus conductors, postmen
Social class V: Unskilled manual; for example, porters, ticket collectors, general labourers

Some version of this classification has been in use since 1921. In those days, more than 35 per cent of work was classified as unskilled or partly-skilled. This proportion has steadily fallen over the intervening period, though by 1970 just under 27 per cent of work still fitted the definitions of social classes IV and V, and that percentage remained relatively stable through the 1981 census. In the 1961 census, social class III, skilled workers, was disaggregated to form two sub-classes: non-manual and manual workers. At the same time, new rules for upgrading supervisors and foremen in social classes IV and V were introduced. This change made a notable difference to the distribution of occupations at the heart of the ranked scale and therefore serves as something of a methodological watershed in time series anaylsis.

is less. The SMR of wives in social class V was 65 per cent in excess of social class I in 1980 (OPCS 1986).

The trend seems to indicate a widening of the mortality gap between the social classes in the UK. In 1921 the SMR of social class I is 82 compared to 125 in social class V; by 1981 the respective SMRs were 66 and 165 respectively. However, interpretation is not straightforward because both the content and size of each occupational class changed over time. This results partly from shifts in the occupational class structure itself, caused by the disappearance of many traditional industrial occupations, and partly from attempts to improve census methodology by shifting specific occupations into different classes.[11]

Changes in the occupational structure have eroded the demand for unskilled workers and increased the demand for better trained employees. Overall, by the time of the 1971 census in the UK, the membership of social class V was drawn from a much smaller and probably more selectively disadvantaged section of the national population. This raises difficulties when the scale is used for the analysis of trends. To compare like with like at two points in time, it is clearly desirable that the proportion of the population classified into each social class remains broadly the same (Illsley 1986). This requirement can be met by the simple expedient of constructing a cumulative frequency of occupational mortality risk and comparing the SMR of the same divisions, for example, population quintiles, at each successive period. The results of such an analysis are given in Table 10.5.

Table 10.5 displays a remarkable stability in the aggregate distribution of mortality risk in England and Wales during the twentieth century. Among men, 1980 breaks the trend. The SMR of the fifth quintile jumps to 161 while the first quintile falls to 58, suggesting that the seventies saw a sharp deterioration in the relative position of the population exposed to the highest risk of mortality (20 per cent of population classified in occupations with the highest mortality risk). Among married women (classified by their husband's occupation) 1970 is the year which breaks the trend, the SMR increases from 134 (in 1960) to 151 (in 1970), rising to 161 a decade later. Increases are matched by reductions in the population recording the lowest occupational mortality. The SMR of the first quintile (the 20 per cent of the population with the lowest SMR) falls from 69 (in 1960), to 59 and then to 58 in 1970 and 1980 respectively. The relative mortality risk of the fifth quintile for men, women, and infants is much greater in 1981 than at any previous point in the twentieth century, suggesting that the increase of health inequality in the last decennial supplement (1979–83) was real and not an artefactual trend. Since there is no theoretical rationale underlying the quintile distribution,[12] the results in Table

Table 10.5. Quintile distribution of mortality risk UK 1921–81 death rate per 1000 live births

Period	1st	2nd	3rd	4th	5th	Ratio 5th as % 1st
Males						
1920	63	85	97	108	141	224
1930	66	87	98	112	141	214
1950	65	83	95	107	145	223
1960	65	89	97	103	140	215
1970	68	90	100	109	145	213
1980	58	78	95	109	161	278
Females						
1950	68	88	96	105	139	204
1960	69	89	99	107	134	194
1970	59	92	103	115	151	256
1980	58	78	95	109	161	278
						SMRS 15–64 years
Infants						
1920	46.4	66.2	77.0	93.4	109.2	235
1930	39.6	51.0	58.3	70.0	89.0	225
1950	17.3	24.0	28.2	32.3	42.9	248
1970	11.0	14.0	16.0	18.2	25.5	232
1980	3.1	6.0	7.0	7.9	12.3	397

Source: Hart 1986*a*.

[11] In the UK, the Registrar General's social classes are constructed out of a data base of occupational titles which has itself changed substantially in the period since 1921. Between 1921 and 1951 the census made use of more than 550 occupational titles. In 1961, following a methodological rationalization, the base number fell to 200. Ten years later, it was increased to 221, and by 1981 a complete reformulation brought it back up to the 550 mark. These changes produce a considerable discontinuity in the composition of each class which undermines efforts at decennial comparison and discourages the use of the time series for the study of trends in longevity. Nevertheless, since the decennial supplements are a unique data base of occupational mortality, many researchers have felt it worth the effort to overcome their methodological problems (Pamuk 1985, Koskinen 1985).

[12] The contrast is striking but caution is necessary in the interpretation of these data. The quintile distribution is based upon a cumulative frequency distribution ordered according to the mortality risk of separate occupations. It is not an occupational class scale. There is no independent conceptual or theoretical rationale for assigning occupations to the quintile distribution, the allocation is a mechanical process based on recorded death rates at each decennial period, aggregated into a cumulative frequency distribution in order to rank the relative mortality risk for equal proportions of the population. The SMRs for each quintile may be artificially inflated and diminished by increases and decreases in the number of occupational titles. The more detailed the occupational classification, the more internally homogenous the population in each category, the greater the range of occupational mortality risk overall. In general, grouping more occupations into a single title can be expected to lower its SMR. In addition the cumulative frequency, particularly the top and bottom quintiles, may be especially prone to numerator/denominator problems.

How might these problems affect the interpretation of Table 10.5? While it is true that the 1980 census more than doubled the number of occupational titles in the census, it does not seem that this fact alone could account for the increase in the SMR of the bottom quintile. In 1950 the number of titles (585) was similar to 1980 (550), yet the SMRs in 1950 are broadly similar to those recorded in 1970 when there were only 220 titles in the classification. Among women, the reduced number of titles in 1970 as compared to 1950 is associated with a large jump in the SMR. In neither of these cases are there any logical grounds for assuming that the movement of the SMR is primarily governed by internal methodological procedures. The same argument applies when the possibility of numerator/denominator difficulties are considered. While inflation and deflation of the SMR for any single occupation can be expected to occur in any one year, there is no reason to assume that this would happen in any one year rather than another; the more reasonable assumption would be that this particular methodological difficulty would exert a random effect in each decennial period, and there is no particular reason why numerator/denominator problems should suddenly become more acute in 1980 (1970 for married women) than they were in the preceding years.

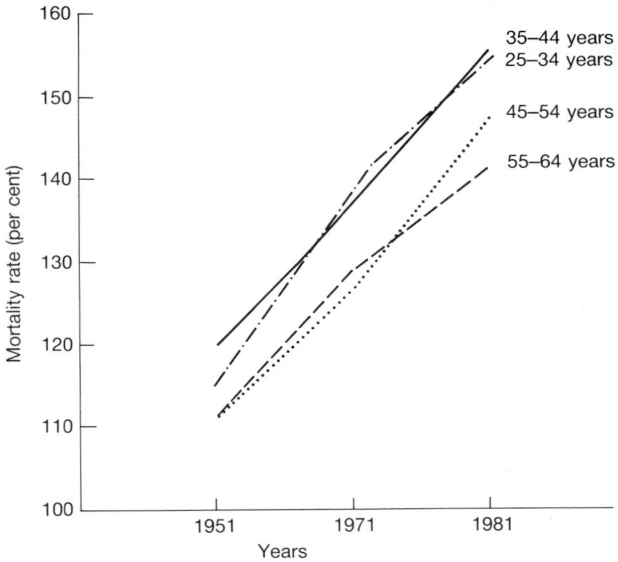

Fig. 10.3. Mortality rates of manual workers expressed as a percentage of non-manual workers (=100%) 1951, 1971, and 1981. (Source: Hart 1990.)

10.5 should be read as a broad indicator of trends in the aggregate level of inequality. They reinforce the pattern revealed in Table 10.4, strengthening the impression that recent decades saw a definite widening of the gap in life chances in British society.

Another way of overcoming problems resulting from the changing size and composition of occupational classes is to combine occupations into larger, easily recognized aggregates. The obvious division is based on the manual/non-manual (blue/white collar) divide. While occupational titles change and new jobs replace old, the manual/non-manual dichotomy remains a fairly visible and therefore valid tool of occupational differentiation. Figure 10.3 displays the trend in mortality differentials for the two broad classes in England and Wales in the post-war era.

The twofold model[13] of occupational class in Figure 10.3 obscures the full extent of inequality in mortality risk by immersing more disadvantaged occupational groups in larger aggregates whose average level of excess mortality is substantially less. The Whitehall Study (Marmott *et al.* 1984) based on a longitudinal survey of British government employees reveals a death-rate among the lowest ranks three times that of the highest. This gap based on longitudinal data of high quality implies that even the Registrar General's full scale (see p. 158) underestimates the extent of health inequality. Figure 10.4 presents an alternative grouping of UK national data for the same period combining the top two and the bottom two occupational classes.[14] Whichever grouping, the conclusion remains the same. The post-war period in Britain witnessed a widening of the gap separating the survival chances of economically advantaged and disadvantaged social groups (Koskinen 1985; Pamuk 1985). The same conclusion emerges from other European societies.

Excess mortality among manual workers is found in every society where accurate and appropriate data are collected and published. Figure 10.5 compares mortality risk between manual and non-manual workers outside of agriculture in five European nations. The gap is widest in France, where the death-rate of manual workers is 56 per cent above that of non-manual workers of the same age. The smallest differential (31 per cent) is found in Denmark, with England and Wales, Hungary, and Finland occupying the middle ground with excess percentages around 40 per cent.

These comparative differentials should be read with caution since they are not based on the same scales of measurement. Each society has its own occupational nomenclature and its own methodological conventions. Even so, the

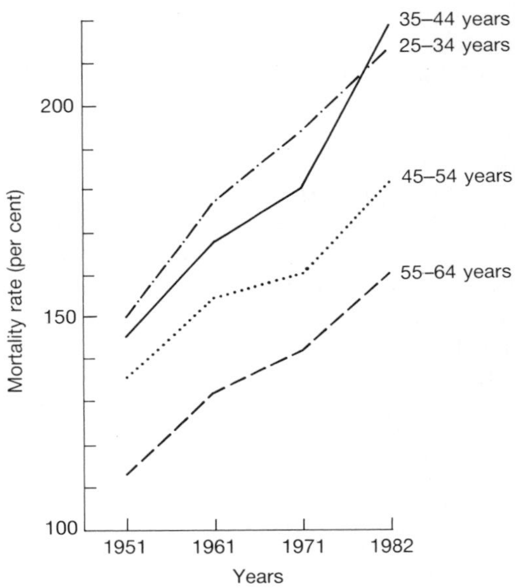

Fig. 10.4. Mortality rates of social class IV and V expressed as a percentage of I and II (England and Wales). (Source: Hart 1990.)

[13] The percentage of the male population classified to this occupational dichotomy was as follows, for England and Wales:

Period	Manual	Non-manual
1951	72.9	27.1
1971	62.5	35.2
1981	61.7	35.9

[14] The amalgamation in this figure is justified because most of the reclassification of occupational titles has taken place between adjacent classes. Added to this is the argument that reclassification is in any case no more than a nominal resorting, that the skill level of new occupations allocated to IV rather than V is indistinguishable. If, as Blackburn and Mann (1979) argue, most workers exercise more skill in driving to work than in carrying out occupational tasks, the combination of classes IV and V would seem to be both a necessary and appropriate means of analysing trends.

general picture for these five nations is clear; the mortality rates of manual workers are consistently above the rates of non-manual workers. Furthermore, there are also clear indications that the gap between occupational classes is widening. In France, for example, between the mid-fifties and the mid-seventies, the probability of death between the ages of 35 and 60 fell 28 per cent among non-manual workers compared with a 12 per cent decline among manual workers. The direction of trends was the same in Finland, Denmark and, as we have seen, in England and Wales (Valkonen 1987). In each of these societies widening differentials do not result from an absolute deterioration in the position of manual workers. Death-rates have declined for everyone, but the scope of the reduction was more favourable among men classified to non-manual occupations. This was not true of Hungary, where the period between 1960 and 1985 witnessed a steady deterioration in the vitality of male manual workers inside and outside agriculture. Figure 10.6 displays the trends.

Decennial supplement estimates are constructed from independent sources of occupational data. The numerator from death certificates, the denominator from the census. Though their combined use assumes consistency in the collection of occupational data, in practice there is considerable discrepancy between the sources.[15]

Linked data overcomes this problem. Since 1970, in the UK, a longitudinal study (hereafter, the LS) has been developed based on one per cent of the census linked to most national systems of registration including birth, marriage, and death, as well as the national cancer registry. The sample can be updated (and augmented) at each future decennial period. The first update was in 1981. The intention is to follow the same population through successive censuses providing a unique data base for the analysis of socio-demographics at the level of individuals and households.

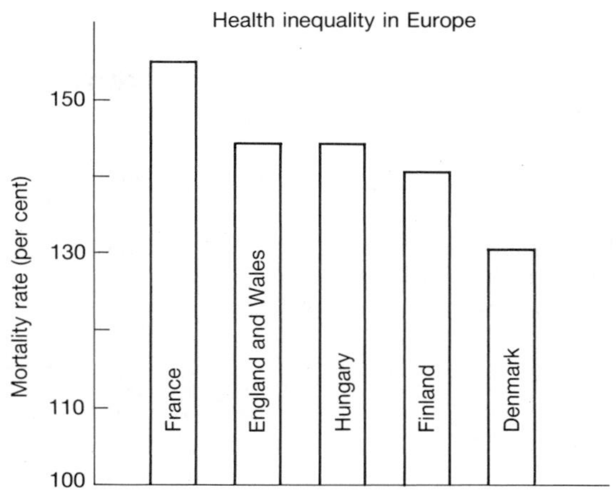

Fig. 10.5. Ratio of mortality of male manual to non-manual workers (outside agriculture). (Source: Valkonen 1987, Table 11, p. 233.)

With linked data, numerator and denominator the same, the occupational characteristics of the deceased (at death registration) can be traced directly to the census return. This also means that all the census variables become available for analysing the distribution of mortality risk. The first results of the LS have already begun to broaden our understanding of the social correlates of human vitality.

A major though not unexpected finding of the earliest LS analysis was the presence of class mortality gradients among the elderly. Under the unlinked method, accurate occupational mortality rates among the retired population could not be obtained because of inconsistencies in the recording of last occupation on the death certificate. The census offers a more complete (though still imperfect) record of occupation, making it possible to calculate the influence of occupational class at retirement on subsequent survival. The continued force of occupational class determining life chances after retirement is revealed in Table 10.6.[16]

[15] The numerator, the recorded incidence of death in a population, may be swelled by the inappropriate allocation of cases where the occupational title is either too broad and encompassing, or too vague. The denominator may be systematically diminished when a sizeable number of people are excluded from the census count. To give some examples. Fishermen record a relatively high SMR. In part this reflects an under-counting of the occupation in the census because a good many may be away at sea when the count is taken. The incidence of death is not so affected, with the result that the recorded risk of death appears higher than it should be because the population at risk is smaller than it really is. The same effect can be produced when the numerator is overestimated. The occupation 'company director' recorded a massive increase in mortality risk in the 1951 decennial supplement. This occurred because a large number of disparate self-employed people were inappropriately assigned to this title on the death certificate. Since the census operated a much stricter definition of this high-ranking occupation, the population at risk was much smaller and the SMR correspondingly raised. The result was a greatly swollen SMR for company directors adding to the mistaken belief that wealth and power may take its toll via 'executive stress' in early death. Numerator/denominator problems are a systematic feature of the unlinked method of calculating occupational mortality rates. At census, occupational data is self-reported but at death, of necessity, it must be someone else (the next of kin, usually the spouse) who reports the occupation of the deceased. Added to this is the possibility that census enumerators and officers registering deaths will not employ either the same conventions or the same degree of rigor in completing the necessary forms. The overall result is that some occupations may be either under- or over-represented in the occupational mortality returns, producing in turn either an inflated or a deflated death rate.

[16] It is worth noting that steep class gradients among the population aged 75+ express the interaction of two opposed processes; selection and material deprivation. Given the effect of material inequalities (differentials in living standards) on survival in infancy, childhood, and before retirement, the population of manual workers (particularly social classes IV and V) is likely to contain a higher proportion of people positively 'selected' on the grounds of health. The converse would be true for non-manual workers who have been exposed to a more 'sheltered' socio-economic environment in their earlier lives. If health selection plays at least some part in determining survival, we must conclude that the observed evidence of health inequalities among the elderly understates the full effect of class, because the population that records the highest mortality risk probably contains a higher proportion of 'robust' individuals who have survived despite highly disadvantaged surroundings earlier in their lives.

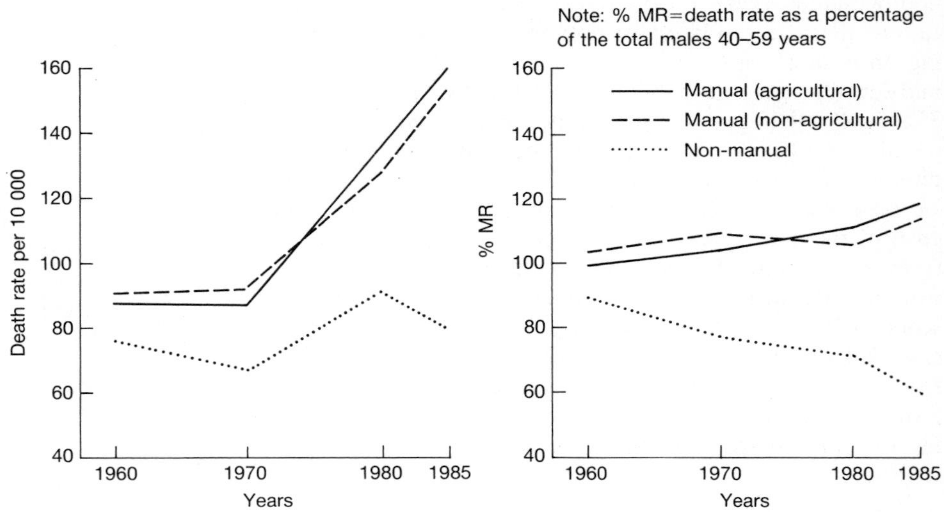

Fig. 10.6. Trends in male mortality by occupational class, Hungary 1960–1985. (Source: Valkonen 1987, based on Table 4, p. 213.)

Employment status

Men

Another important question illuminated by the LS is the effect of employment status on survival. In a controversial series of articles, Brenner (1979, 1983) claimed the existence of a systematic relationship between the mortality rate and the business cycle. Using unlinked cross-sectional data on unemployment and mortality, and lagging the mortality rate by three years, he argued that a quantified increase of unemployment directly causes a proportionate increase in the death-rate. The association is explained by the psychological stress of unemployment. Reasoning that public welfare provision had diminished the material deprivation of job loss, he concluded that its main impact is experienced as a shock to the self-esteem and status of the worker.

Brenner's thesis has been contested by a number of researchers in the UK though the data for or against are ambiguous.[17] The DHSS (Department of Health and Social Security) cohort study of unemployment found no substantial increase in morbidity in a follow-up study of the unemployed (Ramsden and Smee 1981; Stern 1983*b*; Wood 1982). The study also shows that the people at most risk of unemployment do not have previously unbroken work histories, but are unskilled workers often with repeated experience of job loss. Drawn disproportionately from social classes IV and V, they are the same class that records the highest mortality. Increases in the mortality of the unemployed during economic recession may thus indicate that their ranks have been swollen by men whose death rates are routinely higher than average anyway. It is not the single episode of unemployment that constitutes a stressful life event but the unskilled occupational career punctuated by episodes of unemployment which exposes an incumbent to a lifetime of material deprivation and economic insecurity. Any correlation between unemployment and mortality may therefore be an artefact of well-established class mortality differentials, with unemployment recast as real material hardship rather than simply the status degradation of job loss and going on the 'dole'. This interpretation also suggests that material inequalities within the occupational hierarchy act processually on health rather than in sudden episodic bursts.

The LS does reveal excess mortality among the unemployed and it suggests that three possible causal mechanisms may be at work:

1. Unemployment may be an interlude between ill-health and premature death.
2. It may itself cause socio-economic hardship leading to stress, damaging coping responses and hence death.
3. It may be the artefact of excess mortality in the classes most at risk to unemployment.

Table 10.7 classifies men of working age according to their health and employment status in the 1971 census. Four categories are distinguished: currently employed, work seekers, permanently sick, and the temporarily sick. The mortality rates of the four groups are given for the two periods, 1971–5 and 1976–81. In every case the mortality of the employed is well below the rates observed for every category of the 'unemployed'. Among employed men and work seekers in 1971, the standardized mortality ratio increases over the subsequent interval for all causes except accidents and violence.

[17] The relationship between unemployment and health is inherently complex. Though the excess mortality of the unemployed is not in doubt, the causal significance of unemployment is unclear since most deaths before a person ceases work are preceded by a spell of inactivity. As a result it is impossible with ordinary unlinked statistics to determine whether a terminal illness leads to unemployment rather than the other way round.

Table 10.6. Mortality of men by social class and age 1976–81

Social class in 1971	Age (years) at death							
	50–54	55–59	60–64	65–69	70–74	75–79	80–84	85+
I	58	63	79	54	77	67	76	78
II	88	76	75	80	83	81	90	81
IIIN	99	114	103	95	77	88	97	96
IIIM	92	96	100	98	100	107	105	102
IV	114	110	101	102	110	106	108	109
V	124	122	125	121	104	111	118	115

Source: OPCS 1990, Table 6.4 based on the LS.

Among the other two groups, the trend is generally reversed, death rates fall over time.

This pattern reflects the operation of the 'healthy worker effect'.[18] The more vulnerable people in these two groups record higher death-rates in the first period, and in most instances lower death-rates in the second as the selection process takes its toll. The fact that the same process is not observed among employed men and work seekers constitutes powerful evidence that being unemployed undermines health in the subsequent period. In effect, the mortality trend of work seekers corresponds to the population of employed men. Death-rates are lower in the first period suggesting that the sample was not negatively selected for health. Rising mortality in the second period is exactly as predicted by the 'health worker' effect, suggesting that the mortality of men seeking work can be compared legitimately with employed men in order to gauge the health risks of unemployment. The comparison reveals substantial excess mortality for every cause. Accidental and violent causes do not manifest the 'healthy worker' effect for either of these two groups.

How much of the excess mortality of 'work seekers' can be explained by their class composition? The unemployed are disproportionately drawn from social class V, namely unskilled workers, so much so that when the SMRs for all causes in Table 10.7 are standardized for social class, they register a decline from 129 to 121 in 1971–5, and from 146 to 129 in 1976–81. This still leaves a substantial excess mortality among job seekers which cannot be attributed to prior ill-health or to an artefact of class mortality in general. The conclusion is that unemployment does have a corrosive effect on health. This is reinforced by further evidence from the LS showing that the level of excess mortality varies according to the regional level of unemployment. Where it is high, the mortality of job seekers is proportionately higher than it is in areas where unemployment rates are lower. The highest levels of unemployment in the UK in recent years are recorded in the North, the West, and the Midlands, areas of industrial decline where chances of re-employment are bleak. In these areas, the mortality of job seekers is 40 per cent above the same regional average (SMR 140 CI 118–166). In the more affluent south east, where the risk of unemployment is comparatively less, the mortality of men seeking work is not significantly raised (SMR 120 CI 98–145). This squares with the argument that where unemployment is more threatening (in terms of frequency and duration), the negative consequences for the health of the jobless are proportionately greater (OPCS 1990, Chapter 5).

Women

The study of unemployment and health among women is complicated by their uneven labour force participation and by limitations imposed by the census record of their occupational circumstances. The same reasons retard the analysis of occupational class mortality differentials among employed women.

'Unoccupied' women are a mixture of traditional housewives, work seekers, and those too sick to work. However, it is not always clear how each category is treated in the census. Women not classified as 'employed' may be recorded in one of four ways in the British census; unoccupied, inadequately described, permanently sick, other inactive. Since each records a trend of high initial mortality followed by decline, it seems certain that all of them are affected by a reverse 'healthy-worker effect', and it is not possible to identify a distinctive category of female job seekers, that is, unemployed, whose subsequent mortality may be monitored.[19] Nevertheless there is evidence that unemployment has a major impact on women's health. A conclusive finding of the LS is the raised mortality of women married to unemployed men.

A married woman's own employment status makes a major difference to the health threat posed by her husband's joblessness. Where she has her own job, the unemployment

[18] Follow-up studies of occupational mortality have frequently commented on the tendency for age-specific mortality risk to rise over time. This happens because the starting sample tends to be 'health-selected', that is, it excludes people unfit for work whose subsequent death rates will be proportionately above average. This probably exerts a larger influence on samples of manual workers because their work is more physically demanding, and therefore less compatible with debilitating illness. This is likely to diminish the extent of class mortality differentials in the early years of a longitudinal follow-up since the sample ranks of manual workers will contain a disproportionate number of 'healthy workers'. Fox and Collier (1976) noted the operation of the 'healthy worker' effect on the mortality rates of strenuous industrial occupations which can only be performed by relatively 'fit' workers. Because such occupations draw on and retain a more robust selection of the potential work-force, their mortality risk is systematically disguised.

[19] The mortality of 'employed' women is systematically below average for all causes except lung cancer.

Table 10.7. Mortality of men (SMRs) in the LS by health and employment: status, cause, and period of death: standardized mortality ratios; men 15–64

Cause of death	Health and employment status							
	Employed		Seeking work		Permanently sick		Temporarily sick	
	71–5	76–81	71–5	76–81	71–5	76–81	71–5	76–81
All causes	85	93	129	146	408	301	329	249
Malignant neoplasms	91	97	142	148	241	129	227	173
Circulatory disease	87	95	113	121	361	235	297	235
Respiratory disease	63	77	91	222	892	907	604	547
Accidents and violence	92	88	212	187	341	431	378	289

Source: OPCS 1990 forthcoming, Table 5.4.

of her spouse raises her mortality risk only slightly. Where she is not employed, the SMR is 40 per cent above average. Overall, women married to male job seekers record a mortality risk 22 per cent in excess of that of all married women during the decade 1971–81. This excess rises from SMR 111 in the first half of the decade of follow up to SMR 129 during the second (1976–81), duplicating the pattern of male job seekers, that is, exhibiting the 'healthy worker effect', the sign that the population in question (wives of the unemployed) was not negatively selected for health when they were recruited into the LS sample at the time of the 1971 census. This is a further strong indication that unemployment is damaging to the health of both unemployed male breadwinners and their wives.

Apart from suicide, the largest excess recorded by wives of unemployed men in Table 10.8 is for ischaemic heart disease, SMR 152. The comparable SMR among men seeking work is 113, though among the age group 15–44 the SMR for circulatory disease is 186, for ischaemic heart disease it rises to 216 at the same age.

Housing tenure

Housing tenure further differentiates mortality risk among unemployed men and their wives. The earliest LS findings demonstrated the power of housing tenure as an alternative proxy for social class. In 1971 50 per cent of households owned their own homes, 17 per cent rented in the private sector, and 30 per cent were tenants of the state. Table 10.9 compares the distribution of mortality risk between these 'housing classes'.

Death-rates among the population who rent their homes are 15 per cent above average, those of home owners are 16 per cent below average.[20] The effect of unemployment raises the mortality risk of owner occupiers to the national average (SMR of 100) while elevating that of tenants to 152 and 161. This rise is smaller among married women.

There are good reasons why home ownership has emerged as a sensitive indicator of material well-being in UK government statistics. Though owner-occupied homes are generally better furbished than rented tenancies, this is not the primary reason their inhabitants record lower mortality. More important are the economic circumstances that enable a household to purchase its home: property ownership, material security, and creditworthiness. These are core ingredients of material welfare which have helped establish tenure as a powerful proxy for social class. In the LS, owner-occupiers record the lowest unemployment rates, (2 per cent compared with 5 per cent among tenants) and even the mortality of the home-owning jobless is well below that of people in rented accommodation.

Unemployment is not randomly distributed in the population.[21] Job loss tends to be concentrated in unskilled and partly skilled work. It is also concentrated among certain racial and ethnic sections of the population who are also disproportionately represented on the lowest rungs of the occupational class structure. The next section takes up the question of racial inequalities in longevity.

Race

Though a genetic variable, fixed at conception and generally irreversible, race is also a socio-economic variable and a fundamental source of social and health stratification.

The meaning of the word race shifted extensively during the relatively short course of recent history in which people from different parts of the globe have been brought into routine contact with one another (Banton 1977). The era of industrial society and that immediately preceding it witnessed a dramatic increase in the scope of long-range human communication. Peoples of different continents were 'united'

[20] Hart (1985, Fig. 4.3, p. 55) details the interaction of tenure and occupational as determinants of mortality risk in England and Wales. Unskilled manual workers who own their own homes record a lower SMR than professional workers who rent their homes. Indeed, tenure appears to 'neutralize' the effect of class.

[21] Unemployment rates in the UK among manual workers were more than double the rate of non-manual workers in 1985 (see Marmott and McDowell 1986).

Table 10.8. Mortality of women married to men seeking work in 1971. The standard population is married women resident in private households whose husbands are 15–64 years in 1971

A SMRs by cause

Cause of death	SMR
All causes	122
Malignant neoplasms	110
Lung cancer	84
Circulatory disease	134
Ischaemic heart disease	152
Respiratory disease	137
Accidents	93
Suicide	160

B SMRs by women's own employment status

	All married women	Women married to men seeking work
Employed married women	83	86
Non-employed married women	12	140
	100	122

Source: OPCS 1990, Table 5.18 and 5.20.

Table 10.9. Mortality (SMR) by tenure and unemployment status

1971 Housing Tenure	All men	Men seeking work	All married women	Wives of men seeking work
Owner occupiers	84	100	88	101
Private tenants	109	161	101	135
Public tenants	115	152	115	141

Source: adapted from Tables 5.10 and 5.19 OPCS 1989. SMRs in the LS 1971–81.

through voyages of exploration, trade, military conquest, religious conversion, and colonization. This unification of the human species led to material progress over much of the globe, but at the cost of expropriation, enslavement, and even the extinction of some human communities (McNeill 1977).

One modern legacy of these historical experiences is manifested in the distribution of life chances between black and white Americans. Afro-American people did not enter the US via Ellis Island. They came much earlier and not through free choice, but as human cargo, captured in their aboriginal homeland (Africa) and transported like animals to be sold to European settlers in newly colonized parts of North and South America. The ideological justification for the slave trade was a belief in the inherent superiority of the European races, who also happened to possess the means of dominating, coercing, and terrorizing other races. Like sex, belief in racial inequality was given scientific respectability by craniological scientists who purported to show differences in brain size as a basis for species variation (Banton 1977; Sayers 1972). Though such ideas no longer enjoy scientific respectability, and though acts of sexual and racial discrimination have been formally outlawed in most advanced industrial societies, historical memories enshrined in modern beliefs remain a fundamental barrier to the establishment of equality of opportunity in both Europe and America.

Table 10.10. Life expectancy by sex and race: US 1900–45

	Birth	10	60	Birth	10	60
Coloured males				**females**		
1900	32.5	41.9	12.6	35.0	43.0	13.6
1920	46.9	44.9	13.7	48.0	44.9	14.0
1930	50.1	46.6	14.1	52.6	49.0	15.3
1945	56.1	50.1	14.7	59.6	53.8	16.5
White males				**females**		
1900	48.2	48.0	14.4	51.1	52.2	15.1
1920	57.9	54.6	14.8	60.0	56.4	15.7
1930	60.6	55.9	14.9	64.5	60.0	16.4
1945	64.4	57.9	15.4	69.5	62.4	17.8

Source: Dublin and Lotka 1949, p. 50 and p. 54, Tables 12 and 14.

Compared with men and women of twentieth century Africa, black Americans live longer and enjoy higher living standards. Compared with fellow white citizens, they are relatively deprived on both counts (Table 10.10). In 1900, the longevity gap at birth between white and coloured Americans (largely African and mixed descent) was over sixteen years for men and women. More than half was due to infant mortality as the reduced gap at age 10 years reveals (males 6 years, females 9 years). At 60, the gap was only two years for both sexes.

The narrowing of race differentials before 1950 favoured men more than women, and by 1945 the differential at birth had fallen to 8 and 10 years respectively. The greater relative improvement of men is explained by the fact that the sex mortality differential developed faster among whites. In other words, the health of white women (relative to their men) improved faster than that of their black sisters before 1950. The relative position of black women improved noticeably after 1950 and by 1985; the race differential was larger among men (6.7 years) than women (5.2 years).

Race mortality differentials narrowed between 1950–80, though the most recent evidence (1985/6) indicates a slowing down of the process with some demographic reversal in the black population (Table 10.11). This almost certainly reflects the increase of poverty sponsored by 'Reaganite' policies in the eighties, demonstrating the sensitivity of longevity as a barometer of social and economic inequality.

Race differentials by cause offer further clues about the causes of inequalities in life chances in the US. Heart disease and cancer rank first and second as causes of death for both races and sexes. From here the ranks diverge. Two causes

Table 10.11. Life expectancy at birth by race and sex since 1940

Year	White		All other		Black	
	Male	Female	Male	Female	Male	Female
1940	62.1	66.6	51.5	54.9	—	—
1950	66.5	72.2	59.1	62.9	—	—
1960	67.4	74.1	61.1	66.3	—	—
1970	68.0	75.6	61.3	69.4	60.0	68.3
1980	70.7	78.1	65.3	73.6	63.8	72.5
1984	71.8	78.7	67.4	75.0	65.6	73.7
1985	71.9	78.7	67.2	75.0	65.3	73.5
1986	72.0	78.8	67.2	75.1	65.2	73.5

Source: US Department of Health and Human Services (1988).
Data by race other than 'white' not available before 1970.

Table 10.12. Death rates by sex and race—ten leading causes: US 1986

Rank	Males		Females	
	White	Black	White	Black
1	Heart	Heart	Heart	Heart
2	Cancer	Cancer	Cancer	Cancer
3	Stroke	Accidents	Stroke	Stroke
4	Accidents	Stroke	Pneumonia	Diabetes
5	Pulmonary	Homicide	Pulmonary	Accidents
6	Pneumonia	Pneumonia	Accidents	Pneumonia
7	Diabetes	Perinatal	Diabetes	Perinatal
8	Suicide	Pulmonary	Athero-sc	Nephritis
9	Cirrhosis	Cirrhosis	Nephritis	Homicide
10	Athero-sc	Diabetes	Septicaemia	Septicaemia

Source: US Department of Health and Human Services (1988).

which make a major contribution to the race mortality differential are accidents and homicide. Black males record a staggering 55.0 homicides per 100 000 population. Ranking fifth among causes of black male mortality it accounts for one in twenty of all deaths, and makes a major impact on life expectancy because homicide is a 'young man's cause of death'. Among black females, homicide (12.1 per 100 000) ranks nine; it is not among the leading ten causes of death in the white population (Table 10.12).

Another factor dividing white and black life chances is the prevalence of perinatal causes. Ranked seven in the black population for both sexes, it makes a substantial contribution to the gap in life expectancy. The high rate of infant mortality in a nation which leads in the volume of GNP for medical care is the direct outcome of socio-economic inequalities between the races.

In 1986 the risk of death in the first year of life for black infants (19.6 per 1000) was more than double the risk for white infants (8.7 per 1000). Some part of the inequality may be traced to public and private negligence in access and utilization of antenatal care. It is unlikely to be the most important explanation. More significant is the rate of teenage motherhood in the black population.

Maternal age is a critical correlate of infant survival. Infants born to teenage mothers have the lowest prospects of survival and of growing up in propitious material circumstances. Jones (1985) estimates a total pregnancy rate of 90 per 1000 US women aged 15–19 years. This compares with an average of around 40 in Canada, France, Sweden, and the UK. Herein lies one key to understanding the fundamental causes of health inequality between black and white citizens of the US.

The high infant mortality in the black population is heavily influenced by the large proportion of children born to single, adolescent, unsupported mothers. The rate in the black population is more than double the white rate, ensuring that a much larger proportion of black infants begin their lives in materially and socially deprived circumstances. Even those who survive birth grow up in relatively underprivileged circumstances. Many may end up fostered or spending part of childhood in care. Those lucky enough to remain with a natural parent will very likely experience poverty in their formative years.

There is nothing inevitable about either the high fertility or the poverty of teenage mothers. The same infants born to the same category of mothers in Sweden have a much brighter future. Though they may still lose on account of the youth and inexperience of their mothers, they are less likely to be systematically deprived of material resources, health care, and decent housing. They have priority in access to child care while their mothers work. Young unsupported mothers in Sweden cannot expect to live on welfare (they have higher wage employment rates than their married counterparts) but they can expect systematic help from the state towards the cost of maintaining their children while they work or complete their education. The result is a system where teenage motherhood is not a substitute for wage employment, and therefore not an alternative route to independence. Nor is it so crucially associated with extremes of poverty and public neglect (Kindlund 1988).

The most advanced system of market capitalism in the world, the US, has the most primitive system of welfare rights. The casualties are disproportionately black because of the fusion of race and class in American history. Public neglect is exacerbated by highly ambivalent attitudes to abortion. The US records one of the highest abortion rates in the West, yet ironically, those in most need may be least likely to gain access to safe abortion. Up to a quarter of US citizens are not covered by medical insurance, and in many states the use of public funds for the voluntary termination of pregnancy is disputed through moral/religious uncertainty exploited by political expediency. The casualties are disproportionately very young, poor, black, and other minority women whose babies record the highest infant mortality rates in the developed world.

Racial inequalities highlight the importance of maintaining a historical perspective in the analysis of health differentials. Though slavery was abolished in the US after the civil war (1865), discrimination against black Americans (former slaves) has never been extinguished. It is worth remembering that it was only in the sixties that President Kennedy sought to enforce the rights of black students to equality of opportunity in education, that freedom riders insisted on equality of access to public transport, and that Martin Luther King led his followers to Washington to remind the American people of the first article of their constitution. Though racial discrimination is now illegal, apartheid-style attitudes and behaviours persist throughout the US, as they do towards black citizens in many European countries.

As a result, conventional routes to social mobility have been systematically denied to black families. Unlike his white counterpart, the black male could hardly aspire to be a family breadwinner. Discrimination in education and training, exclusion from desirable jobs, exposed to the highest risk of unemployment, he was systematically denied access to career opportunities open to other minority groups. The development of a black middle class to serve as a reference group for sequential generations has been severely retarded. For the same reason, the black woman could never aspire to retreat into the domesticated housebound role idealized for her white counterpart (Matthieu 1980). Their rates of wage employment have always been high, and unlike the white family, ties of economic interdependency have never acted as the cement of domestic life. The necessity for self-reliance, is one reason for the prevalence of single-headed black families, their state of relative poverty is the combined result of racial and sexual discrimination in employment. Both factors shed light on why the sex mortality ratio developed more quickly in the white population, and why the race mortality differential was higher among women before the war.

The legacy of systematic racial discrimination in the labour market, that is, the means of valuing individuals in any advanced society, is also the key to understanding crime and homicide in the black community. Though at a formal level black Americans enjoy the same citizenship rights as do white Americans, they have remained the 'outsiders' of American society. Denial of access to legitimate means of achieving status in a society where material success is prized above all else, rates of illegitimate economic behaviour are disproportionately higher in the black population. Though at first sight rates of homicide appear less appropriate as indicators of human vitality than, let us say, pneumoconiosis, both reflect participation in restricted opportunity structures. The son of a Yorkshire miner who follows his father into the pits becomes exposed to predictable respiratory risk. In the same way the unemployed black teenager who joins an older friend on a criminal exploit enters an 'occupational' realm where violence is routine. In the end, the life of each individual may be brought to a premature end by factors associated with the way they seek income and status in their respective and restricted communities. This is their direct experience of the social and economic environment.

Education

Education is both a measure of intellectual training and an indicator of socio-economic status. Reflecting opportunity and effort in the early decades of life, it is the most basic and important qualification in the labour market. Not surprisingly, it is highly correlated with longevity. The power of education as a predictor of the length of life is revealed in Figure 10.7.

Valkonen (1986), estimates a decrease of male mortality risk of between 8–9 years for each additional year of education in six selected European countries (based on an inequality coefficient, calculated from the slope of the regression line in Fig. 10.7). Among men, the picture is remarkably uniform. Though the mortality rate of Hungary and Finland is much higher, the slope of the gradient is broadly the same in all four countries. Among women there is less uniformity. A clear inverse relationship is found in Sweden, Norway, and England and Wales. Among Danish, Finnish, and Hungarian women there is a hint of increased mortality with the longest educational careers. For all countries except Hungary the data were drawn from linked data sets permitting further analyses of short run trends.[22] Figure 10.8 charts these trends.

For both sexes, Sweden stands out for a small but consistent decline regardless of educational class. In England and Wales, by contrast, we find a remarkable increase in mortality risk by length of education, which Valkonen attributes to increasing inequality in cardiovascular disease. In the first period mortality from this cause declined 4.8 per cent for each additional year of education, by the second (1976–80) the percentage decline increased to 13.7 per cent. While deaths from this cause grew slightly in the lowest educational categories, it fell 20–40 per cent among the better educated.

The potential of education as a means of discriminating short-run changes in mortality risk is revealed strikingly in this research. The number of years of full-time education means broadly the same thing everywhere, and unlike occupation it applies to the whole population, including those not currently in the workforce; the unemployed, retired, and housewives. Furthermore, since education is generally complete by the early twenties, it is a relatively fixed attribute of adult men and women, not subject to what Valkonen (1986) calls 'reverse causal path: weakened health leads to low income especially among men'.

The main disadvantage is the highly skewed distribution of formal educational qualifications caused because the majority everywhere complete the compulsory minimum. Consequently, large and socio-economically diverse sections of the population fall into a single class. Even so, in the Nordic Countries, it reveals a mortality gradient obscured by alternative measures based on occupation (compare Denmark in Figs 10.5 and 10.7). Given its methodological advantages, it is worth exploring the possible mechanisms which translate the number of years in school and college into additional years of life.

[22] In each of the four Nordic countries, the source was census linked death records for the period 1971–80, for England and Wales, the longitudinal study 1971–5 and 1976–80).

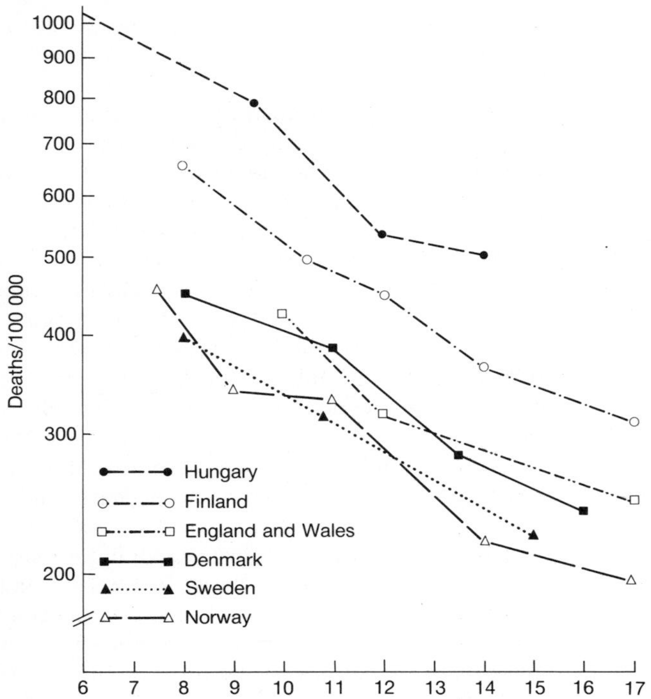

Fig. 10.7. Age-standardized mortality (per 100 000) from all causes of death by years of education and country, males aged 35–54, period 1976–80 (log-scale). (Source: Valkonen 1989, p. 147.)

Education has a dual character; it denotes both intellectual training and vocational preparation. Intellectual growth is concerned with at least one dimension of human development in a full sense. Against this goal, the more superficial business of gaining credentials to get a good job seems more mundane. In this latter sense education operates as an indicator of socio-economic status, that is, a measure of economic welfare or social class.

The links between occupation and longevity are well established. When education substitutes for occupation as an indicator of class, the causal implications are broadly the same. An indicator of lifetime earning capacity, education provides a forceful demonstration of the significance of material welfare on longevity. This does not exhaust the potential of educational influence on vitality. Intellectual training might be thought to lead to a greater capacity for understanding, controlling, and acting upon the social and economic environment. This may be manifested simply through greater exposure to knowledge, or going further, literally providing individuals with greater confidence in their own mental powers and faith in their ability to shape 'health destiny'.

Belief in the causal potential of education as a means to improved personal health care is rooted in 'Enlightenment' beliefs in the power of human reason as the road to social progress. Industrialization involved a massive application of human ideas (scientific knowledge, inventions, and technology) to material production and reproduction. Whether the emergence of industrial capitalism merely coincided with or actually caused a great emancipation and flowering of human intellect, the end result is the same. Science provided a rational account for events and experiences which previously only made sense by recourse to magic, superstition, and religion. The decline of supernatural beliefs (including religion) undermined fatalistic social attitudes (thought to be typical of pre-industrial peasant communities) and strengthened faith in rational planning and purposeful action.

The growth of knowledge and its application to the human condition was not an unfettered social process. With new ideas and techniques rose new social groups, brotherhoods, guilds, and professions claiming privileged access to enlightened thought and technical expertise. This itself created new barriers to the spread of knowledge, encouraging not so much the destruction of fatalism as a transference from a sense of impotence in the face of the gods, to sense of dependency on the esoteric language, skills, and potions of professional interest groups. Illich (1976), the best known

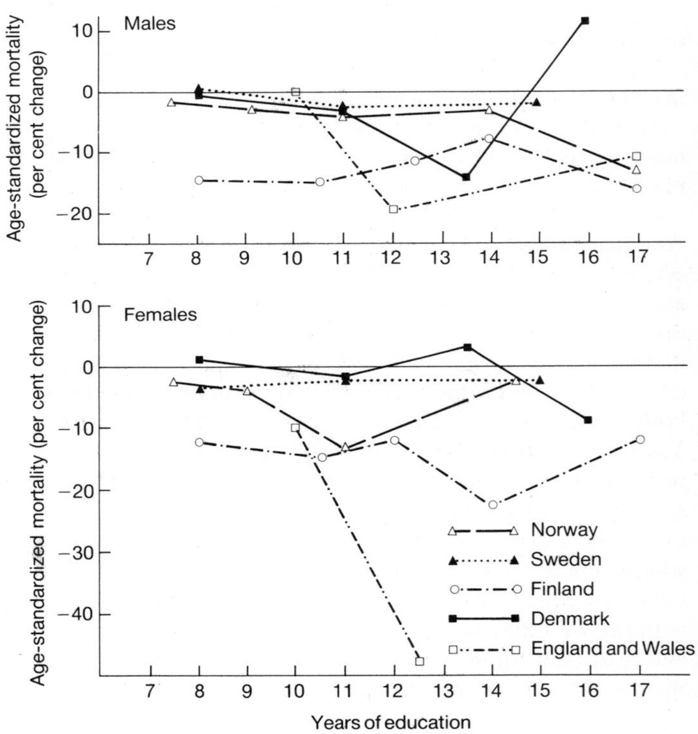

Fig. 10.8. Percentage change in age-standardized mortality from 1971–75 to 1976–80 by years of education, country and sex; all causes of death. (Source: Valkonen 1989, p. 153.)

spokesperson for this pessimistic view of the subversion of reason and knowledge by occupational interest groups, makes clinical medicine a favourite target. His critique of medicine is part of a general attack on professional and bureaucratic organizations which he sees as intrinsic to industrialization, overwhelming individuals, disabling their creativity, and crushing personal initiative. The clinical perspective in medicine, though posing as the guardian of human health, is actually the worst enemy of a salubrious way of life because it induces the modern form of fatalism, dependency on the expert.[23]

Illich identifies a number of real threats to human autonomy in the professionalization of knowledge and technique. He also exaggerates the extent of the monopoly of knowledge and expertise in the late twentieth century. Though the income and prestige of a number of professions depends on the subjugation of lay initiative, as many British general practitioners know only too well, the articulate and well-informed patient may often develop a specialized knowledge which alters the asymmetry of the consulting relationship. More generally, the growing popular appeal of prevention and the elaboration of 'healthy life-styles' testifies to the potential of more preventive 'public health' oriented strategies of health care in which well-informed and committed individuals take greater initiative in their own health care.

Can we identify a direct causal role for education as a determinant of health status? The highly educated have more access to new knowledge and perhaps possess a more sceptical view of the virtues of curative medicine. Could the gradients in Fig. 10.7 be partially the product of personal agency stratified by unequal access to education and technical training. The idea is persuasive, though it must be said that in contrast to the evidence linking economic deprivation and insecurity to health status, there is a dearth of empirical information on the potential of well-informed personal initiative as a determinant of health and longevity. Some modern-day health 'fads' remain shrouded in controversy, and the many precedents that today's wisdom becomes tomorrow's heresy provides ample encouragement for sustaining a more complacent life-style.

The correlation of survival increments beyond the age of sixty with economic welfare in pre-retirement life rests on a firmer empirical base (see p. 161). Bearing this in mind, it perhaps makes more sense to focus on the importance of commitment, as opposed to information, in sustaining what Dubos called '. . . the more difficult task of living wisely' (Dubos 1959). Knowledge about some health damaging behaviour and consumption is so well established and publicized that we must seriously doubt whether the people most at risk live in a state of ignorance. More plausible is the idea that prolonged material deprivation and lack of security encourages coping responses which are themselves either detrimental to, or not optimum for, the promotion of personal vitality.[24] As Orwell (1937, p. 95) remarked on the diet of the unemployed in his tour of unemployment-ridden Wigan in the 1930s:

> Would it not be better if they spent more money on wholesome things like oranges and wholemeal bread or even saved on fuel and ate their carrots raw? . . . the peculiar evil is that the less money you have, the less inclined you feel to spend it on wholesome food. A millionaire may enjoy breakfasting off orange juice and Ryvita biscuits, an unemployed man doesn't. When you are underfed, harassed, bored and miserable, you don't want to eat dull wholesome food.

That the substance of education rather than occupational credentials accounts for the longevity differentials observed in Figure 10.7 is understandably popular with health educationalists. It implies that even in the presence of material scarcity, people can be taught more effective ways of safeguarding their own health. While the goal of removing barriers to the dispersion of well-founded knowledge on means to prevent disease must remain a priority of modern public health, it is vital that its practitioners do not lose sight of the alternative significance of education as occupational credentials, and as a gateway to labour market privilege and the material securities that this confers.

Sex and gender

In pre-industrial societies, the longevity of men and women is roughly equal (Kynch 1985). True today among predominantly agrarian peoples, it was also true of pre-industrial Europe. In 1850 male and female life expectancy at age 15 in England and Wales was 43.2 and 43.9 years respectively, less than a year's difference. A hundred years later, the comparable figures were 56.2 and 62.1, a gain of 41 per cent among women, compared with 30 per cent among men (Hart 1985*a*).

Excess male mortality is universal in advanced societies. Given the importance of material livelihood to the rise of modern life expectancy, the growing divergence of male and female durability suggests that women have benefited most from economic progress over the last two centuries. It further implies that changes in the sexual division of labour associated with industrialism played an important part either in

[23] Friedson (1970) deals with some of the same issues from the perspective of sociological theory.

[24] This observation has relevance to widening class differentials in cigarette consumption which have developed over the last two decades. In the mid-fifties, smoking was a widespread habit among all men; today it is increasingly concentrated in social classes IV and V. Since many of the diseases with the steepest class gradients are smoking-related, cigarette consumption stands out as the obvious 'cause' of health inequalities. Giving up cigarettes involves giving up an addictive drug with renowned tranquillizing properties. The correlation between social and economic deprivation, and smoking, must be placed in this context. Though smoking is probably a dangerous way of coping with stress, it is more appropriate to see it, not as the organic cause of disease, but as an indicator of risk. This interpretation of the role of smoking also makes sense of the discrepancy between smoking and disease-specific mortality differentials. In the Whitehall Study (Marmott *et al.* 1984) the smoking differentials between occupational ranks were much smaller than mortality differentials for lung cancer, indicating that cigarette consumption was not a sufficient explanation for class differentials in this disease.

raising women's health status, or in diminishing the prospects of men achieving their optimum.

The gradients in Fig. 10.9 measure percentage excess male mortality (male/female death rate) throughout the life-cycle for cohorts specifically chosen to estimate the contribution of environmental forces to health. Two of the cohorts, 1891 and 1921, reached early adult life during the First and Second World Wars respectively. The effect is witnessed in the raised mortality of these cohorts in the twenties when a large proportion of their male members were mobilized for military service.[25] During the Great War (1914–18), more than 17 per cent of males in the 1891 cohort lost their lives. Those who died were disproportionately fit and healthy. They had been certified 'medically fit' in a military recruitment examination which rejected half the applicants. This is a cohort which was heavily depleted of a large proportion of its 'fittest' members; it may be thought of as a population negatively selected for vitality in the aftermath of the war. However, the percentage excess male mortality of this cohort in middle age is well below the level of the 1901 cohort, born too late to be sacrificed in the first world war and too early for the second. The 1901 cohort records the largest sex mortality ratio in Fig. 10.9. This fact cannot be explained by selection (see p. 176) and so it must be connected to other developments in social and economic life in the first half of the twentieth century.

The century that witnessed the trend in Fig. 10.9 also saw major changes in male and female roles. Though the sexual division of labour in the family, the maternal female and the breadwinning male, appears to rest on a 'natural' foundation, its historical chronology is short. It emerged in the latter half of the nineteenth century, a response to high marital fertility combined with the growing separation of home and work against a background of legislation designed ostensibly to restrict and thereby protect women and children from exploitation in an industrial labour market. Women gradually gained the opportunity to become specialized in nurturance and to devote a much larger proportion of their time to their children. They also became economically dependent on the earnings of husbands and older children. The health consequences were both positive and negative. On the one hand women could devote more time to caring for their families; on the other, they had to feed and clothe their families on a very small housekeeping allowance which they had little or no opportunity to augment. Whatever the forces removing women from active wage employment after 1850, sufficiency of average household income was not among them and, as the poverty surveyors of the turn of the century testify, a very large proportion of the (especially female) population lived on the margins of subsistence (Hart 1989*b*).

The divergence of the sex mortality ratio coincides with the development of the male breadwinner/female housewife role division.[26] This implies that industrial work, the effort, the risks, and the heavy breadwinner responsibility, might be the source of excess male mortality. Empirical evidence to support this is lacking. Where women maintained a significant presence in the factories, the mortality of 'active' women was below the rate of the 'inactive' (Johansson 1977). Equally, the evidence of real material scarcity born principally by married women in the late nineteenth/early twentieth century family leaves the strong impression that women lost as much as they gained in being relieved of wage employment (Rowntree 1902, Bowley and Burnett-Hurst 1915).[27] Among the cohorts of men born successively from 1870, smoking and the risk of death relative to women have grown hand in hand (Hart 1989*a*).

The circumstantial evidence is strong but, while cigarette consumption has undoubtedly played a major role in the growth of sex differences in survival, it is not a sufficient explanation (Valkonen 1985). Furthermore, too much emphasis on smoking neglects changes affecting women as well as men. Reversing the focus and asking the question why women's mortality has fallen more than men's, attention shifts to the possible contribution of improvements in the material circumstances of women. Differential neglect of the female is a widespread feature of pre-industrial society.

[25] These data are based on civilian registrations. The peaks of male excess occurring in these two cohorts therefore exclude military casualties and deaths occurring outside the UK. They indicate a short term 'healthy soldier' effect.

[26] For most of human history, the activities of production, consumption, and nurturance have been indistinguishable features of everyday family life. This was reflected in the transition to the earliest factories which employed whole families under the jurisdiction of the father who directed the labour of his wife and children and received their wages as out of his own property (Smelser 1959). As factories became larger and more impersonal, they found the tier of paternal authority a cumbersome anachronism. Efficiency demanded that workers be treated as individuals under a single authority figure, the foreman or supervisor. This preference was first manifested as a threat to familial solidarity and patriarchal rights. When employers sought to bypass the male household head and hire his 'dependents' on an individual basis, contemporary moralists rightly recognized that a threat to traditional order was at hand. The problem was not only the threatened loss of the father's right to receive the value of his wife and children's labour, it was also the prospect that employers might prefer to hire women and children (a cheaper and more docile workforce) and that adult men would be without a personal livelihood. This potentiality was averted by a programme of protective legislation which progressively removed adult women and children from wage employment. By the end of the nineteenth century, patriarchal authority was reconstructed in the role model of male breadwinner, and the subservience of wife and children was redefined as a matter of economic dependency. These gender divisions began to dissolve within a few generations of their formation. By the mid-twentieth century, with marital fertility at a fraction of its former rate, housewives past childbearing age were beginning to search for wage employment opportunities, and twenty five years later younger women with dependent children were following suit. Today in many families with young children both parents work, and the housebound wife/mother is beginning to appear as either a luxury for the very rich or sign of poverty among the very poor (see p. 173).

[27] This view is strongly reinforced by Winter's (1982) research on the recruitment of married women into the labour force to fill the jobs vacated by mobilized male workers during the great war. Infant mortality rates fell fastest during the period of the wartime emergency, and specifically in urban areas where the pre-war rates had been high and where women were recruited into the industrial labour force in large numbers. Ironically, the war brought a large reduction in family poverty; it gave women direct access to wage employment and direct control over the family purse.

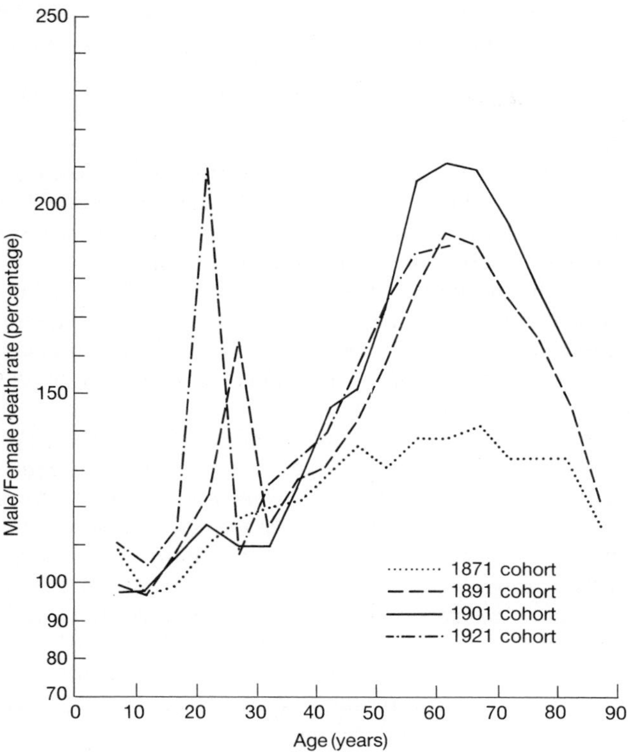

Fig. 10.9. Sex mortality differentials: Male death rates over female death rates by cohort for England and Wales. (Source: Hart 1989*a*, p. 118.)

Whether female infanticide or systematic neglect and underfeeding, the historical conclusion is the same, that the female has been the devalued sex in most historical cultures.

Changing sex mortality differences in Japan illustrates this long standing feature of pre-industrial society. In 1928 male death rates were lower than their female age equivalents from childhood through to age 40. Today female mortality is lower at every age. In the early period the sex mortality differential resulted from social rather than physiological difference, a matter of gender rather than sex. Parents preferred sons and allocated more resources to their growth and development.[28] The cultural devaluation of the female sex was not unique to Japan, it was widespread throughout the pre-industrial era, and its persistence today in some less developed societies explains why more girls die during famine (Sen 1981).

The trend in Japan (Table 10.13) illustrates the flaw in the assumption that the roughly equal longevity of pre-industrial men and women represents a natural base for the measuring of excess male mortality of males in the twentieth century. In pre-war Japan, a similar overall mortality rate for men and women is composed of highly variable age-specific rates. The superior vitality of the female sex appears only after the age of 40, before this males record lower death-rates.[29]

Today, in societies where food scarcity no longer determines survival, young females appear if anything physiologically stronger; they record lower fetal loss, stillbirth, and perinatal mortality. Their lower death-rates later in childhood are associated with reduced risk of accidents as well as some major diseases. In both centuries cultural norms and values interact with economic resources to distribute life chances between the sexes, though today the female is the main beneficiary, or alternatively the male appears to be the principal casualty.

Some evidence points the possible role of sex (that is, physiological difference) in the divergence of male and female vitality. That there is more wastage of male life at all stages of human development, before as well as after birth, suggests at least some causal role for biological selection in the unequal survival of the two sexes. Contrary to immemorial and near universal social belief, the male and not the female may be the weaker sex (Hart 1978). The evidence from Japan suggests that historically any 'natural' advantages possessed by the female were neutralized by norms of gender discrimination. Today it is unclear how much, if any, of the diverging sex mortality ratio can be explained by genetic variation. With cigarette consumption standing out as the major risk factor for much preventable loss of life in men, it seems certain that gender differentials in nicotine addiction are a primary factor in excess male mortality.

Gender differentials in consumption before the war reflected inequalities of spending and norms proscribing smoking as 'unfeminine' behaviour. Recent trends in consumption leave no doubt that smoking is sex-neutral. The entry of married women into the labour force since 1950 is associated with a major increase in smoking among women and signs of higher lung cancer mortality among wage earning women. Women will soon be the major consumers of

[28] The fate of the poor in pre-industrial societies is remarkably similar across different continents and cultures. In pre-war Japan, the children of the poor, like their counterparts in pre-industrial Europe, were unlikely to spend much of their adolescence in their natal homes. From a very early age they were literally sold by their families to other households who needed their labour and who could afford to keep them. The majority became farm servants, some were apprenticed to artisans, and a good many young women ended up in brothels (Robins-Mowry 1983). The preference for male heirs meant that female children were disproportionately among the young people transferred to the human brokers who scoured the Japanese countryside in pursuit of their trade. The same factor produced an interesting reversal of the sexual division of labour in Japan's emerging industrial economy at the turn of the century. Up until 1920 more than 60 per cent of industrial workers were female (Hane 1982; Sievers 1983). They were not free wage labourers but indentured workers sold by their families to the textile factories which dominated Japan's early industrial development. The conditions of work were unspeakable. Young women endured a 12 hour day, seven days a week, usually for about seven years at a stretch. Their living conditions were overcrowded, unsanitary, and their diet was inadequate. No wonder then that the rate of respiratory tuberculosis was high in factory dormitories, and girls rather than boys record the highest risk of death because of the norm of sexual discrimination which pervaded the way parents cared for and used their children.

[29] This pattern was also evident in nineteenth century Europe in the general population, though interestingly not among the upper classes where the female mortality rate only exceeds the male between the ages of 10 and 17 (Ansell 1874, p. 26). Though the risks of childbirth might be thought to make some contribution to excess mortality of women of

Table 10.13. Sex and age-specific mortality in Japan, rates per 1000 population, 1928 and 1986

Age	1928		1986	
	Male	Female	Male	Female
0–4	47.1	42.6	1.5	1.2
5–9	4.1	4.4	0.23	0.14
10–14	2.7	3.8	0.20	0.12
15–19	7.3	8.7	0.66	0.23
20–24	9.2	10.1	0.81	0.34
25–29	7.8	8.9	0.78	0.42
30–34	7.0	8.6	0.92	0.51
35–39	7.9	9.2	1.3	0.76
40–44	10.2	9.7	2.2	1.2
45–49	14.3	10.7	3.5	1.8
50–54	20.0	13.7	6.1	2.8
55–59	28.8	18.7	9.0	4.0
60–64	43.4	28.1	12.7	6.2
65–69	61.9	42.4	21.0	10.4

Source: Japanese government statistics (1988).

cigarettes in the UK and other industrial nations.[30] The price of equality will add increased profits for the tobacco companies, it will also be paid by women themselves in smoking-related disease. Ominous signs are already detectable in very recent years in the decline of the sex mortality differential in the UK (Fig. 10.9). If anti-smoking policies remain relatively ineffective, and if trends in consumption continue on their present path, we may soon be in a better position to evaluate the contribution of sexual physiology to sex mortality differentials of the twentieth century.

The detailed analysis of gender differentials in mortality by other socio-economic variables in the LS is hampered by the fact that less than 60 per cent of women in any age group are classified to an occupation. This reflects the 'discretionary' status of wage employment for women complicated by their marital and maternal status. Thus while the correlation between being employed and good health is probably more marked in women, unoccupied females still contain a substantial number 'fit' for work for whom the concept 'unemployed' is ambiguous. For this reason it is useful to incorporate a wider range of variables to study the effect of occupation and economic status on women's health. Prominent among these is marital status.

Marital status

The preservative effect of marriage was noticed by Durkheim in his late nineteenth century survey of suicide. As Fig. 10.10 indicates, the effect holds for all causes of death: being married is one of the best means of reducing actuarial risk. The greater survival of married people is partially a function of what Durkheim called 'matrimonial selection'. As the 1970–72 decennial supplement (OPCS 1978*b*, p. 34) put it:

> Marriage, like work, is a selection process—some groups like the physically and mentally disadvantaged may never pass the 'fitness hurdle'.

Though selection probably accounts for some of the excess of the unmarried, it is less plausible for the divorced and widowed, whose age-related mortality profiles are very similar to never-married men and women. Part of the excess mortality of the formerly married may be explained by the 'grief' of widowhood or divorce, a substantial part may be explained by more general features of the unattached way of life which unite all categories of single people (Hart 1976; Susser 1981). What do these include?

For Durkheim, the preservative effect of matrimony lay in normative regulation. Marriage restrains excessive egoism, forcing individuals to conduct their lives with a respect for the values, expectations, feelings, and needs of spouse and family. In other words, marriage is a more social, group-oriented way of life, necessarily involving greater co-operation, concern for people other than oneself, as well as having the benefits of mutual support. Though he emphasized the 'social control' dimensions of conjugality, more recent socio-medical research stresses the mutuality (i.e. emotional interdependence) of married life as the primary source of its health dividend (Brown and Harris 1978). Marriage operates primarily as the most intimate and intense social support network.

The value of social support for health preservation is demonstrated in a number of large epidemiological studies (Berkman and Syme 1979; House *et al.* 1982). Researchers report a steady decline in mortality risk with the increasing size of social networks. Good support does not even have to be human, warm relationships with canine, feline, or even feathered friends seem to materially influence the prospects for human survival (Friedman 1980).

The mechanism of affiliation and physical well-being is easier to imagine than to demonstrate. Human beings are social animals; they need to affiliate with other people, their well-being depends upon contact, communication, and co-operation. Social isolation represents a denial of species being and a form of social death. This explains why people reportedly lose the will to live in situations of personal alienation.[31]

Among sources of potential support, marriage, the most intimate, intense, and committed of all relationships is the social bond *par excellence*. The conjugal partner is the indi-

childbearing age, the available evidence suggests that this was not a major reason for the divergence (Hart 1989*a*; Schofield 1984).

[30] Once again Japan is the exception (Hirayama 1987).

[31] The causal link between mind and matter in theories of social support and health is stress. When exposed to intolerable stress—physical, psychological, social—bodily response may itself constitute the symptoms of disease. Selye's graphic image of the bleeding stomach ulcer, spawned by an excessive discharge of adrenalin, is the prototype of stress disease (Selye 1956). In this disease equation, social support operates by diminishing the individual's perception that an event or experience is threatening, thereby containing the body's hormonal responses within tolerable limits.

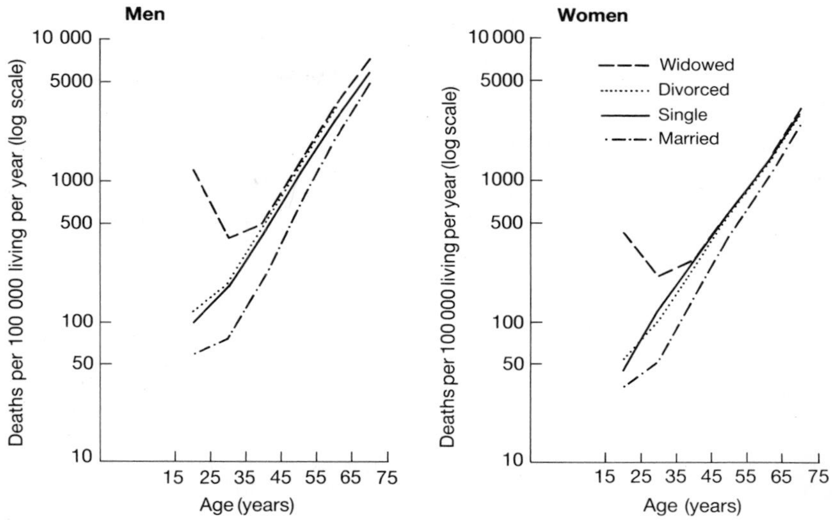

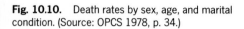

Fig. 10.10. Death rates by sex, age, and marital condition. (Source: OPCS 1978, p. 34.)

vidual's alter ego whose mere presence sustains a sense of normality, security, and permanence in everyday life (Berger and Kellner 1964). This is why the loss of a spouse is so devastating, why the separated and divorced swell the ranks of clients in therapy, and why the widowed are at special risk of literally breaking their hearts (Parkes *et al.* 1969).[32]

The positive emotional benefits of the matrimonial state are reinforced by the material gains of co-habitation, the pooling of resources, and the possibility of higher living standards. What part do these more concrete benefits contribute to the enhanced vitality of married men and women?

Once again, the findings of the LS are illuminating. Table 10.14 cross-classifies women by their marital status, their own and their husband's occupational class, their tenure and car ownership. 'Occupied' married women have below average mortality no matter how they are cross-classified. Even in households with no car, primary evidence of low income, the SMR is below average. Compared to the single 'occupied' (excluding the divorced and widowed), it appears that marriage holds real advantages especially among manual workers.

Among unmarried females in the labour force, each of the 'class' indicators influences mortality risk. The largest differential is by occupational class, those classified in manual work record almost double the mortality of their non-manual sisters (SMR 160 as against 84). The manual excess of married women workers is far less (SMR 97 against 79) suggesting that marriage mediates the influence of class on female mortality risk. How much?

The preservative effect of marriage is much diluted among 'unoccupied' wives. The SMR of the non-working wife of a manual worker is substantially above that of her employed equivalent (SMR 121 as against 96) with even larger differentials for tenure and car ownership. The death-rate of unoccupied wives in non-car households is 50 per cent above average. Figure 10.11 takes the analysis further by charting the SMR (with confidence intervals) for distinctive clusters of these variables. Single women, manually employed with no car, record the highest mortality risk. Their non-manual

Table 10.14. Mortality (SMR) among single and married women by occupational class, tenure, and car ownership: women 15–59 at death 1976–81

	Single	Married occupied	Married unoccupied
Own class			
Non-manual	84 (68)	79 (262)	—
Manual	160 (62)	97 (310)	—
Husband's class			
Non-manual	—	72 (172)	71 (155)
Manual	—	96 (400)	121 (383)
Tenure			
Owner	85 (48)	79 (269)	84 (258)
Tenant (P)	129 (82)	93 (79)	111 (75)
Tenant (S)		99 (223)	130 (208)
Cars			
Yes	80 (44)	83 (394)	83 (312)
No	133 (86)	99 (177)	144 (224)

Source: Moser *et al.* (1988), Table III, p. 1222.
Observed number of deaths in brackets.
Single = never married.

[32] Though the private household is probably unique in its ability to provide the individual with spontaneous love and unwavering commitment, we should not lose sight of the fact that it is also a context where extremes of emotion, passion, and violence are let loose. The statistics of divorce, spouse and child abuse, domestic homicide, and increasing homelessness among the young all point to the fact that marriage and family do not always possess the capacity or resources to preserve the health and vitality of people at risk.

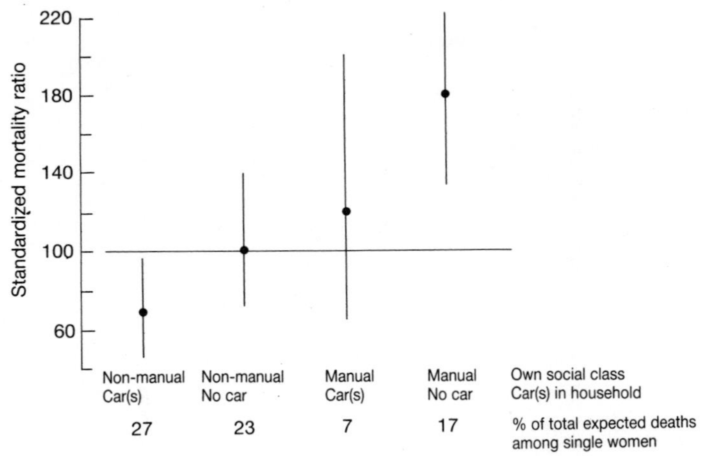

Note: Mortality among single women aged 15–59 at death, 1976–81.
Points are standardized mortality ratios and 95% confidence intervals.

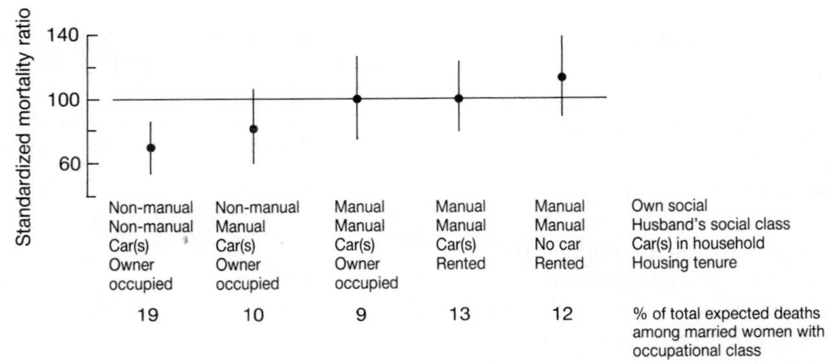

Note: Mortality among married women with an occupational class aged 15–59 at death, 1976–81.
Points are standardized mortality ratios and 95% confidence intervals.

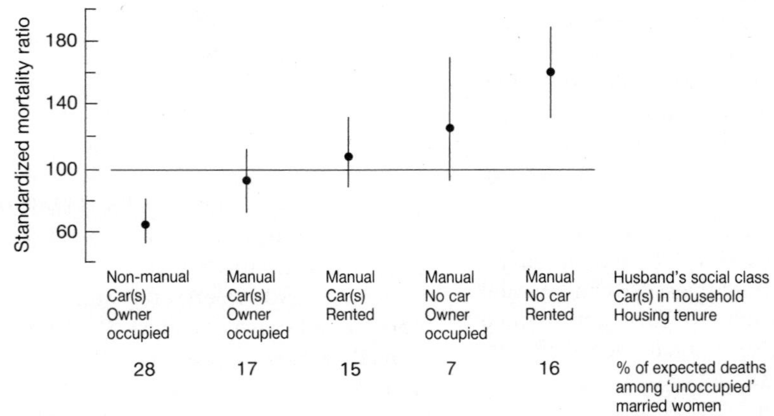

Note: Mortality among 'unoccupied' married women aged 15–59 at death, 1976–81.
Points are standardized mortality ratios and 95% confidence intervals.

Fig. 10.11. The relationship of marital and employment status to mortality in women. (Source: Moser *et al.* 1988, p. 1223.)

equivalents with cars record substantially below average risk (69 based on 30 deaths), suggesting that where economic circumstances are favourable the protective effect of marriage is much less, and even absent.

Again the married 'occupied' stand out as the most advantaged, though there is some differentiation in their ranks. The least at risk are non-manual wives of non-manual husbands, car and home owners. The most at risk are manual wives of manual husbands, owning no car and renting accommodation.

Among 'unoccupied' wives, economic circumstances are all important. The wife of a non-manual worker owning both car and home records the lowest risk of all, an SMR of 65 compared with one of 70 for her employed equivalent. In contrast, the wife of a manual worker in rented accommodation with no car has an SMR of 161 (based on 147 deaths). Car ownership is revealed as a sensitive indicator of inequality among married women. Where a manual husband lives with an 'unoccupied' wife, car ownership is a better indicator of mortality risk than home ownership.

These data suggest that the protective benefits of marriage for women have a substantial economic component. Marriage may be operating as a proxy for dual earner households, raising living standards and reducing mortality risk principally among the employed working class women. In affluent homes (owner occupied, non-manual, and car owning) a wife derives no additional years of life from labour force participation. Indeed, with these material advantages, the woman who works is exposed to a slightly higher mortality risk (SMR 70 based on 90 deaths). In other words, in circumstances of material sufficiency, employment carries no observable health benefit for the married women. In other circumstances, it makes all the difference. In poor homes (rented, manual husband, no car) the working wife records an SMR of 113 as against one of 161 for the non-working equivalent. Among the very poor it seems that the emotional benefits of marriage cannot compensate for economic deprivation.

Though these census-based indicators offer no means of gauging the quality of social support in households in different 'regions' of the social and economic landscape, they do suggest that economic factors play a larger part than has hitherto been recognized in mediating the preservative effect of social support on human health.

It is unclear whether the same arguments would apply to men. So far there are no published data for men cross-classifying their socio-economic and marital status. However, Fig. 10.12 contains a very striking set of class mortality gradients by male marital status, suggesting that matrimony operates in the same way for men. The preservative effect of marriage (the mortality gap between different marital statuses) increases with the descent of social class, suggesting that the benefits of marriage are much greater for unskilled manual workers (social class V) and much less among professionals (social class I). The mix of economic factors and social support in this result remains to be researched.

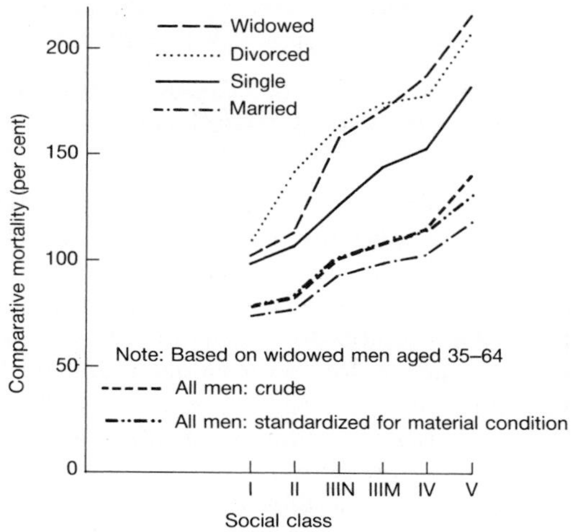

Fig. 10.12. Mortality by social class and marital condition: men aged 15–64. (Source: OPCS 1978, p. 34.)

Living standards and longevity: explaining the link

There is a mass of evidence linking health to the social and economic environment in which the strongest connection is the link of longevity and living standards. The distribution of life chances over the last two centuries have been closely associated with improved living standards in industrial societies. This is witnessed in historical trends in the life table, in the international infant mortality/life expectancy league, and in the distribution of survival chances within national populations. Can we explain these associations in any way other than that health is the product of the social and economic environment?

The alternative to viewing health as the dependent product of social and economic organization is to see it as an independent force in socio-economic development and in social stratification. Could the 'modern rise of population' have come about from spontaneous changes between disease organisms and their human hosts? Did the European human stock suddenly develop a new resistance to all sorts of pathogens in the late nineteenth century, improving their vitality and vigour to such an extent that their work output and standard of livelihood was doubled or trebled? Could we venture so far as to suggest that industrialism itself was stimulated by the improved vitality of the European population? One suspects there are few takers for this line of argument.

The possibility that personal 'fitness' underlies the achievement of material wealth carries the germ of eugenic argument. To propose, counter-intuitively, that improved vitality pre-dates socio-economic development risks the interpretation that European races led the industrial revolution because of their superior health stock (McNeill 1977). The same argument within a national population concludes that the vitality of the rich reflects their inherent fitness, their health and wealth are the dependent consequence of physiological superiority. The converse is that the poor die young because of their weakness, or as Tawney (1931, p. 74) put it:

> The poor, it seems are beloved by the gods, if not by their fellow mortals. They are awarded exceptional opportunities for dying young.

Stern (1983*a*) offers one of the few serious attempts to explain social class mortality differentials as a selection process. He suggests that the evidence of worsening health inequalities between social classes is artefactually produced by improved opportunities for social mobility. If good health status is a 'prerequisite' for upward mobility, the health stock of the recruiting class is raised at the expense of the donor class. The result is a widening gap in life chances between classes. Only where the poor are kept in their place do class mortality gradients keep a stable course.

Stern offered no hard evidence of 'personal fitness' operating as an independent factor in achieved social class; neither did he show any correlation between the movement of class mortality gradients and trends in social mobility (Wilkinson 1986), nor of shifts in age-specific mortality risk and the timing of mobility in the life career (Hart 1986*b*). The absence of supporting data is not unexpected. The concept of 'good health stock or status' is not easy to measure. Given the coincidence of morbidity or mortality and social class, it is well nigh impossible to demonstrate the distinctive contribution of genetic/intrinsic attributes to achieved health status. Even Illsley's (1955) much quoted study associating maternal height, upward mobility, and low infant mortality has proved in recent re-analysis the importance of material welfare in the natal home to achieved female stature (Illsley 1986).

The difficulty of finding data to demonstrate selection is matched by the difficulty of finding theorists who are willing to speak up unambiguously for the theory. Though eugenic theories enjoyed a certain vogue at the beginning of the century, today they hold more dubious overtones (Szreter 1984). Hence, selection, where evoked, tends to be cloaked in ambiguity, or presented as a blended mix of unweighed natural and socio-economic forces. Even Stern leaves the meaning of selection unclear. His argument might be read as illustrating how processes of social mobility selectively recruit young adult men and women who grew up in pockets of material privilege within ostensibly (by statistical measures) deprived socio-economic environments. This interpretation, which fits the re-valued Illsley data, boils down to a materialist argument.

The most striking evidence of the limited contribution of health selection to achieved health status is the relative gains in longevity achieved by men and women of the 'lost generations' (p. 170). The Great War offered the nearest equivalent to a social laboratory for the study of natural selection in longevity. The combatant nations literally selected their 'fittest' young men and despatched them to a battle field where the risk of slaughter was one in five. Though the male cohort born in 1896 lost more than 17 per cent of its 'fittest' members, it recorded a smaller sex mortality differential than the 1901 cohort whose ranks were not depleted by military sacrifice. The same is true of France where the 1896 cohort was even more 'burnt out' (Hart 1989*b*). This shows the limited contribution of selection to life chances as against other forces emanating from the social and economic environment. The same conclusion may be drawn from the continued operation of occupational class on survival in old age (p. 161).

Despite the absence of confirming evidence, the selection thesis continues to play an informal though important part in shaping the explanatory perspective of casual medical observation. This happens because medical education embodies a concept of health as an individual attribute. Critics of biomedicine take the argument even further, accusing doctors of neglecting the whole person by treating patients as impersonal machines whose parts may be removed for service and repair (Mishler *et al.* 1981). These tendencies epitomized in anonymous treatment settings, divorced from any intelligence about the socio-economic identity of the patient, cannot but lead to a focus on the nuts and bolts of symptoms and a diversion from the external forces which, as we have seen, play by far the more important part in shaping the quantity and quality of life.

Here we see the disciplinary perspective shaping scientific understanding even where careful empirical analysis recommends other interpretation. Taken together, the established links between longevity and living standards do not lend themselves with any force to the selection thesis, or to the idea that the health of an individual can be understood without recourse to the social and economic environment of daily life. Neither theoretical logic nor the chronology of industrialization can uphold a plausible account of health as an independent force in the process of social and economic development. The search for evidence of eugenic processes for longevity differentials between societies has proved equally futile. When, as in Japan, the length of the human lifetime virtually doubles in the space of a few decades, in combination with massive economic growth and rising living standards, there must be very strong grounds for preferring a eugenic over a materialist explanation of trends.

The same must be said of differential levels of health inequality between societies. The degree of health inequality varies substantially between Denmark and France, who would argue that this results from national variation in genetic 'fitness'?

The strong causal connection between processes of social and economic development and human health, at the level of demographic history and comparative political economy, suggests that the search for the causes of observed mortality differentials within national populations should be carried out in the social and economic environment. The same choice is reinforced by the detailed catalogue of associations between mortality risk and economic circumstances enumerated in preceding pages.

Historical demography and international trends reaffirm the conclusion of the 'Black' Report (DHSS 1980) which identified materialist explanation as the most persuasive means of comprehending the relationship between the social and economic environment and human health. A materialist explanation locates variation in life chances in access to 'real' tangible resources, property, income, housing, eligibility for government services, resources, and so on. The human body has a material form, it makes sense to explain its durability in the same terms. The generation of health capital is after all not so different from other forms of wealth (Muurinen and Le Grand 1985). However, there are also signs of non-material factors helping to shape the health status of populations. Non-material refers to intangible dimensions of social life: ideas, knowledge, morality, beliefs, and the immeasurable quality of relationships. It is easier to distinguish the material and the non-material in the abstract than to identify ways in which they operate independently. This is evident when we classify the variables correlated with longevity.

The scope for distinguishing the operation of material and non-material forces is limited with all three of the different forms of evidence considered in the body of this chapter. The material progress of the first century of industrial capitalism went hand-in-hand with intellectual revolution, the spread of education, literacy, the growth of personal freedom, and political equality. While the evidence perhaps points in the direction of rising living standards as the primary cause of improved health, it is unlikely that it acted alone, and as the economists would put it, the 'multiplier' effects were probably substantial.

Comparison between societies offers one means of identifying the action of non-material factors in health. The demographic miracle of post-war Japan is difficult to reduce to economic change alone. As noted, the standard of living is relatively modest by international criteria, yet longevity is well above rates achieved in far richer societies such as the US. While part of the gap is explained by the degree of stratification, (that is, more or less extremes of inequality) high levels of cultural cohesion in Japan may also contribute extra increments of longevity. The same might be said of Sweden, though there one might emphasize political culture and the degree of citizenship integration as the important mechanisms.

The contrast between these two countries, very different but each with exemplary national health records, offers suggestive insights on 'social support' as a property of society at large. In Japan, social relationships between men and women, parents and children, workers and employers, are highly stable. Almost unique among advanced industrial societies, Japan's already low divorce rate continues to fall. Illegitimacy is very low, aided no doubt by the fact that more than 90 per cent of children complete high school. Stability of family relationships helps explain the low level of poverty and material deprivation in a market society where welfare provision is only partially developed. In Sweden, divorce rates are much higher, family life is less stable, and childbirth outside marriage is more widespread and socially permissive. The difference between Sweden and other western capitalist democracies lies in the development of welfare legislation to support a heterogeneity of domestic arrangements, and to remove economic sanctions from those which deviate from the traditional male-breadwinner–maternal-female role model (Kindlund 1988). Here are two models of 'social support' as a property of the social and economic environment. The support in question involves a very large material component. Indeed in Sweden it is not so much stable social relationships which underpin human vitality, but state intervention to provide the stable flow of resources which is a critical feature of kinship reciprocity. The Swedish model involves an enlarged role for the state, and appears relatively expensive from a monetarist standpoint. This is why many western European governments having embarked on the same reformist project, have recently invoked the image of the self-sufficient family as a more cost-effective means of providing for individual welfare. The Japanese model represents the survival of a modified form of the feudal family life in an otherwise fully fledged capitalist society. Whether such an arrangement can be re-implanted in Europe and North America, after the mass erosion of traditional kinship norms in the face of market forces and rampant individualism, remains to be seen. So far the experiment has been a dismal failure with rising levels of poverty, homelessness, narcotic abuse on a grand scale, epidemic infection, and real destitution.

The unification of the material and the non-material is also evident in the personal experience of everyday life. Several variables closely associated with the distribution of health within societies have both material and non-material dimensions.

Sex is a fundamental source of health stratification. The material difference between men and women is self-evident at two levels. Biological sex has a material reality independent of the cultural norms which prescribe the appropriate behaviour for men and women. It is also distinct material inequality emanating from the sexual division of labour which designates and values the work performed by men and women. Inequalities of income and property constitute a material difference between the sexes which pre-date the emergence of modern industrial society, yet they stem from an ideological source, from socially constructed norms of gender division. In recent decades the ideological character of sexual difference has become increasingly visible as women transgress the traditional cultural barriers of gender (work, smoking, and so on).

Similar arguments can be made for other variables which stratify health status. Education exerts both a material and a non-material impact on the individual life career (see p. 167). Facilitating access to secure well-paid employment, it is a primary determinant of economic welfare. At the same time it is a source of intellectual enlightenment, broadening ideas, raising expectations, enlarging horizons. Separating the joint influence of these dimensions with any precision is a near impossible analytical task.

Though we cannot exactly weigh the blend of social and

economic forces which shape survival we can, with the aid of the LS, reach some judgment about the relative contribution of material security. Among the complex interaction effects of tenure, employment status, marriage, and occupational class, one conclusion stands out: economic insecurity makes a critical difference to survival. Differences between women by marital status, class, and tenure are most revealing here. We learn that the preservative power of matrimony is corroded by economic scarcity; only in relatively affluent non-manual households can women afford to remain outside of paid employment without reducing their survival chances. This is strong evidence that material welfare is a foundation of longevity and that other non-material—cultural, ideological, or relational—factors exert their preservative effect only against a background of basic material sufficiency and security. This was the judgement of the Black Report and the evidence examined here offers no grounds for doubting its wisdom.

The political environment and human health

Political arithmetic, the original and apt term for public health statistics, is both complex and controversial. Some sources of methodological complexity have been explored in this chapter. These difficulties may exaggerate or disguise the degree of inequality. Evidence from studies which control for possible artefacts suggest, if anything, that official statistics underestimate the full extent of inequality in health in a population (Marmott *et al.* 1984).

The controversial side of data is a different matter. Political arithmetic (public health statistics) carries policy implications. From its earliest origins, it has disturbed the public conscience by bringing to light the human cost of government inaction. In modern societies where public authorities are democratically accountable and where the protection of health is among the rights of citizenship, trends in public health offer a potent means of evaluating the integrity and competence of elected governments. Public authorities may prefer to confine debate to issues of government resources for medical treatment as a means of deflecting attention from the true sources of physical and mental well-being in social and economic life. It is much cheaper to fund high-profile clinical research on AIDS than to underwrite income maintenance programmes for vulnerable groups in the population. Public health policy cannot be confined within a bio-clinical medical perspective. Given the overriding influence of the social and economic environment on human health, public health specialists cannot discharge their responsibilities without advising public authorities of the inseparability of 'medical fitness' from economic and social welfare. In the nineteenth century, the leaders of the sanitary movement could achieve major improvements through engineering projects, though even then it was apparent that the foundations of decent health lay in social, economic, and political reform (Taylor and Rieger 1984). Today the scope for achieving even minor improvements outside the sphere of social and economic policy is negligible, and public health specialists must be simultaneously social and medical scientists to comprehend the ramifications of government policy on the health of the people. To accomplish this, the discipline depends on a full and free flow of socio-medical statistics.

We have examined evidence from a number of advanced societies. Our choice was largely determined by available data. Many societies do not collect or publish mortality statistics by socio-economic variables. The neglect is not unexpected given the fiscal threat imposed by full awareness of the causes of health inequality. This is why governments often prefer not to monitor the health of the people. It also illuminates why government statisticians sometimes insist on extreme caution in the interpretation of official data. Worsening trends in class mortality gradients are the favourite object of the artefact model. Changes of classification may be emphasized to induce extreme caution or even to excuse discarding the data altogether. This was recommended by the latest Decennial supplement of Occupational Mortality in the face of a sharp increase in the gap separating the survival chances of the richest and the poorest sections of the British population (OPCS 1986, p. 42). The published guidance was undermined within a week of publication when the report's author used the official data in a Lancet paper to show that worsening mortality gradients could not be explained as artefacts (Marmott and McDowell 1986).

These comments underline an important, though often overlooked, dimension of the relationship between health and the social and economic environment. The direction and quality of the work of public health professionals in any society depends upon a national commitment to collect political arithmetic. While it is easy to criticize extremes of caution on the part of government statisticians, at least those national governments which make the data available reveal a commitment to public debate and scientific discourse on the health and welfare of the population. In those societies where national data of the socio-economic distribution of mortality are unavailable (for example, USSR and US), the government cannot be held accountable for the health of citizens.

National health statistics of high quality are expensive to collect. They involve the co-ordination of a number of government departments and the nurturance of specialized skills. Their availability is a measure of the commitment of public authorities to the health of the population. This is the 'political environment' at work, shaping awareness and stimulating the will to safeguard public health. It is no less important a determinant of health than the social and economic dimensions of the environment that have formed the focus of this chapter.

References

Adelstein, A. and Mardon, C. (1975). Suicides 1961–74. *Population Trends*, **2** (Winter), 13. Office of Population Censuses and Surveys, HMSO, London.

Ansell, C. (1874). *Statistics of families in the upper and professional classes*. National Life Assurance Society, London.

Banton, B. (1977). *The idea of race*. Tavistock Publications, London.

Berger, P. and Kellner, H. (1964). Marriage and the social construction of reality. *Diogenes*, **46**, 1.

Berkman, L. and Syme, L. (1979). Social networks, host resistance and mortality: A nine year follow up study of Alameda county residents. *American Journal of Epidemiology*, **109**, 186.

Blackburn, R. and Mann, M. (1979). *The working class in the labour market*. Macmillan, London.

Bonfield, L., Smith, R.M., and Wrighton, K. (1984). *The world we have gained: Histories of population and social structure*. Blackwell, Oxford.

Bourdieu, P. and Passeron, J. (1977). *Reproduction in education, society, culture*. Sage, London.

Bowley, A.L. and Burnett-Hurst, A. (1915). *Livelihood and poverty*. G. Bell and Sons, London.

Brenner, M.H. (1979). Mortality and the national economy: A review and the experience of England and Wales 1936–76. *Lancet*, **i**, 568.

Brenner, M.H. and Mooney, A. (1983). Unemployment and health in the context of economic change. *Social Science and Medicine*, **17**, 1125.

Cohen, M.N. and Armelagos, G.J. (1984). *Paleopathology at the origins of agriculture*. Academic Press, New York.

Compton, P.A. (1985). Rising mortality in Hungary. *Population Studies*, **39**, 71.

Department of Health and Social Security (1980). *Inequalities in health: Report of a working party*, (The Black Report). DHSS, London.

Doyal, L. and Pennell, I. (1979). *The political economy of health*. Pluto Press, London.

Dublin, L. and Lotka, A.J. (1949). *Length of life; A study of the life table*. Ronald Press, New York.

Dubos, R. (1959). *Mirage of health*. Harper and Rowe, New York.

Durkheim, E. (1896). *Suicide*, (1952 edition). Routledge & Kegan Paul, London.

Farr, W. (1885). *Vital statistics*, (reprinted 1975). The Scarecrow Press, Metuchen, New Jersey.

Floud, R. (1984). *Measuring the transformation of the European economies: Income, health and welfare*. Discussion paper No. 33, Centre for Economic Policy Research, London.

Fox, A.J. and Adelstein, A.M. (1978). Occupational mortality: Work or way of life. *Journal of Epidemiology and Community Health*, **32**, 73.

Fox, A.J. and Collier, P.F. (1976). Low mortality rates in industrial cohort studies due to selection for work and survival in the industry. *British Journal of Preventive and Social Medicine*, **30**, 225.

Fox, A.J., Goldblatt, P., and Adelstein, A.M. (1982). Selection and mortality differentials. *Journal of Epidemiology and Community Health*, **36**, 69.

Fox, A.J. (ed.) (1989). *Inequalities in health in Europe*. Gower, London.

Freidson, E. (1970). *Profession of medicine*. Dodd Mead, New York.

Friedman, E. (1980). Animal companions: One year survival after discharge from a coronary care unit. *Public Health Reports*, **95**, 307.

Fries, J.F. and Crapo, L.M. (1981). *Aging and vitality*. W.H. Freeman, San Francisco, California.

Gergely, M. (1987). *Ropirat az ongyilkos sagrol*. Medicina, Budapest.

Glendinning, C. and Millar, J., (1987). *Women and poverty in Britain*. Wheatsheaf Books, Harvester Press, Brighton, UK.

Hane, M. (1982). *Peasants, rebels and outcastes: The underside of modern Japan*. Pantheon, New York.

Hart, N. (1976). *When marriage ends: A study in status passage*. Tavistock, London.

Hart, N. (1982). *Which is the weaker sex? Radical Community Medicine*, **11**, 25.

Hart, N. (1985). *The sociology of health and medicine*. Causeway Press, Ormskirk.

Hart, N. (1986*a*). Life chances: Trends in class mortality in the UK. Paper presented to the ESF Workshop, Inequalities in Health in Europe, June 1986 (unpublished).

Hart, N. (1986*b*). Inequalities in health: The individual versus the environment. *Journal of the Royal Statistical Society, Series A*, **149**, 228.

Hart, N. (1986*c*). Is capitalism bad for your health? *British Journal of Sociology*, **33**(3).

Hart, N. (1989*a*). 'Sex, gender and survival: Inequalities of life chances between European men and women. In *Inequalities in Health in Europe* (ed. A.J.Fox). Gower, London.

Hart, N. (1989*b*). Gender and the rise and fall of class politics. *New Left Review*, **175**, 19.

Hart, N. (1990). *Life chances: Love, livelihood and longevity in the twentieth century*. Macmillan, London. (In press.)

Hirayama, T. (1987). The problem of smoking. *Japan Journal of Cancer Research*, **78**, 203.

House, J., Robbins, and C. Metzner, H. (1982). The association of social relationships and activities with mortality: Prospective evidence from the Tecumseh community health study. *American Journal of Epidemiology*, **116**, 123.

Illich, I. (1976). *Limits to medicine*. Marion Boyers, London.

Illsley, R. (1986). Occupational class, selection and the production of inequalities in health. *Journal of Social Affairs*, **2**, 151.

Johansson, S.R. (1977). Sex and death in Victorian England: An examination of age and sex specific death rates, 1840–1910 in *A widening sphere: Changing Victorian women* (ed. M. Vicinus) Methuen, London.

Jones, E. (1985). Teenage pregnancy in developed countries: Determinents and policy implications. *Family Planning Perspectives*, **17**, 53.

Joshi, H. (1987). The cost of caring. In *Women and poverty in Britain* (ed. C. Glendinning and J. Millar). Wheatsheaf Books, Harvester Press, Brighton, UK.

Kindlund, S. (1988). Chapter 4. In *Child support: From debt collection to social policy* (ed. A.J. Kahn and S.B. Kamerman). Sage Publications, London.

Koskinen, S. (1985). Time trends in cause specific mortality by occupational class in England and Wales. Paper presented at the IUSSP General Conference, June 1985, Florence.

Kynch, J. (1985). How many women are enough?: Sex ratios and the right to life. *Third World Affairs*, **2**, 156.

Lewis, R.A. (1952). *Edwin Chadwick and the public health movement 1832–54*. Longmans, London.

Madigan, F.C. (1957). Are sex mortality differentials biologically caused? *Millbank Memorial Fund Quarterly*, **35**, 202.

Marmott, M.G., Shipley, M.J., and Rose, G.A. (1984). Inequalities in death—specific explanations of a general pattern. *Lancet*, **ii**, 1003.

Marmott, M.G. and McDowell, M.E. (1986). Mortality decline and widening social inequalities. *Lancet*, **i**, 274.

Matthieu, J. (1980). *An economic history of women in America*. Harvester Press, Brighton.

McKeown, T. (1976). *The modern rise of population*. Edward Arnold, London.

McKinlay, J.B. (ed.) (1984). *Issues in the political economy of health care*. Tavistock Publications, London.

McNeill, W.H. (1977). *Plagues and peoples*. Blackwell, Oxford.

Mishler, E., AmaraSingham, L.R., Hauser, S.T., Liem, R., Osherson, D., and Waxler, N. (1981). *Social contexts of health, illness and patient care*. Cambridge University Press, Cambridge.

Moser, K., Pugh, H.S., and Goldblatt, P. (1988). Inequalities in women's health: Looking at mortality differentials using an alternative approach. *British Medical Journal*, **296**, 1221.

Muurinen, J. and Le Grand, J. (1985). The economic analysis of inequalities in health. *Social Science and Medicine*, **20**, 1029.

Office of Population Censuses and Surveys (OPCS) (1978*a*). *Social and biological factors in infant mortality*, 1975–6. HMSO, London.

Office of Population Censuses and Surveys (OPCS) (1978*b*). *Occupational mortality 1970–72, Decennial supplement for England and Wales*. HMSO, London.

Office of Population Censuses and Surveys (OPCS) (1986). *Occupational mortality 1979–80, 1982–83. Decennial supplement, pt. 1. Commentary*. HMSO, London.

Office of Population Censuses and Surveys (OPCS) (1990). *Mortality and social organisation*. HMSO, London.

Orwell, G. (1937). *The road to Wigan pier*. Victor Gollanz, London.

Pamuk, E. (1985). Social class inequality in mortality from 1921 to 1972 in England and Wales. *Population Studies*, **39**, 17.

Parkes, C.M., Benjamin, B., and Fitzgerald, B.G. (1969). A broken heart: A statistical study of increased mortality among widows. *British Medical Journal*, **i**, 740.

Ramsden, S. and Smee, C. (1981). The health of unemployed men. Department of Health and Social Security, Cohort Study. *Employment Gazette* (September), 397.

Robins-Mowry, D. (1983). *The hidden sun: Women of modern Japan*. Westview Press, Colorado.

Roosevelt, A.C. (1984). Population, health, and the evolution of subsistence. In *Paleopathology and the origins of agriculture* (ed. M.N. Cohen and G.J. Armelagos). Academic Press, New York.

Rose, G. and Marmott, M. (1981). Social class and coronary heart disease. *British Heart Journal* **45**, 13.

Rowntree, B.S. (1902). *Poverty*. Macmillan, London.

Sayers, J. (1982). *Biological politics*. Tavistock, London.

Sen, A. (1981). *Poverty and famines: An essay on entitlement and deprivation*. Oxford University Press.

Schofield, R. (1984). Did the mothers really die? Three centuries of material mortality in 'The world we have lost'. *The world we have gained* (ed. I. Bonfield, M. Smith, and K. Wrighton). Blackwell, Oxford.

Selye, H. (1956). *The stress of life*. McGraw Hill, New York.

Sievers, S. (1983). *Flowers in salt: The beginnings of feminist consciousness in modern Japan*. Stanford University Press, Stanford, Connecticut.

Smelser, N.J. (1959). Social change in the industrial revolution. Routledge & Kegan Paul, London.

Stern, J. (1983*a*). Social mobility and the interpretation of social class mortality differentials. *Journal of Social Policy*, **12(1)**, 27.

Stern, J. (1983*b*). The relationship between unemployment, morbidity and mortality in Britain. *Population Studies*, **37**, 61.

Susser, M. (1981). Widowhood: A situational life stress of a stressful life event. *American Journal of Public Health*, **71**, 793.

Szreter, S.R.S. (1984). The genesis of the Registrar General's social classification of occupations. *British Journal of Sociology*, **35**, 522.

Tawney, R.H. (1931). *Equality*. George Allen and Unwin, London.

Taylor, R. and Rieger, A. (1984). Rudolf Virchow on the typhoid epidemic in Upper Silesia: An introduction and translation. *Sociology of Health and Illness*, **6**, 2.

United States Department of Health and Human Services (1988). *Advance report of final mortality statistics 1986*. Monthly Vital Statistics Report, Vol. 37, No. 6.

Valkonen, T. (1985). *The mystery of the premature mortality of Finnish men*. Reprint number 128, Department of Sociology, University of Helsinki.

Valkonen, T. (1989). Adult mortality and level of education: A comparison of six countries. In *Inequalities in health in Europe*. (ed. A.J. Fox). Gower, London.

Valkonen, T. (1987). Social inequality in the face of death. Paper presented to the European Population Conference. IUSSP/UIESP, EAPS, Finnco. Tilastockeskus Statistikcen Tracen.

Vicinus, M. (ed.) (1977). *A widening sphere: Changing roles of Victorian women*. Methuen, London.

World Health Organization (1986). *World Health Statistics*. World Health Organization, Geneva.

Whitehead, M. (1987). *The health divide: Inequalities in health in the 1980s*. Health Education Authority, London.

Wilkinson, R. (1986). Socio-economic differences in mortality: Interpreting the data on their size and trends. In *Class and health. Research and longitudinal data* (ed. R. Wilkinson), pp. 4–30. Tavistock, London.

Winter, J. (1982). Aspects of the impact of the 1st World War on infant mortality in Britain. *Journal of European Economic History*, **11**, 713.

Wood, D. (1982). *The Department of Health and Social Security Cohort Study of Unemployed Men*, Working Paper No. 1. Department of Health and Social Security, London.

Zelizer, V. (1985). *Pricing the priceless child: The changing social value of children*. Basic Books, New York.

11

Education and life-style determinants of health and disease

LAWRENCE W. GREEN and DENISE G. SIMONS-MORTON

Introduction

Behaviour has long been appreciated as an inescapable factor in determining health. Technological, engineering, and biomedical remedies have tried to circumvent it; legal and regulatory approaches to public health have sought to control it and to protect people from each other's behaviour. These strategies have alternately declared their victories, only to find behaviour breaking out somewhere else as a cause of ill-health. Educational approaches to behaviour and health have been less spectacular, but perhaps are more dependable. Education of the public provides, at the very least, a palliative solution while technological solutions await development. Education enables people to take personal and collective action to protect themselves and to support the development of technology and the passage of legislation.

This chapter attempts to sort out the ways in which behaviour and its more complex manifestation, called life-style, influence health and disease, and ways in which education shapes or modifies behaviour and life-style. We use the life-style construct broadly to refer to any combination of specific practices and conditions of living reflecting habitual patterns of behaviour that are influenced by family history, culture, and socio-economic circumstances. We know that behaviour can be influenced directly by education targeted to individuals and groups. Life-style changes more slowly and usually requires some combination of educational, organizational, economic, and environmental interventions or changes.

Acquired immune deficiency syndrome (AIDS) presents the obvious contemporary example of both a disease awaiting a technological solution, and a life-style problem responding to some combination of educational, organizational, economic, and environmental interventions and changes. But virtually any public health or medical victory from the past can be traced in part to an educational component that preceded the technological or legislative innovation and another that facilitated the diffusion, adoption, and implementation of the technology or legal solution. Unless and until a human immunodeficiency virus vaccine is developed, society must depend on the control of behaviour to curb the spread of AIDS. Much of the behaviour in question with AIDS cannot be controlled by legal means because of its private nature. We are left, then, with health education as the primary means to control the spread of AIDS. The success of health education in filling this gap has been modest but not insignificant in changing relevant behavioural patterns in high-risk populations (Levy *et al*. 1983; Nelkin 1987; Taylor 1987; Vincent *et al*. 1987). More complex health promotion interventions that include education might be slowly changing some of the more complex life-styles associated with AIDS.

We start with a reductionist look at specific behaviours as they relate to health and disease, then progress to the more complex, socio-cultural aspects inherent in the term life-style. Finally, we examine the functions of education in reducing disease and promoting health in populations or communities.

Specific behaviours and health

Considerable evidence exists that, for many diseases, behaviours increase the risk of developing disease and can be considered causes of disease. In addition, some types of behaviours correlate with and precede better health, increased longevity, and decreased disease risk. Examining the relationships between specific behaviours and specific measures of health and disease status provides the foundation for assessing behavioural and life-style factors as health determinants.

Behaviours and disease—the causal links

In some cases the evidence for a relationship between a specific behaviour and a measure of health status or disease is correlational in nature; in other cases there is stronger evidence of causality. Evidence from observational epidemiological studies, human experimental trials, and animal models along with potential mechanisms of biological action lead one to conclude that many behaviours are, in fact, contributing causes (causal risk factors) of specific diseases.

Studies of the relationship between smoking and lung cancer provide probably the strongest evidence of a behaviour as a cause of a disease. Observational epidemiological studies have found strong and consistent measures of association, the correct temporal sequence, and a dose–response relationship; laboratory studies have demonstrated that components of tobacco smoke cause cancer in animal models (Office on Smoking and Health 1982). A randomized trial of smoking cessation provides preliminary experimental evidence for smoking as a cause of lung cancer. Rose *et al.* (1982) demonstrated a significantly lower rate of smoking in smokers who were provided with a smoking cessation programme compared with smokers in a control group. Resultant lung cancer death-rates were 23 per cent lower in the experimental group, although this difference was not statistically significant.

For other behavioural factors, there are varying degrees of causal evidence for differing diseases. For example, evidence that saturated fat and cholesterol in the diet are contributing causes of coronary heart disease (CHD) comes from numerous ecological studies showing a correlation between dietary fat consumption and CHD mortality and incidence rates (Keys 1970; McGill 1979), and is supported by studies showing that a high serum cholesterol level increases the risk of CHD development (Dawber 1980; Kannel *et al.* 1984; Pooling Project Research Group 1978), that changes in dietary fat lead to changes in serum cholesterol (Brown 1983; Grande 1983), and that lowering the serum cholesterol level decreases the occurrence of CHD (Lipid Research Clinics Program 1984*a*, *b*). Although strong evidence can be found for each step in a causal chain, studies attempting to confirm the causal hypothesis by examining the incidence of CHD in response to modifying the fat composition of the diet have been inconsistent in their findings (Stallones 1983).

For another example, evidence that physical inactivity is a risk factor for CHD comes from biological plausibility and epidemiological studies. Epidemiological evidence includes consistent measures of association, relatively strong measures of association, the correct temporal sequence, and a dose–response relationship (Powell *et al.* 1987). Positive results from experimental studies, however, are lacking.

Many causal risk factors are not themselves behaviours, but have determinants that are behaviours; in these cases the behavioural determinants can be considered indirect risk factors that act earlier in the causal pathway. For example, obesity is a risk factor for type II (adult onset) diabetes mellitus (Bennett 1982), with a combination of high caloric intake and low energy output as a behavioural determinant of obesity (Horton and Danforth 1982).

For many diseases, behaviours contribute to disease prognosis; such behaviours include seeking medical care and complying with medical treatment. For example, the prognosis of type I (insulin dependent) diabetes mellitus is a function of patients' compliance with their insulin prescriptions, and the prognosis of breast cancer is a function of the stage of disease at which the woman obtains medical care. Because of the important relationship between patient behaviours and disease outcome, a large literature on patient compliance with medically prescribed regimens has been catalogued (Haynes *et al.* 1979) and subjected to meta-analysis (Mullen *et al.* 1985).

Behavioural risk factors and the public's health

Behavioural determinants of health and disease status can be found for almost every disease, either through a causal risk factor that is a behaviour, through behavioural factors that influence physiological risk factors, or through behavioural factors that influence treatment and prognosis.

The leading causes of death in developed nations are primarily chronic degenerative diseases and injuries (National Center for Health Statistics 1988*a*). Listed in Table 11.1 are the ten leading causes of death in the US in 1985, their generally accepted risk factors (behavioural and physiological), and some behavioural determinants of the physiological risk factors. As can be seen in Table 11.1, each of the leading causes of death has a behavioural determinant somewhere in the causal chain.

In developing countries, the leading causes of death primarily are infectious diseases (World Health Organization 1986*b*). Although the causes of these diseases are the specific infectious agents, behaviours are contributing determinants of the transmission of infectious agents and of the individual's susceptibility. Such behaviours—which include actions to decrease the spread of the infectious organisms or exposure to vectors, or to receive immunizations—cannot take place in environments that do not enable these actions or provide health care resources. For these reasons, major public health emphases in developing nations are the provision of health care environments, particularly primary care, the use of appropriate technology, and the provision of sanitation and safe water (World Health Organization 1979, 1981, 1986*b*). The role of behavioural risk factors for disease has been emphasized primarily in developed nations. It is clear, however, from the historical trends in morbidity and mortality in developed nations (US Department of Health, Education and Welfare 1979) that attention to behavioural risk factors is warranted early in a nation's development. Perhaps such early attention would help to prevent, or at least to mitigate, the dissemination and social acceptance of those behavioural practices that increase the risk for chronic degenerative diseases and injuries.

When one examines the generally accepted behavioural risk factors for disease, it readily becomes apparent that in developed countries a few categories of behaviours are related to a large proportion of deaths. Smoking, dietary practices, and alcohol misuse alone are causally related to all of the ten leading causes of death in the US, in addition to being related to various causes of morbidity, disability, diminished functional capacity, and quality of life. Smoking is a

Table 11.1. The ten leading causes of death in the US (1985), their generally accepted behavioural and physiological risk factors, and behavioural determinants of the physiological risk factors. (Behaviours are indicated by italics)

Cause of death	Behavioural and physiological risk factors	Behavioural determinants of physiological risk factors
Diseases of the heart	*Smoking*	
	Physical inactivity	
	Oral contraceptive use	
	High serum cholesterol	*High fat diet*
	Obesity	*High calorie diet*
	Hypertension	*High salt diet*
	Diabetes mellitus	*High calorie diet (via obesity)*
Malignant neoplasms	*Smoking*	
	High fat diet	
	Low fibre diet	
Cerebrovascular disease	Hypertension	*High salt diet*
	Atherosclerosis	*High fat diet*
Unintentional injuries (including fires)	*Alcohol use*	
	Unsafe driving	
	Seat-belt non-use	
	Smoking	
Chronic obstructive lung disease and allied conditions	*Smoking*	
Pneumonia and influenza	Influenza immunization status	*Failure to receive immunizations*
	Malnutrition	*Inadequate diet*
Suicide and homicide	*Alcohol use*	
	Handgun use	
Diebetes mellitus	Obesity	*High calorie diet with low energy expenditure*
Chronic liver disease and cirrhosis	*Alcohol use*	
Atherosclerosis	*Smoking*	
	High serum cholesterol	*High fat diet*

risk factor for CHD (Office on Smoking and Health 1983), cancers of the lung, larynx, oral cavity, and bladder (Office on Smoking and Health 1982), chronic obstructive lung disease (Office on Smoking and Health 1984), unintentional injury from fire (Mierley and Baker 1983), and adverse pregnancy outcomes such as low birth-weight, premature rupture of membranes, and abruptio placenta (Office on Smoking and Health 1980). Dietary factors are related to the development of atherosclerosis and CHD (through serum cholesterol level) (Kannel *et al.* 1984), cancer (through fat, fibre, and vitamin intake) (Ames 1983; Willett and MacMahon 1984), and diabetes mellitus (through obesity) (Bennett 1982). Alcohol use is related to a variety of major diseases and health problems including cirrhosis of the liver and other liver diseases, suicide and homicide, unintentional injury, and congenital anomalies (National Institute on Alcohol Abuse and Alcoholism 1983, 1987; Ravenholt 1984).

The population attributable risks (PAR) for causal risk factors—the proportion of the disease in the population that can be attributed to the behavioural factor—can be estimated for many diseases. For such an estimation to be appropriate, strong evidence of causality must exist. The US Office on Smoking and Health (OSH) has used estimates of the relative risk and risk factor prevalence, taken from numerous epidemiological studies, to calculate the proportion of disease-specific deaths in the US attributable to smoking (OSH 1982, 1983, 1984). The proportion of deaths estimated to be attributable to cigarette smoking for all cancers, lung cancer, coronary heart disease, and chronic obstructive lung disease are shown in Table 11.2. These estimates translate into more than 350 000 deaths per year in the US due to cigarette smoking alone (OSH 1984). Clearly this is just a small portion of the deaths in the US and worldwide than can be attributed to behavioural causes.

Table 11.2. Percentage of deaths attributable to cigarette smoking (population attributable risk) for selected leading causes of death in the US, 1980

Disease	Percentage of deaths
All cancers	30
Lung cancer	85
Coronary heart disease	30
Chronic obstructive lung disease	85

Source: Office on Smoking and Health (1982, 1983, 1984).

Based on information about the causal associations between behaviours and health, the importance of the health problems, and the prevelance of those behaviours in the US, the US Government identified five major behavioural areas to emphasize in federal policy for health promotion and disease prevention (US Department of Health, Education and Welfare 1979; US Department of Health and Human Services 1980): smoking, alcohol and drug misuse, nutrition, exercise and fitness, and stress and violent behaviour.

Combinatorial effects of specific behaviours

Various evidence suggests that combinations of behavioural risk factors may act together synergistically. In the Framingham study, a large prospective study of heart disease in the US, glucose intolerance (related to diet via obesity) almost doubled the risk of a CHD event in men with elevated serum cholesterol (related to dietary practices); the addition of cigarette smoking further increased the risk (Dawber 1980; Kannel *et al.* 1984). In the Pooling Project, in which data from a dozen large cohort studies of CHD incidence were pooled, the presence of two risk factors increased the risk of a CHD event more than would be expected if the individual risks from each factor were added, and the presence of three risk factors further increased the risk to a level greater than the additive effect of the individual factors (Kannel *et al.* 1984; Pooling Project Research Group 1978). This evidence suggests that there is more than an additive effect of risk factors. It is likely that causal factors for other diseases also result in a greater-than-additive effect on disease occurrence.

Table 11.3. Some known and suspected influences on four major behavioural risk factors

Cigarette smoking	Dietary practices	Alcohol use/abuse	Physical inactivity
Knowledge of adverse health effects of smoking	Personal food preferences	Expectations of alcohol effects	Beliefs of physical activity benefits
Attitudes about smoking	Cultural food preferences	Child of alcoholic	Attitudes toward physical activity
Skills in smoking cessation/prevention	Perceived social acceptance of foods	Alternatives to alcohol	Self-motivation
Cigarette cost	Social context of eating	Psychological stress	Self-discipline
Availability of cigarettes	Availability and convenience of foods	Low self-esteem	Accessibility of exercise facility
Cigarette advertising	Skills in menu planning	Early drinking experience	Skills in relapse prevention
Peer influences to smoke	Skills in food preparation	Heavy social drinking	Skills in goal setting
Social support for non-smoking	Skills in food selection	Parent and peer influences	Enjoyability of physical activity
	Food advertising	Alcohol advertising	Family support
		Cost of alcohol	
		Availability of alcohol	
		Supervision of drinking	

Complexity and determinants of behaviours

Examination of the generally accepted behavioural risk factors and evidence of their causal relationship to disease, their prevalence in the population, and their potential combined effects lead us to conclude that there is a large effect on the health of the population from just a few major health related behaviours. The most important preventive behaviours appear to be smoking, dietary factors, and alcohol consumption.

Despite the implied simplicity in identifying only a few major preventive behaviours, those behaviours are highly complex. Most behavioural risk factors, and health care behaviours also, are the product of a variety of component behaviours, tasks, or actions. For example, food consumption has been said to be 'the product of a chain of behaviours that includes procuring and selecting foods, planning menus or selecting from a menu, preparing or ordering foods, and eating'. Most people are 'confronted with literally hundreds of food-related choices, including where to shop or eat, what to purchase or prepare, how to season food, and with whom to eat' (Simons-Morton *et al.* 1986). Similar chains of component behaviours and behaviour-related choices are important in other health behaviours.

Not only are health behaviours complex, but each behaviour has numerous influences or determinants. Factors that influence behaviours can be grouped into three major categories (Green *et al.* 1980): predisposing, reinforcing, and enabling. Predisposing factors reside in the individual and include attitudes, values, and beliefs. Reinforcing factors are positive consequences of behaviour, such as peer acceptance, or negative consequences, such as social disapproval. Enabling factors generally are conditions of the environment that allow the behaviour or, alternatively, create barriers to it.

Most behaviours have influences from each category. Some of the known and likely influencing factors for the three most important preventive health behaviours, plus another important health behaviour, physical activity, are shown in Table 11.3.

The influences on smoking initiation and cessation are numerous (Simons-Morton D.G. *et al.* 1990; McCaul *et al.* 1982; Warner 1986*a*,*b*). Predisposing factors include attitudes about smoking and beliefs about and knowledge of the health effects of smoking. Reinforcing social factors include social support, peer influences, and cigarette advertising (providing vicarious reinforcement). Enabling factors include availability and cost of cigarettes.

A variety of factors influences dietary practices (Birch 1980; Birch *et al.* 1980; Rozin 1984; Simons-Morton *et al.* 1986). These include both personal and cultural food preferences, perceived social acceptance, social context, availability and convenience of foods, and skills in menu planning, food purchasing, food selection, and food preparation.

Numerous factors appear to influence alcohol use and abuse (Biddle *et al.* 1980; Breed and DeFoe 1984; Glynn 1981; National Institute on Alcohol Abuse and Alcoholism 1987; Simons-Morton, B.G. *et al.* 1990; Zarek *et al.* 1987). Predisposing factors may include expectations about the effects of alcohol, psychological stress and lower self-esteem, perceptions of insusceptibility to adverse consequences of drinking such as losing one's job, being a child of an alcoholic, and early drinking experiences. Reinforcing factors include parent and peer influences, and may include advertising and modelling in the visual media. Enabling factors and barriers include availability, or unavailability, of non-alcoholic drinks, cost of alcoholic beverages, access to alcohol, and supervision of adolescents.

Besides the three major behavioural risk factors for causes of mortality, every behaviour related to morbidity and well-being also has a variety of influences. Physical inactivity is an important health behaviour and is a good illustrative example. Physical inactivity is not only a risk factor for CHD (Powell *et al.* 1987), but also is related to hypertension (Blair *et al.* 1984; Duncan *et al.* 1985), osteoporosis (Aloia 1981; Smith *et al.* 1981), and mental health (Lichtman and Poser 1983; Taylor *et al.* 1985), all of which are highly prevalent health problems in developed countries. The numerous influences on physical activity (Dishman *et al.* 1985; Oldridge 1982, 1984; Simons-Morton *et al.* 1988*b*) include: beliefs about the importance of physical activity, attitudes about physical activity, motivation and self-discipline, accessibility of an exercise facility, skills in relapse prevention and goal setting, unenjoyability of exercise, and family support.

In addition to the complexity of risk behaviours and their numerous determinants, the performance of each behaviour is interwoven with other behaviours and with socio-economic and cultural factors.

Socio-economic and cultural factors and health

To enable us to understand better the behavioural health determinants, we must examine the context within which behaviour occurs. That context is necessarily social, cultural, and economic. The social context includes contemporary personal interactions with family and others, as well as with complex organizations. The cultural context includes the cumulative weight of these interactions over generations as reflected in values and traditions related to behaviour.

There is substantial evidence that socio-economic conditions are associated with health status—an association that is addressed in considerable detail in Chapter 10 of this volume. Culture plays an intimate role in determination of health status, most clearly through health behaviours, but also apparently through traditional patterns of social support (Berkman and Syme 1979). Here we review representative studies and major reports for an overview of the relationships between socio-economic status and health, socio-economic status and use of health care services, and culture and health, to provide a context for the behavioural factors. This will lead us to the more complex socio-cultural–behavioural construct, life-style.

It may be that education is the most basic aspect of socio-economic status as presented in this review. Duncan (1961) described the relationship between the basic components—income, education, and occupation—of socio-economic status: 'Education qualifies the individual for participation in occupational life, and pursuit of an occupation yields him a return in the form of income'.

Socio-economic status and mortality

In 1967, Antonovsky reviewed literature from the 1600s through to the early 1960s on the relationship between socio-economic status and mortality, including over 30 studies primarily from the European countries and the US (Antonovsky 1967). The reviewed studies measured socio-economic status in a variety of ways, including type of occupation, median rental costs in census tracts, taxpayer status, and indices comprising education, occupation, and median family income. Antonovsky's conclusion was: 'Despite the multiplicity of methods and indices used in the 30-odd studies cited, and despite the variegated populations surveyed, the inescapable conclusion is that [socio-economic] class influences one's chance of staying alive. Almost without exception, the evidence shows that [socio-economic] classes differ on mortality rates'. People with lower socio-economic status have higher mortality rates. He observed that the greatest difference in mortality rates occurred during the middle years of life (30s and 40s), which he conjectured may be due to differences in preventable deaths between those with different socio-economic status.

Since Antonovsky's historical review, subsequent reports and studies have provided additional evidence of a persisting relationship between socio-economic status and all-cause mortality. Table 11.4 shows some representative data. In the US in 1960, socio-economic status as measured both by education level (years of schooling completed) and family income was inversely associated with mortality ratios (Kitagawa and Hauser 1973). In 1982, the Black Report, by the UK Working Group on Inequalities in Health, cited higher 1970–2 mortality rates in occupational groups of lower socio-economic status for both males and females in each of the three major age groups: infants, children, and adults (Black *et al.* 1982). In 1979, Morris reported that in the UK in 1975–6, all-cause mortality ratios in men and infant mortality rates continued to be higher for occupational groups of lower socio-economic status (Morris 1979). In the UK, the Office of Population Censuses and Surveys' (OPCS) Longitudinal Study determined mortality rates for 1976–81 in relation to occupational groups of 1971 to examine the prospective relationship between occupational class and mortality (Fox *et al.* 1986). A mortality gradient persisted, suggesting that selection into lower occupational class due to illness is not the major operating factor for the gradients seen in cross-sectional studies.

A socio-economic differential for infant and childhood mortality has been seen in many studies. A few recent studies are mentioned here. Early childhood death-rates (ages 0–5 years) in Southampton, England, from 1977 to 1982 were higher in districts with high unemployment, poor housing, and single-parent families (Robinson and Pinch 1987). In Kentucky, US, 1982–3 infant mortality rates, adjusted for a variety of variables, were significantly higher in poor than in non-poor infants during the post-neonatal period (Spurlock *et al.* 1987). In Canada, 1971 and 1981 infant mortality rates were 1.3–2.0 times higher among the poor than the non-poor (Shah *et al.* 1987).

Most within-country studies have been conducted in developed nations; however, comparisons between nations provides a global perspective as shown in Tables 11.5 and 11.6. In general, less developed countries with lower per caput incomes exhibit lower life expectancy and higher infant mortality rates than more developed countries with higher per caput incomes.

An inverse socio-economic differential for cause-specific mortality for most diseases has also been seen in many studies. The Whitehall Study of British civil servants revealed gradients of mortality rates for each of two categories of diseases—CHD and other—with lowest rates in administrators (highest grade) and highest rates in 'other' employees (lowest grade) (Marmot *et al.* 1984). Ischaemic heart disease

Table 11.4. Socio-economic status and mortality within developed countries

Year	Country/population	Measure of socio-economic status	Mortality
1960	US/white males, 25–64 years	Years of schooling	Obs./exp. ratio*
		0–4 years	1.15
		5–7 years	1.14
		8 years	1.07
		High school, 1–3 years	1.03
		High school, 4 years	0.91
		College, 1–3 years	0.85
		College, 4+ years	0.70
1971	England and Wales/males, 15–64 years	Occupational class	Rate/1000
		I (professional)	3.98
		II (intermediate)	5.54
		IIIN (skilled/non-manual)	5.80
		IIIM (skilled/manual)	6.08
		IV (partly skilled)	7.96
		V (unskilled)	9.88
1975–6	England and Wales/males, 15–64 years	Occupational class	SMR†
		I (professional)	77
		II (intermediate)	81
		IIIN (skilled/non-manual)	99
		IIIM (skilled/manual)	106
		IV (partly skilled)	114
		V (unskilled)	137
1976–81‡	England and Wales/males, 15–64 years	Occupational class	SMR†
		I (professional)	66
		II (intermediate)	77
		IIIN (skilled/non-manual)	105
		IIIM (skilled/manual)	96
		IV (partly skilled)	109
		V (unskilled)	124

Sources: Black *et al.* (1982), Fox *et al.* (1986), Kitagawa and Hauser (1973), Morris (1979).
* Obs./Exp. ratio = observed to expected ratio, total group = 1.00.
† SMR = standardized mortality ratio, all men = 100.
‡ From the Office of Population Censors and Surveys' longitudinal study; occupational classes excluded men with inadequate occupational descriptions or 'unoccupied', many of whom were permanently sick; occupational class was determined in 1971.

mortality in the UK is higher in manual than in non-manual workers (Marmot and McDowell 1986; Pocock *et al.* 1987), and cancer mortality rates in New York City are higher in lower income groups (Shai 1986). In the US, educational level has been seen to be inversely related to CHD risk factors (blood pressure, cigarette use, relative weight) and to subsequent mortality rates from CHD and all causes (Liu *et al.* 1982). Death due to CHD in the UK, and possibly in other developed countries, has shifted from a higher rate in those in a higher occupational class level to the reverse (Marmot *et al.* 1978).

All of the above patterns for socio-economic status are reflected also in white–non-white differences in mortality. For both males and females in the US, the black death-rate exceeds the white rate by 50 per cent (Andersen *et al.* 1987).

Socio-economic status and morbidity

Prevalence of illness has been seen to be associated with socio-economic status, with greater morbidity in lower socio-economic groups (Table 11.7). In 1971–6 in the UK, 'limiting long-standing illnesses' showed a socio-economic gradient, with such illnesses three times more prevalent in unskilled manual than in professional workers (Black *et al.* 1982); the gradient has persisted into the 1980s (for 1981–2 rates see Arber 1987). In the US, the number of restricted activity days per caput remained about twice as high for poor compared with non-poor from the 1960s through to 1983 (Health Resources and Services Administration 1986). In 1980, the number of bed disability days per caput decreased with higher income status, as did restricted activity days (Health Resources and Services Administration 1986). A clear income gradient was seen for reported health status in the 1985 US Health Survey: five times the proportion of lowest income persons compared with highest reported fair or poor health status, and twice the proportion of highest income persons compared with lowest reported excellent health (Health Resources and Services Administration 1986). Similar patterns of disadvantage are found in black morbidity rates compared with those of whites in the US (Manton *et al.* 1987).

Socio-economic status and utilization of health care services

The Black Report in the UK pointed out the difficulty in evaluating the utilization of health care services by different population groups because of differing needs, but concluded that the level of health care consultation for workers of lower

Table 11.5. Per caput income with life expectancy at birth, infant mortality rate (1985), and health-care personnel (1981) for selected countries, developed and developing

Country	Per caput income (US $)	Life expectancy at birth (years)	Infant mortality rate (per 1000)	Population per	
				Physician	Nurse
Low income economies	**270**	**60**	**72**	**5 770**	**3 880**
Ethiopia	110	45	168	88 120	5 000
Mozambique	160	47	123	37 000	5 610
Nepal	160	47	133	28 770	33 430
Central African Republic	260	49	137	22 430	2 120
India	270	56	89	3 700	4 670
Kenya	290	54	91	10 140	990
China	310	69	35	1 730	1 670
Senegal	370	47	137	14 200	1 990
Middle income economies	**1 290**	**62**	**68**	**5 080**	**1 380**
Liberia	470	50	127	9 400	2 940
Zimbabwe	680	57	77	7 100	1 000
Thailand	800	64	43	6 870	2 140
Turkey	1 080	64	84	1 530	1 240
Tunisia	1 190	63	78	3 900	950
Jordan	1 560	65	49	1 200	1 170
Brazil	1 640	65	67	1 300	1 140
Malaysia	2 000	68	28	3 920	1 390
Yugoslavia	2 070	72	27	700	300
Greece	3 550	68	16	400	370
Israel	4 990	75	14	400	130
High income oil exporters	**9 800**	**63**	**61**	**1 380**	**620**
Libya	7 170	60	90	620	360
Kuwait	19 270	70	22	700	180
Industrial market economies	**11 810**	**76**	**9**	**530**	**180**
Spain	4 290	77	10	360	280
UK	8 460	75	9	680	120
France	9 540	78	8	460	110
Japan	11 300	77	6	740	210
Canada	13 680	76	8	550	120
Switzerland	16 370	77	8	390	130
US	16 690	76	11	500	180

Source: Adapted from World Bank (1987).

Table 11.6. Selected global indicators of health status and conditions for nations in three stages of development, 1986

Indicator	Least developed nations	Other developing nations	Developed nations
Safe water supply (%)	34.7	57.2	95.8
Sanitation (%)	16.8	29.1	76.6
Local health care (%)	48.5	72.8	99.8
Infant mortality rate (per 1000 births)	141.2	78.9	12.8
Life expectancy at birth (years)	47.9	61.5	72.4

Source: World Health Organization (1986*a*).

socio-economic status does not appear to match their needs for heath care (Black *et al.* 1982); inequalities appeared to be greatest for preventive health services.

In the US, the number of physician visits per year in 1964 was greater for higher income families, whereas by 1980 the gradient, although not substantial, was reversed (US Department of Health and Human Services 1986). These data are difficult to interpret because they are not presented in relation to need. Socio-economic differentials for morbidity and mortality suggest that lower income groups have a greater need for health care services than upper income groups, whereas the actual utilization is about the same. The 1983 National Health Interview Survey showed that lower income families were more likely to visit hospital out-patient departments, whereas higher income families were more likely to visit a doctor's office or obtain a telephone consultation (US Department of Health and Human Services 1986). This illustrates the differential sources of care in the US between income groups, where lower income groups lack personal physicians and have less continuity of care. A recent study in the US found that among insured adults of working age, the poor were about four times as likely as the non-poor to have not received the supportive health care services that they needed (Hayward *et al.* 1988). Biases in methods of data collection on health status and health care utilization might make these underestimates of the problems of lower socio-economic groups (Andersen *et al.* 1987).

Between-country comparisons show a striking difference in availability of health care for lower and higher income countries. Table 11.5 shows the population per physician and per nursing person for 1981 arrayed according to 1985 per caput

Table 11.7. Socio-economic status and measures of morbidity within developed countries

Year	Country/population	Type of morbidity	Measure of socio-economic status	Measure of morbidity
1964	US/all	Restricted activity days	Income	Number per person/year
			Poor	24.6
			Non-poor	14.1
1971–6	UK males	Limiting long-standing illness	Occupational class	Average rate per 1000 population
			Professional	79
			Managerial	119
			Intermediate	143
			Skilled manual	141
			Non-skilled manual	168
			Unskilled manual	236
1980	US/all	Bed disability days	Annual income ($)	Number per person/year
			Less than 3000	12.7
			3000–4999	14.5
			5000–6999	10.3
			7000–9999	8.6
			10 000–14 999	7.1
			15 000–24 999	5.5
			25 000 or more	4.5
1983	US/all	Reported health status	Annual family income ($)	Percentage fair or poor
			Less than 10 000	21.1
			10 000–14 999	13.7
			15 000–19 999	10.4
			20 000–34 999	6.9
			35 000 or more	4.6

Sources: Black *et al.* (1982), Health Resources and Services Administration (1986).

income in dollars (World Bank 1987) and Table 11.6 shows the percentage of the population with access to local health care for least developed, other developing, and developed nations (World Health Organization 1986*a*). Most dramatic is the range of one physician per 88 000 population in Ethiopia to one physician per 390 population in Switzerland.

The general conclusion from the available evidence is that poorer population groups have less access to health care services, and may not make good use of services that are available.

Culture and health

Culture appears to play an important role in health status. Culture is intimately related to accepted social practices, many of which are in turn related to health and disease.

Epidemiological evidence for the impact of culture on health comes from the ecological studies of diet and CHD, mentioned earlier, and immigrant studies of cardiovascular disease. The ecological studies show clear cultural differences in both dietary practices and cardiovascular disease consequences. Studies of Japanese men living in Japan and emigrating to California revealed CHD and stroke rates comparable to the country of residence only in subsequent generations, while those emigrating to Hawaii had intermediate rates (Kato *et al.* 1973; Keys *et al.* 1958). The implications are that as the Japanese became acculturated, they assumed both the dietary and cardiovascular patterns of the new country. This is a clear argument against the hypothesis that genetic factors have the dominant influence on heart disease, and for the hypothesis that cultural factors play a prominent role. Dietary factors appear to be powerfully influenced by culture (Rozin 1984).

The Roseta Study in Pennysylvania provides another example of the effects of culture on health. Although in Roseta the diet was more caloric and fatty than the typical American diet, and Rosetans were more obese and hypertensive than residents of a comparison community with similar smoking habits, mortality due to myocardial infarction was significantly lower in Roseta than in the comparison community (Lynn *et al.* 1967; Philips *et al.* 1981; Stout *et al.* 1964). In addition, the prevalence of myocardial infarction increased as native-born Rosetans moved away from the community (Bruhn 1965). The results were attributed to the apparent protective effect of a unique social, ethnic, and family cohesion in the ethnically homogeneous community (Bruhn *et al.* 1982; Philips *et al.* 1981).

Life-style and health—the merging of perspectives

Definition of life-style

Life-style has emerged as a concept in modern discourse to describe in shorthand what Madison Avenue advertisers call market segments—groups or types of people differentiated by a set of consumption patterns related to their income, education, occupation, gender, residence, and geopolitical and ethnic identification. This commercialization of the term is not totally unrelated to the social science origins of the concept. In the health field, however, the term has been used more variously to describe, at one extreme, discrete, narrowly defined behaviour related to chronic diseases or health enhancement with elements of intense individualism (discussed in Coreil and Levin 1984–5). At the other extreme the term has been used to describe the total social milieu including the 'psycho-socio-economic environment' as well as personal health behaviours (Hancock 1986).

As a behavioural concept, life-style generally implies more complex, repetitive (if not habitual) patterns of behaviour conditioned by living standards but still under the control of the individual or family within their economic means. The public health application of this behavioural notion of life-style has tended to associate it with 'health-related' (e.g. food consumption patterns) as distinct from 'health-directed' (e.g. diet) behaviour (Gottlieb and Green 1987; Steuart 1965). As a sociological and political concept, life-style refers more to a set of conditions that surround the social group, including their cultural history and socio-economic circumstances, but it is still the social behaviour of this social group that is the object of interest. The public health application of this notion of life-style has been to seek policies and environmental regulations that would redirect life-style or 'make healthy choices the easier choices' (World Health Organization 1986*b*).

The latter view of life-style is made more explicit in the formal attempt at definition by Nuttbeam (1985), writing for the European Regional Office of the World Health Organization: 'life-style is taken to mean a general way of living based on the interplay between living conditions in the wide sense, and individual patterns of behaviour as determined by socio-cultural factors and personal characteristics'. This definition seems to blend the behavioural perspective with the sociological one, giving prominence to the sociological: 'life-styles are usually considered in the context of both collective and individual experiences' (Nuttbeam 1985).

Relationship between life-style factors

The concept of life-style as the interplay between habitual behavioural patterns and socio-cultural conditions leads one to put into a broader context the reductionist examinations, presented in the first half of this chapter, of specific behaviours and specific measures of socio-economic status as they relate to health and disease. The dynamic interplay between the specific measures creates a complex and intricate system of social, economic, cultural, and behavioural factors, interwoven with disease risk factors and health status, influenced by the health care and physical environments. A simplified scheme of such a system is shown in Figure 11.1.

Although such a complex system is extremely difficult (if not impossible) to study directly, it can, and has, been studied in its parts. Evidence for its existence comes from the relationships presented earlier between specific health behaviours and health, socio-economic factors and health, and culture and health. In addition, there is evidence of relationships between specific health behaviours and socio-economic factors, and, although inconsistent, health behaviours to each other.

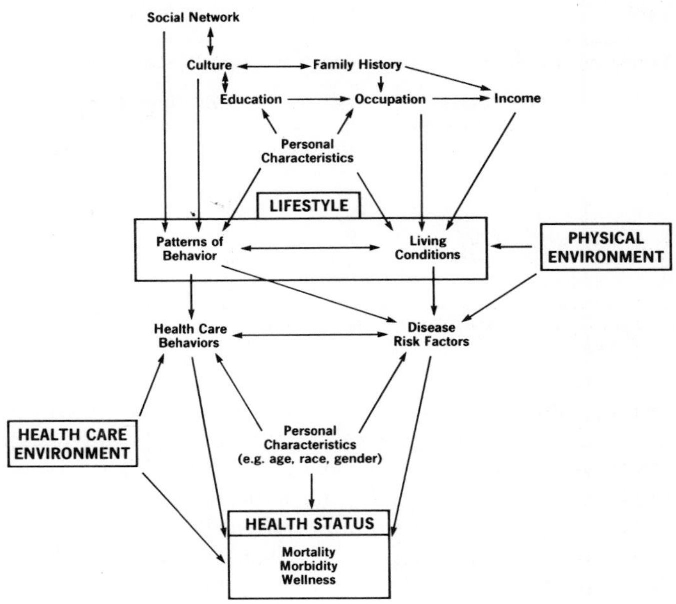

Fig. 11.1. Some interrelationships in the complex system of life-style, environment, and health status.

Numerous surveys have shown socio-economic differentials for health behaviours. One of the most recent, the Health Promotion and Disease Prevention section of the 1985 US National Health Survey (National Center for Health Statistics 1988*b*), showed relationships between numerous specific health behaviours and socio-economic status, measured both by years of education and family income. For 45 reported survey items of health behaviour and knowledge, 18 (40 per cent) showed a trend of increasing prevalence with increasing family income. Only one (2 per cent) showed a negative relationship, and the remainder (58 per cent) showed no obvious relationship (National Center for Health Statistics 1988*b*). Behaviours and knowledge associated with increasing income and higher education level included lower prevalence of obesity and cigarette smoking, greater exercise and sports participation, greater use of preventive health services for PAP smears and breast examinations, and greater use of seat-belts. Some of these gradients are shown in Table 11.8. A negative relationship was seen for increasing perceived stress with increasing income and education.

Other studies also have found associations between different social/educational and behavioural factors in the complex system. For example, the Alameda County Study in California found that the lower the socio-economic level, measured by income and educational level, the more likely the subject was to have high-risk health behaviour, measured by an index of health practices (Berkman *et al.* 1983). (The relationship was stronger among 30–39 year olds than among 60–69 year olds.) An analysis of the 1979 US National Survey of Personal Health Practices and Consequences showed that

Table 11.8. Selected health practices by family income, US, 1985

Family income ($)	Percentage of population aged 18 years and over			
	Current smokers	Regular physical activity	Regularly wear seat belts	Had a PAP smear in past year*
Less than 10 000	32.4	33.7	27.2	37.4
10 000–19 999	33.1	36.7	29.7	42.9
20 000–34 999	30.7	40.9	36.1	48.9
35 000–49 999	27.9	45.9	44.1	51.7
50 000 or more	23.5	52.5	51.8	52.6

Source: National Center for Health Statistics (1988*b*).
* Women.

income, education, and social support were positively related to a life-style health practices index (Gottlieb and Green 1984). In Norway, subjects in the Tromsø Heart Disease Study who had the highest education also were less overweight, smoked less, were more physically active, and had fewer atherogenic dietary habits than subjects with lower educational levels (Jacobsen and Thelle 1988).

Studies of the relationships of health behaviours to each other have been inconsistent in both their conduct and their findings. Studies have varied as to whether they examined preventive medical care (health care) behaviours (Steele and McBroom 1972), other preventive health behaviours as defined by the researchers (Green 1972; Langlie 1979), or health behaviours as defined as by the study subjects (Norman 1985). Some experts have stated that correlations, albeit low, exist between some preventive health care behaviours, although other preventive health actions are independent of each other (Kirscht 1983). Yet, some studies have shown significant positive correlations between various preventive health actions, including both medical care and non-medical care behaviours (Langlie 1979). It appears that the inclusion of income and/or education into the analysis increases the opportunity of finding a relationship. This implies that income and education are highly important components of the system.

It may well be that education is not only the most basic aspect of socio-economic status, but also of the complex system of health-related life-style. Education may be the aspect that can most influence the rest of the system. Although educational level is primarily a function of social, cultural, and economic circumstances, education to influence the life-style system is clearly an important aspect of public health practice.

Education for health

Functions of education

Education generally emerges from epidemiological studies as a powerful and pervasive correlate of health and of health-related behaviour (Green 1972; Pincus *et al.* 1987). It can be seen to influence behaviour in at least four ways: expanding opportunities for the individual; increasing knowledge of the world and the options it offers; building self-confidence; and increasing specific skills and capabilities. As education advances, so does the individual, the family, and the community on each of these dimensions of development. With advancement on these comes the fruits of personal, family, or community development. These fruits include improved health, reduced exposure to environmental threats to health, increased access to health resources, and purchasing power to buy primary health care and advanced medical care.

Channels of educational influence on health

The term education, like the term life-style, takes on various meanings depending on the context of its use. As a descriptive characteristic of individuals, it generally refers to years of formal schooling. As an epidemiological and demographic variable, it most typically serves as a surrogate measure of socio-economic status. As a family variable, education of the main earner often stands as an indicator of family's socio-economic status, although research generally shows that the education of the female head of household is more influential in determining family health and the health behaviour of other family members (Carmelli *et al.* 1986; Davis and Robinson 1988; Green 1972). Figure 11.2 shows the relationship between years of schooling of adults between 18 and 64 years of age in the US, their probability of having at least one chronic health problem, and their relative frequency of health conditions (Pincus *et al.* 1987).

Education is also used as a term to describe organizations, social institutions, and the status of communities. Whole sectors of the community may be broadly identified as educated, as in 'the education establishment', or 'higher education'. Education is also a function of most social institutions and departments of government. It is in this latter context that the term 'health education' is used to describe the educational function of health agencies, but health education also has a place in a variety of different sites such as schools, churches, workplaces, and recreational facilities. The term

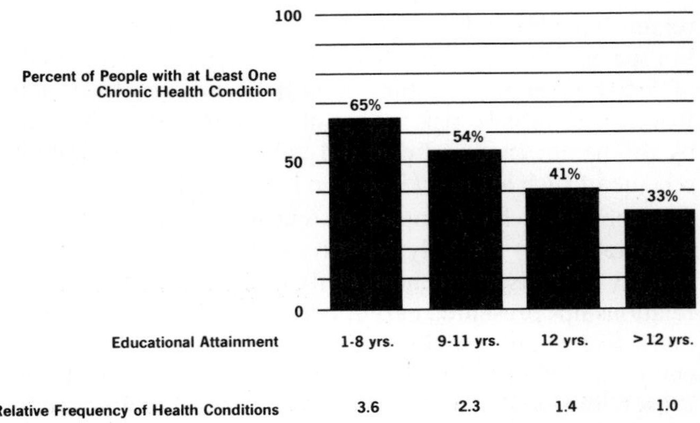

Fig. 11.2. Relationship between education and chronic health problems.

health education applies to a wide range of approaches and topics relevant to health, including, for example, education in basic hygiene for children, safety education for children and youth, parent education for young adults, and chronic disease prevention and management in later adulthood. The various forms of health education occur through a variety of channels, some institutional, some inter-personal.

Health education can be defined as the combination of planned learning experiences to facilitate voluntary actions conducive to health (Green *et al.* 1980). The actions may be of individuals to protect or promote their own health, of families to protect or promote the health of their members, and of organizational or community leaders and influential persons to change environmental conditions affecting health.

Education of those at risk of threats to their health

The assumption, well grounded by decades of experimental research, underlying most health education programmes for people whose health may be at risk is that, by providing people with some combination of new information, skill, and reinforcement, they will take actions to protect themselves or to improve themselves. Unfortunately, many of the behavioural changes people can make to help protect or improve their health are not inherently satisfying or reinforcing. Knowledge and skill in making the change in behaviour are usually easier to introduce through health education than is the reinforcement of the behaviour. In the clinic or the classroom, the health professional or teacher can reward the patient or the student with words of praise and with good marks. This is a critical component of counselling and of formal education. In the community, such direct reinforcement is more difficult to provide or for the individual to find. When community health education programmes depend on educational or communication opportunities through mediated channels, such as radio, television, or the written word, rather than through direct inter-personal contact, they can strengthen knowledge, beliefs, attitudes, and skills, but they are limited in their ability to reinforce behaviour.

Cognitive models of health education emphasize the informational and affective components of the educational process. The Health Belief Model (Becker 1974, 1990), for example, centres on three beliefs that account for most of the variance in predispositions to adopt a recommended health practice: belief in susceptibility (or belief that you could have the disease and not know it in the case of undertaking screening or treatment for conditions like hypertension); belief in severity of the consequences of not taking action; and belief that the benefits will outweigh the costs (including social benefits and costs such as inconvenience, discomfort, or embarrassment). This model has been widely tested and found to have predictive validity (Becker 1974, 1990).

Even the proponents of the Health Belief Model and other cognitive models that have proved useful in the development of health education interventions acknowledge that the task of behavioural change requires more than changing beliefs, attitudes, and perceptions. These factors may produce strong desires to change, but without skills and resources the highly motivated individual will only be frustrated. Frustration leads to a need to deny the prior motivation, which leads to rationalization or other defence mechanisms that erect a barrier to future attempts to convince the individual to change. Simplistic approaches to health education based entirely on information transfer can therefore backfire in two ways. First, a result worse than being ineffective in changing behaviour for many people is aroused expectations and motivation that lead to frustration and disappointment when the target behaviour proves out of reach. Second, they may set up defence mechanisms in the disappointed individuals that make subsequent attempts to reach them with health messages more difficult.

A more promising alternative to information-only approaches to health education is the combination approach referred to in the definition cited above. This approach recognizes that behaviour is complex and has multiple causes and sources of influence variously impinging on it. Motivation must be backed with skills and resources to enable the behavioural change and with rewards or social support to reinforce the behavioural change. These additional elements require community organization, environmental changes, and training of professionals, family members, employers, or others to provide social support for the behaviour.

Health behaviour in the home tends to be correlated with some other behaviour, and behaviours tend to change in patterned ways in relation to each other (Langlie 1979). Exercise, for example, tends to be correlated with other health habits, and when a group of people begin an exercise programme a proportion of them also take up changes in their diet and smoking and drinking habits. Such patterns suggest that knowledge, attitudes, and values associated with the education on one health behaviour generalize for many people to other health practices (Green *et al.* 1986).

Besides the rational model of education influencing cognitive predispositions, and enabling or reinforcing behaviour, an understanding of education influencing health through alternative routes that do not necessarily involve consistent behaviour changes has been suggested (Lorig and Laurin 1985). Education may influence the process of social support, which can have a direct influence on health without changing health behaviour (Berkman and Syme 1979; Nuckolls *et al.* 1972). Education can also increase self-confidence, self-image, or self-efficacy, any of which might have an independent effect on health with or without behavioural change (Ewart *et al.* 1983; Kaplan *et al.* 1984).

Education for community development

Education can serve health at yet another level through the community development process. Here the function of education can be seen as more circuitous. Education arouses interest and increases the consciousness of the public about local issues or problems. The public then seeks more information, participates more actively in debating the priority that should be given to the problem and to optional solutions, and watches more vigilantly the process of governmental or institutional response. All these ways in which the public becomes more active in participating in community affairs has a ripple effect on the community's ability to solve other problems more effectively (Green 1986). With an active population, public agencies tend to be more responsive, elected and appointed officials tend to be more sensitive to public needs, and community organizations tend to be more co-operative in working with each other than in communities where the public waits for governmental and other organizations to provide all the leadership on health matters (Cottrell 1976; Goeppinger and Baglioni 1985). Community development leads to better schooling, which results in improved levels of education in the community, which comes full circle to the functions of education for health described above.

Education to influence environmental conditions

In addition to education of people to influence their health behaviours and their socio-economic status, education also can be directed toward those people who have the power and authority to change the physical and health care environments—environments that can influence health directly or that can enable health behaviours.

Organizations, communities, and governments establish more or less healthful environmental conditions through policies, practices, facilities, and resources. Those conditions are influenced by decision-makers such as managers, department heads, and administrators within organizations; legislators, regulators, enforcers, and agency administrators within local, state, and federal government; and community leaders. Education of the decision-makers is a crucial avenue for facilitating healthful environmental change (Simons-Morton *et al.* 1988*a*).

Approaches to influencing organizations, communities, and governments are often called something else, but always contain educational components. Such approaches include organizational change, consulting, social and political action, community organization, persuasive communication, and political process (e.g. lobbying) (Greene and Simons-Morton 1984; Simons-Morton *et al.* 1988*a*).

Summary

Behaviour and life-style are crucial determinants to health, illness, disability, and premature mortality. Socio-economic status, particularly education, is a crucial determinant of behaviour and life-style. Education can be augmented through planned health educational interventions and, together with organizational, economic, and environmental supports, can contribute significantly to the protection and promotion of health.

Acknowledgement

We are indebted to Carol Holt for bibliographic assistance.

References

Aloia, J.F. (1981). Exercise and skeletal health. *Journal of the American Geriatrics Society* **29**, 104.

Ames, G. (1983). Dietary carcinogens and anticarcinogens. *Science* **221**, 1258.

Andersen, R.M., Mullner, R.M., and Cornelius, L.J. (1987). Black–white differences in health status: Method or substance? *Milbank Memorial Fund Quarterly* **65** (supplement 1), 72.

Antonovsky, A. (1967). Social class, life expectancy and overall mortality. *Milbank Memorial Fund Quarterly* **45**, 31.

Arber, S. (1987). Social class, non-employment, and chronic illness: Continuing the inequalities in health debate. *British Medical Journal* **294**, 1069.

Becker, M.H. (1974). The Health Belief Model and personal health behaviour. *Health Education Monographs* **2**, 324.

Becker, M.H. (1990). Theoretical models of adherence and strategies for improving adherence. In *The adoption and maintenance of behaviours for optimal health* (ed. S.A. Shumaker, E. Schron, J. Okene). Springer, New York. (In Press.)

Bennett, P.H. (1982). The epidemiology of diabetes mellitus. In *Diabetes mellitus and obesity* (ed. B.N. Brodoff and S.J. Bleicher), p. 387. Williams and Wilkins, Baltimore, Maryland.

Berkman, L.F. and Syme, S.L. (1979). Social networks, host resistance, and mortality: A nine year follow-up of Alameda County residents. *American Journal of Epidemiology* **109**, 186.

Berkman, L.F., Breslow, L., and Wingard, D. (1983). Health practices and mortality risk. In *Health and ways of living: The Alameda County study* (ed. L.F. Berkman and L. Breslow), p. 61. Oxford University Press, New York.

Biddle, B.J., Bank B.J., and Marlin, M.M. (1980). Social determinants of adolescent drinking. *Journal of Studies on Alcohol* **41**, 215.

Birch, L.L. (1980). Effects of peer models' food choices and eating behaviors on preschooler's food preferences. *Child Development* **51,** 489.

Birch, L.L., Zimmerman, S.I., and Hind, H. (1980). The influence of social–affective context on the formation of children's food preferences. *Child Development* **51,** 856.

Black, D., Morris, J.N., Smith C., and Townsend, P. (1982). *Inequalities in health: The Black Report* (ed. P. Townsend and N. Davidson). Penguin Books, Harmondsworth, Middlesex, England.

Blair, S.N., Goodyear, N.N., Gibbons, L.W., and Cooper, K.H. (1984). Physical fitness and incidence of hypertension in healthy normotensive men and women. *Journal of the American Medical Association* **252,** 487.

Breed, W. and DeFoe, J.R. (1984). Drinking and smoking on television, 1950–1982. *Journal of Public Health Policy* **5,** 257.

Brown, H.B. (1983). Diet and serum lipids: Controlled studies in the U.S. *Preventive Medicine* **12,** 103.

Bruhn, J.F. (1965). An epidemiological study of myocardial infarctions in an Italian-American community: A preliminary sociological study. *Journal of Chronic Disease* **18,** 353.

Bruhn, J.G., Philips, B.U., and Wolf, S. (1982). Lessons from Roseto 20 years later: A community study of heart disease. *Southern Medical Journal* **75**, 575.

Carmelli, D., Swan, G.E., and Rosenman, R.H. (1986). The relationship between wives' social and psychologic status and their husbands' coronary heart disease. *American Journal of Epidemiology* **122,** 90.

Coreil, J. and Levin, J. (1984–5). A critique of the life style concept in public health education. *International Quarterly of Community Health Education* **5,** 103.

Cottrell, L.S. (1976). The competent community. In *Further explorations in social psychiatry* (ed. B.H. Kaplan, R.N. Wilson, and A.H. Leighton), p. 195. Basic Books, New York.

Davis, N.J. and Robinson, R.V. (1988). Class identification of men and women in the 1970s and 1980s. *American Sociological Review* **53**, 103.

Dawber, T.R. (1980). *The Framingham Study: The epidemiology of atherosclerotic disease*. Harvard University Press, Cambridge, Massachusetts.

Dishman, R.K., Sallis, J.F., and Orenstein, D.R. (1985). The determinants of physical activity and exercise. *Public Health Reports* **100**, 158.

Duncan, J.J., Farr, J.E., Upton, J., Hagan, R.D., Oglesby, M.E., and Blair, S.N. (1985). The effects of aerobic exercise on plasma catecholamines and blood pressure in patients with mild essential hypertension. *Journal of the American Medical Association* **254**, 2609.

Duncan, O.D. (1961). Occupational components of educational differences in income. *Journal of the American Statistical Association* **56,** 783.

Ewart, C.K., Taylor, C.B., Reese, L.B., and DeBusk, R.F. (1983). The effects of early myocardial infarction exercise testing on self-perception and subsequent physical activity. *American Journal of Cardiology* **51,** 1076.

Fox, A.J., Goldblatt, P.O., and Jones, D.R. (1986). Social class mortality differentials: Artefact, selection, or life circumstances? In *Class and health: Research and longitudinal data* (ed. R.G. Wilkinson), p. 35. Tavistock Publications, London and New York.

Glynn, T.J. (1981). From family to peer: A review of transitions of influence among drug-using youth. *Journal of Youth and Adolescence* **10,** 363.

Goeppinger, J. and Baglioni Jr., A.J. (1985). Community competence: A positive approach to needs assessment. *American Journal of Community Psychology* **13,** 507.

Gottlieb, N.H. and Green, L.W. (1984). Life events, social network, life-style, and health: An analysis of the 1979 national survey of personal health practices and consequences. *Health Educationtion Quarterly* **11,** 91.

Gottlieb, N. and Green, L.W. (1987). Ethnicity and lifestyle health risk: Some possible mechanisms. *American Journal of Health Promotion* **2,** 37.

Grande, F. (1983). Diet and serum lipids–lipoproteins: Controlled studies in Europe. *Preventive Medicine* **12,** 110.

Green, L.W. (1972). *Status identity and preventive health behaviour*. Pacific Health Education Reports No. 1. University of California School of Public Health, Berkeley.

Green, L.W. (1986). The theory of participation: A qualitative analysis of its expression in national and international policies. In *Advances in health education and promotion* (ed. W.B. Ward, Z.T. Salisbury, S.B. Kar, and J.G. Zapka). Vol. 1, Part A, p. 211. JAI Press, Greenwich, Connecticut.

Green, L.W., Wilson, A., and Lovato, C.Y. (1986). What changes can health promotion produce and how long will they last? Trade-offs between expediency and durability. *Preventive Medicine* **15,** 508.

Green, L.W., Kreuter, M., Deeds, S.G., and Partridge, K.B. (1980). *Health education planning: A diagnostic approach*. Mayfield, Palo Alto, California.

Greene, W.H. and Simons-Morton B.G. (1984). Social change. In *Introduction to Health Education*, p. 193. Macmillan, New York.

Hancock, T. (1986). Lalonde and beyond: Looking back at 'A new perspective on the health of Canadians'. *Health Promotion: An International Journal* **1,** 93.

Haynes, R.B., Taylor, D.W., and Sackett, D.L. (1979). *Compliance in Health Care*. Johns Hopkins University Press, Baltimore, Maryland.

Hayward, R.A., Shapiro, M.F., Freeman, H.E., and Corey, C.R. (1988). Inequities in health services among insured Americans: Do working-age adults have less access to medical care than the elderly? *New England Journal of Medicine* **318**, 1507.

Health Resources and Services Administration (1986). *Health status of the disadvantaged: Chartbook 1986*. US Department of Health and Human Services, Public Health Services, DHHS Pub. No. HRS–P–DV86–2. US Government Printing Office, Washington, DC.

Horton, E.S. and Danforth, E. (1982). Energy metabolism and obesity. In *Diabetes mellitus and obesity* (ed. B.N. Brodoff and S.J. Bleicher), p. 261. Williams and Wilkins, Baltimore, Maryland.

Jacobsen, B.K. and Thelle, D.S. (1988). Risk factors of coronary heart disease and level of education: The Tromsø Heart Study. *American Journal of Epidemiology* **127**, 923.

Kannel, W.B., Doyle, J.F., Ostfeld, A.M., Jenkins, C.D., Kuller, L., and Podell, R.N. (1984). Optimal resources for primary prevention of atherosclerotic diseases: Report of the Inter-Society Commission for Heart Disease Resources. *Circulation* **70,** 155A.

Kaplan, R.M., Atkins, C.J., and Reinsch, S. (1984). Specific efficacy expectations mediate compliance in patients with COPD. *Health Psychology* **3,** 223.

Kato, H., Tillotson, J., Nichaman, M.Z., Rhoads, C.G., and Hamilton, H.B. (1973). Epidemiological studies of coronary heart disease and stroke in Japanese men living in Japan, Hawaii, and California. Serum lipids and diet. *American Journal of Epidemiology* **97**, 372.

Keys, A. (1970). Coronary heart disease in seven countries. *Circulation* **41** (Supplement 1), 1.

Keys, A., Kimura, N., Kusukawa, A., Bronte-Stewart, B., Larsen, N., and Keys, M.H. (1958). Lessons from serum cholesterol studies in Japan, Hawaii, and Los Angeles. *Annals of Internal Medicine* **48,** 83.

Kirscht, J.P. (1983). Preventive health behaviour: A review of research and issues. *Health Psychology* **2,** 277.

Kitagawa, E.M. and Hauser, P.M. (1973). *Differential mortality in the United States: A study in socioeconomic epidemiology*. Harvard University Press, Cambridge, Massachusetts.

Langlie, J.K. (1979). Interrelationships among preventive health behaviors: A test of competing hypotheses. *Public Health Reports* **94,** 216.

Levy, S.R., Iverson, B.K., and Walberg, H.J. (1983). Adolescent pregnancy programs and educational interventions: A research synthesis and review. *Journal of the Royal Society of Health* **3,** 99.

Lichtman, S. and Poser, E.G. (1983). The effects of exercise on mood and cognitive functioning. *Journal of Psychosomatic Research* **27,** 43.

Lipid Research Clinics Program (1984*a*). The Lipid Research Clinics Coronary Primary Prevention Trial results: I. Reduction in incidence of coronary heart disease. *Journal of the American Medical Association* **251**, 351.

Lipid Research Clinics Program (1984*b*). The Lipid Research Clinics Coronary Primary Prevention Trial results: II. The relationship of reduction in incidence of coronary heart disease to cholesterol lowering. *Journal of the American Medical Association* **251,** 356.

Liu, K., Cedres, L.B., Stamler, J., *et al.* (1982). Relationship of education to major risk factors and death from coronary heart disease, cardiovascular diseases and all causes: Findings of three Chicago epidemiologic studies. *Circulation* **66,** 1308.

Lorig, K. and Laurin, J. (1985). Some notions about assumptions underlying health education. *Health Education Quarterly* **12,** 231.

Lynn, T., Duncan, R., Naughton, J., Brandt, E.N., Wulff, J., and Wolf, S. (1967). Prevalence of evidence of prior myocardial infarction, hypertension, diabetes, and obesity in three neighboring communities in Pennsylvania. *American Journal of Medical Science* **254,** 385.

McCaul, K.D., Glasgow, R.E., O'Neill, H.K., Freeborn, V., and Rump, B.S. (1982). Predicting adolescent smoking. *Journal of School Health* **52,** 342.

McGill, H.C. (1979). The relationship of dietary cholesterol to serum cholesterol concentration and to atherosclerosis in man. *American Journal of Clinical Nutrition* **32,** 2664.

Manton, K.G., Patrick, C.H., and Johnson, K.W. (1987). Health differentials between Blacks and Whites: Recent trends in mortality and morbidity. *Milbank Quarterly* **65** (Supplement 1), 129.

Marmot, M.G. and McDowall, M.E. (1986). Mortality decline and widening social inequalities. *Lancet* **ii,** 274.

Marmot, M.G., Adelstein, A.M., Robinson, N., and Rose, G.A. (1978). Changing social-class distribution of heart disease. *British Medical Journal* **2,** 1109.

Marmot, M.G., Shipley, M.J., and Rose, G.A. (1984). Inequalities in death—Specific explanations of a general pattern. *Lancet* **i**: 1003.

Mierley, M.C. and Baker, S.P. (1983). Fatal house fires in an urban population. *Journal of the American Medical Association* **249,** 1466.

Morris, J.N. (1979). Social inequalities undiminished. *Lancet* **i**, 87.

Mullen, P.D., Green, L.W., and Persinger, G.S. (1985). Clinical trials of patient education for chronic conditions: A comparative meta-analysis of intervention types. *Preventive Medicine* **14,** 753.

National Center for Health Statistics (1988*a*). *Vital Statistics of the U.S., 1985, Volume II—Mortality, Part A.* DHHS Pub. No. (PHS) 88–1101. US Government Printing Office, Washington, DC.

National Center for Health Statistics (1988*b*). *Health promotion and disease prevention: United States 1985.* Public Health Service, Vital and Health Statistics Series 10, No. 163, DHHS Pub. No. (PHS) 88–1591. US Government Printing Office, Washington, DC.

National Institute on Alcohol Abuse and Alcoholism (1983). *Fifth special report to the US Congress on alcohol and health.* DHHS Pub. No. (ADM) 84–1291. US Government Printing Office, Washington, DC.

National Institute on Alcohol Abuse and Alcoholism (1986). Safety update: Drunk driver fatalities declining. *Alcohol Health and Research World* **10,** 78.

National Institute on Alcohol Abuse and Alcoholism (1987). *Sixth report to the US Congress on alcohol and health.* DHHS Pub. No. (ADM) 87–1519. US Dept. of Health and Human Services, Rockville, Maryland.

Nelkin, D. (1987). AIDS and the social sciences: Review of useful knowledge and research needs. *Review of Infectious Diseases* **4,** 980.

Norman, R.M.G. (1985). Studies of the interrelationships amongst health behaviors. *Canadian Journal of Public Health* **76**, 407.

Nuckolls, K.B., Cassels, J., and Kaplan, B.H. (1972). Psychosocial assets, life crisis and the prognosis of pregnancy. *American Journal of Epidemiology* **95,** 431.

Nuttbeam, D. (1985). *Health promotion glossary*. World Health Organization Regional Office for Europe, Copenhagen.

Office on Smoking and Health. US Department of Health and Human Services (Public Health Service) (1980). *The health consequences of smoking for women.* US Government Printing Office, Washington, DC.

Office on Smoking and Health. US Department of Health and Human Services (Public Health Service) (1982). *The health consequences of smoking: Cancer*. DHHS Pub. No. (PHS) 82–50179. US Government Printing Office, Washington, DC.

Office on Smoking and Health. US Department of Health and Human Services (Public Health Service) (1983). *The health consequences of smoking: Cardiovascular disease.* DHHS Pub. No. (PHS) 84–50204. US Government Printing Office, Washington, DC.

Office on Smoking and Health. US Department of Health and Human Services (Public Health Service) (1984). *The health consequences of smoking: Chronic obstructive lung disease.* DHHS Pub. No. (PHS) 84–50205. US Government Printing Office, Washington, DC.

Oldridge, N.B. (1982). Compliance and exercise in primary and secondary prevention of coronary heart disease: A review. *Preventive Medicine* **11,** 56.

Oldridge, N.B. (1984). Adherence to adult exercise fitness programs. In *Behavioral health: A handbook of health enhancement and disease prevention* (ed. J.D. Matarazzo S.M. Weiss, J.A. Herd, N.E. Miller, and S.M. Weiss), p. 467. John Wiley and Sons, New York.

Philips, B.U., Bruhn, J.F., and Wolf, S. (1981). Smoking habits and reported illness in two communities with different systems of social support. *Social Science and Medicine* **15A**, 625.

Pincus, T., Callahan, L.F., and Burkhauser, R.V. (1987). Most chronic diseases are reported more frequently by individuals with fewer than 12 years of formal education in the age 18–64 United States population. *Journal of Chronic Disease* **40,** 865.

Pocock, S.J., Shaper, A.G., Cook, D.G., Phillips, A.N, and Walker, M. (1987). Social class differences in ischaemic heart disease in British men. *Lancet* **ii**, 197.

Pooling Project Research Group (1978). Relationship of blood pressure, serum cholesterol, smoking habit, relative weight and ECG abnormalities to incidence of major coronary events: Final report of the Pooling Project. *Journal of Chronic Disease* **31,** 201.

Powell, K.E., Thompson, P.D., Caspersen, C.J., and Kendrick, T.S. (1987). Physical activity and the incidence of coronary heart disease. *Annual Review of Public Health* **8,** 253.

Ravenholt, R.T. (1984). Addiction mortality in the United States, 1980: Tobacco, alcohol, and other substances. *Population and Development Review* **10,** 687.

Robinson, D. and Pinch, S. (1987). A geographical analysis of the relationship between early childhood death and socio-economic environment in an English city. *Social Science and Medicine* **25,** 9.

Rose, G., Hamilton, P.J.S., Colwell, L., and Shipley, M.J. (1982). A randomized controlled trial of anti-smoking advice: Ten-year results. *Journal of Epidemiology and Community Health* **36,** 102.

Rouwenhorst, W. (1983). *Leren Gezond te Zijn?* (Learning to be Healthy?). Walters-Noordhoff, Gronigen, The Netherlands.

Rozin, P. (1984). The acquisition of food habits and preferences. In *Behavioral health: A handbook of health enhancement and disease prevention* (ed. J.D. Matarazzo, S.M. Weiss, J.A. Herd, N.E. Miller, and S.M. Weiss), p. 590. Wiley, New York.

Shah, C.P., Kahan, M., and Krauser, J. (1987). The health of children of low-income families. *Canadian Medical Association Journal* **137,** 485.

Shai, D. (1986). Cancer mortality, ethnicity, and socioeconomic status: Two New York City groups. *Public Health Reports* **101,** 547.

Simons-Morton, B.G., O'Hara, N.M., and Simons-Morton, D.G. (1986). Promoting healthful diet and exercise behaviors in communities, schools, and families. *Family and Community Health* **9,** 1.

Simons-Morton, B.G., Brink, S.G., Parcel, G.S., *et al.* (1990). *Preventing acute alcohol-related health problems in adolescents and young adults*. Centers for Disease Control, Atlanta, Georgia.

Simons-Morton, D.G., Simons-Morton, B.G., Parcel, G.S., and Bunker, J.F. (1988*a*). Influencing personal and environmental conditions for community health: A multilevel intervention model. *Family and Community Health*, **11,** 25.

Simons-Morton, D.G., Brink, S.G., Parcel, G.S., Tiernan, K.M., Harvey, C.M., and Longoria, J.M. (1988*b*). *Promoting physical activity among adults: A CDC community intervention handbook*. Centers for Disease Control, Atlanta, Georgia.

Simons-Morton, D.G., Parcel, G.S., Brink, S.G., Tiernan, K.M., and Harvey, C.M. (1990). Smoking control among women: Needs assessment and intervention strategies. *Advances in health education and promotion*. Jessica Kingsley Publishers, London. (In press.)

Smith, E.L., Reddan, W., and Smith, P.E. (1981). Physical activity and calcium modalities for bone mineral increase in aged women. *Medicine and Science in Sports and Exercise* **13,** 60.

Spurlock, C.W., Hinds, M.W., Skaggs, J.W., and Hernandez, C.E. (1987). Infant death rates among the poor and nonpoor in Kentucky, 1982 to 1983. *Pediatrics* **80**, 262.

Stallones, R.A. (1983). Ischemic heart disease and lipids in blood and diet. *Annual Review of Nutrition* **3,** 155.

Steele, J. and McBroom, W. (1972). Conceptual and empirical dimensions of health behavior. *Journal of Health and Social Behavior* **13,** 382.

Steuart, G.W. (1965). Health, behavior, and planned change. *Health Education Monographs* **20,** 3.

Stout, C., Morrow, J., Brandt, E.N., and Wolf, S. (1964). Unusually low incidence of death from myocardial infarction in an Italian-American community in Pennsylvania. *Journal of the American Medical Association* **188,** 845.

Taylor, C.B., Sallis, J.F., and Needle, R. (1985). The relation of physical activity and exercise to mental health. *Public Health Reports* **100,** 195.

Taylor, W.T.L. (1987). *Market research for Australia's national AIDS education campaign.* Paper presented at the Third International Conference on AIDS, June 1987, Washington, DC.

US Department of Health, Education and Welfare (1979). *Healthy people: The Surgeon General's report on health promotion and disease prevention.* US Government Printing Office, Washington, DC.

US Department of Health and Human Services (1980). *Promoting health/preventing disease: Objectives for the nation.* US Government Printing Office, Washington, DC.

US Department of Health and Human Services (1986). *Health status of the disadvantaged: Chartbook 1986.* DHHS Pub. No. (HRSA) HRS–PD–V86–2. US Government Printing Office, Washington, DC.

Vincent, M., Clearie, A., and Schluchter, M. (1987). Reducing adolescent pregnancy through school and community-based education. *Journal of the American Medical Association* **257,** 3382.

Warner, K.E. (1986*a*). *Selling smoke: Cigarette advertising and public health.* American Public Health Association, Washington, DC.

Warner K.E. (1986*b*). Smoking and health implications of a change in federal cigarette excise tax. *Journal of the American Medical Association* **255,** 1028.

Willett, W.C. and MacMahon, B. (1984). Diet and cancer: An overview. *New England Journal of Medicine* **310,** 633 and 697.

World Bank (1987). *World development report 1987.* Oxford University Press, New York.

World Health Organization, Executive Board (1979). *Formulating strategies for health for all by the year 2000.* World Health Organization, Geneva.

World Health Organization (1981). *Global strategy for health for all by the year 2000.* World Health Organization, Geneva.

World Health Organization (1986*a*). *1986 world health statistics annual.* World Health Organization, Geneva.

World Health Organization, European Regional Office (1986*b*). *Health promotion concepts and principles in action. A policy framework.* WHO Regional Office for Europe, Copenhagen.

Zarek, D., Hawkins, J.D., and Rogers, P.D. (1987). Risk factors for adolescent substance abuse. *Pediatric Clinics of North America* **34,** 481.

12

Medical care and public health

M. HOBBS and K. JAMROZIK

Introduction

This chapter examines the role of medical care services as a determinant of the level of health experienced in a particular community. The discussion is deliberately limited to personal medical services that are delivered in a one-to-one setting. As well as emergency medical or surgical care, rehabilitation, and palliative measures, such personal medical services can include a range of preventive activities such as health education, immunization, and family planning. In considering the entire range of activities designed to maintain or improve the health of a given population, a distinction is drawn between personal medical care services of the type described and, on one hand, measures for protecting health such as provision of clean water, waste disposal, quarantine, and food standards, and, on the other hand, community-wide health promotion activities such as advertising campaigns or creation of public facilities for exercise. Clearly, proper control of certain health problems, such as infectious diseases, requires a co-ordinated approach containing elements of each of health protection, health promotion, and personal medical care services. In practice, however, these three different activities are often overseen by separate parts of the health service or even by separate governmental departments, with this division of responsibility being maintained all the way up from the local to the national level. While activities in one sector can have a profound effect on the need for services provided by the other two, their budgets, priorities, and programmes are usually planned independently and guarded jealously. It is therefore meaningful to examine the effect on the health of the population of personal medical services, even if this fragmented approach is somewhat remote from the ideal one.

Since medical care aims to improve the health of individuals and populations, how best to measure health is a matter of central importance. The basic tools and a number of approaches to the measurement at least of ill-health are reviewed before the discussion moves to the related and often equally difficult task of evaluating the effects even of single, readily identifiable components within complex systems of medical care. Different conceptual paths to solving this puzzle are identified, after which we embark upon a series of case studies of particular medical care activities. Most of the examples are drawn from Western Australia but they are intended to illustrate general issues that are common to many settings. Thus, the activities that we have chosen range from preventive to palliative in therapeutic intent, and from prenatal to post-retirement in target age group. Some appear to pay an obvious and direct dividend in terms of benefit on health, but it is less easy to be certain whether others do justify the corresponding investments.

Historical background

Western industrialized nations invested heavily in health protection long before medical care services began to develop into the highly organized and elaborate systems that we see today. While it is true that hospitals and various kinds of individual practitioners have been providing personal medical care services in these countries for many years, quarantine regulations date back at least to the fifteenth century (Burnet and White 1972) and the state became embroiled in the campaign to provide basic water supplies and sanitation at the beginning of the nineteenth century. In the UK, governments had provided hospitals for the poor but state involvement in provision of universal personal medical care services is usually dated to the inception of systematic antenatal care in the early 1900s.

The relationship between declining mortality in the nineteenth and early twentieth centuries and medical care measures has been extensively reviewed by McKeown (1965, 1979). He provides convincing evidence that improvements in health as measured by falling mortality rates during that time had little to do with personal medical care. We have therefore restricted our examination of the effects of medical care on health to the last 50 years, the era of 'scientific medicine'.

Throughout this period, there have been progressive improvements in a number of key areas of medical care. Immunization has led to the eradication of smallpox worldwide (Fenner 1980) and, in the developed countries, the virtual control of poliomyelitis, diphtheria, and tetanus.

Substantial progress has been made towards reduction in morbidity from measles and there is hope that rubella and consequent congenital abnormalities will be controlled. Through the exigencies of two World Wars and several other major conflicts substantial improvements have been made in the management of injury, and in restorative surgery and rehabilitation. In the field of reproduction, large advances have occurred in the control of fertility and in obstetric and neonatal care; it is now possible to contemplate radical programmes for population control as part of public health. Modern anaesthesia and parenteral therapy have greatly increased the safety of emergency surgery and have extended the scope of corrective surgery. New drugs have led to the successful treatment of many bacterial conditions and have also contributed significantly to improved prognosis in surgery for accidents and emergencies. Certain malignant tumours of childhood are now curable and substantial relief, if not cure, can be provided for many of the common cancers of adults and for psychoses, epilepsy, diabetes, ischaemic heart disease, and some forms of arthritis. New prosthetic devices have opened the door to corrective surgery for conditions previously associated with major disability and handicap. Advances in methods for diagnosis have supported many of these developments in treatment.

In parallel with technical advances there have been major changes in the philosophy of provision of health care and in the structure of health systems. In most developed countries the concept of universal access to essential medical care, regardless of means, has been accepted. Even though systems of payment differ, most health care is heavily subsidized by governments and the principle of charity no longer applies (Kohn and White 1976; Maxwell 1981; Roemer and Roemer 1981).

Hospital and professional structures have changed to reflect increases in specialized knowledge in medicine. This has led to improvement in technical care but at the expense of holistic care. On the other hand, the shortcomings of the past in relation to the care of the chronic sick are now becoming recognized and are slowly being redressed. For example, geriatric medicine, pioneered in the UK, has greatly increased insights into the care of the disabled aged (Brocklehurst 1975). New concepts for multi-disciplinary services have been developed and put into practice in both hospital and community settings. The hospice movement has drawn attention to the relief of pain and discomfort as legitimate primary objectives in health care. Not only has the scope of medical care been broadened, but access to medical services has been greatly increased.

Evaluation of the contribution of medical care services to health

That the advances in medical care described above have made an important contribution to health would appear to be self-evident to many people. And yet this has been seriously questioned by some writers. The most radical views have been expressed by Illich (1975), who suggests that apart from the direct iatrogenic consequences of some medical treatments, the overall approach of the medical and allied professions to health issues—what they define as health 'problems' and the ways they think about and discuss them—has broader negative repercussions for both individual patients and society at large. These views are probably shared by only a small minority, but other observers, including McKeown (1979) and Cochrane (1971), have also drawn attention to the fallacies inherent in some beliefs about the contribution of medical care to health. McKeown has argued that much of the fall in mortality during the nineteenth century was due to social changes and, therefore, that any further improvements in health seen now should not automatically be attributed to modern medical care. Cochrane was more concerned to show that the benefit of much medical care activity is unproven and should be carefully reappraised, partly to ensure that unsafe procedures are discontinued and partly to make certain that resources for medical care are employed in the most effective and efficient ways. What these writers do is to broaden our focus from direct benefits of health care to consider also indirect effects, adverse effects as well as beneficial effects, and social effects in groups as well as biological effects in individual patients.

Health services which are at all effective can have profound effects at the population level. For example, slowing population growth in underdeveloped countries by successful control of fertility may lead to major improvements in the social environment in favour of health. In contrast, declining mortality without an associated reduction in fertility can lead to rapid population growth, which by overtaxing the social environment is eventually counter-productive as far as health is concerned. Concomitant falls in fertility and mortality, while stabilizing population growth, will also produce the type of demographic transition observed in industrial countries which will in turn produce entirely new patterns of morbidity, with relative increases in the dependent aged and reductions in the dependent young.

As far as direct adverse effects are concerned, the rapid development of new methods of treatment has produced many dramatic and salutatory lessons which must caution us against unqualified enthusiasm for the presumed benefits of modern medicine. While thalidomide is perhaps the landmark example, there are unresolved questions pertaining to many other drugs including oral contraceptives, oral hypoglycaemics and diethylstilboestrol, drugs to which many people world-wide are or have been exposed, and to procedures such as electroconvulsive therapy.

Conceptual problems in evaluation of health care

There are major conceptual and technical difficulties in determining the contribution of medical care to health. We have noted that the task of medical care is to contribute to health through prevention, cure or mitigation of disease. When discussing the contribution of medical care to health we need to specify which of the above functions is in question. The most fundamental problems, however, are difficulties in defining and measuring need for medical care and in devising suitable measures of its effects. First, it is obvious that benefits of health care can be determined only if the size and nature of the health problem in question can be clearly defined. In other words, a baseline must be established against which changes in frequency or severity of disease can be measured. Much has been written about the need for more satisfactory and comprehensive measures of health status that could be used to define specific needs for health care and to determine the impact of various types of care at the population level (Bergner and Rothman 1987; Fingerhut *et al.* 1980; Hunt *et al.* 1986; Murnaghan 1981). The volume of this literature is in itself testimony to the difficulties involved and to the fact that answers to the problem remain elusive. Second, because of the diversity of the medical task, there can be no single index or convenient set of indices that can adequately summarize health needs or encompass all health outcomes. Third, while it is possible to examine disease outcomes in biological terms it is difficult to measure the social outcomes or benefits.

Biological states that can be measured and classified as disease do not necessarily correspond to a subjective experience of illness or to a socially acknowledged sick role (Hunt *et al.* 1986). Thus an important issue in assessing health outcomes is the assessment of personal benefit—the extent to which the change produced by treatment is valued by the subject. Here it is essential that we recognize the difference between extension of life and its quality. This is necessarily a matter of subjective judgement, but whose judgement?

We must also consider social outcomes. As discussed by Butler and Vaile (1984) there are again several aspects to this. For example, do we consider benefits in terms of the productive capacity of the individual, the impact of disability on families, or the opportunity costs of expending resources on this type of care rather than on some other service which might yield greater benefit overall to the community? Thus, although we might regard the case for particular preventive and acute care programmes of proven effectiveness beyond doubt, this may not be so if the opportunity costs are high.

Another difficulty is that medical care is usually provided by generic services. The objectives of such services are not explicitly defined in terms of prevention, cure, or care and it is therefore difficult to identify, let alone quantify, the components of medical care that are devoted to each of these functions, or, as health economists would wish, to apportion costs to them. Quite different measures are required to assess the impact of a preventive programme (measured by the absence of disease), of acute care as in the management of accidents (not only reduction of mortality but also absence of residual disability), or the management of chronic illness (prolongation of life in some instances and also minimization of pain and restriction of function).

Problems are also encountered in moving from a consideration of the benefits of specific treatments for specific diseases to the aggregate benefit of medical services applied to the whole population. Thus it is common experience that many diseases have been reduced or eliminated through immunization, that persons who are injured benefit from injury services, that obstetric care has minimized the likelihood of maternal and perinatal mortality or morbidity, that surgery for accidents and particular abdominal emergencies may be life-saving, and that drugs developed in the past 50 years offer the prospect of a cure or substantial relief of symptoms in many illnesses. However, the lack of valid measures of outcome and difficulty in controlling for confounding frequently make the formal evaluation of medical care services exceedingly difficult in practice. Yet without some guide to desirable levels of service provision or indicators of performance, the rational management of health services would be impossible. Proxy measures of outcome are therefore frequently used. Arising from the classic paper of Donabedian (1966), structural and process measures have also become widely accepted as alternatives to measures of outcome to assess the performance of health services. It is assumed that the 'better' or 'worse' their performance, the greater or smaller the contribution of health services to health is likely to be.

Ultimately, the use of structural and process variables to assess medical care is likely to yield misleading conclusions if some nexus between them and the effectiveness of care is not established. But this has rarely been done, and various service norms in common use have been determined mainly by historical precedent rather than through explicit rational processes. Uncritical acceptance of measures of structure and process carries dangers because of the complacency that it may engender in removing the apparent need for true measures of outcome. The use of such measures is reviewed in detail in Vol. 2, Chapters 1–6 and Vol. 3, Chapter 9.

At the population level it is difficult to sum the benefits (and in some instances the adverse effects) of all aspects of medical care. Apart from the magnitude of this task there are other factors to consider. For example, it may not be possible to generalize from the benefits of specific treatments evident through randomized controlled trials or well-conducted observational studies because the subjects of such research may be more compliant with treatment than are patients in normal practice. In addition, the treatment regimens used in trials are necessarily restricted and may not reflect everyday methods. There are also structural reasons why the benefit of therapy obtained under optimum conditions of supervision (termed by some writers as 'efficacy') is not achieved in practice. For various reasons access to medical care may be restricted, or follow-up and discussion with patients may not be optimal. Moreover, in many instances the relevant clinical trials simply have not been done. Instead, attempts have been made to use routinely collected data on health and disease to assess the impact of various medical care activities.

But we must acknowledge that the readily available data that have been used in this way were not specifically intended for this function and may therefore be misleading.

Data used to assess benefits of medical care

Any rational approach to the assessment of the impact of medical care on health requires that we have basic knowledge about the incidence, prevalence, and natural history of the problems that collectively comprise ill-health. Unfortunately our understanding of these aspects of ill-health is often rudimentary.

We have already suggested that one of the problems of measuring the benefits of medical care lies in the fact that most readily available data relating to health or medical services were intended for other purposes. The type of data generally available for monitoring health and the performance of health services are reviewed in Vol. 2, Chapters 1–6 and Vol. 3 Chapter 9. Here we review briefly the main classes of data in relation to their possible use in assessing the contribution of medical care to health.

Mortality data

Mortality data have the great advantages that they are collected in a relatively uniform way in different countries and are available for the whole period in which effective medical care has evolved. Such data are therefore of central importance in making broad comparisons of health status between countries and between different periods of this century. However, mortality data do have some shortcomings as indicators of the state of health of populations. To begin with, they do not reflect the need for medical care associated with conditions, such as mental illness, which cause significant morbidity but are infrequent causes of death. Mortality data are also subject to variations caused through changes in diagnostic practice and through changes in coding and classification procedures so that, at the level of specific diseases, it is often not possible to provide valid trends over time or to make comparisons between countries.

The use of mortality data to assess the impact of changes in medical care is subject to special difficulties in the area of confounding. As the general economic situation of a country improves and progressively more money per head of population can be invested in health services, so it is likely that the average standard of living, including housing and nutrition, also improves. It then becomes impossible, *post factum*, to ascribe any improvement in overall mortality rates to increases in total expenditure on health because changes in the other likely determinants of general health usually have not been monitored and because their actual effects on health and mortality are indirect and very difficult to measure.

The potential importance of confounding is exemplified by comparisons of trends in perinatal mortality and in mortality rates from ischaemic heart disease in developed countries over the last two to three decades as discussed below.

Morbidity data

The universality, historical continuity, and comparability across populations of morbidity data are all limited. Compared with mortality data, the quality of morbidity data is suspect as it is dependent firstly on the quality of medical records, and, secondly, the records themselves tend to be coded and processed at several locations, whereas in most countries the coding of mortality data is relatively centralized and subject to greater quality control. Even so, morbidity data have the potential to fill in some of the gaps in mortality data, although, like mortality data, they do present some difficulties in use.

The most commonly used measures of morbidity such as hospital statistical collections or data generated from general practice are in fact measures of service utilization rather than measures of the level of illness in the community. As such they may obviously be affected by factors that are quite independent of true need, such as changes in clinical practice, availability of resources, and general factors that affect access to medical care of all kinds.

Morbidity data frequently measure episodes of care or occasions of consultation rather than persons treated. Record linkage provides a means of overcoming this, providing person-based statistics relating to hospital use, while cross-linkage with mortality data allows study of survival following admission to hospital for selected conditions or surgical procedures (Acheson 1967; Goldacre 1987; Golding *et al.* 1987). Opportunities for this type of analysis are extremely limited but the use of morbidity data as one source of information for disease registers, and as a starting point for *ad hoc* studies of the effect of medical care should be kept in mind. In Perth, Australia, for example, linked morbidity and mortality data have been used to study trends in incidence and survival in acute coronary disease, parameters which reveal much about the medical care received both by persons at risk of developing ischaemic heart disease and by those who have developed symptoms (Hobbs *et al.* 1984; Martin *et al.* 1984, 1987*a*, 1987*b*).

The technical problems relating to consistent capture and coding of data from general practice are even greater than in the case of hospital morbidity. In general it is difficult to recruit general practitioners to continuing studies, and compliance with collection of data is likely to be low. Most studies of illness or treatment using general practice records have therefore been based on selected practices which may not be fully representative. It follows that general practice data of most value derive from special broad-based but time-limited studies which have specific goals such as the study of adverse effects of oral contraceptives undertaken by the Royal College of General Practitioners (Royal College of General Practitioners 1974). General practice data have other important potential uses in assessing the effects of and in improving health care. Given appropriate records it is at least possible to measure the proportion of the target group which has received a particular service and this is an important step on the way to determining the impact on health of that medical care activity.

Community surveys

Intensive community studies such as Framingham (Dawber 1980; Dawber *et al.* 1951), North Karelia (Puska *et al.* 1985) and Busselton (Curnow *et al.* 1969), which measure the incidence and prevalence of specific health problems, provide a basis for detailed epidemiological studies of various forms of health care intervention. However, most studies of this type are necessarily restricted to a narrow range of causes of ill-health and are therefore of limited use in assessing the impact of medical care on a wider front.

More general sample surveys of illness and disability in the community such as the health component of the General Household Survey in the UK and National Health and Nutrition Examination Survey in the US have provided insight into levels of treated and untreated sickness and disability in the community as well as into illness behaviour. However, the types of questions that can be asked in surveys using non-medical interviewers are limited, with the result that the information obtained may be too general for the purpose of relating the effects of medical care to the health of the community. Changes in the survey instruments over time and variations in response according to various demographic characteristics, including age, ethnic group, and economic and educational status, may make interpretation of trends and geographical differences difficult. The problems of using various measures of health status in the US for this purpose have been extensively reviewed by Wilson and Drury (1984).

Disease registers

Properly designed, operated, and exploited disease registers are far more than elaborate counting exercises (Clemmesen 1978; Horwitz 1978; Keil 1978; Tunstall-Pedoe 1985). They can be used for evaluation of both process and outcome, and to monitor standards of care. Kessner *et al.* (1973) proposed that suitable sentinel or tracer conditions would be selected on grounds of the magnitude of their functional impact, ease of definition and diagnosis, and prevalence. Further factors to be considered might include demonstrated responsiveness to medical care, general consensus on methods of prevention or treatment, and understanding of the effects of non-medical treatment on the natural history of the particular disease. Possible tracer conditions suggested by the authors included middle ear infections and hearing loss, iron deficiency anaemia, hypertension, urinary tract infection, and cervical cancer. Trends in incidence and response to treatment of such diseases can provide important pointers to risk factors and changes in exposure to risk factors, and to the impact of health services designed either to prevent the disease developing in those who are at risk or to limit its effects once the disease has become clinically evident.

By extension, a great deal can be learnt about the operation and performance of medical care services in general, particularly if a suitable range of sentinel conditions is included in a network of disease registers. A combination of registers for congenital malformations, selected injuries, cardiovascular disease, and cancer, for instance, would yield insights into antenatal and perinatal care, the management of chronic conditions in early and middle life, the treatment and long-term outcomes following trauma, environmental hygiene, and diagnostic, treatment, and palliative services for patients with malignant disease. If coupled with surveillance of suitable infectious diseases such as one or more immunizable diseases of childhood, sexually transmitted diseases including acquired immune deficiency syndrome, and tropical diseases that are rarely seen in temperate climates, one would have a system of registers that not only encompassed many of the major causes of mortality and morbidity in developed countries, but also allowed the performance of various preventive activities to be monitored. The list of conditions to be covered obviously could be extended. But it does serve to show that an imaginative combination of disease registers could provide much information on the overall performance of a wide range of health services as well as producing intrinsically interesting descriptive information about each of the individual health problems.

Approaches used in assessing the contribution of medical care to health

As we have indicated, there are major conceptual and technical problems in assessing the health of populations and hence the contribution of medical care services to this. Nevertheless, a number of different approaches which provide partial answers is possible.

Global approaches using mortality data

The work of McKeown (1965, 1979), which examined the possible reasons for falling mortality rates in the UK since the early nineteenth century, has greatly increased awareness of the issues surrounding the use of mortality data to assess effects of medical care on health. He has demonstrated that the major proportion of the improvement in mortality was probably due to improvements in nutrition, water supply, and housing with relatively minor contributions from specific medical interventions, including immunization. On the basis of earlier trends, McKeown suggests that at least some part of the continuing decline in mortality following the development of effective medical care would have still occurred without the advent of scientific medicine. The fact that mortality rates in middle aged and elderly men demonstrated little improvement during most of the twentieth century, even after effective medical measures became available, has also been cited as evidence that medical care has contributed little to the decline in mortality.

However, as suggested by Levine *et al.* (1983), further sharp declines in mortality rates in males and females of all ages which have occurred recently in some countries point to a need to review this interpretation. First, it must be noted that mortality in females in several countries including Australia has continued to improve throughout most of the twentieth century. Since 1900, expectation of life in Australian females has increased at all ages but the trend has accelerated in the past 15 years. Expectation of life at birth also increased steadily in Australian males, albeit at a slower rate than in females. Moreover between 1932–4 and 1974, as in other developed countries, there was no improvement in male life expectancy at the age of 60 years. Since then, expectancy of life at all ages in males has increased at a rate comparable with that in females (Australian Bureau of Statistics Life Tables). Figures 12.1 and 12.2 summarize the percentage decline in the risk of dying between selected ages in Australia since the 1930s, a period generally recognized for increases in the effectiveness of medical care.

Second, Morris (1967) has shown that mortality in men in England and Wales would have declined continuously if there had not been increases in mortality from lung cancer and ischaemic heart disease of such proportions as to swamp an underlying downward trend in mortality from other causes. While some writers have suggested that the apparent increases in lung cancer and ischaemic heart disease were due to statistical artefact (Stehbens 1987), the subsequent marked decline in ischaemic heart disease in some countries, and downturn of mortality from lung cancer in others (Doll and Peto 1981), attest to the epidemic nature of these diseases. Although the fall in mortality from ischaemic heart disease in countries such as Australia, Finland, New Zealand, and the US, and the more recent commencement of the same phenomenon in most countries in western Europe (Uemura and Pisa 1985), are not necessarily related to medical care, by the same argument, the earlier intractability in male mortality rates cannot be used as evidence against benefits from medical care.

Data from Western Australia in Tables 12.1 and 12.2 show that, apart from recent improvements in mortality from ischaemic heart disease, impressive reductions in mortality rates have occurred in the past 30 years across most broad disease categories as well as across all age and sex groups. With the exception of deaths due to malignant neoplasms, the mortality rate fell in virtually all major categories of disease subsumed in the chapters of the International Classification of Diseases. In both males and females, the greatest proportion of this improvement by far was attributable to

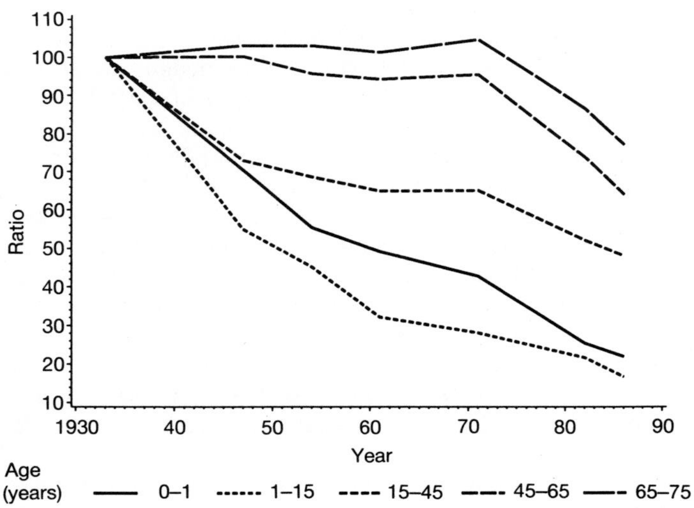

Fig. 12.1. Australian males: changes since 1930 in risk of dying between certain ages. (Source: Australian Bureau of Statistics—life tables)

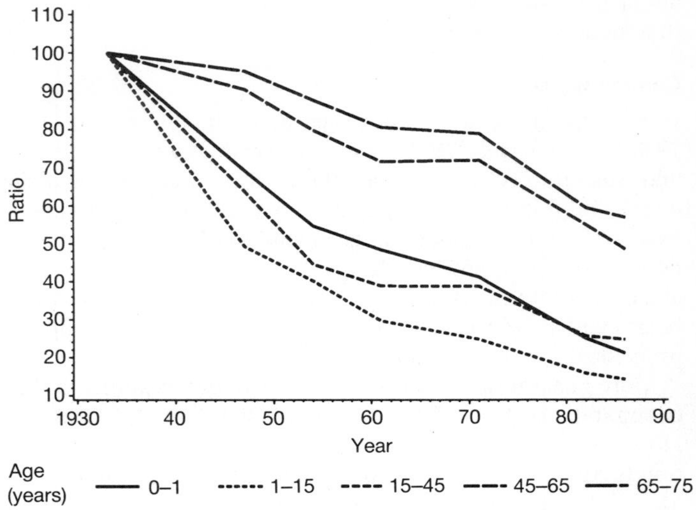

Fig. 12.2. Australian females: changes since 1930 in risk of dying between certain ages. (Source: Australian Bureau of Statistics—life tables)

Table 12.1. Changes in mortality* from selected diseases in males in Western Australia between 1953–7 and 1979–82

Condition/system	1953–7	1979–82	Relative change (%)	Proportion of decline† (%)
Circulatory	467.6	318.9	−31.8	53.9
Injury	110.4	67.8	−38.5	15.4
Perinatal	39.2	12.9	−67.1	9.5
Genito-urinary	32.1	9.1	−71.7	8.3
Infections	20.1	4.4	−78.1	5.7
Digestive	33.5	24.3	−27.5	3.3
Congenital	14.4	9.4	−35.0	1.8
Musculoskeletal	3.1	1.4	−64.6	0.6
Skin and subcutaneous	1.3	0.1	−92.4	0.4
Neoplasms	134.3	162.3	+20.8	—
All other causes	102.0	107.6	+ 5.5	—
All causes	958.0	718.2	−25.0	—

Source: Holman and Brooks (1987).

* Annual mortality rates per 100 000 for conditions subdivided according to chapters of the International Classification of Diseases and age-standardized to the 'world' population.

† Contribution of the particular group of conditions to the overall decline in mortality among those groups for which mortality fell.

Table 12.2. Changes in mortality* from selected diseases in females in Western Australia between 1953–7 and 1979–82

Condition/system	1953–7	1979–82	Relative change (%)	Proportion of decline† (%)
Circulatory	321.8	182.2	−43.4	66.6
Perinatal	29.9	9.4	−68.1	9.8
Respiratory	34.8	24.4	−29.9	4.9
Digestive	21.5	13.2	−38.6	4.0
Injury	33.8	25.6	−24.3	3.9
Genito-urinary	13.2	7.5	−43.1	2.7
Infections	8.0	3.4	−58.2	2.2
Endocrine	15.2	10.7	−29.7	2.2
Pregnancy	3.4	0.4	−88.3	1.4
Haematological	2.5	1.2	−52.7	0.6
Skin and subcutaneous	1.5	0.3	−81.8	0.6
Neoplasms	101.3	98.9	− 2.4	1.1
All other causes	26.3	25.8	− 1.9	0.2
All causes	613.2	403.0	−34.3	100.0

Source: Holman and Brooks (1987).

* Annual mortality rates per 100 000 for conditions subdivided according to chapters of the International Classification of Diseases and age-standardized to the 'world' population.

† Contribution of the particular group of conditions to the overall decline in mortality among those groups for which mortality fell.

falls in mortality from circulatory diseases (see Tables 12.1 and 12.2). However, reductions in mortality rate from injuries, perinatal disorders, genito-urinary diseases, infectious diseases, and disorders of the digestive system each made substantial contributions to the decline in males, while in females, perinatal disorders, respiratory diseases, disorders of the digestive system, and genito-urinary diseases contributed much of the remaining improvement. Less frequent causes of death which showed marked relative improvements include diseases of the skin and subcutaneous tissues in both sexes, congenital and musculo skeletal disorders in males, and infectious diseases, disorders of pregnancy, labour and the puerperium, endocrine disorders, and diseases of the blood and blood-forming tissues in females.

Substantial though these changes may be, trends in overall age-standardized mortality rates obscure the complexity of different rates of change in specific age groups or specific diseases. More detailed examination of the latter is helpful in understanding some of the multiple factors that may be contributing to general improvements in mortality. For example, modest changes in age-standardized total mortality rates for respiratory diseases conceal larger improvements in persons under the age of 45 years and rising rates or mixed trends at older ages (Fig. 12.3). One explanation for these findings is that mortality from acute respiratory infections has declined at younger ages, due at least partly to the use of antibiotics, while mortality due to chronic, smoking related, and industrial respiratory diseases in older age groups has proved less tractable. Similarly, in overall mortality from disorders of the digestive system, marked improvements in persons aged under 45 and over 64 years are partially obscured by smaller decreases between these ages due to rising mortality from cirrhosis of the liver. Thus, if we discount the latter, it is clear that mortality from digestive disorders has fallen at all ages.

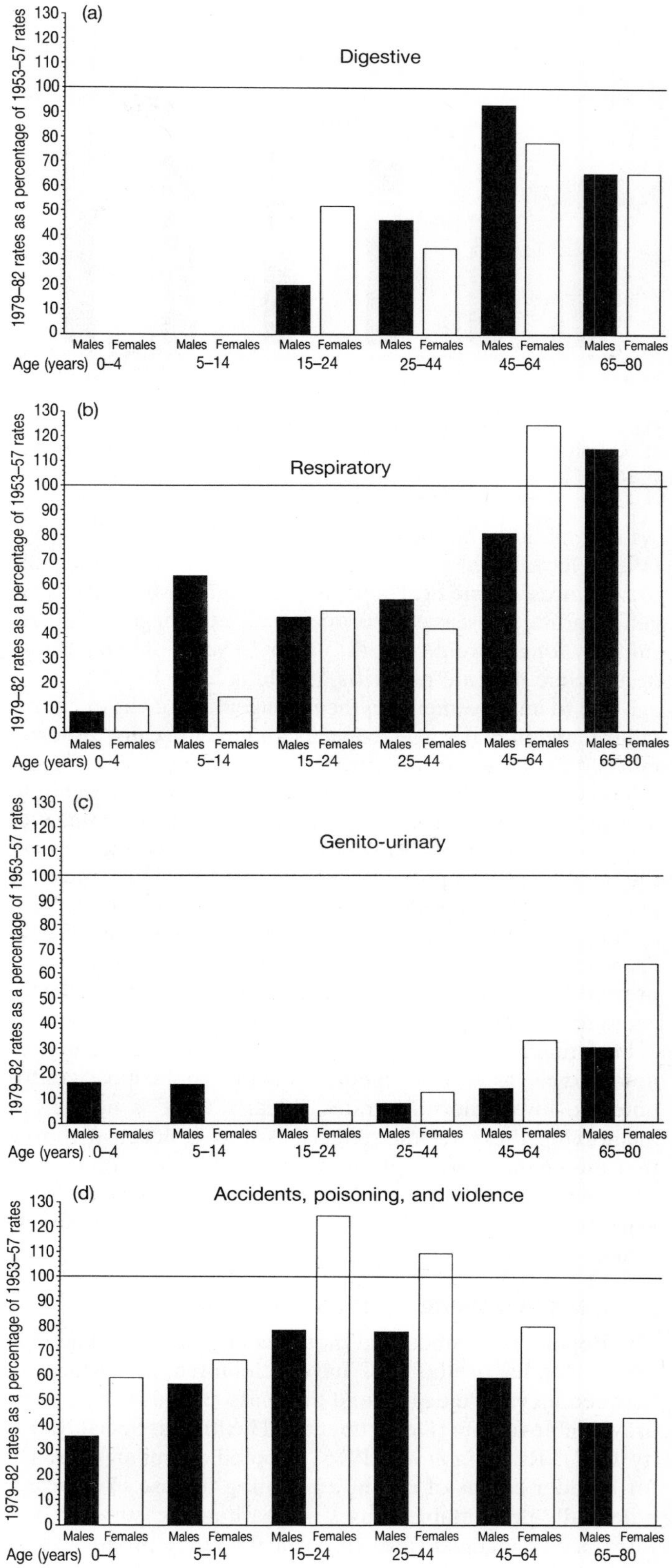

Fig. 12.3. Caption overleaf.

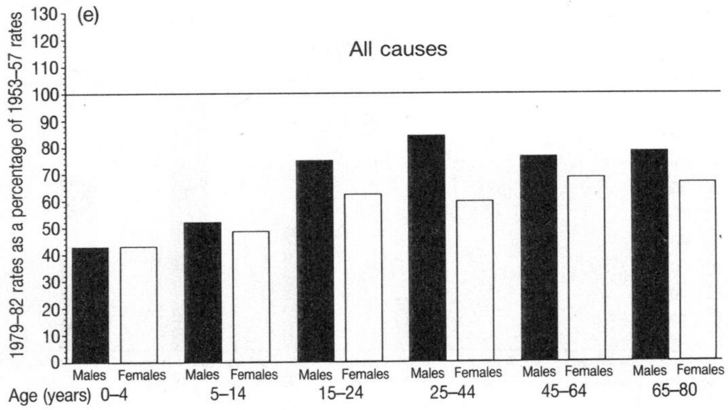

Fig. 12.3. Mortality rates in Western Australia in 1979–82 as a percentage of mortality rates in 1953–7 for selected conditions by age. (Source: Holman 1984)

As several authors, including McKeown (1979) and Beeson (1980), have suggested, it is likely that this is due principally to advances in medical and surgical care. Mortality from genito-urinary diseases also decreased at all ages, but presumably for different reasons: under 15 years of age because of a decline in acute nephritis, in adults from 15–64 years of age due to improvements in the management of chronic renal failure, and in males aged 65 years and over because of lower mortality associated with obstructive uropathy.

Trends in mortality from injuries are more difficult to interpret. Because of the disproportionately heavy impact of transport accidents in younger adults, and to a lesser extent, the middle aged, improvement in mortality from injuries was greater in children under the age of 15 years and in persons over 64 years of age. Reduced mortality from injuries in children may be due to a number of improvements in accident prevention. Lower rates in the elderly, on the other hand, could reflect improvements in medical care.

In virtually all of the examples discussed above, a range of non-medical as well as medical factors could have contributed to lower mortality rates. Clearly there is no way of quantifying these various effects. Nor is it necessarily true that the changes are replicated elsewhere. Nevertheless, as Kleinman (1982) suggested, there may still be lessons to learn from closer scrutiny of mortality rates for the past 40 years.

An approach using residual mortality

The Reports of Confidential Inquiries into Maternal and Perinatal Deaths in the UK have attempted to distinguish between inevitable deaths and those associated with possibly preventable factors (Department of Health and Social Security 1982). Rutstein *et al.* (1976) adopted a similar approach for broader causes of death, advocating the use of selected, potentially preventable causes of death as measures of performance of medical care. In particular, they proposed that single deaths from their list should be used in audit enquiries, although rates based on aggregated data could be used to monitor overall performance of health care. Charlton *et al.* (1983) developed this method further and examined variation among Area Health Authorities in mortality from 14 conditions selected from Rutstein's original list and representing 6.5 per cent of all deaths. They found differences among health areas in standardized mortality ratios for deaths from preventable causes after taking into account a number of social indicators associated with variation in deaths from all causes and in non-preventable mortality. While this residual variation might provide a measure of relative differences in medical care, it does not address the question of absolute effects. Nevertheless, it may be possible to use trends in such sentinel conditions as well as disease registers to assess the impact of medical care on selected aspects of health.

The eclectic approach

One alternative to the method of relating trends in mortality to medical care is the eclectic approach whereby the possible benefits of specific treatments for specific diseases are identified and then summed together. Reid and Evans (1970), for example, assessed the possible effects of changes in drug therapy on mortality rates of selected non-communicable diseases. While this approach has not so far been applied over a wide range of diseases, Beeson (1980) used subjective judgement to compare the efficacy of medical treatments for a large range of diseases in 1975 with the impact of measures in use in 1927. Emphasizing also the increased scope and safety of surgery and improved care of severe injuries, including burns, resulting from antibacterial drugs and other technical improvements, he concluded that 'substantial advances have been made along the whole frontier of medical treatment' and that 'a patient today is likely to be treated more effectively, to be returned to normal activity and to have a better chance of survival than fifty years ago'.

While this qualitative assessment falls far short of what we require to know about the effects of medical care on health, it is an approach that might be furthered by the combination of meta-analyses of the effects of specific medical interventions with good epidemiological data relating to incidence, prevalence, and the natural history of the corresponding diseases. Louis *et al.* (1985) have, for example, reviewed the potential of meta-analysis for assessment of selected public health measures.

A functional approach

Rather than assess the performance and impact of the health system as a whole, or at the other extreme, the benefit from treatment of individual diseases, it is possible to identify and examine specific functions or 'programmes' performed by particular sectors of the health care system (Mooney *et al.* 1986). These may be further divided into subprogrammes, each with their own explicit goals relating to particular problems of client groups. The assumption is made that all of the individual medical care activities undertaken by such a programme contribute, either directly or indirectly, to the achievement of those goals, this assumption obviating the requirement to identify the contribution of each individual activity to reaching the goals.

Notwithstanding the apparent merits of such an approach, the task rapidly becomes complicated because it is possible to subdivide medical care activities in many different ways. For example, the various medical care services within the health system can be grouped together according to their broad function or 'therapeutic intent'. Activities with a principal aim of preventing ill-health are clearly different from those that provide emergency care or otherwise aim to cure established disease. Rehabilitation services designed to maximize remaining function and limit the progression of chronic disease are different again, and palliative care, which aims to control symptoms, whether major or minor, but without any curative intent whatsoever, represents a fourth broad function of the health system. That this subdivision of medical care activity has some basis in reality is suggested by the fact that it is possible to identify doctors and other health personnel who have specialized in each of the four areas.

At the same time, the whole complex range of activities within the health care system can be subdivided in a quite different way by considering the 'life stages' with which they are concerned. For example, obstetricians and midwives deal with the events of pregnancy, labour, and the perinatal period, paediatricians deal with the problems of infancy and childhood, and other specialists have emerged for the adolescent, adult, and elderly age groups. At each life-stage, separate individuals and sometimes separate institutions serve the broad functions of prevention, emergency/curative care, rehabilitation, and palliation defined above. Thus the individual medical care activities can be fitted into a two-dimensional matrix as shown in Table 12.3. One or more individual activities fall into each 'cell' of the matrix and, while it is sometimes possible to evaluate each component in isolation, as shown by the example of rubella vaccination discussed below, the overall effect of medical care services on health will require 'summation' across all cells of the matrix.

Medical care activities can be classified in yet other ways. For example, the services delivered in a primary care or community setting are qualitatively different from those delivered on a worksite or in a hospital. Again, a number of medical specialties such as neurology, urology, gastroenterology, and cardiology correspond to particular organs or organ systems within the body, while others such as oncology and venereology focus on particular disease processes, regardless of the anatomical site affected. Even the scope of the health system as a whole varies from one country to the next; 'boundary issues' such as residential care for the aged or domiciliary support services for the frail or disabled fall within the province of the health system in some jurisdictions but are the responsibility of welfare services in others.

Whatever criteria are selected to define programmes and subprogrammes, the resulting framework must be both logically consistent and comprehensive; there should be no areas of overlap and all activities should be included.

Case studies

The following case studies illustrate some of the cells in the matrix of health services defined by stage of life and therapeutic intent. In western countries at least, the services most commonly used tend to lie along the diagonal of the matrix because as an individual gets older, that is, moves down the matrix, common health problems change from those that are potentially totally preventable to the other end of the spectrum where, once started, the pathological process cannot be halted or reversed and the only need that can be met effectively is that for relief of symptoms. In some ways, the health of whole communities follows the same cycle, with preventable problems of infancy and childhood predominating in developing countries and chronic, degenerative, incurable diseases constituting the greatest health problems of developed nations.

For these reasons our five case studies are drawn from cells lying along the diagonal of the matrix. Beginning with the relatively straightforward example of a specific preventive measure for which the outcome relates directly to the explicit objective of the programme, the case studies illustrate situations that are progressively more complex in terms of both the health problem and the medical response to it. Correspondingly, it becomes increasingly difficult to demonstrate conclusively that the medical care does have an overall beneficial effect.

Table 12.3. Subdivision of some medical care activities according to broad function and age group of the target population

Target group	Broad function			
	Prevention	Cure	Rehabilitation	Palliation
Pregnancy/perinatal	Rubella vaccination	Caesarean section	Closure of meningomyelocele	Care of lethal malformations
Infancy and childhood	Vaccination, developmental screening	Treatment of infections, acute leukaemia	Surgery for squints	Treatment of solid tumours, muscular dystrophy
Young adults	Contraception	Treatment of injury	Meniscectomy	Treatment of spinal injuries
Middle life	Mammography, cervical cytology	Acute coronary care	Renal transplant, coronary surgery	Treatment of cancer, coronary surgery
Elderly	Chiropody	Treatment of non-melanocytic skin cancer	Cataract surgery	Hospice services

The services included in this matrix are examples only. Many of them, such as coronary surgery, span more than one function and more than one age group as they can be applied in a number of clinical situations.

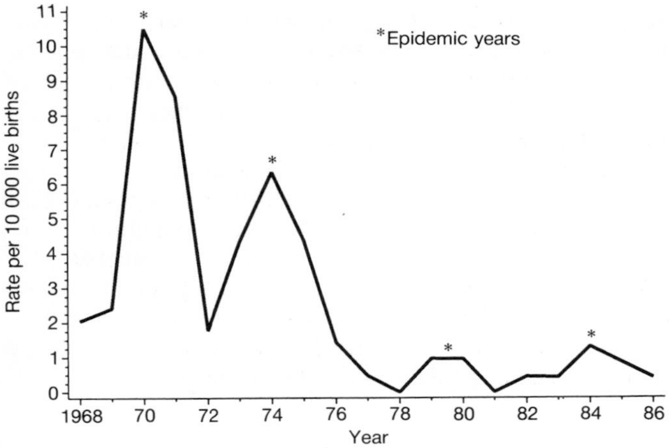

Fig. 12.4. Frequency of congenital rubella syndrome in Western Australia. (Source: Bower *et al.* 1987; Stanley *et al.* 1986)

Control of infectious disease by immunization

Modern developments in immunization have seen the global eradication of smallpox and the virtual elimination of poliomyelitis in some populations. Promising results have been achieved with both measles and rubella immunization, although both present problems in relation to vaccine failure and hence optimal vaccination routines.

While world-wide eradication of measles and rubella is not yet a realistic goal, we may judge the efficacy of immunization in terms of the major sequelae of infection. For example, Stanley *et al.* (1986) demonstrated that the inception of a universal immunization programme for young teenage girls is the most likely explanation for an observed decline in the frequency of congenital rubella syndrome in Western Australia. Vaccination of schoolgirls aged 12 or 13 years began in 1971 (when the proportion immunized was 75 per cent), and acceptance had increased to 88 per cent by 1983. The observed 'background' frequency of congenital rubella in Western Australia was approximately 2 per 10 000 live births from 1968 onwards, but rose sharply in association with epidemics of rubella in 1970 and 1974 (Fig. 12.4). At the latter time, the first girls to have participated in the programme would have been 15 years old. Over the subsequent decade, the frequency of congenital rubella syndrome in Western Australia fell to approximately 0.5 per 10 000 live births and none of the 19 reported cases was born to mothers who would have been eligible for the programme. Furthermore, the low frequency of congenital rubella syndrome has continued in the years following the period covered by the original report. There is therefore strong presumptive evidence that a specific medical care activity has virtually eliminated one source of lifelong and at times life-threatening disability in the target community. However, the examples provided by the Rubella Vaccination Programme in Western Australia and the Measles Eradication Campaign in the US (Mitchell and Balfour 1985) are relatively straightforward compared with the problems encountered in assessing the impact of more complex medical care activities where the relationship between the intervention and the outcome is less specific.

Obstetric and neonatal care

Although antenatal care is usually thought of as serving a purely preventive function, in practice it operates as a screening system to identify pregnancies at high risk of poor fetal or maternal outcome in order that the 'full force' of the obstetric services can be brought to bear early, before irreversible problems develop. Most obstetric activity is therefore fundamentally curative in intent, in that it seeks to recognize and terminate established disease processes rather than to prevent them from ever starting.

Historically, standards of obstetric and neonatal care, together with levels of reproductive fitness, have been assessed by monitoring maternal and perinatal mortality. Maternal mortality is now so low in most developed countries that it makes a negligible contribution to total mortality in females and is of doubtful value as a measure of outcome in obstetric care. Through the Confidential Inquiries into Maternal Mortality it is apparent that most of the deaths that still occur under the general rubric of 'complications of childbirth, pregnancy, and the puerperium' are related more to social and maternal factors or to abortion than to obstetric care *per se*. There is much evidence to suggest that the virtual elimination of maternal mortality has been due to factors such as improved maternal fitness, stature, and nutrition in early life, together with better education, control of fertility, and, in many countries, more liberal attitudes to termination of pregnancy with a subsequent reduction in illegal abortions. But there is equally strong evidence, from the Confidential Inquiries into Maternal Mortality, that certain avoidable causes of maternal death have been eliminated through the disciplined application of modern obstetric and anaesthetic techniques.

With the marked decline in maternal mortality, attention has turned to the assessment of preventable factors in perinatal deaths. The relatively high frequency of these compared with maternal deaths dictates the need to focus attention on deaths which are likely, *a priori*, to be associated with potentially avoidable factors. For example the Perinatal Mortality Committee in Western Australia has excluded perinatal deaths associated with lethal malformations and concentrated its attention on singleton still births weighing 2000 g or more or neonatal deaths weighing at least 1500 g at birth, and, for multiple births, still births weighing 1500 g or more and neonatal deaths weighing 1000 g or more (Health Department of Western Australia 1987).

In 1986 approximately 20 per cent of perinatal and infant deaths met the above criteria for full investigation. Of these cases 14.8 per cent, representing 3.4 per cent of total perinatal deaths, were associated with potentially avoidable factors. Thus it appears likely that in most developed countries with rates of perinatal mortality similar to Western Australia

Table 12.4. Changes in still births, neonatal, and post-neonatal infant mortality rates in Western Australia 1970–86

	Stillbirth rate (per 1000 total births)		Infant mortality rate (per 1000 live births)	
	Antepartum	Intra-partum	Neonatal	Post-neonatal
1970–5	10.3	2.2	12.8	5.5
1976–81	7.8	1.3	8.0	3.8
1982–6	5.6	0.8	5.1	3.4
Decline (%)	45.6	63.6	60.2	38.3

Source: Stanley and Hartfield (1979); F.J. Stanley, unpublished data.

Table 12.5. Changes in birth-weight-specific neonatal mortality rates per 1000 live births in Western Australia (excluding lethal malformations) from 1971–4 to 1980–4

Birth-weight (g)	1971–4	1975–8*	1980–4
1000–1499	526	252	137
1500–1999	141	79	37
2000–2499	35	18	17
2500–2999	7.5	6.0	2.3
3000–3499	3.3	3.7	2.3
3500+	2.4	1.8	1.0

Sources: Stanley and Hartfield (1979); F.J. Stanley, unpublished data.
* Data not available for 1979.

(12 per 1000) perinatal mortality, like maternal mortality previously, has declined to something approaching an irremediable minimum.

However, concepts of preventability must clearly be considered in the context of medical practices of the day. Thus the birth-weight limits for preventability now used are considerably lower than those used by Baird and his colleagues (1954) in their studies of perinatal mortality in Aberdeen in the 1950s. This change is emphasized particularly by the accelerated decline in neonatal mortality that has occurred in developed countries in the past two decades in association with the development of neonatal intensive care and, more recently, intra-uterine and intra-partum fetal monitoring.

In Western Australia, perinatal mortality fell from 28 per 1000 births in 1970–5 to 12 per 1000 in 1982–6 (Stanley and Waddell 1985). This improvement occurred in all components of perinatal mortality, but particularly in intra-partum still births and neonatal deaths (Table 12.4). Approximately 20 per cent of the improvement after 1970 can be attributed to improvement in the distribution of birth-weights but, as shown in Table 12.5, the greatest part is due to improved survival in lower birth-weight babies, during a period in which there have been marked changes in neonatal intensive care. Similarly, the virtual elimination of intra-partum still births has occurred in association with new methods for monitoring intra-uterine growth in late pregnancy and fetal distress during labour.

A concern relating to the development of neonatal intensive care and consequent survival of very low birth-weight babies who would previously have died, is the potential for increases in long-term morbidity and handicap, particularly resulting from cerebral damage. Stanley and Watson (1988) monitored the incidence of cerebral palsy in Western Australia over the period of falling neonatal mortality described above. They found no increase in the birth-weight-specific incidence of cerebral palsy, but observed an overall increase in cases of approximately 5 per cent due to the combined effects of the relatively high frequency of cerebral palsy in low birth-weight babies and their improved survival.

In summary, maternal mortality has become obsolete as a measure of obstetric outcome in developed countries probably due to a combination of general improvements in maternal health and better medical care. Furthermore, perinatal mortality is now also approaching what appears to be an irreducible minimum, and the contribution of improved obstetric and paediatric care to this trend would appear to be beyond question. However, there is some evidence that this reduction in mortality has been associated with an overall increase in the prevalence of long-term disability in children of low birth-weight, underlining the necessity for taking into account a number of different measures of outcome, including those relating to adverse as well as beneficial effects.

Orthopaedic and ophthalmic surgery in the aged

In the past 20 years greatly improved anaesthesia and the development of prosthetic devices have led to a wide range of surgical procedures for the prevention or correction of disability and handicap. These include improved methods for internal fixation of fractures, arthroplasty, ophthalmic procedures, valvular surgery, arterial grafting, cardiac pacing, and organ transplantation.

Two of the principal objectives in aged care which can be readily related to quality of life are the maintenance of mobility and independence in activities of daily living. Particular conditions which threaten these and are amenable to surgery include fractures of the proximal femur, osteoarthrosis in the hip, and impairment of vision.

An important factor in the care of the aged is the frequent occurrence in the same individual of multiple diseases and impairments. Thus, people suffering from proximal femoral fractures may already be infirm; in the case of Western Australia approximately one-third of persons admitted to hospitals with such fractures are residents of nursing homes. From this it is clear that the collaboration of orthopaedic surgeons and geriatric physicians in the care of these cases is highly desirable (Devas 1974). A combined orthopaedic geriatric unit has existed in Perth, Australia, for the past 10 years and provides integrated medical and orthopaedic care for patients with femoral neck fractures from admission to eventual discharge following rehabilitation (Lefroy 1980). Continuous monitoring of the work of the unit has shown that among patients who are admitted from the community and are alive 6 months after discharge, approximately 80 per cent have

Table 12.6. Cumulative incidence (%) of hip arthroplasty at selected ages in Western Australia, 1982–4

	Males	Females
By age		
75 years	1.77	1.69
80 years	2.46	2.46
85 years	3.03	3.09
All ages	3.37	3.41

Source: Health Department of Western Australia Inpatient Hospital Morbidity Statistics, 1984–6 (unpublished data).

Figures represent the cumulative incidence (%) assuming only unilateral procedures are performed. If only bilateral procedures were performed, each figure would be halved.

regained the same level of mobility as they had before the fracture and a similar proportion continues to live independently in their own homes. Although these results are impressive there has not been a formal comparison of the combined approach with routine surgical management alone.

Data relating to rates of elective surgical procedures show substantially higher rates for many operations in countries such as the US, Canada, and Australia, compared with the UK (Holman and Brooks 1987; Rutkow and Starfield 1984; Schacht and Pemberton 1985). An exception to this appears to be arthroplasty of the hip which is performed with equal frequency in the UK and Australia. It is therefore likely that there is general agreement across the two countries about the indications for this procedure, which is recognized as being highly successful in reducing the pain and disability associated with osteoarthrosis of the hip (Attenborough 1977). Thus arthroplasty of the hip may play an important role in maintaining mobility and independence in old age.

In Western Australia rates for this operation are approximately equal in males and females and rise with age to reach a peak at 75–79 years (Table 12.6). The cumulative incidence of operations by the age of 75 years is approximately 2 per cent, and by 80 years is approximately 2.5 per cent. Allowing for the need for bilateral prostheses in some individuals, it appears that over 1 per cent of 75-year-olds and nearly 2 per cent of 80-year-olds benefit from the procedure.

In contrast to elective operations on the hip, rates for surgical treatment of cataract vary between countries and this points to differences in the views of ophthalmic surgeons concerning indications for the procedure. In the past this may have been due to concerns about intra-ocular complications and problems in some patients with visual adjustment following operation (Jaffe 1978). Such difficulties have been greatly reduced by refinements in surgical technique and the development of new prosthetic lenses which are inserted directly into the lens envelope. Improved post-operative function and the ability to perform the operation under local anaesthesia as a day case have led to a more liberal attitude to cataract surgery. Thus, having been relatively stable from 1974 to 1980, rates for lens surgery in Western Australia increased rapidly between 1980 and 1984 and now appear to have stabilized again, but at a higher level (Holman and Brooks 1987) (Fig. 12.5).

Restoration of mobility and vision in the elderly are clearly matters of considerable importance. While the data shown here fall mainly into the category of process measures, the problems of relating them to true measures of outcome in at least samples of the population are not insurmountable. It therefore should be possible to develop models to assess the contribution of various restorative procedures to the well-being of the disabled aged.

Management of cardiovascular disease

Ischaemic heart disease

Several chronic diseases, including peptic ulcer (Morris 1967) and many neoplasms (Doll 1967; Doll and Peto 1981), have shown remarkable changes in incidence in the twentieth century. However, none has had the same impact on mortality rates as deaths related to diseases of the circulatory system. Within this broad group ischaemic heart disease and cerebrovascular disease have attracted the most interest. While there are difficulties in assessing long-term trends in these diseases because of changes in diagnostic and classification practices, ischaemic heart disease appears to have increased in most countries throughout the twentieth century. Moreover, major differences in mortality rates exist between developed countries (Epstein and Pisa 1979; Pisa and Uemura 1982; Thom *et al.* 1985; Uemura and Pisa 1985). Of even greater interest has been the demonstration of divergent trends in mortality from ischaemic heart disease which are unlikely to be due to classification artefact. As noted previously, mortality from ischaemic heart disease has declined markedly since the late 1960s in a number of countries including the US, Finland, Australia, New Zealand, Canada, and Japan.

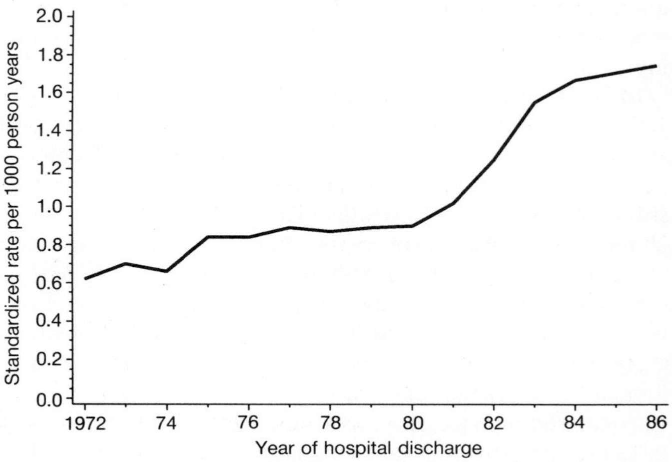

Fig. 12.5. Trends in surgery for cataract in Western Australia. (Source: Holman and Brooks 1987)

In most European countries, rates either continued to rise or remained stable during the 1970s and early 1980s. More recently mortality from ischaemic heart disease has declined in virtually all western European countries but not to the same extent as in the countries which first showed this trend. In Sweden, Italy, and several eastern European countries mortality rates are still increasing (Uemura and Pisa 1985).

The only explanation for the spectacular decreases in mortality from ischaemic heart disease in the US, Australia, and other selected countries that can reasonably be excluded is that of classification artefact, as there is no evidence of transference between major disease groups, and there have been commensurate declines in mortality from all causes combined. That aside, other reasons for the decline, including the possible role of medical care, remain the subject of speculation, but any explanation or set of explanations must be consistent with the following observations:

(1) other cardiovascular diseases, particularly stroke, have undergone equally large and even longer declines;
(2) the decline commenced in all adult age groups at approximately the same time;
(3) the decline has been greater in females than males and, in some countries, greater in younger age groups;
(4) in countries outside Europe, the decline commenced at about the same time (in the mid or late 1960s in males but earlier in females);
(5) divergent trends have occurred in countries with comparable health care systems and common methods of treating vascular disease.

It is difficult to identify any single environmental, behavioural, or medical care factor which could satisfy all of these conditions, but a number of studies now suggest that the decline in mortality has been associated with a decline in incidence of acute coronary events (sudden cardiac death or acute myocardial infarction) rather than improved survival after infarction (Elveback *et al.* 1981; Goldberg *et al.* 1979, 1986; Stewart *et al.* 1984). In Perth, Australia, a consistent fall in both fatal and non-fatal events occurred between 1971 and 1982 (Martin *et al.* 1984; Thompson *et al.* 1988), whereas there was no improvement in survival at 28 days or at 1 year after infarction in patients reaching hospital alive. Approximately 80 per cent of the decline in fatal events occurred in deaths out of hospital, while the fall in numbers of deaths in hospital was due mainly to reduction in admission rates rather than an improvement in case fatality. There is thus little evidence to support a direct or substantial contribution from acute coronary care. While these findings point towards changes in the primary factors responsible for ichaemic heart disease and underlying atherosclerosis, they do not exclude the possibility that medical care of persons with clinically recognizable, subacute ischaemic heart disease may have also contributed to the decline in incidence of acute coronary events. The interventions that could be beneficial include counselling about smoking, drug therapy, and coronary artery surgery, measures that range in therapeutic intent from preventive to palliative.

Some authors such as Dwyer and Hetzel (1980), Marmot (1985) and Dobson (1987) have linked the decline in mortality from ischaemic heart disease to changes in life-style factors such as diet and smoking or to changes in the treatment of hypertension. Others have attributed at least part of the decline to direct treatment of ischaemic heart disease. Goldman and Cook (1984), for example, suggested that 60 per cent of the decline was due to changes in life-style and the remainder to medical care. Beaglehole (1986), using a similar approach, estimated that 42 per cent of the decline in ischaemic heart disease in New Zealand could be attributed to medical care.

Cerebrovascular disease

Mortality from cerebrovascular disease has fallen in many countries over the past 30 years (Bonita and Beaglehole 1986; Levy 1979; Prineas 1971; Uemura and Pisa 1985), but the causes for this trend are not yet completely clear. Although the relationship between hypertension and the incidence of stroke is well established, as is the protective effect of treatment of hypertension (Australian National Blood Pressure Study 1980; Hypertension Detection and Follow-up Program 1982*a*; Medical Research Council 1985), the issue is not perfectly straightforward. First, mortality from stroke appeared to be decreasing before improvements in hypotensive drugs had reached the stage where it was possible to treat mild to moderate hypertension easily and without unacceptable side-effects (Lovell and Prineas 1971). Second, while hypertension carries a high relative risk of stroke, the attributable risk is less than is generally believed. For example, application of the incidence rates for stroke observed in the Framingham Study (Dawber 1980) to the distribution of blood pressure observed in the Australian population (National Heart Foundation (1983) shows that more than one-half of all strokes occur in individuals with a diastolic blood pressure below 90 mmHg. Third, using data from the Australian National Blood Pressure Study (1880) and the Hypertension Detection and Follow-up Program (1982*b*), Bonita and Beaglehole (1986) estimated that better treatment of hypertension would explain less than one-tenth of the decline in mortality from stroke during the 1970s. Fourth, there remains some uncertainty about the safety of treating hypertension in the elderly.

The problems in determining the effect of medical care on the incidence and outcome of stroke are even greater than those encountered in the case of ischaemic heart disease because a sizeable number of both fatal and non-fatal episodes of stroke never come to the notice of the hospital system. As Ward *et al.* (1988) have noted,

> It is therefore impossible to determine [from hospital statistics] whether a decline in mortality from stroke is due to a decline in incidence or in severity of new cases, either of which might reflect better control of risk factors, or to improved care of new cases, resulting in a lower case-fatality ratio. Moreover, a change in the mixture of the

diverse pathological entities encompassed by the term 'stroke' might also bring about a fall in mortality in the absence of a change in overall incidence of the condition. Clearly, all of these factors might have contributed to the downwards trend in mortality from stroke that has been observed [in several countries], and all of the factors must be examined if we are to explain this trend.

Despite the fact that both stroke and acute myocardial infarction are readily recognizable as distinct clinical and pathological entities, the epidemiology of each is complex and it is extremely difficult to determine the true impact of medical care activities designed to alleviate the burden of each that is borne by a particular community. The difficulties stem from several sources. First, individual episodes of illness defined as 'stroke' or 'myocardial infarction' represent only single points in the spectra of the underlying disease processes, processes whose manifestations range from infrequent and transient symptoms to sudden catastrophic events that can be rapidly fatal or followed by chronic disability. Second, the medical care response to each disease addresses numbers of different points in the spectrum and therefore crosses boundaries between preventive, curative, and palliative functions and between services delivered in hospitals, nursing homes, and general practices. The corollary is that each disease process potentially can be influenced by a number of different medical care activities. Since these medical influences are often applied concurrently it becomes difficult to ascribe changes in the behaviour of the disease in either an individual or a community to any one particular medical intervention. Third, the wide spectrum of clinical manifestations means that a number of different outcome measures are relevant. There is no simple way of combining measures of successful prevention, quality of life, continuing morbidity and disability, and mortality or case-fatality, among others, to determine whether the overall medical management of cardiovascular disease or cerebrovascular disease is 'adequate' or whether its adequacy is changing over time.

Palliative care

Broadly defined, palliative care means alleviation of symptoms in chronic or intractable illness. It is important to recognize that the medical system attempts to respond to the burden of discomfort caused by highly prevalent 'nuisance' disorders, such as headache or allergic rhinitis, as well as to the severe pain or loss of function associated with less common but more serious conditions such as rheumatoid arthritis or terminal cancer. Through detailed studies, medical care can be shown to provide very significant relief from many such disorders individually, but it is difficult to find an appropriate single index of benefit that can be applied to the whole range of such problems as they affect entire communities.

Nevertheless it is both possible and important to determine the level of resources applied to what is essentially palliative care. Studies of the cost of medical care at the end of life have been reviewed by Scitovsky and Capron (1986). These authors concluded that medical care in the last year of life accounts for 20–30 per cent of total expenditure on health care. The question then arises as to whether this expenditure represents unjustified use of technology and other costly measures. The evidence suggests that this is not the case, one reason being that it is difficult to distinguish, from among patients with similar problems, those who are likely to die and those who will not. While such gross studies confirm that a large proportion of our health resources are defrayed in what is judged retrospectively as 'palliative care', they do not throw much light on the social and health benefits that dying patients and their relatives might obtain from such care. Even in the specific and rapidly increasing area of hospice care, there has been little formal evaluation of benefits, and in those studies that have been conducted under properly controlled conditions using objective assessment, the evidence to show that the hospice movement has provided more satisfactory care than might be received from generic services is equivocal (Torrens 1985).

Conclusion

Just as health has many dimensions—long life expectancy, mental and social well-being, absence of physical impairment or illness—so modern medical care services constitute a complex system serving functions ranging from prevention to palliation. Yet history shows that marked improvements in health, as reflected in life expectancy at birth, occurred in the century *preceding* the 'scientific' practice of medicine characteristic of developed nations now. This observation suggests that measures in the realms of health protection and perhaps health promotion were responsible for most of the advances seen soon after the industrial revolution, and begs the question as to what has been the contribution of personal medical care services to the continuing, but mostly slower, improvements seen this century. To be in a position to answer this question requires both a single overall measure of health and some way of pooling the disparate processes and effects that make up modern medical care.

As indicators of health in developed countries mortality data have the advantages of universality and long historical continuity, but they do not encompass major but non-fatal illnesses, are subject to changes in diagnostic and classification practices, and may provide only a poor reflection of changes in medical care services due to confounding by alterations in more general economic and environmental factors. Systematic collection of morbidity data, on the other hand, has only a short tradition, is far from complete, and is probably even more prone to problems in diagnosis and coding. General community surveys, with application of special measurement techniques to subgroups with conditions of particular interest, may be required to obtain a true picture of the prevalence and level of ill-health in a population as opposed to the rate at which that population 'consumes' medical services. But even if the present level of health can be established, the need for medical care may remain undefined because an 'ideal' state of health has subjective and social dimensions beyond the absence of physical impairment.

For the present purposes we are concerned mainly with the direct effects of medical care services rather than the indirect effects that become apparent over generations as changes in fertility and mortality mould the demographic, social, and economic features of a community. But even the direct effects of medical care are difficult to assess since we need to take both beneficial and adverse effects into account and these might properly be measured in ways that are very difficult to reconcile; for example, how is the social cost of labelling asymptomatic individuals with hypertension as 'ill' to be weighed against a reduction in mortality rate resulting from treatment of the raised blood pressure?

Over and above these conceptual difficulties, there are intellectual and attitudinal problems arising from the fact that the tradition of evaluation of medical care is not strong and that structures and processes are so much easier to describe than are outcomes. Coupled with these are the facts that much of medical care is provided by generic services, which often leads to preventive, curative, restorative, and palliative interventions being applied simultaneously, and that many of the prominent diseases in our communities are complex ones for which a wide range of outcome measures are relevant. While there are isolated examples of specific medical interventions for which single, specific outcomes can be defined, in many instances the problem is truly 'multifactorial'.

In considering whether medical care services in their totality improve the level of health in a community, neither an exclusive focus on the 'bottom line' of the 'balance sheet', nor becoming enmeshed in the complicated algebra of individual 'transactions', seems workable in practice. Potentially useful alternatives include meta-analyses which pool the results of those interventions which have been submitted to randomized controlled trials under conditions closely resembling day-to-day practice, and networks of registers measuring the incidence and outcome of carefully selected sentinel conditions. If assessments of the relative cost-effectiveness of different medical care activities are built into such studies, then it might be possible not only to determine which interventions 'pay their way' with regard to health, but also to plan future investments in medical care rationally.

References

Acheson, E.D. (1967). *Medical record linkage*. Nuffield Provincial Hospitals Trust and Oxford University Press, London.

Attenborough, C.G. (1977). *Arthritis in the elderly*. In *Geriatric orthopaedics* (ed.) M. B. Devas p. 77. Academic Press, London.

Australian Bureau of Statistics. *Life tables 1932–34, 1946–48, 1965–67, 1970–72, 1975–77, 1982, 1986*. Canberra.

Australian National Blood Pressure Study: Report by the Management Committee. (1980). The Australian therapeutic trial in mild hypertension. *Lancet* **i**, 1261.

Baird, D., Walker, J., and Thomsen A.M. (1954). The causes and prevention of stillbirths and first week deaths. *Journal of Obstetrics and Gynaecology of the British Empire* **61**, 433.

Beaglehole, R. (1986). Medical management and the decline in mortality from coronary heart disease. *British Medical Journal* **292**, 33.

Beeson, P.B. (1980). Changes in medical therapy during the past half century. *Medicine* **49**, 79.

Bergner, M. and Rothman, R.L. (1987). Health status measures: An overview and guide for selection. *Annual Review of Public Health* **8**, 191.

Bonita, R. and Beaglehole, R. (1986). Does treatment of hypertension explain the decline in mortality from stroke? *British Medical Journal* **292**, 191.

Bower, C., Stanley, F., Forbes, R., and Rudy, E. (1987). *Report of the Congenital Malformations Register of Western Australia 1980–86*. Health Department of Western Australia, Perth.

Brocklehurst, J.C. (ed.) (1975). *Geriatric care in advanced societies*. MTP, Lancaster.

Burnet, M. and White D.O. (1972). *Natural history of infectious disease*. Cambridge University Press, Cambridge.

Butler, J.R. and Vaile, S.B. (1984). *Health and health services*. Routledge and Kegan Paul, London.

Charlton, J.R.H., Silver, R., Hantly R.M., and Holland W.W. (1983). Geographical variations in mortality from conditions amenable to medical interventions in England and Wales. *Lancet* **i**, 691.

Clemmeson, J. (1978). Registration in the study of cancer. In *Health care and epidemiology* (ed. W. W. Holland and L. Karhausen) p. 153. Henry Kimpton, London.

Cochrane, A.L. (1971). *Effectiveness and efficiency: Random reflections on health services*. The Nuffield Provincial Hospitals Trust, London.

Curnow, D.H., Cullen, K.J., McCall, M.G., Stenhouse, N.S., and Welborn, T.A. (1969). Health and disease in a rural community. *Australian Journal of Science* **31**, 281.

Dawber, T.R. (1980). *The Framingham Study: The epidemiology of atherosclerotic disease*. Harvard University Press, Cambridge.

Dawber, T.R., Gilcin, M.F., and Moore, F.E. (1951). Epidemiological approaches to heart disease: The Framingham Study. *American Journal of Public Health* **41**, 279.

Department of Health and Social Security (1982). *Report on confidential enquiries into maternal deaths in England and Wales 1976–78*. HMSO, London. (These are a long-standing series of reports of which this is one example.)

Devas, M.B. (1974). Geriatric orthopaedics. *British Medical Journal* **i**, 190.

Dobson, A.J. (1987). Trends in cardiovascular risk factors in Australia, 1986–1983: Evidence from prevalence surveys. *Community Health Studies* **11**, 2.

Doll, R. (1967). *Prevention of cancer: Pointers from epidemiology*. The Nuffield Provincial Hospitals Trust, London.

Doll, R. and Peto, R. (1981). *The causes of cancer: Quantitative risk of cancer in the United States today*. Oxford University Press.

Donabedian, A. (1966). Evaluating the quality of medical care. *Millbank Memorial Fund Quarterly* **44**, 169.

Dwyer, T. and Hetzel, B.S. (1980). A comparison of trends in coronary heart disease in Australia, USA, England and Wales with reference to three major risk factors—hypertension, cigarette smoking and diet. *International Journal of Epidemiology* **9**, 65.

Elveback, L.R., Connolly, D.C., and Kurland, L. (1981). Coronary heart disease in residents of Rochester, Minnesota. II. Mortality, incidence and survivorship, 1950–1975. *Mayo Clinic Proceedings* **56**, 665.

Epstein, F. and Pisa, Z. (1979). International comparisons in ischaemic heart disease mortality. In *Proceedings of the conference on the decline in coronary heart disease mortality* (ed. R. J. Havlik and M. Feinleib), NIH Publication No 79–1610, p. 58. US Department of Health Education and Welfare, Bethesda, Maryland.

Fenner, R. (1980). Smallpox and its eradication. In *Changing disease patterns and human behaviour*. (ed. N. F. Stanley and R. A. Joske), p. 215. Academic Press, London.

Fingerhut, L.A., Wilson, R.W., and Feldman J.J. (1980). Health and disease in the United States. *Annual Review of Public Health* **1**, 1.

Goldacre, M.J. (1987). Implications of record linkage for health services management. In *Textbook of medical record linkage* (ed. J. A. Baldwin, E. D. Acheson, and W. J. Graham) p. 305. Oxford University Press.

Goldberg, R.J., Gore, J.M., Alpert, J.S., and Dalen, J.E. (1986). Recent changes in attack and survival rates of acute myocardial infarction (1975 through 1981): The Worcester Heart Attack Study. *Journal of the American Medical Association* **255**, 2774.

Goldberg, R., Szklo, M., Tonascia, J.A., and Kennedy, H.L. (1979). Time trends in prognosis of patients with myocardial infarction; a population-based study. *Johns Hopkins Medical Journal* **144**, 73.

Golding, J., Vivian, S.P., Baines, C.J., and Baldwin, J.A. (1987). Analytical methods for time-sequenced linked records. In *Textbook of medical record linkage* (ed. J. A. Baldwin, E. D. Acheson, and W. J. Graham) p. 55. Oxford University Press.

Goldman, L. and Cook, E.F. (1984). The decline in ischaemic heart disease mortality rates: An analysis of the comparative effects of medical interventions and changes in lifestyle. *Annals of Internal Medicine* **101**, 825.

Health Department of Western Australia (1987). *The Seventh Annual Report of the Perinatal and Infant Mortality Committee of Western Australia and the Annual Report of the Maternal Mortality Committee for 1986*. Statistical Series No. 7. Health Department of Western Australia, Perth.

Hobbs, M.S.T., Hockey, R.L., Martin, C.A., Armstrong, B.K., and Thompson, P.L. (1984). Trends in ischaemic heart disease mortality and morbidity in Perth Statistical Division. *Australian and New Zealand Journal of Medicine*. **14**, 381.

Holman, C.D.J. (1984). *Mortality in Western Australia 1953–84. An analysis of age, sex and disease specific rates, trends in rates, and causes of premature death*. Position Paper No. 1. Steering Committee on the Review of Health Promotion and Health Education in Western Australia. Health Department of Western Australia, Perth.

Holman, C.D.J. and Brooks, B.H. (1987). *Surgical procedures in Western Australia. An analysis of surgery type in 1985 and trends in surgical procedure rates 1972–1985*. Occasional paper No. 15. Health Department of Western Australia, Perth.

Horwitz, O. (1978). Epidemiological parameters for the public health evaluation of chronic disease. In *Health care and epidemiology*. (ed. W. W. Holland and L. Karhausen). p. 143. Henry Kimpton, London.

Hunt, S.J., McEwen, J., and McKenna, S.P. (1986). *Measuring health status*. Croom Helm, London.

Hypertension Detection and Follow-up Program Co-operative Group (1982*a*). The effect of treatment on mortality in 'mild' hypertension: Results of the Hypertension Detection and Follow-up Program. *New England Journal of Medicine* **307**, 976.

Hypertension Detection and Follow-up Program Co-operative Group (1982*b*). Five-year findings of the Hypertension Detection and Follow-up program. III. Reduction in stroke incidence among persons with high blood pressure. *Journal of the American Medical Association* **247**, 633.

Illich, I. (1975). *Medical nemesis: The expropriation of health*. Marian Boyers, London.

Jaffe, N.S. (1978). Cataract surgery—a modern attitude toward a technological explosion. *New England Journal of Medicine* **299**, 235.

Keil, U. (1978). Community registers of myocardial infarction as an example of epidemiological register studies. In *Health care and epidemiology* (ed. W. W. Holland and L. Karhausen), p. 171. Henry Kimpton, London.

Kessner, D.M., Kalk, C.E., and Singer, J. (1973). Assessing healing quality: The case for tracers. *New England Journal of Medicine* **288**, 189.

Kleinman, J.C. (1982). The continued vitality of vital statistics. *American Journal of Public Health* **72**, 125.

Kohn, R. and White, K.L. (ed.) (1976). *Health Care. An International Study. Report of the World Health Organization/International Collaborative Study of Medical Care Utilization*, p. 125. Oxford University Press.

Lefroy, R.B. (1980). Treatment of patients with fracture of the neck of the femur in a combined unit. *Medical Journal of Australia* **ii**, 669.

Levine, S., Feldman, J.J., and Elinson J. (1983). Does medical care do any good? In *Handbook of health care and the health professions* p. 394. (ed. D. Mechanic). The Free Press, New York.

Levy, R.I. (1979). Stroke decline: Implications and prospects. *New England Journal of Medicine* **300**, 490.

Louis, T.A., Fineberg H.V., and Mosteller F. (1985). Findings for public health from meta-analyses. *Annual Review of Public Health* **6**, 1.

Lovell, R.H.H. and Prineas, R.J. (1971). Trends in prescribing hypotensive drugs and in mortality from stroke in Australia. *Medical Journal of Australia* **ii**, 509.

McKeown, T. (1965). *Medicine in modern society*. George Allen and Unwin, London.

McKeown, T. (1979). *The role of medicine: Dream, mirage or nemesis?* The Nuffield Provincial Hospitals Trust, London.

Marmot, M.G. (1985). Interpretation of trends in coronary heart disease mortality. *Acta Medica Scandinavia* (Supplement) **701**, 58.

Martin, C.A., Hobbs, M.S.T., and Armstrong, B.K. (1984). The fall in mortality from ischaemic heart disease in Australia: Has survival after myocardial infarction improved? *Australian and New Zealand Journal of Medicine* (Supplement) **14**, 435.

Martin, C.A., Hobbs, M.S.T., and Armstrong, B.K. (1987*a*). Estimation of myocardial infarction mortality from routinely collected data in Western Australia. *Journal of Chronic Diseases* **40**, 661.

Martin, C.A., Hobbs, M.S.T., and Armstrong, B.K. (1987*b*). Identification of non-fatal myocardial infarction from hospital discharge data in Western Australia. *Journal of Chronic Diseases* **40**, 1111.

Maxwell, R.J. (1981). *Health and wealth. An international study of health care spending*. Lexington Books, Toronto.

Medical Research Council Working Party (1985). MRC trial of treatment of mild hypertension: Principal results. *British Medical Journal* **291**, 97.

Mitchell, C.D. and Balfour, H.H. (1985). Measles control: So near and yet so far. *Progress in Medical Virology* **31**, 1.

Mooney, G.H., Russell, E.M., and Weir, R.D. (1986). *Choices for health care*. MacMillan, London.

Morris, J.N. (1967). *Uses of epidemiology*. Livingstone, Edinburgh.

Murnaghan, J.H. (1981). Health indicators and information systems for the year 2000. *Annual Review of Public Health* **2**, 299.

National Heart Foundation of Australia (1983). *Risk factor prevalence survey, Number 2–1983*. National Heart Foundation, Canberra.

Pisa, Z. and Uemura, K. (1982). Trends of mortality from ischaemic heart disease and other cardiovascular diseases in 27 countries, 1968–1977. *World Health Statistics Quarterly* **35**, 11.

Prineas, R.J. (1971). Cerebrovascular disease occurrence in Australia. *Medical Journal of Australia* **ii**, 557.

Puska, P., Tuomilehto, J., and Salonen, J.T. (1985). Ten years of the North Karelia project. *Acta Medica Scandinavia* (Supplement) **701**, 66.

Reid, D.D. and Evans, J.G. (1970). New drugs and changing mortality from non-infectious disease in England and Wales. *British Medical Bulletin* **3**, 191.

Roemer, M.T. and Roemer R.J. (1981). *Health care systems and comparative manpower policies*. Marcel Dekker, New York.

Royal College of General Practitioners (1974). *Oral contraceptives and health*. Pitman Medical, London.

Rutkow, I.M. and Starfield, B.H. (1984). Surgical decision making and operative rates. *Archives of Surgery* **119**, 899.

Rutstein, D.D., Berenberg, W., Chalmers, T.C., Child, C.G., Fishman, A.P., and Perrin, E.B. (1976). Measuring the quality of medical care. A clinical method. *New England Journal of Medicine* **294**, 582.

Schacht, P.J. and Pemberton, A. (1985). What is unnecessary surgery? Who shall decide? Issues of consumer sovereignty, conflict and self-regulation. *Social Science and Medicine* **20**, 199.

Scitovsky, A.A. and Capron, A.M. (1986). Medical care at the end of life: The interaction of economics and ethics. *Annual Review of Public Health* **7**, 59.

Stanley, F.J. and Hartfield, M.J. (1979). *Livebirths and perinatal mortality in Western Australia 1976–78*. Report of NH and MRC Research Unit in Epidemiology and Preventive Medicine and Department of Health and Medical Services, Western Australia.

Stanley, F.J. and Waddell, V. (1985). Changing patterns of perinatal and infant mortality in Western Australia: Implications for prevention. *Medical Journal of Australia* **143**, 379.

Stanley, F.J. and Watson, L. (1988). The cerebral palsies in Western Australia. *American Journal of Obstetrics and Gynecology* **158**, 89.

Stanley, F.J., Sim, M., Wilson, G., and Worthington, S. (1986). The decline in congenital rubella syndrome in Western Australia: An impact of the school girl vaccination programme? *American Journal of Public Health* **76**, 35.

Stehbens, W.E. (1987). An appraisal of the epidemic rise of coronary heart disease and its decline. *Lancet* **i**, 606.

Stewart, A.W., Beaglehole, R., Fraser, G.E., and Sharpe, D.N. (1984). Trends in survival after myocardial infarction in New Zealand, 1974–81. *Lancet* **ii**, 444.

Thom, T.J., Epstein, F.H., Feldman, J.J., and Leaverton, P.E. (1985). Trends in total morbidity and mortality from heart disease in 26 countries from 1950 to 1978. *International Journal of Epidemiology* **14**, 510.

Thompson, P.L., Hobbs, M.S.T., and Martin C.A. (1988). The rise and fall of ischaemic heart disease in Australia. *Australian and New Zealand Journal of Medicine* **18**, 327.

Torrens, P. (1985). Hospice care: What have we learned? *Annual Review of Public Health* **6**, 65.

Tunstall-Pedoe, H. (1985). Monitoring trends in cardiovascular disease and risk factors: The WHO MONICA project. *WHO Chronicle* **39**, 3.

Uemura, K. and Pisa, Z. (1985). Recent trends in cardiovascular disease mortality in 27 industrialized countries. *World Health Statistics Quarterly* **38**, 142.

Ward, G.R., Jamrozik, K.D., and Stewart-Wynne, E.G. (1988). Incidence and outcome of cerebrovascular disease in Perth, Western Australia. *Stroke* **19**, 1501.

Wilson, R.W. and Drury T.F. (1984). Interpreting trends in illness and disability: Health statistics and health status. *Annual Review of Public Health* **5**, 83.

C

Public Health Policies and Strategies

13

Governmental and legislative control and direction of health services in the United States

PHILIP R. LEE and A. E. BENJAMIN

Introduction

Consideration of governmental and legislative control and direction of health services in the US requires an examination of the health care system and the political context within which health care policies are established. It also demands an understanding of the respective roles of government and the private sector in the planning, provision, financing, and regulation of health care. It is the purpose of this chapter to provide the context for understanding the evolution of health care policies in the US, particularly the role of the private sector, pluralism, and inter-governmental relations (federalism) in those policies. After presenting an overview of the American health care system, we will discuss a political framework for understanding the process and substance of health policy in the US. Following that, the bulk of this chapter will describe the evolution of contemporary health policy, with particular attention to themes involving public–private sector roles and shifting conceptions of federalism. We conclude with a brief commentary on the future of health policy in the US.

An expanding private health care system based on a growing public subsidy

Governmental policies and programmes play a major role in the planning, regulation, and financing of health care. The major function of governmental policies related to health care is to support services in the private sector, particularly those provided by physicians, hospitals, and nursing homes. The complicated system of health services in the US is composed of a large private sector that includes thousands of non-profit institutions; increasing public subsidy or direct financing of the system; multiple federal, state, and local governmental programmes and agencies; and over 238 million consumers threading their way through the maze.

Health workers totalled over 6 million people in 1986, including approximately 1.59 million registered nurses, 569 000 physicians, 180 000 pharmacists, 158 000 dentists (US Department of Commerce, Bureau of the Census 1989), and more than three million workers in 150 other health service and support categories. More than 5.5 million of these people are engaged directly in providing health care services. There are approximately 6900 hospitals in the country, including non-profit community hospitals, public hospitals (federal, state, and local), and proprietary hospitals discharging nearly 35 million patients and providing over 270 million days of in-patient care annually. Over 18 000 nursing homes provide care for more than 1.6 million patients, most of them elderly. There are a variety of different arrangements for ambulatory medical care: solo practitioners and small partnerships, including primary care physicians, specialists, or sub-specialists; large and small multi-specialty and single specialty group practices, including health maintenance organizations; community health centres, mental health centres, and public health clinics, usually subsidized by government funds; hospital out-patient clinics; emergency rooms; adult day care centres; and a host of other arrangements. Any of these may serve as a point of entry into the health care system.

National health expenditures for 1987 totalled approximately $500 billion, with $443 billion spent for personal health care, including $195 billion for hospitals, $103 billion for physicians' services, $41 billion for nursing homes, and $104 billion for all other services (Levit and Freeland 1988). Health care financing is decentralized and includes more than 1200 private health insurance companies; 50 state Medicaid programmes (supported by federal, state, and, in some states, local government funds); the federal Medicare programme for the elderly; thousands of large companies that self-insure; the Veterans' Administration, the Army, Navy, and Air Force; the US Public Health Service (the Indian Health Service and Merchant Seamen); state governments providing in-patient psychiatric care for the indigent mentally

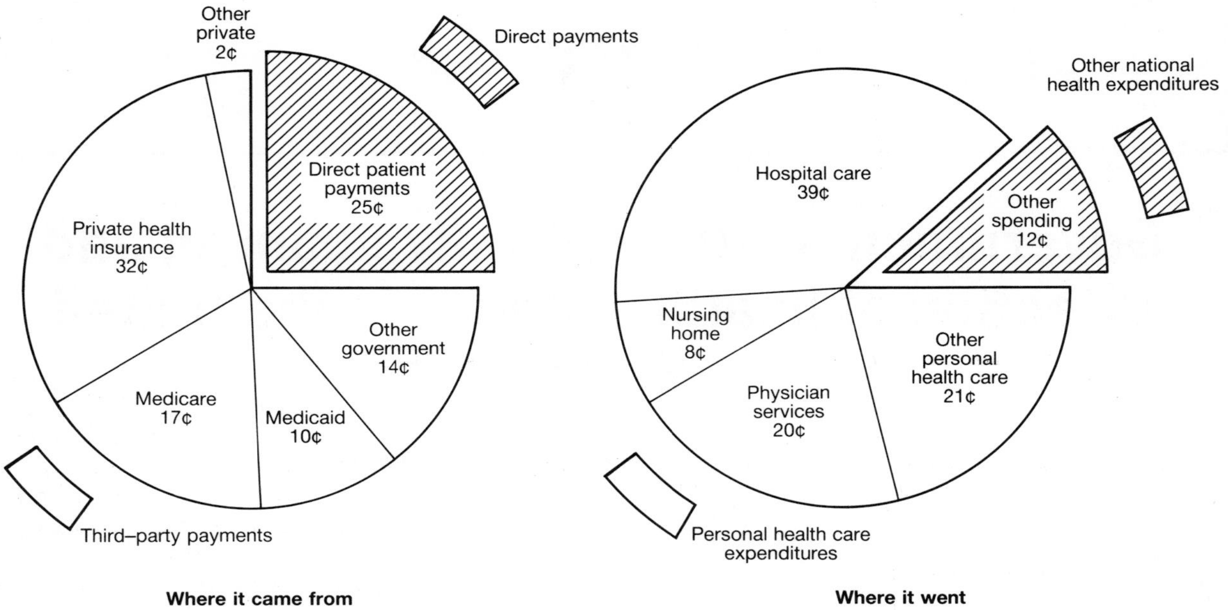

Fig. 13.1. The US health dollar, 1987. Almost three-quarters of national health expenditures were channelled through third parties; nearly two-thirds were channelled through private hands. The bulk of that expenditure was for patient care, and the remaining 12 per cent was spent on research, construction, administration, and government public health activity. (Source: Health Care Financing Administration, Office of the Actuary, Division of National Cost Estimates).

ill; federal, state, and local governmental programmes for the mentally retarded and mentally ill, as well as for alcoholics and drug abusers; local government (county or municipal) programmes providing or paying for care of the indigent; private foundations, voluntary organizations, and charities; and millions of individuals who pay directly out of pocket for all or a portion of their health care.

The sources of funds for health expenditures in 1986 were: private insurance (32 per cent), federal Medicare (17 per cent), federal–state Medicaid (10 per cent), other governmental programmes (14 per cent), direct payments by patients (25 per cent), and other private sources (2 per cent). Almost three-quarters of national health expenditures (e.g. those for health services) are channelled through third parties (e.g. private health insurance companies). Approximately 88 per cent of total expenditure is for patient care, while the remaining 12 per cent is spent on research, construction, administration, and governmental public health activity (Fig. 13.1) (Levit and Freeland 1988).

Based on the assumptions that present trends will continue and that there will be no major changes in prices related to health care financing, the federal Health Care Financing Administration has projected that national health expenditures will reach $1.5 trillion in the year 2000 (15 per cent of the gross national product). Total spending per caput is projected to rise from $1837 in 1986 to $5550 in 2000 (Health Care Financing Administration 1987, p. 18).

Despite billions of dollars spent annually and a vast array of programmes and providers, there remain millions of people with limited access to personal health care in the US. These are largely the working poor and those with chronic illness who do not have adequate public or private third party coverage, often called the uninsured and underinsured. Although Medicare and Medicaid dramatically improved access to care for the elderly and the poor after 1965, in recent years federal, state, and local cutbacks, particularly in Medicaid, have increased the number of individuals dependent on local governments for needed medical care. This is also true for the growing number of people without health insurance. The most recent data suggest that between 31 and 37 million Americans (depending on the population base used), mostly the working poor and their families, are without health insurance (Wilensky 1988).

A political framework for health policy: the federal system and the role of government

It has been observed that each nation's health care system is a reflection of its own particular traditions, values, organizations, and institutions (Litman and Robins 1971). Certainly, the experience of the US is no exception. In order to understand health policy and the politics of health in the US, it is important to know something of the factors that influence the system. We view the health care policy process as shaped by a variety of political, economic, and social factors. In this review we focus on three: (1) the dominance of the private sector; (2) federalism as the basis for the system of government, with an emphasis on minority rights, majority rule, and the preservation of individual liberty; and (3) interest-group pluralism.

Private sector dominance

It has long been recognized that the role of government in the US, particularly in relation to domestic social programmes, has often been to stimulate or support the private sector. The basic economic justification for governmental intervention is as a remedy for market failure (Blumstein and Zubkoff 1973), and despite the enormous change in the role of government in the past 200 years, the basic idea persists.

Health policies and programmes of the US government have evolved piecemeal. Most, such as federal support for biomedical research, developed in response to needs that were not being met by the private sector or by state and local governments. Others, such as federal support for hospital construction, arose because results of the 'free market' were grossly inequitable across communities and states. Still other programmes, such as Medicare and Medicaid, developed because health care was so costly that many elderly and the poor could not afford the private option, nor could hospitals and physicians meet their needs through charity. The result has been a proliferation of categorical programmes that are administered by more than a dozen federal government departments. Over the years new programmes have been added, old ones redirected, and numerous efforts made to integrate and co-ordinate services. Functions of the public and private sectors have become increasingly interrelated, and roles are often blurred. Public programmes pay over 40 per cent of the nation's personal health care bill; most physicians and other health care personnel are trained at public expense; almost 65 per cent of all biomedical and health-related research and development funds are provided by the federal government; and most non-profit community, municipal (public), and university hospitals have been built or modernized with governmental subsidies. The bulk of governmental expenditures are federal, with state and local government contributing significant but much smaller amounts.

Federalism

Federalism, which is the term commonly used to describe the relationship between the US government and the 50 states, is the basic foundation of public policy in the US. Originally, federalism was a legal concept, emphasizing the constitutional division of authority and function between the national government and the states. Not only did the US constitution define broadly the relationship between the federal government and the states, it also established separate executive and judicial branches to place adequate checks and balances upon the federal executive branch. The concept of federalism, however, had its roots not in the US, but in ancient Greece and in even earlier times (Hale and Palley 1981).

Federalism represents a form of governance different from that in a unitary state, where regional and local authority derive legally from the central government, and a confederation, in which the national government has limited authority and does not reach individual citizens directly (Reagan and Sanzone 1981). Federalism initially stressed the independence of each level of government from the other and incorporated the idea that some functions, such as foreign policy, were the exclusive province of the central government, while others, such as education and police protection, were the responsibility of regional units—state and local government. Both public health and medical care were long thought to be primarily the responsibility of state and local government and the private sector, and responsibility for payment for health services rested with the individual. Although the concept of separation of functions continues today, there is greater interdependence among federal, state, and local governments, particularly with respect to major national issues such as income maintenance (e.g. unemployment insurance), health care, housing, transportation, and education.

The question of the proper role of government, as well as the balance of power among the federal, state, and local governments, has been the subject of debate throughout the history of the US, with the final decision in any given controversy usually drawn by the courts. Over time the courts, particularly the Supreme Court, have reached different conclusions on key issues. The first Supreme Court under Chief Justice John Marshall held to the doctrine of national supremacy, while the Supreme Court that followed under Chief Justice Taney subscribed to the doctrine of dual federalism, with the states having concurrent powers in matters considered to be truly local in character (Litman 1984, p. 6). This view of dual federalism prevailed until the Great Depression in the 1930s and the advent of the New Deal, when the concept of co-operative federalism emerged, ultimately with the support of the Supreme Court in 1937.

The bases for federal and state government involvement in public health and health care rest on quite different constitutional principles. Federal involvement rests on the interpretation of the commerce clause and the welfare clause of the Constitution. It was not until the enactment of the Social Security Act of 1935 that the welfare clause was interpreted broadly to permit a wide range of federal intervention in areas previously deemed the responsibility of the states. The commerce clause has also been more broadly interpreted since the mid-1930s. State governments, in contrast to the federal government, have drawn on the police powers '. . . to enact and enforce laws to protect the health, safety, . . . and general welfare' (Grad 1973).

The enactment of Medicare and Medicaid in 1965, as amendments to the Social Security Act, represented a watershed in the role of federal and state governments in financing and regulating personal health care. Since then, public support has risen from 21.6 per cent of total personal health care expenditures to over 40 per cent. Medicare and Medicaid represent the largest, most important, and most expensive federal health programmes. In addition, hundreds of other federal and state programmes affect health care. These range from federal grants that support biomedical research in university health science centres to block grants to the states for maternal and child health. Well over 100 separate federal programmes provide support for research, training, services, or construction of health facilities.

With the exception of Medicare and social security payments to individuals, most federal assistance has been in the form of categorical grants-in-aid made to state and local governments and other public and private organizations or institutions for specific purposes. Such grants-in-aid enabled state and local governments, as well as non-profit institutions, to maintain their autonomy within a framework of federal assistance.

With the rapid growth of categorical grants-in-aid during the 1960s, there was growing concern on the part of state and local government officials about excessive federal direction and control. This concern resulted in two separate approaches—block grants and revenue-sharing. Block grants were given for broader purposes than categorical grants-in-aid, but were limited to a particular area, such as mental health, child health, or public health. Revenue sharing, in contrast, simply turned over to state and local governments federal revenues that could be used for broad purposes at the discretion of state and local governments. Revenue-sharing disappeared in the 1980s because of the rapid increase in the federal deficit.

Pluralism

The political context in which US health policy is debated, developed, and implemented has greatly influenced its scope and effectiveness. Most federal law designed to affect the health of the American people, their access to health care, and the resources available for such care reflect the public's faith in the ideology of pluralism, the strength of special interests, and the primary role accorded the private sector (Alford 1975; Cater and Lee 1972; Estes 1979). Ginzberg (1977) identified four power centres in the health care industry that influence the nature of health care and the role of government: (1) physicians; (2) large insurance organizations; (3) hospitals; and (4) a highly diversified group of participants in profit-making activities within the health care arena. These groups are major actors, with governmental executives and congressional committee members, in what Cater described as 'a new form of federalism'. He noted:

> In the politics of modern America, a new form of federalism has emerged, more relevant to the distribution of power than the old. The old federalism which ordered power according to geographic hegemonies—national, state, and local—no longer adequately describes the governing arrangement. New subgovernmental arrangements have grown up, by which much of the pressing domestic business is ordered. (Cater 1972, p. 4)

Health has become one of the largest and most powerful subgovernments, and the growing influence of the power centres identified by Ginzberg (1977) is evident in governmental policies at the federal, state, and local levels. Medicare and Medicaid policies reflect the powerful influence of physicians and hospitals, as well as their allies in the health insurance industry. In enacting Medicare, Congress adopted policies that permitted physicians to be paid on the basis of their usual charges, not on the basis of a fee schedule negotiated with the government. This system of physician reimbursement is highly inflationary because it provides incentives to physicians to raise prices and to provide ancillary services, such as laboratory tests, electrocardiograms, and radiographs. Hospital reimbursement was initially based on costs incurred by hospitals in providing care, creating strong incentives to provide more and more services. In spite of the impact of rising costs on Medicare expenditures, Congress steadfastly refused to alter Medicare's methods of payment to physicians and hospitals until the budget crisis that followed the major recession of 1981–2. In 1983, Congress adopted a prospective payment system for hospitals, based on the estimated cost per case, but did not modify physician payment. In 1989, Congress adopted reforms for physician payment in the Medicare programme, including establishment of a resource-based relative value fee schedule, limits on balance billing by physicians, and volume performance standards.

An historical framework: the development of health policy from 1798 to 1988

The evolution of US health policy has not been as lengthy as it has been complex. In order to simplify the following discussion, we have divided the history of American Government into four historical periods and attempted to highlight the major political and health policy developments that give each period its character. Perhaps not surprisingly, the most detailed attention is given to the most recent period of policy evolution.

Dual federalism and the role of the private sector: from the early days of the republic to the Great Depression

With the major changes in health care and public health that have occurred over the past 200 years, particularly the changes in the past 50 years, has come a transformation in the role of government. Well before the American Revolution, the colonies were adopting policies related to health care. In 1730, American seamen were taxed by the British government to pay for hospital care. In 1760, New York City adopted licensure requirements for physicians to practise medicine. In 1772, New Jersey adopted the first act in the colonies to regulate the licensing of physicians. After the Revolution, Virginia became the first state to tax seamen for hospital care.

During the early years of the republic, the federal government played a limited role in health care activities, which were largely within the jurisdiction of the states and the private sector. Private charity shouldered the responsibility of care for the poor. The federal role in providing health care began in 1798, when Congress passed the Act for the Relief of Sick and Disabled Seamen, which imposed a 20 cents per month tax on seamen's wages for their medical care. The federal government later provided direct medical care for merchant seamen through clinics and hospitals in port cities, a policy that continues to this day. The federal government also played a limited role in imposing quarantines on ships entering US ports, in order to prevent epidemics (Lee and Silver 1972). It did little or nothing, however, about the spread of communicable diseases within the US.

The states, within the federal system, have always played a role in protecting the health of the public. Indeed, the health of every citizen is protected under the authority of state law (Miller *et al.* 1977). Massachusetts passed the first compulsory immunization law in 1809, requiring cowpox immunization to prevent smallpox. Most states first exercised their public health authority through committees or commissions, with actions to deal with epidemics. The first state health department was established by Louisiana in 1855, and Massachusetts followed in 1869. Gradually, other states followed suit.

Initially, the most significant role played by state health departments in personal health care was in the establishment of state mental hospitals. These first developed as a result of a reform movement in the mid-nineteenth century led by Dorothea Dix. Over the next century, state mental institutions evolved into isolated facilities for custodial care of the chronically mentally ill. The development of these asylums reinforced the stigma attached to mental illness and placed the care of the severely mentally ill outside the mainstream of medicine for more than a century (Foley 1975).

Local health departments were established in some large cities even before state departments of health. Later, local health departments were set up in rural areas, particularly in the South, to counteract hookworm, malaria, and other infectious diseases that were widespread in the nineteenth and early twentieth centuries. Originally, the emphasis of local health departments in the cities was on environmental sanitation and epidemic diseases (Miller *et al.* 1981).

Local hospitals, public and private, began to play an important role in medical care for the poor early in the eighteenth century. Hospital sponsorship at the local level was either public (local government) or through a variety of religious, fraternal, or other community groups. Thus the non-profit, community hospital was born in the eighteenth century. The nonprofit, voluntary hospital, rather than the local public hospital, gradually became the primary locus of care. Physicians provided voluntary care for the sick poor in order to earn the privilege of caring for their paying patients in the hospital (Silver 1976). Charity was the major source of care for the poor, and publicly provided or supported services also began to grow in the nineteenth century. Gradually, the public sector at the local level assumed responsibility for indigent care.

The Civil War brought about a dramatic change in the role of the federal government. Not only did the federal government engage in a war to preserve the union, but it began to expand its role in other ways that were to alter forever the nature of federalism in the US. The changing federal role that began with the Civil War was reflected in congressional passage of the first programme of federal aid to the states. In 1862, Congress passed the Morrill Act, which granted federal lands to each state. The profits from the sale of these lands supported public institutions of higher education, known as land grant colleges (Hale and Palley 1981). Towards the end of the nineteenth century, the federal government began to provide cash grants to states for the establishment of agricultural experiment stations. Although the federal role was expanding, the change had little impact on health care. One step was made, however, in 1878, when the Surgeon General of the Marine Hospital Service (later to become the US Public Health Service) was given congressional authorization to impose quarantines within the US (The National Quarantine Act of 1878). This was the first time that the federal government assumed a public health responsibility in an area where the states had previously held jurisdiction.

The next major change in the role of the federal government in public health was in the regulation of food and drugs. This occurred because of public outcry about adulteration and misbranding of foods. Initially, money, not health, was the primary concern. After 20 years of debate, Congress enacted the Federal Food and Drug Act in 1906 to regulate the adulteration and misbranding of food and drugs. The law was designed primarily to protect the pocket-book of the consumer, not the consumer's health. However, it represented a major change in the role of the federal government, which assumed a responsibility previously exercised exclusively by the states. The commerce clause of the Constitution was used to justify a federal role in the regulation of inter-state commerce. The legislation had some impact on fraudulent practices, but far less than was hoped by its advocates.

Children began to be the focus of national concern following the first White House Conference on Children called by President Theodore Roosevelt in 1909. In 1912, Congress established the Children's Bureau, which initially turned its attention to child labour and infant mortality and stimulated a great deal of action at the state and local levels. In 1921, Congress passed the Maternity and Infancy Act (Sheppard–Towner Act), which provided federal grants to states to develop health services for mothers and children. Due to the virulent opposition of the American Medical Association, Congress permitted this programme to lapse in 1929. The 1930 White House Conference on Children issued its famous Children's Charter, which still stands as a powerful policy statement on behalf of children (Silver 1988), but it was not until 1935 that Congress enacted new categorical grant-in-aid programmes for maternal and child health services.

From dual federalism to co-operative federalism: from the New Deal to the Great Society

Federalism in the US evolved from a pattern of dual federalism, with a limited role in domestic affairs for the federal government, to co-operative federalism, with a strong federal role in the 1930s. The Great Depression in the US brought action by the federal government to save banks, support small business, provide direct public employment, stimulate public works, regulate banks and business, restore consumer confidence, and provide social security in old age. As a result, the role of the federal government was transformed in the period of a few years.

The Social Security Act of 1935 was, without doubt, the most significant domestic social programme ever enacted by the US Congress. This marked the real beginning of co-operative federalism. This act established the principle of federal aid to the states for public health and welfare assistance. It provided federal grants to states for maternal and child health and crippled children's services and public health. It also provided for cash assistance grants to the aged, the blind, and destitute families with dependent children. This cash assistance programme provided the basis for the current federal–state programme of medical care for the poor, first as Medical Assistance for the Aged in 1960 and then as Medicaid in 1965. Both programmes linked eligibility for medical care to eligibility for cash assistance. More important, however, the Social Security Act of 1935 established the Old Age, Survivors' and Disability Insurance (OASDI) programmes that were to provide the philosophical and fiscal basis for Medicare, a direct programme for federal health insurance for the aged, also enacted in 1965.

In 1938, following the death of a number of children due to the use of elixir of sulphonamide, drug safety became a real issue in drug regulation. The result of this disaster was the Food, Drug and Cosmetic Act of 1938, which required drug manufacturers to demonstrate the safety of drugs before marketing. Amendments in the Food, Drug and Cosmetic Act in 1962 enhanced the law by specifying that drugs be effective, as well as safe. Advertising was also strictly regulated and more effective provisions included the removal of unsafe drugs from the market.

Federal concern for maternal and child health, particularly for the poor, was reflected in a temporary programme instituted during the Second World War to pay for the maternity care of wives of men enlisted in the army and navy. This means-tested programme successfully demonstrated the capacity of the federal government to administer a national health insurance programme. With the rapid demobilization after the war and opposition by organized medicine, the programme was terminated, but it was often cited by advocates of national health insurance, particularly those who accorded first priority to mothers and infants.

The introduction of the scientific method into medical research at the turn of the century and its gradual acceptance had a profound effect on national health policy and health care. The first clear health policy impact of the growing importance of research was the establishment of the National Institutes of Health (NIH) in 1930, followed by the enactment of the National Cancer Act of 1937 and the establishment of the National Cancer Institute within the framework of NIH. There followed multiple legislative enactments after the Second World War that created the present-day institutes, primarily focused on broad classes of disease, such as heart disease, cancer, arthritis, neurological diseases, and blindness, within the NIH. In the 15 years immediately after the Second World War, NIH grew from a small governmental laboratory to the most significant biomedical research institute in the world. The NIH became the principal supporter of biomedical research in the US, surpassing industry and leaving only a limited role for private foundations. Federal support for biomedical research was then one of the few areas of health policy activity, because the federal government avoided involvement in medical care for the civilian population and it did little to support health professions education in the 1950s and early 1960s. The influence of organized medicine was a critical factor in limiting the federal role during this period.

After the Second World War, it was evident that many of America's hospitals were woefully inadequate, and the Hill–Burton federal–state programme of hospital planning and construction was launched. This was a major federal initiative. Its initial purpose was to provide funds to states to survey hospital bed supply and develop plans to overcome the hospital bed shortage, particularly in rural areas. The Hill–Burton Act was amended numerous times as its initial goals were met. This legislation provided the stimulus for a massive hospital construction programme, with federal and state subsidies primarily for community, non-profit, voluntary hospitals. Public hospitals, supported largely by local tax funds to provide care for the poor, received little or no federal support until the needs of private institutions were met. The programme was a model of federal–state–private sector co-operation. It was a prime example of co-operative federalism and the major force—until enactment of Medicare and Medicaid—behind modernization of America's voluntary (non-profit) community hospital system.

During the period from 1945 to 1952, repeated attempts by President Harry S. Truman to have the US Congress enact a programme of national health insurance, funded through federal taxes, were thwarted, largely due to the efforts of the American Medical Association. No progress was made until the 1960s in extending the federal role into medical care financing.

By 1953, when the Department of Health, Education, and Welfare (now the Department of Health and Human Services) was created, the federal government's role in the nation's health care system, although limited, was firmly established. Biomedical research, research training, and hospital construction were the major pathways for federal support. Traditional public health programmes, such as those for venereal disease control, tuberculosis control, and maternal and child health, were supported at minimal levels through categorical grants to the states. Federal support for medical care was restricted to military personnel, veterans, merchant seamen, and native Americans (Indians) until 1960, when enactment of the Kerr–Mills law authorized limited federal grants to states for medical assistance for the aged. This programme was short lived, but it highlighted the need for a far broader federal effort in medical care for the poor and the aged.

From co-operative federalism to creative federalism: the transformation of health policy from 1961–9

A number of major federal health policy developments took place between 1961 and 1969, during the presidencies of John F. Kennedy and Lyndon B. Johnson. The bulk of the health legislation during that period was enacted in 1965–6 by the 89th Congress. Its record of legislative accomplishment in health was unparalleled. Never before or since had one session of Congress produced more legislation of such far-reaching implications for the health of the American people. Federal support to universities, hospitals, and non-profit institutes conducting research was expanded during this period, and federal aid in public health channelled through the states was also increased; but the largest increase in federal spending came as a result of the Medicare and Medicaid programmes which paid for medical care previously provided largely by the private sector. The term 'creative federalism' was applied to policies developed during the Johnson presidency that extended the traditional federal–state relationship to include direct federal support for local governments (cities and counties), non-profit organizations, and private businesses and corporations to carry out health, education, training, social services, and community development programmes (Reagan and Sanzone 1981).

The primary means used to forward the goals of creative federalism were grants-in-aid. Over 200 grant programmes were enacted during the 5 years fo the Johnson presidency. Among the many programmes initiated, only Medicare was directly administered by the federal government. Among the more important new laws enacted during the Johnson presidency were the Health Professions Educational Assistance Act 1963, which opened the door for direct federal aid to medical, dental, pharmacy, and other professional schools, as well as to students in these schools; the Maternal and Child Health and Mental Retardation Planning Amendments of 1963, which initiated comprehensive maternal and infant care projects and centres serving the mentally retarded; the Civil Rights Act of 1964, which barred racial discrimination, including segregated schools and hospitals; the Economic Opportunity Act of 1964, which provided authority and funds to establish neighbourhood health centres serving low income populations; the Social Security Amendments of 1965, particularly Medicare and Medicaid, which financed medical care for the aged and the poor receiving cash assistance; the Heart Disease, Cancer and Stroke Act of 1965, which launched a national attack on the major killers through regional medical programmes; the Drug Abuse Control Amendments of 1965; the Health Research Facilities Amendments of 1965; the Water Quality Act of 1965; the Clean Air Act Amendments and Solid Waste Disposal Act of 1965; the Health Professions Educational Assistance Amendments of 1965; the Medical Library Assistance Act of 1965; the Comprehensive Health Planning and Public Health Service Amendments of 1966, and the Partnership for Health Act of 1967, which re-established the principle of block grants for state public health services (reversing a 30-year trend of categorical federal grants in health). This legislation created the first nation-wide health planning system, which was dramatically changed in the 1970s to focus on regulation of health care as well as health planning.

The programmes of the Johnson presidency had a profound effect on inter-governmental relationships, the concept of federalism, and federal expenditures for domestic social programmes. Grants-in-aid programmes alone (this excludes social security and Medicare) grew from $7 billion at the beginning of the Kennedy-Johnson presidencies in 1961 to $24 billion in 1970 at the end of that era. In the next decade the impact was to be even more dramatic as federal grants-in-aid expenditures for these programmes grew to $82.9 billion in 1980. 'Grants-in-aid', noted Reagan and Sanzone (1981), 'constitute a major social invention of our time and are the prototypical, although not statistically dominant (they now constitute over 20 per cent of domestic federal outlays), form of federal domestic involvement.'

Not only did the programmes of the Johnson presidency have a profound effect on the nature and scope of the federal role in domestic social programmes, they also had an impact on health care itself. Federal funds for biomedical research and training, health staff development, hospital construction, health care financing, and a variety of categorical programmes were designed primarily to improve access to health care and, secondarily, to improve its quality. During the Kennedy–Johnson presidencies (1961–9) the primary agenda was one of assuring greater equity in access to health care (Lewis 1982).

The consensus reached in the 1960s that resulted in the flood of health legislation and a rapid increase in federal expenditures had its root in what Fox has called 'hierarchical regionalism'. This phrase was used by Fox to summarize three assumptions that he found to be the basis for organizing health policies in industrialized nations. These assumptions were:

> The causes and cures for most diseases are usually discovered in the laboratories of teaching hospitals and medical schools. These discoveries are then disseminated down hierarchies of investigators, institutions, and practitioners, which serve particular geographic areas. A central goal of health policy is stimulating the creation of hierarchies in regions which lack them and making existing ones more efficient. (Fox 1986, p. 76)

In the 1970s and 1980s, the success of the policies began to erode because of the rapidly growing expenditures for medical care and the growing competition among providers within regions (Fox 1986). The result was a growing emphasis on competition, decentralization, deregulation, consumer choice, and the free market (Havighurst 1986).

From creative federalism to new federalism in an era of limited resources

During the 1970s, President Richard M. Nixon coined the term 'new federalism' to describe his efforts to move away from the categorical programmes of the Johnson years toward general revenue sharing (transferring federal revenues to state and local governments with as few federal strings as possible) and, later, toward block grants (grants to state and local governments for broad general purposes), which fall between the no-strings-attached approach of general revenue-sharing and the detailed restrictions characterized by categorical grants. During the Nixon and Ford presidencies (1969–77), continuing conflict raged between the executive branch controlled by the Republican Party and the Congress controlled by the Democratic Party with respect to domestic social policy, including the new federalism strategy originally advocated by President Nixon. Congress strongly favoured categorical grants and was opposed to both revenue-sharing and block grants.

President Nixon also differed sharply with President Johnson in his explicit support for private rather than public efforts to solve the nation's health problems. On this fundamental issue the Nixon administration made its position clear:

> Preference for action in the private sector is based on the fundamentals of our political economy—capitalistic, pluralistic and competitive—as well as upon the desire to strengthen the capability of our private institutions in their effort to provide health services, to finance such services, and to produce the resources that will be needed in the years ahead. (Richardson 1971, p. i)

Although the Nixon administration attempted to implement its new federalism policies across a broad front, progress was made primarily in the fields of community development, staff training, and social services. Categorical grant programmes in health continued to expand despite attempts by the Nixon and Ford administrations to transfer programme authority and responsibility to the States (US Department of Health, Education, and Welfare 1976).

In the 1970s the expansion of two programmes—Medicare and Medicaid—dwarfed all other health programmes. The growth of Medicare and Medicaid was due largely to the rapidly rising costs of medical care in the 1970s, not to an expanded scope of benefits. The federal government basically bought into a system that had few cost-constraining elements, and the staggering expenditures had profound effects on health policy.

The federal government's response to skyrocketing medical care costs (and thus governmental expenditures) took a variety of forms. Federal subsidies of hospitals and other health facility construction were ended and replaced by planning and regulatory mechanisms designed to limit their growth. In the mid-1970s health staffing policies focused on specialty and geographical maldistribution of physicians rather than physician shortage, and by the late 1970s concern was expressed about an over-supply of physicians and other health professionals (Lee *et al.* 1976). Direct subsidies to expand enrolment in health professions schools were cut back and then eliminated. Funding for biomedical research began to decline in real dollar terms when an NIH-directed 'war on cancer' launched by President Nixon appeared to produce few concrete results and when Medicare and Medicaid pre-empted most federal health dollars.

More important than the constraints placed on resources allocated for health care were regulations instituted to slow the growth of medical care costs. Two direct actions were taken by the federal government. First, a limit was placed on federal and state payments to hospitals and physicians under Medicare and Medicaid (included in the 1972 Social Security Amendments). Second, wage and price controls were applied to the economy generally and continued for health care providers in the early 1970s. After wage and price controls on hospitals and physicians were lifted in 1974, costs rose rapidly.

Another regulatory initiative was designed to control costs through limiting the utilization of hospital care by Medicare and Medicaid beneficiaries. Although the original Medicare and Medicaid legislation mandated hospital utilization review committees, these appeared to have little effect on hospital utilization of costs. The Social Security Amendments of 1972 required the establishment of professional standards review organizations (PSROs) to review the quality and appropriateness of hospital services provided to beneficiaries of Medicare, Medicaid, and maternal and child health and crippled children programmes (paid for under authority of Title V of the Social Security Act). The PSROs were composed of groups of physicians who reviewed hospital records in order to determine if length of stay and services provided were appropriate. Results of these efforts were mixed. In only a few areas where PSROs were in operation was there evidence that cost increases were restrained, and in these areas it is not clear that the PSRO was a critical factor.

An attempt was also made to control costs through major changes in the organization of health care. Efforts were made to stimulate the growth of group practice prepayment plans, which provide comprehensive services for a fixed annual fee. Later, capitated prepayment systems developed that included more loosely organized groups of physicians, many engaged primarily in fee-for-service practice. These capitation-based prepayment organizations were defined in federal legislation enacted in 1973 as health maintenance organizations, or HMOs. Studies have demonstrated that certain HMOs (e.g group practices) provide comprehensive care at significantly less cost than fee-for-service providers, primarily because of lower rates of hospitalization (Luft

1980). Predictably, the federal stimulus for development of HMOs ran up against strong resistance from organized medicine. Nevertheless, the programme successfully enhanced professional and public awareness of HMOs and assisted in the development of a number of small prepaid group practices. While the impact on costs at the national level was minimal by the early 1980s, the plans were growing rapidly. By the late 1980s, after a few years of slow growth, HMO enrolment began to increase rapidly. Prospects for future growth, at a somewhat slower pace than originally predicted, seem excellent.

Additional steps were taken during the Nixon, Ford, and Carter administrations to strengthen health planning and make it a more effective tool to slow the rate of increase in health care expenditures. These efforts were not particularly successful. With few exceptions, local health planning agencies (Health Systems Agencies) were not part of local government, but rather non-profit private agencies strongly influenced by health care providers, particularly physicians and hospitals represented on their boards of directors. Hampered by broad and ambiguous mandates, with pressure to do something about the rising costs of health care, inexperienced staffs, limited resources, and the lack of direct links to third party payment (e.g. Medicare, Medicaid, private health insurance), it is not surprising that local health planning agencies were not effective in curbing the rapidly rising costs of health care (Salkever and Bice 1976).

In the Carter years (1977–81) there were few new successful health policy initiatives, in part because of the influence of special interests, such as the hospitals. The Carter administration tried, without success, to get Congress to enact comprehensive hospital cost-containment legislation. The hospitals and physicians were able to convince Congress that a voluntary effort would work more effectively. The rising costs of health care did moderate during the debate in Congress, but when mandatory controls were rejected by Congress, health care costs shot up at a rapid rate.

One major health programme initiated in the Carter years to stimulate more effective programmes in health promotion and disease prevention was to have a lasting impact. A national effort initiated by the Assistant Secretary for Health, Dr Julius Richmond, established national goals for health promotion and disease prevention. It also developed a process for involving state and local government officials and leaders in the private sector to improve knowledge, disseminate information, expand programme efforts, and identify barriers to progress related to health promotion and disease prevention. Concern about problems associated with cigarette smoking, sexually transmitted diseases, injuries, low birth-weight, adolescent pregnancy, and diet (especially related to excess fat) stimulated a wide range of national, state, and local programmes. Much was accomplished by federal leadership with a relatively small investment of federal funds.

The Reagan administration, which took office in 1981, accelerated the degree and pace of change in policy that had been developing since the early years of the Nixon presidency. Three major shifts in federal policy advanced by the Reagan administration have directly affected health care: (1) significant reduction in federal expenditures for domestic social programmes; (2) decentralization of programme authority and responsibility to the states, particularly through block grants; and (3) deregulation and greater emphasis on market forces and competition to address the problem of continuing increases in the costs of medical care.

Another policy development of major importance, the Tax Equity and Fiscal Responsibility Act (TEFRA) of 1982, has had direct and indirect effects on health care. First, by reducing taxes it has severely limited the fiscal capacity of the federal government to fund programmes; second, by requiring the development of a prospective payment system for hospitals it initiated a major reform in the Medicare programme. In the 1983 Social Security Amendments, Congress established a new hospital prospective payment system based on the use of diagnosis-related groups (DRGs). On 1 September 1983, the Health Care Financing Administration ushered in a new era in hospital payment with the publication of the regulations implementing the DRG-based prospective payment system. The payment system is now in effect for all hospitals accepting Medicare patients.

Although the Reagan administration strongly favours deregulation and the stimulation of pro-competition market forces, this has not always been translated into policy, except in the elimination of federal support for health planning. Indeed, in the Medicare programme, regulations to limit hospital reimbursement and physicians' fees have increased dramatically. Medicare's prospective payment system based on DRGs has been described as 'an exceedingly sophisticated, highly regulatory form of administered prices that changes the incentives facing hospitals' (Marmor 1987).

Although President Reagan's new federalism and pro-competition/deregulation policies attracted attention, it is the dramatic reduction in federal fiscal capacity due to tax cuts and the record increase in the federal deficit that have had the most significant effect on health services. In addition, in the early 1980s a number of states moved to restrict expenditures for Medicaid beneficiaries because of the continued impact of the rising costs of medical care and the increased Medicaid expenditures at the state and federal levels. Several states enacted policy changes that restricted patients' freedom to choose providers, reduced levels of hospital and physician reimbursement, and shifted the burden of care for large numbers of poor back to local government.

The severe recession of 1981–2 reduced revenue growth and had a profound effect on health policy at all levels of government. Even with the subsequent recovery and continued, but slow, economic growth, millions of Americans were left with limited or no health insurance. By the mid-1980s approximately 31 million Americans were without public or private health insurance (Wilensky 1988), with the number of uninsured increasing by about ten million between 1977 and 1985 (Mundinger 1985). The growing number of uninsured and underinsured has translated into diminished access to care for the poor and minorities, and this appears to be having some impact on the health status of these groups. Recent research has brought to light evidence of a levelling off of the 20-year decline in infant mortality; an increasing number of low birth-weight babies; and the deteriorating health status of children, women of childbearing age, and adults with chronic diseases (Mundinger 1985). The Robert Wood Johnson Foundation supported three independent studies on access to care in 1976, 1982, and 1986. In the 1986 survey a number of important findings were reported: (1) between 1982 and 1986 overall use of medical care declined; (2) access to physicians' services for individuals who were poor, black, and uninsured decreased between 1982 and 1986; (3) while hospital care declined for these groups, the decline in hospital use was comparable to that for the general population, but the uninsured, blacks, and Hispanics received less hospital care that might be appropriate given their burden of illness; (4) under-use of services was found among key population groups; (5) the urban–rural differences in access to care have disappeared; and (6) most Americans continue to be highly satisfied with physician and in-patient hospital care (Freeman *et al.* 1987).

Medicaid cuts reduced the number of beneficiaries by 600 000 between 1981 and 1983. Currently, Medicaid covers only 38 per cent of the poor, as compared with 65 per cent in 1976 (*AHA News* 1988), and in some states fewer than 20 per cent of the poor are covered (Mundinger 1985). Pregnant women and children have been the hardest hit by the Medicaid cuts. In 1984 Congress began to initiate Medicaid policy changes to redress some of the inequities that had occurred as a result of the cuts in 1981–3. These policy changes emphasized greater access to prenatal care for poor women and improved access to care for poor children. By the late 1980s a number of states were beginning to expand their Medicaid programmes to reach these groups, but millions remain dependent on local governments for their medical care.

The problem of the uninsured and uncompensated care is receiving more and more attention. In a study by Blendon and associates (1986) it was found that access to care for the poor in states with limited Medicaid programmes was markedly restricted when compared with states with more generous programmes. In contrast to the marked inter-state variation found in access to care for Medicaid recipients, the elderly, regardless of economic status, were found to have comparable access to care. The difference was the federal Medicare programme's uniform benefits and provider reimbursement policies that assured more equitable access to care. Others (Davis and Rowland 1983) have also found significant differences in access for the insured and uninsured.

Growing recognition of the problem of the uninsured and uncompensated care (for hospitals and physicians) has stimulated a variety of proposals to deal with the problem (Cohodes 1986; Lewin and Lewin 1987; Wilensky 1984). Basically, four approaches are available: (1) direct funding of providers, either through an all-payer system or by granting additional revenues from a fund of pooled resources; (2) targeting individuals, through a variety of insurance mechanisms—catastrophic insurance, state-sponsored insurance for high-risk individuals, insurance for the unemployed and mandated employer-funded private health insurance; (3) grants to states and local governments to permit state or local discretion in the type of approach adopted; and (4) doing nothing and letting the costs fall increasingly on local government and providers (Wilensky 1984).

Even more important than issues related to access and the uninsured have been the issues related to the rapid increase in expenditures for medical care. National health expenditures rose from $42 billion (5.9 per cent of gross national product) in 1965 to approximately $500 billion (11.1 per cent of gross national product) in 1987. In constant dollars the increase from 1965 to 1986 was from $46.8 billion to $107 billion. In dollars per caput the increase for the same period was from $229 to $431 (Table 13.1) (Ginzberg 1987).

Although the Reagan administration did not develop a comprehensive pro-competitive strategy, several factors have led to an increase in competition in the 1980s: (1) a dramatic increase in physician supply in the past 20 years; (2) Medicare's prospective payment for hospitals; (3) new forms of prepaid health care, often referred to as health maintenance organizations or preferred provider organizations; and (4) the growth of for-profit health care enterprise (Ginzberg 1987). Fuchs would add to this list increased consumer cost sharing and actions by state regulatory agencies (Fuchs 1986). Although there has been some amelioration in the rate of increase in hospital costs in the 1980s, expenditures for physicians' services continue to increase at a rapid rate. The basic problems identified by Ginzberg (1987) include the fragmented system of financing, with many third party payers; the continuing expansion of technology in the provision of care; and resistance to domination by the federal government.

The three basic approaches to cost containment—increased production efficiency, reduced prices, and reduced number of services per caput or per patient—have been described and analysed by Fuchs (1986). Countries such as Canada that have managed to contain costs more effectively than the US during the past decade have used all three approaches, with an emphasis on the last. Fuchs (1988) believes that this is the only approach likely to be effective in the long run.

Fuchs has also done much to clarify the current debate on the role of competition in health care. While discussing the changes in the 1980s, he found little evidence of increased competition, except for the growth of organized health plans. He noted that most of the pressures perceived by physicians and hospitals as increased competition are in reality due to the changing policies of buyers of care (employers, private health insurance, Medicare, Medicaid). He also noted that the supply-side factors that increase costs—hospitals, phys-

Table 13.1. National health care expenditures, 1960 through 1987*

Year	In current dollars			In constant dollars†	
	Billions of dollars	Percentage of gross national product	Average change from previous year shown (%)	Billions of dollars	Dollars per caput
1960	26.9	5.3	—	34.0	185
1965	41.9	5.9	9.3	46.8	229
1970	75.0	7.4	12.3	62.2	292
1975	132.7	8.3	12.1	78.5	349
1980	248.1	9.1	13.3	93.3	397
1981	287.0	9.4	15.7	97.5	410
1982	323.6	10.2	12.8	98.5	410
1983	357.2	10.5	10.4	100.0	412
1984	390.2	10.3	9.2	102.8	420
1985	425.0	10.7	8.9	105.4	427
1986	465.4	10.8	9.5	107.4	431
1987	500.3	11.1	9.8	—	—

Source: Ginzberg (1987).
* Data are for selected calendar years and are from the US Department of Commerce and the Health Care Financing Administration (Office of the Actuary, Division of National Cost Estimates).
† The Medical Care Price Index has been used to convert current to constant dollars.

icians, and technology—have not declined in the 1980s. Indeed, the number of physicians has continued to increase, as has the supply of new, often expensive, technologies. Once a new technology is in place, it tends to be used (Fuchs 1988).

A major problem that has arisen in the 1980s is the epidemic of human immunodeficiency virus (HIV) including acquired immune deficiency syndrome (AIDS), which is posing an unprecedented challenge to policy-makers at the federal, state, and local levels. The initial problem facing policy-makers was to determine the nature and the magnitude of the HIV/AIDS epidemic. AIDS was first recognized in Los Angeles and New York City in 1981 when a small number of homosexual men were diagnosed with a rare form of cancer (Kaposi's sarcoma) and rare forms of pneumonia (e.g. pneumocystis carinii pneumonia). It has since appeared in individuals who are not homosexual (e.g. intravenous drug users, transfusion recipients) and has spread to communities in all 50 states and the District of Columbia. Over 136 000 cases of AIDS have been reported in the US as of July 1990 and it is expected that the cumulative total of AIDS cases will exceed 270 000, with 170 000 deaths, before the end of 1991. The US Public Health Service estimates that 74 000 AIDS patients will be diagnosed in 1991 and an additional 71 000 previously diagnosed patients will be alive at the beginning of that year and will require treatment (Coolfont report 1986). Estimates of the number of HIV-infected individuals in the US have ranged from 500 000 (Osmond and Moss 1988) to 1.5 million (Curran *et al.* 1988). Recent studies have suggested that virtually all of those infected will develop AIDS unless new approaches to treatment prove effective.

Although the epidemic of AIDS has spread to all 50 states, some states (e.g. California, New York, Texas, Florida) and cities (e.g. New York, San Francisco, Los Angeles, Houston, Miami) have been more heavily affected than others. In addition, neighbourhoods within cities such as the Castro neighbourhood in San Francisco, the lower west side of Manhattan, and areas of West Los Angeles, have suffered devastating personal, social, and economic consequences of the AIDS epidemic.

Currently, the direct costs of health care for persons with AIDS exceeds $1.5 billion annually. The direct costs of health care for persons with AIDS will range from $3.5 billion to $9.4 billion in 1991 (Scitovsky and Rice 1987). Recent estimates, based on revised Centers for Disease Control projections of numbers of cases and the likely impact of treatment with azidothymidine (AZT), the first antiretroviral drug approved by the Food and Drug Administration, forecast a cost of $4.5 billion in 1991, excluding non-personal (e.g. HIV testing) costs (Hellinger 1988). Although these costs will represent less than 2 per cent of national health care expenditures in 1991, the epidemic will pose considerable strain not only on the financing of care, but also on the organization and delivery of health care in communities with large numbers of person with AIDS. Except for Hellinger's 1988 forecast, past cost estimates have been based on data developed before the use of AZT in treatment of persons with AIDS and the treatment with AZT and other antiviral drugs of persons with HIV infection who have not yet developed AIDS. These changes in treatment will certainly significantly prolong the lives of persons with HIV infection and may add substantially to the costs of their treatment.

The policy response at the federal level has stressed biomedical research, epidemiological surveillance, protection of the blood supply, and, increasingly, education and information. The tax subsidy for private health insurance and Medicaid funding for the care of persons with AIDS have received very little attention. Three goals established by the Secretary of Health and Human Services in 1983 have been achieved: the causative agent was identified (in France and the US); the mode of transmission of HIV infection was clarified; and an antibody test was developed and applied to protect the nation's blood supply. In addition, a great deal has been learned about HIV infection and AIDS, and several promising drugs have been developed.

Until very recently the most serious deficiency of federal policy had been the lack of an overall plan to combat the epidemic, to prevent the spread of HIV infection, and to care for those afflicted. Strong federal leadership, beyond the Surgeon General's call for education and prevention, is clearly needed, and appears to be gradually developing.

The second, and equally serious, problem has been the lack of adequate federal funding for research; education, information, and prevention; training; and the financing of acute medical care and long-term care, particularly subacute hospital care, skilled nursing care, and home care.

State policies, with few exceptions, have been very limited in view of the magnitude of the epidemic and heavy costs already borne by communities such as New York City, Los Angeles, West Hollywood, San Francisco, Miami, Houston, Baltimore, and more than a dozen other cities or countries. Less than half a dozen states have responded in any substantial way to the epidemic.

Local government responses, with the exception of San Francisco and New York, have been very limited. Recently, with funding provided by the Robert Wood Johnson Foundation and the US Public Health Service, programmes are being strengthened and expanded in more than a dozen cities and counties. In the communities hardest hit, the private sector has responded, particularly through community-based volunteer organizations often organized under strong gay male leadership.

The essential elements of a national HIV/AIDS programme have been outlined in the National Academy of Sciences Institute of Medicine reports (1986, 1988), the Public Health Service's Coolfont report (1986), the Surgeon General's Report (US Department of Health and Human Services 1987*a*), the Public Health Service's information/education plan (US Department of Health and Human Services 1987*b*), and the Report of the Presidential Commission on the Human Immunodeficiency Virus Epidemic (1988). It is essential that the federal, state, and local governments collectively and co-operatively with the major groups and institutions in the private sector (the insurance industry, hospitals, nursing homes, physicians, nurses, community-based home care agencies, foundations, the pharmaceutical industry, and academic medical centres) move forward to propose national policies, a comprehensive plan, and specific programmes to deal effectively with the HIV/AIDS epidemic.

The future of health care: pluralistic, bureaucratic, or radical change

In examining the health policy developments from 1965 to 1980, one of the nation's most able health policy scholars, Lewis (1982, p. 17) noted:

> The period of the 1960s and the 1970s was the age of special-interest liberalism. What we did was denigrate government and public service to the point where we regarded them as just another special interest instead of the essential broad framework for the processes of choice and effective decisions in the public interest. It is very easy to be smart when there is a lot of money, but it is not so easy to be smart when money is tight, especially when there is no solid political framework for choice.

It is the lack of solid political framework for choice that represents the most important barrier to addressing the issue of health care costs, their impact on governmental expenditures, patients' access to care, and equity. To achieve equity, for example, some of the liberties currently accorded physicians, dentists, pharmacists, hospitals, nursing homes, and other health professionals and institutions will have to be restrained. Pluralism is a prime characteristic of the US health care system, and it also characterizes the process by which domestic social policy decisions are made. As Lewis noted, 'Pluralism has come to mean a system of government where everybody is in charge, and nobody is in charge' (Lewis 1982). In this pluralistic system of interest group influence, government has become only another actor at the bargaining table.

Another astute analyst of health care politics, Alford (1975), described three models or theories of the causes of existing arrangements in health care: (1) the pluralist or market perspective; (2) the bureaucratic or planning perspective; and (3) the institutional or class perspective. Much of the description and analysis of health care policies has been based on either the pluralist or the bureaucratic perspective. While the pluralists hold that the present system is appropriate for our time and society, the bureaucrats see a more rational planned system in which resources are effectively co-ordinated (as opposed to the fragmentation accepted by the pluralists) and more appropriately allocated. Perceiving the primary obstacle to rational planning and resource allocation as the professional monopoly of physicians over medical education and practice, bureaucratic reformers would basically adjust the existing system to achieve agreed upon goals, such as equity of access and cost containment.

There has been relatively little research in the US examining the class basis of health policies. Those who hold that the defects in health care are deeply rooted in the structure of a class society would radically alter the present health care system, creating a national health service, with decentralization of administration and community control over health care institutions and health professionals. Those who view the defects in health care as having a class basis do not believe that tinkering with the health care system can achieve the desired outcomes. They call for major structural changes in society. The advocates of the class perspective have had little impact on health policy.

Policy developments of the coming decades will depend on which of these views of health care politics—pluralist, bureaucratic, or class—predominates. To date, the pluralists have played the most influential role in health care politics and policies.

Acknowledgements

This chapter was prepared while Dr Lee was a fellow at the Center for Advanced Study in the Behavioral Sciences. He is grateful for the financial support provided by the Henry J. Kaiser Family Foundation and the John D. and Catherine T. MacArthur Foundation.

References

Alford, R.R. (1975). *Health care politics, ideological and interest group barriers to reform*. University of Chicago Press, Chicago, Illinois.

American Hospital Association suggests plan to change Medicaid for people with AIDS. *AHA News* **24**(30), 1&5 (25 July, 1988).

Blendon, R.J., Aiken, L.H., Freeman, H.E., *et al.* (1986). Uncompensated care by hospitals or public insurance for the poor: Does it make a difference? *New England Journal of Medicine* **314**(18), 1160.

Blumstein, J.W. and Zubkoff, M. (1973). Perspectives on government policy in the health sector. *Milbank Memorial Fund Quarterly* **51** (summer), 395.

Cater, D. (1972). An overview. In *Politics of health* (ed. D. Cater and P.R. Lee), pp. 1–7. Medcom Press, New York.

Cater, D. and Lee, P.R. (ed.) (1972). *Politics of health*. Medcom Press, New York.

Cohodes, D.R. (1986). America: The home of the free, the land of the uninsured. *Inquiry* **23**, 227.

Coofont report (1986). A PHS plan for prevention and control of AIDS and the AIDS virus. *Public Health Reports* **101**, 341.

Curran, J.W., Jaffe, H.W., Hardy, A.M., *et al.* (1988). Epidemiology of HIV infection and AIDS in the United States. *Science* **239**, 610.

Davis, K. and Rowland, D. (1983). Uninsured and underserved: Inequities in health care in the United States. *Milbank Memorial Fund Quarterly* **61**(2), 149.

Estes, C.L. (1979). *The aging enterprise*. Jossey-Bass, San Francisco, California.

Foley, H.A. (1975). *Community mental health legislation*. Lexington Books, Lexington, Massachusetts.

Fox, D.M. (1986). The consequences of consensus: American health policy in the twentieth century. *Milbank Quarterly* **64**(1), 76.

Freeman, H.E., Blendon, R.J., Aiken, L.H., *et al.* (1987). Americans report on their access to health care. *Health Affairs* **6**(1), 6.

Fuchs, V.R. (1986). Has cost containment gone too far? *Milbank Quarterly* **64**(3), 479.

Fuchs, V.R. (1988). The 'competition revolution' in health care. *Health Affairs* **7**(3), 5.

Ginzberg, E. (ed.) (1977). *Regionalization and health policy*. US Government Printing Office, Washington, DC.

Ginzberg, E. (1987). A hard look at cost containment, *New England Journal of Medicine* **316**(18). 1151.

Grad, F. (1973). *Public Health Manual* (3rd ed.). American Public Health Association, Washington, DC.

Hale, G.E. and Palley, M.L. (1981). *The politics of federal grants*. Congressional Quarterly Press, Washington, DC.

Havighurst, C.C. (1986). The changing locus of decision making in the health care sector. *Journal of Health Politics, Policy and Law* **11**(4), 697.

Health Care Financing Administration, Office of the Actuary, Division of National Cost Estimates (1987). National Health expenditures, 1986–2000. *Health Care Financing Review* **8**(4), 1.

Hellinger, F.J. (1988). Forecasting the personal medical care costs of AIDS from 1988 through 1991. *Public Health Reports* **103**(3), 309.

Lee, P.R. and Silver, G.A. (1972). Health planning—a view from the top with specific reference to the USA. In *International medical care* (ed. J. Fry and W.A.J. Farndale) p. 284. Medical and Technical Publishing, Oxford.

Lee, P.R., Le Roy, L., Stalcup, J., *et al.* (1976). *Primary care in a specialized world*. Ballinger, Cambridge.

Levit, K.R. and Freeland, M.S. (1988). National medical care spending. *Health Affairs* **7**(5), 124.

Lewin, L.S. and Lewin, M.E. (1987). Financing charity care in an era of competition. *Health Affairs* **6**(2), 47.

Lewis, I.J. (1983). *Evolution of federal policy on access to health care, 1965–80. Bulletin of the New York Academy of Medicine* **59**(1), 9.

Litman, T.J. (1984). Government and health: the political aspects of health care—a sociopolitical overview. In *Health politics and policy in perspective* (ed. T.J. Litman and L.S. Robins), pp. 3–4. Wiley Interscience, New York.

Litman, T.J. and Robins, L.S. (1971). Comparative analysis of health care systems: A sociopolitical approach. *Social Science and Medicine* **5**, 573.

Luft, H.S. (1980). *Health maintenance organizations: Dimensions of performance*. Wiley Interscience, New York.

Marmor, T.R. (1987). *American health politics, 1970 to the present: Some comments*. Presentation at Health Care Policy Conference, College of William and Mary, Williamsburg, Virginia, 12–14 November.

Miller, C.A., Brooks, E.F., DeFriese, G.H., *et al.* (1977). Statutory authorizations for the work of local health departments. *American Journal of Public Health* **67**, 940.

Miller, C.A., Moos, M., Kotch, J.B., *et al.* (1981). Role of local health departments in the delivery of ambulatory care. *American Journal of Public Health* **71** (supplement), 15

Mundinger, M.O. (1985). Health service funding cuts and the declining health of the poor. *New England Journal of Medicine* **313**(1), 44.

National Academy of Science, Institute of Medicine (1986). *Confronting AIDS: Directions for public health, health care, and research*. National Academy Press, Washington, D.C.

National Academy of Sciences, Institute of Medicine (1988). *Confronting AIDS: Update 1988*. National Academy Press, Washington, D.C.

Osmond, D. and Moss, A. (1988). The prevalence of HIV infection in the United States: A reappraisal of the Public Health Service estimate. *1988 AIDS Clinical Review* **1**, 1.

Presidential Commission on the Human Immunodeficiency Virus Epidemic (1988). Final Report (July).

Reagan, M.D. and Sanzone, J.G. (1981). *The new federalism*. Oxford University Press.

Richardson, E.L. (1971). *Towards a comprehensive health policy in the 1970s*. Department of Health, Education, and Welfare, Washington, D.C.

Salkever, D.S. and Bice, T.W. (1976). The impact of certificate-of-need controls on hospital investment. *Milbank Memorial Fund Quarterly* **54**(2), 185.

Scitovsky, A.A. and Rice, D.P. (1987). Estimates of the direct and indirect costs of acquired immunodeficiency syndrome in the United States, 1985, 1986, and 1991. *Public Health Reports* **102**(1), 5.

Silver, G.A. (1976). *A spy in the house of medicine*. Aspen Systems Corporation, Germantown.

Silver, G.A. (1989). Ending the reign of dogma: Designing child health policy for America. *Bulletin of the New York Academy of Medicine* **65**(3), 255.

US Department of Commerce, Bureau of the Census (1989). *Statistical abstract of the United States: 1989* (109th edn). US Government Printing Office, Washington, DC.

US Department of Health and Human Services (1987*a*). *Surgeon General's report on acquired immune deficiency syndrome*. US Government Printing Office, Washington, DC.

US Department of Health and Human Services, Public Health Service (1987*b*). *Information/education plan to prevent and control AIDS in the United States*. US Government Printing Office, Washington, DC.

US Department of Health, Education, and Welfare (1976). *Health in America: 1776–1976*. DHEW Publication No. (HRA) 76–616. US Government Printing Office, Washington, DC.

Wilensky, G.R. (1984). Solving uncompensated hospital care: Targeting the indigent and the uninsured. *Health Affairs* **3**(4), 50.

Wilensky, G.R. (1988). Filling the gaps in health insurance: Impacts on competition. *Health Affairs* **7**(3), 133.

14

Public health policies and strategies in Europe

JAN E. BLANPAIN and MIA DEFEVER

Introduction

Before the early 1960s, public health policies and strategies in Europe were conceived and formulated predominantly at the national level. In retrospect, those policies seem to differ from country to country when considered at any given moment. However, when analysed over a longer period of time, a rather consistent pattern emerges of a similar sequence of dominant policies and policy shifts in Europe since 1883 when Bismarck made sickness insurance compulsory for part of the working population of Imperial Germany.

Looking back over the period since 1883, a sequence of seven dominant policies can be distinguished in most European countries: (1) provision of access to physicians; (2) priority for hospitals; (3) neglected areas as new priorities: mental health and care for the aged; (4) promotion of comprehensive integrated health care systems; (5) development of health care personnel as the crucial resource; (6) more equitably distributed and more efficient health care; (7) and cost containment (Blanpain *et al.* 1978).

Individual European countries were rarely synchronized with respect to the dominant policy of the period. Although similar policies were eventually adopted, there was no systematic effort, apart from anecdotal exceptions, to benefit from the experience in another country where a major shift in policy had occurred.

Since the beginning of the 1960s, European public health policies and strategies gradually began to develop. They redirected, complemented, or replaced national policies. These 'European' policies emanate from various supranational agencies and, depending on the agency concerned, relate to different European frameworks. The Organization for Economic Co-operation and Development (OECD), the Council of Europe, and even the North Atlantic Treaty Organization (NATO) occasionally address public health policies.

The periodic OECD health expenditure surveys and related policy considerations are, so far, the most comprehensive ones available at European level (OECD 1985, 1987). The Council of Europe, within its endeavours for human rights, has been active in the area of patients' rights, and it inspired the European Parliament of the European Economic Community (EEC) in this respect. A series of resolutions, recommendations, and charters regarding patients' rights has been issued: the Resolution of the Committee of Ministers of the Council of Europe on sterilization (1975), and on removal, grafting, and transplantation of human substances (1978); the Recommendation of the Parliamentary Assembly of the Council of Europe on the rights of the sick and the dying (1976); the Recommendation of the Council of Europe on the participation of the sick in their own treatment (1980), and on the legal protection of persons suffering from mental disorders placed as involuntary patients (1983); the Recommendation of the Council of Europe on automated medical data banks (1981), and on the legal duties of doctors regarding their patients (1985); and the Recommendation of the Council of Europe on the use of human embryos and fetuses for diagnostic, therapeutic, scientific, industrial, and commercial purposes (1986). The influence of the European international efforts to promote patients' rights is limited. National legislative practice is hardly affected beyond the moral value of the resolutions and recommendations (Leenen 1987).

Two agencies, the World Health Organization (WHO) and The EEC, are more engaged in public health policies and strategies than the other supra-national agencies, and further discussion is therefore limited to WHO- and EEC-related policies.

The WHO has developed a coherent and comprehensive set of health policies for Europe: 'Targets for health for all' (WHO 1985). In contrast, EEC health policies are rather fragmentary and are often the indirect result of economic policies, However, EEC health policies, to the extent that they are part of a directive, can be enforced whereas the implementation of WHO policies depends on the degree to which national health authorities honour the pledge made when they endorsed the regional strategy. There is a tendency at both national and supra-national level to pursue alignment of WHO and EEC health policies (House of Lords 1985), although the WHO and the EEC represent different sets of Member States. The WHO European Region comprises over 850 million people in 32 Member States, and includes Israel, Morocco, Tunisia, Turkey, and The Asian part of the USSR. The EEC which, since its creation in 1958, has gradually increased its membership from six to the current 12 Member States (Belgium, Denmark, the German Federal Republic, Greece, France, Ireland, Italy, Luxembourg, Spain, Portugal, The Netherlands, the UK), has a total population of around 320 million.

The WHO European Health For All Policy

History

At the thirteenth session of the Regional Committee of the Member States of the WHO European Region held in Fez in September 1980, a first common health policy for Europe was approved. This policy was developed in line with the WHO global strategy for Health For All By The Year 2000 launched at the 1979 World Health Assembly. The European Health For All regional strategy, approved at Fez, stresses: health promotion and disease prevention; inter-sectoral collaboration of all sectors with impact on health; and mobilization of individual and community resources to improve and maintain health. Primary care is emphasized as the major approach to attain, Health For All. At Fez, it was also decided to formulate specific regional targets to support the implementation of the strategy. Eventually, 38 specific regional targets were approved in September 1984. At the same time, a group of regional indicators was proposed to be tested and used for periodic evaluations of the progress made towards the realization of the targets. A first evaluation was made in 1985 (WHO 1986).

The targets strategy

The Health For All By The Year 2000 statement has occasionally been met with the criticism of being utopian. It was therefore defined in more precise terms as 'improved health status' which could realistically be reached in Europe by the year 2000. The pursuit of realistic and attainable improvement in health status led to the selection of health problems on the basis of the magnitude of the problem together with important social implications and the feasibility of effective intervention.

To achieve improved health status the targets strategy recognizes, moreover, that health in society is profoundly influenced by such basic factors as food, housing, education, employment, and natural or human-made disasters. Therefore, the scope of health policy must be broader than health services proper and must encompass the provision of the basic commodities and the freedom from the fear of disaster.

The goal of 'improving health status' is directed at four main objectives:

1. To ensure equity in health by reducing by 25 per cent the present inequalities in health status within countries and between countries.
2. To add 'life' to years by improving the quality of life of individuals primarily by assisting them to develop and maintain healthier lifestyles.
3. To add health to life by reducing morbidity and disability.
4. To add years to life by reducing avoidable death.

Of the 38 European regional targets (see Appendix) targets 1–12 cover these four main objectives in terms of health outcomes. Targets 6–12 relate to reduction in mortality. These are expressed in terms of increased life expectancy, reduction in infant and maternal mortality, or decreased disease-specific mortality (e.g. diseases of the circulatory system, cancer, accidents, and suicide). The remaining targets (13–38) deal with specific activities needed to reach the objectives. These activities address four major areas of concern.

First, healthy lifestyles are stressed as a central focus of health policy as well as the related required changes in the social, cultural, economic, and physical environment to enable individuals to pursue healthier lifestyles. Second, risk factors in the environment are emphasized. Third, the re-orientation of the health care system is stressed with a number of major issues: the strengthening of primary care and disease prevention; better care for the elderly; and appropriate use of health technology. The fourth major focus outlines the required staffing, research, and technological, political, management, and other support necessary to bring about the desired changes in the first three areas.

Evaluation of the European Health For All strategy

The first major evaluation of progress in implementing the WHO regional targets strategy was completed in 1985. It is intended that such an evaluation will be repeated every 6 years. The main purpose of the evaluation was to ensure that the agreed strategy has been followed by action. Moreover, countries were provided with a health bench-mark for future assessment and policy development.

Historical trends for 65 essential regional indicators or group of indicators incorporating the 12 global indicators, developed to assess the global Health For All strategy, were used to measure progress towards regional targets. Table 14.1 gives some examples of indicators used in the 1985 evaluation.

In general, percentage change between a 'historical year' (usually around 1970) and a 'latest available year' (usually around 1980) has been measured. For non-quantitative indicators, the evaluation is based on the number of countries that implemented a given type of intervention. An important handicap in conducting the evaluation was the lack of essential data regarding a number of targets and related indicators.

A significant assessment of the progress towards reducing health-related inequalities was not possible since data were available only for a few countries. Likewise, morbidity data and, in particular, disability data were in general either fragmentary or absent. Data on quality of life were virtually non-

Table 14.1. Examples of regional indicators used to measure progress towards regional targets

Adding years to life
Life expectancy at birth
Life expectancy at various ages
All malignant neoplasms
External injury and poisoning
Suicide
Adding health to life and life to years
Intestinal infections
Sexually transmitted diseases
Disability days per person per year
Long-term disability
Working days lost
Lifestyles
Tobacco consumption per adult
Alcohol consumption per adult
Intake of calories
Birth-weight
Adult literacy
Environmental health
Control of trans-frontier pollution
System for control of environmental hazards
Water and sanitation
Food safety
Work-related risks
Health care
Percentage of gross national product spent on health
Expenditure on local care
Quality of care
Technology assessment
Expenditure on research

existent and the lack of adequate indicators made it difficult to assess progress in this field. Measurement of trends with respect to lifestyle-related targets and assessment of the impact and success or failure of efforts to influence positively lifestyle-related behaviour proved difficult, given the available data. Quantitative measures of environmental hazards were scarce. Finally, the measurement of the situation and trends regarding the reorientation of the health care system proved to be fraught with problems of definition and standardization. In particular, direct information on the movement towards primary health care was generally unavailable. A crude estimation was derived from the trends in expenditure on institutional care developed by the OECD (1985).

Ideally, a national information system for health policy purposes should comprise the following elements:

(1) population data: age, gender, race, geographical distribution, literacy, income distribution, family structure, housing, and employment;

(2) epidemiological data: mortality, prevalence and incidence of morbidity including data on health status, nutritional status, access to safe drinking water, and exposure to environmental hazards;

(3) health resources data: health care personnel, health technology, and health facilities;

(4) utilization data: e.g. visits to physicians or dentists, hospital admissions, in-patients days, drug utilization, and diagnostic tests;

(5) health expenditure data: as a share of national income and expenditure, by subpopulation, and by type of care, e.g. health promotion, primary care, in-patient care, long-term care, pharmaceuticals;

(6) outcome data: effects on health status of health resource utilization.

The evaluation with respect to target 35 (health information system) indicates that countries are making progress toward developing timely and relevant information for health policy and health management purposes.

Implementation of the strategy

The following findings on the implementation of the Health For All strategy have been abstracted from various WHO reports and OECD studies (Asvall 1986; Asvall *et al.* 1986; Brzezinski 1986; OECD 1985, 1987; WHO 1986). Although the commitment of the WHO Member States to a common health policy was apparent in the formal adoption of the Health For All strategy and in the quality of most of the national evaluation reports, only a limited number of countries had formulated detailed Health For All strategies by 1985. Also, the provision of information to the public and their involvement in Health For All policies was very slight and information systems concerning the needs of the strategies needed substantial improvement.

The 1985 evaluation showed that the regional targets are realistic and feasible. Many countries have already, or will by the year 2000, reach many of the targets. Yet, great differences continue to exist between countries of similar levels of development.

Prerequisites for health

Demographic changes, in particularly the aging of the population and the marked increase in the numbers of the very old, contribute to a continuing rise in demand for health care and social services. Furthermore, the negative effects on health of unemployment, particulary long-term unemployment, are increasingly being recognized. These increasing demands occur within an environment of prolonged economic crisis with its concomitant policies to contain health care costs.

Equity in health

The progress towards reducing by 25 per cent differences in health status between countries and between groups within countries has been poor. For six of 12 mortality-related regional indicators of health status, the gap between countries has been widening. For the few countries where data allowed the assessment of differences between groups, no decreases of the existing gap was found on average; occasionally a widening of the gap has been noted. These differences seem unresponsive to significant increases in health expenditure and deployment of health resources.

Adding years to life

Increase in life expectancy has resulted from both a significant improvement in infant mortality and from increasing the longevity of people aged over 65 years. Around 1982, 26 countries had achieved an infant mortality rate of less than 20 deaths per 1000 live births compared with only 12 countries in 1970. By the early 1980s, six countries had already reached the regional target of 75 years' expectancy at birth and it is expected that most countries will reach the target by the year 2000. Maternal death dropped to less than 15 per 100 000 live births in 22 countries.

Among the three leading causes of death (cardiovascular disease, cancer, and injury), decreasing trends in the mortality rate due to diseases of the circulatory system and from accidents were noted. Mortality from cerebrovascular diseases showed a pronounced decline whereas the trend for ischaemic heart disease was mixed: a decline in mortality among younger men in several countries and an increase in adults aged under 65 years in most countries. Mortality due to all cancers and suicide, in particular among young adults, also increased. Progress in the area of reducing avoidable deaths resulting from smoking and nutritional patterns has been limited. The reduction of tobacco and alcohol consumption and desirable changes in food habits will require stronger and more multi-sectoral action to reach the targets.

Adding health to life and life to years

New infectious diseases, such as legionellosis and in particular AIDS, have appeared on the scene. The incidence of gonorrhoea and primary syphilis has increased since the 1970s. Progress in the reduction of infectious diseases, such as tuberculosis, diphtheria, acute poliomyelitis, and tetanus, is encouraging but the situation for measles and pertussis is less satisfactory. By 1984, routine rubella vaccination had been introduced in eight countries. A resurgence of malaria occurred in Turkey. Information on other communicable disease such as influenza, infectious intestinal diseases, and viral hepatitis proved to be unreliable and incomplete.

Data on disability and impairment of activities of daily living are fragmentary and do not allow to region-wide trends to be determined.

Lifestyles

Nutritional data indicate that fat intake stabilized recently after an increase during the period 1970–82. However, in 17 countries (Compared with 14 in 1970) the amount of fat in the diet still exceeded the recommended level of 35 per cent despite the mounting epidemiological evidence that excess fat intake is an important risk factor for cardiovascular disease and certain types of cancer.

Eleven countries out of 26 recorded a decline in cigarette consumption, although the overall consumption in the region increased slightly between 1976 and 1983. Among the countries witnessing a decrease in cigarette smoking, five are predicted to reach a 50 per cent reduction by the year 2000. Legislation aimed at reducing tobacco consumption has increased throughout the region. Measures against advertising, fiscal measures to increase the cost of smoking practices, smoking bans in public areas, and health warnings on cigarette packets have been enforced in a rapidly growing number of countries.

The consumption of alcohol decreased by 7 per cent between 1970 and 1983. In particular, those countries with previously high levels showed a decline while several countries showed a marked increase up until the mid-1970s, followed by a stabilization or decline.

The limited data available indicate that illicit substance abuse and licit use of psychotropic drugs is increasing at an alarming rate.

Mortality resulting from homicide and deliberate injury increased by some 20 per cent from 1972 to 1981. Various forms of violent social behaviour—child abuse, wife-beating, sexual abuse, and rape—were stressed as areas of great concern in many Member States.

Traditional health promotion campaigns aimed at the avoidance of given risk factors, such as traffic accidents, smoking, alcoholism, and substance abuse, have recently been complemented by a trend towards active support for activities focused on positive health promotion, such as increased physical activity, balanced nutrition, and the promotion of breast-feeding. Available data so far do not allow an evaluation of the effectiveness of the various health promotion campaigns.

Environmental health

While region-wide quantitative data on environmental hazards are scarce, most Member States in the WHO European region have developed comprehensive regulations to monitor and regulate key areas of environmental health. The areas of greatest concern are housing, food safety, water pollution, air pollution, dangerous consumer goods, occupational health, hazardous waste, and radiation. Interest in the latter has been boosted by the Chernobyl nuclear accident.

The housing situation in the European Region remains far from satisfactory, and relevant legislation in general lags behind known health requirements. The proportion of the population covered by appropriate water supplies is 92 per cent, but only 66 per cent of the regional population has adequate sanitary facilities. Some countries will probably not meet the targets of the International Drinking Water Supply and Sanitation Decade which aim to provide all people with a continuous supply of safe drinking water and appropriate means of sanitation by 1990.

The need for international co-operation and agreements to monitor, assess, and control the main environmental hazards, which basically have a trans-frontier character, has been recognized. International co-operation is being implemented through the development of sophisticated national monitoring systems linked to region-wide, and even global, monitoring networks, through increased involvement of international agencies and through various agreements regarding air pollution, as well as storage, treatment, and disposal of hazardous waste. Environmental health issues have become the rallying cry of new political formations, the so-called 'Green parties', in a growing number of countries. Together with well organized and determined activist groups, such as Greenpeace, they basically pursue the protection and improvement of the environment. Recent political developments in Scandinavia, the German Federal Republic, Belgium, and France indicate the growing influence of Green parties on both the political process and on specific environmental issues.

Changes in health care systems

Health expenditure, expressed in terms of percentage of gross national product, continued to rise until the beginning of the 1980s when the regional average was about 5.5 per cent. Since then, the rise in health expenditure has levelled off in a growing number of countries, with some actually witnessing a decrease in percentage of gross national product devoted to health. Similarly, public expenditure as a percentage of total health spending, which had been rising since the 1960s, levelled off at the beginning of the 1980s.

Cost containment policies are apparently having an effect. However, governments continue to address health expenditure in terms of an 'out-of-control cost explosion'. This attitude leads to arbitrary budgetary restrictions and cost-sharing by patients, and sometimes to outright rationing of health resources. It seems to be dictated more by the imperatives of diminishing public deficits than by the goal of bringing the rise in health expenditure in line with the growth of gross national product.

Health care personnel increased substantially, to the extent that an over-supply of doctors is acknowledged in several countries such as Belgium, Italy, and Spain.

Hospital facilities in terms of the bed:population ratio in 30 countries increased only slightly between 1970 and the early 1980s compared with the marked increase during the period 1960–70, indicating the effect of national policies of hospital containment.

Information from 25 countries reports availability of primary health care for the total population. However, the available information does not permit determination of the content of this primary care or assessment of whether it provides the full range of health promotion, curative, rehabilitative, and supportive services in response to the basic needs of the population. This information gap is to a substantial degree caused by the different meanings attached to the term 'primary care' in different countries and by the resulting confusion reflected in policy statements and analysis.

Programmes for the systematic evaluation of the quality of health care are seldom available (Blanpain 1985). Likewise, systematic efforts to assess technology are only sporadic, apart from well established national programmes for the assessment of the efficacy and safety of new drugs, and the rapid development of post-marketing drug safety surveillance.

Prospects for the European Health For All strategy

The production of the WHO European target document has gradually been emulated by similar health policy formulations at national, sub-national, and even metropolitan area level using the WHO target strategy as a guiding framework. At the end of 1986, the WHO reported that five countries had already completed large-scale planning exercises to adapt national policy to fall in line with WHO policy and that eight more were in the process of doing so while, in 1987, the remaining countries were planning events to initiate a national health policy discussion (Steering Committee on Future Health Scenarios 1988).

The health authorities in The Netherlands were among the first in Europe to initiate the formulation of a coherent health policy memorandum on the future developments of the health situation of the Dutch population along the lines of the WHO strategy. After wide consultation with all interested parties, the 'Health 2000' memorandum was presented to the national parliament in 1986 (Gunning-Schepers 1986). In conjunction with the formulation of the 'Health 2000' memorandum, the Dutch health authorities started a pioneering effort in 1983 with the systematic development of a series of future scenarios in the field of public health and health care. These 'health scenarios' are sets of possible and desirable 'futures' based on quantitative and qualitative analyses of future health needs and anticipated health technology. These insights into future needs and resources can be used to increase the anticipatory ability of policy-making and strategic planning in the field of health and health care. Scenarios on a wide range of topics continue to be published in The Netherlands: (a) growing old in the future; (b) the heart of the future—the future of the heart; (c) anticipating and assessing health care technology; (d) accidents and traumatology; (e) cancer in The Netherlands; (f) appropriate health technology at home; and (g) the hospital of the future. Scenario approaches for national Health For All strategies are also being used in Finland, Poland, and Sweden (Steering Committee on Future Health Scenarios 1988).

At the WHO European level, a priority research policy to promote Health For All has also been formulated in view of discussions throughout the region, and developmental work has started to define an educational strategy in support of Health For All (Asvall 1988).

Public health policies and strategies in the EEC

The EEC, created in 1958, has its main thrust in the economic sphere: removing barriers to trade, capital, and services, and introducing common policies first in agriculture and later in other economic fields. Following the ratification and implementation of the Luxembourg 1987 Single European Act, the 12 Member States are fully committed to the creation, by 31 December 1992, of a single market without internal frontiers in which the movement of goods, persons, services, and capital is ensured (EEC 1987).

Although the original treaty governing the EEC did not explicitly include health policies, these were, step by step, given attention at various levels in the institutional structure of the Community, and specific policies and programmes gradually emerged in several areas: environmental health, industrial health, food sanitation, pharmaceuticals and veterinary medicines, health personnel deployment, medical education, medical research and health services research, control of substance abuse, acquired immune deficiency syndrome (AIDS), cancer, and medical informatics. Some understanding of the structure and functioning of the EEC is needed to grasp the decision-making process regarding EEC health policies and strategies.

The EEC has an Assembly or European Parliament which, since it was constituted on the basis of direct European elections, increasingly reflects the daily preoccupations of the European citizens. Health policy has become an important issue on the agenda of the European Parliament and, in recent years, several resolutions and recommendations have been issued. As the function of the European Parliament is predominantly advisory, the real power is exerted by the Council of Ministers and by the Commission.

Each country is represented in the Council of Ministers by one Minister. Depending on the business before the Council, the appropriate Minister, be it finance or health, will represent his or her country. Since 1969, summit meetings have been held by the Heads of State and lately they have included health-related items on their agenda. Yet, the Council of Ministers of Health interrupted its formal meetings between 1978 and 1985 primarily because some Member States considered health affairs to be largely a national prerogative.

The Commission is the executive body of the EEC and is designed to counterbalance the Council of Ministers, where nationalistic viewpoints tend to prevail. The Commission is a body of permanent and full-time commissioners chosen from each of the Member States 'for their general competence and indisputable independence'. The Commission represents the Community as a whole and is responsible for carrying out the provisions of the Treaties, the Single European Act, and the decisions of the Council. The Commission not only implements decisions but also drafts policy, and thus has the power to propose ways in which the general objectives of the Treaties can be implemented. The implementation of the 1987 Single European Act has led the Commission to draft more than 300 measures to realize the objectives of the Act. The Commission is assisted by civil servants and by a number of committees among which are several health-related committees, e.g. the Committee on the Environment, Public Health and Consumer Protection, the Committee for Proprietary Medicinal Products, and the Advisory Committee on Medical Training. The main vehicles for implementation of Community policy are directives which are binding upon Member Governments as to their ends but allow the Member States to make appropriate arrangements to reach these ends.

Industrial health and radiation health policies

Industrial health was among the first health concerns addressed by the Community, and action was initiated under both the Euratom Treaty and the Treaty of Rome (1957).

Several directives have been issued regarding the principles of surveillance and the control of exposure to toxic substances. Maximum limits of exposure have been issued and are regularly being adjusted and updated. Certain specific agents and industrial processes have been banned.

In the field of radiation health, the protection of the workers and the population at large in the 12 Member States is based on the same principles of prevention, monitoring, and exposure limits. The Chernobyl accident has led to expanded responsibilities of the Commission regarding the analysis and interpretation of the consequences of the accident and in determining new reference limits with respect to contaminated food.

Health personnel policies

Within the perspective of free movement of labour and the related mutual recognition of diplomas envisaged by articles 48 and 57 of the Treaty of Rome, directives were issued in 1975 providing for mutual recognition of professional qualifications and freedom of movement of health professionals throughout the Community (physicians, veterinarians, pharmacists, midwives, nurses, and dentists). In reality, this free movement has not materialized to the degree that some predicted: in 1988, migration had hardly increased from the level before the directives.

Within the framework of harmonizing medical education and specialist training a directive was adopted in 1986 whereby a specific complementary training for general practitioners became obligatory. From 1988, Member States must have introduced a 2-year training after 6 years of basic training along the lines stipulated in the directive. Starting on 1 January 1995, a certificate of special training will be required for generalists to be eligible for social security-related reimbursement of medical fees.

Towards a single European market for medicinal products

A substantial body of Community legislation has already harmonized the rules for authorizing the marketing of new medicines within the Community. As long ago as 1965, it was laid down that the sole criteria for authorizing new medicines should be the quality, safety, and efficacy of the medicinal products concerned. In 1975, two further directives dealt in detail with the types of tests and trials necessary to demonstrate these characteristics, with labelling and packaging requirements, and with the obligations of manufacturers. In addition, a series of notes for guidance on how to conduct the various types of tests and trials has been adopted.

Although the criteria for registration are harmonized, the actual assessment of applications and the final decision on individual products in 1988 still remains the responsibility of Member States. Two procedures for co-ordinating national decisions at Community level are available. The first procedure, a decentralized one, enables a company which has previously obtained authorization in one Member State to apply for the extension of the authorization to cover two or more other Member States. If one or more of these countries objects to the product, the application is referred to a Community level committee, the Committee for Proprietary Medicinal Products (CPMP), for an opinion. The second procedure, which is reserved for medicinal products derived from biotechnology and other high technology medicinal products, involves the concerted assessment of new products from the moment an application is first made to any Member State. At the end of the procedure, the CPMP gives an opinion to the Member States on the acceptability of the drug concerned. In neither case is the opinion binding but experience shows that opinions of the CPMP can have strong influence. A central feature of the Single Act internal market programme is the establishment of a definitive system for free circulation (mutual recognition, single authorization, or a combination of both) regarding new drugs, which will be valid throughout the 12 Member States. The objective is to achieve this situation by August 1989. It is also intended to formalize the arrangements for post-marketing drug safety monitoring and the withdrawal of unsafe or ineffective products which have evolved within the Community.

Within the countries of Europe, pharmaceutical consumption is covered by compulsory health insurance schemes, and a substantial proportion of the cost is met by the health insurance schemes. Price controls or restrictions on the range and type of medicinal products covered by the health insurance scheme of a given country are common practice. This practice results in different prices for the same drug in different countries and leads to parallel imports of drugs between countries. A Commission proposal to increase transparency of these national price control and reimbursement policies has been issued to help solve the complex problems posed by the realization of the internal market in view of existing price and reimbursement controls.

By May 1990, Member States are bound to have reviewed the quality, safety, and efficacy of older medicinal products, after which officially approved scientific information should be available for all medicinal products on the Community market.

Proposals have been introduced by the Commission to extend the scope of the Community directives to cover immunological products, medicinal products derived from human blood, and radiopharmaceuticals. In addition, it is intended to rationalize the information given to patients about over-the-counter products, to improve guarantees of the quality of manufacturing of pharmaceutical products by requiring compliance with a Community guide to Good Manufacturing Practice, and to improve the provision of information about medicines to Third World countries. In this regard, the Community supports the drug strategy of the WHO and recognizes the importance of the WHO quality certification scheme and provisions for product information to Third World countries.

The authorization of veterinary medicinal products within the Community is also covered by directives which contain provisions to ensure that food derived from animals does not contain hazardous residues for the consumers.

Environmental health

Since the adoption in 1972 of a first Community policy on the environment, several directives have been adopted by the Council aimed at the prevention of environmental risks and at the safeguarding of the natural environment. Common norms and objectives regarding the quality of the respective environments (water, air, and soil) have been agreed upon.

The fight against disease

A European cancer programme has been started. It focuses on research and prevention. Contrary to the mere co-ordination of concerted actions which characterize most of the Community-initiated medical research, this programme is directly financed by the Community.

AIDS has been the subject of an inter-governmental resolution aimed at co-ordinated action.

Following resolutions of the European Parliament, the Commission is developing plans for controlling substance abuse, in particular illicit drugs, alcohol, and tobacco. Close consultation is taking place to avoid duplication with similar efforts by the WHO.

Medical and health research

Four streams of health-related research programmes have been developed in the Community with different resources and modalities. The oldest one originated within the framework of the European Coal and Steel Community and included ergonomic research in the coal and steel industries and studies on industrial hazards (pneumoconiosis, noise levels, etc.).

A second series of programmes, started in 1960, focused, via contract research (300 projects in the period 1985–9), on radiation protection in support of the normative activities of the Commission in the fields of radiation health and radiation protection. Current research is centred on cellular changes in radiobiology, effects of radiation on the nervous system, skin lesions due to radiation, and methodologies to evaluate the radiation consequences of a nuclear disaster.

A third set of research projects started in 1978 as concerted actions whereby the Commission co-ordinates research by national teams around themes chosen in consultation. The main goal is to address research activities with pooled national financial resources and skills. The programme has grown from three concerted actions in 1978–81 involving 100 national teams, into 34 concerted actions, during 1982–6, with 1400 national teams participating. It is estimated that about 4000 scientists were involved. Non-Member States have been given the opportunity of participating in selected parts of the programme. Formal agreements of co-operation have thus been concluded with Sweden, Switzerland, and Canada, and an informal exchange of expertise exists with Austria, Finland, Norway, and Yugoslavia, as well as with the National Institutes of Health, US, and occasionally with Japan.

The co-ordination programme of medical and health research for the period 1987–91 is divided into two subprogrammes: (a) major health problems; and (b) health resources. New Targets (cancer research, as a follow-up of the European Councils since Milan in 1985, and research into AIDS) are included in the subprogramme major health problems. Continuation of research included in previous programmes is provided for to the extent that they are referring either to age-related (including disabilities) or to environment and lifestyle-related health problems, respectively. The second subprogamme is oriented to the improvement and joint use of those health resources required to ensure optimization of the cost:effectiveness ration in the health care field. Particular attention will be given to strengthening medical technology development on the one hand and to developing health services research on the other.

The fourth set of health-related research is the Community programme in Advanced Informatics in Medicine in Europe (AIM) which started its pilot phase in 1988. In its pilot phase, AIM intends to analyse and evaluate future information needs of medicine and biotechnology and to define precise objectives and optimal approaches for concerted action and collaboration in the field of medical informatics and bioinformatics in order to re-enforce public and private efforts in this respect.

Towards a European health policy

The growing importance of health issues in Community law has not reached the point where the EEC can be considered as a space governed by a common health policy (Massart-Pierard *et al.* 1988). Although an expanding set of health policies has been issued by the Community there is still a long way to go before the fragmentation will be overcome and comprehensiveness and coherence can be attained.

Appendix: the 38 regional targets for Health For All

1. By the year 2000, the actual differences in health status between countries and between groups within countries should be reduced by at least 25 per cent, by improving the level of health of disadvantaged nations and groups.

2. By the year 2000, people should have the basic opportunity to develop and use their health potential to live socially and economically fulfilling lives.

3. By the year 2000, disabled persons should have the physical, social, and economic opportunities that allow at least for a socially and economically fulfilling and mentally creative life.

4. By the year 2000, the average number of years that people live free from major disease and disability should be increased by at least 10 per cent.

5. By the year 2000, there should be no indigenous measles, poliomyelitis, neonatal tetanus, congenital rubella, diphtheria, congenital syphilis or indigenous malaria in the Region.

6. By the year 2000, life expectancy at birth in the Region should be at least 75 years.

7. By the year 2000, infant mortality in the Region should be less than 20 per 1000 live births.

8. By the year 2000, maternal mortality in the Region should be less than 15 per cent 100 000 births.

9. By the year 2000, mortality in the Region from diseases of the circulatory system in people aged under 65 years should be reduced by at least 15 per cent.

10. By the year 2000, mortality in the Region from cancer in people aged under 65 years should be reduced by at least 15 per cent.

11. By the year 2000, deaths from accidents in the Region should be reduced by at least 25 per cent through an intensified effort to reduce traffic, home, and occupational accidents.

12. By the year 2000, the current rising trends in suicides and attempted suicides in the Region should be reversed.

13. By 1990, national policies in all Member States should ensure that legislative, administrative, and economic mechanisms provide broad inter-sectoral support and resources for the promotion of healthy lifestyles and ensure effective participation of the people at all levels of such policy-making.

14. By 1990, all Member States should have specific programmes which enhance the major roles of the family and other social groups in developing and supporting healthy lifestyles.

15. By 1990, educational programmes in all Member States should enhance the knowledge, motivation, and skills of people to acquire and maintain health.

16. By 1995, in all Member States, there should be significant increases in positive health behaviour, such as balanced nutrition, non-smoking, appropriate physical activity, and good stress management.

17. By 1995, in all Member States, there should be significant decreases in health-damaging behaviour, such as overuse of alcohol and pharmaceutical products, use of illicit drugs and dangerous chemical substances, and dangerous driving and violent social behaviour.

18. By 1990, Member States should have multi-sectoral policies that effectively protect the environment from health hazards, ensure community awareness and involvement, and support international efforts to curb such hazards affecting more than one country.

19. By 1990, all Member States should have adequate machinery for the monitoring, assessment, and control of

environmental hazards which pose a threat to human health, including potentially toxic chemicals, radiation, harmful consumer goods, and biological agents.

20. By 1990, all people of the Region should have adequate supplies of safe drinking water, and, by the year 1995, pollution of rivers, lakes, and seas should no longer pose a threat to human health.

21. By 1995, all people of the Region should be effectively protected against recognized health risks from air pollution.

22. By 1990, all Member States should have significantly reduced health risks from food contamination and implemented measures to protect consumers from harmful additives.

23. By 1995, all Member States should have eliminated major known health risks associated with the disposal of hazardous wastes.

24. By the year 2000, all people of the Region should have a better opportunity of living in houses and settlements which provide a healthy and safe living.

25. By 1995, people of the Region should be effectively protected against work-related health risks.

26. By 1990, all Member States, through effective community participation, should have developed health-care systems that are based on primary health care and supported by secondary and tertiary care as outlined at the Alma-Ata Conference.

27. By 1990, in all Member States, the infrastructures of the delivery systems should be organized so that resources are distributed according to need, and so that services ensure physical and economic accessibility and cultural acceptability to the population.

28. By 1990, the primary care system of all Member States should provide a wide range of health-promotive, curative, rehabilitative and supportive services to meet the basic health needs of the population and should give special attention to high-risk, vulnerable, and underserved individuals and groups.

29. By 1990, in all Member States, primary health systems should be based on co-operation and team-work between health care personnel, individuals, families, and community groups.

30. By 1990, all Member States should have mechanisms by which the services provided by all sectors relating to health are co-ordinated at the community level in a primary health care system.

31. By 1990, all Member States should have built effective mechanisms for ensuring quality of patient care within their health care systems.

32. Before 1990, all Member States should have formulated research strategies to stimulate investigations which improve the application and expansion of knowledge needed to support their Health For All developments.

33. Before 1990, all Member States should ensure that their health policies and strategies are in line with Health For All priniciples and that their legislation and regulations make their implementation effective in all sectors of society.

34. Before 1990, Member States should have managerial processes for health development geared to the attainment of Health For All, actively involving communities and all sectors relevant to health and, accordingly, ensuring preferential allocation of resources to health development priorities.

35. Before 1990, Member States should have health information systems capable of supporting their national strategies for Health For All.

36. Before 1990, in all Member States, the planning, training, and use of health personnel should be in accordance with Health For All policies, with emphasis on the primary health care approach.

37. Before 1990, in all Member States, education should provide personnel in sectors related to health with adequate information on the country's Health For All policies and programmes and their practical application to their own sectors.

38. Before 1990, all Member States should have established a formal mechanism for the systematic assessment of the appropriate use of health technologies and of their effectiveness, efficiency, safety, and acceptability, as well as reflecting national health policies and economic restraints.

Source: WHO (1986), with permission.

References

Asvall, J.E. (1986). Towards a European policy in health *Health Policy* **6**, 221.

Asvall, J.E. (1988). The healthy public policy—stimulating health for all development. In *Proceedings of the Second International Conference on Health Promotion,* Adelaide (in press).

Asvall, J.E., Jardell, J.–P., and Nanda, A., (1986). Evaluation of the European strategy for health for all by the year 2000. *Health Policy* **6**, 239.

Blanpain, J.E. (1985). The role of medical associations in quality assurance. *Health Policy* **4**, 291.

Blanpain, J.E., Nys, H., and Delesie L. (1978). *National health insurance and health resources: The European experience.* Harvard University Press, Cambridge, Massachusetts.

Brzezinski, Z.J. (1986). Mortality indicators and health-for-all strategies in the WHO European region. *World Health Statistical Quarterly* **39**, 365.

European Economic Community (1987). *Europe without frontiers—completing the internal market.* EEC Official Publications, Luxembourg.

Gunning-Schepers, L.J. (1986). 'Health for All by the Year 2000': A mere slogan or a workable formula? *Health Policy* **6**, 227.

House of Lords Select Committee on the European Communities (1985). *Co-operation at Community level on health-related problems.* HMSO, London.

Leenen, H.J.J. (1987). Patient's rights in Europe. *Health Policy* **8**, 33.

Massart-Pierard, F., Anrys, H., Belanger, M., *et al.* (1988). *L'Europe de la santé: Hasard et/ou nécessité?* Academia, Louvain-la-Neuve.

Organization for Economic Co-operation and Development (1985). *Measuring health care 1960–1983.* OECD, Paris.

Organization for Economic Co-operation and Development (1987). *Financing and delivering health care*. OECD, Paris.

Steering Committee on Future Health Scenarios (1988). *Scenarios and other methods to support long-term health planning*. Uitgevery Jan van Arkel, Utrecht.

World Health Organization (1985). *Targets for health for all*. WHO Regional Office for Europe, Copenhagen.

World Health Organization (1986). *Evaluation of the strategy for health for all by the year 2000*. WHO Regional Office for Europe, Copenhagen.

15

Public health policies and strategies in England

CLIFFORD GRAHAM*

Introduction

In observing developments over the 40 years' life of the National Health Service in the UK, the outside observer is likely to note the apparent absence of clearly defined measures for the control and direction of health services by Government and Parliament. This could be interpreted as a demonstration of the real difficulty of exercising such control and direction. The Department of Health (DH), which is charged with exercising this general responsibility on behalf of Government and Parliament, essentially must rely on persuasion and exhortation in dealing with the many other central government departments that have an indirect interest in health services, and its direct operations with the National Health Service (NHS) are based on an agency relationship not a chain of command. But this apparent absence of control and direction could also be explained by reference to the fact that, after the early years of centralization required to establish the NHS, it has been the will of recent Governments and Parliaments that responsibility for the control and direction of health services should be devolved to the local administrative unit most closely involved in the actual delivery of services. In England, with effect from 1 April 1982, this responsibility falls on the District Health Authorities (DHAs), working within general guidelines produced by the DH and Regional Health Authorities (RHAs).

In the final analysis, it may simply be that health is too complex and local an issue to be generalized about or subjected to central control and direction. But, so long as Parliament pays the piper, in meeting almost all the massive costs of the NHS through general taxation, the Government will continue to call the tune, at least in terms of the general strategy to be adopted towards the provision of health services, if not the operational activities required for day-to-day direction, management, and control.

The constitutional framework in England

The planning, management, and delivery of health services in England depend on an agency relationship between the Secretary of State (Minister) for Health and the DH on the one hand, operating at the national level, and, on the other hand, the NHS, operating sub-nationally through 14 RHAs and 190 DHAs. It is the duty of the Minister to 'promote the establishment . . . of a comprehensive health service designed to secure improvement in the physical and mental health of the people of England . . . and the prevention, diagnosis and treatment of illness, and for that purpose to provide or secure the effective provision of services' (National Health Service Act 1977).

The Minister has wide general powers to provide health services, and specific duties to provide services, including hospital and other accommodation; medical, dental, nursing, and ambulance services; facilities for the care of expectant and nursing mothers and young children; facilities for the prevention of illness; other facilities required for the diagnosis and treatment of illness; and facilities for family planning. But the Minister's specific duties to provide services are qualified 'to such extent as he considers necessary to meet all reasonable requirements'.

The Minister's responsibilities are for providing services and facilities, not for providing diagnosis and treatment. It is for the patient's doctor (or other professionals with clinical responsibility) and not the Minister to decide whether and what treatment is appropriate. For services provided by general medical, dental, and ophthalmic practitioners and the pharmaceutical service the Minister's duty is to 'make arrangements' with practitioners who are responsible for providing services as independent contractors. An illustration of the organization and management of the NHS and DH is given in Figure 15.1.

* This chapter contains the views of the author alone: it should not be read as an expression of the views of the Department of Health.

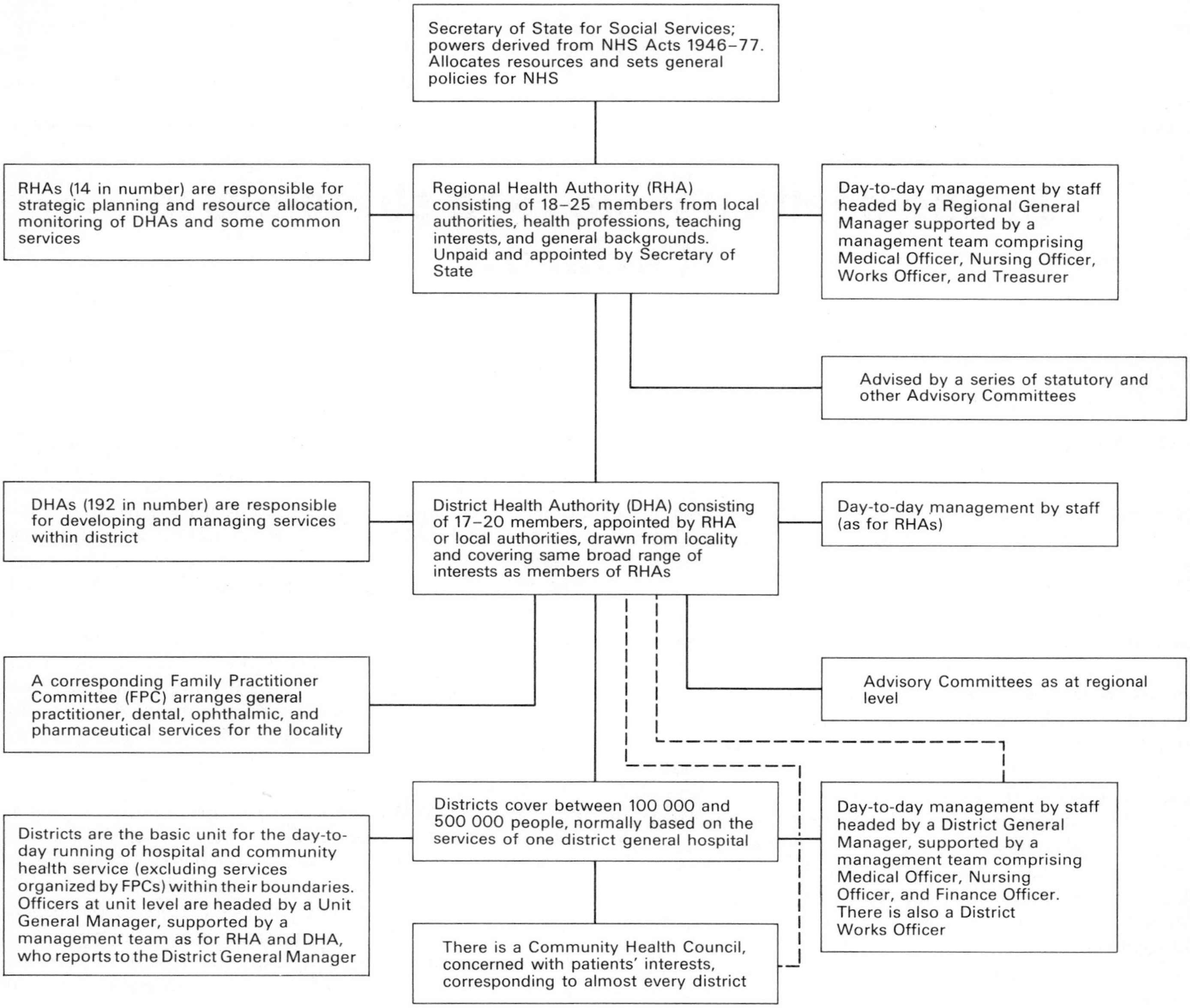

Fig. 15.1. Structure of the NHS and DH.

Underlying political philosophy

The NHS in England has been accepted by all political parties as the best way of meeting the needs of the people for a health service funded from general taxation which is free of charge at the point of need and use, subject to prescription charges for a highly selective and limited group of medicines, drugs, and a few items included in the family practitioner service. This bipartisan approach has been sustained for over 40 years, following the Beaverbook (Liberal) Report calling for a state-funded comprehensive health service in 1942, the Winston Churchill (Conservative) White Paper proposing 'A National Health Service' in 1944 and the Aneurin Bevan (Labour) National Health Service Act 1946. It was a Labour Government (Gaitskell) in 1951 which first introduced prescription charges (for teeth and spectacles); it has been Conservative Governments, from 1970 to date, which have reorganized the management structure of the NHS (in 1973, 1982, and 1984) and introduced measures to improve the efficiency and effectiveness of the NHS; and, it was a Labour Government (Dr David Owen) in 1976 which redistributed NHS resources on the basis of geographical need and attempted to separate privately funded and publicly funded medicine in NHS hospitals. The arrangements to remove private medicine from NHS hospitals were overturned by the incoming Conservative administration in 1979. A Conservative Government, in 1984, introduced the general management function at all levels of the NHS as recommended in the report of the 1983 NHS Management Inquiry headed by Mr (now Sir) Roy Griffiths, who subsequently became the prime minister's adviser on health service management.

General management

For the first 30 years of the NHS, top management (at regional level) was undertaken through a process of consensus management including a doctor, a nurse, an administrator, and a finance officer, and, on occasions, a works officer. In 1974, this process was extended to area, district and hospital level. In 1984, the Government set up, within the DH, a Health Services Supervisory Board (HSSB) and an NHS Management Board (NHS MB), and required the Minister to appoint general managers at unit, DHA and RHA level. The membership and role of the two Boards and the main items required in the job description for the general managers are described below.

The Health Services Supervisory Board

The role of the HSSB is to strengthen existing arrangements for the oversight of the NHS and to be concerned with:

(1) determination of purpose, objectives, and direction for the health service;
(2) approval of the overall budget and resource allocations;
(3) strategic decisions;
(4) receiving reports on performance and other evaluations from within the health service.

The Board is chaired by the Minister and comprises his or her ministerial colleagues in the DH, the Permanent Secretary of the DH, the Chief Medical Officer, the chairperson of the NHS Management Board, the President of the Royal College of Physicians, and the chairpersons of the RHAs. It relates to the statutory and professional bodies in exactly the same way as do Ministers and the DH.

The NHS Management Board

The role of the NHS MB is to plan implementation of the policies approved by the HSSB and, in particular, to:

(1) give leadership to the management of the NHS;
(2) to control the performance of the NHS;
(3) to achieve consistency and drive, throughout the NHS, over the longer term.

The NHS MB is chaired by a businessperson from outside Government, and its membership contains people from private business, the civil service, and the NHS, covering functions such as personnel, operations, finance, medicine, and nursing. The chairperson of the NHS MB is graded as a civil servant (Second Permanent Secretary) and, as such, is also a member of the top management team of DH; he or she is the Treasury Accounting Officer for all public expenditure on the hospital and community health services, and he also provides the DH voice and presence on NHS management on all the various gatherings and platforms across the NHS at large. (An Accounting Officer is personally accountable to Parliament for the proper and efficient use of funds allocated by Parliament. This personal accountability is increasingly being interpreted by Parliament as including the policies which give rise to expenditure as well as financial propriety.)

The general manager

The *general manager's* broad areas of responsibility must include as a minimum:

(1) direct accountability to the authority, or in the case of units to the district general manager, for the general management function within the undertaking;
(2) direct responsibility and accountability for the managerial performance within the authority or unit;
(3) leadership of the authority's management team, or unit equivalent, and accountability for the performance of the team as a whole in developing policies and possible courses of action and ensuring the provision of proper advice;
(4) ensuring that management and administrative practices enable the care of patients to be constantly to the fore;

and to these ends he or she should:

(5) ensure that the authority or unit is provided with the range of advice and information it needs to formulate policies, decide priorities, set objectives, and monitor progress;
(6) ensure that full weight is given to clinical priorities in the light of advice from nurses and doctors:
(7) ensure that timely decisions are reached;
(8) ensure that objectives are achieved;
(9) provide the necessary leadership to stimulate initiative, urgency, and vitality in management, for example in ensuring a constant search for constructive change and cost-improvement;
(10) co-ordinate activities, functions, and personnel as necessary;
(11) ensure that responsibility, including the management budgeting responsibility, is delegated to the point where action can be taken effectively;
(12) secure effective motivation of staff.

The essence of the general management function is the bringing together, at each management level of the NHS, responsibility for the planning, implementation, and control of the authority's or the unit's performance. The general manager carries personal and visible responsibility for this, and he or she is personally accountable to the authority for its discharge. The authority in turn must be seen clearly to give full support and backing to the general manager at all times. This officer can be appointed from any discipline and can be drawn from the public, private, or voluntary sectors, inside or outside the health sector, at home or abroad. The sole criterion is fitness to undertake all aspects of the general management function on a full-time basis (except only where practising clinicians wish to undertake the job, and the authority wishes to make special, and perhaps part-time, employment arrangements to cover such a candidate). Against this background, the key objectives to be achieved over time are as follows: (Department of Health and Social Security 1984)

1. To act on the Government's overriding concern, to see that the NHS provides the best possible service to patients, within the resources made available by the general taxpayer to meet the health service requirements of Parliament, Government and the public at large.
2. To ensure that the massive public expenditure devoted to the health service—currently running at almost £20 billion per annum in England—does reach its target: improvement in the physical and mental health of the people and in the prevention, diagnosis, and treatment of illness, and the promotion of good health and well-being.
3. To ensure that systems for service plans, resource allocation and budgeting, and performance monitoring and review directly relate to the Government's broader aim of producing a well-managed health service which is geared primarily to the interests of patients: 'it cannot be said too often that the NHS is about delivering service to people. It is not about organising systems for their own sake . . . the driving force . . . is the concern to secure the best motivation for staff. As a caring, quality service, the NHS has to balance the interests of the patient, the community, the taxpayer and the employees'.

These overriding objectives naturally lead on to more specific objectives for general management, including the following:

(1) to assert the importance of achieving visible improvements in the service provided to users of the NHS;

(2) to pinpoint where the personal responsibility for taking decisions actually lies;

(3) to give management authority to take decisions (as mandated by the health authority);

(4) to reduce the 'fudging' encouraged by the pre-existing 'consensus management' approach;

(5) to achieve significant cost improvements in order to release resources for the wider benefit of health service users.

The Department of Health

The Department of Health is headed by the Secretary of State for Health, aided by a Minister of State for Health and two Joint Parliamentary Under Secretaries of State (usually drawn from both the House of Commons and the House of Lords). Five groups of administrative Divisions within the DH have functions related to the NHS, as follows:

1. *The Health and Personal Social Services Policy Group* is responsible for the development of national policies and priorities and a strategy for health.
2. *The NHS Management Board* is responsible for three main groups as follows:

   (a) *The Regional Group* is responsible for the operation of the health services by health authorities, the arrangements for allocating resources to health authorities, the procurement directorate, medical supplies and industries, and medical exports;

   (b) *The Health Authorities Personnel Group* is responsible for staffing policies, and pay and conditions, and the education and training of nurses, midwives, and health visitors;

   (c) *The Finance Group* is responsible for the overall administration of financial resources including accounting and control systems, statistics, management services, and computers and research, and liaison with the media.
3. *The Departmental Personnel Management Group*, whose responsibilities include headquarters establishment and personnel and central resource management.

The staff of the DH includes members of the main professional groups serving in the NHS.

The functions of these groups of Divisions include those related to the NHS, services provided direct to the public, and a number of other functions not mainly related to the NHS. The main functions and services provided under these headings are:

1. NHS-related functions
   Development and explanation of policy
   Allocation of resources
   Monitoring and control
   e.g. Financial controls (including audit)
   Quality controls (including a Health Advisory Service for elderly and mentally ill people, and a National Development Team for mentally handicapped people)
   Information including statistics
   Planning system
   Provision of central services for NHS
   e.g. Building development and guidance
   Central pay negotiations
   Procurement
   Advisory functions such as NHS computing; technical advice; personnel development
   Management services
   Research and development
2. Provision of services to the public
   Special hospitals for the criminally insane
   Disablement services

3. Functions not related exclusively to the NHS public health
   Medicines legislation
   International relations
   Industries and exports
   Control of abortion and inspection of private abortion clinics
   Work on smoking and health including voluntary and statutory agreements with the tobacco industry.

Control and direction of health services in theory

The NHS is one of the largest users of the nation's resources. In 1988 it spent some £20 billion a year in England, and received about 85 per cent of its money from general taxation. About 11 per cent came from the health element in the national insurance contribution, and the rest mainly from charges to patients. Therefore, the main method of control and direction lies in the allocation of resources and the monitoring of expenditure.

Allocation

Resources are distributed between hospital and community health services, family practitioner services, and central services. About 70 per cent of the NHS's total money is spent on the hospital and community health service through cash-limited budgets administered by RHAs and DHAs. This has recently grown in real terms by about 1.5 per cent a year. The family practitioner services, i.e. family doctors, general dental, and ophthalmic practitioners, and the drugs and appliances they prescribe, use about 20 per cent of the NHS's money for current services. They are funded so that they can meet the needs of patients as they arise. Current spending on these services has grown in real terms by 2.3 per cent a year on average. The remainder of NHS current spending is on a small group of services provided centrally. These include support for the Health Education Authority and for training of some health and personal social services staff, central grants to voluntary bodies, the provision of some welfare milk, the special hospitals for the criminally insane at Broadmoor and elsewhere, wheelchairs and other aids for disabled people, and the public health laboratory service.

Distribution of resources to health authorities

Present policy for public health care accepts that the most appropriate unit for direct administration of health services is a local one, the DHA. The most recent reorganization, through the introduction of a general management function and the appointment of general managers at RHA, DHA and unit level, has made local management stronger and more efficient. The main employers of NHS staff are the RHAs and DHAs, and to a large extent they are free to decide what mix of staffing and other resources they should use to produce the services they seek to provide, within the overall constraint of their cash limits.

The current approach to national control and direction of NHS resources seeks to balance the decentralization in decision-making by systematic performance monitoring and review, using the chain of accountability from the DHA through the RHA to the NHS Management Board, the Health Services Supervisory Board, DH Ministers, Government, and Parliament. It also seeks to help RHAs and DHAs improve their use of resources in various specific ways.

Once cash limits have been set for current and capital expenditure on hospital and community health services, this money is allocated by the DH to RHAs. Allocations are determined on the basis of a formula which broadly follows the recommendations of the Resource Allocation Working Party (RAWP) (1976). It is intended to ensure that each year's 'new' money is distributed to those parts of England which are short of services or where the population is growing, so as to secure as far as possible equal opportunity of access to health care for people at equal risk. The RAWP formula means that RHAs receive different proportionate increases in their cash allocation. RHAs are then responsible for the allocation of funds to DHAs and for monitoring their use. In allocating funds, RHAs and DHAs are expected to take account of the need to remedy local shortages of services and to implement Government priorities, such as improving services for the elderly, mentally ill, and mentally handicapped, and expanding community care. Where a hospital has been built, the RHA is expected to take account of the need to provide new services, staff, and other running costs.

In addition the DH gives RHAs, and RHAs give DHAs, planning guidelines which include specific resource assumptions one year ahead and very broad, long-term resource assumptions as a basis for strategic planning.

The money allocated to DHAs becomes their cash-limited budget. Districts in turn allocate funds to the different services and budget-holders, and this in practice determines the level of 'funded establishment' (the level to which the budget-holder for any particular function or department can afford to recruit). It ensures, broadly speaking, that levels of staffing in each authority are within its revenue capabilities, and that future budgets are not over committed in advance by unplanned recruitment. The system also means that the first responsibility for ensuring that services are provided and that staff are deployed as efficiently as possible lies with health authorities and, within them, with managers at all levels.

Monitoring

RHAs and the DH have a responsibility to ensure that the resources distributed to DHAs are used effectively and in accordance with Government priorities. Spending is monitored in a number of ways:

1. There is a fully developed financial control system at DHA, RHA, and DH levels to ensure that cash limits are observed. The DH determines the RHAs' cash limits,

and the RHAs and DHAs are subject to their strict discipline, now based on statute through the Health Services Act 1980.

2. During the course of the financial year, the DH monitors RHA spending—and RHAs monitor DHA spending—on a cash basis through a financial information system which in turn enables the DH to supply information required by HM Treasury's financial information system.
3. DHAs are responsible for producing strategic plans fitting within outline RHA strategies, which look ahead about 10 years and deal with major shifts in resources and services. Examples of strategic decisions are building new acute hospitals to replace old ones and/or providing new services, and reshaping services for the elderly or mentally ill or handicapped so that large isolated long-stay hospitals are gradually replaced by small local units and domiciliary and day care. Changes such as these require long-term planning to build and close hospitals and to recruit, train, and redeploy staff.
4. The RHA is responsible for pulling the DHA plans together into a regional strategic plan, and for ensuring that this observes national priorities and is compatible with the financial, staff, and capital resources likely to be available. It must also be reasonably flexible so that it can be modified if resource expectations change. DHA plans may need to be modified when they are brought together in this way. The DH then examines RHA plans to ensure that they observe national priorities and are feasible in terms of the resources likely to be available to the RHA: it considers not only overall financial constraints but also key staffing constraints, for example the shortage of geriatricians. Strategic plans are discussed with the RHAs and may be modified as a result.
5. In the shorter term, DHAs are responsible for producing operational programmes to implement their strategy, looking ahead firmly for one year and tentatively for a second. These short-term programmes are monitored by the DH to ensure that they fulfil the strategy and are actually implemented.
6. DHAs and RHAs also monitor what services are provided in order to identify deficiencies, as a starting point for preparing their plans. The DH has developed a comprehensive system of monitoring the development of services historically, looking at trends over a long period. This monitoring is based on national activity and staffing statistics and the health authority accounts and hospital costing returns. These are drawn together to provide an overall picture of how services have developed and resources have been used.

The DH also carries out manpower planning of a strategic kind, principally for doctors and for other groups of staff where the NHS is a major employer. The DH sets the target level of intake into medical schools. It can also indirectly influence the teaching capacity for nurses taking basic nurse training courses through the central funding of Regional Nurse Training Committees, which administer the funding of the salaries of tutors and certain running costs of schools of nursing. Training for occupational therapists is funded centrally, but for other professional groups the main role of DH is to provide a national overview to help local decision-makers. The DH exercises a number of other strategic controls over staff. New consultant posts (which are bid for by RHAs on the basis of their assessment of the availability of the resources to support them) cannot be established without DH approval; and the DH also exercises close control over the number and distribution of training posts. Where there is a national shortage, new posts are rationed on a basis of need. There are also direct controls over certain top management posts.

There are other specific ways in which the DH and RHAs seek to direct control and improve the use of resources:

1. At the national level, both the Health Advisory Service, which advises on long-stay services and services for children, and the National Development Team for the Mentally Handicapped provide regular reports on services they have visited. These reports are copied to the appropriate RHA(s) and are available within the DH. Staffing levels are also monitored on a regular basis by statutory auditors, who identify high-cost areas by comparison with costs in other authorities and through periodic examination of costs under specific heads of expenditure. Through this process both excessive staffing and excessive or unwarranted use of overtime are drawn to the attention of NHS Chief Officers and to the RHA and DHA formally by way of an audit report in the worst cases. The DH also promulgates good practices in the use of resources in many ways, for example by encouraging cost–consciousness in clinical services and by providing direct advice to the NHS.
2. Within regions, RHAs are responsible for ensuring that suitable information is available for line managers to set and monitor staffing levels. Where comparisons suggest that more economic and efficient use of resources may be possible, the authority in question is responsible for ensuring that managers take appropriate action. RHAs provide management services such as operational research and 'organization and method' (O & M) studies. Several RHAs also issue guidelines on staffing levels (or other indicators) which managers can use in planning or reviewing their use and development of manpower locally. For example, several RHAs (including Trent) have devised formulae to assess staffing requirements in particular areas, and the South East Thames RHA manpower information system produces routine reports for each constituent authority. The region follows up matters such as abnormally high overtime levels or exceptionally high proportions of staff employed in higher grades.

Alongside the services provided by the NHS are those provided by local authority social services departments. For groups such as the elderly, the disabled, the mentally ill, and the mentally handicapped, effective care requires joint planning of health and social services. To encourage provision of services across administrative barriers, health authorities have since 1976/7 been given special allocations by DH to use on projects planned and funded jointly with local authorities. From a modest beginning of £8 million in 1976/7, the money available for 'joint finance' projects had risen to £113 million in 1988/9.

The planning and resource allocation processes of the NHS and DH were designed to operate on the basis of the agency relationship described above, in which most of the responsibilities are devolved to health authorities but with the chain of accountability stretching back through the Minister to Parliament and the people. These processes therefore form a most important part of the organization of the NHS and DH and secure the Government and legislative control and direction of health services required by Parliament.

In terms of organization and management these requirements have been summarized as follows:

1. An essential feature of the management arrangements which the Minister requires the NHS to adopt is a system of control in which performance is monitored against plans and budgets.
2. Planning systems are seen as a principal means of achieving a clear line of responsibility for the whole NHS from the Minister down to and within DHAs, with corresponding accountability from DHAs back to the Minister through the DH.
3. The NHS planning, resource allocation, and performance review systems are intended to provide the main management control to be exercised by the Minister and health authorities in undertaking the functions covered by the NHS legislation.

These processes provide the main means by which the requirements of the law are made explicit and can be monitored in accordance with the law.

The resource allocation process has a clear relationship to, and a very considerable influence on, the organization of the NHS and DH and on their planning processes. In particular, the resource allocation process provides:

(1) a realistic framework in which planning options can be considered and a balance of priorities agreed;

(2) a need-related baseline, showing the share of the available resources that should be consumed by a given population;

(3) an important first stage in planning the provision of services for defined populations;

(4) An organizational tool for the monitoring of performance.

Accordingly, the DH has been operating a comprehensive planning process since 1974, drawing on a Programme Budget devised by the DH Finance Division and a balance of care model devised by the DH Operational Research Service. The NHS has been operating a complementary planning process since 1976, and in 1976 both the NHS and DH introduced an improved method of resource allocation based on the recommendations contained in the Report of the Resource Allocation Working Party.

Control and direction of health services in practice

So what has actually been achieved in practice? Since 1979, government has set broad objectives for the NHS. A number of documents, published by the DH, have set out these objectives. The main objectives of the hospital and community health services in recent years have been to:

(1) expand services sufficiently to keep pace with the increasing numbers of old and very old people, and to meet the demands placed on the NHS since 1978 due to the rising birth-rate;

(2) develop services generally, to make use of new technology (such as dialysis and hip replacement), and to reduce waiting times;

(3) reduce perinatal mortality and morbidity;

(4) improve standards of care for the mentally ill, mentally handicapped, the disabled, and other long-stay patients, and to care for patients in the community rather than in hospital whenever possible;

(5) develop and extend preventive measures to decrease the incidence of disease (for example immunization and cervical cytology) and promote good health.

(6) improve services in deprived regions and districts;

(7) improve efficiency and obtain better value for money.

In response, the NHS has many achievements to show for its performance since the 1974 reorganization:

1. With the assumption of community health responsibilities, RHAs have greatly improved the integration of hospital and community services. There has also been much progress in integrating health services with personal social services, particularly in the case of the elderly, the mentally ill, the mentally handicapped, and the disabled.
2. Services for the elderly, the mentally ill, the mentally handicapped, and the disabled have been improved. In hospitals the nurse:patient ratios have risen, and the balance of care has shifted towards local provision. Day and domiciliary care have also been increased, e.g. home nursing has risen by about 4 per cent per annum since 1976 and day hospital attendances by about 3 per cent per year.
3. There has been a substantial increase in the level of activity in the acute sector, more than enough to keep

pace with the increasing number of old people, but still not enough to make full use of treatments such as hip replacement and renal dialysis.

4. Resources have been redeployed to relatively deprived regions: at the end of 1986/7 all but 2 of the 14 RHAs were within 4 per cent of their RAWP revenue target; while in 1977/8 the most deprived region was 27 per cent below that of the best-off region.
5. The new cash limits system has been introduced successfully.
6. Unit costs have in general steadied or fallen, whereas in the period 1964–74 they were rising rapidly in acute and maternity services. This has been achieved in the face of a rising number of elderly people being treated in acute hospitals.

There are, however, still serious weaknesses in the service. In particular:

1. There are still major deficiencies in care of the elderly, the mentally ill, the mentally handicapped, and the disabled, and long waiting lists persist for some specialties in some places.
2. There is still some mismatch between capital developments and the ability to finance their running costs, although this is partly due to the fact that hospitals planned in the early 1970s, when growth was expected to be twice the post-1975 level, are still coming on-stream.
3. The planning dialogue between the NHS and the DH has been devoted mainly to longer-term strategies, and there has not been sufficient emphasis on what should be achieved in the short term, and whether short-term plans have in fact been implemented.
4. Systematic planning and monitoring has been concerned with deployment of resources between services—by client group, the balance of care, geographical distribution—and less with efficiency; there has been a lot of work on efficiency, but it has been *ad hoc*.
5. The information systems have been inadequate in certain respects, such as the difficulty in relating financial and staffing information to information about activity or output, and the lateness and inaccuracy of some returns.
6. There has been an excessively long chain of command, which results in duplication of effort and diffusion of responsibility; this has now been remedied by the reorganizations in the health service structure.

In recognition of these difficulties, the DH is both stepping up existing efforts to encourage efficient management and effective resource use in the NHS and is developing new initiatives.

Apart from the general channels of control and direction identified above, these impressive results have been achieved by the other means of general advice and exhortation through which the DH is able to help promote efficient ways of working in the NHS. These include:

(1) publication of building and engineering standards and cost allowances for health authorities, and advice on operating and maintenance costs;

(2) advice on catering and domestic service matters;

(3) publication of papers and reports produced by the NHS/DH Joint Manpower Planning Advisory Group (MAPLIN); joint conferences at the NHS Training Authority and other educational institutions in the private and voluntary sector;

(4) advice, by visits of medical officers to general practitioners, on better practice organization and premises, and improved co-operation with other primary health care workers;

(5) information about central promotion and development of transferable computer systems and information technology, intended to produce economies and greater efficiency;

(6) provision of national data for use in incentive bonus schemes for maintenance craftsmen and some groups of ancillary staff;

(7) advice to the professions on ways of achieving the most efficient use of available skills;

(8) reports of working parties and research groups, dealing with better use of resources in many fields.

These long-standing and regular ways of advising the NHS are now augmented by several new initiatives designed to help the NHS cope with the problems which still need solving.

Regional reviews and the planning system

The maximum delegation of responsibility to health authorities must be accompanied by systematic monitoring to ensure that DHAs are properly accountable through RHAs to the Minister and to Parliament. The planning system—seen as an essential tool for monitoring the performance of health authorities in meeting Ministers' broad policy objectives and strategies for the NHS since the 1974 reorganization—has been refined considerably since it was introduced in 1976. The Government has now introduced annual Regional Reviews to strengthen the monitoring of strategic planning and to broaden it to include an assessment of overall performance. Starting from 1982, Ministers now lead a Departmental review on the long-term plans, objectives, and effectiveness of each region with the chairperson and chief officers of RHAs. Each Regional Review aims:

(1) to ensure that each region is using the resources allocated to it in accordance with the Government's policies; to agree with the chairperson the progress and development which the regions will aim to achieve in the ensuing year; and, to review progress against previously agreed plans and objectives;

(2) to assess the performance of RHAs and DHAs in using

staff and other resources effectively. Ministers hold RHAs to account for the ways in which resources are used in their regions, and RHAs in turn hold their constituent DHAs to account.

Performance indicators and review

A DH Group was set up in 1981 to explore the indicators at RHA and DHA level that could be used to help assess performance, with particular emphasis on the efficiency with which resources are used and the effectiveness of the services provided. The Group proposed a range of indicators, including cost per case and average length of stay, which could be used in annual review meetings between the DH and RHAs and the RHAs and DHAs. A range of NHS, RHA, and DHA indicators has now been developed.

Rayner reviews

Following the success of the scrutinies of particular functions, operations, and activities of Government departments carried out under the auspices of Lord Rayner (chairman of Marks and Spencer plc and the prime minister's, then, personal adviser on efficiency in Government) similar techniques have been applied to the NHS. Routine administrative and managerial functions, systems, and processes are subjected to intensive investigation, starting from the initial justification for the continued existence of the system or function and continuing with a detailed examination of the cost-effectiveness of the way in which it is carried out. Scrutinies are conducted by senior NHS officials, operating jointly with DH officials, and reporting directly to the appropriate RHA chairperson followed by a report to the Minister. The advice and support of Lord Rayner's 'efficiency unit' and the DH is made available to those responsible for conducting the scrutinies.

The NHS information system

Following a fundamental review, started in 1980 under the direction of a steering group chaired by Mrs E. Körner (then vice-chairperson of South Western RHA), an NHS data base has produced information which is useful to local management. This links, where possible, information on activity or output with information on financial and staffing resources, so that the resource cost of activities is known. This provides a common core of information so that DHAs can be compared and information aggregated at RHA and DH level. It is intended that information used locally will be more accurate and will be provided more quickly. The computerization of staffing data together with the implementation of the Körner Group's recommendations has already had a marked effect on the provision of statistics.

NHS procurement

The Government has set up a Procurement Directorate, within the NHS Management Board, to develop policies and introduce arrangements that enable health authorities to make the best use of their resources, and to demonstrate how the NHS supplies services should be organized, including computer-based information systems to improve procurement decisions. The Directorate also advises health authorities on the application to the NHS of the Government's policy for using public purchasing to improve the competitiveness of UK suppliers, and on the introduction of a quality assurance scheme designed to help the NHS to purchase goods manufactured to an acceptable standard. Each of these initiatives is subject to continuing surveillance.

The Directorate is now focusing more on specific areas where improved value for money can be achieved. These include the identification of new purchasing arrangements which will result in savings; the development of a means of monitoring the effectiveness of procurement policies; a review of training for supplies staff; and a review of storage and distribution arrangements.

Land sales

The Government has introduced greater flexibility into the rules governing the disposal of land and property by the NHS in order to increase the efficiency and effectiveness of their estate management, and to speed up the process, and to produce a greater return on sales for the benefit of the exchequer and individual health authorities.

Promoting more cost-effective practice in the National Health Service

Clinical practice and procedure

The DH, the Medical Research Council (MRC), the Royal Colleges (professional medical associations concerned with maintenance of standards), the NHS, and universities carry out a significant programme of evaluation and research into the cost-effectiveness of clinical procedures, both of existing and new technology. Britain probably leads the world in the per capita number of clinical trials, but this should prompt the question: do such trials lead to more effective treatment or just to more blind alleys? To an increasing extent, economic evaluation is built into many of these. Examples of such research sponsored by the DH include: the breast cancer screening trials, co-ordinated at the Institute of Cancer Research; evaluation of cardiac transplantation at Papworth and Harefield Hospitals; and the multi-centre study of the cost-effectiveness of diagnostic radiology co-ordinated at the Welsh National School of Medicine. In the psychiatric field the DH sponsored a survey of electro convulsive therapy (ECT), based at the Royal College of Psychiatrists, which uncovered failings in some ECT departments, requiring changes in practice.

Although Britain has led the way with clinical trials, research into the economic consequences of clinical practice has been hampered by the comparative crudeness of available costing data. To help to remedy this, a method of specialty costing has been developed recently, and tested successfully in seven hospitals.

An attempt is also being made to develop a practical method of patient, or disease, costing within a financial information project based in the West Midlands RHA.

It has often been said that clinicians in the UK (and, indeed, in other countries) are given few incentives to economize. Providing clinicians with better cost information alone may not lead to significant improvements in resource allocation. The DH is now sponsoring a series of trials of clinical budgeting and resource management by consultants and their clinical teams. These will allow clinical managers to assume financial responsibility for the considerable resources the use of which they dictate and direct.

International comparisons

It is well known that England devotes a lower proportion of its gross national product to health expenditure than most other industrialized countries. In 1985, the share in this country was about 6 per cent (including private as well as public spending), compared with shares which are typically in the range 7–11 per cent in the major European and North American countries. But such aggregate comparisons say little about the comparative standards of health care between countries, as allowance needs to be made for other factors such as differences in relative costs and in the efficiency with which health services are organized.

There are a number of reasons for thinking that the NHS provides relatively good value for money. For example, as a national organization it can achieve economies in the purchase of drugs and other supplies; the building of new hospitals is planned and organized so as to avoid any duplication of expensive facilities; the strength of the primary care system, and in particular the 'gatekeeper' role of the general practitioner, probably reduces the extent of any unnecessary hospitalization; the administrative costs of financing the NHS are very low because of the absence of insurance and reimbursement arrangements; in contrast to the more open-ended financing mechanisms commonly found elsewhere, the cash limits system provides a continuing incentive for service managers to seek efficiency savings; and the method of paying doctors and other professionals mainly by salary or capitation payments minimizes incentives for the over-provision of services such as laboratory tests and X-rays.

The impact of these various factors on standards of health care and on the health status of the population cannot easily be assessed. It is, however, worth noting that according to conventional indicators of health, such as mortality rates, there is little difference between England and other leading industrialized countries, despite our lower level of health spending. For example, life expectancy at birth is somewhat lower in England than in Sweden, the Netherlands and Denmark, but broadly the same as in France and Canada, and higher than in West Germany, Australia, and the US.

Conclusion

Although in theory the legislature has not chosen to provide itself with too many direct and explicit measures for the direction and control of health services in England, in practice the NHS is under reasonably tight governmental control without recourse to detailed legislation. But there can be no doubt that if the NHS were seen to be getting out of control Parliament would take urgent steps to arm itself with the necessary means of direction and control, as it has in the past in the case of cash limits (Health Services Act 1980). At present, this characteristically English system operates on the basis of an unwritten constitution and relies heavily on general advice and exhortation, coupled with a carefully conceived structure for the organization and management of health services and well-ordered systems for the planning, funding, and monitoring of the delivery of services. This reflects the general constitutional approach in England and the workings of the English democracy. But the apparent reluctance of Government and Parliament to be drawn into the development and implementation of more detailed measures of direction and control could also reflect the clear understanding of all concerned that the delivery of health services in England is far too complex a matter to be directed and controlled through legislative means. The complexity of these issues is reflected in the model presented in Figure 15.2.

This suggests that, although attempts are being made through the delivery of health services to affect the health status of the population in England, at present health status cannot be defined too well, e.g. see the quotation from the NHS Act 1977 under 'The Constitutional Framework . . . ' above (which, itself, repeats the 1946 legislation), and so far no useable indicators to this end have been devised, collected, or displayed. Instead, improvements in health status have been attempted by devising policies for health services, without assessing first the full significance of hereditary, industrial, environmental, and social factors, and individual freedom. In retrospect these latter sets of factors, which are in the main outside the control of the DH, might be judged to have a much greater impact on health status than DH policies. In any event, these factors need to be taken more fully into account in devising and implementing health service policies. This could suggest a broader central strategy for health services and related activities, not only to take account of all the different factors affecting the health status of the population but also of the pressures for or against change from politicians, the professions, and the community, and of the capacity for change in the existing system, facilities, money, and staffing.

But the real key to success in determining approaches to the direction and control of health services in England, as described above, lies in the careful blending of the essential interests in pursuit of a clearly defined and accepted goal. In England, we have chosen to blend the contributions of the statistician, economist, and operational researcher; the doctor, the nurse, and the social worker; the treasurer and

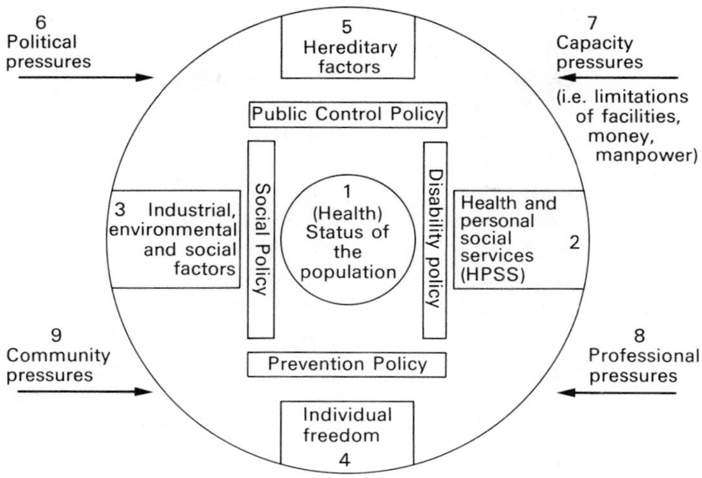

1 **Health status** – includes direct measures such as prevalence and incidence of a particular disease, or disability, functional levels of individuals or population groups, and behaviour and attitude towards health-related activities; and indirect or proxy measures such as marital status, use of health services, risk factors, and outcome from particular services or programmes

2 **HPSS** – covers the full range of policies and services as at present developed by DH and delivered by NHS and local authorities

3 **Industrial, environmental, and social factors** – includes health consequences flowing from policies developed by industry (presence or absence of factories), environment (the living and working environment of the population), employment (increases in unemployment and leisure), education (falling school population, redundant buildings, health education, and prevention) and Home Office (mentally disordered offenders, children, fire precautions)

4 **Individual freedom** – covers all individual aspects of human behaviour at which the prevention policy is directed, plus wider health and social issues controlled by the individual in his or her chosen environment

5 **Hereditary factors** – embrace both the inescapable consequences of human inheritance and possible control concepts, such as screening, cloning, euthanasia, abortion.

6 **Political pressures** – the expressed demands of national and local politicians within the general political process as it affects the HPSS. This includes pressure from the trade unions

7 **Capacity pressures** – the limitations on change presented by the physical features and location of the HPSS estate, the money allocated to or assumed to be available for HPSS, and the manpower actually or potentially available for employment or use within the HPSS

8 **Professional pressures** – the expressed demands of the national and local professionals engaged in delivering the HPSS, within both the NHS and local government

9 **Community pressures** – the organized (Community Health Council, Parliamentary Commissioner for Administration, i.e. the Ombudsman), and spontaneous ('Campaign Against the Cuts') expression of community reaction to the existing pattern of services, or proposals for change, in the HPSS

Fig. 15.2. Limited model to assist the development of a strategy for the HPSS before the twenty-first century.

administrator; and the epidemiologist. Equally important is the blend of general disciplines; the political and policy-making requirement, the academic approach, the professional discipline, and the administrative and management action. But an overriding and invaluable requirement is a clearly stated and accepted general objective, strongly backed by central government and set against a reasonably tight deadline. There are many possible permutations in deciding on the provision of the academic approach and the professional discipline, depending upon the country and health care system in question and the desired objectives, but to make this contribution relevant and effective it must be related to the political and policy-making requirement, and its implementation must be secured through general management action. In short, for any such method to be sensibly developed and successfully applied there must be multi-professional teamwork at all levels.

A further test for the English system of providing health care came early in 1989 when the Government produced legislation on three important areas of the NHS as follows:

1. A White Paper containing the Government's proposals, for consultation prior to legislation, on the Minister's 'Review of the NHS' to make the NHS more businesslike in keeping with the 'enterprise economy' of the UK (but not turn it into 'a business'); requires the NHS to match the private sector in terms of efficiency, customer choice, standards, good management, and clear decision-taking;

develop the right relationship between the NHS and the growing private health care industry in the UK, with the maximum of buying and selling services between the public and private sectors; and develop much greater consumer choice and voice—in the patient's choice of a general practitioner, in the GP's choice of a hospital, in the consultation between GPs and hospitals within the NHS, in the choice of appointments systems, and in the effectiveness of complaints procedures; and, in the full involvement of doctors, nurses, and patients in the administration and management decisions about how the available resources can best be used for the benefit of patients.

2. The Health and Medicines Act 1988 is designed to achieve important improvements in the provision of primary care in the UK. These aim to benefit the consumer through the provision of better services, the professions by relating financial rewards more closely to performance, and the taxpayer by more cost-effective targeting of the resources available.
3. The implementation of the 'Overview of Community Care', already submitted to the Government by Sir Roy Griffiths, in his role as personal adviser to the prime minister, requires major changes in the organization and funding of community care, particularly for the elderly, the mentally ill, the mentally handicapped, and the disabled; and in the importance of the 'health service' elements of such care from the 'local government' support for the individual in the community: a Government White Paper will contain an agenda for action on the second 'Griffiths' report, for consultation prior to legislation.

Apart from an examination of the merits of each of these proposed changes, individually and collectively, in the context of the NHS for the 1990s (and, indeed, for the twenty-first century), these proposals will provide a further, stern, test for the bipartisan approach to the politics of health care in the UK. For perhaps the first time in the 40 years life of the NHS, the Government will try to secure an all-party agreement to a much fuller process of effective collaboration between a continued, stronger, NHS and the growing private health care industry, within the wider, mixed 'enterprise economy' of England in the 1990s and the next century. Some challenge, some test, some human prize: watch this space! The Minister (presently Mr William Waldegrave) has already made the first bold step in this direction by assuming the chairmanship of the NHS MB, with Sir Roy Griffiths continuing as his deputy chairman, and by recruiting a new Chief Executive for the NHS MB [and nailing his personal colours firmly to the mast, as in 1–3 above, at his first Conservative Party conference as Minister]. What follows may make history equal to the very beginnings of the NHS itself.

References

Department of Health and Social Security (1971). *Better services for the mentally ill*. Cmnd. 4683. HMSO, London.

Department of Health and Social Security (1975). *Better services for the mentally handicapped*. Cmnd. 6233. HMSO, London.

Department of Health and Social Security (1976*a*). *The National Health Service planning system*. HMSO, London.

Department of Health and Social Security (1976*b*). *Priorities for health and personal social services in England—a consultative document*. HMSO, London.

Resource Allocation Working Party (1976). *Sharing resources for health in England*. HMSO, London.

Department of Health and Social Security (1983). *Report of the NHS management inquiry*. DH, London.

Department of Health and Social Security (1984). *Implementation of the NHS management inquiry report* HC(84) 13 June 1984. DH, London.

Department of Health (1989). *Working for patients: the health service: caring for the 1990s*. CM 555. HMSO, London.

16

Public health policies and strategies in Japan

NOBUO ONODERA

Introduction

During the 40 years since the end of the Second World War, public health has made dramatic advances in Japan. This course of development can be divided into the first 20 years and the last 20 years, with great changes being observed not only in the nature of the problems of public health but also in the policies to cope with these problems. In particular, from the 1960s and early 1970s when there was rapid economic growth followed by a period of low economic growth, there was an improvement in the health of the people and their standard of living together with a marked extension in average life-span and a decrease in the mortality rate. Despite the drastic change in socioeconomic structure, the improvement of health care in Japan was remarkable compared with the rest of the world. In 1947 the average life-span of the Japanese was 50.1 years for males and 54.0 years for females, about 15 years less than the average life expectancy in advanced western countries, but in 1965 the average life expectancy of males reached the level of western countries—70 years for males and 75 years for females. In 1987, the average life expectancy of males was 75.6 years and that of females was 81.3 years, making Japan the country with the highest life expectancy in the world.

This chapter provides a systematic review of this successful and effective development of public health from the scientific standpoint of policies and strategies. Further, it attempts to answer the following questions. What was the impact of changes in policy? What kinds of health service organizations proved effective? What kinds of future problems are envisaged in public health? What kinds of actions should be taken for these?

Background of health policy-making

The basic policies for the recovery of Japan from the Second World War included (1) recovery from war damage, (2) control of inflation, (3) increased food production, (4) promotion of key industries, and (5) improvement of public health. For the first 20 years after the Second World War public health administrators used traditional public health approaches with emphasis placed both on control of infectious diseases, including tuberculosis and parasitic diseases, and on improvement in nutrition and environmental health. Factors which probably played a key role in improving public health during this period included the following.

1. Public health was developed under a democratic form of government.
2. Constitutionally, public health was recognized as a fundamental human right and an obligation of the government.
3. A national network of health centres was established, which served as the regional nuclei of public health activities.
4. A medical care delivery system was developed which recognized both unrestrained private practice and a system of public medical care institutions.
5. Arrangements were made by the government to bear expenses from public funds for patients with tuberculosis, mental disorders etc. who were faced with difficulty in meeting medical care expenses.
6. A social insurance system of medical care was formulated with complete coverage of all citizens.
7. Community activities for public health were initiated including campaigns for control of human disease transmitted by arthropods, mosquitoes, and flies.
8. Public health surveys, such as the All Japan Tuberculosis Prevalence Survey, were conducted using epidemiological techniques.

These policies were enforced in a manner commensurate with the existing consciousness of the people towards health and the existing social and economic trends of the nation. Dramatic changes were accomplished in post-war Japan through the dedicated efforts of public health administrators with the strong support of the Supreme Commander for the Allied Powers during the occupation of Japan. Advances in public health were also due to the self-sufficient attitude of the Japanese, the introduction of new laws, systems, and structures, and more adequate budgets to cope with the problems facing public health.

The development and promotion of statistically sound health information systems for identification of community health problems and requirements for care also contributed to improved public health. Public health administrators were obliged to tread a difficult path, however, because the first priority of the government was economic development.

From the 1960s remarkable development and advances were made in elevating the standard of living of the Japanese people. The valuable experience gained during the recovery from post-war difficulties coupled with the introduction of scientific methodology in public health, improvements made in legal systems, and systematic training of human resources resulted in significant improvements in public health in this period.

Though the policy for rapid economic development which gave priority to industrial development was influential in elevating the standard of living of the people, it also became necessary to develop strategies to cope with the resulting drastic changes in the social environment arising from over-population in some urban areas and under-population in rural areas, increases in industrial pollution, the problems of stress and the different health problems of an ageing population, and increases in the incidence of intractable diseases. The formulation of a health policy for the twenty-first century will need to take into account these issues as well as the increased demand for comprehensive health care, the increases in disease associated with ageing such as Alzheimer's disease, changes in the industrial structure, social support of workers, and changes in the Japanese political structure.

Structure of health policy-making

Decision-making structure of the central health administration

The central focus for the formulation of health policies in Japan is the Diet (House of Representatives and House of Councillors), the highest legislative body for the enactment of laws in the nation. Within the Diet the Social Affairs and Labour Committee of both houses deliberates on draft bills, and after acquiring national consensus the bills are presented to the Diet for approval.

In this process, it is, in abstract terms, the people themselves who participate in the formulation and drafting of the health policies, but in practice it is the government officials serving in the central health administrative bodies who are authorized to compile the budget, present draft bills to the Diet, and collect and collate the information necessary for decision-making witihn the legal, technical, and administrative structure.

Within the administrative structure of the government, the Ministry of Health and Welfare is responsible for health administration. The Ministry of Education is responsible for school health, the Ministry of Labour is responsible for industrial health, and the Environment Agency, which was established in 1971, is responsible for public health administration related to pollution and other environmental hazards. Japan has a parliamentary system under which the majority of the Cabinet ministers must be chosen from members of the Diet. The ministers of state are authorized to function as decision-makers according to the stipulations of the law. For example, on 7 June 1983 the Cancer Control Specialist Council, a cabinet council, was created under the guidance of the Prime Minister to formulate the Ten-Year Cancer Control Strategy.

In the process of making health policies, a judgement must be made as to whether the policy is just and proper and whether the fundamental human rights of the people have been considered. In drafting policies pertaining to general health services, the Population Problems Council and Public Health Council (based on Article 7 of the Establishment Law of the Ministry of Health and Welfare), the Health and Welfare Statistics Council and Living Environment Council (based on the Organization Regulations of the Ministry of Health and Welfare), the Medical Services Council (based on the Medical Services Law), and the Council on Health and Medical Services for the Aged (based on the Law on Health and Medical Services for the Aged) each have a vital function in the decision-making process.

The Public Health Council, in response to requests from the Minister of Health and Welfare, conducts investigations into public health issues and presents its opinions to the administrative bodies concerned. In 1978 the Public Health Council was expanded to include the former Infectious Disease Prevention and Investigation Council, the Tuberculosis Prevention Council, the Nutrition Council, and others. This enlarged council with its expert subgroups is composed of representatives of related agencies, scholars, and experts in the field.

In addition, in order to respond to the remarkable socio-economic changes in recent years and to the needs of the people, temporary private advisory and investigative discussion groups reporting directly to the minister or to bureau directors have been established for a given period, but they have no legal basis. However, these *ad hoc* groups play an active role in policy formulation. Examples of some of these are the Discussion Group on Life and Ethics, the Discussion Group on Social Problems of People over 80 Years of Age, the Discussion Group on Family Doctors, the Study Panel on Health Insurance of Remotely Located Communities, the Discussion Group on Stress and Health, and the Group to Review Future Plans for Regional Health. Futhermore, in November 1986 the Health and Welfare Science Conference was established as a discussion group under the Minister for Health and Welfare for the purpose of formulating policies which would take into account recent advances in scientific and technological studies related to health and welfare. This conference was designed to review (1) the middle- and long-term basic strategies for the promotion of scientific technology for health and welfare, (2) priority research themes for the future, (3) methods of evaluating research in health and welfare science, and (4) guidelines for the smooth promotion of research.

Structure of decision-making for local health administration

Local government in Japan is composed of 47 *do-fu-ken* (prefectures) and 3276 cities, towns and villages. Local public entities are based on the provision that regulations may be enacted within the scope of the law stipulated in Article 94 of the Constitution of Japan. Based on Article 14 of the Local Autonomy Law, regulations may be enacted through decisions reached in the local assembly. In local assemblies, committees are established to examine various policies on health and welfare and on the budget draft. In recent years, local public health agencies have made greater use of their regulation enactment power for the advancement of the health and welfare of the local residents.

Prefectural governors, mayors of cities, heads of towns and villages, and members of the local assemblies are elected by direct vote of the residents. Their term of office is four years with no restriction on re-election. The offices of the prefectures, cities, towns, and villages assume a wide range of administrative responsibilities of direct relevance to the residents. A Department of Health Administration exists and plays an important role in budget formulation and local health administration and management. Various types of local advisory councils and other expert groups also exist. These include the Medical Care Advisory Council, the Health Centre Management Council, and the Council for the Promotion of Health in Cities, Towns and Villages.

Health centres are established in the 47 prefectures for the promotion of community health services based on the Health Centre Law, and also are established in 32 government ordinance-designated cities and 23 special wards of Tokyo based on the enforcement ordinance of the Health Centre Law. Health centres are established, as a rule, at a rate of 1 per 100 000 population, but they may be established taking into consideration transportation conditions, relation to other government offices, local health conditions, and population distribution.

As of March 1988, there were 851 health centres in Japan, composed of 638 prefectural health centres, 160 city health centres, and 53 ward health centres. The personnel strength as of March 1987 was 35 253. The director of a health centre is a permanent local technical officer who is a medical doctor with three or more years of experience in public health work, who has successfully completed the training course provided by the Institute of Public Health, or a medical doctor recognized by the Minister of Health and Welfare to have technical skill or experience equivalent to or greater than the foregoing. Duties and responsibilities of health centres are stipulated in the Health Centre Law and the specific nature of its operations is based on the Tuberculosis Prevention Law, the Maternal and Child Health Law, the Mental Health Law, the Law on Health and Medical Services for the Aged, and the Foods Sanitation Law so that uniform services can be provided throughout the nation. In addition, the health centres engage in work specific to the problems of each locality, such as measures of endemic diseases and health checks related to pollution. The budget for health centres is provided from various sources including national funds in the form of management subsidy, operational grants, grants for facilities and equipment, and grants from the local authority concerned, Flexible budgetary management is made possible within the framework of the Regional Finance Law and with the application of the stipulations of the Special Measures Law on the Rationalization of Administration Related to Operational Expenses of Projects of Health Centres.

At the prefectural level, the collection and analysis of data related to investigation, research, and guidance made at local health laboratories, environmental pollution research centres, mental health centres, and hospitals plays an important role in public health decision-making.

With the aim of promoting the health of the residents in local cities, towns and villages, a ten year plan to establish 4000 city, town, and village community health centres where health consultation, health education, and health checks could be provided was initiated in 1978 (as of March 1989 there were 979 centres).

Various types of statistical information are extremely important factors in deciding health policies. The major health statistics compiled by the Statistics and Information Department of the Ministry of Health and Welfare can be broadly classified as follows:

(1) a vital statistics survey;
(2) a comprehensive survey of living conditions, initiated in 1986 by the integration of four basic household surveys on structure (family size, housing and age), income, savings, and health of household members;
(3) a survey of the numbers and distribution of medical personnel, medical institutions, and social welfare institutions;
(4) statistical reports on the activities of health administrative agencies such as activities carried out by health centres, medical services for the aged, and approval of foods.

Health centres play an important role in the implementation of these surveys and also have a major function in decision-making in the health centres, at both the national and regional level.

Development of public health activities, special scientific organizations, and civilian groups

It is evident that effective public health activities can be developed by better defining the responsibilities assumed by national and regional public entities and by strengthening the activities and defining the role and responsibilities of such

Table 16.1. A comparison of traditional and modern concepts of public health services

Traditional concepts	Modern concepts
Strict vertical-order systems	Mutual understanding systems
Strict regulation	Emphasis on functions
Centralization of authority	Decentralization of authority
Social defence against diseases	Individual response

professional organizations as the medical, dental, pharmaceutical, and nursing associations.

Voluntary health organizations have played a vital role in the development of public health in Japan. Soon after the end of the Second World War, organized community health activities commenced under the positive guidance of national and local public entities. The voluntary activities included the promotion of environmental health activities, tuberculosis prevention activities (by the Japan Anti-Tuberculosis Association and the Regional Women's League for Anti-tuberculosis), maternal and child health activities (*Aiiku-han*—a women's community for better health and welfare of mothers and children, the name of which was derived from *Aiiku-kai*, a non-profit organization founded in 1934), grass-roots activities for measures against parasites (the Japan Association for Parasite Control), and nutrition improvement activities through groups initiated by recipients of nutrition courses organized by health centres. The last of these proved particularly effective in rural agricultural communities.

In 1960, with major changes in the socio-economic environment, there was a turning point in organized community activities in both rural and urban areas. In the changing society, with greater employment opportunities and increased dependence on nuclear families, community solidarity within society began to decrease. Futhermore, the community response to pollution, intractable disease, and problems regarding regional differences in medical service resources resulted in a general lack of confidence and a number of disputes between public administrators and the community. As a result community-organized groups became more active in providing direct public services. A systematic evaluation of community health problems emphasized a greater involvement of the members of the community themselves. It is thus evident that there was a transition from the traditional to the modern concepts as shown in Table 16.1.

Changes in the attitudes of local communities and civilian organizations were greatly influenced not only by changes in public health structure and dramatic technological advances in medical care but also by developments in fields related to public health such as computer science and biotechnology, development of new materials, modes of transportation and communication, retrenchment in the national budget, and changes in the international economic environment.

In the light of these various changes in environmental conditions, scientific groups and civilian organizations, including existing professional groups and single-purpose organizations, took urgent measures to improve and reform their constitution and structure in order to permit integrated interdisciplinary but autonomous activities so that they would be able to cope adequately with the changes of the new age. For example, the Japanese Medical Association, to strengthen its policy formulation function, created within its own framework a number of specialist scientific panels such as the Medical Care Policy Council and the Health Education Commission. A large number civilian organizations in promoting their studies on policy formulation took steps to activate their branch organizations so that they could actively participate in and support regional health activities. For example, they actively participated in various advisory and other councils and developed and reinforced political power for the formulation of health policies. As a consequence, there was a transition from policy based on the previously bureaucratic leadership to one based on community leadership. Concurrently, people tended to utilize the political power of pressure groups less and the results of policy research more. There is a tendency for the formulation of health policies in Japan either to increase or to decrease depending on the relative relationship between bureaucratic and scientific groups. The more complicated the problem, the greater is the importance of the political power of Diet members, in particular the members of the Social Affairs and Labour Committees.

Special concerns of public health

The environment surrounding public health in Japan has changed substantially with the ageing of the population, advances in science and technology, improvement in the national standard of living, and developments toward institutionalization. For the maintenance and promotion of national health in response to these changes, it is necessary both to ascertain the needs of the people and to promote appropriate scientific measures.

Public health measures for an ageing society

In recent years there has been a rapid change in the population structure in Japan; it has been estimated that the growth in the elderly population of Japan is far higher than that of other developed countries of the West. The proportion of the population aged 65 years or more in Japan was 7 per cent in 1970, but it is predicted that this will reach 14 per cent by 1996, i.e. within a period of 26 years. In contrast, in France, where the elderly population reached 14 per cent in 1975, it required 85 years, in the UK and the FRG it required 45 years, and in the US, where the proportion is expected to reach 14 per cent in 2020, it will have required 75 years.

Public health measures for an ageing society should be directed towards services which enable people to live their entire life in good health. In February 1983, the Law on Health and Medical Services for the Aged was enforced to promote health and medical service programmes such as health examinations, health education, and health consultation with emphasis on the prevention of adult diseases so that people would be able to enjoy old age in good health. In order to introduce this programme smoothly, based on the opinions formulated by the Panel on the Health and Medical Services for the Aged under the Public Health Council, guidelines for activities to promote health in the elderly were presented in the First and Second Five-Year Project for the Health and Medical Services for the Aged (1982–6 and 1987–91 respectively). In preparation for the increased elderly population in the early part of the twenty-first century, on 6 June 1986 a cabinet decision was made on the General Measures for the Aged Community as a guideline for the projects for the elderly to be promoted by the Government. The basic policy was as follows.

(1) to promote economic prosperity and an active elderly community;
(2) to promote the formation of communities based on the spirit of social solidarity and support for the active health of the aged community;
(3) to provide relief from economic concerns so that the elderly can have a healthy and satisfactory life;

The conditions of the policy were as follows:

(1) to establish a system permitting stable life during one's life span;
(2) to establish to basic conditions for healthy and satisfactory life.

Apart from satisfying these policies, consideration must be given for a satisfactory system of employment, income, health, welfare, education, social participation, housing, and living environment, all of which have important implications and roles.

Measures for mental health

In the rapidly changing contemporary society, people face a multitude of stresses. Thus maintenance and promotion of mental health, respect of human rights, proper medical care, protection of the mentally handicapped, and promotion of social rehabilitation are very important public health issues. It was only after 1950 when the Mental Health Law was enacted in Japan that mental health became an indispensable field of contemporary public health. A remarkable improvement in medical care in this field was subsequently made through greater availability of hospitals, but as in-patient care alone is not sufficient, the need for after-care as outpatients has been strongly emphasized. At the same time, with the development of mental medical care and enhancement of social consciousness came the desire for community mental health activities. in 1975, with the strengthening of the activities of mental health centres, a consultation service was initiated at health centres.

In 1986 the Mental Health Panel of the Public Health Advisory Council submitted to the Minister of Health and Welfare 'Opinions regarding social rehabilitation of the mentally handicapped' in which the promotion of social rehabilitation and social participation of the mentally handicapped was sought. Furthermore, following the revision of the Mental Health Law, a new direction was set for the improvement of mental care and in particular for the promotion of the human rights of the mentally handicapped and for the maintenance and promotion of the mental health of the people. It is estimated by the *Ad Hoc* Headquarters Office that, accompanying the increase in the elderly population in recent years, the prevalence rate for senile dementia in the entire nation as of 1987 was 4.8 per cent in the population aged over 65 years.

With the increase in the number of cases of senile dementia an urgent need arose for the establishment of comprehensive measures, and in August 1986 the Ministry of Health and Welfare established the *Ad Hoc* Headquarters Office for Promoting Measures for Senile Dementia. This office has been engaged in the study of senile dementia cases, supportive measures for their home care, and improvements in facilities for treating them.

To cope with the increase in social problems such as truancy, domestic violence, suicide, bullying, delinquency, school violence, refusal to eat, and vocational inaptitude, programmes are being promoted to create mental health in accordance with the individual's life-cycle. Furthermore, as part of the comprehensive labour health programme in the work-place, mental health care had been incorporated in the programmes for maintaining and creating health in both mind and body. In addition to health consultation and health guidance, efforts are being made to improve the working environment and work content aimed at maintaining mental health.

Measures against alcoholism

With the increase in national income and changes in life-style in recent years, there has been an increase in alcohol consumption and thus a strong need has arisen for measures to cope with the resulting family collapse and social problems. According to the annual statistical reports of the National Tax Agency, the per capital consumption of alcohol in Japan, computed as pure alcohol, was 2.13 litres in 1955, which increased to 4.79 litres in 1970, 5.80 litres in 1980, and 5.97 litres in 1984. The 1984 patient survey conducted by the Ministry of Health and Welfare has estimated the number of alcohol dependents to be about 1.8 million, 2.9 per cent of the estimated total of people consuming alcohol.

From 1979, measures were strengthened for the prevention of alcoholism with the Mental Health Centres serving as the core. Education has been provided to the public on alcoholic beverages and how should they be consumed. Treatment and guidance regarding abstinence has been given to alcoholics. In 1980 the Japan Health and Alcohol Incorporated Association was established for the purpose of disseminating knowledge regarding alcoholic beverages and of strengthening the related research system. The Public Health Council in 1985 submitted to the Minister of Health and Welfare its opinions on measures to be taken for alcohol-related problems, such as strengthening measures for prevention, medical care, social rehabilitation of alcoholic cases, and related research.

Measures against smoking

There has been mounting global concern over the adverse health effects of smoking. According to the National Survey of the Smoking Population conducted by the Japanese Tobacco Industry Corporation, the smoking population had declined from 83.7 per cent in males and 18.0 per cent in females in 1966 to 62.5 per cent in males and 12.6 per cent in females in 1986. As early as 1900 a law prohibiting minors from smoking was enacted in Japan, and in recent years public health measures against smoking have gradually been strengthened. In 1964 a notification was issued by the Public Health Director of the Ministry of Health and Welfare that effective health education should be given on the problem of smoking and health. Together with the Medical Affairs Director of the same ministry they advised that smoking should be restricted to specific places within medical treatment institutions. There has also been an increase in the view that non-smoking rights be recognized. From April 1985, the tobacco monopoly system was abolished. Tobacco manufacturing was placed under private management and the import of foreign tobacco was liberalized. As a result the sales of tobacco have become increasingly competitive. Consequently, to maintain the health of the people, positive measures had to be taken against smoking. An Expert Committee on Smoking and Health Problems was established within the Public Health Council of the Ministry of Health and Welfare, and, after making a study and literature review on the present status of smoking, the effects of smoking on health, knowledge and attitudes regarding smoking and health problems, and present status of measures against smoking, they released their report in October 1987. According to this report, the trend in smoking by age group showed that in males the decrease in the smoking rate in those in their twenties and thirties was relatively small, but the decrease in those in their forties and over was substantial. In females, however, the decrease in the smoking rate was marked in those aged 50 and over, although the rate in those in their twenties presented a problem in that the rate increased from 9.8 per cent in 1970 to 12.7 per cent in 1975 and to 16.2 per cent in 1980. To reinforce public health policy towards smoking, in November 1987 the Sixth World Congress on Smoking and Health was held in Tokyo to develop measures against smoking with the concerted effort of the people and the government.

Measures for the promotion of health

Maintenance and promotion of national health is indispensable for the creation of a rich and active society. The promotion of health with emphasis on primary prevention, in particular, is vital. Since 1967, Nutrition and Health Exhibitions have been held in regional cities of Japan and, through the activities of public health and nutrition classes in health centres, guidance has been given on nutrition, exercise, and recreation. From 1978, national integrated health measures have been enforced with the aim of providing a healthy life for the entire nation. These have included the following:

(1) promotion of health throughout the entire life-span, encompassing maternal and child health, school health, industrial health, and geriatric health;
(2) establishment of the bases for health such as health promotion centres and training of human resources including public health nurses;
(3) strengthening of the Health and Physical Strength Institute, establishment of Health Promotion Councils in cities, towns, and villages, and intensified activity of diet improvement promoters aimed at education and dissemination of health activities.

Measures for environmental health

Important public health policies were necessary to cope with environmental policies in Japan which differed from locality to locality with natural conditions and socio-economic conditions attributable to industrial development. It was recognized that the water supply and sanitation required improvement nationally. As a result plans for a waterworks administration to ensure an extensive water supply were released in March 1984 by the Living Environment Council. It stipulated that the local public authorities were responsible for the disposal of their own garbage and domestic waste. Since 1963 a series of five consecutive long-term garbage and waste disposal plans have been developed and work has begun for the establishment of waste disposal facilities and the effective use of waste. Regional co-ordination and co-operation were needed in selecting disposal sites. The responsibility for the disposal of industrial waste rests with the enterprise which produces the waste and the details regarding the methods of disposal and standards for the disposal facilities have been established. As an administrative measure, the prefectural governor seeks the opinion of the Prefectural Pollution Control Council in advance and must establish an industrial waste disposal plan in accordance with the law.

Pollution in Japan has been extremely serious in this period of rapid economic growth. Minamata disease, Itaiitai disease, Yokkaichi pollution and other pollution-related diseases have been targets of public concern, and in lawsuits the injured party has invariably won the case in court. The government has come to grips with the pollution as a national problem and in a 1967 the Basic Law for Environmental Pollution Control was enacted. Thereafter efforts have been made to prevent environmental pollution and to maintain a satisfactory environment.

Medical care delivery system and medical care security

The medical care system in Japan is characterized by the freedom of doctors to open private practice, the freedom of the people to choose the medical care of their own choice, and security from medical expenses through insurance systems. The medical care delivery system has shown a desirable trend in recent years in increased accessibility and availability with the establishment of modern medical facilities. However, high-quality effective medical care must focus on the qualitative rather than the quantitative aspects of health care. In December 1985 the medical services law was revised and a regional medical care plan was establishment for each prefecture which incorporated (1) the establishment of medical care zones, (2) estimates of the required number of beds, (3) targets for hospital renovation in accordance with medical care priorities, (4) a system for the co-ordination of function, and (5) recruitment of medical care personnel. Furthermore, in January 1987 the Ministry of Health and Welfare, in order to respond adequately to the medical care needs of the people, established the *Ad Hoc* Headquarters Office for Comprehensive Measures for National Medical Care.

For medical care security, a universal health insurance system was established in April 1961 as a basic policy. The characteristic feature of the medical care security system in Japan is that medical care expenses are borne by public sources (national government and/or local public authorities) legally based either on the Tuberculosis Prevention Law, the Mental Health Law, the Law on Health and Medical Services for the Aged, the Atomic Bomb Survivors Medical Treatment Law, the Child Welfare Law, and the Livelihood Protection Law, or on budgetary measures in accordance with Guideline Rules on Measures for Intractable Diseases established by the Ministry of Health and Welfare in 1962 for patients with specific diseases and for patients economically unable to bear medical care expenses. The national medical care expenses for the year 1987 totalled 18 trillion 75.9 billion yen (US$ 11.7 billion), of which 55.1 per cent was met by health insurance, 32.4 per cent by public funds (6.4 per cent by public health and welfare services, and 26.0 per cent under the Law on Health and Medical Services for the Aged), and 12.3 per cent by others (12.0 per cent by the patient). The proportion borne by public funds was 15.9 per cent in 1955, but this had increased to 27.6 per cent in 1970 and 36.4 per cent in 1983. In preparation for the increased proportion of elderly people in society, a stable base for the medical care insurance system is needed for adequate cover of the needs of the expected life-span of 80 years. The proportion of gross national product devoted to medical care was 2.7 per cent in 1955, 3.32 per cent in 1975, 4.98 per cent in 1985, and 5.15 per cent in 1987. This proportion is gradually stabilizing as a result of recent policies on medical expenses.

Industrial health

The labour force in Japan is approximately 57 million. Japan entered a period of rapid economic growth in the 1960s, and with it arose developments in industrial activities and changes in working patterns and working environment accompanied by a mounting concern for prevention of pollution. New technologies and materials were rapidly introduced to the industrial sector and with the increase in occupational diseases came the need to promote measures for safety and health which would be able to cope with the rapid changes in industrial activities. The Ministry of Labour had developed measures for industrial health cased on the Labour Standards Law, but in 1972 the Labour Safety and Health Law was enacted and enforced. Through this new legislation, the Ministry of Labour, in addition to maintaining the previous minimum standards, worked positively towards developing measures for preventing health hazards in the changing labour environment and towards the formation of more comfortable places of work. In contemporary society, efforts are being made to formulate a comprehensive industrial health system. The three pillars of a comprehensive system are the management of the working environment, labour force management, and management of health factors which incorporate advances made in medicine and health engineering.

The future of public health in Japan

Through the great efforts made to restore the means of production and national economy of Japan following the catastrophic effects of the Second World War, there has been continued socio-economic development, promotion of science and technology, and formulation of effective measures for public health. In looking ahead to the twenty-first century, we can expect further ageing of the population, development of better health information systems, changes in national health consciousness with greater involvement of the public in formulating health activities in the economy, advancement in industrial structure, sophisticated municipalization and an internationalized society. New public health problems, which must be tackled with comprehensive policies and strategies not confined to the realm of traditional measures, can also be expected.

The increases in cancers, heart disease, and mental disorders are examples of changes in disease patterns in Japan. The increase in traffic casualties is attributable to increase in the volume of transportation and has become a matter a vital concern. The increasing demand for medical care with rapid ageing of the population (with over 2 million individuals over the age of 80 in 1984) poses a need for improving care systems for the aged. The elevation of popular health consciousness and behaviour with advances made in public health has brought an increase in health demands, while specialization and sophistication of medical care have introduced at the same time the problem of effective distribution of medical care resources and medical ethics. The levelling off of economic growth to a stable level accompanying the changes in economy has indicated a need to arrest the present increasing trend of medical care expenditure. Accompanying the trend toward internationalization, many problems have to be addressed, including the problems of more rapid introduction of infectious diseases with international travel and the health management of foreigners residing in Japan. The problems of environmental pollution resulting from chemical substances introduced by industrial innovation and of anti-biotics and anti-bacterial substances remaining in foodstuffs must be addressed. The promotion of the Radiation Effects Research Foundation is another problem in public health in Japan which is the only country in the world to have experienced the atomic bomb in warfare.

The future of public health in Japan is dependent on how health, medical care and welfare policies will be integrated and developed in the community. The scope of public health is dramatically evolving, and thus public health is in need not only of public policies but also of the establishment of a system to promote both community and private sector participation.

Moveover, the development and application of sophisticated information systems in Japanese society, such as surveillance systems for tuberculosis and infectious diseases, information systems for emergency medical care, and medical care information systems for remote areas, should be positively promoted. In this endeavour, it is important that consideration be exercised to safeguard privacy.

In the formulation of new health policies studies should be made of examples of creative public health activities in the community. The formulation of health policy in Japan is characterized by legislation being initiated by politicians, government officials, scholars, experts on learning and experience, and representatives of the people, followed by enactment through comprehensive and systematic actions of political adjustment, policy drafting, council activities, public opinion surveys, community participation, activities by residents, petitions, and mass campaigns. In this process, an interdisciplinary consensus-seeking approach lies at the core of policy formulation. Such action is possible at both the national and the local level.

In the development of future policies and strategies of public health, it is important to appreciate that the essence of public health is the concept of respect for life and the philosophy of peaceful coexistence. At the same time philosophical advancements can be achieved by integrating the universal and creative ideas with analytical logic through the pursuit of scientific truth, changes in social structure, and appreciation of the inconsistent nature of human beings and societies in the development, application, and practice of public health. From these standpoints, it may be expected that further progress will be made by incorporating the results of future international studies on health policy science to the course, characteristics, and future of public health policies and strategies in Japan.

Further reading

National Land Agency of Japan (1987). *The Fourth Comprehensive National Development Plan*. National Land Agency of Japan, Tokyo.

Japan International Cooperation Agency, Japan International Medical Foundation. National Health Administration of Japan (1987). *Textbook for Seminar on National Administration*. Ministry of Health and Welfare, Tokyo.

Health and Welfare Statistics Association (1987). *Trends of National Health Services in Japan*. Vol. 34, No. 9, Tokyo.

17

Public health policies and strategies in China

CARL E. TAYLOR, ROBERT L. PARKER, and ZENG DONG-LU

Few countries have experienced as dramatic changes in health policy and strategy in recent years as China. This chapter describes how recent political changes have influenced health care.

China has one of the most decentralized health systems in the world. This is contrary to the perception that most people in the west have of health care in a communist country. Since liberation in 1949, health services have been organized and paid for locally, presumably because officials in Beijing logically realized that they could not take responsibility for trying to pay for free care for a quarter of the world's people. Since 1979, Deng Xiao Ping's open policy and modernization reforms have accelerated decentralization of all governmental functions.

In 1979, a long discussion with the head of the planning bureau in Beijing described the new pragmatic process by which the planning bureau was translating high level official policies to practical implementation. He ended the discussion with the summary statement, 'in Beijing we decide what should be done, but people at local level decide how it should be done'. This is a perceptive definition of the distinction between policy and strategy. In accordance with general directives from national leaders, the ministry proposes policies on specific issues which are sent to the periphery for comment. Responses from the various parts of the country are summarized and policies are revised with repeated feedback to the periphery for clarification depending on local conditions, priorities, and constraints. Wording of policy statements is made sufficiently general to accommodate differences in the various parts of the country. Once agreement is reached there is a meeting of responsible officials with the level of attendance depending on the priority given to the issue. The policy is announced by the highest official responsible for that policy. It is publicized in the media and from that point every unit is expected to work out the local adaptations needed for implementation.

This description of the health planning process in China came at an important transition point and reflects the evolving system. Before 1979 the process of exchange with the periphery was probably minimal and depended mainly on officials from Beijing following Chairman Mao's example and going to the countryside to hear for themselves what people were saying. Especially during the erratic and wildly uncertain period of the cultural revolution, all policies depended on interpretations of Chairman Mao's sayings. Policy reversals were abrupt and had to be implemented immediately with little opportunity to raise questions or make adaptations. Some theoretical positions that were widely admired internationally actually caused great hardship and later backlash and resistance. Other dramatic innovations met longstanding public demand and have gradually helped to evolve the general principles on which the present health system is built (Hillier and Jewel 1983).

Since 1979, the continuing decentralization has systematized procedures for obtaining feedback from the periphery. Regular opportunities are provided through bureaucratic channels as well as the party to report on how various strategies are working and how policies might be improved. An important Party Congress in July 1988 increased the role of administrators and health professionals in decision-making.

In the shift from a hierarchically centralized to demand-based setting of policy there was a period during which China's unique achievements in health care seemed to be slipping (Sidel and Sidel 1982). Families were required to arrange their own financing and this seemed to diminish the priority that had been given to prevention and equity. The earlier pattern had been portrayed around the world as 'the model' of communities providing their own care with an idealized image of dedicated barefoot doctors from the community growing their own herbs in order to provide immediate and culturally appropriate care to neighbours at very low cost. This image was based on visits to selected model services. Under the communes, care depended on workpoints rather than on money, so local officials were able to distribute the limited care in their units relatively equitably at commune level even though there was great regional variation. This changed in 1979 when the Economic Responsibility System took financial control for agricultural and business activities from communal units run by party cadres and gave it to families (Schell 1985; Wang Guichen and Zhoun Qicen 1985).

The transition has been difficult for all social services because economic gain now dominates the thinking of officials and families, rather than the previous emphasis on social welfare. On the other hand there has been more money available for family discretionary spending because the economy was growing at the rate of 10–15 per cent per year. This is now being eroded by inflation, especially for urban areas. Care standards in health institutions are improving as services are being professionalized. There has now presumably been something lost in community participation and personalized relationships, but even so the emphasis on 'serve the people' permeates services more than in most countries. There are also compensating gains because care has become more scientific and effective. Even with the dramatic changes in recent years, official policy continues to insist that equity in the distribution of care should not be jeopardized. A point that is beginning to be recognized is that the previous regional disparities need to be corrected by greater central government subsidy to the poorest areas. It may be possible to improve regional equity because data gathering and analysis can be focused on more accurate definitions of conditions in the poorer areas.

In this chapter the general transitions described above will be traced as they influence particular policies and strategies. In each section a general category of public health activity will be described. Because our own experience in China has been mainly in maternal and child health services, those areas will receive particular attention. We have learned that the legendary patience that comes from being one of the world's oldest continuing civilizations causes the Chinese to plan in the perspective of centuries rather than years. This does not mean, however, that decisions are delayed: we have been frequently surprised by the willingness of officials to try out new ideas. The general attitude is that if something works it will be continued, if it doesn't work it will be changed. This pragmatic emphasis on scientific modernization means that decision-making follows Deng Xiao Ping's principle of gathering data and 'seeking truth from facts'.

Three-tiered health system

The health system is based on a clear hierarchy of organization and all units and workers know where they fit. The three-tiered organization follows the political/administrative structure and is dependent on it. Where population units are particularly large a fourth tier is introduced so that the size of peripheral units can be equalized.

The basic organizational unit is the county or *Xian*. The population of a county ranges from less than 100 000 where people are dispersed to over one million in crowded eastern provinces. Each county is expected to be self-sufficient in decision-making and financing, and there is great diversity between counties. Neighbouring counties have very different patterns of organization and use of personnel. But the counties seem to be learning from each other's experience and these differences may prove to be transitional as greater uniformity gradually evolves in the various regions. In addition to hospital services, there are public health bureaux for anti-epidemic services, maternal and child health services, laboratories, etc. Both public health and hospital services in the county are controlled by local county officials. Most qualified doctors are financed from provincial subsidies to local units, and salaries continue to be standardized at a relatively low level, but there is great flexibility in financing for other local staff.

Within the county the second tier of organization is the township or *Xiang*. In the reorganization that accompanied the economic responsibility system the old township structure was re-established. The townships are approximately the size of the communes that were started as part of the Great Leap Forward in the late 1950s but have now been abolished. Most townships have a population of from 20 000 to 70 000, and contain one or more towns which are undergoing rapid industrial development. The many small-scale industries provide much of the money used by the administration of both *Xians* and *Xiangs* for development activities. The economic responsibility system is encouraging entrepreneurs, and the private sector is taking over new industrial and service enterprises in affluent areas but not in conservative provinces such as Hunan.

Township or *Xiang* health services reflect county level organization but with considerable streamlining. There is usually a small hospital and several clinics, including maternal and child health services, but these are usually not staffed with specialists.

The village is the third tier or most peripheral administrative unit. Under the communes, production brigades and production teams organized communal work. In 1970 about 70 per cent of China was covered by health co-operatives (Hillier and Jewell 1983). When agricultural control was returned to families these production units no longer had a function and the administrative structure returned to what were called natural villages with 1000–3000 population. Clusters of hamlets are sometimes combined in administrative villages. The health posts where barefoot doctors had worked were upgraded. The village had to decide how these health posts would be supported. The more affluent villages have tended to maintain health co-operatives that were functioning well, and either require per caput payments from each family or fund them from profits from small industries.

In much of China, however, the feeling was that since private enterprise for farmers was encouraged under the contract responsibility system the same rules should apply to health workers. At first many barefoot doctors simply shifted to farming because that seemed to be where more money could be made (Sidel and Sidel 1982). When fee-for-service private practice was permitted many of them were willing to take the special courses and exams to obtain the credentials of a village mid-level doctor. The role relationships are still

Table 17.1. Numbers of professional health personnel, 1949–1985

	Number (1000s)				
	1949	1957	1965	1975	1985
Total	505.0	1039.2	1531.6	2057.1	3410.9
Doctors and assistant doctors of traditional Chinese medicine	276.0	337.0	321.4	228.6	336.2
Pharmacy workers of traditional Chinese medicine	*	53.5	71.8	86.2	151.2
Doctors of western medicine	38.0	73.6	188.7	293.0	602.2
Feldschers	49.4	135.7	252.7	356.1	472.8
Pharmacists	0.5	2.4	8.3	12.8	33.0
Assistant pharmacists	2.9	18.4	37.2	57.2	89.7
Nurses	32.8	128.2	234.5	379.5	637.0
Midwives	13.9	35.8	45.6	64.9	75.5

* Data not available.
Source: Adapted from Cui Yueli (1986).

evolving and tensions will probably increase between salaried professionals and health workers in private practice.

Above the county, administratively, is the prefecture. Several counties are included, with population size ranging from half a million to about five million. Referral hospitals with specialists are usually available, and some have medical schools. Supportive services for public health activities include laboratory facilities and the peripheral branches of national special programmes.

Above the prefecture is the province or autonomous region. There are 23 provinces (excluding Taiwan), five autonomous regions (Xinjiang, Ningxia, Tibet, Guizhou, and Guangxi), and three separate municipalities (Beijing, Shanghai, and Tianjin). One province, Sichuan, has a population of more than 100 million, which makes it the equivalent of the eighth largest country in the world. The fantastic diversity of China includes geographical conditions that range from some of the world's most intensively farmed areas in the east to the sparsely populated deserts and mountains in the western half of the country. Economic and social development levels also vary greatly. Health conditions in the eastern half of the country, with 41 per cent of the geographical area but 89 per cent of the population, reflect the types of problems found in developed countries (Chen Haifang and Zhu Chao 1984). In the western three provinces and four autonomous regions, with 59 per cent of the land area and 11 per cent of the population, the illness patterns are still similar to those of developing countries.

Provincial health services often have to care for a larger population than most countries in the world. They are expected to be self-sufficient in producing their own personnel and financial support. Each province and autonomous region has at least one medical school and multi-specialty centres and research institutions. They set their own priorities and make their own decisions except for a few national programmes such as the expanded programme for immunization which is controlled by the central antiepidemic bureau.

Health workers

China needs massive numbers of health workers to care for its population of over one billion. Table 17.1 shows that at the time of liberation in 1949 it was estimated that there were only 38 000 doctors of western medicine but 276 000 practitioners of traditional Chinese medicine (Cui Yueli 1986). By 1965 the number of doctors had increased to almost 200 000, and by 1985 to over 600 000. The number of traditional practitioners increased more slowly: 320 000 in 1965, and 336 000 in 1985. Mid-level assistant doctors were almost 50 000 in number by 1949, over 250 000 in 1965, and 472 000 in 1985. The number of nurses increased from 33 000 in 1949, to 234 000 in 1965, and 637 000 in 1985. The number of midwives went up from 14 000 in 1949, to 45 000 in 1965, and 75 000 in 1985.

This phenomenal rate of increase was because of the great priority given to health by the national leaders. Under the influence of Russian advisers, the main emphasis during the 1950s was to develop medical schools and hospitals. Extensive building programmes provided facilities for later development. The priority given to educating doctors and the rapid expansion of clinical services in urban areas meant that rural populations were still not being served.

Mao's roots in rural China and experiences during the guerrilla fighting led to an abrupt reorientation about the time of the Great Leap Forward in 1956. When Russian assistance was terminated, China shifted to radical policies of promoting bottom-up efforts to obtain quantitative expansion of simple services to reach the people in greatest need.

The origins of the barefoot doctor system, as well as most of the principles of Primary Health Care as enunciated at Alma Ata, were strongly influenced by a rural project in Ting Hsien (Ding Xian) in the early 1930s. This rural county about 100 miles south of Beijing was the experimental area for a remarkable rural reconstruction movement organized by the reformer Jimmy Yen. The health service component was run by Drs C. C. Chen and John B. Grant of the Peking Union Medical College and the Rockefeller Foundation (Seipp

1963). Rural health projects around the world since that time have been trying to implement principles discovered at Ting Hsien. They showed that through community participation and simplification of scientific procedures it was possible to provide comprehensive care for about twelve cents per caput per year (at 1930s prices). Most of the work was done by village health workers who were trained to provide simple care for their communities. In communist-controlled areas, similar experiences led to the formation of people's schools and the training of medical assistants for the guerrilla armies assisted by Drs Norman Bethune and George Hatem (Ma Haide) (Chen Haifang and Zhu Chao 1984). One reason for the high morale of the People's Liberation Army was their good medical care: 80 per cent of battle wounds were treated in 15 minutes, and 70 per cent of casualties were transported to a medical facility in 30 minutes by relay stretcher teams.

In two great surges of training in the last half of the 1950s and the last half of the 1960s, all parts of China were covered by services provided by barefoot doctors. Recent medical graduates in the 1950s and essentially all doctors in the 1960s were required to go to the countryside for extended periods of service. There were many problems and much ruthless coercion of professionals but one good thing that happened was that they trained one and a half million barefoot doctors (Hillier and Jewell 1983). Joshua Horn (1969) described the pain and exhilaration of being sent with a team from an orthopaedic hospital in Beijing to remote villages in northern China. It was so cold that in order to build their own quarters and the training facilities for barefoot doctors these orthopaedic specialists had to heat the water to make cement.

The results were remarkable. Shared struggle generated a sense of service that was transmitted to the barefoot doctors with a value orientation that fused with community concerns. The specialist teams learned what conditions were really like in the countryside—Mao had said they should learn from the people. They greatly improved the quality of local services. Because the coverage by hospital teams was limited, most of the training of local personnel was done by county health staff. The role of the antiepidemic services became especially important because they eventually took on the main responsibility for supervising the work of the barefoot doctors. They learned to appreciate the usefulness of these village workers in promoting preventive programmes.

For curative care a major limitation was the lack of drugs and equipment. Barefoot doctors were trained to collect or grow their own herbs in accordance with the ancient practices of traditional medicine. Every health post had its own herb garden, stacks of plants drying, and equipment to grind or otherwise prepare the traditional medicines. About one-third of the barefoot doctors' work time went into preparing their own medicines. Growing medicinal plants was not too different from the agricultural responsibilities assigned to them by the production brigades. Some barefoot doctors specialized according to their own interests and skills in areas such as acupuncture, massage, dentistry, or midwifery. In addition, each production team had health assistants with even less training, for routine preventive services.

Table 17.2. Numbers of barefoot doctors, village doctors, and birth attendants in China, 1970–86

	Barefoot doctors	Village doctors (total certified)*	Health assistants	Village birth attendants
1970	1 218 266			
1975	1 559 214			615 184
1980	1 463 406			634 858
1984†		1 396 452 (117 236)	2 006 560	584 565
1985		1 293 094 (341 972)		513 977
1986		1 279 935 (341 783)		507 538

* Figures in parentheses indicate the number of village doctors who have been certified by passing credentialling examinations. The others continue to practise without certification.
† Adapted from Chen Haifang and Zhu Chao (1984).
Source: Adapted from Ministry of Public Health (1986). *Chinese health statistical digest.* Beijing.

When the economic responsibility system was started families were encouraged to take communal land on contract. They had to sell to the Government at official rates what the land would have produced under communal farming, but they could sell anything more that they could grow in the rapidly expanding free market. Agricultural production almost doubled in 5 years. Contracts for land have now been extended from 1 year to as much as 15 years.

Social support activities such as day care centres for children collapsed under the contract responsibility system because there were no longer workpoints to pay for services. Parents decided to keep their children at home rather than turn them over to public facilities. This became especially true after the one-child policy was announced because families wanted as much contact as possible with their one child.

After the modernization reforms the emphasis in health care shifted mainly to curative activities because that was what families were willing to pay for under the new financial arrangements (Young 1984). The Government set up an extensive programme of retraining; after about 6 months' work in township or county hospitals, the barefoot doctors could take a certification exam. If they passed, they became certified village doctors. Those that didn't pass continued to be uncertified village doctors. In most places both groups could charge fees that were fixed by local authorities. Table 17.2 shows the village health personnel situation from 1970 to 1986, during the period of transition when some barefoot doctors went back to farming and the certification process was getting started. Of the 1.5 million barefoot doctors, about 1.3 million had, by 1986, become village doctors, of whom about one-quarter were fully certified (Cui Yueli 1986).

It was no longer efficient for village doctors to grow their own herbs. Government pharmaceutical factories flooded local markets with medicines which were subsidized and very cheap. While most barefoot doctors have shifted to using commercially available medicines there is also a growing market for commercially produced traditional medicines which people tend to buy directly.

Preventive services suffered in the transition, especially maternal and child health services. The United Nations Children's Fund (UNICEF) was able to help as the Government re-emphasized preventive services for children and their mothers. National programmes such as the immunization of 85 per cent of children in each province by 1988 and 85 per cent in each county by 1990, and the control of iodine deficiency diseases, became part of the world-wide child survival and development revolution. The maternal and child health Model County Programme in 30 counties around the country promoted collaboration between county health services and local medical and public health schools. They developed new approaches, adapted to local conditions through demonstration and research activities. Village workers specializing in maternal and child health services were trained to provide comprehensive services. This Model County Project was extended to an additional 95 of the poorest counties in the country to ensure coverage where the needs are greatest, and is now being extended further to the 300 poorest counties in the country.

Local financing of health services has been recognized as a ubiquitous problem. Trials are being conducted of various forms of health insurance. For instance an innovative approach that is spreading rapidly arose from concerns about the immunization programme. In 1986, village doctors' charges to families for each injection varied around the Government's recommendation of 10 fen (3 cents US). Depending on local affluence, actual charges ranged from 5 to 30 fen. Immunization coverage was found to be directly correlated with charges, and ranged from 45 to 95 per cent. One county tried an insurance contract between the health system and family of 6 yuan ($2 US). Complete immunization included the commitment that if the child developed any of the six EPI diseases included in the expanded programme for immunization (EPI) the local health service would not only pay all medical expenses, but would also compensate the family as much as 100 yuan. This contract arrangement has spread rapidly to many counties. Now it is being extended to other maternal and child health activities. In some places families can take out a contract for all prenatal, natal, and infant care at a fixed price of about 30 yuan. This seems to be accepted because of good faith established by the guaranteed compensation for immunization failure.

Training

In 1985, China had 117 university level medical schools (Cui Yueli 1986) of which 24 were for Chinese traditional medicine. This had increased from 38 in 1949, with 15 234 students enrolled. In 1985, the number of students enrolled was 157 388 and there were 40 030 new admissions. The intensity of training varies greatly. In the approximately ten national or 'key' medical schools under the ministries of health and education, the curriculum is for 7 years and in PUMC (Peking Union Medical College) it is 8 years. Most of the regular medical schools under provincial or municipal authority have curricula of 5 or 6 years. The older medical universities tend to still follow the Russian system of having five faculties—medicine, paediatrics, public health, dentistry, and pharmacy.

There are 19 schools of public health (Chen Haifang and Zhu Chao 1984). These are part of medical universities and have their own dean, faculties, and facilities. Students take the national entrance exam for medical schools and their grade determines where they will be admitted. Those with the highest grades go to the key medical colleges, and otherwise to their local colleges. Within a medical university there is the expected ranking of preferences, with public health falling considerably below the clinical faculties. The public health curriculum provides basic training in clinical medicine but with public health courses distributed through the 5 year programme instead of the advanced clinical work taught in medical faculties. Graduate training in some public health schools provides certain master's degrees in specialties of public health. A new union school of public health in Beijing is being started to provide comprehensive graduate public health training for leaders.

Nursing education and training for mid-level doctors and technicians in secondary medical colleges are being gradually extended but with great variation by province (Chen Haifang and Zhu Chao 1984). There were 515 secondary medical schools in 1985 with 221 441 students. Students take an average of 3 years to train in medicine, public health, maternal and child health, dentistry, laboratory, biological products, maintenance of medical apparatus, health statistics, and health management. In the autonomous regions and prefectures there were 83 of these secondary medical schools. Traditional medicine is taught exclusively in 19 secondary medical schools and is integrated into the teaching of 79 more. One of the areas in which the ministry in Beijing is increasingly exerting control is the development of national standards to define the roles of health personnel. This is being done partly by standardizing entrance examination and those required for licensure.

Educational methods are being intensively discussed in all schools and universities. Awareness is growing that the previous patterns of didactic presentation and memorization do not produce the kinds of problem-solvers needed in modern professional work. Emphasis is being placed in the educational reforms on promotion of creativity and innovations that encourage learning by doing. Some of the greatest needs are in county health schools which train peripheral field workers, including those for maternal and child health.

An important training activity is in-service training which has always been greatly emphasized. Current priority given to modernization has included efforts to upgrade all categories of personnel. Numerous short courses have been run in health institutions and public health departments. Many

Table 17.3. Summary measures of estimated fertility and mortality rates for 5-year periods, 1940–80

Period	Crude birth rate	Crude death rate*	Rate of natural increase	Total fertility rate	Infant mortality rate	Expectation of life at birth (years)
1940–45	38.0	38.6	−0.6	5.3	290	27.7
1945–50	41.7	35.7	6.0	5.9	265	30.5
1950–55	44.7	30.9	13.8	6.5	236	34.1
1955–60	39.8	27.3	12.5	5.8	229	34.8
1960–65	38.3	21.9	18.4	5.9	208	37.7
1965–70	38.2	13.3	24.9	5.5	137	49.0
1970–75	28.6	9.5	19.1	4.1	96	57.3
1975–80	19.5	8.5	11.0	2.7	65	64.2

* Rates per 1000 population per year.
Estimates derived from intercensal extrapolations.
Source: Adapted from Jamison *et al.* (1984).

innovative programmes are run for self-learning on the job. These include correspondence courses, formal courses on television or radio, and various combinations such as practical demonstrations during periodic one-day workshops and examinations by mail. The numbers of people involved run into several millions.

Traditional Chinese medicine still occupies an important place in the beliefs of common people, especially the elderly (Unschuld 1985). Most hospitals and pharmacies have separate sections where traditional practitioners work. Efforts to integrate traditional and modern scientific medicine were most active in the training of barefoot doctors but now the two systems tend merely to coexist. Certain traditional practices such as acupuncture for chronic pain are included in the training of several categories of personnel. Acupuncture is, however, decreasingly used for anaesthesia (one reason given is that 'surgeons are too impatient to wait'). The strength of traditional medicine lies not only in the large numbers of practitioners carried over from the past, but also in its strong influence on medical ethics and values (Unschuld 1979).

Put prevention first

China's success in improving health status to levels approaching those of developed countries is due largely to the emphasis given to prevention. Mao's famous slogan of 'put prevention first' set priorities during the period of communal control. Speculation continues about what forces were most important in producing the dramatic reduction in mortality and morbidity in the 1950s and 1960s. Probably a combination of factors had varying influence according to time and place. There is no doubt about the importance of the relative political stability after many decades of civil war. This brought a stop to the frequent famines and disruption from waves of fighting. Everyone shared in the distribution of food, which was adequate though minimal.

Much publicity has been given to the dramatic mass patriotic health campaigns. Seldom in world history has there been equivalent participation of millions of people sharing the hard work of changing their environment for public good. A hierarchy of committees chaired at the top personally by Premier Zhou Enlai mobilized the people to attack the major epidemic and endemic diseases (Horn 1969). Under the four pests campaign, people used the simplest methods to destroy flies, rats, mosquitoes, and bedbugs. For instance, everyone had to meet a community quota of flies that they had swatted. At first sparrows were included but then it was realized that the grain they ate was not as damaging as the insects they destroyed. Some of the most publicized mass campaigns were in eastern provinces where schistosomiasis was a massive problem. The people were mobilized to destroy snails, which they collected with long chopsticks or buried by diverting streams and canals to new channels (Hillier and Jewell 1983). Perhaps most dramatic was the eradication of venereal diseases through careful contact tracing and treatment. The head of the national malaria programme on one occasion explained the slower response in malaria control by saying 'you can tell people what not to do but you can't tell mosquitoes'.

Although the mass campaigns were dramatic, the control of epidemic and endemic infections was probably successful mainly because of the rapid expansion of infrastructure for continuing services (Hillier and Jewell 1983). The key organizations were the antiepidemic bureaux, which became the best staffed and organized of the public health services. They had units at all organizational levels and were the main point of contact and supervision for barefoot doctors. They were responsible for monitoring epidemics and for the prompt use of control measures. They organized the immunization programmes. They also promoted better sanitation through improved water quality and excreta disposal. Many different designs of latrines were developed to meet local conditions since human excreta are considered very precious for use as fertilizer. In some areas successful use has been made of family biogas units which also provide methane gas for cooking.

Table 17.3 (Jamison *et al.* 1984) shows the dramatic decline that has occurred since liberation in mortality rates, especially infant mortality. In 20 years the country moved from developing to developed country rates. This was in spite of a major disaster from 1958 to 1961 when massive famine occurred when drought was added to the earlier social trauma of taking the land away from families as part of forced rural industrialization. In the dramatic demographic impact shown in Figure 17.1, it is estimated that 20 million people

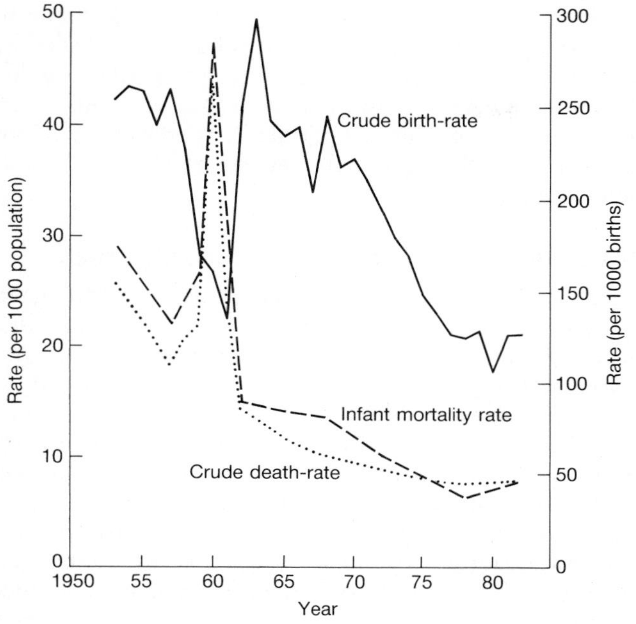

Fig. 17.1. Estimated birth and death-rates, and infant mortality rates, China, 1953–82

died. But as soon as the trouble was over the secular decline in mortality returned to the earlier trend lines.

For the 800 million people living in the crowded eastern provinces, improved health status is reflected in markedly changed patterns of illness. The causes of death now are similar to these of any developed country. Cardiovascular diseases have become the leading cause of death, particularly problems related to hypertension which may be associated with very high salt intake, especially among northerners. Coronary heart disease is still not as common as in western countries, perhaps because of better dietary and exercise patterns. Cancer is not only very common but is pocketed in geographical areas where studies are being done on epidemiological associations. The production of a national cancer atlas was a remarkable example of scientific collaboration in all parts of the country.

The maternal and child health services were poorly developed until recently. One beneficial impact of the one-child family planning programme has been that much greater value is now placed on the care of children. Parents say that if they can have only one child they want it to be a perfect child. Both families and the Government give high priority to improving the health and development of children. The best developed parts of maternal and child health services are those for maternity care, especially delivery services. Child care is less systematic although there are some efforts to monitor growth. Of special importance in local services is early diagnosis and treatment of health problems through surveillance and prompt care locally. A remarkable achievement is the low mortality rate from diarrhoea even though morbidity is still high. This seems to be associated with the use of salt and sugar solution and traditional liquid medicines by barefoot doctors (Taylor and Xu 1986).

Among children, the acute infections are still important. The leading cause of death is pneumonia, especially during the neonatal period. A major effort is being made to introduce the World Health Organization/UNICEF case management protocol to improve early diagnosis and appropriate treatment. Neonatal mortality is generally the second cause of death. Although infant mortality in the country is now down to 35 per 1000 live births, in the western provinces and autonomous regions there are still significant areas where it is over 100. Although basic nutritional needs are being met, about one-third of children are stunted by international standards, about one-third have anaemia, and in the north about one-third have rickets. For all of these problems a combination of preventive and curative approaches is needed.

Perhaps the most urgent public health problem of the immediate future is very heavy pollution of air and water. The efforts to achieve rapid industrialization at low cost carry health costs that are beginning to be recognized. Cities such as Beijing still rely on stoves burning soft coal for home heating, and in the winter air pollution is as bad as has ever been seen anywhere. Occupational health measures are being introduced. Epidemiological studies are tracing the causes of pockets of problems such as high prevalence of neural tube congenital defects in counties north of Beijing. The diversity of the country and the relative stability of the population makes such geographical epidemiological studies particularly promising.

Curative services

After three decades during which curative services had lower priority than prevention there has been a steady effort in the 1980s to improve clinical care. Public demand has focused on better medical facilities and personnel capability. People no longer seem satisfied with earlier successes in providing coverage with simple primary care. Now that people have more choice of where they will obtain care, because most people have to pay for each service, they tend to bypass the health post in their own village and go directly to a hospital or urban facility. As in the rest of the world, people in China are increasingly impressed by high-technology equipment and specialists. Competition between practitioners is leading to the over-use of drugs and growing problems of individuals attempting procedures that are beyond their training and capability. Massive use of the most powerful and potentially toxic medications, especially a wide range of antibiotics, has raised growing concerns about iatrogenic health problems. For instance, studies of deafness in Shanghai showed that a major cause has been the uncontrolled use of streptomycin. Gradually the health system is developing more controls and considering what can be done to prevent over-use of potent drugs.

The shift to chronic diseases rather than acute infections carries with it a necessary change in the types of control measures to be used. Simple preventive measures are no longer sufficient. The care of chronic illnesses requires expanded curative services. Preventive approaches depend on changes in behaviour and lifestyle. For instance there has been extremely heavy smoking among men which is now reflected in high lung cancer rates. Intensive health education has been started to try to control smoking.

The expansion of hospitals and institutional care is now being given high priority. An indication of the continuing emphasis on equity is the priority given to building small rural hospitals at local level.

Population growth

China's experience shows that population growth can be controlled, although few countries will want to use the draconian approach that was applied in the 1970s and early 1980s. Coale has used the analogy that population growth in China was like a train travelling at increasingly high speed where nothing was done for a long time about applying the brakes and eventually they put up a brick wall as a crisis measure (Coale 1981). Mao had relied on the Marxian concept that overpopulation was a capitalist phenomenon that would disappear when distribution was corrected. Nothing was done until the 1970 census showed that population growth had gone from less than 500 million to over 800 million since liberation in 1949. China was beginning to be dependent on large food imports. Controls were then applied abruptly and, because the cultural revolution was raging and all policies were coercive, a two-child policy was rigidly enforced (Bannister 1984; Coale 1984). Birth-rates declined precipitously in just 10 years, as shown in Table 17.3 (Jamison *et al.* 1984) and Figure 17.1. After the worldwide 1970 censuses, predictions were made that world population would be over 9 billion by the year 2000. After the 1980 censuses, the projection dropped to between 6 and 7 billion, largely because of what China had done in cutting population growth rate by one-half in one decade. The number of babies born each year dropped from about 30 million to less than 20 million. China has moved from having one-quarter of the world's population to slightly more than one-fifth, and India now has more babies born each year than China.

In 1979 the one child per family policy was introduced because the birth-rate was again going up because the baby boom generation of the 1950s and 1960s was entering its period of peak reproduction. This was only a temporary policy to prevent the social trauma of cyclic surges in demand on education, employment, etc. It took some time to adjust implementation policies from the earlier coercive measures, but in 1984 the Central Committee's document seven established new principles that the programme should be based on public education and be made voluntary. The number of exceptions permitted when the one-child policy will not be applied has been greatly expanded and decentralized. The one-child policy had never been applied to minorities or to people living in remote areas with difficult living or working conditions. Some provinces now permit a second child if the first was a girl. Public compliance has been remarkable, but even so birth-rates have risen sufficiently to push projections of world population by the year 2000 to over 7 billion. The average number of children per family is now about 1.7 and the goal of stabilizing population size at about 1.3 billion seems reasonably within reach.

Health services in remote and minority areas

The deserts of the north-west and the Tibet/Qinghai plateau have some of the most rugged living conditions in the world. Health conditions have not yet been improved but these areas are being given special priority by the ministry of health. For instance, Tibet is the only part of the country where health services are free because of central government subsidies. A particular effort is being made to train local health personnel because Han settlers from crowded provinces have refused to stay in these difficult areas. There are 55 ethnic minorities in China, with their own cultures and languages. Most of them live in the western mountains under difficult conditions. The new open policy of the present administration is giving priority to minority rights, especially in education and health. Rapid improvements are occurring in accordance with genuine commitment to providing equity of opportunity.

Summary

The unpredictability of results from massive social change permits only tentative conclusions about the new directions of health policy and strategy in China. So far it does seem that a successful transition is being made in moving away from the idealized health care system under the communal structure that provided an important model for the Alma Ata definition of primary health care. In fact, such new developments as the decentralized structure and mixed financing evolving from current changes will be a more relevant model for other countries. Many public health leaders around the world shrugged off what China was doing earlier because they said their people would not put up with so much coercion. The present achievements come much closer to fitting conditions in most countries. The Chinese model is, therefore, extremely important for public health leaders to understand as they develop their own health systems.

References

Bannister, J. (1984). Analysis of recent data on the population of China. *Population and Development Review* **10**, 241.

Chen Haifang and Zhu Chao (1984). *Modern Chinese medicine, Vol. 3, Chinese health care*. MTP Press, London, and People's Medical Publishing House, Beijing.

Coale, A.J. (1981). Population trends, population policy, and population studies in China. *Population and Development Review* **7**, 85.

Coale, A.J. (1984). *Rapid population change in China, 1952–1982*. National Academy of Sciences Report, Washington, DC.

Cui Yueli (1986). *Public health in the People's Republic of China*. People's Medical Publishing House, Beijing.

Hillier, S.M. and Jewell, J.A. (1983). *Health care and traditional medicine in China, 1800–1982*. Routledge and Kegan Paul, London.

Horn, J. (1969). *Away with all pests: An English surgeon in People's China 1954–1969*. Monthly Review Press, New York.

Jamison, D., Evans, J.R., King, T., Porter, I., Prescott, N., and Prost, A. (1984). *China, the health sector*. World Bank, Washington, DC.

Schell, O. (1985). *To get rich is glorious—China in the 1980s*. Mentor-Pantheon, New York.

Seipp, C. (ed.) (1963). Health care for the community—Selected papers of Dr John B. Grant. The Johns Hopkins University Press, Baltimore, Maryland.

Sidel, R. and Sidel, V.W. (1982). *The health of China*. Beacon Press, Boston, Massachusetts.

Taylor, C.E. and Xu Zhao Yu (1986). Oral rehydration in China. *American Journal of Public Health* **76**, 187.

Unschuld, P.U. (1979). *Medical ethics in Imperial China*. University of California Press, Berkeley, California.

Unschuld, P.U. (1985). *Medicine in China—A history of ideas*. University of California Press, Berkeley, California.

Wang Guichen and Zhou Qicen (1985). *Smashing the communal pot—formulation and development of China's rural responsibility system*. New World Press, Beijing.

Young, M. (1984). A study of barefoot doctors' activities in China. PhD Thesis. Department of International Health, The Johns Hopkins School of Public Health, Baltimore, Maryland.

18

Public health policies and strategies in developing countries

ADETOKUNBO O. LUCAS

Introduction

The historic International Conference on Primary Health Care at Alma Ata has had a profound effect on the health policies of governments throughout the world. By projecting 'Health For All By The Year 2000' as the common goal and by adopting primary health care as the basic strategy, the Alma Ata declaration provided all nations with a coherent framework for planning, implementing, and evaluating their health services. Over the past decade, ministries of health have reviewed and revised their health policies to comply with the spirit of the Alma Ata declaration.

The consensus about goals and strategies cannot obscure the large differences in the health situation in the various countries. At one end of the spectrum are developed countries with highly sophisticated health care systems, with access to a variety of powerful modern technologies, with abundant resources, and with large reserves of capacity which are immediately available for further growth and expansion. At the other end of the spectrum are the poorest nations which cannot as yet provide health, water supply and sanitation, education, modern transportation, and other basic services to their entire population, and which are struggling to maintain progress in spite of severe economic constraints (United Nations 1985).

Developing countries, the group of nations defined by their average low level of per caput income, share some common features but there are striking differences among them. The low income economies are defined as having 1984 gross national product (GNP) below US $400 per person. The middle income countries have 1984 GNP per person of $400 or more. It is widely recognized that GNP by itself, is an imperfect measure of development (World Bank 1986). More penetrating indicators of development can be obtained by including other variables such as the distribution of incomes, level of education, especially female literacy, and the provision of basic human needs. The United Nations Children's Fund (UNICEF) has proposed that the under-fives mortality rate (U5MR) be used as the single most important indicator of the state of a nation's children, and that this together with average life expectancy and adult literacy rates provide good indicators of the social development of nations (Grant 1989). U5MR reflects food availability and the nutritional health of children, the level of immunization and control of diarrhoeal diseases, the quality of maternal and child health services, the availability of safe water, and the general safety of the environment in which children live and grow. The annual rate of reduction of U5MR is also a sensitive measure of social development. Some developing countries have achieved low U5MR (below 30 per 1000 population), e.g. Jamaica, Costa Rica, and Cuba, but it is above 170 in some 33 developing countries.

Table 18.1 shows the distribution of the percentage of children who survive to the age of 5 years in different developing countries. Some developing countries have achieved such low U5MR that over 95 per cent of their children survive to their fifth birthday.

A similar case can be made in using maternal mortality rate as an additional indicator of social and health development in a nation. Whereas maternal deaths occur rarely in developed countries, many developing countries still experience very high death-rates in association with pregnancy and childbirth (Table 18.2). Maternal death-rates show a higher differential between developing and developed countries than any other major public health problem. According to a recent estimate, for the African woman the life-time risk of dying from pregnancy-related causes is 1 in 21 as compared with a risk of 1 in 6000 or less for women in developed countries. The high rates in developing countries reflect the general state of the health services but more specifically indicate women's reproductive health and their general status within the community (Starrs 1987).

The pattern of health and disease in developing countries shows wide differences. In some countries, parasitic and infectious diseases, acute respiratory-tract diseases and malnutrition occur frequently as major causes of morbidity and

Table 18.1. Child survival rates in developing countries: percentage of children born who survive to the age of 5 years

Percentage surviving to age of 5 years	Americas	Middle East and North Africa	Sub-Saharan Africa	Asia
95–	Argentina Chile Costa Rica Cuba Jamaica Panama Trinidad and Tobago	Kuwait United Arab Emirates	Mauritius	China North Korea South Korea Malaysia Singapore Sri Lanka
90–	Brazil Colombia Dominican Republic Ecuador El Salvador Mexico Nicaragua Paraguay	Iran Iraq Jordan Lebanon Syria Tunisia Turkey	Botswana	Burma Mongolia Papua New Guinea Philippines Thailand
85–	Guatemala Honduras Peru	Algeria Egypt Libya Morocco Oman Saudi Arabia	Congo Ghana Lesotho Zambia Ivory Coast Kenya Liberia Zimbabwe	India Indonesia
80–	Bolivia Haiti	Yemen	Benin Cameroon Madagascar Sudan Togo Zaire Burundi Gabon Nigeria Tanzania Uganda	Bangladesh Bhutan Laos Nepal Pakistan
75–		Democratic Yemen	Burkina Faso Togo Mauritania Rwanda Somalia Central African Republic Uganda Niger	Cambodia
< 75			Angola Guinea Mali Sierra Leone Ethiopia Malawi Mozambique	Afghanistan

mortality; infant and maternal mortality rates are high; fertility rate is high; and expectation of life at birth is low. Other developing countries in a transitional phase are undergoing rapid demographic and epidemiological change; infant, child, and maternal mortality rates are declining, fertility rates are high but falling, life expectancy is rising; parasitic and infectious diseases are still prevalent but chronic degenerative diseases and non-communicable diseases associated with modern life-styles and aging populations are increasing. The more advanced developing countries have acquired the epidemiological pattern which is typical of the developed countries: fertility rates and infant, child, and maternal mortality rates are low; life expectancy at birth is high; cancer, cardiovascular, neurological and mental disorders, degenerative diseases, and problems associated with life-styles and behaviour are as common as in developed countries.

Table 18.2. Estimates of maternal mortality

Region	MMR*	Maternal deaths†
Africa	640	150
Asia	500	308
Latin America	270	34
Oceania	100	2
Developing countries	450	494
Developed countries	30	6
World	390	500

Source: World Health Organization statistics, from Starrs (1987).
* Maternal mortality rate per 100 000 live births.
† Estimated number of deaths in thousands.

The Alma Ata accord

In spite of major differences in their social, political, and other characteristics, all nations, both developed and developing, reached an accord at Alma Ata and they endorsed a declaration which is widely recognized as the fundamental basis of national health policies throughout the world. This chapter examines the translation of this universal accord into national policies and strategies in developing countries.

Historical perspective

The recent history of the developing countries which gained political independence after the end of The Second World War includes a period of colonial government. The current situation in these countries is historically linked to their colonial past which has determined and still influences contemporary attitudes, their systems of health care, the training and use of health personnel, and other aspects of their health services. The developing countries which were not recently under colonial rule have also been influenced by the particular developed countries with which they had preferential cultural, economic, and professional links.

The colonial era

In many developing countries, the foundation of modern health care was established during the colonial period (Sabben-Clare *et al.* 1971). Although the various imperial governments did not adopt uniform systems of care, colonial medical services had some features in common.

Coverage and access

On the whole, colonial health services provided uneven coverage of the population. In some colonies, large sections of the population did not have regular access to modern health care. In many cases, the health services were developed strategically to assign priority to military, governmental and commercial priorities rather than to the needs of the general population. This determined the geographical distribution of hospitals and other health care facilities and influenced the operations of the services in public institutions which gave priority access to defined groups of the population such as civil servants and soldiers. At independence, the health services were still relatively underdeveloped. This was particularly true in Africa. For example, at independence, there were only 1000 doctors serving a population of over 30 million in Nigeria, and half of them were expatriates (Schram 1967).

Innovative approaches to health care

Colonial governments introduced strategies for health care delivery which were adapted to local circumstances. Some of these approaches have survived and have been incorporated into the health care systems after independence. Such innovations included the training and use of a variety of health personnel whose education and skills differed significantly from the traditional professional workers who serve in developed countries. Persons with limited education and, sometimes, even illiterate candidates were trained to deliver a limited range of health care services to the population. These health auxiliaries and paramedical personnel usually served under the supervision of qualified professional personnel.

Mobile health services were used to deliver health care in many colonial territories. Sometimes, the mobile teams provided a broad range of basic health services to populations which were not regularly served by local health institutions. They were also used for the control of endemic diseases or for dealing with epidemics.

Confronted by endemic diseases which were major causes of disease and disability, colonial medical services devised and used special campaigns as an important strategy for dealing with health problems. Classical examples of problems tackled in this way included the mass treatment of yaws, the control of African trypanosomiasis and the management of leprosy. Some of these campaigns were later integrated into the basic health services. For example, the French Government used a mobile force, the Organisation de Coordination et Cooperation pour la Lutte Contre Grandes Endemies (OCCGE), for disease control in their West African colonies.

Modern health care in part replaced traditional systems of medical care but generally the two systems operated side by side. No uniform policies were developed with regard to traditional medicine but some effort was made to involve some traditional medical practitioners in the delivery of modern medical care (Last and Chavunduka 1986).

Research

The colonial medical services made important contributions to knowledge of the local diseases in various parts of the world. This was the era during which major discoveries were made about the aetiology, pathology, epidemiology, and control of major tropical diseases in developing countries.

Apart from studies on the wide variety of infective conditions, the colonial research institutes contributed new knowledge on nutritional disorders, genetic, and other diseases which were common in the colonies or which presented unusual features. Institutes of tropical medicine, located in developed countries, chiefly in western Europe, provided technical back-up for the research being conducted in colonial territories.

Private health care

Private organizations and individuals made significant contributions to the delivery of health care in many colonies. Private voluntary organizations, especially missionary bodies, provided the bulk of the health services available to populations in some parts of developing countries. Missionary bodies have been particularly prominent in providing care for groups and problems that receive low priority from the official health services as, for example, the care of leprosy patients and handicapped children.

These and other historical features of colonial medical services shaped the pattern of health care at independence, and some of the traditions continue to influence the evolution of health systems in developing countries.

The post-independence period

Certain trends characterize the evolution of health services during the post-independence period and before the Alma Ata declaration. As in other sectors, political independence brought with it a rising tide of increased expectations by the general public with a strong demand for health care, especially for curative services. In some countries this led to increased investment in large urban hospitals. Expensive specialist hospitals were constructed and some of these served as teaching hospitals for the newly established medical schools. In many cases, these urban-based curative services consumed increasingly higher proportions of the national health budget leaving preventive services, especially in rural areas, relatively starved of funds.

Some of the experiences gained in colonial times were extended during the post-independence period, some of which led to key experiments in health care. For example, some of the important studies which led to new strategies for the care of pre-school children in developing countries were carried out at this time.

Changes outside the health sector, especially those associated with accelerated technological and social development after independence, also influenced the health status in developing countries. For example, many countries in sub-Saharan Africa substantially increased their investment in education during this period. New schools were established in rural areas which previously were mainly populated by illiterate persons. Industrial developments and new commercial enterprises attracted rural populations into the urban areas leading to an explosive growth of some cities.

Pre Alma Ata

Before Alma Ata, health policies in developing countries followed different paths which, in some cases, were related to their recent colonial past. Mostly, the health services showed heavy urban concentration, they were heavily dominated by curative services, and the population did not have uniform access to care. Efforts were made to correct some of these imbalances through the establishment of basic health services; packages which included curative and preventive services were delivered through health centres, dispensaries and other small peripheral units. The services were usually centrally planned and managed with little or no involvement of the community in the process of making decisions about health policies.

Post Alma Ata

In May 1977, the thirtieth World Health Assembly passed a resolution which adopted as the main goal of all member states of the organization, the attainment by all citizens of the world by the year 2000 of a level of health that would permit them to lead a socially and economically productive life. This policy was further developed and analysed at the International Conference on Primary Health Care in Alma Ata, Soviet Union, which was jointly sponsored by the World Health Organization and UNICEF. Delegations from 134 governments and representatives of 67 organizations of the United Nations system and other organizations attended the conference. The thirty-second World Health Assembly in 1979 endorsed the report of the conference and adopted a Global Strategy for its implementation. As the lead international agency on health matters, the World Health Organization has been promoting the Alma Ata declaration as the central policy for global action on health. It has adopted a Plan of Action and, with the collaboration of member states, it is monitoring progress in implementing the strategy and evaluating its effectiveness.

Since its promulgation, the Alma Ata declaration has been analysed, extensively debated, and subjected to varying interpretations and to some controversy. Even with their well-developed health services and relatively large resources of personnel and technology, developed countries find the ambitious goal a challenging task. Developing countries have also bravely accepted the challenge of working towards the common goal of Health For All.

Primary health care strategy

As defined in the Alma Ata declaration, primary health care is: 'essential health care based on practical, scientifically sound, and socially acceptable methods and technology made universally acceptable to individuals and families in the community through their full participation and at a cost that the community and country can afford to maintain at every stage of their development in the spirit of self-reliance and self-determination'. These services are to be 'sustained by integrated functional and mutually-supportive referral systems, leading to the progressive improvement of comprehensive health care for all and giving priority to those most in need'. (World Health Organization 1978).

Each developing country is seeking appropriate solutions for the issues which arise in the process of translating the concepts of primary health care into practical programmes. The range of questions and problems can be illustrated by examining seven major issues which commonly demand attention and resolution:

(1) equity and coverage;
(2) inter-sectoral action;
(3) organization and management of services;
(4) selection and use of technologies;
(5) financing of health care;
(6) health information;
(7) research.

Equity and coverage

The goal of primary health care is to ensure for all persons a level of health that will permit them to achieve their full potential, both social and economic. It emphasizes the concept of equity and social justice, and it denounces as unacceptable gross inequalities in health status and access to health care (Montoya-Aguilar and Marin-Lira 1986). The mechanisms used to overcome inequities in health systems vary with political and economic systems of different nations but in all cases must involve social justice as a fundamental principle for health care. It has been rightly said that Health For All By The Year 2000 requires social justice for all by the year 1999.

Equity in the context of primary health care implies both fairness in the distribution of health services and also social justice in aspects of development which have an impact on health. Health expenditures must be fairly distributed to the entire population, with particular attention to such vulnerable and disadvantaged groups as rural communities, urban poor, women and children, the elderly, and the handicapped, as well as to ethnic and religious minorities. Poverty, illiteracy, and other social-economic factors are underlying causes of inequities in health. Development policies must be reformed to ensure equitable growth within nations and to empower the disadvantaged groups, enabling them to overcome social and economic barriers to their development. All the people should have access to a fair share of health resources—staff, drugs, and other facilities. It is important that the most vociferous groups within the population and the most politically powerful do not arrogate to themselves an unfair share of health resources.

The concept of equity also applies between, as well as within, nations. The issue has become more urgent with the deterioration of the economies of some African and Latin American countries. The mounting external debts of some of these nations has increased poverty and is having a deleterious effect on the health of the populations. UNICEF has drawn attention to the damaging effects of economic adjustment on the health of children and has proposed modifications to current debt recovery programmes (Grant 1989; Jolly 1988). 'Adjustment with a human face', as proposed by UNICEF would include a concern for basic human welfare and a commitment to protect the minimal levels of nutrition of children and other vulnerable groups. The revised programme should include a minimum floor for nutrition and basic human needs; it should restructure the productive sectors with emphasis on small-scale producers in the informal sector; it should seek to maintain growth momentum and ensure maximum coverage within health, education, and other social sectors by adopting low-cost technologies; and it should seek more international support for these aspects of the adjustment programme.

Inter-sectoral action

Although it is well known that socio-economic factors have a significant effect on health, mechanisms for translating this knowledge into practical programmes have not been well developed. The goal of Health For All can be achieved only through health development which includes the participation of all relevant sectors. In the central government, this requires the involvement of several key ministries—agriculture, education, finance, planning and development, public works and engineering, housing, and communications. At the local community level, mechanisms are being developed to ensure appropriate inputs from each sector. For example, in a rural community, the primary health care programme may work closely with agricultural extension, adult literacy, water and sanitation, and similar projects. The best mechanisms for co-ordinating these elements and the specific list of inter-sectoral activities are best determined locally. Health care givers should provide leadership and guidance in identifying elements of development that can most effectively promote health as well as drawing attention to dangers that may be associated with development projects.

Inter-sectoral action can also be used to promote equity in health by tackling poverty, malnutrition, poor housing, and other factors which put segments of the population at a disadvantage. The promotion of health then becomes the goal of government as a whole and not merely the preoccupation of the ministry of health. Each government can give concrete expression to its commitment to the goal of Health For All by its policies in the health sector but also through all other relevant sectors. In practice, this would involve a greater allocation of resources to the health sector and a reorientation of the programmes of other sectors to promote the health and welfare of the entire population in an equitable manner (Royal College of Physicians 1987).

Organization and management of health services

Many questions need to be answered in the process of converting traditional health services in developing countries to fit in with the concept of primary health care. The issues include the content of the services, organization and management, the role of health care givers, and community involvement.

Content of primary health care

The challenge of primary health care is to ensure that all citizens have access to the eight basic elements of service at the point of first contact:

(1) health education;
(2) promotion of food supply and proper nutrition;
(3) adequate supply of safe water and basic sanitation;
(4) maternal and child health care, including family planning;
(5) immunization against major infectious diseases;
(6) prevention and control of endemic diseases;
(7) treatment of common diseases and injuries;
(8) provision of essential drugs.

These eight elements represent the minimum package of services that should be available in every community; each individual should have access to those services which are relevant to the person's needs. Within the package, the priority given to each element will vary from place to place and will also shift with time.

Some countries have not achieved a wide coverage of their populations with these basic services and it is unlikely that they can expand the health sector rapidly to achieve this objective in the near future. This observation has prompted the suggestion of an alternative strategy which would consist of selectively tackling high priority diseases for which effective technologies are available (Walsh and Warren 1979). This 'selective primary health care' strategy has been presented as an interim option. This approach has attractive features in that it could rapidly reduce morbidity and mortality due to a few major problems. The danger is that it could postpone the systematic development of broadly based health care. Furthermore, there is the danger that priorities selected at the national level may not fit well with the local realities in each community. The argument between primary health care and selective primary health care can easily be resolved in practice. In selecting priorities for primary health care, emphasis should be given to problems which are of major public health importance and for which effective technologies are available (Walsh 1988). Specific activities such as the child survival programme including immunization, diarrhoeal disease control, the promotion of breast-feeding, etc. can be given special emphasis in the development of the eight primary care elements. These special projects represent peaks in a broadly based programme of health development. Taken at the extreme, selective primary health care could become another vertical programme, a fire-fighting operation which generates initial enthusiasm that may dissipate in time having contributed little to the development of basic health services.

Organization of services

Orientation of health services toward primary health care extends beyond the physical arrangements for providing the list of services. It should involve a redefinition of the roles of the institutions and personnel, and of the objectives of the health services. The peripheral and community services form the broad base of the health care pyramid with appropriate back-up by referral units. A typical pyramidal structure includes three main levels:

I *Primary health care* serving populations in the range 2000–30 000;
II *Secondary referral level*—usually a hospital providing the main specialist services for a population of 200 000–500 000;
III *Tertiary referral level*–providing highly specialized services to a population in the region of 1 000 000 or more.

The system should be integrated such that all the health service units in a geographical area form a functional unit linking promotive, preventive, and curative services as well as rehabilitation. Although the central government must provide leadership, commitment, and drive for the primary health care movement, the organization of services should be decentralized so as to ensure that the allocation of health resources can be made more relevant to local needs. Successful decentralization depends on making resources available at the local level, providing guide-lines to ensure compliance with overall national policies and strategies, and delegating authority to make decisions.

The district health system

The district has been identified as the focus for effective management of primary health care. The district health system, as defined by the World Health Organization:

(1) comprises a segment of the population living within a clearly delineated administrative and geographical area;
(2) includes all institutions, both public and private, providing health care in the area;
(3) consists of a variety of interrelated elements that contribute to health in homes, schools, offices, factories, and on the farm, through health and related sectors;
(4) extends from the most peripheral units to the hospital at the first referral level and to the appropriate diagnostic and logistic support services including laboratories;
(5) requires co-ordination by an officer who can draw together the elements into a fully comprehensive range of promotive, preventive, curative, and rehabilitative health activities.

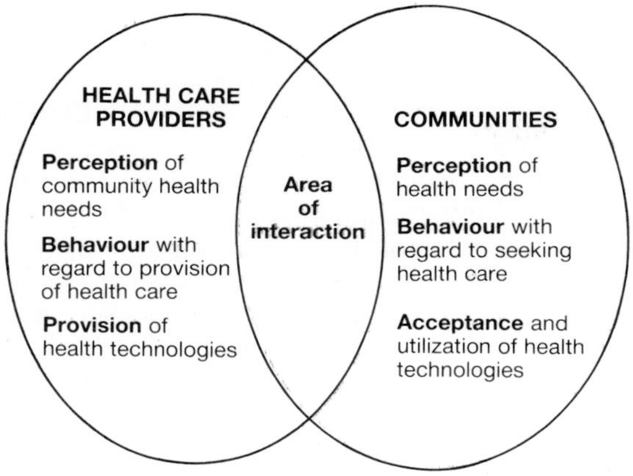

Fig. 18.1. A new approach to health education. From: World Health Organization 1983*b*.

Table 18.3. Some quantitative health personnel indicators

	Population per physician		Population per nurse	
	1960	**1980**	**1960**	**1980**
Low income countries	12 222	5 785	7 217	4 668
China	8 330	1 920	4 020	1 890
Ethiopia	100 470	58 490	14 920	5 440
India	4 850	3 640	10 980	5 380
Nepal	73 800	30 060		33 420
Pakistan	5 400	3 480	16 960	5 820
Zaire	79 620	14 780	3 510	1 920
Lower-middle income countries	28 870	7 751	4 925	2 261
Brazil	2 670	1 700		740
Egypt	2 550	970	1 930	1 500
Guatemala	4 420	8 000	9 040	1 620
Indonesia	46 780	11 530	4 510	2 300
Ivory Coast	29 190	21 040	2 920	1 590
Nigeria	73 710	12 550	4 040	3 010

Source: World Health Organization (1985*a*).

Community involvement

The Alma Ata declaration notes that it is both the duty and a right of people as individuals, families, and communities to take responsibility for their own health. In the primary health care model, people cannot be treated as passive recipients of health care on terms dictated by a professional élite. Rather, they must be involved in the process of identifying priorities, designing programmes, mobilizing resources, delivering care, in identifying problems and in seeking feasible solutions, in monitoring progress, and in evaluating outcomes.

This emphasis on community involvement calls for a redefinition of the role of health personnel, especially in their educational function. The old-fashioned approach to health education was based on the downward communication of health knowledge from health personnel to an ignorant public. Little effort was made in the past to understand traditional cultures, beliefs, religions, and practices. The new relationship is based on the appreciation of the value of the conventional wisdom in every community. The health care giver then acts in the dual role of being a learner as well as a teacher and facilitator. The community members also serve as learners as well as teachers. Figure 18.1 illustrates the contrasting perspectives of health care givers and the communities which they serve. The health care givers perceive community health needs and respond with the provision of health technologies. The community members have a different perception of their own health needs and respond by seeking health care and utilizing health technologies. The area of interaction, representing the common ground between the community and the health personnel, can be expanded through dialogue between health personnel and the community.

Private voluntary organizations are important resources which can make valuable contributions to primary health care. Some of the associations—social clubs, market women and other trade groups, religious bodies—provide the useful mechanism for promoting dialogue with the people and for mobilizing community action.

The effectiveness of community involvement is dependent on the quality and dedication of its leadership. Potential leaders who are committed to social justice should be identified and supported. They can be found in various sections of the community—professional persons, labour leaders, market women and other trade groups, religious leaders, traditional healers, etc. They need information which would guide their actions and the opportunity for meaningful participation to give them credibility within the community.

Training of health personnel

In the past much of the emphasis about staffing development was placed on the numbers in the various categories, their distribution in various parts of the country, and their technical skills (Table 18.3). Since Alma Ata, attention has been paid to other qualities of health personnel, especially in preparing them for their social functions in relation to the communities that they serve.

The training programmes for various categories of health personnel need to be revised in the light of their new functions and the reorientation of the health care system (World Health Organization 1985*a*, 1986). The training of doctors has come under particular scrutiny because the traditional medical school programmes did not adequately prepare the graduates for functioning effectively in the primary health care system. Following extensive consultations and analysis, representatives of academic institutions and governments, meeting in Edinburgh, produced new guide-lines for medical training (*Lancet*, 1988). This Edinburgh declaration provides guide-lines that were unanimously accepted by representatives of both developed and developing countries. It recommends changes in the curriculum and teaching methods which would provide doctors with skills and attitudes that they need to function more effectively in working within the primary health care system. Other professional groups are similarly engaged in reviewing their training programmes and proposing appropriate modifications.

Selection and use of health technologies

Primary health care implies the use of technologies that are 'scientifically sound and socially acceptable . . . at a cost that the community can afford to maintain . . . ' This definition has given rise to the concept of appropriate technology for health. Ideally, technologies for use in primary health care in developing countries should meet five criteria. They should be:

1. *Effective*: All technologies, including traditional medical practices, should be evaluated in terms of their effectiveness—their ability to achieve the desired results in the circumstances of their use.
2. *Safe*: The technology should have minimal undesirable effects on the users, operators, and the environment.
3. *Simple to apply and maintain*: For drugs, vaccines, and other therapeutic procedures, the treatment schedules should be as simple as possible and, where feasible, they should be formulated for single-dose treatments. For technologies requiring special equipment, primary health care programmes should select models that are easy to operate, sturdy in design, and easily maintained by local technicians.
4. *Culturally and socially acceptable*: The selection of technologies should take note of cultural and religious beliefs of the population and the social circumstances that could affect the acceptability of the technology.
5. *Affordable*: Every effort should be made to minimize the cost of health care in developing countries by selecting cheaper alternatives of proven effectiveness and safety. Programmes that acquire and operate expensive technologies which are donated by external agencies sometimes find that they are unable to maintain the service once the external source of support is withdrawn.

Appropriate technologies for use in primary health care in developing countries include the child survival package popularly known as GOBI-FFF, representing growth monitoring, oral rehydration, breast-feeding, immunization, family planning, supplementary feeding, and female education.

Essential drugs

Drugs and other pharmaceutical substances play a central role in health care and account for a high proportion of its cost. In support of primary health care, the World Health Organization proposed a programme on essential drugs. The programme generated much controversy at its inception but many of the arguments against it were based on misunderstanding of the objectives and the proposed administration of the scheme. There was fear that the World Health Organization would define a restricted list of drugs and would use its influence to coerce the compliance of developing countries. The problem was defused when it was made clear that the decision to establish a list of essential drugs and to implement a scheme is an issue of national policy. Once these misconceptions were corrected, many countries have been designing and implementing essential drug programmes to meet their need according to the principles enunciated by successive expert committees of the World Health Organization (1988).

The main recommendations and guide-lines produced by WHO committees can be briefly summarized. As defined, essential drugs are those that satisfy the health care needs of the majority of the population. The selection of the drugs should be based on the pattern of prevalent diseases, the treatment facilities, the training and experience of the available personnel, the financial resources, and genetic, demographic, and environmental factors. A local committee, which includes competence in the fields of medicine, pharmacology, and pharmacy, and peripheral health workers should oversee the management of the programme. The World Health Organization advises that international non-proprietary (generic) names for drugs be used and that the list should be accompanied by concise, accurate, and comprehensive information on each drug. The quality of each drug, including stability and bioavailability, should be assured by appropriate tests. Decisions should be made as to the particular level of expertise required for authority to prescribe each drug. The drugs on the essential list should always be available and in the appropriate dose forms. For efficient administration of the programme, careful attention must be paid to the procurement of the drugs, storage and distribution, and careful management of the stock to eliminate waste and to ensure continuity of supplies.

Equipment

Some initiatives have been taken in recent years to examine and deal with the problem of selecting equipment and maintaining it in good order (Perry and Chu 1988; World Health Organization 1985*b*). For example, the World Health Organization has promoted the concept of rational development of radiological diagnosis, proposing three levels of services: a basic radiological system, a general purpose radiological service, and a specialized radiological service. The basic radiological system includes the use of simple radiological equipment, some of which can operate on batteries, and the training of staff in the use and maintenance of the equipment and in interpretation of the findings.

Each country should establish mechanisms for the assessment of health technologies which should lead to national policies on their acquisition and use. Many savings can be achieved and efficiency improved by standardizing equipment for use in primary health care.

Financing health care

Progress with primary health care is being constrained by serious economic problems facing many developing countries. This has accentuated interest in exploring the question of how to finance health care. The specific issues relate to the allocation of public funds to the health sector and to other activities relevant to health, e.g. to education, water supply, and sanitation. The economic situation has also stimulated debate and discussion on other sources of funds for the health sector, including the question of user fees.

In the global strategy for primary health care, each country would aim at a target of spending at least 5 per cent of GNP on health and also would redirect a higher proportion of the national budget to primary health care. The issue has been complicated by the severe economic crisis that has affected some developing countries especially in Africa where, in recent years, the gross domestic product has fallen in many countries. The debt burden and the economic readjustment required to cope with it have put serious strains on development in these countries, with severe impact on health, education, and social services.

The economic crisis has stimulated interest in looking for other sources of funding for the health sector in developing countries. In some developing countries, sections of the population are covered by social security, and public and private insurance schemes. In other countries, the state assumes most of the responsibility for providing health care from public funds which subsidize the cost of the services. Additional funds are being sought and, in particular, there is increasing interest in generating income by charging fees for services that were previously offered free (Shephard and Benjamin 1988). This is a difficult issue as the burden of user fees may fall heavily on the poor who are already suffering other effects of economic recession. User fees may also discourage the use of cost-effective preventive services such as vaccination.

One new initiative is designed to generate funds by charging fees for drugs which are provided at primary health clinics. Known as the Bamako initiative, the scheme aims to create at the district level, a sustainable self-financing mechanism to support primary health care. The initial input of funds will be obtained from external sources which will be sufficient to provide essential drugs for five years. Revenue derived from the sale of the drugs will provide a revolving fund to finance the cost of primary health care in the district, including the running costs of the health centre, peripheral units, and the community health workers.

It would be unrealistic to expect that substantially large funds will soon become available to the economically depressed countries of the developing world. Health resources must therefore be prudently managed to gain as much efficiency as is possible. Although there is a broad correlation between the economic strength of a nation and the health of its population, there are examples of countries with low to modest GNP which have achieved a high level of health. In a recent review, case studies of China, Costa Rica, Kerala State in India, and Sri Lanka gave interesting insights into the process of achieving 'good health at low cost' (Halstead *et al.* 1985). Although other nations cannot reproduce all the conditions in these four examples, much can be learnt on how health can be improved through the application of simple cost-effective measures. The case studies also emphasized the value of education, especially female literacy.

Health information systems

Efficient management of primary health care demands systematic and analytical information which can be used for setting priorities, designing strategies, and for monitoring progress. It is important to determine what data are to be collected, how they will be organized, and how the information will be used to improve health.

Health managers in many developing countries are severely handicapped because they do not have at their disposal timely information on which to base decisions. In some cases, basic demographic data about the population are lacking and plans for the delivery of health care have to be based on estimates and projections. They also lack epidemiogical data about the pattern of diseases and the risk factors associated with major health problems. Apart from the demographic and epidemiological data, health planners also require relevant information from health-related sectors.

Most developing countries will have difficulty in providing sophisticated data collection and analysis at the primary health level. Efforts are being made to increase the quality and the quantity of health statistics. The availability of affordable but powerful microcomputers facilitates the handling of data at least at the district level and some demonstration projects are exploring their use.

Research

The value of research in strengthening primary health care cannot be overemphasized. Well-designed studies can generate findings that will provide a sound objective basis for making decisions about primary health care. The objectives of research at the primary health care level include the definition of the pattern of health and disease and the identification of determinants and of risk factors; the testing of alternative strategies for the delivery of health care; the evaluation of technologies in the local environment; and the monitoring of health trends and assessment of the impact of health and other interventions (World Health Organization 1983*a*).

In keeping with the basic concepts of primary health care, the research programme will not be conducted exclusively by biomedical scientists. It should involve other disciplines, in particular social scientists whose skills will contribute to the study of non-medical issues that have a bearing on health. Such studies should extend beyond the standard surveys of knowledge, attitudes, and practice of selected samples of the population, and should probe, in broad imaginative studies, various aspects of life, disease, and death within the community.

Evaluation of primary health care

The World Health Organization has selected 12 global indicators for monitoring progress towards Health For All (Table 18.4). Member states have used this protocol in reporting to the organization. The first major analysis was reported in the

Table 18.4. Twelve global indicators for monitoring progress towards Health For All

The number of countries in which:

(1) Health For All has received endorsement as policy at the highest official level;
(2) mechanisms for involving people in the implementation of strategies have been formed or strengthened;
(3) at least 5 per cent of the gross national product is devoted to health care;
(4) a reasonable percentage of the national health expenditure is devoted to local health care;
(5) resources are equitably distributed;
(6) needs for external resources are receiving sustained support from more affluent countries (developing countries only);
(7) primary health care is available to the whole population;
(8) the nutritional status of children is adequate;
(9) the infant mortality rate for all identifiable subgroups is below 50 per 1000 live births;
(10) life expectancy at birth is over 60 years;
(11) the adult literacy rate for both men and women exceeds 70 per cent;
(12) the gross national product per head exceeds US$ 500.

Source: World Health Organization (1985*c*).

Table 18.5. Some major initiatives related to health during the past decade

Year	Activity
1976–85	United Nations decade for women: equality, development, and peace
1981–90	International drinking water supply and sanitation decade
1979	International year of the child
1984	Task Force on Child Survival (World Bank, WHO, UNICEF, and Rockefeller Foundation)
1987–90	Commission on Health Research for Development
1987	International safe motherhood conference, Nairobi, Kenya
1988	World conference on medical education

Seventh Report on the World Health Situation, covering the period from 1977 to 1984. The report card shows a mixed picture with some significant gains but also some worrying setbacks (World Health Organization 1987).

On the positive side, the consensus achieved at Alma Ata has survived and all governments have repeatedly ratified both their commitment to the goal of Health For All and to the strategy of primary health care as the key to its achievement. Heads of states and parliaments have demonstrated their commitment though a variety of actions and significant policy changes. Furthermore, the primary health care programme has stimulated many other complementary initiatives on water supply and sanitation, on the welfare of women and children, and on new approaches to the training of doctors (Table 18.5). Both the World Health Organization and UNICEF, the original sponsors of the Alma Ata conference, have kept the issue at a high level of visibility.

Developing countries have reviewed and reorganized their health services to bring them closer to the primary health care concept. There is objective evidence of successful expansion of services to reach large populations in some countries. UNICEF notes significantly higher immunization rates and wider availability of oral rehydration salt packages. Although it is early to assess the impact of these changes on health indicators, some significant gains have been recorded. There are, however, some worrying signals. For example, the malaria situation has deteriorated in many endemic areas; and there is much concern about the rising incidence of infection with the human immunodeficiency virus in many developing countries, especially in some parts of Africa.

The most important threat to progress during the past decade has been the deteriorating economies of many developing countries in Latin America and in Africa. The debt burden has imposed severe restrictions on government spending in many of these nations. The economic readjustment programmes have had adverse effect on public spending on health and education; in some countries, UNICEF noted that the allocations for health and education were more severely cut than was the defence budget. The economic crisis has been compounded in Africa by droughts, famine, civil wars, and political strife. Per caput food production and consumption has declined in some countries. In spite of these reverses, most observers are convinced of the value of the new approach to global health. In fact, without the guiding principles enunciated at Alma Ata, the current economic crisis would probably have had a more damaging effect on the population.

References

Grant, J.P. (1989). *The state of the world's children*. Oxford University Press, Oxford.

Halstead, S.B., Walsh, J.A., and Warren, K.S. (1985). *Good health at low cost*. Rockefeller Foundation, New York.

Jolly, R. (1988). A UNICEF perspective on the effects of economic crises and what can be done. In *Health, nutrition and economic crises: Approaches to Third World policy* (ed. D.E. Bell and M.R. Reich) p. 81. Auburn House Publishing Company, Dover, Massachusetts.

Lancet (1988). *World conference on medical education, Edinburgh. Lancet* **ii**, 464.

Last, M. and Chavunduka, G.L. (1986). *The professionalisation of African Medicine*. International African Seminar. Manchester University Press, Manchester.

Montoya-Aguilar, C. and Marin-Lira, M.A. (1986). International equity in coverage of primary health care: Examples from developing countries. *World Health Statistics Quarterly* **39**, 336.

Perry, S. and Chu, F. (1988). Selecting medical technologies in developing countries. In *Health, nutrition and economic crises: Approaches to Third World policy* (ed. D.E. Bell and M.R. Reich) p. 379. Auburn House Publishing Company, Dover, Massachusetts.

Royal Colleges of Physicians of the UK (1987). *Equity—A prerequisite for health: Intersectoral challenges for health for all by year 2000*. Royal College of Physicians, London.

Sabben-Clare, E.E., Bradley, D.J., and Kirkwood, K. (1971). *Health in tropical Africa during the colonial period*. Clarendon Press, Oxford.

Schram, R. (1967). *The history of Nigerian medical services*. University of Ibadan Press, Ibadan, Nigeria.

Shephard, D.S. and Benjamin, E.R. (1988). User fees and health financing in developing countries. In *Health, nutrition and econ-*

omic crises: Approaches to Third World policy (ed. D.E. Bell and M.R. Reich) p. 401. Auburn House Publishing Company, Dover, Massachusetts.

Starrs, A. (1987). *Preventing the tragedy of maternal deaths.* A report on the International Safe Motherhood Conference, Nairobi, Kenya.

United Nations (1985). *Socio-economic differentials in child mortality*. United Nations, New York.

Walsh, J.A. (1988). *Establishing health priorities in the developing world*. United Nations Development Programme, New York.

Walsh, J. and Warren, K.S. (1979). Selective primary health care *New England Journal of Medicine* **301**, 967.

World Bank (1986). *World development report, 1986.* Oxford University Press, Oxford.

World Health Organization (1978). *Primary health care*. World Health Organization, Geneva.

World Health Organization (1983*a*). *Research for the reorientation of national health systems*. Technical Report Series No. 694. World Health Organization, Geneva.

World Health Organization (1983*b*). *New approaches to health education: report of an Expert Committee*. Technical Report Series No. 690. World Health Organization, Geneva.

World Health Organization (1985*a*). *Health manpower requirements for the achievement of health for all by the year 2000 through primary health care*. Technical Report Series No. 717. World Health Organization, Geneva.

World Health Organization (1985*b*). *Future use of imaging techniques in developing countries.* Technical Report Series No. 723. World Health Organization, Geneva.

World Health Organization (1985*c*). *World health statistics, 1985*. World Health Organization, Geneva.

World Health Organization (1986). *Regulatory mechanisms for nursing training and practice: Meeting primary health care needs in developing countries*. Technical Report Series No. 738. World Health Organization, Geneva.

World Health Organization (1987). *Evaluation of the strategy for health for all by the year 2000*. Seventh Report on the World Health Situation. World Health Organization, Geneva.

World Health Organization (1988). *The use of essential drugs*. Technical Report Series No. 770. World Health Organization, Geneva.

19

Comparative analysis of public health approaches

ALBERT VAN DER WERFF

Advantages and limitations of comparative analysis

People enjoy travelling, and while doing so they like to compare the situation at home to that in other countries. Politicians, health administrators, managers, and health policy analysts also like to discuss common problems and share views with colleagues in other countries, and look abroad for public policy approaches which might be copied and adapted for their own purposes.

Advantages of a comparative analysis are that it creates awareness of the existence of common health problems across countries, it may provide new and fresh ideas to policy-makers for public health approaches which could be useful for adaptation, it prevents them from re-inventing the wheel, and it contributes to the sharing of public health knowledge internationally which is essential in today's world.

Comparative studies also have limitations. Comparisons between nations and geographical regions have the practical disadvantage that the differences observed have to be explained, which in turn demands consideration of a very large number of factors. Important factors influencing health are those of the growth, composition, and distribution of populations, their genetic and biological determinants, and their life-styles, as well as socio-economic and cultural determinants, physical environmental factors, and living conditions. Futhermore, due account must be taken of the availability of or the willingness to provide resources in the form of personnel, facilities, and finance. These factors are, of course, interdependent. Thus public health approaches alone do not determine health.

It may be better to say that external factors determine health. The complexity and the dynamics of these interdependent variables are such that comparative studies may easily get no further than a 'description' and 'explanation' of the similarities and the differences; the effectiveness and efficiency of particular public health approaches may not be identified and measured.

In international comparisons the health statistics and the terms and definitions which are used at an aggregate level should be considered with caution. With respect to cost it is important not to overemphasize the relationship between health expenditures and gross domestic product (GDP). The scope and range of health activities may be different in different countries because of various restrictions of rights of patients and health professionals, and the existence of public and private co-payments. As may be the case with respect to health statistics, the definitions of costs and GDP may differ between countries. Percentages may reflect fluctuations in growth rates of GDP rather than improvements in cost effectiveness. Thus, generally, no wholly valid conclusions can be derived from comparing the data.

In addition to the practical disadvantages, there is a theoretical objection to comparative studies. The comparability of health systems depends on certain health systems are only comparable if they share:

(1) the same social goals, such as the value placed on health, the balance between collectivism and individualism, and distributional responsibility (WHO 1974);

(2) the same strategic output objectives, expressed in terms of improvement of levels of health, and the same way of carrying out these functions;

(3) common operational objectives directed towards the same specific health problems.

If all these requirements are met, the health effectiveness and the efficiency of public health approaches can be measured to serve as a basis for choosing between them (WHO 1970). In practice, it is impossible to meet these requirements in full, not even with respect to having the same overall social goals, let alone having the same strategic and operational objectives (Van der Werff 1976).

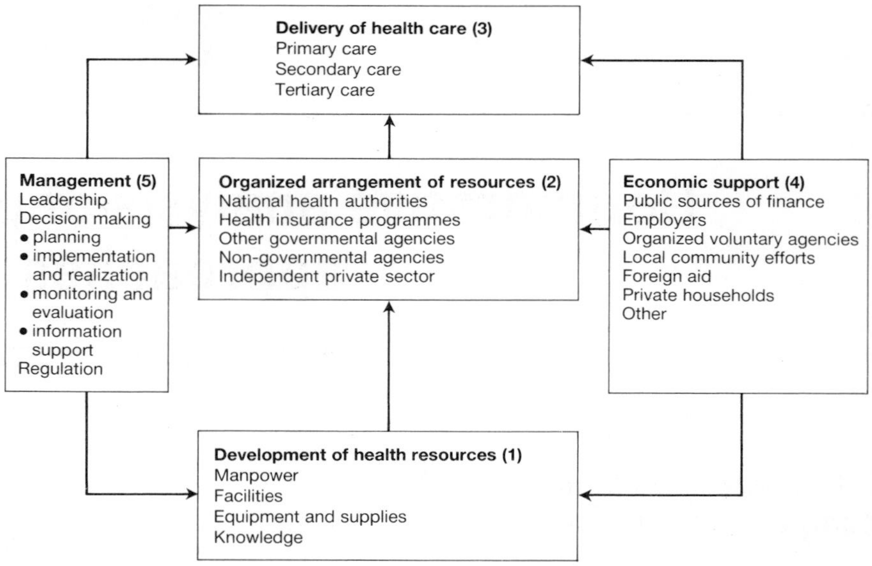

Fig. 19.1. Public health policies and strategies.

The situational context of application of public health approaches

The need for context analysis

To reduce the implications of these practical and theoretical disadvantages of comparative analysis, and to avoid incorrect conclusions, it is important to recognize the different factors that are characterizing the context in which public health approaches are being applied. This set of different, mutually dependent, and interrelated factors determines the 'situational environment' for application of public health approaches. Commonly, a different situational environment requires a different approach. This relationship is clearly a dynamic one which means that generally applied public health approaches will have to be adapted to a changing environment.

This line of reasoning is of practical value for actual policy-making and planning, and for comparative analysis at the same time. Decision-makers select certain public health approaches to achieve their objectives. The feasibility of implementing the selected approach has also to be taken into account. This means that before a particular approach can be selected, the characteristics of the situational environment have to be considered. To express this differently: decision-makers may accept a particular approach, or parts of it, but refuse others on the basis of their judgement that under the circumstances the chances of the successful application of one particular approach, or parts of it, will be estimated as 'high', and of other approaches as 'low'. The basic idea is that there is no single solution which is 'best' under all circumstances. Whether one approach is the best one cannot be decided beforehand, but is largely dependent of the objectives which have to be achieved, and on the situation in a certain place at a certain point in time (Blum 1974; Van der Werff 1986; Vught 1982).

In summary, when public health approaches are compared the situational context should be carefully analysed and understood before conclusions can be drawn.

Public health approaches

A wide variety of public health approaches are applied to improve and maintain health and well-being by preventing and curing disease, correcting the genetic and biological make-up of people where possible, changing life-styles, and improving the environmental conditions. There are many different ways of describing and analysing public health approaches, depending on the degree of thoroughness intended. Five main clusters can be distinguished (Fig. 19.1) each of which is directly or indirectly related to the others; these are public health policies and strategies directed at:

(1) development of health resources;
(2) organized arrangement of resources;
(3) delivery of health care;
(4) economic support;
(5) administration and management (Roemer 1977; Kleczkowski *et al.* 1984).

This set of public health approaches varies widely from country to country, in particular between affluent developed countries, on one hand, and the poor, least developed countries on the other.

Development of health resources

An early stage in the operation of any health system involves the development of the human and physical resources necessary to provide health care and perform supportive functions in the system. Many different types of resource are required, and their development entails diverse actions. In their simplest form, these health resources may be classified into four principal categories:

(1) health manpower;

(2) health facilities;

(3) health equipment and supplies;

(4) health knowledge.

Organized arrangement of resources

To translate the various resources of health systems into health activities and enable them to function properly requires social organization of some type. Organized arrangements are necessary to bring health resources into effective relationships with each other, and also to bring individual patients or community groups into contact with the resources through health care delivery mechanisms. The degree of formality in these organized arrangements and relationships varies greatly in different types of national health systems.

In any national health system these arrangements and relationships may be promoted in several ways, some through the actions of government (at various levels) and others outside government. The major groupings or organized arrangements of health resources may be classified in five categories:

(1) national health authorities;

(2) health insurance programmes (public);

(3) other government agencies;

(4) nongovernmental agencies (voluntary);

(5) the independent private sector.

Delivery of health care

The third major approach to be considered is the analysis of the variety of processes by which various health care services are provided. In different countries or in different parts of the same country these processes may vary greatly.

Health care delivery may be classified in different ways. Most often it is categorized according to the objective of the service delivered. This separates health activities into promotional, preventive, curative, rehabilitative, and the socio-medical care of the profoundly disabled and incurable.

In the context of national health systems, it is more customary to consider health care according to its level of complexity, or the sequential order in which the health needs of populations are served. Thus the services to be delivered comprise primary, secondary, and tertiary health care. The manner of delivery of all three levels of care differs among national health systems, but the differences are probably greatest for both the curative and the preventive aspects of primary health care.

Economic support

All the health resources and health care delivery mechanisms discussed above require economic support in any society. Since there are obviously many competing needs in a country, there must be procedures for channelling money into the health system. Unlike food and shelter, the need for therapeutic health care often cannot be predicted by the individual, and the need for many valuable preventive services may not even be recognized. Moreover, the ability of the various sections of the population to pay for health services in relation to their needs is dependent on income level. For these reasons, all national health systems have established certain mechanisms of economic support outside the operation of the free market.

The methods of economic support may be categorized in various ways. WHO classified the sources of finance as follows:

(1) public (all levels of government, including ministries of health, health insurance schemes, and other ministries);

(2) employers (industrial and agricultural enterprises);

(3) organized voluntary agencies (charity, voluntary insurance, and so on);

(4) local community efforts (financial contributions and unremunerated services);

(5) foreign aid (both governmental and philanthropic, the latter often from religious agencies)

(6) private households (both for payments to organized programmes and for purely private purchases);

(7) other possible sources (such as lotteries and donations).

Administration and management

The role of administrative or managerial processes has been implied in much of the above discussion: their importance for the proper functioning of a health system is so great that together they are regarded as a distinct factor in effective organization and operation. Ultimately, the pattern of management applied depends on the history, culture, and social values of a country. It also depends on the structure of authority (i.e. centralized, federal, or decentralized) of each country's government.

Framework for context analysis of public health approaches

As pointed out before when public health approaches are compared the situational context should be analysed carefully and understood before conclusions can be drawn. A comparative situational context analysis can be carried out on the basis of the following framework (see Fig. 19.2 and Table 19.1) which is designed in three, interdependent, concentric layers, namely:

(1) health and well-being;

(2) health context;

(3) socio-economic political–administrative context.

The health context

Following Blum (1974), Lalonde (1974), and WHO (1985) the health context can be described using three main clusters

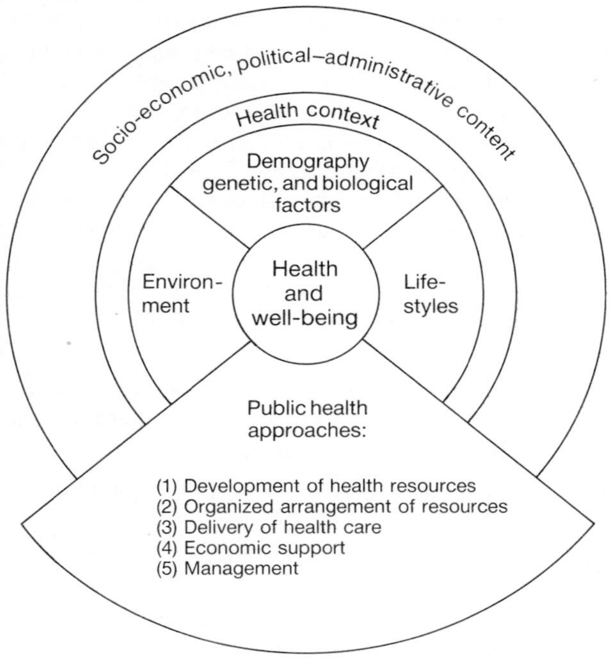

Fig. 19.2. Framework for comparative situational context analysis.

of mutually dependent and interrelated variables which directly determine health; that is

(1) populations: size, distribution and growth rate and genetic and biological factors;

(2) life-styles;

(3) health environment.

Population

Populations vary considerably from country to country with respect to size, distribution, and growth rates. The genetic and biological element includes all those aspects of health, both physical and mental, which are developed within the human body as a consequence of the basic biology of man and the organic make-up of the individual, the processes of maturation and ageing, and the many complex internal systems in the body. This element contributes to all kinds of ill health and mortality, including chronic diseases and others (genetic disorders, congenital malformations, mental retardation), but taken proportionally it is not very different in man across countries. In relation to the size, distribution, and growth of population, however, its effect on international differences can be substantial.

Life-style

Health may be enhanced by some life-styles, but damaged by others. Harmful practices include insufficient and unbalanced nutrition, overnutrition, smoking, alcohol and drug abuse, dangerous driving, and violence. These are all common and urgent problems in all relationships between health and the components of particular life-styles. To be effective, health programmes concerned with life-styles must deal with 'structural' influences on behaviour, as well as with its 'specific individual' components. Ethical issues have also to be considered.

Health environment

In all countries there is a fast growing, widespread interest in the quality of the environment and its influence on human health. Environmental policy and management should aim not only at safeguarding human health from the potential adverse effects of biological, chemical, and physical factors, but also at enhancing the quality of life by, for example, providing people with clean water and air, adequate living and working conditions, and facilites for leisure pursuits. All are factors which can have substantial positive effects on health. The circumstances are very different around the world. The developing countries in particular have few resources to protect their populations.

Socio-economic and political–administrative context

Economic context

An important background influence is the size and strength of a country's economy, which is usually expressed in terms of one of the national accounting aggregates, such as gross national product (GNP), or gross domestic product (GDP). These aggregates measure the total volume of national econ-

Table 19.1. Framework for context analysis of public health approaches

Health and well-being:
The main focus of the analysis concerns the health (health status) and well-being of a population (group) in a particular geographic area (central layer).

Health context:
The set of interdependent factors that directly determine health and well-being (intermediate layer); that is,

(1) population: size, distribution and growth rate, and genetic and biological factors;
(2) life-styles, conducive to health;
(3) environment;
(4) development of health resources, organized arrangement of resources, delivery of health care, economic support, and administration and management (public health approaches).

Socio-economic political–administrative context:
The set of factors which are indirectly influencing health, and largely determine the actual choice and application of public health approaches (outer layer); that is,

(1) the present economic situation and future outlook;
(2) the socio-political context of planning; that is, ideologies, norms and values, the political context, and attitudes towards planning;
(3) the political–administrative setting; the extent of government control, the institutional framework of the health administration, multi-sectoral co-ordination, and the balance between centralization and decentralization;
(4) the main characteristics of decision-making processes in health; that is, the complexity of decision-making, the degree of legalization, the role and power of the actors involved, and styles of management.

omic activity valued at current or constant prices. By dividing the GNP or GDP by the total population, one arrives at per caput GNP or GDP, which are common general purpose indicators of national wealth.

Per caput GNP may thus serve as a general measure of human welfare, that is, of health in a very broad sense. In practice, many health variables are indeed correlated with per caput GNP or GDP. Countries with a high per caput GNP are predominantly industrially developed, while those with a low one are predominantly agricultural, or developing.

Health problems in industrialized countries have passed through various evolutionary stages, each characterized by different challenges to public health and personal health care. In the initial stage, infectious diseases, malnutrition, and poor housing were combated by socio-economic improvements in combination with public health measures such as the provision of a pure water supply and sewage disposal facilities. As scientific advances were made, broader control of acute bacterial and viral diseases was achieved by means of immunization and chemotherapy, as well as increased health care for individuals.

The second evolutionary stage has been dominated by chronic diseases, particularly cardiovascular, cerebrovascular diseases, and cancer. Scientific and technological progress has produced a wide array of medical interventions for diagnosis and cure, higher levels of specialization in medical practice, and transfer of much of the care previously rendered in doctors' offices and patients' homes to increasingly elaborate and expensive hospitals. The cost of health care has risen dramatically, and in most countries it has become a matter of substantial public concern.

There is evidence in some industrialized countries of a third stage, which might be described as social and environmental pathology. Threats to health arise not only from intrinsic disorders of bodily structure and function, but from environmental hazards related to urban development and exposure to toxic substances, as well as from changes in social behaviour associated with violence, alcohol, and drug abuse of epidemic proportions.

Industrialized countries have passed through these three stages over the course of more than a century. Developing countries, on the other hand, face the challenge of coping with all three stages simultaneously, with just a fraction of the human and material resources available to their industrialized counterparts (Kleczkowski *et al.* 1984).

The socio-political context

Undeniably, the health system in any country is part of its social and political setting. Different ideologies, norms and values are at the roots of the position placed by society on health, the choices made with respect to the balance between collectivism and individualism, and the decisions taken on the degree of equity in the distribution of services and resources. These socio-political foundations of health systems are deeply rooted in society and do not change rapidly.

Fundamental changes in the socio-political foundations of health systems usually require changes in the power structure of the country. Major changes in health policy need political will and the support of a group which is politically large and strong enough to carry out the designed plans in spite of resistance. Generally speaking, health is not an important political issue, and it rarely provokes a major political debate. Most recently the economic recession has changed the picture in a number of countries, as the health sector has been found to be one of the main causes for the disproportionate increase of public expenditure and budgetary deficits. For the first time the political pressure is there, but ironically, it is coming from sources outside the health service.

The differences in attitudes towards planning are reflected in the classical debate on the relations between 'planning' and 'freedom'. Even today these discussions about the possibilities or impossibilites of planning influence the literature. 'Orthodox' or 'classical' planning, for instance, has been rejected by some authors because it is said to limit freedom, lead to rigidity and require technical methods. For these critics, 'modern' planning is considered to be better because it is arguably more open, flexible, and takes political aspects into account. Although there is not much proof of their validity, these beliefs still have great influence on the acceptance and application of planning. In periods of economic stringency and instability, these conflicts of opinions, or beliefs, are becoming more explicit. Whereas in some political groups in society trust remains in planning, in other parts opinions are dominated by the conviction that the extent of planning should be more restricted.

The political–administrative context

The political–administrative context in which national health systems are built and function is an important influence upon the situational characteristics of public health approaches. The main issues discussed below are the extent of government regulation and intervention, the institutional framework of health administration, multi-sectoral co-ordination, and the balance between centralization and decentralization.

The extent of government regulation and intervention ranges from 'maximal' when society takes full financial and organizational responsibility for the provision of health services, through 'moderate', when financial and organizational responsibility is shared between society and the individual, to 'low', when society assumes little financial and organizational responsibility. The precise point on the continuum depends on what is considered the 'collective optimum', and the extent of individual tolerance of government regulation and intervention. In the past the public has come to demand access to health care as an inviolable human right. Political acceptance of this demand has enforced a shift from private to public financing, with corresponding government control. However, under the pressure of economic circumstances the 'market model' has been advocated again, which may lead to a re-balancing act for the collective optimum. If the State fails in its efforts to regulate the public sector's provision and finance of health care, it can divest itself of these functions, that is, they can be 'privatized'. On balance, none the less, the scope and range of tasks and responsibilities of health authorities has broadened over the years, along with the increasing role of government.

While health planning may have gained in its concept of comprehensiveness, there are marked differences between countries in the scope and range of tasks and responsibilities which are conceived to belong to the health sector. In consequence, health policy seems to offer itself readily to a wide variety of organizational combinations with other sectors. But the health sector is not free to organize its multi-sectoral co-ordination. The political–administrative context of a country usually settles the matter. Also, the prospects for multi-sectoral co-ordination fluctuate in the course of time.

Finally, of critical importance is that the degree of decentralization in the health sector differs greatly among countries, and it is an issue continuously under debate as the balance has to be struck between social and geographical equity, on the one hand, and responsiveness to local need on the other. In the recent past, in a majority of countries, decentralization has been fostered by extending local autonony and community participation. Decentralization does imply a new division of labour between the health authorities, new definitions of staff responsibilities, increasing managerial competence, and probably an increase in staffing levels away from the centre.

Decision-making processes

Next to the political–administrative context, decision-making processes also influence the situational context of public health approaches. The main features are the complexity of decision-making, the degree of legalization, and the role and power of the actors involved.

A distinguishing feature of the health planning environment is the complexity of health problems and decision-making processes. The first characteristic of complexity is the existence of a great number of different personnel with heterogeneous interests. The second feature of complexity can be found in the nature of health problems. In solving health problems a great variety of factors need to be taken into account. Moreover, between these variables exist interrelationships and interdependencies which also have to be considered. Not only are health problems complex by themselves, but so are the decision-making processes in health. Health systems are social systems by which and within which decisions are taken, and these decisions are interrelated. The more decisions there are to be made, the more complex these systems will become. The shift towards governmental control, the extension of the scope and range of tasks and responsibilities of health authorities, multi-sectoral co-ordination, decentralization, increasing democratization in a 'multi-factor' society, and last but not least the rationalization of decision-making processes are factors which have increased the number of decisions to be made, and so the complexity of decision-making. The fourth element of complexity can be described as uncertainty. This means that information, or knowledge, is less than certain about the outcome of decisions and the result of subsequent activities. Political changes, demographic trends, shifts in morbidity patterns, rapid development of technology and economic circumstances are the main causes of uncertainty in health matters. The possibilities of eliminating uncertainties are limited. Besides, the cost of reduction of uncertainty may easily exceed available budgets.

Heterogeneity, variety, quantity, and uncertainty are inherent factors of decision-making in health. The purpose of planning is to simplify this complexity by regulating these features.

The increasing role of the State in regulating health has led to the increasing intervention of the legislator in the health field. Initially, the legislator entered the health field to fight communicable diseases and to guarantee public order through food hygiene and environmental measures. Also, professions have been regulated to a greater extent. From these problem-oriented approaches, the approach has become more global, covering the entire health field with legislation on the organization of health systems. Most recently these approaches have been exemplified the appearance of the legislative guarantee of the right to health, and the entry of the legislator into the field of promotion of healthier life-styles. Along with these developments, a shift has taken place from an emphasis on 'codifying' towards 'modifying' legislation. Codifying legislation confirms generally accepted practice by law, whereas modifying legislation provides policy and planning for the future with the necessary regulatory framework. Over the years, the roles and power of the different personnel involved in decision-making in health have undergone great changes. In the earliest phases of development, when the problems were merely of a medical– technical character, the 'health professionals' were undeniably in a dominant position. Along with development came the problems of large scale organization and financing which required different or additional types of personnel actors. 'Public administration and management' became much more important. This trend has been observed throughout the world after the Second World War, but it is probably most apparent in Europe. In this whole spectrum of personnel the 'consumers' have continuously been under-represented. In principle, the consumers are involved via their (elected) representatives in the different political bodies. Depending on the type of political–administrative system, many countries are now developing additional organizational structures and mechanisms for consumer participation.

Styles of health management

Some individuals are risk takers and some are risk averters, some are cautious and some are daring, some believe in slow adaptation and some in constant change. Similarly, some organizations demonstrate risk-taking, entrepreneurial, innovative characteristics; whereas others show a cautiously adaptive, conservative view. While both accept change, the former's tolerance for it is much greater than the latter's.

One believes in quantum changes, the other in incremental adjustments. The differences between the two modes are anchored in different ideologies. The ideology of conservative management emphasizes stability, evolutionary rather than revolutionary change, a view that the status quo is good unless proven otherwise. Management styles are different combinations of various dimensions, such as risk taking, optimization, flexibility, coercion, and participation (Khandwalla 1977).

Naturally, each country has a unique management style which is a combination of risk taking, optimization, flexibility, coercion, and participation. The same is true for each organization. However, some management styles in general are more in tune with characteristics of changing environments than others. When the situational context of public health approaches is likely to be marked by change, fluctuation, and uncertainty, entrepreneurial, risk-taking decision-making, with an emphasis on research and innovation, is required.

Yet, economic conditions have brought individuals who are mainly interested in cost containment to the centres of power. Representatives of this type of management tend to remain on the conservative side, prefer cautious, pragmatic, stability-oriented decision-making and a proclivity to low risk. These adaptors characteristically produce ideas on existing agreed definitions of problems and likely solutions. They look at these in detail. Much of their effort in the management of change is in improving and doing better. This satisfying style is not fully consistent with strategic long-term health planning, but prefers short-term resource input planning. Naturally, adaptors are interested in cost estimates and budgeting and tend to rely on personnel with experience and common sense in decision-making. They also tend to use structured channels of communication and restricted flows of information. Essentially, this style of management is marked by a conservative, risk-avoidance orientation.

Types of health system according to social, economic, and political characteristics of society

According to their socio-economic and political–administrative characteristics health systems vary widely across countries. No two seem alike. Nevertheless, it can be useful to apply a typology of national health systems. On the basis of estimates of a country's socio-economic and political characteristics the health system of any country can be placed on a theoretical matrix, as suggested by Kleczkowski *et al.* (1984).

An economic classification can be easily reached using a country's per caput GNP, as discussed above (although this measurement cannot reflect the distribution of income within a country). The socio-political characteristics of a country, or even of its health system, are not so easy to quantify: more qualitative judgements are required which can be based on the system's embodiment of the characteristics discussed above. Health systems may also be graded according to their administrative structure; that is, from governmental to private, from centralized to decentralized, and from pluralistic to unified. Thus a socio-political framework that vests great power in government and leaves little room for private enterprise would be expected similarly to vest all or nearly all health responsibilities in government, and have little private medical practice; while, at the other end of the scale, private medical and hospital services would be strong, and governmental health programmes weak. Centralization in the general socio-political framework would yield centralized controls, standards, and management in the health system; decentralized policies would generate similarly decentralized financing and control of hospitals, health insurance, and other health programmes. Pluralistic political ideology would undoubtedly be associated with a multiplicity of health programmes and numerous challenges to the achievement of co-ordination, whereas a generally unified political structure would almost certainly lead to the merging of all or nearly all health responsibilities in a single agency, typically a ministry of health. All the above socio-political aspects of health systems may be condensed into a scale of organization ranging from modestly, to moderately, to highly organized. Roemer based his conceptual matrix of the characteristics of national health systems on two factors; namely, the national economic level and the health system's degree of organization. His typology is presented in Table 19.2.

Table 19.2. National health systems: typology based on national economic levels and degree of health system organization

National economic level	Degree of health system organization		
	Modestly organized	Moderately organized	Highly organized
Developed (affluent)	1	2	3
Developing (transitional)	4	5	6
Least developed (poor)	7	8	9

Theoretically, every national health system in the world could be placed in one of these nine conceptual categories. In some categories there would be many health systems, and in others only a few. Moreover, the economic, socio-political, and administrative characteristics of countries and their health systems are continually changing, so that a system might be in one category now and in another five years hence. Subject to this possibility, Table 19.2 may offer general guidance for the comparative analysis of public health approaches. Of course, no type of health system can be singled out as the best: each type reflects a different sort of emphasis. Value judgements, based on the category into which a national health system falls, must also be avoided. In almost all categories, certain systems may rank high or low on criteria of relevance, coverage, effectiveness, and efficiency. National health systems which are rather loosely structured (modestly organized) but highly efficient may have the most favourable effect on the health of their populations. On the other hand, some highly organized but resource- or structure-oriented systems could have a less favourable effect on health, because of organizational rigidity or inadequate financial support.

Common framework and format for comparing public health approaches

Comparing public health approaches on a world-wide scale

Taking into account the complexity of the situational context within which comparative analyses of public health approaches have to be carried out, these exercises are necessarily multi-dimensional and complex processes. There are many areas for which comparative studies are extremely useful: international comparative analyses of health care spending and cost containment policies, the epidemiological assessment of the status of malaria across countries, the world prevalence and distribution of AIDS and the policies and strategies to prevent the spread of this disease from the high-risk groups that have been defined, international comparison of organ transplant policies, the study of the changes in the effectiveness and efficiency of planning along with shifts in the political–administrative context of countries, comparative analyses of the differences in organization of primary health care between countries, and so on. Some studies are restricted to a comparative study between a small number of countries, others are extended to the entire world. Therefore, international comparative analyses of public health approaches may differ in geographical scope, object, level of detail, and method. Within the context of this textbook it may be most useful, however, to concentrate on the work that has been implemented in the recent past by the World Health Organization (WHO). With a view to monitoring and evaluating the progress made in implementing the strategies for 'Health for All by the Year 2000', the WHO has developed a common framework and format that provides a suitable tool for comparative analysis of public health approaches at global, regional, and national level (WHO 1986*a*).

International comparisons by the World Health Organization

In 1979, the Thirty-second World Health Assembly launched the 'Global Strategy for Health for All by Year 2000' by adopting resolution WHA32.30. In this resolution the Health Assembly endorsed the Report and Declaration of the International Conference on Primary Health Care, held in Alma-Ata, USSR, in 1978. In the same resolution, the Health Assembly invited the Member States of WHO to formulate national policies, strategies, and plans of action for attaining this goal. Subsequently the World Health Assembly decided that progress in the implementation of these strategies should be monitored and evaluated at regular intervals so as to provide for a collective appraisal of the situation and its trends at national, regional, and global levels. The first monitoring round took place in late 1982/early 1983, and the first evaluation round in late 1984/early 1985 at national level; subsequent monitoring and evaluation took place in 1987–8 and for 1990–1.

Yet these formal monitoring and evaluation rounds which are first reviewed by the regional committees and the executive board, and then by the World Health Assembly, should be no more than the visible part of the ongoing monitoring and evaluation processes developed by each individual member state to ensure the sound management of its national health system. To facilitate international comparisons WHO has developed a set of global indicators for monitoring progress towards health for all, while additional sets of indicators have been adopted in certain parts of the world for use at the regional level. Some member states have also developed and adopted supplementary indicators for use at the national level (WHO 1981, 1986*b*). In addition, great efforts have been made with respect to the international standardization of terms and to the unification of data for categories of disease and disability, patients, health professionals, services, and resources.

Because of the differences of socio-economic development between countries, not all countries can afford information systems that can be used for international comparisons at an equal level. Even the short list of only 12 basic, global indicators selected by the WHO still exceeds what national health systems can handle in the least developed countries. On the other hand, in some of the most developed countries, any general list of indicators is too limited to provide the sensitivity and specificity required for the management of national health systems.

Indicators for international comparative analysis

European experience

The Member States of the WHO European Region adopted 38 regional targets for achieving their common health strategy and approved a tentative list of 65 groups of essential regional indicators (1980). The European targets have been formulated within a framework developed by the Member States and the WHO Regional Office to meet the needs of the European Region, where the aims and principles of Health for All have formed part of the health policies and strategies of most Member States for many years (WHO 1985). The targets and indicators reflect the overall high level of development in the Region and are more detailed than the basic global indicators. The evaluation itself for the Region as a whole and for individual Member States of the Region is based on two essential elements: the first is the strengthening of health and reduction of disease and its consequences (adding years to life, health to life, and life to years); the second is equity. Indicators measure changes in relation to the selected regional targets, and must be valid, objective, specific, sensitive, and relevant. In order to be useful for health managers, the indicators are also relatively simple, clearly expressed, and not too numerous. On the basis of the European experience a set of indicators can be described briefly that can be used for international comparative analysis.

Health status indicators

Indicators for health status include quantitative indicators on life expectancy at different ages, infant mortality, maternal mortality rates, standardized mortality ratios for all age groups, diseases, and indicators on morbidity and disability.

Demographic indicators

Demographic factors, that is, changes in the size of population, its age and sex structure, geographic distribution, and increase rates, are the obvious indicators and basic to all forms of policy-making and planning.

Life-style indicators

At the non-quantitative level, progress in this area can be monitored through changes in legislation, policies, and programmes for protection against health damaging risks and behaviours (smoking, alcohol and drug abuse, toxic products, violence), for health education, for the promotion of healthy life-styles (nutrition, physical exercise, and so on), and of community participation in health (including the enhancement of family and social support). In this area, surveys are particulary necessary to establish whether appropriate conditions exist for the development of healthy life-styles. Some quantitative indicators exist (for example, tobacco and alcohol consumption), but they are not always easily available in a relevant form.

Indicators for environmental health

Coverage by water supply and sanitation is an important global indicator, and indicators linked to human settlements, housing, and working conditions may help monitor changes in these areas. This is complemented by indicators for other environmental hazards, together with an assessment of surveillance measures of environmental aggression (food safety, air pollution, water pollution, hazardous waste disposal, and so on) and an assessment of the existence and quality of public information on environmental health.

Indicators for health care

Indicators for health care refer to all categories of public health approaches, i.e. development of health resources, organized arrangement of resources, delivery of health care, economic support, and administration and management, as discussed before (see Fig. 19.1).

The indicators for health care include the non-quantitative indicators such as the existence (or absence) of policies, programmes and activities on promotional, preventive, curative, rehabilitative, and socio-medical care for the improvement of health status and of equity in health. Also various non-quantitative indicators are applied to describe the organization of national health systems and of administration and management, as well as of the administrative and managerial processes and mechanisms. In this respect the coverage, access and utilization of the health services are primary, secondary, and tertiary care level can be expressed in quantitative terms. Indicators for health care provision also include the classical quantitative ratios of health resources to population, and cover the distribution of these sources within the population.

Socio-economic and political indicators

Socio-economic factors are easier to characterize in quantitative terms than political factors. As pointed out before, the size of the national economy is an important background influence. This is usually expressed in terms of one of the national counting aggregates such as the gross national product (GNP) or gross domestic product (GDP). The per caput GNP or GDP is a very common general purpose indicator of the average income level and of income distribution. On this basis the proportion of health spending can be calculated per geographical area, population group, disease and disability category, service sector, and resource category. The ratios from the different services of economic support to health expenditures are useful indicators too. They show, among other things, the share which is under the control of government. For international comparison the World Bank Atlas is the most widely used aid. Moreover, indicators are in use describing the stages of development of countries from a social and cultural point of view, as on education, for example.

Summary

Comparative analysis has proven to be a creative source of information on the application of public health approaches across countries. It is widely used and it is generally accepted as a helpful tool for priority setting, policy development, planning, monitoring, and evaluation for international and local purposes. When public health approaches are compared, it is essential to make sure that the situational context of each of the compared approaches is carefully analysed and well understood before conclusions are drawn. A common framework and format for international comparison on a world-wide scale has been developed by the World Health Organization in recent years. This is a concentric approach in three steps, considering health in its context and taking systematically into account the background characteristics of the socio-economic and political–administrative environment. The socio-economic, political–administrative factors are important background influences which tend to be neglected. In particular, for comparing public health approaches, it is also important to analyse the differences and similarities in decision-making processes and styles of management between countries. International comparative studies may differ in geographic scope, subject and objectives, level of detail, and method. Such studies can be restricted to a comparison at a certain point in time. As the relations between the different determining factors are clearly dynamic, comparative analyses over periods of time seem to be more relevant. Considering the complexity of the situational context, comparative analyses of public health approaches are necessarily multi-dimensional and complex processes. These exercises can be facilitated by the application of a common framework and format, use of international indicators, and standardization of terms and data. Not all information can be expressed in quantities, and most of the work will remain descriptive.

References

Blum, H.L. (1974). *Planning for health, development and applications of social change theory*. Human Sciences Press, New York.

Khandwalla, P.N. (1977). *The design of organizations*. Harcourt Brace Jovanovich Inc. New York.

Kleczkowski, B.M., Roemer, M.I., and van der Werff, A. (1984). *National health systems and reorientation towards health for all*, Public Health Paper number 77. World Health Organization, Geneva.

Lalonde, M. (1974). *A new perspective on the health of the Canadians*. Government of Canada, Ottawa.

Roemer, M.I. (1977). *Comparative national policies on health care*. Dekker, New York and Basel.

Vught, F.A. van (1982). *Experimentele beleidsplanning*. Vuga, 's-Gravenhage.

Werff, A. van der (1976). *Organizing health care systems, a developmental approach*. Doctorate thesis, National University of Utrecht, Utrecht.

Werff, A. van der (1986). Planning and management for health in periods of economic stringency and instability: a contingency approach. *International Journal of Health Planning and Management*, **1**, 227.

World Health Organization (1970). Technical Report number 472, pp. 12 and 14. WHO, Geneva.

World Health Organization (1974). Public Health Paper number 55, p. 65. WHO, Geneva.

World Health Organization (1981). *Development of indicators for monitoring progress towards health for all by the year 2000*. WHO, Geneva.

World Health Organization (1985). *Targets for health for all*. World Health Organization Regional Office for Europe, Copenhagen.

World Health Organization (1986*a*). *Evaluation of the strategy for health for all by the year 2000*. Seventh report of the World Health Situation, Vol. 1, Global Review, WHO, Geneva, 1987, Vol. 5, European Region, WHO, Copenhagen, 1986.

World Health Organization Statistics (1986*b*). *Quarterly Report, Special Issue: Indicators for HFA-strategies*. Volume 39, Number 4. World Health Organization, Geneva.

D

Provision and Financing of Health Care

20

Provision and financing of health care in the US*

BARBARA STARFIELD

Introduction

The US is almost unique among industrialized nations in not having a national health system that sets the bounds for the way in which services are organized, paid for, and delivered. The lack of commitment to health care as a 'right' results in a situation where a substantial proportion of the population is unable to pay for necessary care, and where even those who are able to afford services often lack the information and ability to identify the most appropriate care for their needs. Although a commitment to considering health care a 'right' was made two decades ago through the passage of certain federal legislation, philosophical and ideological developments since this time have eroded the concept. Moreover, the system of health care is highly volatile and subject to marked change in nature and degree, even within short periods of time.

Virtually all systems of care in other nations are organized according to the principle of levels of care, in which individuals who need services first seek those services from a generalist physician (a 'primary care' physician). That physician either takes care of the problem or, if it is beyond the scope of his or her expertise, refers the patient to a consultant for short-term diagnosis and/or management, or for longer term care in the case of particularly specialized needs which can be provided only by a specialist trained for these unusual problems. Thus, in most countries with organized health systems, physicians are trained to be primary physicians, consultants, or specialist physicians. Primary physicians work in local offices or health centres; consultants work in area hospitals or large centres; and specialists work in regional hospitals.

The system of health services in the US has developed piecemeal, and is largely bipartite although with considerable overlap of its parts. The system is mainly within the 'private sector', in which physicians generally are self-employed and hospitals may be organized on a profit or non-profit basis. However, some care is provided directly by various levels of government; such is the case for the military and its dependants, and for certain segments of the population (such as the very poor, or American Indians) who would otherwise face extraordinary barriers to entry into the private system.

Moreover, a substantial amount of care provided in the private sector is paid for by public funds. Until recently, accountability for the expenditure of public funds by the private sector has been relatively loose.

The coalescence of three phenomena is responsible for recent changes in organization and financing. First, the percentage of the gross national product contributed by health services is far outstripping all other components of national expenditures, and is rising at a rate far higher than elsewhere. Thus, the costs of care are perceived to be on an ever-increasing spiral upward, without any clear bounds. Second, there is increasing evidence of diminishing access to appropriate care, not only for the poor who lack the means to pay for services (Freeman *et al.* 1987) but also for the rest of the population, which has increasing difficulty in identifying the most appropriate source of care for problems. Third, there is mounting evidence of wide differences in the use of diagnostic and therapeutic procedures across geographical regions and among hospitals, without any apparent reason for the differences or any evidence of better health resulting from it.

This chapter reviews the existing state of affairs with regard to the organization and financing of health services and provides a basis for understanding changes that may occur in the near future.

Organization and financing of health services

Until a few years ago, all but a small proportion (less than 5 per cent) of the civilian population received their medical care from a physician who worked alone or in a small group

* This chapter is adapted from and based upon Starfield, B. and Flint, S. (1989). Economics of medicine. In *Economics of medicine, in principles and practice of pediatrics*, Chapter 4. (ed. F. A. Oski) J. B. Lippincott Company, Philadelphia.

and charged a fee for each service rendered. The availability of private insurance has been increasing since the 1920s, and localized experiments with prepaid group practices have been a feature since the 1940s. But it is only since the mid-1970s that a plethora of new organizational formats began to assume prominence. The major distinguishing characteristics have to do with the degree to which responsibility for a defined panel of patients is assumed and the method of reimbursement for services.

Responsibility for a defined panel of patients

The traditional method of delivering services in the US involves the provision of care by the physician when patients present at the physician's office. In the newer formats, responsibility for care involves a contractual agreement in which both the physician (usually as part of a physician group) and the patient recognize that all care over a defined period of time (usually a year) will be obtained from the same source. This agreement extends to all levels of care; if a patient requires a referral to a specialist, or hospitalization, the physician of record (the 'primary care' physician) will designate the specialist or hospital. In this form of care, patients forego the ability to seek care from other sources unless they wish to pay separately for these services.

Method of reimbursement for services

In the traditional mode of reimbursement, physicians are paid for seeing the patient and for diagnostic and therapeutic procedures. The payment can be paid directly by the patient, either immediately or in response to subsequent billing, or can be reimbursed by insurance companies or other third party payers (such as state agencies who reimburse for services to certain of their employees or the poor who qualify for public assistance). In most cases, insurance fails to cover a substantial proportion of the costs of care, especially for preventive care, so that patients either have additional out-of-pocket expenses or do not obtain the services.

In the past 40 years, this traditional method of paying for care has been gradually eroded in favour of prepayment for services. In this form of reimbursement, third party payers contract with physicians or physician groups, so that these providers of care are at some financial risk. In return for this prepayment, the providers agree to provide a defined package of services according to a contract. In some forms of prepayment, these physicians are also at financial risk of costs of care that result from referrals or hospitalizations, although most arrangements exempt certain 'catastrophic' costs over which the physician has no control.

The new organizational arrangements that are assuming increasing importance in the US health care scene are characterized by various combinations and permutations of organizational format and reimbursement. In the following section, the most important of these forms of care are briefly described.

Health maintenance organizations (HMOs)

The HMO is an organizational format that is unique to the US. In 1932 the Commission on the Costs of Medical Care recommended that health care be provided by organized groups of health professionals, preferably in a hospital setting, on a prepayment basis. Although care provided by such organized groups is a feature of many health care systems, it is only in the US that they are organized within the private sector and reimbursed largely through private insurance mechanisms.

HMOs are distinguished by prepayment for care. The HMO's payment is fixed in advance, depending on the number (and sometimes type) of patients who are enrolled to receive their care in the HMO, regardless of whether the individual needs care or how much care is needed. The burden is on physicians who work in the HMO to control health care expenditures; if more care is provided than the HMO anticipates, the HMO will run a deficit. On the other hand, if too little care is provided, the HMO could be suspected of not providing necessary care. HMOs are characterized as being one of several types described below.

1. *Staff model*: the physicians work directly for the HMO, on salary but sometimes with a bonus depending on the HMO's earnings.
2. *Group model*: the HMO contracts with a separate physician group to provide its services. These are also known as prepaid group practices (PPGPs), and physicians may be paid a salary or a 'capitation' depending upon how many individuals the physician has on his or her panel of patients. The *network model* HMO contracts with several independent group practices instead of just one.
3. *Individual practice association (IPA)*: the HMO contracts with individual physicians who are in independent practice, or with a group of physicians each of whom works independently usually in solo practice or single specialty groups. IPA plans typically require primary care physicians to act as case managers for prepaid patients. This role requires them to approve referrals, admissions, and high cost procedures and tests. A capitation fee is paid to the HMO for each person enrolled. Physicians may be paid a capitation for primary care services or could be reimbursed according to a fixed fee schedule for services rendered; the fees are typically based on a percentage of the physician's usual and customary fee. In a network model IPA, primary care physicians could receive additional capitation for referral and ancillary services which would be arranged by the primary care physician and paid for on a fee-for-service basis or a pre-arranged payment schedule. Usually a portion of the fee or the capitation payment is withheld by the IPA to cover losses at the end of the year. If utilization and costs are in line with expectations, the withheld amount is redistributed to the physicians.

In both the group model and IPA, the physicians are usually

Table 20.1. Payment for care

Source of payment	Pre-negotiated fee			Reimbursed fee
	No enrolment*		Enrolment (managed care)	
	Capitation/salary	Fee-for-service†		
Out-of-pocket			Individual contracts for specified types of care (e.g. obstetric, dental)	Traditional ambulatory services Free-standing emergicenters
Private insurance	Corporate health centres	Individual practice associations Preferred provider organizations Managed indemnity plans	Health maintenance organizations	Most inpatient services Certain ambulatory services
Tax revenues	Some governmental health centres, community health centres	Preferred provider organizations	Pre-paid Medicaid programmes	Medicaid Champus
Federal insurance	Veterans administration facilities	Preferred provider organizations (through Medicare)	Health maintenance organizations (special arrangements)	Medicare

* For the purposes of this table, enrolment means that individual names are provided to physicians, and, in the aggregate, constitute a practice roster.
† Individual physicians are paid fee-for-service, although subject to utilization review.

free to see patients outside of the HMO in addition to the patients enrolled in the HMO.

(Another type of HMO, the *social* HMO or S/HMO has been developed and tested by the Federal Government as a way of integrating the long-term care and acute care services to the chronically ill elderly.)

Managed indemnity plans and preferred provider organizations

A managed indemnity plan (MIP) is a conventional (indemnity) insured fee-for-service arrangement, but the use of services and procedures is carefully monitored and there is significant cost-sharing on the part of the patient.

A preferred provider organization (PPO) is a newer arrangement by which associations of physicians or hospitals contract with employers or insurers to provide services to individuals for a negotiated, generally discounted, fee. In a PPO, patients may use a physician of their choice, regardless of whether the physician participates in the PPO. However, the patient must pay part of the fee (co-insurance or a deductible) out-of-pocket when a non-PPO physician is consulted. In an exclusive provider organization (EPO) use of non-PPO physicians is not covered except in emergency situations. The EPO, therefore, is similar to the IPA, except that EPO is paid a fee-for-service by the third party payer whereas the IPA is prepaid. In both cases, physicians rendering the services are paid a pre-negotiated fee-for-service.

Integral to PPOs are a 'managing' organization that provides certain administrative functions which monitor the use of services. These include pre-admission certification (wherein patients slated for elective admissions to hospitals are reviewed for appropriate admission), second opinions before surgery, certification of treatment plans for certain non-emergency services (such as mental health services), and review of the medical care that is provided.

Free-standing emergicenters

Free-standing emergicenters (FECs) are walk-in facilities for the treatment of non-emergency medical conditions. Most are equipped to provide only minor and non-life-threatening care and some routine health services. They do not require an appointment and are open during times when most physicians' offices and hospital clinics are closed. Most require immediate payment by cash or credit card. Between 1978 and 1984, the average annual growth of FECs was 71 per cent; by the end of 1983, estimates of the number nationwide ranged from 900 to 1300. However, since 1983, the growth in the number of FECs and in-patient volume have abated; many FECs are not profitable and are restructuring or closing. Originally about three of every four FECs were owned by physicians, largely for profit. Currently they are being taken over by large national firms that are involved in other types of organizations for managing care.

Table 20.1 summarizes the different ways in which medical services are organized and financed.

Trends in the development of new organizational formats

Before the late 1920s, most people paid directly for medical care services; those unable to pay either did not receive needed care or else relied upon charitable organizations and local health department clinics, where they existed. Private health insurance became increasingly popular during the late 1920s and 1930s for those who could afford to purchase it; physicians were overwhelmingly paid a fee for their services. During the 1930s and especially 1940s, major societal changes made access to medical care a societal priority. Labour–management negotiations generally included bargaining over health insurance as part of employee benefits; an increasing proportion of the benefit package has consisted of employee and family coverage for health care costs. Moreover, the Federal Government assumed an increasing responsibility to provide financing for care of population groups without the means to provide their own. During this period the precursors of PPOs were developed (the first prototype was in 1934) and the forerunners of the IPA were initiated (generally by county medical societies in the form of 'foundations' for medical care). Kaiser Industries, faced with the need for medical services to care for the influx of workers in defence industries during the Second World War, developed the first large-scale prepaid group practice HMO. These alternative delivery systems grew very slowly; in the 1970s, fewer than 5 per cent of the population and physicians were involved in them. Rapidly exploding medical care costs, fuelled by a growth in technology in the form of diagnostic and therapeutic tests and procedures and a system of reimbursement (fee-for-service) that simply passed the costs on to patients and their insurers, provided impetus for new thinking about prepayment and cost-controls. Federal legislation was passed. The HMO Act of 1973 (implemented in 1976) encouraged the growth of prepaid organizations; in 1982, the Tax Equity and Fiscal Responsibility Act (TEFRA) encouraged the enrolment of the elderly, who had been covered since 1965 for most of their medical services as part of the Social Security Insurance system, in HMOs. (The Act also eliminated the federal loan programme of 1973, thus encouraging HMOs to seek private sector investment capital for start-up cost.)

Between 1981 and 1986, the number of HMOs increased 145 per cent (from 243 to 595). HMO enrolment jumped by 130 per cent from 10.3 to 23.7 million people, or about 10 per cent of the population (Owens 1987). From 1984 to mid-1987, enrolment grew at an annual rate of more than 20 per cent. The IPA type of HMO experienced the largest growth. During the period from 1981 to 1985, aggregate HMO enrolment doubled, while IPA enrolment quadrupled (Dalton 1987). As of the end of 1985, 51 per cent of the HMOs were IPAs, 19 per cent were network HMOs, 17 per cent were group HMOs, and 13 per cent were staff HMOs. By late 1987, however, there was evidence of a marked slowing in the rate of increase in enrolments in HMOs. Many corporations eliminated health insurance coverage for care in HMOs. At least 16 HMOs disappeared in 1987 because of mergers or business failures; the HMO arena is increasingly becoming dominated by a few large national organizations, some of which are the original non-profit predecessors (such as the Kaiser Health Plan and the Health Insurance Plan of New York) and some of which are the newer profit-seeking plans such as Maxicare, United Healthcare, and US Health Care.

Before 1983, there were 20 operational PPOs in the US, whereas there were 413, in 41 States, by 1986. From June 1985 to June 1986 alone there was an increase of 59 per cent in the number of PPOs (Dalton 1987). The total number of individuals served by PPOs is unknown, but various estimates have been made. The 197 out of about 325 existing PPOs in a study conducted in November 1985 had an enrolled population of 18 million (Logsdon *et al.* 1987). Telephone surveys indicated a growth from about 6 million enrolled in July of 1985 to 16.5 million enrolled in the summer of 1986. Although hospitals and Blue Cross/Blue Shield Plans had the largest market share in 1986, commercial insurers and investor-owned PPOs were growing at the most rapid rate (deLissovoy *et al.* 1987). Three States (California, Colorado, and Florida) accounted for 65 per cent of PPO enrollees. As more States strike down statutes prohibiting selective contracting, PPOs are likely to spread to other parts of the country (deLissovoy *et al.* 1987). By mid-1987, the American Association of Preferred Provider Organizations (AAPPO) reported that there were 674 PPOs, involving 32 million people with PPO 'options'. Thus, by 1987, roughly one-quarter of the US population had their health care financed and delivered through an HMO or PPO.

This marked growth in alternative delivery formats has been accompanied by notable shifts in the ownership and management of medical care organizations. Initially, HMOs were established as non-profit organizations. The federal Health Maintenance Organization Act of 1973 provided impetus for the development of HMOs by providing loans; most of the applications were from non-profit organizations. Early in the 1980s, the national Administration encouraged a movement towards 'market-place competition'; this resulted in the entry into the health field of entrepreneurs and venture capitalists. The rapid growth of for-profit national health care corporations also spurred the entry into the field of insurance companies, who had to protect their own indemnity plans from the new competition. From 1981 to the end of 1985, the percentage of HMOs that were for-profit had grown from 18 to 51 per cent. Although the majority (54 per cent as of 1986) of PPOs are operated by hospitals and physician groups, it is likely that insurance companies will capture an increasing share of the market so that these organizations, too, will be dominated by a profit motive (Dalton 1987).

The increasing prominence of new organizational formats for providing health services does not bode well for individuals with the greatest health needs. Health care organizations that are competing for enrolled populations while they attempt to maximize profits cannot be expected to welcome individuals who are likely to require relatively large amounts of services. Competition encourages non-profit organizations into the same situation. Experience with a variety of programmes to encourage the enrolment of employers in health care plans shows clearly that they generally adopt techniques to minimize the possibility of adverse risk selection. Such techniques are the exclusion of certain categories of employment where risks are relatively high (such as construction firms) and pre-enrolment health screening to detect existing medical conditions. As marginal workers and those working for low wages (and their families) are more likely than others to be excluded for these reasons, low-income children will be at an even greater disadvantage in the future than in the recent past. Chronically ill individuals who are poor risks for pre-budgeted health care organizations might also be at a disadvantage. In the past, public agencies have attempted to overcome some of the barriers in the way of access to care for those in high risk groups. Whether or not state legislatures will develop new strategies (such as insurance pools supported by a tax on employers) to pay for care of the uninsured remains to be seen; such options are currently under consideration in several states.

The importance of unimpeded access to health services cannot be overemphasized. An extensive review of the evidence of effectiveness of medical care for children showed how health improved when barriers to access were reduced (Starfield 1985). Conversely, health status worsens when families are required to pay a co-insurance fee for their medical services. In a large-scale insurance experiment, families were randomly assigned to one of a variety of plans differing in the rate of co-insurance for medical visits; one plan required no fee at all. Both health status and use of services were monitored for several years. As compared with the free plan, both adults and children in the co-payment groups were less likely to receive care for conditions for which medical care is effective. The adverse effect of co-insurance was even more striking for poor individuals than for others (Lohr *et al.* 1986, p. S36). Poor individuals in the co-insurance groups also had poorer health status at the end of the experiment than comparable individuals in the free care group. In all groups, requirements for co-payment reduce equally the use of needed as well as discretionary or inappropriate services.

Financing of health services

Sources of financing of services in the US are remarkably diverse. In 1986, 15 per cent of the population had no external source ('third party coverage') of funding. The proportion without third party coverage was greatest for children aged under 5 years, 17.5 per cent of whom had no source of funding other than 'out-of-pocket coverage'. By the age of 45–64 years, about one in ten adults (10.3 per cent) had no third party coverage. In those aged over 65 years, however, more than 99 per cent of individuals had some source of help with payment of medical charges. This difference by age is largely a result of the 'Medicare' programme, which came into being as a result of legislation in the mid-1960s. Medicare is a two-part programme designed to help the elderly pay part of their medical bills. Part A is supported by a trust fund that is replenished each year by taxes on employee wages and employer payrolls. Individuals pay no premium for this governmental insurance other than the tax on their wages, but there are substantial deductibles and co-insurance after the first 60 days of hospitalization. In mid-1988, new legislation ('the catastrophic care bill') expanded Medicare to provide free hospital care after the annual deductible. The new legislation provides for up to 150 days of care in nursing homes and for 38 days of continuous home health care. These benefits would be financed by premiums paid by Medicare enrollees. Part B of Medicare, which covers ambulatory care services, is financed by enrollee premiums (25 per cent) and by general tax revenues. It requires a monthly premium paid by enrollees and requires co-payments as well as an annual deductible. The 'catastrophic care bill' expanded coverage to include a ceiling on out-of-pocket payments for doctor bills, as long as the service is a Medicare-approved service. (The original programme had no limit on co-payments and deductibles.) The programme was also expanded to include half-payment for prescription drugs in excess of US $600 in 1991, with the Medicare share rising to 60 per cent in 1992 and 80 per cent thereafter. Payment for preventive care is not covered by Medicare, although the 1988 legislation created a breast-cancer screening benefit for routine examinations.

Except for the elderly, insurance coverage is inequitably distributed in the population. Although males are only slightly less likely to have no coverage than females, there are marked differences according to family income: 37 per cent of those in the lowest income group and 31 per cent in the second lowest income group have no coverage at all as compared with 2.9 per cent in the highest income group and 8.4 per cent in the next highest group. The source of coverage varies by family income as well; over 93 and 88 per cent of the top two highest income groups, but only 31 and 58 per cent of the two lowest income groups, have some private (non-governmental) insurance.

Very few indemnity insurance plans cover all types of service. As a result, a large proportion of the population must pay out-of-pocket for important aspects of health care. Whereas 91 per cent of the population has some coverage for in-hospital care, only about 80 per cent have at least some coverage for visits to physicians' offices, 88 per cent for visits to hospital clinics, 80 per cent for in-patient mental health visits and 72 per cent for out-patient mental health visits, 55 per cent for skilled nursing facilities, 36 per cent for home health care, 83 per cent for maternity care, and 73 per cent for prescription medicines. Only a minority of the population have even some coverage for dental care (25 per cent), vision care (15 per cent), routine preventive care (10 per cent), and hearing problems (9 per cent). Individuals whose coverage is through a governmental programme (such as Medicare) are more likely to have coverage for the various types of service than individuals whose only coverage is through private insurance. (These data are from 1977, the latest available at the time of this writing.)

Tax-supported health programmes

Although Government has always assumed some responsibility for the medical care of some individuals with no ability to pay for them, it was the Depression of the 1930s that resulted in a federal commitment on a large scale. Title v of the Social Security Act of 1935 and various amendments provided funds to states to support services for children with certain chronic conditions, maternal and child health clinics, family planning, regionalized prenatal care, and dental care. In 1965, the legislation was amended in a major way. For the first time, federal dollars were appropriated from general tax revenues to pay for care, through the states, to families with dependent children whose incomes were below a certain amount. By 1978 (the latest year for which information is known), public programmes contributed 29 per cent of the total expenditure for health care of children under age 19 years; two-thirds of the amount came from the Federal

Government and one-third from states and localities. Just over half (55 per cent) were in the form of reimbursement directly to physicians for services rendered and the rest went to federally supported service programmes (6 per cent), the Department of Defense (14 per cent), state and local hospitals (2 per cent), and various other programmes (3 per cent) (Budetti *et al.* 1982).

Medicaid

The Medicaid programme was initiated by federal legislation in 1965. Its intent was to assist states in paying for basic medical services for individuals who could not afford to pay for care or for insurance against medical expenses. States determine the scope and design of their programmes, based on federal guide-lines. The Federal Government matches state expenditures based on a formula depending on the state's wealth. States are permitted to exceed their matching requirement but must provide for a minimum of care including in-patient and out-patient hospital care, physician services, laboratory and X-ray services, and nursing home care. States are permitted to establish their own income cut-offs for determining eligibility, with the exception of recently mandated coverage of poor pregnant women and infants to the age of 1 year.

In the first decade of its existence, Medicaid eligibility and expenditures grew; since 1976, the process has reversed. Despite the economic recessions in the mid-1970s and early 1980s, high levels of unemployment, and increasing poverty, the Medicaid population has not grown. States have responded to budget pressure at the federal level by limiting income eligibility, reducing coverage of optional groups, and reducing the scope of covered services. Half of Medicaid recipients are children, although children (individuals under 21 years of age) account for only 20 per cent of Medicaid expenditures. From 1980 to 1985, Medicaid coverage of the poor and near-poor population of all ages dropped from 53 to 46 per cent. The 1985 average income eligibility standard for a family of four was US \$4992, or 47 per cent of the federal poverty level, with state to state variability ranging from US \$1776 per year or 21 per cent of the federal poverty level in Alabama to US \$7800 or 93 per cent of the federal poverty level in California. Legislation in the early 1980s (the Omnibus Reconciliation Act 1981, or OBRA) reduced federal funding and gave states increased flexibility to reduce eligibility standards; the non-elderly poor experienced the greatest loss of coverage, with the ratio of Medicaid enrollees to the poverty population declining by almost 13 per cent from 52.1 to 45.5 per cent. Among those families rendered ineligible for Medicaid, approximately one-half were left without any insurance coverage. A national survey of access to medical care showed that use of medical services declined markedly between 1982 and 1985, with poor children, especially those in poor health, suffering the greatest declines (Freeman *et al.* 1987).

In the early 1980s, other changes in the Medicaid programme influenced its organizational format. In OBRA, Congress modified existing Medicaid procedures to permit prepayment for care of recipients without requesting special exemption from the provisions of the original Medicaid legislation that guaranteed freedom of choice of providers. By late 1985, there were a total of 59 waivered programmes in 28 states. There are three main types of programme: (1) mandatory choice in which the recipient must choose a case manager from a list of specified providers; (2) mandatory enrolment in which the state contracts with an intermediary who is responsible for all covered services and who may or may not allow the individual to choose a provider; and (3) voluntary enrolment in which the eligible individual may select a case manager or may choose to remain in the fee-for-service system.

From the physician's viewpoint, there are several alternative formats: primary care provider, who assumes the responsibilities of a case manager ('gatekeeper') and is paid a fee-for-service plus a small additional amount for management; contracted health insuring organization (HIO) in which primary care is arranged through an intermediary which negotiates payment rates (by law, in 1985, HIOs must maintain a non-Medicaid and Medicare (the elderly) enrolment of at least 25 per cent, although some states have already received waivers to avoid doing so); prepayment contracts with primary care organizations in which states contract with organizations or physicians who become the case manager paid by capitation; partially capitated services with physicians or organizations similar to the prepayment contracts except that the capitation covers a more limited range of services and the case manager is not at risk for specialty services; and voluntary HMO enrolment in which the state maintains traditional Medicaid but encourages HMO development for Medicaid recipients.

HMOs serving Medicare and Medicaid patients have also been exempted from certain standards for operation of HMOs. These HMOs are called Competitive Medical Plans (CMPs). The relaxation of standards for these HMOs is an attempt by the Federal Government to increase the number of prepaid plans eligible to compete for publicly insured patients.

Although the experimentation in the financing and delivery of Medicaid services continues to expand, there has been a roll-back in the eligibility restrictions imposed by OBRA and, in the case of some sub-populations, certain expansions have taken place. The Deficit Reduction Act of 1984 (DEFRA) and the Consolidated Budget Reconciliation Act of 1985 (COBRA) required all states to provide Medicaid coverage to pregnant women and children to the age of 5 years who reside in families with incomes below the Aid to Families with Dependent Children (the 'Welfare' programme) payment level for a family of a comparable size. This expansion of Medicaid eligibility required states to extend benefits to pregnant women and young children in two-parent families. Before this time, nearly half of the states excluded pregnant women and children regardless of their income if they did not meet certain family composition requirements.

The Omnibus Budget Reconciliation Act 1986 (OBRA-86) permitted (did not mandate) states to cover pregnant women, infants (initially and as they grew to age 5 years), and the elderly and disabled with incomes less than the federal poverty level. Within a year, most states elected this option. The 1988 catastrophic care legislation mandated Medicaid coverage of pregnant women and infants in families with incomes under the federal poverty level. Other legislation allows states to cover pregnant women and children to the age of 1 year with incomes up to 185 per cent of the poverty level, which is the level for eligibility for WIC services (see below), and increases the age of optional eligibility for poor children to 8 years.

Local health department programmes

The role of local health departments has waxed and waned. In the early decades of the century, public health efforts to provide both environmental and direct maternal and child health services had a major impact in reducing infant and early childhood mortality. In the 1950s, population migration, particularly from southern rural areas to large municipal areas, found local health departments unprepared; as a result, the earlier declines in infant mortality ceased. During the 1960s, the domestic War on Poverty rekindled the commitment to the public's health in the form of community health clinics which were supported both by direct federal grants as well as state and local dollars. The wave of privatization and market-place competition in the early 1980s was accompanied by a declining role for local health departments in the provision of direct health services. Efforts are concentrated largely on environmental health matters and communicable disease control. All states now require children to have basic immunizations as a condition of entry to school and to organized day-care programmes. Some also require a physical examination for entry to school and periodically thereafter, and some jurisdictions provide school health services where children are not otherwise able to obtain them.

Neighbourhood health centres

Initiated in the mid-1960s as an effort to provide comprehensive health services for underserved populations, these centres have grown into a network of about 800 facilities serving nearly six million poor and underserved individuals in 50 states, Puerto Rico, and the District of Columbia. The centres are funded primarily through grants from the Federal Government through the programme for Community Health Centres, Migrant Health Centres, National Health Service Corps, Maternal and Child Health Block grant to States, and the Urban Indian Health Programme, although many also try to attract patients with Medicaid or private insurance. However, fewer than one-quarter of the country's 25 million medically underserved are reached by existing facilities.

Programmes for mothers and children

Maternal and child health services block grants

Although maternal and child health (MCH) services have been provided through federal grants for over five decades, the New Federalism policies in the early 1980s resulted in the formation of block grants that consolidated funding to the states. The programme contains grants for MCH services, Supplemental Security Income for Disabled Children (for children with chronic health problems), lead-based paint poisoning prevention programmes, sudden infant death programmes, haemophilia treatment centres, and adolescent pregnancy and genetics services grants. In most southern, midwestern, and western states that have large networks of county and local health departments, services are often provided directly through governmental agencies in schools, local health clinics, well child clinics, antenatal clinics, health screening, and immunization services. In other areas, particularly the northeast, the state MCH agency often contracts with private health centres to provide services.

Special education

In 1975, Congress passed the Education for the Handicapped Act (PL 94–142) to provide federal funding for state education departments to assure 'free public education' for handicapped children aged 3–21 years. More recent legislation (PL 99–457) extended the programme to children from birth to 3 years of age. Approximately 10 per cent of children, including those with deficits such as hearing problems, are served across the country but, as yet, there is little co-ordination between the educational services and health services for these children.

Supplemental programme for women, infants, and children (WIC)

This programme was initiated in 1972 to provide supplemental food and nutrition education to pregnant and postpartum women, nursing mothers, and children from birth to 5 years whose family incomes are below 185 per cent of the federal poverty level. The programme is administered by the Department of Agriculture through grants to state health departments.

Head Start

This programme resulted from legislation in the mid-1960s to improve the developmental level of low income pre-school children. It is administered by the Federal Administration for Children, Youth and Families, with support given directly to local Head Start groups.

Family planning

Support for family planning comes from a variety of public and private sources including planned parenthood funds, the MCH block grant, and several other sources deriving from federal legislation. Services are provided through a network of unco-ordinated centres located in health clinics, hospitals, and private physicians' offices.

Access to health services

In 1985, there was a total of 22.0 active physicians per 10 000 population in the US. Of these there were 20.7 non-federal physicians for the civilian population. Almost 90 per cent of these civilian physicians were involved in patient care, with the remainder in medical administration, teaching, research, and other activities. About one-third of all physicians and two-fifths of all office-based physicians are involved in primary care which, in the US, consists of family and general practitioners, internal medicine physicians, and paediatricians. (Note that some of the latter two types of physician actually restrict their practice to a sub-specialty of adult or paediatric medicine, so that the proportion of physicians actually practising primary care is lower than the figures given.) There were 56.9 dentists, 9.9 optometrists, 66.3 pharmacists, 4.2 chiropodists, and 641.4 registered nurses per 100 000 population. Of the non-federal physicians in patient care, 76.4 per cent were in office-based practice and the remainder practised in hospitals. Five of every six (83 per cent) generalist physicians in patient care practised outside of hospitals, as compared with 77 per cent of surgical specialists, 72 per cent of medical specialists (including internists and paediatricians), and 66 per cent of other specialists (such as anaesthesiologists, psychiatrists, and neurologists).

The proportion of physicians who are women will increase in the future, as women are entering medical schools in increasing numbers. In 1971–2, the proportion of women in the entering class of medical schools was 10.9 per cent; in 1977–8 it was 23.7 per cent, and in 1985–6 it was 32.5 per cent.

There were slightly more than 6000 short-term hospitals in the country with a total of 4.6 beds per 1000 population. About 8 per cent of beds were owned by the Federal Government; of the 92 per cent of non-Federal beds, 70 per cent were private, 'not for profit', 10 per cent were for profit, and 20 per cent were owned by state and local governments. Occupancy rates have been falling since the early 1970s, largely as a result of efforts to reduce costs of care by reducing the most expensive component of costs (in-patient care). Average occupancy rates in 1985 were 65.5 per cent, down from 75.7 per cent in 1970. The percentage of individuals hospitalized in a year has also fallen, from 9 per cent in 1982 to 7 per cent in 1986 (Freeman *et al.* 1987).

In 1986, the average number of contacts with physicians per person in the US was 5.3, with over half (55.5 per cent) taking place in a doctor's office, 15 per cent in a hospital outpatient department, and 13.2 per cent by telephone, with the remainder in out-of-hospital clinics, laboratories, or the home (NCHS 1986). Females had more contacts, on average, than males: 6.0 versus 4.6. The number of contacts per person was greater the lower the family income (6.6 for those in the lowest income groups as compared with 5.4 in the highest group), and the type of care differed. For those in the lowest income group, 44 per cent were in a doctor's office, 22 per cent were in a hospital clinic, and 11 per cent were by phone, as compared with 60, 12, and 14 per cent, respectively, for those in the highest income group. Over the last several years, an increasing proportion of contacts by individuals in the lowest income groups are occurring in hospital outpatient departments and other types of facilities, with a lower proportion in doctors' offices and by telephone.

Although the overall mean number of physician visits per person has decreased over the most recent several years, visits by individuals who are poor or near poor have declined even more than the average. The disadvantage has been even greater for those who are in fair or poor health; the gap in mean number of visits between the ill poor or near-poor and the non-poor widened from +12 per cent in 1982 to −27 per cent in 1986 (Freeman *et al.* 1987).

Another trend is the decreasing proportion of the population who have a regular source of care; in 1986, 18 per cent of the population (as compared with 11 per cent in 1982) reported no single usual source of care (Freeman *et al.* 1987).

As a result of a periodic survey of practice in physicians' offices (although not in hospital clinics), information is available on various characteristics of care, distributions of diagnoses, therapies, and dispositions. In 1985, 17.7 per cent of all visits of office-based physicians were first visits, 42.6 per cent took 10 minutes or less, and 58.8 per cent had a return visit scheduled.

How are physicians paid for the care they provide?

In 1986, almost half (48 per cent) of all physicians participated in at least one prepayment plan (HMO). Nearly one-fifth (18 per cent) of office-based physicians worked with three or more such organizations. However, only a small percentage of these doctors' visits were prepaid, even for physicians affiliated with HMOs. Physicians with HMO affiliations typically received 10 per cent of their practice earnings from prepaid plans, with the remainder from traditional fee-for-service. Paediatricians, obstetricians/gynaecologists, and family physicians had the highest proportion (12–13 per cent), surgeons (5–8 per cent, depending on type), internists and psychiatrists the lowest proportion (9 per cent) of earnings from prepayment (Owens 1987). Therefore, despite the dramatic increase in HMOs and prepayment plans, fee-for-service continues to be the primary way that physicians are paid in the US.

Arrangements by which physicians are paid by a variety of mechanisms are becoming the norm in the US. In 1987, about 40 per cent of all physicians who were engaged in patient care but not employed by the national Government had at least one contract with a preferred provider organization. This proportion has been rising: in 1983 it was about 11 per cent (Emmons 1988). Twenty-five per cent were members of an independent practice association, and 35 per cent had at least one contract with an HMO. About 6 per cent of physicians received at least 90 per cent of their income from an IPA or HMO; 12 per cent received none of their income from such a source. The remainder of physicians received varying proportions of their income from IPAs or HMOs, the plurality (31 per cent) receiving between 1 and 5 per cent of their income from these types of source. Physicians in the western region of the country, those just beginning in practice (under 35 years of age), and paediatricians were more likely to receive larger proportions of their income from IPAs or HMOs. Few physicians (less than 1 per cent) received most of their income from PPOs; about 12 per cent received none of their income from a PPO and about 45 per cent received 1–5 per cent of their income from a PPO, with little difference by geographical area, age, setting, or specialty (Emmons 1988). The most common combinations of sources of income, at least among paediatricians, were non-contract fee-for-service and contract fee-for-service (11 per cent) and non-contract fee-for-service and salary (8 per cent). None of the other combinations (contract fee-for-service, non-contract fee-for-service, salary, or some other source) characterized the income sources of more than 4 per cent of paediatricians (LeBailly 1985, pp. 16–19).

This general overview indicates the wide variety of types of arrangement for providing care and the diversity in sources of payment and coverage for various types of service. This extraordinary organizational and financial disarray of a health services system is virtually unique in the industrialized world. As a result of the disarray, people who need care often lack the information they need to decide where to obtain care and how to pay for it. Moreover, administrative costs of the system are much higher than is the case for more organized systems of care, even those that are still primarily in the private sector (such as Canada) (Himmelstein and Woolhandler 1986). Will more rationality be brought into the system, and costs lowered as a result? The answer depends upon whether or not the prevailing emphasis on 'market-place competition' can be tempered by systems more like the national health programmes of comparable nations elsewhere in the world.

Quality and costs of care

One of the problems resulting from the great fragmentation of care is the difficulty in assigning responsibility for the quality of care while controlling costs. In most other countries, the vast majority of the population has a regular source of care, that is, a particular doctor or place where care is centralized and co-ordinated. This source of care is known as a 'primary care provider', who is usually a physician. In the US, family physicians and general practitioners, general internists, and paediatricians are also known as 'primary care physicians', but in contrast to the situation in other countries, the majority of patients do not have to enrol on the panel of a physician or physician group, and the majority of physicians do not have a panel of persons for whom they explicitly maintain ongoing responsibility for general care and for referral to specialists when needed.

In the 1970s and 1980s, and largely as a result of rapidly increasing costs of care, both federal and state governments have initiated a variety of mechanisms to assign responsibility to physicians both for the costs of care and for its quality. In this section, the basis for the efforts and the most important of the mechanisms for controlling costs and quality are reviewed.

Physician payment reform

As of early 1989, Congress is actively considering physician payment reforms. Although these reforms will apply only to the Medicare programme, they may be adopted by other third party payers and managed-care organizations as well. Congress mandated the development of a payment system that more adequately reflected the degree of input of resources; this development was reported by Hsiao and colleagues in late 1988. The major impact of this work is likely to be increased payment levels for those services that are provided by primary care physicians, and reduced payment levels for physicians providing highly technological services.

Medical practice variations

A series of studies have provided clear evidence that rates of hospitalization and surgery vary widely from one area to another, even within areas as small as states; similar variability is apparent across larger areas, such as countries. Why should this be the case? In some cases, the differences can be accounted for by differences in population needs. For example, hospitalization rates for children with asthma vary according to the geographical area in which the children live; in poorer areas, where asthma is more severe, hospitalization rates are greater than in non-poor areas. Another possible reason concerns differences in the availability of a mechanism to pay for certain services; where a procedure is reimbursed by a third party payer, it is more likely to be performed. The main reason, however, appears to be physician uncertainty about the effectiveness of diagnostic and therapeutic modalities. Where there is lack of knowledge about effectiveness, or where there is knowledge but it is not widely disseminated, certain practice patterns become established in particular geographical areas and are reinforced by common practice within those areas. Most of the mechanisms to control costs and quality of care are based upon the recognition that there is considerable uncertainty about appropriate care, and that rates of performance that are consistently outside the range of common practice are at least questionable.

Although most of the documentation and research on medical practice variations has focused on hospitalization and in-patient procedures, the decreasing hospitalization rates and rapidly increasing out-patient costs per visit (which rose 88 per cent compared to 77 per cent in cost per hospitalization from 1981 to 1987) are likely to generate considerable interest in medical practice in the out-patient setting.

One way to reduce the variability in costs of care is to pay for care ahead of time, based upon the projected demand for services, so that the hospital or physician is at risk of financial loss if costs are greater than projected, an approach that has major implications for quality of care. Many of the following mechanisms to control costs and quality are based on this principle.

Prospective payment

Prospective payment differs from the traditional retrospective reimbursement in establishing a per-diem or per-case payment ahead of time based on the diagnosis, instead of after care has been provided. Such an approach has primarily been applied to hospitalizations. The earliest of the approaches to be used on a wide scale were those resulting from state cost review commissions or insurance agencies. In one form of prospective reimbursement, a state-wide formula is established to determine the percentage increase in hospital costs from one year to the next. The starting point is the hospital's costs in the previous year; the costs for the current year are those from the previous year, adjusted by the formula for projected increases in labour costs, supplies, technology, and other relatively fixed expenditures. In another type of state rate-setting, negotiations are conducted with each hospital individually.

A third form of prospective reimbursement further decouples the payment for care from actual costs incurred. In this form of reimbursement, which began in New Jersey in 1979, hospitals are paid according to their 'case-mix', on the assumption that costs should primarily be determined by the problems that are treated. The basis for this approach derives from research on 'diagnosis-related groups' (DRGs). In this approach, all diagnoses that were responsible for hospitalizations were grouped into about 450 categories that were relatively homogeneous with regard to clinical type and their lengths of stay. As costs of care are highly associated with the duration of hospital stay, grouping of diagnoses that are similar in their length of stay are presumably similar in their costs of care. In 1983, the Amendments to the Social Security Act mandated the use of DRGs as the basis for reimbursement of federally reimbursed care for Medicare. The law called for a gradual phasing in of the DRG system, so that within 4 years national average prices for each DRG would be the basis for reimbursement of all hospitals. Hospitals are paid according to their case-mix, that is, their mix of DRGs. A hospital's compensation is determined by the number of patients in each diagnostic category times the payment rate for the category summed over all diagnostic categories after adjusting for various other factors such as residency programmes and rural/urban locations. Provisions are made for outliers, that is, patients who have extraordinarily high costs because of unusual circumstances; it is expected that no more than 5 per cent of patients would be outliers. Provisions are also made for increases in costs due to inflation.

Although DRGs initially applied only to patients whose care was reimbursed by Medicare, some private insurers and state Medicaid programmes have adopted DRGs or similar systems in an attempt to reduce variability in expenditures across hospitals and to reduce overall increases in costs. By the end of 1986, at least ten Medicaid programmes, a Blue Shield programme and an unknown number of commercial insurers were using DRG-based reimbursement. In anticipation of possible extension of DRGs to children, a separate case-mix classification has been developed for the paediatric age group.

Furthermore, a prospective approach to care provided in ambulatory (out-patient) settings is likely to be adopted in an attempt to control the variability in out-patient physician practices. As noted earlier in this chapter, prospective payment is already a feature of prepaid group practice HMOs. Physicians in the most rapidly growing new forms of organization of services, the IPAs and the PPOs, however, are still paid a fee-for-service, albeit a negotiated fee. As price competition continues to increase among competing health care financing and delivery systems, the impetus towards greater cost containment will make case-mix adjusted ambulatory care payments a likely prospect.

When control of the costs of care involves procedures that reduce use of services, there is always the possibility that the reduction will include indicated as well as unnecessary services. For example, the DRG system of reimbursement provides powerful incentives to discharge patients too early because the hospital will be paid the same regardless of how long the patient stays. Thus, patients may be discharged before they should be and fail to recover adequately, perhaps even requiring readmission. There is also an incentive to admit patients who may not really require admission, because the hospital will be reimbursed for whatever DRG the patient is diagnosed as having. As a result of these concerns, there have been periodic attempts to develop procedures to monitor the quality of care. The Federal Government, which was first to implement cost-control on a large scale, was also the first to devise large-scale efforts to deal with the issue of quality of practice.

Peer review

In the Social Security Amendments of 1972, Congress established the Professional Standards Review Organizations (PSROs). The legislation mandated that professional organizations assure the quality of care provided in programmes that received federal funding. (These included the Medicare programme for the elderly, the Medicaid programme for the indigent, and the maternal and child health programmes.) These organizations were to be non-profit and comprised of physicians. They might be set up on a state level or be local in character; they were funded by grants. The PSRO programme was intended to use three mechanisms in pursuit of its mandate. The first and most well developed was utilization review wherein admissions to hospitals were reviewed for necessity and where a maximum length of stay was determined. If the admission lasted longer than the set length of stay, the responsible physician had to provide justification. The second function was medical care review, in which the PSROs were required to choose certain diagnoses and review medical records to determine whether the standards of care were met. Re-review after the institution of educational programmes to remedy deficits was required. The third function, which never was well developed, was the collection of data to compare care across institutions and to develop programmes to determine reasons for differences in the 'profiles of care' where they were found. Evaluations of the effectiveness of the PSROs generally concluded that the costs of the efforts were approximately equivalent to the dollars that were saved as a result of them.

Disenchantment with the programme, and its failure to reduce costs or capture the imagination of the medical profession, led to its demise and replacement, through the Tax Equity and Fiscal Responsibility Act 1982 (TEFRA), by the Utilization and Quality Control Peer Review Organizations (PROs). This legislation, prompted by concern that the DRG prospective reimbursement programme might compromise the quality of care provided to Medicare recipients, mandated that PROs review the validity of diagnostic information to make sure that the assigned DRG was correct, and to review the completeness, accuracy, and quality of care to assure that underservice was not occurring. The major difference between the PSRO and PRO programmes is the aegis under which they operate. PSROs were non-profit physician organizations funded by grants; PROs operate on contract (and hence under greater surveillance and control) with the Federal Government and can be for-profit and operated by non-physician organizations such as fiscal intermediaries and insurers if no satisfactory physician group applies. As of the time of writing, little is known of the operations of these agencies or their effectiveness in detecting inadequate patterns of care.

In 1986, Congress passed additional legislation to strengthen peer review by offering immunity from prosecution of peer review groups by physicians whose practices were found deficient. The legislation also mandated the establishment of a national data bank to contain information on the names of physicians who have had successful malpractice judgements against them and descriptions of the acts that led to the claims. The law also mandated the reporting to the data bank of the names of all physicians who have had their licences revoked or suspended for reasons related to professional incompetence, and the names of physicians who have had their clinical privileges suspended by hospitals, health maintenance organizations, or other professional groups. Hospitals must search the data base when physicians are considered for medical staff privileges. At the time of writing, the keeper of the data bank has not been decided. Leading contenders are the American Medical Association, which already maintains a data bank containing demographic information on each physician in the country, or some other organization in the private sector which successfully wins in a competitive bidding process.

Procedures for disclosure to the public of information about practice patterns has also undergone change in the past decade. In creating the PRO programme, Congress altered the ground rules under which PSROs had been operating, in the direction of increased disclosure. The new regulations, which were published in April 1986, permit PROs to disclose interpretations and generalizations concerning the quality of care in a particular institution (such as a hospital) so long as no individuals are identifiable. Increasingly, information about differences in practice patterns, and about differences in performance as reflected in mortality and morbidity rates, are being made available not only to third party payers but also to the actual and potential users of health services. Since 1986 the US Department of Health and Human Services has released lists of the nation's hospitals that have mortality rates that are significantly higher and lower than the national average. The Department cited the Federal Freedom of Information Act as the justification for releasing the list to the public. Although the rates involved only patients on the Medicare programme (those aged over 65 years), the action sets a precedent for public disclosure of information about the practices of health services facilities.

The future of medical practice

Medical practice in the US is likely to undergo radical change in the near future. Recent directions suggest an increasing corporatization of medicine, with the ownership of facilities and control of medical groups being assumed by large for-profit companies that contract with physicians and physician groups to provide care. Concern about expenditures for care are resulting in increasing scrutiny over medical practice and the imposition of various types of control over it. Both community-based and hospital-based physicians will be affected; assumption by corporations of ownership of hospitals, including teaching hospitals, is occurring at an increasing rate (Whiteis and Salmon 1987).

The trend towards 'managed care' has several implications for the practice of medicine. As employers (and especially large corporations) and insurance companies increasingly assume responsibility for 'managing' care, the practice of medicine will be increasingly a matter of renegotiated contracts. Managers, whose emphasis is likely to be on cost-control, will have a propensity to seek new sources of care that appear to be less costly; individuals insured under these 'managed care plans' may be subjected to periodic changes of physicians from whom they may seek care. These managed systems of care also often substitute non-physicians for physicians for certain types of services. Reductions in utilization will reduce the number of physicians that are required. For example, some estimates indicate that insured children in 'managed care' plans require 20 per cent fewer primary care paediatricians than would be the case under ordinary circumstances; adults in these plans require 50 per cent fewer primary care physicians than would otherwise be involved in providing care (Steinwachs *et al.* 1986). One advantage of managed plans, for individuals who are enrolled in them, however, is that they generally include preventive services, which have not been part of the benefits of most indemnity insurance plans.

Managed care systems provide another opportunity for physicians to participate in efforts to improve the accuracy and completeness of information systems in practice. Little is known about the extent to which commonly employed procedures actually achieve the purposes for which they are intended. In managed care systems, the imperative is towards cost-control. Clinicians can contribute in a major way towards developing techniques to measure, document, and assess effectiveness of health services.

But, as this chapter has shown, the revolution in organizational formats for providing care is gradually disenfranchising a growing segment of the population; those who are poor, unemployed, or working in marginal employment and their families are most at risk. There are many areas where medical care is deficient: most office-based physicians provide relatively little care to the socio-economically disadvantaged and to individuals who need care other than strictly 'medical care' (for example, psychosocial problems). Physicians, as individuals and through their organizations, need to find new ways to provide care to those who now lack access to it, to be more readily available to provide care for new problems as they arise, to increase the scope of services they provide, and to work towards reducing barriers to care and improving the health status of all of the country's population.

References

Budetti, P., Butler, J., and McManus, P. (1982). Federal health program reforms: Implications for child health care. *Milbank Memorial Fund Quarterly: Health and Society* **60**, 155.

Dalton, J.J. (1987). HMOs and PPOs: similarities and differences. *Topics in Health Care Financing* **13**, 8.

deLissovoy, G., Bice, T., Gabel, J., and Gelzer, H. (1987). Preferred provider organizations one year later. *Inquiry* **24**, 127.

Emmons, D.E. (1988). Changing dimensions of medical practice arrangements. *Medical Care Review* **45**, 101.

Freeman, H., Blendon, R., Aiken, L., Sudman, S., Mullinex, C., and Corey, C. (1987). Americans report on their access to care. *Health Affairs* **6**, 6.

Himmelstein, D., and Woolhandler, S. (1986). Cost without benefit: Administrative waste in U.S. health care. *New England Journal of Medicine* **314**, 441.

Hsiao, W., Braun, P., Yntema, D., and Becker, E. (1988). Estimating physicians' work for a resource-based relative-value scale. *New England Journal of Medicine* **319**, 835.

LeBailly, S.A. (1985). *A profile of the American Academy of Pediatrics' Fellows: Descriptive data from the AAP membership census*, Statistical Note No. 3. Division of Health Services Research and Information, American Academy of Pediatrics.

Logsdon, D., Rosen, M., Thaddeus, S., and Lazaro, C. (1987). Coverage of preventive services by preferred provider organizations. *Journal of Ambulatory Care Management* **10**, 25.

Lohr, K.N., Brook, R.H., Kamberg, C.J., *et al.* (1986). Use of medical care in the Rand health insurance experiment. Diagnosis and service-specific analyses in a randomized controlled trial. *Medical Care* **24** (Supplement 9), S1.

National Center for Health Statistics (NCHS), Moss, A.J., and Parsons, V.L. (1986). *Current estimate from the National Health Interview Survey, United States, 1985. Vital and health statistics*, Series 10, No. 160, DHHS Pub. No. 86–1588. Public Health Service, US Government Printing Office, Washington.

Owens, A. (1987). What's prepaid care worth to doctors? *Medical Economics* **March 2**, 202.

Starfield, B. (1985). *Effectiveness of medical care: Validating clinical wisdom*. Johns Hopkins University Press, Baltimore, Maryland.

Steinwachs, D., Weiner, J., Shapiro, S., Batalden, P., Coltin, K., and Wasserman, F. (1986). A comparison of the requirements for primary care physicians in HMOs with projections made by the GMENAC. *New England Journal of Medicine* **314**, 217.

Whiteis, D., and Salmon, J. (1987). The proprietarization of health care and the underdevelopment of the public sector. *International Journal of Health Services* **17**, 47.

21

Health care in Europe

C. SPREEUWENBERG and A.J.P. SCHRIJVERS

Introduction

Europe consists of countries with totally different cultural backgrounds, political systems, and economic situations. A number of its countries are members of supra-national organizations, such as the European Economic Community (EEC), the Council for Mutual Economic Assistance (COMECON), and the European Free Trade Association (EFTA). The main aim of those organizations is to improve the economic level of the member states by creating an internal market. These supra-national organizations are increasingly influencing health care systems throughout Europe.

The EEC intends to remove its internal borders at the end of the year 1992. This means a forcing into line of educational systems, the rights of medical professionals to practise, the rules regarding health care facilities, drugs, and medical devices, and quality control. After 1992, medical professionals qualified in one of the member states of the EEC will have, in order to practise in one of the other EEC countries, the same rights and duties as their colleagues qualified in that country.

A quarter of the 300 'European' directives which must be adapted to the new situation, affect the health care systems. A so-called Permanent Committee in which physicians of all 12 member states are represented, advises the European Council of Ministers and the European Committee, the executive board of the EEC. The Permanent Committee is concerned with all professional problems such as education, staffing, ethics, and structure and financing of the health care system.

In all European countries, the level of health has improved during the last century. Health indicators, such as life expectancy and infant mortality, are comparable for many European countries (Table 21.1). The improvement of the health care systems has required a rapid increase in health care expenditure. In comparison with the US, relatively more people are insured by social insurances or by a nationalized health care system.

The proportion of the gross national product (GNP) spent on health care is greater in western and northern European countries than in the UK (Table 21.2).

All governments are concerned about the increase of costs of their health care systems and question its efficiency. All countries face a demographic change as the population aged over 65 years increases disproportionately and the birth-rate stabilizes more or less. Many countries therefore face an aging of their population. Some demographic indicators are shown in Table 21.3.

Across Europe there is a great difference in the number of doctors and hospital beds per 10 000 population. Most governments have tried to reshape their health care system by promoting primary care. In the northern, eastern, and southern parts of Europe, governments have promoted community care by introducing health centres and group practices. Conversely, in western Europe one can observe a tendency to promote competition both between health care deliverers and medical professionals, and a desire to reduce the role of governments.

Table 21.1. Health indicators of some European countries

Country	Life expectancy at birth (years)			Infant mortality in 1986 (per 1000 live births)
	Year	Males	Females	
Austria	1986	71.0	77.8	10.3
Belgium	1984	70.8	77.8	9.7
Czechoslovakia	1985	67.3	74.8	14.0*
Denmark	1985	71.7	77.7	7.9
Federal Republic of Germany	1986	71.9	78.5	8.5
Finland	1986	70.6	78.9	5.9
France	1985	71.8	80.1	7.9
German Democratic Republic	1985	69.5	75.4	9.2
Greece	1985	73.5	78.5	12.3
Hungary	1986	65.3	73.3	19.0
Italy	1983	71.3	77.9	10.9
The Netherlands	1985	73.1	79.9	8.1
Norway	1985	72.6	79.6	8.5
Portugal	1986	70.2	77.1	15.9
Spain	1981	72.6	78.6	10.5
Sweden	1985	73.8	79.9	5.9
Switzerland	1986	73.8	80.8	6.8
UK	1985	71.9	77.6	9.4*
Yugoslavia	1983	67.1	73.0	27.3

Source: World Health Organization (1987).
* In 1985.

Table 21.2. Health care expenditure in selected countries in 1985

Country	Expenditure on health care	
	£ (million)	Percentage of GNP
Austria	451	6.7
Belgium	413	6.6
Denmark	577	7.0
Federal Republic of Germany	722	9.2
Finland	561	6.9
France	616	8.6
Ireland	272	7.7
Italy	332	6.9
The Netherlands	564	8.5
Sweden	796	8.8
Switzerland	843	7.3
UK	364	5.9
Japan	421	4.9
US	1388	10.5

Source: Office of Health Economics (1987).

In this chapter, we outline the health care systems in the different European regions. We will select one or more countries per region to illustrate each region.

Scandinavia

Introduction

The Scandinavian countries cover the northern parts of Europe. Sweden has eight million inhabitants, Norway has four million, and Denmark has five million, and each country is very thinly populated (Sweden has 18 inhabitants per square kilometre). Sweden and Norway are characterized by extensive rural and mountainous areas; in Sweden about 85 per cent of the population lives in towns. This, together with the great differences in climate between the northern and southern parts of the country, causes great problems in the infrastructure of the country. Large areas have been depopulated and the people who are left are relatively old. Despite this, all countries are economicaly well developed. Norway has a well-developed farming and fishing industry, although the exploitation of oil and gas has changed its traditional economic structure. Denmark is composed of a peninsula, and two relatively large and hundreds of small islands.

Politically, all countries have a long democratic tradition. Denmark is a member of the EEC.

Denmark

In 1973, Denmark instituted a system of national health care that was financed by taxation. All Danes who pay taxes are insured by the national health insurance which provides free access to the primary health care system, hospital care, and specialists.

At state level, the responsibility is essentially to initiate, co-ordinate, and supervise the health system. The Ministry of Health is the principal health authority and is responsible for legislation on health insurance, medical and non-medical personnel, hospitals, and pharmicists. The Government supervises the health care system, mainly by issuing general rules and economic guide-lines to areas that are formally under the control of the local authorities.

The regions are responsible for running and planning the major health care services, such as hospital services and primary care.

The districts (275) are responsible for running and planning most of the social services and some of the local health services, such as home nurses, infant health visitors, and school health and dental services.

The majority, about 93 per cent, of public health expenditure is met by the regions and districts, mainly through taxes and, to some extent, through block grants from the state.

The public health insurance reimburses part of the cost of dental care and physiotherapy. In 1985, total expenditure on health care was 35 829 000 D.Kr. and on disability was 13 420 000 D.Kr. During recent years, expenditure on health as a percentage of the GNP had tended to fall (7.3 per cent in 1983 compared with 6.3 per cent in 1986). Of total health expenditures, exclusive of care of the mentally handicapped, 63.2 per cent was spent on the hospital sector and 20.8 per cent on primary health care in 1986.

Citizens may choose between absolutely free medical care paid for by the public health insurance and a co-payment arrangement. Three to five per cent of citizens choose the latter method, which implies that they can consult a specialist without a referral from a general practitioner.

All hospitals, except the University Hospital in Copenhagen, are run by the regions in accordance with the Hospital Act. A list system operates, which means that patients must be referred by a general practitioner if they are to receive

Table 21.3. Demographic indicators for some European countries

Country	Size of population (million)*	Percentage over 65 years†	Live births (per 1000 population)‡
Austria	7.5	14.1	12.5
Belgium	9.9	13.4	12.5
Czechoslovakia	15.6	11.0	15.4
Denmark	5.1	14.9	10.7
Federal Republic of Germany	60.7	14.5	10.1
Finland	4.9	12.3	13.3
France	54.8	12.4	14.5
German Democratic Republic	16.8	13.3	14.4
Greece	9.9	13.1	14.9
Hungary	10.7	12.5	12.9
Italy	57.3	13.0	11.2
The Netherlands	14.6	11.8	12.5
Norway	4.2	15.5	12.5
Portugal	10.3	10.5	17.1
Spain	38.8	11.1	14.6
Sweden	8.3	16.9	11.3
Switzerland	6.4	14.0	11.5
UK	56.1	15.1	13.6
USSR	281.3	9.3	19.0
Yugoslavia	23.3	8.2	16.4

Source: World Health Organization (1987).
* In 1986.
† In 1985.
‡ In the period 1980–5.

specialist or hospital care free of charge. People who choose absolutely free medical care can change their general practitioner only once a year. The others have unrestricted choice. General practitioners in Copenhagen are not allowed to work in a health centre or as associates in a group practice and are paid on a capitation basis such that their income depends on the number of patients. In the rest of the country, 66 per cent of Danish general practitioners work together in a health centre or group practice; they are paid by a mixed system in which they receive a fixed fee per patient plus a fee for certain services. On average, general practitioners receive half their income from the capitation fees and half from other services. The idea is that this system stimulates general practitioners to refer fewer patients. The danger is the encouragement of work that is not strictly necessary. The number of contacts between general practitioner and each patient is high: an average of seven in 1986.

The three Danish universities have each integrated primary care into their medical curriculum. A chair in general practice has been established at two universities.

Interestingly, general practitioner training takes the same time as specialist training. The trainees must spend 4 years in hospital departments and 1 year in general practice.

Denmark has introduced a maximum working time of 30 hours a week for all physicians.

Heart diseases account for more than 33 per cent of all deaths. On average, every Dane is admitted to hospital once every 5 years and is treated in an outpatient clinic once every 18 months.

Sweden

Sweden has a long tradition of governmental responsibility for the health care system. All Swedish citizens and foreign residents are included in the public sickness insurance and are insured for medical care, which covers the so-called 'open care' or non-institutionalized extramural care. All hospital care is financed through taxation. Almost all adults are also insured for a cash benefit in the event of sickness.

Since 1965 regions have been responsible for health care provision in their district. There are less than 1000 private physicians in Sweden; since 1970 most physicians have worked as employees of the state. Patients have to pay for part of the health care they receive.

Hospital medicine dominates the health care system following a period of rapid expansion in the 1950s, and 1970s. Employment as a specialist became so attractive that it caused a great shortage of district doctors for the rural areas. This shortage was a stimulus to the development of the independent profession of district nurses with responsibility for primary care. A great discrepancy was seen between the well-equipped and up-to-date hospitals in the larger cities and the much less sophisticated hospitals in the rural areas. Only from the beginning of the 1970s was general practice again promoted in the Swedish system. Outside the hospitals, health centres were created. However, the problems are still that people can go to the hospitals without a referral from a general practitioner, that district nurses and general practitioners do not co-operate adequately, and that most general practitioners do not work outside office hours. Some regions have less than one general practitioner per 15 000 inhabitants; others have one per 1500 inhabitants (the average is 3100 inhabitants per physician). It is striking that the general practitioners' consultation time per patient is about half an hour and the result is that the public and the politicians are disappointed by the general practitioners' efforts to make the health care system more effective and less expensive. Sweden has more than 12 000 doctors, of whom more than 80 per cent are specialists.

Expenditure on health care peaked in 1982 at 9.7 per cent of GNP and then declined to 9.1 per cent of GNP in 1986.

Health For All and the Swedish health care system

Sweden is one of the countries with the best record of health care outcome. In 1986, life expectancy at birth was 76.9 years, 74.0 for men and 80.0 for women.

In 1985 the Swedish Parliament accepted a bill that recommended the publication of a Public Health Report monitoring the strategies for Health For All By The Year 2000 every 3 years. Strategic goals were formulated and progress will be evaluated.

In the first report (1988), it was mentioned that primary health centres were co-operating to a growing extent with the social welfare services in order to give old, long-term ill, and physically and/or mentally handicapped persons the potential to live on their own for as long as possible with supplementary help from the community in the form of home-help services, medical care at home, and day-care centres.

The numbers of places in old people's homes and beds in psychiatric hospitals have been significantly reduced during recent years. In 1985, 92.4 per cent of the population aged over 65 years and 76.6 per cent over 80 years lived outside institutions. Between 15 and 20 per cent of the people receiving home help are considered to be people in need of extensive care.

Special attention has been paid to programmes for education and for the dissemination of accurate, relevant, and clear information to promote healthy behaviour. Children and teenagers receive information about alcohol, narcotics, tobacco, and safe sexual behaviour. Industrial workers and other groups with higher than average mortality and morbidity rates are informed with the help of the trade unions. In 1985, a bill was proposed to reduce the intake of fat and sugar. The National Institute of Environmental Medicine has decided to give priority to environmental hazards with a project entitled 'Environmental medical warning based on biological indicators'.

Norway

Norway has a separate health care system for rural areas and for towns and larger places. In all rural districts, there are so-called district doctors with responsibility for public health and who function as primary care physicians and district nurses, working mostly independently from the district doctor. In the larger places, there are private physicians—general practitioners as well as specialists. People may choose their doctor freely, and there is no list system as in Denmark and the UK.

Patients need a referral to be treated in a hospital. This treatment is free of charge; all costs are refunded.

Finland

Introduction

In 1988, the population in Finland was 4.8 million, of whom 60 per cent lived in urban and 40 per cent in rural areas. An important feature of the country's political constitution is its long history of local government. Local government has the right to levy income tax and has the main responsibility for education, social welfare, public works, housing, town planning, and health services.

The Finnish health care system

Before the Second World War maternity and child health services were developed by voluntary bodies and local authorities. In 1945, these services were extended to cover the entire population regardless of place of residence or economic status. This can be seen as the first step in the development of present-day primary health care. In the 1950s and 1960s, this service received less attention and continued to be limited to the above-mentioned care, school health care, and some district nursing. Many factors contributed to the decision to implement a new system of primary care in Finland: a great shortage of doctors; an overemphasis on hospital care; the rapid growth of total expenditure on health; a standstill in most health indicators; and a perception that changes in health services could be achieved only by political intervention.

The new philosophy resulted in the Primary Health Care Act 1972. Its guiding principles were to stress community participation in local political processes; to include decisions about health care policies; to cover the entire population; to ensure comprehensive care with a wide variety of primary services; equitable geographical distribution of care; and stimulation of close contacts between patients and care providers.

Based on the Act of 1972, primary care is delivered by health centres run by local government. Such centres may include several buildings and service units located in different parts of the community or groups of communities. The minimum population served by a health centre has been fixed at 8000. This forced numerous communities to amalgamate for this specific purpose. There are now more than 200 health centres in Finland, of which about a half serve a single community. The centre's staff is responsible for different health care activities such as;

(1) promotion of health and prevention of diseases;
(2) primary medical services, including most common radiographic laboratory services;
(3) district nursing;
(4) mother and child care;
(5) school, student, and occupational health services;
(6) ambulance services;
(7) dental care;
(8) hospital services for those in need of long-term care.

The last-mentioned service is provided in so-called 'health centre wards', i.e. small local in-patient units. Depending on the financial position of a community, the state subsidizes investments and running costs of the health centres by 39–70 per cent. This provides a strong incentive for local authorities to allocate their local tax revenues to these services. Each community or group of communities develops its services by means of annually rotating five-year plans, which must be approved by the Government.

In the 1960s, an obligatory social sickness insurance was introduced, recovering a part of the medical services of health centres and private practitioners.

Local government also plays a major role in medical care and hospital services. Federations of communities were formed for the purpose of running one or more specific hospitals. Sometimes the federation borders for different tuberculosis, acute, and mental hospitals overlap one another. Hospitals are financed on an approximately fifty-fifty basis by the state and the local authorities.

Some statistical data

Traditional health indicators of the Finnish population are comparable with other European countries. Life expectancy at birth is 70.6 years for men and 78.9 years for women (Table 21.1). The perinatal mortality rate in 1981 was about 6.5 per 1000 life births. There are about 220 health centres in Finland. The number of doctors working in them is about 2000. During the 1970s the number of posts in these centres tripled, reaching 36 000 in 1980. Investments in their buildings are roughly equal to those in hospitals. About 5 per cent of all doctors work privately and up to 45 per cent of doctors in public hospitals have a private practice.

The number of beds in 1976 was 4.2 per cent per 1000 inhabitants in general hospitals, 4.1 in mental hospitals, and more than 2.2 in health centres. About 7 per cent of GNP is spent on health services, which is substantially lower than the 9 per cent spent in Sweden. This difference may be attributed to the greater influence of the Government on the financing system and the stronger development of primary health care in Finland than in Sweden.

Western Europe

Introduction

Many characteristics can be found across western Europe. With the exception of Switzerland and Austria, all countries are founder members of the EEC. All countries have a pluralistic health care system characterized by the co-operation of many protagonists with different responsibilities and at different levels to provide health services to the public.

In West Germany, the Netherlands, Belgium, Luxembourg, France, Switzerland, and Austria three main groups may be distinguished. The first group includes public authorities, such as national parliament, national government, ministries, regional governments, and municipalities. The most important are the regional governments: in West Germany called *Bundesländer*; in the Netherlands *Provincies*; in Belgium *Communités*; in France *Départements*; and in Switzerland *Cantons*.

The second group entails the health insurance agencies, both public and private. At the beginning of the century, social health insurance organizations or sick funds were founded on a non-profit basis by labour unions, doctors' organizations, and other non-governmental groups. In German-speaking countries they are called *Krankenkassen*, in the Dutch-speaking countries *Ziekenfonndsen*, and the French synonym is *Mutualitées*. Nowadays, the majority, or in some countries (Belgium, Luxenbourg) the whole, of the population is compulsorily covered by the social insurance agencies. The private health insurance companies work for a small proportion of the population or cover additional health services which are not insured by the social insurances.

The third group consists of health care providers, divided into 'own risk' working professionals (for instance, general practitioners, independent medical specialists, dentists, and physical therapists) and institutions, e.g. acute hospitals, nursing homes, and ambulatory care agencies. The health care systems of the Netherlands, the Federal Republic of Germany, and France are discussed as examples. Belgium, Luxembourg, Austria, and Switzerland have comparable systems, although size and density of the population may influence their respective organizational structures.

The Netherlands

Introduction

The Netherlands is one of the most densely populated countries in the world, with 14.5 million inhabitants on the 1 January 1986 including some 560 000 foreigners. The number of live births in 1986 was 177 000, or 12.2 per 1000 inhabitants. The infant mortality rate was 8.1 per 1000 live births. The life expectancy of newly born children is now 79 years for girls and 72 for boys. Politically, the Netherlands has a long democratic tradition. It is a member of the EEC.

The health care system

Health care in the Netherlands originated largely from private initiative, often on a charitable basis. Roman Catholic and Protestant hospitals, services, and nursing homes still exist today. However, such divisions have become considerably less marked and, generally speaking, health services no longer tend to be organized on a sectarian basis.

The history of health care—from charity to Poor Law, from the National Assistance Act to the Health Insurance Act and the Exceptional Medical Expenses Act—reflects the changing relationship between the Government and the private sector. Public health is part of the portfolio of the Minister for Welfare, Health, and Culture.

Administrative and executive powers in respect of health care are vested by legislation in both local and provincial authorities.

One of the features of the Dutch health care system is the strict separation between basic, primary care and secondary/tertiary care. The idea is that the various facilities within a particular sector may be more effectively co-ordinated in this way. People can only enter the services of hospital and specialists after being referred by their general practitioner.

The costs of the health care system have stabilized since 1982, at a level of about 8.7 per cent GNP.

Health care facilities

On 1 January 1986 there were 32 193 doctors, among whom 6189 were general practitioners and 11 206 were specialists, 7118 dentists, and 971 midwives.

The public health sector is responsible for individual and collective preventive measures. There are special doctors and nurses for monitoring the health status and applying preventive measures such as immunizations for infants and schoolchildren.

About 60 per cent of the general practitioners work single handed, and of the remaining 40 per cent, half work in group practices and the other half in health centres. Besides the general practitioners and their assistants, nurses of a 'cross organization' deliver primary care in the community; they are responsible for prevention and nursing of people at home. (A cross organization is an organization responsible for certain public health services such as nursing at home, child care, and other types of home care.)

General practitioners receive a fixed fee of about £30 per year per patient insured under the Health Insurance Act. The other patients (or their insurance company) pay a fee for each service. The services of the cross organizations are free for the whole population; the organization is paid under the Exceptional Medical Expenses Act.

The same Act entitles every inhabitant to free assistance (after referral by the general practitioner) from one of the 59 Institutes for Out-Patient Mental Health Care (RIAGGs). Each RIAGG has a catchment area of 150 000–300 000 inhabitants. These RIAGGs are organizations with their own management, independent from the state, church, or local authorities. They serve people of all ages.

In the Netherlands there are approximately 220 general hospitals with 70 000 beds, 82 psychiatric hospitals with 24 000 beds, 120 hospitals for mentally handicapped persons with 30 000 beds, and 330 nursing homes with 50 000 beds. Almost all specialists work in a hospital. People insured under the Health Insurance Act have to pay a certain amount for drugs and the services of specialists; for them the services of midwives, physiotherapists, and the hospital are free. Specialists, physiotherapists, pharmacists, and midwives are paid on a fee-for-service basis.

The state administers and runs six of the eight teaching (university) hospitals.

Health policy

Until 1987, the concepts of health policy were based on:

(1) regionalization, by which is meant the co-ordination of facilities to form a structural and functional whole, and the administration of the whole within a specific geographical area;

(2) separation of the basic, primary, and secondary health care sectors;

(3) a balanced system of charges and fees by means of the Health Care Charges Act;

(4) compulsory health insurance for people on low income.

Approximately 68 per cent of the Dutch population is insured against medical expenses under the Health Insurance Act. They are entitled to free services of a general practitioner, and care in a general or psychiatric hospital, and to most expenses of specialists, dentists, drugs, and nursing homes. The compulsory health insurance (sick fund) applies to employees whose income does not exceed an annually determined limit. The premium in 1986 was 9.6 per cent of the insured person's wage up to a fixed limit. The employer and the insured person each pay half of the premium. This insurance also applies to those persons aged under 65 years who are receiving benefits and allowances, e.g. disablement and unemployment benefit, or national assistance. A person who is not insured under the health insurance scheme may take out private insurance with an insurance company. Persons insured by such a company for the first time must be accepted if they opt for the standard type of coverage. The companies are not allowed to charge age-related premium supplements if the policy-holder has been insured against medical expenses for a minimum of six months immediately prior to taking out the policy.

Everyone, regardless of income, is insured under the Exceptional Medical Expenses Act for expenses incurred due to long-term illness or serious disability which cannot be borne by the patient personally. The RIAGGs and care in each hospital and institutions for the mentally handicapped for longer than 365 days are included.

In 1987/8 the Government adopted a report of a committee chaired by the former executive director of Philips, Dekker, in which the concepts have changed dramatically:

1. Regionalization and separation of different sectors as a leading principle have been replaced by functional units in which different sectors must co-operate.
2. Market mechanisms will be introduced by the stimulation of competion in price and quality between the different health care facilities.
3. A mixed insurance system will be introduced in which essential medical facilities will be available for all inhabitants; it is assumed that insurance companies will insure for other medical risks, and the sick funds will disappear.
4. The price of health care facilities will be determined by negotiations between insurance companies and health care deliverers.
5. The influence of provincial and local authorities will diminish strongly.

The Government expects that the introduction of these market elements will reduce the costs of the health care system and will increase efficiency and quality.

Some special problems

Outside the common problems of all European countries, such as demographic changes and the costs of the health care system, attention has been paid internationally to the liberal attitude of the Dutch government towards euthanasia and the use of drugs. Under strictly defined conditions it is tolerated that a doctor may terminate the life of a severely ill patient.

Another central objective of Dutch Government policy has been the prevention of the risks to drug addicts, to their immediate environment, and the prevention of risks to society as a whole that arise from drugs. The basic aim is not to combat drug use itself or to prosecute the drug users, but to reduce the risks. Improving social and physical functioning takes a higher priority than curing the addiction as such.

Germany

Some features of the health care system

Within the Federal Government, two ministries have regulatory power over the health services. The Federal Department of Labour and Social Affairs (*Bundesministerium für Arbeit und Sozialordnung*) is responsible for the regulation and the preparation of laws concerning the social health insurances and their agencies. The Federal Department for Youth, Family, and Health (*Bundesministerium für Jugend, Familie und Gesundheit*) is the regulatory body for education of health care providers, for public professional jurisdiction, for prevention, and for quality control of medical drugs. The 11 regional governments (*Bundesländer*) have, under different names, their own ministries for health affairs. They finance the hospital building investments directly, which are, in this way, excluded from the costs of social insurances. Other responsibilities of the *Bundesländer* are the planning of hospital beds, running of public health agencies, protection of the physical environment, labour inspection, and food and water control.

In The Federal Republic of Germany, about 1300 sick funds exist, which cover about 90 per cent of the population and fund more than 50 per cent of total health care expenditure. Some social health insurance agencies cover a special geographical area, and others cover a special branch of medicine or a special type of worker. The insurance premiums and policies are comparable for all sick funds. Of the premium, 50 per cent is paid by the employer and the other half by the employee. Sick funds are allowed to cover additional services for which an extra premium may be asked of their members.

The health care providers work as independent contractors (doctors outside hospitals, dentists) or as salaried employees of, mostly, private, non-profit organizations, often with a religious background. The doctors who work for sick funds belong to special unions (*Kassenärztlichen Gesellschaften*), which are geographically organized and have a legal status. These unions negotiate with the sick funds about charges and other working conditions.

The Federal Republic of West Germany developed, in 1977, a planning strategy with which to integrate these three autonomous groups—the Government, sick funds, and health care providers—through co-operation and consultation on a voluntary basis. The German name for this strategy is *Konzertierte Aktion im Gesundheitswesen*. The role of the Government is to defend the general interests of the population and the national economy and to stimulate the two other parties (sick funds and health care providers) to co-operate. Because of the interests of the providers, the main topics of the German health policy are related to charges and other financial affairs.

An important difference from the UK, the Netherlands, and Denmark is that both general practitioners and specialists may deliver primary care. It is possible for a patient to consult a specialist without referral by a general practitioner. This situation is reflected in the education system. There are only a few chairs in general practice and the contribution of general practice to the basic medical education is much less than in the other European countries.

Activities and expenditures

The Federal Republic of Germany spent, in 1986, 12 per cent of its GNP on health care. In absolute terms, the expenditure was 9.5 million DM in 1960, 25.2 million DM in 1970, and 108.7 million DM in 1984. Most of the financial resources were spent on hospitals. The number of hospital admissions increased from 9.7 million in 1971 to 11.6 million in 1982, while the average length of stay decreased over the same period from 24.3 to 19.4 days. The number of posts in hospitals grew from approximately 525 000 to 700 000 during this period. Each German person insured under the social insurance acts is allowed to go immediately to a medical specialist without first consulting a general practitioner. Nevertheless, many Germans with a health complaint visit a general practitioner, who may work alone, in a group practice, or in a health centre (*Sozialstation*) with other primary health care providers.

The German planning strategy described above has not had much influence on the growth of health care expenditure. This is why, in 1981, the Federal Parliament passed two new laws concerning cost control. Later, in the 1980s, the introduction of more competition and for-profit organizations in the health system was debated as a better cost-control mechanism than further regulation. But the last word on this has not yet been spoken. At present, other health concerns are care of the elderly, promotion of primary health care, the epidemiology of acquired immune deficiency syndrome, and closure of acute hospitals.

France

The French health care system

In the French health care system the private and public sectors co-exist. Most of the costs of both sectors are covered by a public health insurance. The central government has a role in health care planning, financial regulation, education, and quality assurance. The three main parties in French health services are the same as in the Federal Republic of Germany: health care providers, public insurance agencies, and the Government.

As in the Federal Republic of Germany, primary health care is delivered by general practitioners (*omnipracticiens*) and specialists. Almost all of them work as independent contractors on a fee-for-service basis, both for patients insured by public insurance and for private patients. Most of the doctors work independently. Each citizen has the right to consult a medical specialist without referral by a family doctor. In 1982, there was one general practitioner per 1140 inhabitants.

Other primary health services, such as district nursing, are provided by non-profit organizations, partly financed by local government and partly by public insurances. Physiotherapists, pharmacists, dentists, and other primary health care professionals have a comparable position with that of the general practitioners: independent and paid on a fee-for-service basis.

The acute, mental, and long-stay hospitals, with a total of 600 000 beds (11 beds per 1000 population) are divided between the public and private sectors. The former houses approximately 70 per cent of the beds. Most larger and all teaching hospitals are public institutions. Their boards are mostly appointed by local or regional governments who try to create multi-unit systems. The *assistance publique*, *Hôpitaux de Paris*, for instance, encompasses all public hospitals in Paris. The private sector owns about 30 per cent of the hospital beds. Of these, the non-profit (mostly Roman Catholic) organizations, with roughly 12 per cent of all beds, are comparable with the public ones. The for-profit hospitals, with approximately 18 per cent of all beds, are mostly owned by a group of doctors and offer fewer specialties. Many of them are found in the richer parts of the country: Paris and the Côte d'Azur. The doctors in the private hospitals are paid on a fee-for-service basis; their colleagues in public institutions are salaried. Public insurance patients are allowed to visit private clinics, but sometimes their maximal reimbursement is lower than the charges of a private hospital.

The compulsory public insurance financed, in 1983, 72 per cent of the French health care costs, the consumer paid 20 per cent, and state agencies 5 per cent. About 70 per cent of the population subscribes voluntarily to a complementary health insurance, which covers much of the consumers' contribution (23 per cent). Not insured by the compulsory public insurance are 30 per cent of the costs of ambulatory medical consultation, 30–60 per cent of the costs of prescribed drugs, and about 11 per cent of the costs of hospital admission. The Social Security Department, which runs the public insurances, is a non-governmental organization and is managed by employers' representatives and the major trade unions. The resources of the Social Security Department are derived from compulsory public contributions which are a percentage of the revenue or salary of the employer (12.6 per cent in 1984) and the employee (5.5 in 1984) respectively. Only these percentages and not the total amount of public expenditures, are discussed by Parliament. In 1984, the funding system of public hospitals was completely changed. Until that year, these hospitals received income per admission day. Since 1984, the public clinics have received an operating budget, allocated at the beginning of the year and paid in 12 monthly instalments. This system had been further improved: presently, the budget depends on the number of admissions that are diagnosis related according to the Diagnosis Related Groups Systems of the American Medicare Program. The private hospitals still receive their income per admission day. In their prices for a day's admission the cost of the hospital stay, the medical fees, and the drugs are included.

The state plays a minor role in France. The government gives a guideline as to the yearly growth of the national budget for the public hospitals, which, in recent years, has been less than the rate of inflation. For hospital investments, authorization from the Secretary of State for Health and from the Minister of Social Affairs is needed. This permission is given only when the investment is in accordance with the health map of the regional public authority (*le Département*) and with the national plan to close 16 000 acute hospital beds. The state finances 40 per cent of hospital investments.

Middle and Eastern European countries

Introduction

Europe has been divided by the iron curtain. Most countries were communist and their health care systems have been influenced to a large extent by the political system.

The health care system

In Eastern European countries, the health care systems are nationalized. This means that the organization and running of the health care system, including research, is seen as a duty of the state. Citizens of these countries are insured for all health care facilities by public insurance, paid for by the state. The communist governments earmarked the health care system as an important instrument with which to influence the welfare of the entire population.

A characteristic feature of the systems in these countries is the fusion of health care facilities into regions and a hierarchy of interlinking institutions of various categories, types, and levels, from specialized centre up to health district at a higher level. The health care of workers is often delivered by facilities attached to a factory, farm, or other workplace. Czechoslovakia, for instance, is divided into more than 5000 medical districts, in addition to which health care is delivered by about 2400 occupational (works) doctors. The District and Regional Institutes of National Health are the principal organizational units in which all health facilities of a certain area are included. These Institutes are always administered by the respective National Committee body which is elected and controlled by the people. This means that citizens have the potential and right to participate in the running and control of the health service.

Also typical for a country such as Czechoslovakia is the extensive use of medicinal and hot springs. The Czech Republic makes use of 150 hot springs in 35 spas at which there are 93 health institutions with nearly 24 000 beds, whereas the Slovak Republic counts 1160 registered springs, 23 spas, and two hot spring companies. In 1980, 448 000 people underwent a treatment in spas. Roughly 379 000 patients received it free of charge, whereas 42 000 foreigners and 27 000 citizens of Czechoslovakia payed for it themselves. A doctor can prescribe a spa treatment as a medicament.

Education

In all Eastern European countries, primary care is well developed. Continuous medical education is compulsory for all practitioners even after qualification. In Hungary, there are departments of general practice at all four universities which give a clinical training. After 24 months of vocational training doctors can be registered as general practitioners.

In the German Democratic Republic there is less involvement of general practitioners in the basic medical curriculum. On the other hand there is compulsory vocational training of from 48 to 60 months.

In Czechoslovakia, general practitioners are involved in the medical curriculum at three universities. Vocational training is compulsory for 30 months with no theoretical teaching.

Extension of health care facilities

While governments in other parts of Europe try to stabilize the number of doctors and to reduce the number of hospital beds, the governments in the Eastern European countries try to extend health care facilities, particularly those of primary care. In Hungary, Czechoslovakia and the German Democratic Republic, the number of general practitioners has increased more than in the western countries over recent years (53 per cent between 1970 and 1987 in the GDR), and a vocational training with which a doctor can qualify as a specialist in general practice has been established.

In most Eastern European countries, much attention is paid to prevention and immunization. In the GDR, with these measures, measles decreased from an incidence of 1.71 per 1000 population in 1970 to an incidence of 0.01 per 1000 in 1987, and hepatitis from 1.09 to 0.16 per 1000 over the same period. Trends similar to those of other countries can be demonstrated for occupied hospital beds (decreasing), admissions to hospital (increasing), costs (rising), perinatal and infant mortality (decreasing) and life expectancy (increasing).

Comparison of Eastern European countries

It is hard to compare the Eastern European countries. Although they all have a nationalized health care system, past influences are still recognizable.

There is significent contact between doctors in Hungary, Czechoslovakia, and the German Democratic Republic and doctors in western European countries. Although the former have problems with the supply of instruments and drugs, they are oriented to the western (German) way of practising medicine. Life expectancy and infant mortality rates in these countries are comparable with those in western European countries. In Hungary, there is a booming market for unofficial private health care. Future changes in the economic laws will surely influence the health care system. The other Eastern European countries are more isolated. Cultural and language differences play an important role.

Recently, Michael Ryan (1988) wrote a comment on recent developments in the Soviet Union. He demonstrated that the average life expectancy at birth had improved substantially from 46.9 years in 1938–9 to 67.0 years in 1955–6. But, during the 1970s, contrary to the experience of most countries, this key indicator showed a decline, from 69.5 years in 1971–2 to 67.9 years in 1978–9. During that period, the gap between men and women also widened to the extent that, on average, men's lives were shorter by as much as 10.1 years. Recently, however, the picture is more favourable. In particular, life expectancy rates for men improved from 62.5 years in 1978–9 to 65.0 years in 1986, so that, with a life expectancy for women of 73.6 years, the overall life expectancy for the population in 1986 was 69.6 years. Typically, for such a large country, infant mortality varies enormously within the different Soviet Republics.

Leading cadres are clearly aware of the position of Soviet health status and they recognize its seriousness. For example, Fedorov, a famous eye surgeon, posed the question of why it is that the average life expectancy still ranks thirty-fifth in the world (Fedorov 1987). In 1987, the Soviet leaders published a document (Ts K KPSS 1987) which sets out a programme of truly radical changes in health care. One passage summarizes the main problems: 'More than two-thirds of the population are not involved in the systematic pursuit of physical culture and sport, up to 30 per cent are overweight, and about 70 million persons (a quarter of the population) smoke. Drunkenness and alcoholism are widespread, and the number of persons who have recourse to narcotics is increasing'. An interesting question is if, and how, *perestroika* will influence Eastern European health care and status.

The Iberian Peninsula

Introduction

Spain and Portugal are vast, mountainous low-density countries compared with the rest of Europe. Together, both countries take up approximately 13 per cent of Europe's surface area. The population is concentrated in urban zones. About a quarter of the population is 14 years old or younger, while about 11–12 per cent is aged over 65 years. During the 1980s, both the birth-rate and the marriage rate dropped sharply. Economically speaking, both countries have the lowest per caput incomes of Europe, as well as the highest unemployment.

In the 1980s both Iberian countries changed their political system radically, and their health care systems have undergone various reforms during this century. As in other countries, the health care systems have shown technical improvement, specialization, and increase in expenditure.

Portugal

Some features of the health care system

Nowadays, Portugal has a national health service financed by the state. There are private practices and private health services, but they play a subsidiary role. Public expenditure on health care facilities in 1986, around 70 per cent of all expenditure on health care was 3.8 per cent of GNP. Total health expenditure was 5.7 per cent of GNP; this is comparable with the UK. In 1986, the health care system covered almost 80 per cent of the population.

The emphasis of the Government lies with stimulating primary care. General practice, or family medicine, was until recently very poorly developed in Portugal. Primary health care was delivered by nurses. The Government exerted itself to set up health centres and to use the health centre as the basic functional unit of the health care system. It promoted the implementation of general practice as a discipline. With the help of enthusiastic young doctors and international help, a college of general practitioners was founded in a short time, a system for vocational training was built up with a curriculum comparable with that operating in the UK, and postgraduate courses were organized. Nowadays, there are 335 health centres, 1802 ambulatory units attached to health centres, and, in the rural areas, some community hospitals attached to the health centres, with a total of 4294 beds.

New developments

Until recently, it was difficult to identify and develop the market components of the Portuguese health care system. However, the health policy-makers have become increasingly receptive to the introduction of market elements. The present Government programme explicitly states that the Government will take initiatives to change the National Health System Law, especially to find feasible alternatives to that Service, including private health insurance. Private, non-profit institutions, such as the organization *Misericórdias*, are being encouraged to promote health activities, to manage community hospitals, and, at least in one case, to regain control of central, specialized hospitals. In the future, measures necessary to increase market participation will be discussed again.

Spain

Some features of the health care system

Spain has a long tradition of a centrally organized public health care system. In 1967, the government decided to reform the public health care system to make it more comprehensive—*Assistencia Sanitaria de la Seguridad Social*. The population covered increased from 53 per cent in 1967 to 84 per cent in 1973, and the new system gained control of 23 per cent of the beds. In 1973, the system was reorganized into the National Health System INSALUD. By 1982, this covered 86 per cent of the population and maintained 26 per cent of hospital beds. This social security insurance is compulsory for all employees as well as the self-employed. The insurance is paid by insured persons, employers, and, for a small part, the state tax revenues. On the other hand, the private sector covers 30 per cent of all health care expenditure. A recent law sought to achieve the total protection of the population by transforming the health care system into a truly national service. A diagram of the present Spanish health care system is shown in Figure 21.1. At present, the Spanish health service system can be characterized as a public service, side by side with a sizeable private sector. There is a tendency to decentralize health care resources. Some say that the Spanish health care system follows a precarious pattern with an excessive number of physicians and a general lack of hospital beds. Compared with the British NHS, Spain has less than half the number of people per physician, which is also less than the ratio in Italy. Existing hospital beds fail to cover population needs adequately because a significant proportion is devoted to emergency services and to chronic and mentally ill patients. Patients stay in the hospital for a long time (average 14.6 days). The average number of hospital days per person per year is 1.3. In Spain, there are more than 55 000 physicians (21 000 general practitioners), 18 000 pharmacists, 3000 dentists, 31 000 district nurses, and 4000 midwives. Many physicians divide their time between the public and private sectors.

Patients can choose a general practitioner and can be enrolled in his or her list. They cannot choose specialists themselves. Physicians are paid by the government according to the size of their list and not by the number and type of service performed. Professionally, primary care and general practice are less well developed than in Portugal. The process

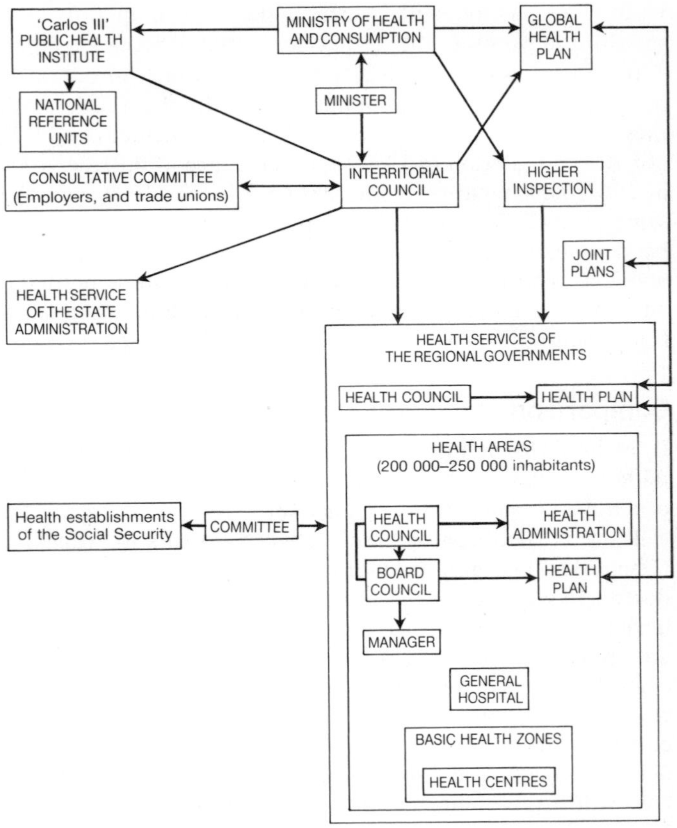

Fig. 21.1. Diagram of the Spanish national health system. (Source: Ley 14/1986.)

of establishing a vocational training has lagged behind. The fact that the EEC has ordered that each member state must have a compulsory training scheme for general practitioners of at least two years in length by 1990, will accelerate this process.

Medical education in the peninsula

In Spain, both medical education and medical expertise have become more international. There are strong links with the US. An important action was the creation of a system in 1979, that allowed a reduction of the number of medical students from almost 80 000 in 1979 to 60 000 three years later. Almost half of all medical students are women. Portuguese medical schools are more comparable with those in the UK. An interesting fact is that the deans of the Portuguese medical schools participated in a workshop with other European and Portuguese educationalists and general practitioners on ways to meet the requirements of the World Health Organization's Alma Ata declaration.

Some health indicators

Despite the reported ineffectiveness of the Spanish health care system, the important health indicators show figures comparable with those of Northern European countries (Table 21.1). In the past, the situation in Portugal was poor, but the health status of that population is improving rapidly.

Inequality in health

Spain, as well as Portugal, has large social and health inequalities. There is some criticism that the Spanish system, INSALUD, generates new health imbalances, rather than eliminating existing ones. Certain social class differences are worth mentioning. The upper classes seek health care at private hospitals at a rate seven times greater than that of the lower classes, but they go to private physicians only three times more frequently than the latter. In part, these differences may be explained by the fact that a private physician is relatively inexpensive compared with a bed in a private hospital. The actual difference, however, is more likely to stem from the upper classes in Spain knowing how to utilize the best of both private and public facilities.

The problem is not a discrepancy in access to health care resources by social class, but the poor quality and low utilization of those resources. Spain's middle and upper classes seek physician consultations and are admitted to hospitals more frequently than the lower classes. Surveys reveal that most Spaniards (64 per cent) believe private hospitals to be the best (20 per cent prefer public hospitals). Physicians, however, do not agree with public opinion; only 33 per cent of them feel that private hospitals provide better health care.

In Portugal and Spain, regional differences in economic status and health care facilities lead to health discrepancies.

The South-East European region

Introduction

Italy is a founder member of the EEC; Greece entered the EEC only recently. Italy has a major land mass and only few islands; Greece is a mountainous country with many isolated islands. Italy is very densely populated with 57 million inhabitants, while Greece has only 10.5 million inhabitants.

Italy

The health care system

In Italy, the economy in the north is much stronger than in the south. This results in greater unemployment and a lower level of health parameters in the south.

In 1978, the so-called Reform Act 833 introduced a national health service which replaced all former structures and regulations. After a debate lasting for ten years, a regionalized system was selected, with public finances and planning, as well as private, non-profit health care providers. The system includes health protection, disease prevention, primary care, hospital care, and rehabilitation facilities. Local health authorities, that are comparable with the British district health authorities and appointed by municipalities, manage these services for a specific area. They are also responsible for health inspections of the physical environment and of farms and factories. Regional health authorities, appointed by regional governments, operate as a link between the local health authorities and the state. They have a legislative autonomy within limits fixed by national laws and, in this way, manage the supply structure within which the local bodies administer the service itself. The Reform Act 1978 defines the health service, leaving the municipalities to deal directly with social services.

The public service is funded from taxation via a National Health Fund. Government and Parliament decide the size of this fund yearly by fixing a percentage of the GNP as the permissible limit of expenditure. Co-payments with patients exist for certain services. The National Health Fund is allocated among the regions by the Ministry of Health, taking into account parameters that aim to reduce the financial inequality between regions and guarantee a minimum level of predetermined care for all citizens (particularly important for the poorer regions in southern Italy). Other allocation criteria are the size of the population and the level of past expenditure. The aggregated financial deficits of previous years are weighting factor for resource allocations to regions: thrifty regions are punished and overspending ones are rewarded. Cost control is not as good as it could be. National decisions are also made about the functioning of the planning and programming system. The regions allocate resources according to previously approved local plans. Some allocations are non-specific; others are earmarked, for instance, for specific services, for research, or for specific projects such as innovations in care delivery.

The Reform Act recognizes the role of private institutions with their long tradition of caring for the sick and old. There are two categories of such organizatons: the religious ones, mostly Catholic, and the for-profit enterprises. With certain conditions, the local health authorities are allowed to buy services from the latter institutions. In this way, some competition exists within the national health service.

The organizational framework within a region and within a local health authority is not outlined and prescribed by the Reform Law, nor is the time fixed in which new frameworks must be established. This is why the building of new frameworks depends on the influence of local and regional politicians and the activities of pressure groups. In 1988, ten years after the introduction of the Reform Law, many old structures still exist at local and regional level and are in a survival struggle with the new authorities. This creates a weak, unclear, and unsure management system. So, health professionals find themselves in organizations with unfavourable conditions for carrying out their jobs, and with inadequate and too many guide-lines.

The debate before 1978 did not focus only on structural reforms. Another main issue was the social reintegration of chronically ill or handicapped people: bedridden patients at home, drugs addicts, handicapped persons, and elderly and psychiatric patients. Well-known in Europe are the Italian efforts to reintegrate psychiatric patients into the community by closing the big mental hospitals and by setting up community based, ambulatory, psychiatric care units. This attempt was not too successful, amongst other reasons because health and social services, as stated above, are not controlled by the same law. The organizational reforms required too much attention and energy from local and regional policy-makers to leave much over for the reintegration of policies and innovation of care delivery.

Some statistical data

Most of Italian health expenditure is publicly financed (87 per cent) with the remaining 13 per cent coming from patient co-payments for some public services and private health insurances. Approximately 6 per cent of the adult population is privately insured. In 1986, 5.2 per cent of GNP was spent on health services, a percentage that had been constant since 1978. The richer regions around Milan and Rome spend, per caput, more than the national average, while the poor areas spend below the mean value. In 1986, 58.5 per cent of all health expenditure was spent on hospitals (48.4 per cent on public hospitals and 10.2 per cent on private ones) and 6.3 per cent on primary health care.

In 1985, the total number of beds in acute, long-stay and mental hospitals was 450 000 or 7.8 per 1000 population. The number of general practitioners was 1.12 per 1000 population. In 1985, the national health service employed 542 000 persons of whom 59 per cent worked as care providers, 30 per cent had technical/professional positions, and 11 per cent performed management roles.

Today, Italian health concerns are the redistribution of care between the south and north of the country, the dissatisfaction of professional health workers because of the unsure local and regional structures and because of underpayment, the lack of nursing staff, and the surplus of doctors.

Greece

The health care system

Greece has experienced badly the detriments of the European wars and conflicts. These events, obviously, have slowed the country's efforts to reform social and health care organization.

After the Civil War in 1953, the Greek health care system was regionally organized and there was a sharp division between public and private services. A progressive law in 1953 did not give the expected impetus to the organization and subsequent developments of the health services in Greece. The public expenditure has always been relatively low (in 1982 it was 4.1 per cent of GNP) while the private sector was extensive and well developed (3.5 per cent in 1982). So the total expenditure in 1982 was 7.6 per cent of GNP.

Health care is financed by three sources:

1. A direct public provision by the Ministry of Health subsidizes most hospital services as well as the regional services for health and hygiene.
2. Health insurance funds provide hospital and ambulatory care. It is estimated that 99 per cent of the population is insured for primary and secondary (hospital) care. Employers and employees each pay a contribution to the fund. There are very large insurance organizations (e.g. the Rural Insurance Organization and the Social Security Organization) and a multitude of small insurance funds. Of the smaller funds, only a few are able to provide high standards of medical services. The rest, including the two largest, are considered to provide a low quality of services. The provision of services by insurance funds is arranged mainly by contracting-out hospital and primary care to private physicians on a fee-for-service basis and through fixed payments per patient to public hospitals.
3. Practices in the private sector are unregulated and difficult to assess. It is estimated that the private sector consumes more than 40 per cent of the total health care expenditure. It plays an important role in the development of the health services.

Up until 1983, almost all doctors worked in both the public and private sectors.

Recent developments

In 1981, the Socialist Party (PASOK) came into power. In 1983, a law for the development of a national health system was passed.

The system is organized on the basis of the following principles:

(1) primary care forms the centre of planning and medical practice;

(2) decentralization and regionalization of all health services;

(3) community participation at all levels of society: central, regional, and local;

(4) creation of a common fund for the financing of health care, which ensures a uniformity in the financing and allocation of resources to the services.

Important features of the service are:

1. The creation of a Central Health Council which sets out the national strategies and priorities for health care and advises the Government on health planning and development. The Council is composed of health professionals, lay members nominated by local authorities, and trade unions. Regional Health Councils will be formed along the same pattern.
2. Full-time and exclusive employment for doctors. The removal of multiple employments is expected to eliminate conflicts of interests.
3. The introduction and development of primary health care, with the creation of health centres in rural and urban areas, and the promotion of general practice and community medicine. Geographical accessibility to and equitable distribution of health centres will bring primary health care to all parts of Greece.

4. The training and development of medical and nursing personnel.

Under the National Health Law the creation of new hospitals and medical enterprises is strictly forbidden. However, the Public Insurance Organization and health insurance funds are allowed to contract-out hospital and ambulatory services to the private sector.

Staffing of the health services

In 1986, Greece had 30 196 doctors, 6000 trained nurses, 2000 midwives, 5000 ward assistants, and 10 000 partly trained nurses. The problem is that doctors are unevenly distributed over the country. Most of the doctors work in Athens and Saloniki (there is one doctor per 257 inhabitants in Athens).

While the number of doctors is relatively high, amongst European countries Greece has the lowest ratio with respect to nurses (only 1 nurse per 1898 population). Also, there is a great shortage of qualified administrative and managerial staff.

Public–Private mix

The problem of the Greek health care system was the uneven growth of the public and private sectors. The free market practices deregulated the health care system and great inequalities in health care, both social and geographical, were manifested. The creation of the national health service in 1983 has curtailed the growth of the private sector. However, the private clinics, the medical diagnostic centres, and the pharmaceutical and medical instrumentation industries are still operating under free market conditions. One might conclude that the public health sector will continue to develop until it comes to maturity. Only then is it expected to be able to compete with the private sector.

Conclusions

In this chapter, a number of examples are given of differences in structure, organization, financing, staffing and indicators of the outcome of health care in Europe. Also, the varying expenditure on health care is discussed.

The structure varies from strictly nationalized systems in the Eastern European countries to the mixed system in continental western Europe. The health indicators in the west are better than in Eastern Europe. It is easy to say that private initiatives facilitate this and that a strong governmental influence diminishes the effectiveness of the health care system, but we also saw examples of countries with health care systems with a clear governmental influence, such as Sweden and Finland, where the quality of the health care system may compete with those in western Europe.

The organization differs from bureaucratic, hierarchically organized systems in the Eastern Europe, to less well organized systems as in Greece. In Scandinavia and Finland, local, municipal authorities have a large influence. In western Europe a balance is sought between governmental responsibility, traditional charitable organizations, and professional bodies. In general, local authorities are of little importance in the organization of the health care system. It is interesting to see how a country such as the Netherlands, in which the Government tried to create a hierarchical and regional system, has changed its policy and now relies on market mechanisms to create an efficient health care system.

The financing of health care varies with structure and organization. Expenditure varies from 6–10 per cent of GNP. It is not easy to gain a clear understanding of the expenditures. GNP is not always comparable, and the expenditure includes both public and private sector health care. It is not always clear what is included in those sectors. For instance, extensive expenditures on alternative medicine are included in some countries in central Europe but are excluded in northern and western Europe.

A country such as Finland, where the health care system has been nationalized and in which the expenditure on health care is 7.0 per cent of GNP, has better indicators of health than the mixed system of Greece with an expenditure of 7.6 per cent.

All countries have to face the problem that the costs of health care rise. In western Europe, for example the Federal Republic of Germany and the Netherlands, governments expect that competition between insurance companies and health care deliverers may reduce the costs and increase the efficiency of the system.

All countries face the problem of a surplus of doctors. Such a surplus easily creates an increase in consumption of health care. The health indicators in northern and western parts of Europe are better than in eastern and southern parts. It is likely that political, cultural, and economic factors are responsible for these differences. The overall impression is that the effectiveness of the system depends on the extent to which government and health care workers succeed in creating a well organized, non-bureaucratic system.

It may be expected that the difference between southern and western European countries will diminish, with the advent of an EEC internal market in 1992, with the removal of internal borders and the bringing into line of social systems and rules concerning health care systems.

Whether or not the economic situation improves, all governments have to choose whether to spend their finances on improvement of the curative sector, on more preventive activities, on improved care of disabled people, or on a reduction of air and water pollution. It is likely that without clear choices only the curative sector will grow, although such one-sided growth is not necessarily in the interest of the health status of the European population.

Bibliography

Baumgarten, J. (1988). Hospital financing by self administrated institutions. Proposal for a new form of organization and financing of hospital care in West Germany. *Acta Hospitalisa* **28**, 67.

Bentzen, N., Chriansen, T., and Pederson, K.M. (1988). *The Danish Health Study*. Odense University, Odense.

Bocconi, L. (1989). Structure, problems and changes in the Italian National Health Service. Paper to the European Association of Health Management Conference, Milan.

Daten des Gesundheitswesens (1988). Band 157, Schriftenreihe des Bundesministers für Jugend, Familie, Frauen und Gesundheit. Verlag W. Kohlhammer, Stuttgart.

Fack, W.G. (1986). Konzertiert Aktien im Gesundheitswesen der Bundesrepublik Deutschland. *Acta Hospitalisa* **26**, 29.

Fedorov, S.N. (1987). Sluzhba zdorov'ya trebuet rekonstruksii. *Meditsinkaya Gazeta* 27 February 1987, p. 3.

Femmer, H.J. (1986). Das Krankenhaus und seine Finanzierung. *Das Oeffentliche Gesuntheitswesen* **48**, 241.

Hoffmann, A. (1986). Das Krankenhaus im Jahre 2000: Entwicklung der Medizin. *Das Krankenhaus* **11**, 4430.

Institut für Medizinische Statistik und Datenverarbeitung (1986). *Das Gesundheitswesen* DDR, Berlin.

Law, N. (1983). 1397: National health system. *Government Gazette of the Hellenic Republic*, No. 143. English translation by the Ministry of Health, Athens.

Ley 14/1986 (1986). de 25 de abril, General de Sanidad. *Boletin Oficial del Estado* 29 April 1986.

Maynard, A. (1987). Logic in medicine. *British Medical Journal* **295**, 1537.

Melkas, T. (1988). Health for all by the year 2000—National Finnish impact on the global development programme. *Scandinavian Journal of Social Medicine* **16**, 1.

Ministry of Social Affairs and Health and the National Board of Health (1978). *Primary health care in Finland*. Government Printing Centre. Helsinki.

Ministry of Social Affairs and Health and the National Board of Health (1987). *Health care in Finland*. Government Printing Centre, Nordic Medico-Statistics in the Nordic Countries 1986. Eloni Tryk, Copenhagen.

Neues Deutschland (1987). Zahlen und Fakten zum Thema soziale Sicherheit: Politik sum Wohle des Volkes. 19–20 March, p. 9.

Neuhaus, R. and Schräder, W.F. (1985). Planning and management of public health in the Federal Republic of Germany. *Health Policy* **5**, 99.

Office of Health Economics (1987). *Compendium of health statistics, 1987*. Office of Health Economics, London.

Publikace na Počest 30, Výročí sjednocení Československého Socialistického Zdravotnictví (1982). *Health for 15 million*. Avidenumo.

Ryan, M. (1988). Life expectancy and mortality data from the Soviet Union. *British Medical Journal* **296**, 1513.

Sissouras, A. (1988). *Background and developments in health care in Greece*. Paper presented at the symposium 'A future for competitive health care for Europe?' Rotterdam, June 1988.

Socialstyrelsen (The National Board of Health and Welfare) (1988). *Monitoring the strategies for health for all by the year 2000 in Sweden*. Socialstyrensen, Stockholm.

Ts K KPSS i Soveta Ministrov SSSR (1987). Osnovnie napravleniya razviya okhrany zdorov'ya naseleniya i perestroiki zdravookhraneneiya SSSR v dvenadtstatoi pyatiletke i na period do 2000 goda. *Izvestiya* 27 November 1987, p. 1.

World Health Organization (1987). *World health statistics, 1987*. World Health Organization, Geneva.

22

Provision and financing of health care in the UK

JOHN FORBES, JOHN HOWIE, MIKE PORTER, and ROSEMARY RUE

Introduction

Since its introduction in 1948 the National Health Service, for which all UK citizens are eligible, has dominated the provision of health care and its financing in the UK. Some companies with large numbers of employees additionally provide occupational health services and some pay private health insurance subscriptions as a condition of employment. There is a small private sector based on individual insurance subscription, and the NHS itself obtains income from private patients (UK and overseas) for whom it can provide services. The Secretary of State for Health is, however, responsible for providing a comprehensive health service, including diagnosis, treatment, and prevention of illness, and the success and popularity of the NHS has resulted in the virtual exclusion of the private sector from many areas of health care. Public funds are made available for NHS purposes: over £23 billion in 1989. It is the accountability for these funds and the quality of services provided with them that necessitates a structured NHS through which the Secretary of State can delegate his powers and remain answerable to Parliament.

Projections for expenditure on health and for all other public services are made in the Autumn Statement by the Chancellor of the Exchequer following submissions made by the various Departments of State as to their needs. The Spring Budget results in cash allocations being made for the forthcoming year. Health obtains funds for England, Wales, Scotland, and Northern Ireland and other subdivisions are made by the Departments concerned. Revenue (current) expenditure and capital (for investment in buildings and equipment) are separately identified, although there is some flexibility between these allocations. Some amounts are retained for the provision of central services and some allocations for specific purposes are made; for example, 'joint finance' for expenditure in collaboration with Local Authorities. The main division is between Family Practitioner Services—the funds needed to pay for the services of independently contracted practitioners who provide primary care, and Hospital and Community Services—funds distributed to Health Authorities who provide hospital and specialist services and community health staff in support of primary care teams.

Each year the Government estimates the likely rate of inflation and allows for this in the allocations made. To the extent that this may be underestimated there is a risk of shortfall for the NHS. A further potential cause of financial difficulty is a failure to predict the level at which settlements will be reached in the annual centrally-conducted salaries and wages negotiations or awards made to professional staff and practitioners through independent Review Bodies. Notwithstanding these difficulties and the financial effect of many pressures on the services provided, the NHS has to date ended each financial year in balance, so that a fresh start is made annually. Local debts have to be made good by the local services incurring them, thus there are incentives to financial control. Local savings may be carried forward. Strenuous efforts have been made over many years to improve efficiency in the NHS, to derive additional funds for services by making cost improvements and to ensure that value for money is obtained. A good deal of cash is regularly released from within the system by these means. In comparison with other countries NHS salaries, fees, and wages are low, administrative costs are low, and there has been limited capital investment over the period of existence of the NHS. These features of expenditure account for much of the UK's low position in international league tables of proportions of gross domestic product spent on health. It would appear that the UK is moving upwards (around 6 per cent) following some recent salary adjustments. In view of the comprehensive nature of the NHS and its accessibility throughout the UK, there is ample evidence that the funds available are used with relative efficiency. The planning, organization, and delivery of services are important considerations in obtaining such results.

Primary medical care in the UK

Primary health care is everything that is beyond the scope of self-care and the social support networks of families and communities, but is not within the remit of the secondary and tertiary care services provided in hospitals. Primary health care involves the two separately financed sectors of the NHS: (i) the family practitioner services which include general medical services, general dental services, ophthalmic, and pharmaceutical services; and (ii) the hospital and community services which, apart from hospital services, include community nursing services, health visiting services, family planning clinics, and health education.

General medical practitioners see themselves providing 'comprehensive primary and continuing health care to patients and families registered with them'. In contractual terms, general practitioners are required to be continuously available to patients registered with them and to make decisions about the care of all and any problems presented to them.

In reality, general practitioners provide a buffer between potentially infinite demand and limited specialist resource. This way general practice has evolved to interpret needs, wants, and expectations and has balanced the challenge to define its quantitative and qualitative role.

Evolution of primary medical care

The early years of the National Health Service were characterized by an imbalance between the 'wants' and 'expectations' of the public and the ability of general practice and general practitioners to meet them. What was particularly serious was that neither public nor profession felt that 'needs' were being met either; dissatisfaction with the service rose as morale fell and the discipline faced problems of maintaining and recruiting able doctors. The Collings report of 1951 highlighted the deficiencies; the foundation of the College of General Practitioners in 1952 indicated the intention of the profession to fight for a structure compatible with a quality service; and the synergy of the College with the political wing of the profession (the General Medical Services Committee of the BMA) resulted in the 'Charter' of 1966 which marked the beginning of the modern pattern of general practice in the UK.

The 'Charter' introduced contractual and financial measures to encourage doctors to work together in groups, to work from better premises, to employ nursing staff, to organize the provision of their services (through appointments systems and better record keeping, for example), and to take part in continuing medical education. The changes which followed were rapid and dramatic. General practice was at once transferred from being a 'cottage industry' into a discipline which began to promote a positive clinical ideology (the balancing of the physical, psychological, and social elements of health and illness), which encouraged a professional approach to management, which appreciated the need for training and steadily attracted recruits of increasingly high calibre. The introduction of mandatory vocational training for new entrants to the discipline in 1982 was a confirmation of a trend towards comparability with other specialisms which had been evident for a decade, and the evolution of its Royal College and of a positive undergraduate role within universities confirms that the discipline is developing positively.

Change and evolution are concepts which need constant stimulation and the Government's 1986 discussion paper entitled *Primary Health Care—An Agenda for Discussion* represented a statement of intent to raise the standard of all general practice and general practitioners to that of the best. The mechanisms proposed centred round the creation of a 'good practice allowance' which would recognize the provision of a package of services weighted towards preventive and screening services and health education, alongside a greater sensitivity to patient needs in the provision of information and doctor availability. The subsequent 1987 White Paper (*Promoting Better Health*) indicated the Government's intention to implement its proposals despite professional disquiet that allocation of time and resource to promote the measurable dimensions of cure and servicing might be at the cost of that available for more abstract but equally important elements of caring and comforting.

The 1989 White Paper *Working for Patients*, and the 1990 proposals for a 'New Contract' for general practitioners, centre on three stated aims: to give patients more choice and more information on which to base their decisions, value for money, and devolving of decision-making. In primary care, practices will be required to produce informative practice leaflets and annual reports, and the mechanism for changing doctors has been simplified to make it easier for patients to change. Value for money is being promoted by setting practices to achieve pay-related targets for immunizations and certain 'screening' procedures, rather than around the previously proposed mechanism of a 'good practice allowance'. More radically, some practices of 9000 or more patients will be able to become 'budget holders', which the Government hopes will lead to greater value for money through the development of an internal market between the primary and hospital sectors, and to innovative ways for organizing practices and for meeting patients' needs.

Finance and expenditure of primary care

The NHS is financed predominantly from general taxation. In 1989/90 it is estimated that 81 per cent of the gross cost of the NHS will be met by general taxation on incomes, expenditure, and capital. A further 16 per cent is financed by the NHS component of the National Insurance (Social Security) contributions paid jointly by employees and employers. The remaining finance is raised from statutory patient charges for health services and other receipts (for example, the sale of land and miscellaneous income generation schemes). Throughout the 1980s there has been a steady shift away from general taxation towards national insurance contributions and patient payments for prescriptions, dental care, private accommodation, and treatment. This trend, along with more reliance on the private sector, is likely to continue as the government considers the potential of alternative sources of finance for health care in the UK.

About 25 per cent of central government expenditure on health and personal social services is spent on the Family Practitioner Services (FPS) which comprise the general medical, pharmaceutical, dental, and ophthalmic services. For every £1 spent on the FPS just over £3 is spent on the Hospital and Community Health Services which account for the bulk of health service expenditure. In 1989–90 *per capita* public spending on FPS amounted to around £100 per year compared to around £400 for all health services.

About one-third of FPS expenditure is allocated to general medical services (GMS). This represents around 8 per cent of total government spending on health services. Although in the 1980s there has been a slightly higher rate of expenditure growth in the FPS compared to overall growth in health service spending this proportion has remained virtually static for over two decades. In proportional terms, spending on GMS was actually higher during the 1950s but was gradually overtaken by a steady shift of resources in favour of the hospital sector.

Total GMS expenditure per capita is about £32 per year, giving an overall average cost per consultation of just under £8. Alternatively, public spending on GMS amounted to about £58 000 per practitioner per year, of which £31 000 (54 per cent) represents remuneration for doctors. The remainder comprises directly reimbursed expenses by the government for running costs (for example, employment of ancillary staff) rent, and capital investment in practice premises. Since 1966 the government has subsidized the employment of certain categories of ancillary staff (nurses, dispensers, secretaries, and receptionists) up to a maximum of two whole-time equivalent staff per GP principal. From 1990 restrictions on categories and numbers of ancillary staff will be removed but a cash limit will be imposed on this component of GMS spending.

Unlike the hospital service where the net amount of cash spent each financial year is subjected to firm budgetary control via the imposition of cash limits, FPS expenditure has until now been demand-led. In practice, this reflects an amalgam of expressed patient demand and decisions taken by individual doctors when caring for patients. Some degree of control over expenditure is indirectly exercised by restricting the numbers of practitioners and controlling the price of drugs. The first move towards the imposition of an explicit budget for FPS has emerged under powers in the Health and Medicines Act 1988 which ensured that government expenditure on ancillary staff and capital investment in the general medical services were cash-limited from April 1990.

The costs of the pharmaceutical service (drugs and appliances prescribed by general practitioners and remuneration and expenses of retail pharmacists, appliance contractors, and dispensing doctors) account for half of FPS spending. The total gross cost of prescribed drugs and appliances represents 84 per cent of spending on the pharmaceutical services. The gross cost per prescription is currently £5.27, which in per capita terms amounts to around £42 per year, or some 30 per cent more than is spent on General Medical Services. Not surprisingly, given the importance of drugs within overall spending on primary medical care, the Government has considered ways of controlling the rate of growth in drug costs and expenditure. Examples of initiatives in this area include the introduction in 1985 of the Selected List Schemes limiting NHS prescribing in seven therapeutic categories to those drugs which are believed to meet all clinical needs at lowest possible cost, and a new computer system PACT (Prescribing Analyses and Cost) designed to provide the doctor with quarterly information on prescribing patterns and cost. Recent estimates suggest that overall public expenditure on drugs has declined by about 4 per cent due to the operation of the Selected List Schemes. Current policy proposals have raised the prospect of introducing cash-limited drug budgets for general practitioners which will undoubtedly strengthen the process of central government budgetary control of the FPS.

Prescription charges were first introduced in 1951 and have been in continuous use ever since, except for a two-year period (1966/67) when they were temporarily abolished. They represent about one-quarter of the total revenue raised from patient charges for health services. Prescription charges increased nominally over time but tended to rise by less than the rate of inflation. Beginning in 1979, however, this trend reversed as prescription charges per item have steadily increased from 20 pence to £2.80 in 1989, representing a rise of over 200 per cent in real terms. These increases have meant that those patients who are not exempt from charges now contribute about half of the gross average cost of NHS prescription drugs, compared to less than 10 per cent in 1978. The additional government revenue generated by the increased prescription charges is about £100 million per year. A further £300 million is 'saved' due to reduced patient/doctor demand for prescriptions.

Paying the general practitioner

In the UK general medical practitioners under contract with the NHS are currently remunerated using a system of allowances, capitation fees, and miscellaneous fees for specific items of service. The number and level of allowances and fees, and hence remuneration, is set by government. Since 1971 the government has been advised by an independent 'Review Body on Doctors' and Dentists' Remuneration' which annually recommends what it believes are appropriate levels of pay (and corresponding allowances and fees). The government, however, retains the power to implement any or all of the Review Body's recommendations.

In 1989/90 the average annual gross income per principal was £45 761. Gross income should not be confused with net pay. Included within gross income is reimbursement in whole or part for certain practice running costs which on average amount to £14 656 per principal or around half of the total cost of practice expenses incurred by the average doctor. Average net pay (before tax) therefore is just over £31 000.

The most important single component of pay is capitation fees, which currently account for about 47 per cent of total receipts from allowances and fees. Annual capitation fees are paid for each patient registered with a GP. The standard capitation fee varies from £8.95 for patients aged less than 65 years to £14.25 for patients aged 75 and over. The patient's sex has no impact on the capitation payment, despite the fact that women account for about 60 per cent of consultations. If the doctor contracts to provide an out-of-hours service a supplementary capitation fee, representing around 20 per cent of the standard fee, is paid for each patient in excess of 1000 on the registered list.

Along with capitation fees, GPs receive a variety of annual allowances or lump sum payments. Collectively, allowances account for about 40 per cent of total fees and allowances. The largest single allowance is the Basic Practice Allowance (£9250 in 1989) payable to all full-time GPs with at least 1000 registered patients. Part-time GPs or those with smaller list sizes are paid on a pro-rata basis. Doctors who have completed a vocational training scheme for general practitioners receive a special allowance. Seniority (years of experience in general practice) is also rewarded. A group practice allowance is payable for working in a partnership of three or more doctors. If out-of-hours service is provided a supplementary practice allowance is paid. A doctor may also qualify for a 'designated area' allowance if he or she is working in a part of the country where average patient list sizes are considerably greater than the national average. Allowances are available for employing a doctor working as a GP assistant or providing post-graduate training for doctors in the vocational training scheme.

A relatively minor but growing component of remuneration is fee for service (13 per cent of total pay). Specific items of service include night visits (between 11 p.m. and 7 a.m.), vaccinations and immunizations, cervical cytology tests, and contraceptive services. Doctors on an 'obstetric list' receive a fee for providing maternity medical care. Income from item-of-service fees averages around £2.60 per patient per year, or nearly one-third of the standard capitation fee. GPs employed to work in hospital on a sessional basis also receive additional payments. Finally, drug and appliance dispensing provides an additional source of income for some GPs.

This hybrid remuneration system has remained virtually unchanged and unchallenged for over two decades. Following on from their plans contained in the 1989 NHS Review White Paper *Working for Patients*, the government has proposed changes to the GP's contract which, if implemented, would radically alter the balance of fees and allowances. The basic thrust of the revised contract is to relate doctors' income more closely to the number of patients on their lists and to the services that they provide. By placing more emphasis on capitation payments as a source of income, many of the existing allowances would be abolished or modified. Capitation payments would represent at least 60 per cent of income from fees and allowances. Target payments to encourage GPs to achieve minimum levels of population coverage for immunization and cervical cytology will replace item of service payments. Sessional fees are to be introduced for providing minor surgery and health promotion clinics. Special capitation supplements for children under five years of age are also under consideration for GPs who routinely conduct paediatric surveillance and developmental screening.

General medical services structure and organization

Numbers and distribution

General practitioners work as 'individual contractos' to the National Health Service. In England and Wales, GP services are administered through Family Practitioner Committees (FPCs). Since 1985, FPCs have been independent of District Health Authorities (DHAs), and have been expected to play a significant role in the planning, organization, and monitoring of primary health care services. The 1989 White Paper proposed that FPCs be made accountable to Regional Health Authorities (as are DHAs) and that they (the FPCs) should be staffed by chief executives and stronger management teams. The 1989 White Paper also proposed that the government should reserve powers to control the number of doctors entering into contract with the FPCs. In Scotland and Northern Ireland, GP services are administered by sub-committees of the Health Boards (HBs).

In 1986, about 32 800 doctors were working in general practice in the UK, an increase of about 20 per cent since 1975. About 90 per cent of them were 'unrestricted' principals who held individual contracts with FPCs or HBs, and who provided primary medical care to patients registered with them. About 7 per cent were qualified doctors undergoing a year 'traineeship' in general practice as part of the mandatory three years of vocational training for general practice. The remaining 2 per cent were either restricted principals (doctors with unusually small lists, or providing only limited services), or assistants employed by unrestricted principals but with their salaries supported by special allowances).

Throughout the 1980s there has been an increase in the proportion of women general practitioners (from 13 to 19 per cent), and a decrease in the proportion of unrestricted principals born in Britain.

The 1946 NHS Act established a Medical Practices Committee (MPC) with the function of securing an even distribution of general practitioners throughout the country. Since 1966 there has been a steady decline in the number of patients living in 'under-doctored' areas, but there is still particular concern both for the availability of doctors in inner-city areas (and peripheral housing estates), and for homeless people and some members of certain ethnic groups in inner-city areas who are not registered with, and do not use, general practitioner services. The 1987 White Paper proposed a special allowance for doctors working in areas of deprivation, and this has been incorporated in the 1990 New Contract.

List size, practice size, and premises

As the number of unrestricted principals practising in the UK has increased, so the number of people registered on their lists has fallen (from 2291 in 1976 to 1988 in 1986). Scottish doctors have the lowest average lists (1644 per unrestricted principal, 1986), followed by Northern Ireland (1840), Wales (1880), and England (2032).

When the NHS was founded, the majority of doctors worked singly or in twos usually from converted shop fronts or their own homes, seldom from purpose-built premises. By 1986, only 11 per cent of unresticted general practitioner principals were single-handed and 52 per cent worked in groups of three or more.

From 1966 onwards, there was an increase in the number of practitioners working from specially adapted premises, or from purpose-built health centres or medical centres. In England, in 1965, there were 28 health centres with 215 practitioners working from them. By 1977 there were 731 health centres with about 3800 practitioners working from them. England has fewer patients registered with doctors working from health centres (17 per cent) as compared with Wales (21 per cent), Scotland (22 per cent), and Northern Ireland (54 per cent). These trends have continued throughout the 1980s, but at a slightly reduced rate, and the 1987 White Paper has proposed particular assistance to improve premises in deprived areas.

The trend away from single-handed practice to group practice and health centres has been accompanied by an increase in the proportion of patients who reckoned that there was no other practice within a reasonable distance. Most patients find it easy or fairly easy to reach their doctors' surgeries, but studies have found that it is not so much the distance as the availability of transport that affects ease of access and women, older people and people in social classes IV and V all found access more difficult than did the young, middle class, and male. The implications of this are important for the organization and use of group practice given that it is young mothers with children, elderly people, and people from social classes IV and V who have the greatest need for health care services.

Team work

Closely associated with the development of health centres and group practice has been the concept of team care. Initially, this involved the employment of ancillary personnel like receptionists and secretarial staff, but has expanded to include practice-employed nurses, DHA/HB-employed 'attached' treatment-room nurses, and 'attached' health visitors and district nurses. More recently, there have been a number of experiments involving the attachment of social workers and clinical psychologists; other, but less common members of health care teams have been midwives, dentists, physiotherapists, radiographers, dieticians, pharmacists, occupational therapists, chiropodists, and marriage guidance counsellors. Teamwork has not always proved easy, because different perceptions of each other's roles, abilities, and suitable case material has created friction and ill-feeling—particularly between general practitioners, health visitors, and social workers. Integration of and access to patient records has sometimes proved difficult, and all too often there has been no suitable or effective mechanism for discussing or resolving difficulties.

Patients' responses to team care in general practice appears to have been mostly favourable. However, although patients are fairly well aware of the role and duties of practice/treatment-room nurses and attached district nurses, they are less certain of the work of health visitors. This is important because the development of preventive services is closely associated with the role of health visitors and with the development of teamwork in primary care. (For further discussion of the work of the community health services in relation to health promotion, see Chapter 34, Volume 3.)

Administration

About three-quarters of patients attend practices with full or partial appointment systems, and doctors working from large group practices, are more likely to consult 'by appointment' than doctors in small group practices or single-handed practices. Doctors in health centres, even if in small practices, are also more likely to consult 'by appointment'.

Appointment systems do save patients from waiting as long as they do when seeing a doctor without an appointment, but some systems are not as effective as they might be, partly because doctors often start late, and often because doctors book patients at quicker intervals than they actually consult. Some patients also report difficulties making an appointment to see a doctor on the same day.

One disadvantage of appointments systems is that people have to rely on the telephone or on a visit to the surgery to make an appointment—two factors which make appointment systems less easy for elderly people and people from manual working classes to use. Generally, elderly people are not overtly critical of appointments systems but a greater proportion of people from manual working classes prefer an open (non-appointment) consulting system.

Until recently, the least progress in practice organization had been in the development of medical records. The medical record envelope is still the standard NHS record in general practice and has been in use since 1920. Into this small 7″ × 5″ envelope is stuffed the clinical record card and those letters and laboratory reports that the GP chooses to keep. Not only is the system impractical for the storage of all this paper, but the opportunity to use the record in a systematic and problem-solving fashion is severely limited. Its only advantage is its low cost to the general practitioner. Far more suited to the ongoing needs of primary care is the A4 size record, particularly if organized on a 'problem orientated' basis or with a list of current acute and ongoing chronic problems prefacing the clinical notes. The number of practitioners using A4 notes is increasing, particularly because of recommendations for becoming a training practice, but progress has been retarded by the reluctance of successive governments to fund both the cost of converting to A4 and also the cost of storing the larger records.

With the development of computer technology, many practices have invested in micro-computers which handle computerized age/sex registers linked to a disease or morbidity register. These systems print repeat prescriptions, and enable the practices to carry out audit of selected groups of patients. Most recently, a few enthusiastic practices have experimented with the use of computer-held medical records with access to information through desk-top terminals in their consulting rooms. The costs of such systems are high because of the difficulties of creating files which are sufficiently well structured to allow for fast search, display, and editing but which also allow for individual patient peculiarities. Most of these expensive interactive systems are provided free to practices provided they provide regular information on prescribing back to the computer company which, in turn, sells this information to interested third parties.

With the increased volume of administration, most practices (90 per cent) now employ reception and secretarial staff, and practice administrators are becoming increasingly common in larger group practices. There are recognized training courses for medical secretaries and practice managers, but there is still a need for better training of reception staff, and for joint sessions involving all practice staff—administrative, medical, and other. The 1989 White Paper and the 1990 New Contract put considerable emphasis on the need for better information systems and practice management. It is likely that there will be a considerable increase in the resources devoted to general practice administration.

General medical services patterns of work

GP/patient interface

The 'average' doctor carries out around 10 000 items of service per year, about 80 per cent of these being 'direct' (face-to-face) consultations. The national average figure of consultations per patient per year (known as the 'consultation rate') is four per year, with some practices reporting figures of only half of that, while others exceed the figure quite substantially. One factor which contributes to doctors' workload is doctors' own behaviour. About 40 per cent of doctor–patient contacts are brought about by doctors bringing patients back to see them.

The average general practitioner spends about 38 hours each week on general medical services activities, but there are wide variations between individual doctors, with one study suggesting that about 15 per cent work 50 hours while 24 per cent work fewer than 30 hours. At the time of writing, unrestricted principals are expected to devote an average of 20 hours per week on direct services to patients in surgery consultations and home visits, and to have at least 1000 patients on their lists. The 1987 White Paper proposes that doctors increase the average number of hours per week spent in surgery sessions, and increase the minimum list size.

The issue of list size and its relationship to the process and quality of care has long been a matter of debate and argument. Although the 1986 Discussion Paper on Primary Care observed that there was no obvious relationship between consultation rates and list size, there is some evidence of a weak but positive association between consultation rates and list size. However, there are important confounding factors when looking at these associations, the most important of which is the apportioning of patients to a GP's list in a group practice where GPs do not see 'their own' patients.

The temptation to equate higher consultation rates with higher quality of care must be resisted—although not necessarily discarded. All studies of workload have described more similarities than differences in consultation rates for *new* episodes of illness and the wide fluctuations appear to reflect doctor-initiated work (return consultations) which do not necessarily reflect patient need.

The amount of time spend with the patient on a home visit probably differs remarkably little with that spent on a surgery consultation (on average, about 7 minutes), but the time the doctor, as against the patient, spends on travelling is significant and can be substantial if visits are not geographically contained. The trend in home-visiting rates is downwards, to the point almost of disappearance in some practices. An estimated figure of six home visits per day out of an average of some 35 consultations per doctor per day, gives a 15 per cent home-visiting rate and available literature seems to confirm this as an average figure.

As far as out-of-hours work is concerned, the use of deputizing services has grown since 1964 when about 9 per cent of practitioners said that they sometimes used 'emergency care' services. By 1977, this had risen to 26 per cent who used a deputizing service regularly and another 18 per cent who used one occasionally. There was little evidence that the increase in the service had led to an increase in its acceptability by patients—63 per cent would have preferred an alternative service. Although a visit by a deputy might be preferable to a telephone conversation, there is clearly a need for a more acceptable form of deputizing service than that currently provided in many areas.

Serious illness, in terms of threat to life, forms a relatively small proportion of total work. Perhaps six patients in an average list practice present with a new malignant condition in any one year, and half of these will be lung cancer. Five patients will have appendicitis and five will have a stroke. Rather more (perhaps 8–10) will have a myocardial infarction

Table 22.1. Analysis of common symptoms and their management in a practice of 4500 patients for one year

Symptoms	No. of consultations for 4500 patients	Management		
		% investigated	% referred to hospital	% receiving prescription
Cough	527	2.9	0.4	97
Rashes	292	0.3	1.0	79
Sore throat	287	1.7	0	95
Abdominal pain	197	5.6	7.6	73
Disturbances of bowel function	195	2.7	1.1	89
Spots, sores, ulcers	181	0	0.6	82
Back pain	172	8.1	0	88
Chest pain	168	9.5	4.8	81
Headaches	149	3.8	1.5	80
Disturbances of gastric function	146	0.7	1.4	79
Joint pain	145	13.5	5.0	70
Changes in balance	65	8.1	2.7	66
Disturbances in breathing	59	22.4	1.7	69
Changes in energy	58	3.3	6.6	90

Source: Adapted from Morrell, D.L. (1972). Symptom interpretation in general practice. *Journal of the Royal College of General Practitioners*, **22**, 297.

but perforated ulcers and meningitis occur less than once per doctor per year. The prevalence of chronic illness is predictable (epilepsy, diabetes, and thyroid disease all affect around 2 per cent of the population) and births and deaths (an average 25 each per doctor per year) will vary from practice to practice. The most useful summary of the incidence of consultations for minor illness is probably to be found in a paper by Morrell (1972) in which he lists the 14 commonest symptoms seen in a practice of 4500 patients over the course of one year. Table 22.1 shows a summary of his results together with a summary of how these illnesses were handled in terms of investigation, referral to hospital, and the prescribing of drugs.

Prescribing

Some 60–70 per cent of consultations between general practitioners and patients are concluded by the writing of a prescription. Although the majority of these prescriptions are intended to be therapeutically helpful, there are many occasions where the act is a token of sympathy or a sign that the consultation is being concluded. The average patient receives some 6–7 prescription items each year. Half of these are issued through repeat prescription systems without the doctor and patient meeting face-to-face.

The issue of 'prescribing' is often used as a proxy measure for discussing problems of general practitioner work more generally. There is wide variation between doctors in how frequently and how expensively they prescribe, and this variation is both difficult to explain and difficult to reduce. Prescribing is seen as a target for cost-saving (drug costs are some 7 per cent of the total NHS bill), as a focus for professional accountability (audit), and as an appropriate area for education and research. It is, however, difficult to define either 'goodness' or 'badness' in this area of activity and the uneasy overlap between clinical, academic, and political priorities has often hindered rather than promoted progress.

Patients in the UK pay a flat rate charge towards the cost of each prescription (presently reflecting about half the average cost) but exemption is given for some two-thirds of prescriptions on grounds of age, of hardship, or in certain clinical conditions (mainly endocrine). A minority of doctors working some distance from pharmacies dispense as well as prescribe for their patients. Relatively few active drugs are available 'over the counter'; antibiotics, oral contraceptives, almost all NSAIDs (non-steroidal anti-inflammatory drugs), and all steroid creams (except 1 per cent hydrocortisone) have to be prescribed by a doctor. Recent changes have, however, restricted general practitioners' freedom to prescribe a range of preparations which were formerly available (notably remedies for coughs and colds). There has been considerable effort put into attempting to promote 'generic' prescribing of drugs in which there is a significant cost difference between 'generic' and 'brand' costs.

Particular emphasis in the research field has been given to prescribing in the areas of psychotropic drugs (the commonest area of prescribing overall) and antibiotics (the drug most commonly prescribed at face-to-face consultation).

Investigation and referral

We have already described the likelihood of investigation and referral for a variety of symptoms seen by general practitioners (Table 22.1). Access to laboratory and diagnostic facilities have improved considerably for general practitioners throughout the 1970s and 1980s. Most general practitioners now have open access to diagnostic facilities like X-ray, biochemistry, bacteriology, and haematology, though general practitioners working in rural areas do find that the turn-round is often rather slow and that this requires extra administrative arrangements.

Referral of patients from general practice to hospital raises many of the same issues which apply to prescribing of medicines. Many papers have been published in this field and these highlight the difficulty of getting good statistical information in an area where the numerator data is often inconsistent (for example, referral may include out-patient and/or in-patient referral) and denominator characteristics may include significant uncontrolled variables (for example, age of patients, list sizes of doctors, availability of services). It seems that around 10 per cent of patients are referred to hospital each year and that referral is decided on at around 5 per cent of consultations. However, very wide variations between doctors have been demonstrated (a 24-fold variation in one study) and these have attracted the interest of administrators and politicians despite warnings from the researchers that this variation may be a statistical 'small numbers' artefact.

From the patient's point of view, the availability of a cost-free second opinion is known to exist and in general, a referral can be arranged within days or weeks rather than months. In a few specialties (particularly orthopaedics and ophthalmology) delays are, however, often longer. Referral may also be made to the private sector and again this is normally routed through the general practitioner thus ensuring that proper background information should be available to the specialist. The co-ordinating role of general practitioners is emphasized by their also having access to community nursing.

From the general practitioner's angle, the advantages of the system lie in his or her ability to decide which line of investigation to pursue and to whom he or she would like to refer. The disadvantage is that—particularly in the larger urban areas—GPs cannot guarantee that their patient will be seen by a senior as against a junior clinician. Communication between doctor and doctor is often a problem due to clerical delay and this may lead to confusion and worry for patients. There are also complaints that patients are not discharged back to their general practitioner's care at an appropriate time.

Secondary and tertiary care

Structural framework

National Health Service hospital and specialist services are provided through local (district) health authorities (or Boards in Scotland and Northern Ireland), covering the populations of the UK by geographical areas. These authorities also provide the community health services—a term used to describe the employed staff and support services for primary health care, i.e. district nursing, health visiting, and community midwifery services. The Boards in Northern Ireland also provide personal social services which elsewhere are the responsibility of local government. The authorities are accountable to Ministers directly in Scotland, Wales, and Northern Ireland, but in England there is an intermediate regional tier of health authorities. There are 19 'special' health authorities directly responsible to Ministers for providing some specialized services of national importance—postgraduate teaching hospitals in London, hospitals for the criminally insane, public health laboratory services, blood products, health education. All authorities are expected to inspect and collaborate with the private sector, which provides acute beds mainly in small hospitals, and nursing home beds in a variety of establishments throughout the UK. A good deal of support and assistance for the NHS is obtained from voluntary organizations who may raise funds or supply volunteers for activities (for example, transport to hospitals) which improve services locally. Each local health authority or board has its complementary community (or local) health council which has the role of representing the consumer viewpoint on both health authority and family practitioner services.

Since 1976, Regional Health Authorities in England have had their funding allocated according to a formula based on population size weighted for a variety of factors designed to account for different demographic factors and health care needs and service costs (Resource Allocation Working Party, RAWP). Similar arrangements were introduced in Scotland, Wales, and Northern Ireland. The direct effect of RAWP has been to shift resources from the relatively well endowed regions to the less well endowed regions.

Regional health authorities in England are directly responsible for a few services (e.g. blood transfusion), but their major responsibility lies in preparing plans and policies for health services in the region and issuing guide-lines and allocating resources to their district health authorities (DHAs). DHAs are responsible for the actual provision of hospital and community health services in their district, and for collaboration between health service and related local authority services (joint planning). DHAs also prepare plans and priorities for their districts. Both DHAs and RHAs have some scope to adapt DHSS guide-lines to suit local requirements.

At the same time that RAWP was reallocating resources between regions, the DHSS issued priorities for the reallocation of funds to programmes directed to particular client groups: elderly people, people with mental illness, and people with mental handicap or physical disability. (Again, similar sets of priorities were also drawn up for the Scottish, Welsh, and Northern Ireland Health Boards.) Guide-lines were also issued which exhorted health authorities to develop community care (as opposed to hospital) services.

For some regions, the implications of 'no growth' funding arising from the RAWP formula, meant that the only way not to overspend was to cut services. For various reasons, small community hospitals were the targets most commonly chosen for closure. More recently, attempts to meet continuing budget constraints at a time of pressure to develop community care have led to the closure of hospital wards prior to the establishment of appropriate accommodation and support services in the community. The setting up of 'joint planning' and of 'joint finance' has generally been too haphazard and too inadequate to compensate for the lack of growth money available to many health authorities. Two reports on community care have highlighted the present organizational, managerial and financial problems, and a recent White Paper (*Caring for People*) has attempted to address the issues raised.

The 1989 White Paper proposes the discontinuation of RAWP, and recommends the use of a simpler formula based on population size adjusted for demographic characteristics only (age/sex). This proposal should lead to an increase in growth money for those regions which had suffered under RAWP.

Authorities and boards

Members of authorities, boards, and councils are appointed by Ministers (via regional authorities in England). They are predominantly lay people from all walks of life, recruited through wide canvassing and advertisement and appointed not as representatives but for the personal contribution each can make. There is a specialist (consultant) doctor, a general practitioner, a nurse, a university and a trades union nominee, and local government nominees, among the members of each local authority with some variations; for example, for teaching district authorities, and specialist health authorities. This group of unpaid volunteers, acting corporately under their ministerially appointed chairman, discharges locally the hospital and community health services (HCHS) functions of the NHS through delegated authority from the Secretary of State. The chairman receives a small honorarium for his services which in a commercially run organization would be very highly rewarded. The authorities meet and take decisions in public, their papers being available to the press, and in this way there is public accountability for funds distributed, services provided and proposals made. The extent to which such a system can maintain freedom from party political influence or reflect the best interests of the local community is a subject of continuing controversy, but the achievements of these unelected amateur bodies during the first 40 years of the NHS have been considerable.

The 1989 White Paper proposes important changes in the management structure of the NHS. The membership size of health authorities will be reduced to ten, and there will be no place for professional, trade union, or local authority representatives.

NHS management

The functions of the authorities are concerned with maintaining and improving services, planning, and the allocation of resources. These functions are fulfilled by the appointment of professionals. Since health services are by their nature extremely labour-intensive, the health authorities have a major role as employing authorities for NHS salaried staff. Each authority appoints a general manager, and there are general managers, accountable to him, for each of the units which are the managed components of the authorities' services—there is much variation in the substructure within authorities' areas. All staff are managerially accountable to the appropriate general manager but professional accountability is organized by each of the health professions, taking particular account of the need for qualified, senior supervision for health professionals in training. Senior medical and dental staff are employed directly by authorities and some levels of training for medical and other professions are organized on a multi-authority basis. The wide personnel functions of authorities include a substantial element of training and career development for staff, the interpretation of terms of employment, conditions of service, grading and appeals machinery, and local and statutory appointments procedures. National review bodies advise the Government directly on salaries for doctors, dentists, nurses, midwives, and professionals allied to medicine. Councils established under Whitley agreements fulfil the management side role in negotiations for other staff. The health authorities make their input to the evidence and briefing considered by these important bodies. At every level there is some kind of interface with staff organizations to their employment.

Participation and professional advice

Concerning the general range of issues affecting the health service, participation in policy- and decision-making by staff most knowledgeable of the problems is encouraged and means for obtaining advice is incorporated into the structure and management style of the service. Each authority is required to make arrangements for receiving advice from representatives of the medical, dental, nursing and midwifery, ophthalmological, and pharmaceutical professions locally. Professional advisory committees are normally elected for this purpose and their chairmen offer advice formally, often publicly, on topics which the health authority is considering, or which the committee feels should be drawn to the authority's attention. There are senior officers of the main health service professions and managers of the authorities' major functions who form the senior management team in each authority's headquarters and who have individual responsibility for a specialist contribution to the health service at that level. Such teams usually comprise, with the general manager: a doctor; a nurse; finance, personnel, and planning officers, and other heads of services for which the authority has responsibility. Such multi-disciplinary teams with appropriate variations are a common feature of NHS management at every level. They assist with the resolution of complex issues where there are several specialist considerations and enable a wide range of experience and advice to be brought to bear on such matters. The necessity for these teams to reach consensus which was a feature of management for the decade 1974–84 has been superseded by the concept of individual accountability for decisions taken by managers in the light of multi-disciplinary advice.

Consultation

Advisory committees, specialist officers, and staff representatives are also involved in the procedure of consultation which is evoked in connection with planning, proposals for change, major decisions concerning priorities or the development of policy and with the implementation of specialized advice. Consultations may be quite limited among those with specialized knowledge, but, since health issues can be of far-reaching concern, consultation is commonly extremely wide, attempting to reach the local and neighbouring populations through local government and parliamentary representatives, interested organizations, the press, and public meetings in addition to the internal health service network. At such times the Community (Local) Health Councils have a particular role and responsibility to promulgate the proposals and obtain views and comments on behalf of the local community. In cases of closure of services or major changes, the Councils can effectively object and request that the matter be referred to Ministers for review.

University liaison

The relationship of the NHS with universities having medical schools, i.e. those teaching medical students, is the subject of special arrangements. A general responsibility is laid upon health authorities in the relevant districts to make provision in hospitals and to facilitate the teaching function. Special revenue funding (service increment for teaching) is identified in central funding and special cost allowances are made for buildings and equipment. A spirit of mutual support and co-operation has been developed and a knock-for-knock system of financial flexibility has avoided the cost and irritations of a cross-charging system for minor sums. The informal relationship is strengthened by representation of the university at each management level, by formal liaison committees, and by the exchange of honorary NHS contracts for all university medical staff undertaking clinical work in hospitals. Additionally, many NHS authorities support academic departments with staff and research costs, much teaching is undertaken by NHS staff, and there is great freedom in the way NHS and university staff work side by side in clinical departments and in the many advisory roles which fall upon the medical profession. There is shared concern that bio-medical and health services research is not obtaining the resources which could be justified by this very active partnership, unique in the NHS.

Postgraduate training and medical manpower

While the numbers of medical students are determined for the UK as a matter of governmental policy, the NHS has the responsibility for the clinical employment of doctors during their postgraduate training years and has to collaborate with universities and with the medical Royal Colleges in making suitable posts and contracts available in appropriate numbers. There have been great problems in attempting to achieve the right mix of doctors trained for general practice and the various specialities so that excesses and deficiencies do not occur. Very substantial efforts in this field of medical manpower planning have had only limited success and the controls currently in force, as a result of voluntary agreements, are fiercely monitored.

Hospital activity, standards, and outcomes

The performance of health authorities in fulfilling their strategic and annual plans, in managing their resources in accordance with national priorities and in meeting national and local objectives is monitored through a formal Ministerial Review process on an annual basis. By agreement, some regional and supra-regional specialities are identified and provided in a limited number of hospitals catering mainly for tertiary referrals with the intention of concentrating rare skills and maintaining high standards. Much attention is, however, given to the activity in a range of specialist services provided in most districts and about which a great deal of routine information is collected. The statistics are formidable. Numbers of hospital specialist consultations are equivalent to half the population each year with a further equivalent of one-fifth of the population having consultations occurring in specialist Accident and Emergency Departments. In each year the equivalent of one in ten of the population is admitted to hospital. Many of these events relate, of course, to the same individuals, but they give some indication of the scale of NHS activity. This kind of information relates to the process of health care and includes great detail, e.g. on waiting times for operations. It is available on a standard basis which allows comparison between districts and specialties, raising legitimate questions for analysts and managers. It is not, however, information which answers questions about quality of care, as attempts to develop a range of 'performance indicators' have shown. Interest is focusing more at the present time on quality issues through medical auditing, a method by which doctors critically review their own work with their peers, in the light of good medical practice. All medical Royal Colleges have a commitment to medical audit and some already require evidence of its operation in their approval of training posts. While the ultimate goal of medical care is the best outcome for the patient, the determination and measurement of outcomes is a difficult area in which it is agreed more research is urgently required. For much of medicine the traditional short-term results which are commonly used are clearly inadequate. There is scope for longer-term studies of the effects of modern medical intervention and the quality of care offered under medical supervision.

Prospects

The NHS has, in common with other health care systems world-wide, run into difficulties in finding resources for medical advances and for the care of the increasing proportion of elderly persons in the population. The Government, apparently in the hope of retaining control and limiting public expenditure on the NHS, has put forward some proposals for management reforms which are described as radical. These consist of the introduction of an 'internal market', increased autonomy and concurrent accountability for some hospitals and general practitioners, and the extension of medical management of resources and medical audit. The proposals depend heavily on a rapid expansion of management information systems and have met with substantial professional and political opposition because of their cost, philosophy, and the time-scale and mode of introduction.

While the proposals for general practitioner services are criticized for their orientation towards financial incentives, wrongly conceived, the proposals for specialist services appear to relate mainly to non-emergency surgery and to have disregarded the interdependence of such activity with emergency and long-term care and with teaching, training, and research. Public health and prevention have not been included, other than through a reiteration of the Government's acceptance of the Report of an earlier inquiry into the public health function. This places a clear responsibility on health authorities to survey and report annually on the health status and problems of the population for which they are accountable and to take account of their findings when allocating resources. This should ensure that the NHS remains comprehensive, locally accessible, and concerned with long-term health issues on behalf of all citizens. Advice concerning the long-term care of the elderly, the handicapped and the mentally ill and the use of resources by health authorities and local government for these purposes has also recently been partially accepted by the Government, signalling some acknowledgement of the reality of demand facing the NHS now and for the future. It is to be hoped that the detail and manner of eventual implementation of the various changes being considered will lead to improvement and strengthening of the NHS which enjoys great popularity in the UK.

Suggested reading

Audit Commission (1986). *Making a reality of community care*. HMSO, London.

British Medical Association (1983). *General practice: A British success*. British Medical Association, London.

British Medical Journal (1989). *The NHS review—What it means*. British Medical Journal, London.

Committee of Inquiry into the Future Development of the Public Health Function (1988). *Public Health in England (The Acheson Report)*. Cm 289. HMSO, London.

Department of Health and Social Security (1976*a*). *Sharing resources for health in England*. Report of the Resource Allocation Working Party (RAWP). HMSO, London.

Department of Health and Social Security (1976*b*). *Priorities for health and personal social services in England*. HMSO, London.

Department of Health and Social Security (1984). *Care in action. A handbook of policies and priorities for the health and personal social services in England*. HMSO, London.

Department of Health and Social Security (1987). *Hospital medical staffing: achieving a balance: Plan for action*. HMSO, London.

Department of Health and Social Security: (1983) Circular (83) 6; (1985) Circular HN (85) 3; (1986) Circular HN (86) 34.

Department of Health and Welsh Office (1989). *General practice in the National Health Service. A New Contract*. DHSS, London.

Griffiths, R. (1988). *Community care: Agenda for action*. HMSO, London.

Körner, E. (1982). *Steering Group on Health Services Information: First report*. HMSO, London.

Office of Health Economics (1989). *Compendium of health statistics*. (7th edn). Office of Health Economics, London.

Secretaries of State for Health, Wales, Northern Ireland and Scotland (1986). *Primary health care: An agenda for discussion*. Cmnd 9771, HMSO, London.

Secretaries of State for Health, Wales, Northern Ireland and Scotland (1987). *Promoting better health*. Cm 249, HMSO, London.

Secretaries of State for Health, Wales, Northern Ireland and Scotland (1989). *Working for patients* (plus associated working papers). HMSO, London.

Secretaries of State for Health, Social Security, Wales & Scotland (1989). *Caring for people*. Cm 849. HMSO, London.

Social Services Committee (1987). *Primary health care*. Vols I and II. HMSO, London.

23

Provision and financing of health care services in Japan

N. MARUCHI and M. MATSUDA

Introduction

The health status of the Japanese has been greatly affected by the improvement in the standard of living, extensive progress of medical science, and rapid socio-economic growth since the Second World War. In 1986, life expectancy in Japan was ranked highest in the world, and Japan's infant mortality rate in 1985 was among the lowest (5.5 per 1000 live births). However, the aging of the Japanese population has been accelerated by the increasing life expectancy and decreasing birthrate. Thus, the ageing population should be tackled not only as a health problem but also as a social one.

Japan has almost attained the goals of the Primary Health Care Declaration of Alma Ata in 1978, but efforts are still required to raise the level of its health and medical care services to meet increasing and diversifying needs. The major problems are mental health, geriatric health, environmental pollution, adverse drug reactions, and financing health insurance and the pension system for the increasingly aged population.

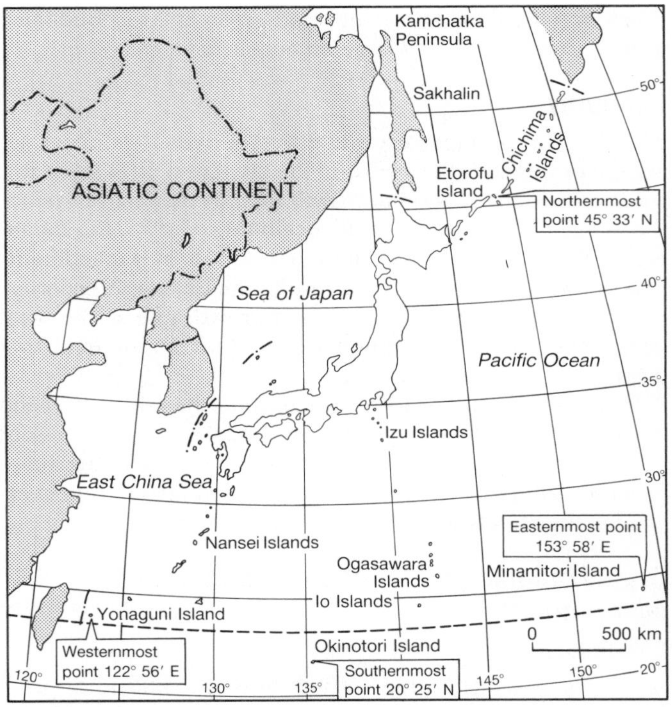

Fig. 23.1. Geographical location of Japan.

The country and its people

Geography

Japan consists of four main islands—Hokkaido, Honshu, Shikoku, and Kyushu—as well as a number of island chains and thousands of smaller islands and islets. This Japanese archipelago, lying off the east coast of the Asian continent, stretches in an arc 3800 km long (Fig 23.1). It covers an area of 377 384 km^2. The total land area is about one-twenty-fifth that of the US, and one-ninth that of India. In terms of world area, Japan occupies less than 0.3 per cent.

The islands of Japan lie in the temperate zone and at the north-east end of the monsoon area. The climate is generally mild, although it varies considerably from place to place. The four seasons are clearly distinct. Summer, which is warm and humid, contains a rainy season that usually lasts for about a month. Except in northern Japan, the winter is mild with many sunny days. The month of September brings typhoons which may strike inland, bringing torrential rains and violent winds.

Much of the land is mountainous, 66 per cent of the total area is forest, 14 per cent is cultivated land, 3 per cent is wilderness, and 16 per cent is other. The proportion of land that is cultivated is low compared with that of European countries.

Social and cultural background

Apart from a very small number of *Ainu*, a people who exhibit certain Caucasian characteristics, the Japanese population has been, since early times, ethnically and linguistically uniform. The racial origins of the Japanese are still obscure, but both Mongolian and southern Pacific traits can be seen in today's population.

The national language is Japanese, used and understood by all. A single spoken language provides the basis for Japanese cultural homogeneity. Regional differences in speech can identify a person's geographical origin, but they do not seriously impede communication, since standard speech is taught and spoken everywhere. The level of literacy among Japanese, although it varies with economic circumstances and distance from urban centres, is among the highest in the world.

The traditional religions in Japan are Shintoism and Buddhism. Neither is exclusive, and many Japanese subscribe, at least nominally, to both. Since the war, a number of religions based on an amalgamation of Shinto, Buddhist, Taoist, Confucian, and Christian beliefs have grown up.

Economy

The economy of Japan grew at a rate of over 10 per cent per annum throughout the 1960s. In the fiscal year 1974 there was a decrease in the economic growth rate for the first time since the Second World War. At the beginning of 1975, the Japanese Government tried to stimulate business by expanding its public investment, and, as a result, the Japanese economy recovered, achieving a 3.6 per cent growth rate in the fiscal year 1975, while the economies of other major industrial countries were still contracting.

In the second half of 1978, just as domestic private demand began to show a strong recovery after a long process of adjustment, the second oil crisis occurred. The Japanese economy in the fiscal year 1986 performed favourably, overcoming the effects of this oil crisis.

Organizational and functional structure of government

Constitution

In the new constitution, which was promulgated on 3 November 1946 and came into effect on 3 May 1947, the Japanese people pledged to uphold the ideals of peace and order. The preamble of the constitution states: 'We, the Japanese people, desire peace for all time. We wish to occupy an honoured place in an international society striving for the preservation of peace, and the banishment of tyranny and slavery, oppression and intolerance for all time from the earth'.

Legislature

The *Diet* is the highest body of state power and the sole law-making body. It consists of the House of Representatives with 511 seats and the House of Councillors with 252 seats.

Executive

Executive power is vested in the Cabinet, which consists of the Prime Minister and not more than 18 State Ministers, and is collectively responsible to the Diet.

Judiciary

The Judiciary is completely independent of the executive and legislative branches of government.

Local administration

For the purpose of local administration, Japan is divided into 47 prefectures, including the Metropolis of Tokyo. Local administration is conducted at the levels of prefectural city, town, and village governments, each with its own assembly.

Education system

The level of literacy among the Japanese has been almost 100 per cent since before the turn of the century. Immediately after the Second World War, with the introduction of democratic ideas into Japanese education, the educational system and policies for it underwent executive reforms, including the adoption of the 6–3–3–4 system. In the 6–3–3–4 system, the former 6–3 indicates the years of compulsory schooling in respect to primary school and junior high school respectively, and the latter 3–4 indicates the years of voluntary schooling with regard to senior high school and college or university respectively. The Fundamental Law of Education of 1947 sets forth the central aims of education as follows: the 'upbringing of self-reliant citizens with respect for human values and equality of educational opportunity based on ability'.

There are three types of institutions of higher education: universities, junior colleges, and technical colleges. There are also kindergartens for pre-school children and special education schools for the physically handicapped and/or mentally retarded, special training schools, and miscellaneous schools for vocational and practical training.

One of the striking features of education in Japan today is the increased competition for places at good universities. The general standards of education in Japan are very high, especially in mathematics and foreign languages.

Steadily increasing numbers of young people from Asian countries are coming to Japan for technical training at scientific and technological institutions and at factories. In 1980, a total of 6543 students from Asian countries and other areas were enrolled in Japanese colleges and universities.

Apart from school education, social education has been promoted by the enactment of relevant laws, the establishment and improvement of facilities, and the education of leaders. Social education facilities include public halls, libraries, various kinds of museum, youth centres, and facilities for women. Public halls provide opportunities for educational classes and lectures, cultural gatherings, sports, and recreational activities.

Population

The population of Japan at 1 October 1986 was 121.7 million, which ranks seventh in the world, and makes up about 2.5 per cent of the world's population. The population density in 1986 was 326 persons per square kilometre; Japan is the fifth most densely populated country among the countries with more than 5 million people. Since Japan has a limited area of flat land, its population density in terms of habitable land area is actually far higher.

From the seventeenth century to the middle of the nineteenth century, the population of Japan remained stable at approximately 30 million. After the Meiji Restoration in 1868, the population began to increase, and reached 60 million in 1926 and exceeded 100 million in 1967. The average annual growth rate of the population throughout these years was a little over 1 per cent. In 1986, the growth rate of the population was at a low of 0.5 per cent.

Population growth in Japan is caused mainly by the excess of births over deaths, since the net migration to or from abroad is negligible. The birth-rate declined sharply from its highest level during the post-war baby-boom period of 1947–9, and was 11.4 per 1000 population in 1986. The death-rate decreased significantly, and was 6.2 per 1000 in the same year. As a result, the population increase for the year was 5.2 per 1000.

The life expectancy of the Japanese population has shown a remarkable improvement, and had reached 75.23 years for males and 80.93 years for females in 1986.

Currently, the productive proportion of the population (15–64 years of age) is larger than in any other major industrial country. However, the elderly population (65 years old and over) exceeded 12.8 million in 1986, demonstrating the remarkable speed and the extent of the aging of population structure (Fig. 23.2).

The population of Japan is expected to reach 131.2 million in 2000. On the basis of recent population trends and the increasing life expectancy, one person in every six Japanese will be 65 years of age or older. This means that Japan is changing faster demographically than many other countries where population aging is already well under way (Table 23.1).

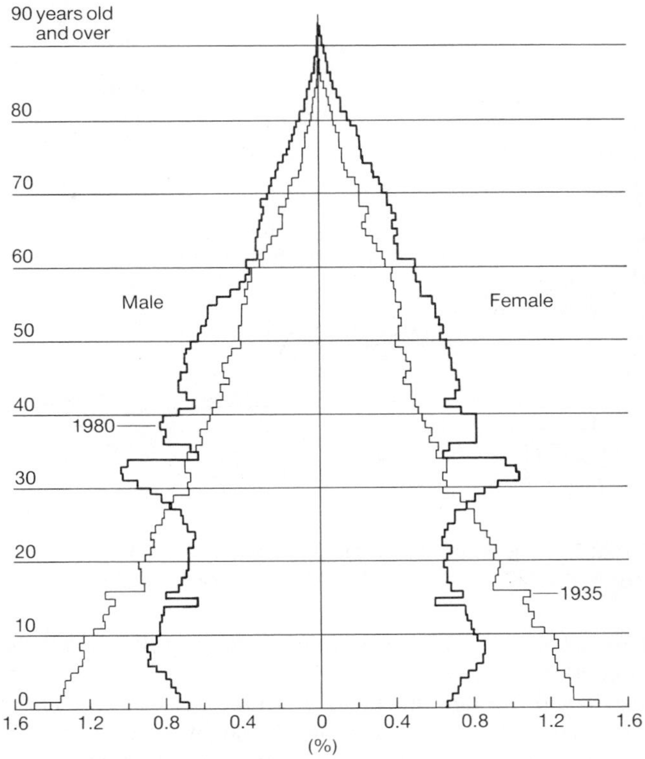

Fig. 23.2. Population pyramid in Japan, 1935 and 1980. (Source: National Census 1935, 1980.)

Table 23.1. Time trends in the age structure of the Japanese population

Year	Percentage of population		
	Age (years)		
	≤ 14	15–64	≥ 65
1950	35.4	59.6	4.9
1986	20.9	68.5	10.6
2000 (estimated)	18.0	65.8	16.2

Source: Kokumin Eisei no Dohko (1988).

Development of health care in Japan

The Government decided to adopt the German system as the official medical policy of Japan in 1870, and its strong influence lasted until the end of the Second World War. The First World War brought tremendous progress to Japanese industry. In those years, spinning was the key industry and was plagued by a high incidence of tuberculosis among its female workers. The Government started a health insurance programme in 1927 covering manual workers and in 1937 started a programme for farmers and fishermen.

The Second World War almost completely destroyed the old system of health care. The post-war years may be described as the period of American influence.

Vital statistics

Whereas Japan suffered from a high death-rate and serious health problems in the pre-war period, health conditions in post-war Japan have improved remarkably along with the development of the socio-economic situation. The infant mortality rate in Japan dropped rapidly in the post-war decades from over 150 per 1000 live births in the 1920s down to 5.2 in 1986. This rate is lower than in most western countries. The perinatal death-rate in Japan was 27.4 per 1000 births in 1950 and only 3.1 in 1986. The crude death-rate in

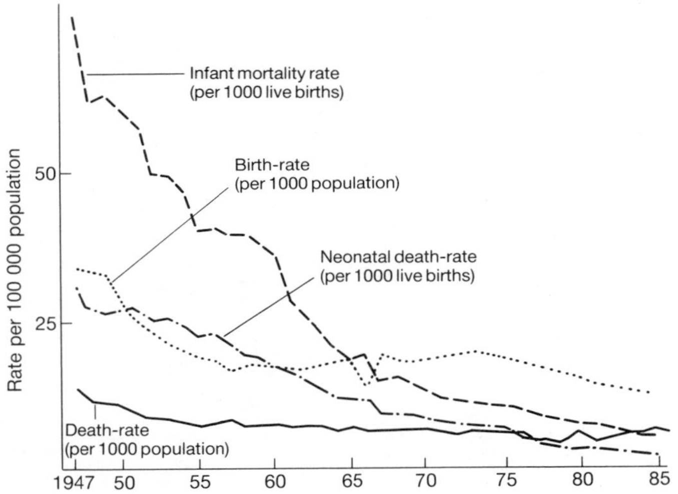

Fig. 23.3. Time trends in vital statistics, Japan, 1947–1985. (Sources: Ministry of Health and Welfare, Statistics and Information Department 1986.)

Japan continued to decrease in the 1950s at a slower rate than soon after the Second World War, but has remained almost unchanged through the 1960s up to the present day. On the other hand, the birth-rate showed a slight increase for some 10 years after 1961, and in 1986 the rate reached 11.4 per 1000 (Fig. 23.3).

Major causes of death

A recent trend has been the gradual increase in deaths from degenerative disease, which comprise a large proportion of the total number of deaths. There has been an increase in the proportion of so-called adult diseases, such as cancer, heart diseases, vascular lesions affecting the central nervous system, and some others, while the number of deaths from infectious diseases, such as tuberculosis, pneumonia, typhoid fever, and other gastro-intestinal infections has decreased greatly (Fig. 23.4).

Prevalence and morbidity of diseases

According to the National Patient Survey, an annual, nationwide, one-day sampling of the number of in-patients and out-patients who utilize hospitals and clinics, there has been a notable increase in the morbidity rate (medical care receiving rate) during the past three decades. The rate per 100 000 population in 1948 was 1306, and the figure in 1984 was 6403. After the second post-war decade, as concomitants of urbanization, industrialization, aging of the population, and a fast increase of the number of the aged, geriatric diseases become prevalent, as is shown in Fig. 23.5. The five leading diseases in the National Patient Survey in 1984 were hypertension, mental disorders, cerebrovascular diseases, heart diseases, and liver diseases. There is also a high prevalence of diabetes mellitus among the Japanese population.

Other measures of health status

From the beginning of the third post-war decade, new types of health problems such as those related to environmental pollution, mental disorders, health problems of the aged and handicapped, and 'Nanbyo' (intractable diseases) have emerged.

According to the National Nutrition Survey of 1985, while the level of nutrition has shown a remarkable improvement generally, there has been the problem of over-eating, possibly leading to degenerative diseases, and of imbalanced or inappropriate nutrition.

Health status has improved in terms of growth in height and body-weight of younger people, and the prolongation of life expectancy. The physique of Japan's youth has shown a sharp improvement since the Second World War. In 1986, the average height of 17-year old boys and girls were 170.3 cm and 157.7 cm, and the average weights were 61.8 kg and 52.8 kg. A recent problem, however, is the increasing number of overweight children. Japan is facing a noticeable increase in degenerative diseases which is possibly connected with the rise in the level of obesity.

Present system of health care in Japan

Government health administration and organization

The modern national health administration is highly centralized, and is divided into four main areas: general health

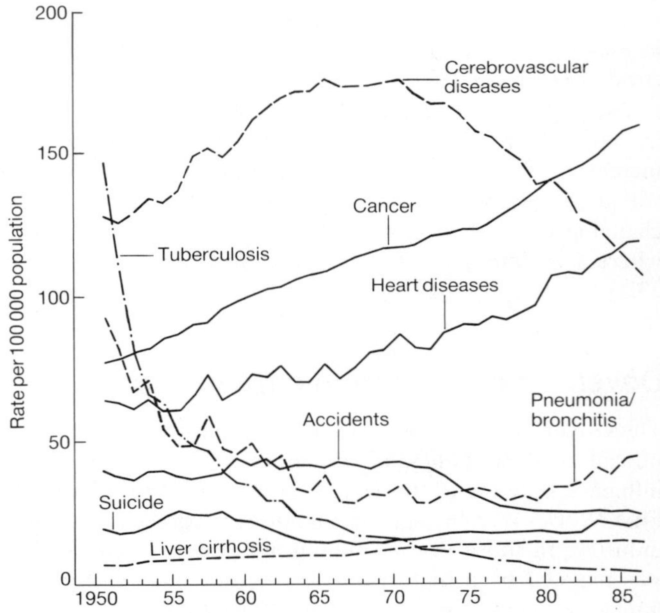

Fig. 23.4. Time trends of the major causes of death in Japan, 1950–1985. (Sources: Ministry of Health and Welfare, Statistics and Information Department 1986.)

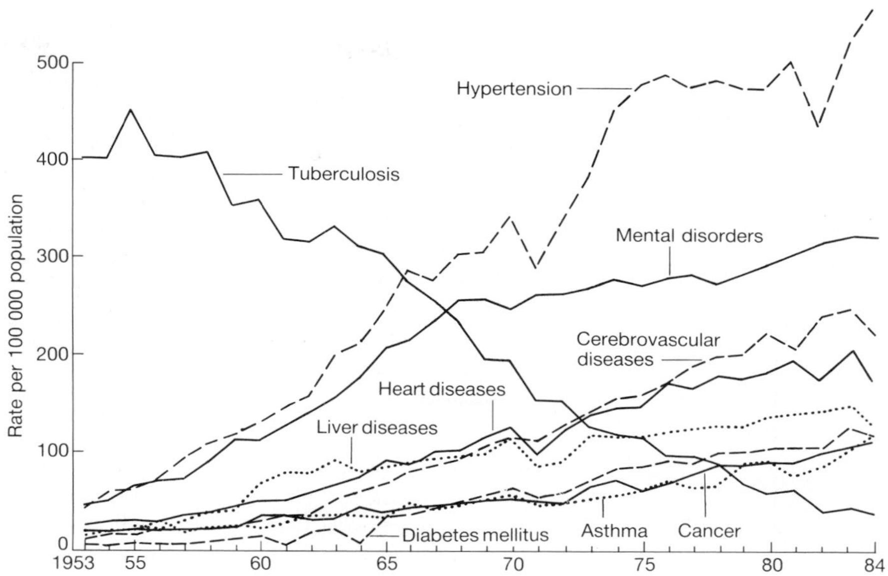

Fig. 23.5. Time trends of the prevalence of major diseases in Japan, 1953–1984. (Sources: Ministry of Health and Welfare, Statistics and Information Department 1985.)

administration, school health administration, industrial health administration, and other agencies. The organizational structure of each, especially for general health administration, is briefly summarized below.

General health administration

General health administration forms a nation-wide system at four levels: central government, prefectural government, health centres, and city, town, and village governments.

Central government

In the central government, the Ministry of Health and Welfare (Koseisho) is the principal body responsible for general health administration, including welfare administration. The main functions of the Ministry of Health and Welfare are:

1. Health and medical services—taking measures for the prevention and treatment of various diseases, improvement of environmental health, adequate provisions for medical care facilities, training of health and medical personnel, quality control of pharmaceutical preparations, and control of narcotics, etc.
2. Social welfare services—taking measures to provide a minimum standard of living for indigent people, giving aid to disabled people and widowed families that will eventually be able to support themselves, and adequate care and protection of children.
3. Social insurance—administration and implementation of various national health insurance and pension schemes to relieve the economic burden of the sick, aged, and disabled.

In addition to the above, the Ministry also gives aid to war-wounded veterans and war-bereaved families as well as those affected by the atomic bomb. National health and medical research institutions also directly administer public hospitals and sanatoria that have been established throughout the country.

Prefectural government

Organization of the 47 prefectural governments is established by the Local Autonomy Law, under which a department charged with health care administration is established in each prefectural government and administrated by elected governor.

Health centres

Health centres have been established, based on the Health Centre Law, in all 47 prefectures, in 23 Tokyo special districts, and in 31 cities designated by a Cabinet order. In 1987, there was a total of 850 (637, 53, and 160 respectively) health centres in operation across the country.

Health centres are responsible for improvement and promotion of public health in the districts. Their major projects are conducting health education, keeping vital health statistics, improving nutrition and food sanitation, environmental sanitation, public health nursing, medical social services, laboratory services, mental health and its social rehabilitation, prevention and control of tuberculosis, venereal diseases and other communicable diseases, maternal and child health, dental hygiene, and other health programmes as required, such as endemic disease control. The general organizational chart for health centres is shown in Fig. 23.6.

From the viewpoint of local health administration and health services, a nation-wide network of health centres is the principal framework for these activities, and each health centre is the focal point for community health.

During the first decade after the Second World War,

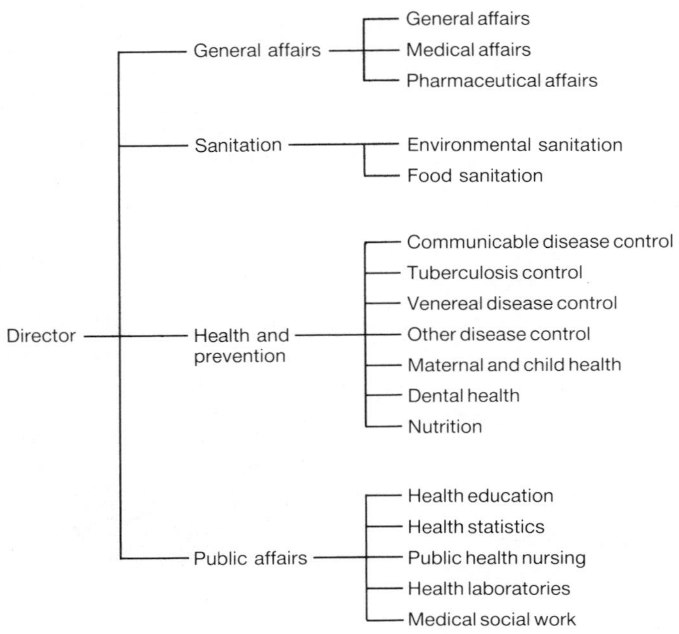

Fig. 23.6. General organization of health centres in Japan. (Source: Kokumin Eisei no Dohko 1988.)

health activities in cities, towns, and villages were very weak, and health centres contributed significantly to the improvement of community health. This was especially true for communicable disease control, maternal and child health, nutrition, tuberculosis control, food and sanitary inspection, and health education.

Particularly from the 1960s, however, the country underwent dramatic socio-economic changes, and there have been notable changes in community health needs due to the changes in disease patterns and the people's mode of living. To cope with these changing community needs, the Ministry of Health and Welfare promoted 'integrated health planning' at the local authority level. Then the ministry developed a plan to reform the existing health centre network.

Even though some reorganization of the health centre system was carried out in the 1980s, it is generally observed that the present function of health centres cannot necessarily cope with the changing needs of different communities; while many of the personnel concerned with health centre activities talk about 'community-based care', most of their ways of thinking and acting are rather technology oriented. They rarely include the viewpoint of members of the community in their planning activities.

City, town, and village government

As of 1980, there was a total of 3255 municipalities (646 cities, 1991 towns, and 618 villages) and in each of these municipalities a section or a subsection charged with health care administration was established. From 1978, city, town, and village government started to establish 'city, town, and village health centres' which are managed by public health nurses, etc., and in 1987 there were 861 such centres in Japan.

Therefore, on average, each prefectural health centre has three or four local authorities in its district. In the case of the health centres established by large municipalities, each situation is different and each health centre is in charge of a part of the respective large city.

At the city, town, and village governmental level there is a division or section responsible for health, and public health nurses are usually employed. The local governments carry out various health programmes, such as health screening against tuberculosis, medical examinations for those over 65 years of age, immunization, and basic environmental services, such as water supply, refuse and night soil disposal, and so forth. Furthermore, in recent years, most local authorities developed programmes for screening for cardiovascular disease and cancer.

All local governments are required by law to establish and carry out a national health insurance programme. In many cases, medical care facilities, established by the national health insurance body, provide medical services directly to citizens.

As mentioned above, integrated health planning by city, town, and village governments has been encouraged by the Ministry of Health and Welfare since 1960, and this kind of community health planning has become popular among the local authorities in recent years.

School health administration

School health is the responsibility of the Ministry of Education, and its Physical Education Bureau maintains a health section. The school health programme at the local level is delegated to the Education Board of the prefectures and municipalities.

Industrial health administration

Industrial health is under the jurisdiction of the Ministry of Labour and is administrated by the Labour Health Section in the Labour Standards Bureau of the Ministry.

Other local agencies and institutions involved with health

In addition to the above, there are various local agencies, institutions, and facilities involved with the health of the people, such as hospitals, general clinics, and dental clinics. Welfare-oriented agencies and institutions, such as welfare offices, social welfare councils, and maternal and child welfare institutions, are also important for health. This is especially the case for the aged, mothers, children, and the handicapped. Mental health centres must also be mentioned here.

Welfare offices

These offices are the comprehensive, front-line institutions for social welfare administration and service provision. They carry out various aid programmes for the needy, care activities, fostering, and rehabilitation, as stipulated by the relevant laws. The total number of welfare offices was 1175 in 1985.

Child guidance centres

These centres form the nucleus of all child welfare services, and one is established in each prefecture and in larger municipalities. There were 645 such centres in 1986. In recent years, day-care facilities and ambulatory rehabilitation facilities have gradually been established by local authorities, but the number of these is still rather small.

Mental health centres

According to the Mental Health Law, the prefectural government is requested to establish a mental health centre. As of 1986, there were 43 such centres in the country. The major roles of these centres are: (a) to disseminate relevant knowledge about mental health; (b) to conduct investigations and study mental health; and (c) to provide counselling services including guidance which is particularly complicated and difficult.

Levels of care

Self-care—grass roots organization and community participation

It should be noted that the traditional grass roots system of organization is highly popular in both urban and rural communities in Japan, as it makes daily life easier and more convenient.

From the viewpoint of community health activities, it is worth noting that community participation in various health programmes increased after the war. The level of community participation has grown steadily, and has diversified to include mental and child health, family planning, parasite control, tuberculosis control, nutrition, and so forth. At present, a number of local authorities nominate voluntary health leaders or spokespersons for the promotion of community health.

In regard to community participation, we must mention the existence and significance of so-called 'patients' groups'. Although more than 100 patients' groups can be identified in Japan, some of the groups have recently tried to develop 'community-based care' in collaboration with medical/health professional groups. An example of such a group concerned with community care for intractable diseases can be seen in some suburban areas of Tokyo. The name of the patient's group is 'Tokinkyo' (muscular dystrophy patients and their families). The group was the promoter and supporter of the development of community care for muscular dystrophy in the community. Its experiences and outcome activities should serve as a boost for the establishment of other types of community care activities, such as care for bedridden senior citizens, mental health care (for example, 'Yadokari-no-sato'), and care for mentally retarded persons.

Primary care—the contribution of local medical practitioners/associations to the development of community health activities

In Japan, community physicians are not the same as general practitioners in western countries. They basically work alone in private practices. Most are also engaged in public health activities in the community as part-time physicians and they also play a role in the prevention and control of community medical/health problems.

Despite the establishment of a national health insurance scheme in Japan, which meets all the demand for medical services, private practitioners continue to control the supply of medical services.

In Japan, private practitioners spend five or more years at a university hospital after completing six years of medical education to obtain the higher academic degree of Doctor of Medical Sciences, which roughly corresponds to a Doctor of Philosophy degree. This is a valuable status symbol for private practitioners in Japan. They are the most influential doctors in the community. As a result, the national health policy tends to favour local private practitioners.

Local medical associations are composed of private practitioners, and combine to form the Japanese Medical Association which currently includes about 75 per cent of the licensed physicians and 90 per cent of the practising physicians in Japan. The Japanese Medical Association is a powerful national group and has an important influence on health policy in Japan.

In Japan, national, prefectural, and local health planning at the administrative level have been developed as described above. On the other hand, for the past 15 years, the Japanese Medical Association has been trying to develop local health planning under the title of 'community health planning' from the viewpoint of community medicine. It should be remembered that the primary medical care system in Japan has basically been developed and supported by the collective efforts of private practitioners all over Japan. This might be one of the reasons why each local association of the Japanese Medical Association, has voluntarily been trying to develop 'community health planning' based on the comprehensive medical care system or community medicine.

One of the main elements of community medicine should be 'primary medical care'. During patients' first contact with community physicians, primary medical care plays an important role in the total process of their medical care. Since 1970, the Japanese Medical Association has been stressing the importance of the role of primary medical care in the community, because primary medical care in Japan has a strong effect on secondary and tertiary care.

To meet the above idea of primary medical care in Japan, the Japanese Medical Association has, so far, also tried to establish community medicine association-operated hospitals and biomedical examination centres.

Community care and hospital—health personnel and hospitals/clinics

Table 23.2 summarizes the current status of health personnel showing their number by category, at the end of 1986.

With regard to physicians, in 1986, 95.8 per cent of the total number were actively engaged in clinical practice, and 4.2 per cent were engaged in non-clinical work, such as research and health administration. The geographical distribution of physicians is rather uneven. For example, the average number of physicians per 100 000 population in 1986 was 157.3, with several prefectures as high as 216.3 (Tokyo) and others as few as 96.6 (Saitama).

At the end of 1986, Japan had 9699 hospitals with 1 533 887 beds, compared with 7974 hospitals and 1 062 553 beds in 1970. Recently, in response to the changing disease pattern, there has been a considerable increase in number of general and mental hospitals, and an expected drop in tuberculosis and leprosy hospitals. At the end of 1986, there were 79 369 general clinics and 47 174 dental clinics.

Table 23.2. Number and population ratio of health personnel, Japan 1986

Category	No.	No. per 100 000 population
Physician	191 346	157.3
Dentist	66 797	54.9
Pharmacist	135 990	111.8
Public health nurse	22 050*	18.1
Midwife	24 056*	19.8
Registered/assistant nurse	639 936*	526.0
Dental hygienist	32 666*	26.8
Dental technician	31 139*	25.6

Source: Kokumin Eisei no Dohko (1988).
* Active number.

Thus, in 1986 there were 12.6 beds per 1000 population, of which 9.2 were general hospital beds, 2.8 were mental health-care beds, and 0.4 were tuberculosis-care beds. When the number of clinic beds is added to this, the bed:population ratio becomes 14.9 per 1000, which is more than double the ratio in the US. The major difference is caused by the traditional concept of hospitalization in Japan. Japan has no 'acute hospitals' so that chronic patients who would usually be in nursing homes in western countries are always treated as in-patients in hospitals or clinics. The average length of stay in general hospitals in Japan, 40 days in 1986, indicates this peculiarity (Table 23.3). In addition, most hospitals in Japan are based on a closed pattern with full-time, salaried medical staff. Group practice is still at an experimental stage.

With regard to the ownership of medical facilities in Japan 76.2 per cent of all hospitals and 92.5 per cent of all general clinics were conducted on a private basis in 1986.

There are still a number of small and remote communities without a permanent physician. However, from the viewpoint of each health centre district, the levels of social resources in these areas seem to be good in terms of their number and capacity.

It should also be noted that, at the prefectural and large municipal levels, a number of regional health-related institutions and facilities exist. There are, for instance, 80 medical schools or medical colleges with teaching hospitals.

Financing and provision of care—medical care security system

Japan has three major components to its medical care security system, which is a critical element of the entire social security system.

The first, and the core of the system, is the social health insurance programme covering the entire population.

The second is the medical assistance scheme based on the Daily Life Security Law 1950. The law provides for public assistance to all needy people on the basis of a means test. This scheme consists of aid for living expenses, educational aid, housing aid, medical aid, maternity aid, occupational aid, and funeral aid. All assistance except for medical care is provided in cash.

The third component is the public medical care programme for selected diseases and disorders. At present, there are more than ten programmes based on such laws as the Tuberculosis Control Law 1951, the Mental Health Law 1950, amended in 1987, the Maternal and Child Health Law 1965, the Old People's Welfare Law 1963, and Elderly Health Law 1982, among others.

In addition, there are several programmes established by budgetary measures, rather than through legislation, which deal with so-called intractable diseases (Nanbyo), selected chronic child diseases, and others.

Health insurance scheme—background and classification of the programme

In 1922, the Health Insurance Law, the first social insurance programme in Japan and in Asia, was enacted to cover workers covered by the Factory Law or the Mining Law. The enforcement of the law was delayed until 1927, and when the programme started the insurants numbered only 1.9 million. In 1938, the National Health Insurance Law, the first community health insurance programme covering farmers, foresters, and fishermen was enacted. National coverage was achieved in 1961.

In Japan, medical services are supported by health insurance. As for the employees' health insurance system, there are two programmes. One is the Government-managed health insurance (Seikan-kempo), whose insurance carrier is the Government, and the other is the society-managed health insurance (Kumiai-kempo), which is established for larger enterprises by a specific health insurance society. There is also the seamen's health insurance programme, the workers' day health insurance programme, and mutual aid association insurance programmes such as the one for civil servants.

Table 23.3. The average bed occupancy rate and the average length of stay by type of hospital, 1986

Type of hospital	Total	Psychiatric hospitals	Infectious disease hospitals	Tuberculosis sanatoria	Leprosaria	General hospitals
Bed occupany rate (%)	85.7	100.6*	1.4	55.9	72.1	83.8
Length of stay (days)	54	533	19	200	10 736	40

Source: Kokumin Eisei no Dohko (1988).

* Theoretically, the value should be less than 100 per cent, but in Japan when a patient discharges and a new patient admits on the same day for the same bed, two patients are counted in a statistical sense.

Table 23.4. Coverage of health insurance, 1986

Insurance programme	Total (thousands)	Insurants (thousands)	Dependants (thousands)	Percentage
Employees' health insurance (total)	75 705	33 931	41 774	62.4
Government-managed	32 624	15 430	17 194	26.9
Society-managed	30 052	13 023	17 029	24.8
Day workers' health insurance	227	146	81	0.2
Seamen's insurance	542	173	369	0.4
Mutual aid association insurance	12 260	5 159	7 101	10.1
National health insurance	45 536	—	—	37.6
Total population insured	121 241	—	—	100.0

Source: Kokumin Eisei no Dohko (1988).

The national health insurance programme is managed by the local authorities (cities, towns, and villages). Table 23.4 shows the coverage of social health insurance, by programme as of the 1 April 1986.

Health insurance benefit system

The benefits of health insurance include physician and dentistry services in hospitals and clinics, diagnostic procedures and treatment including medication, surgical operations, and hospitalization. Screening services and normal child delivery are not included in the benefits. There are also cash benefits for absences from work due to sickness, injury, or childbirth.

The benefit rate for the employees' health insurance programme is 90 per cent of the medical care costs. The costs of a private room in hospital, special denture material, or a private nurse are excluded from the benefits. The benefit rate for dependants covered by the Government-managed employees' health insurance programme was raised from 50 to 70 per cent of medical care expenditure in 1975. In the case of the national health insurance programme, the benefit rate has been 70 per cent for both householders and dependants since 1968.

Since the fiscal year 1973, the partial payment by the patient and dependants has been subsidized by the Government in the case of tuberculosis and mental diseases, and where the dependant's payment exceeds 54 000 yen a month it is reimbursed as a 'special payment for expensive medical care' (Kogaku-ryoyohi). In addition, patients aged over 70 years contribute 800 yen per month for out-patient services, and 400 yen per month for in-patient services, but otherwise they receive free medical services. When an insured person wants to receive medical service, he or she is free to go to any hospital or clinic upon presentation of an insurance scheme membership card.

Finance and payment methods of health insurance

Health insurance is financed by the contributions of members and the subsidies granted by national Government.

Contributions to health insurance are shared on a 50:50 basis by employee and employer, and the amount is determined by the wage of the insurant. The association-managed health insurance is generally the lower rate of contribution compared with the government-managed scheme.

In national health insurance, most insurers collect premiums from their insurants as tax under the Local Tax Law. In the government-managed health insurance system, on account of continuing deficits, the National Treasury was made liable under a Law in 1973. At present, the national subsidy rate is 16.4 per cent of the medical care and the selected cash benefits (Fig. 23.7).

In 1943, the present fee-for-service payment system was adopted. At the end of each month, each participating hospital or clinic issues a bill for the services it has provided. Each act of medical practice is given a certain point score. The bill is submitted to the Social Insurance Medical Care Fee Payment Fund, which pools the money from each programme,and is checked by medical consultants before payment of the bills. Unnecessary treatment is checked here and deleted from the bill.

To determine the fee schedule, the Minister of Health and Welfare requests the recommendation of the Central Social Insurance Medical Council, which consists of insurers and

Category			Insurer (as of March 1987)	Insurance benefits: Medical care*: Medical benefits	Insurance benefits: Medical care*: Dependants' medical benefits	Insurance benefits: Cash payment	Financial Resources: Contribution Rates	Financial Resources: Government subsidy for insurance benefits cost
Employees' health insurance	Regular employees	Government-managed health insurance	Government	90%(80% from the day following the date of public notification by the Health and Welfare Minister pursuant to the Diet approval)	80% for in-patients 70% for out-patients	Injury and sickness allowance/ Maternity allowance/ Delivery expenses etc.	8.3% 1.0% Special insurance contribution (from March 1986)	16.4% of paid benefits etc.
		Society-managed health insurance	Health insurance societies 1777		The same as above (additional benefits available)	The same as above (additional benefits available)	8.103% (average of all insurance societies) as of March 1987	¥7.05 billion in benefits disbursed
National health insurance		Farmers, self-employed, etc.	Municipalities 3270 National health insurance associations 167	70%		Midwifery expenses Funeral expenses Nursing allowance etc. (Optional)	Contribution based on individual income, assets etc.	50% of paid benefits 32–52% of paid benefits
		Retirees from employees' health insurance schemes	Municipalities	Insured person 80% Dependants 80% for in-patients, 70% for out-patients		(as above)	(as above)	None

*Benefit for high-cost medical care; maximum amount to be paid by patients ¥ 54 000 (¥ 30 000 for low-income patients)

Health and medical services for the aged (end of fiscal 1987)	Municipalities	100% Partial payment ¥800 per month for out-patients, ¥400 per day for in-patients (¥300 per day up to 2 months for low-income patients)	Paid by the insurers of each medical insurance scheme	Financing shares Government 20% Prefectures 5% Municipalities 5% Insurers of each medical insurance 70%

Fig. 23.7. Medical insurance schemes in Japan. (Source: Ministry of Health and Welfare 1988.)

providers (physicians, dentists, and pharmacists), and representatives of the public interest. Treatment under health insurance is almost free from restrictions.

Utilization of health care services

In Japan, patients have different modes of entry into the health care net, and they may visit a specialist or enter a hospital without being referred by general practitioner. There is no linear form of referral system in which a general practitioner may refer an individual to a consultant or specialist.

In 1986, 40 per cent of patients visited a general practitioner (clinic), 21 per cent a hospital and 26 per cent a general hospital, 5 per cent of patients visited a university hospital for specialized care, and the remainder (6 per cent) visited an acupuncture or a massage clinic (Table 23.5).

Individual patients in Japan see their doctor, on average, on about five or six occasions each year. In 1986, 30 per cent of patients changed their clinic or hospital for treatment for the same disease either by their own will or were referred by their own doctor. Of the drugs that are ordered by medical doctors, 90 per cent are dispensed in clinics or hospitals and 10 per cent are dispensed at drug stores under the prescription of a medical doctor. All these drugs are covered by health insurance, in which benefit rates range from 70 to 100 per cent. In 1987 there were 35 915 drug stores which were operated by pharmacists.

Besides those treated under western medicine, 6 per cent of patients go to eastern medicine or traditional medicine, such as Anma massage and acupuncture, which were provided by 86 806 and 108 782 specialists, respectively, in 1986.

Role of health insurance on the development of health care services

The rapidity of Japanese reduction of mortality might be explained by the combination of socio-economic development, political commitment to health, educational–human development, and increasing medical resources.

After the Second World War, the number of physicians in Japan increased from 92 442 in 1954 to 191 346 in 1986, and the number of nurses increased from 119 428 in 1954 to 639 936 in 1986. Those increases in medical resources are due to the economic development of Japan and the country's political commitment to health.

Of physicians in Japan, 80 per cent belong to private practice. They are connected with private enterprise and the governmental health insurance mechanism. This motivates medical practitioners and consumers to maintain the accessibility and appropriate utilization of medical resources.

Based on the high levels of education of the people, which are rooted in the influence of 1400 years of Chinese culture, 30 per cent of young people now go to western-style university. This has been the key to the expansion of self-care and mutual help. For example, there are over four million members of the women's Anti-tuberculosis Association.

Present problems—an alternative approach to provision and financing of health care services in Japan

Japanese approaches to the provision and financing of health care services (based on the general network approach)

'General networking' is the discipline for the 'Japanese approach to the provision and financing of health care services'. This approach is applicable to all complicated problem-solving that cannot be solved by conventional approaches alone. This approach has been developed over ten years, based on the concept of primary health care (World Health Organization/United Nations Children's Fund (UNICEF) 1978). (The basic understanding of the general network approach differs from person to person because of practical differences in human nature and attitude or behaviour which are discussed in detail at the end of this section.)

Table 23.5. Estimate of out-patient behaviour in Japan (based on the one-day prevalence survey on 4 September 1986)

Place of out-patient attendance	Percentage
Clinic (general practice)	40.0
Hospital (general practice, secondary care)	20.5
General hospital (general practice, secondary care, tertiary care)	26.2
University hospital (specialized care)	4.7
Acupuncture, massage (traditional medicine)	5.9

Source: Kokumin Eisei no Dohko (1988).

In order to evaluate not only the past status but also the present problems and future perspectives of health care services in Japan a theory is required to contain all aspects of the health care services. It should be qualitative and total in nature, and should be applicable to all levels—national, provincial, and municipal. It should not be so idealistic that it could be termed a philosophy and not a science.

A well-known example is the Asahi village project in central Japan, which combines idealistic and realistic activities. The village activities are deeply rooted in the needs of the general population, individuals, and patients, and lead to public health policy and services, and their co-ordination and development.

At present, Japan combines the free enterprise US-style health system with private practice, and the UK National Health Service-style system of nation-wide health insurance. An alternative method of health care provision in Japan is demonstrated by the Asahi project.

The unique experience of Asahi village, its achievements and perspectives as a 'healthy village' over the past 20 years, is applicable both to developed and developing countries, because it was initially a farming village which has become increasingly urbanized, with its own electricity and water supplies, cars, and even a village-owned television station.

The Asahi 'healthy village' project

Asahi village is located in central Japan, 18 km from Matsumoto city, Nagano prefecture, and has a land area of 70.39 km^2, of which 88 per cent is forested and mountainous. The population has been between 4000 and 5000 for the past 60 years, and is made up of 1000 households.

Asahi village embarked on a policy of the construction of a healthy village in 1965 in order to integrate all the existing health activities, such as parasite control, maternal and child health, and environmental health, and to establish a comprehensive approach to rural health problems. This involved setting priorities and selecting measures for each problem-solving process, with the co-ordination of the local village government, community organizations, and the other health-related agencies outside the village, such as health centres and the local medical school (Fig. 23.8).

In 1965, 60 per cent of the employees in Asahi village were engaged mainly in farming, and 10 per cent of the total population was over the age of 65 years; in 1985, however, these figures had changed to 33 per cent (36 per cent in industry, 31 per cent in the service sector), and 16 per cent respectively. Although people now have electricity, water supply, cars, and even a cable television network in Asahi village, 60 per cent of households are still partially engaged in farming. Asahi village is thus classified as a rural area, as the proportion of workers in farming in 1985 was at the same level as the whole of Japan in 1960.

In 1985, Asahi village was awarded the 37th 'Health Culture' in Japan, which is given for the most excellent local health care activity in Japan each year. In 1965, when Asahi

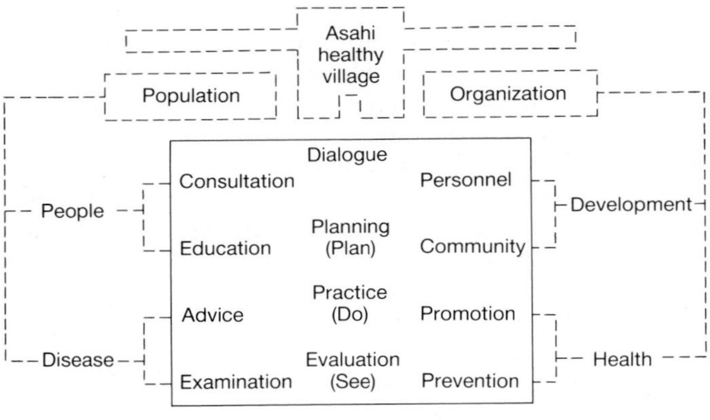

Fig. 23.8. Ideal scheme for the Asahi healthy village project.

village began the healthy village project, one of the main health problems in the village was cerebrovascular disease, which was the leading cause of death in Japan in the 1950s, 1960s, and 1970s. In the village, the local government and people organized the 'health promotion council', which is the main body for implementing the processes necessary for a healthy village, i.e. community research and diagnosis, evaluation of health services problems, and then appropriate health education and mass screening for the relevant diseases, with the collaboration of village people and the local health professions.

As a result of the Asahi healthy village project, the average age of the episode of cerebral apoplexy was delayed from 69 years in 1965 to 75 years in 1980 (a gain of 6 years), and the age at death from apoplexy was increased from 73 years in 1965 to 77 years in 1980 (a gain of 4 years). On the other hand, medical/health costs per person in the village were 14 per cent lower than the national average in 1984.

Provision and financing of health care services in Asahi village

The health promotion activities of Asahi village over its 24 years of existence have recently been reviewed, and it is considered that its financing has been well maintained in comparison with that of most other local governments in Japan.

Problem finding

In 1965, Asahi village had problems with both the provision and financing of health care services, namely cerebrovascular disease and a deficit in health insurance, which were common health problems in Japan at that time (Fig. 23.9).

Problem solving

Both the provision and financing problems were combined, and a clear statement of village policy was issued. After 24 years of practice, both problems were solved. The incidence of cerebrovascular disease was reduced, and the lowest medical/health costs in Nagano prefecture were achieved.

Evaluation

The difference between Asahi village and other places lies in the 'two-in-one' process of solving provision/financing problems. Conventional local governments usually tried to solve these problems separately, with few mechanisms for co-ordination.

The Asahi healthy village project is unique because of its community participatory approach to the problem-solving process, which is usually, in contrast, confined to health specialists in Japan.

In Asahi village, two public health nurses and three medical practitioners are provided in collaboration with the prefectural government, health centre, and the local medical school. The healthy village council co-ordinates them, which gives the people concerned the opportunity to learn about the problem-solving process. Once the people have learned problem-solving for a health problem such as cerebral apoplexy, they can apply the same process to other health problems, such as cancer detection, heart disease, and health promotion.

This problem-solving process is a managerial/psychological process, comprised of problem finding, solving, and evaluation. In medical science, this process is called the 'clinical course', i.e. diagnosis, treatment, and prognosis. In nursing, it is called 'nursing process', and in public health the public health process, i.e. policy, planning, practice (methodology), and programme evaluation. Asahi village is unique in the common understanding of problem-solving between the people in the village and health professionals. This co-operation has strongly supported the construction of a healthy village.

Future needs

Provision/financing of health care services in Japan

The main difference between the approach of Asahi village and elsewhere in Japan centres on differences in output in terms of provision and financing. Asahi has achieved a good provision of health services at low cost; Japan, in general, has achieved a good health standard in terms of mortality rates, but has many new health problems, such as urban mental

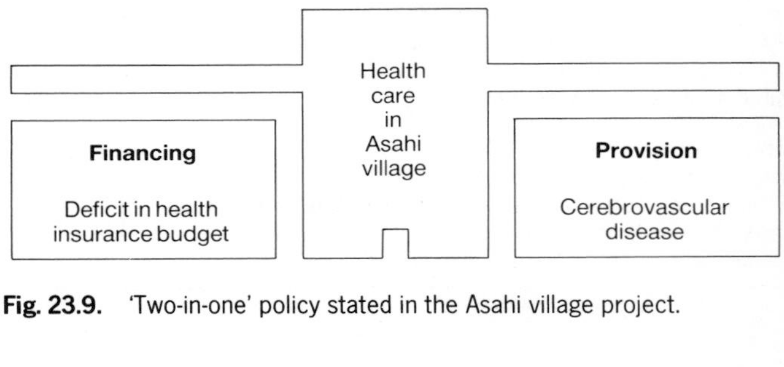

Fig. 23.9. 'Two-in-one' policy stated in the Asahi village project.

health, problems of the elderly, and the increasing cost of financing health care.

One of the reasons for differences in achievement between Asahi and Japan in general, in terms of provision and financing of health care services, is the existence of a clear statement of health policy which is achievable if supported and understood by the people concerned.

Psychological processes for problem-solving

Based on the above Asahi experience, we propose that the general networking problem-solving process could be applicable for general use in all problem-solving. (The process of writing this paper itself also follows the general network problem-solving approach through the development of the idea and a variety of exercises.)

Application of the general network approach to the other current health problems in Japan

The Asahi experience could be applied to other health problems such as cancer, thyroid diseases, glaucoma, and acquired immune deficiency syndrome studies which are oriented to prevention on a daily basis, with special emphasis on aetiological and epidemiological studies.

Provision and financing of health care services in Japan—the general network approach, advantages and disadvantages

The general network approach could be validated through team approaches for solving common problems. Alternatively, the approach could be seen simply as a philosophy to be discussed in an ivory tower.

The main prerequisite for the general network approach is that human behaviour has a 'two-in-one' horizontal nature; 'two' being social and individual attitudes, and 'one' being human independence. The so-called 'four major principles of human independence' are listed in the middle of Figure 23.10—dialogue, empathy, autonomy, and learning. Social and individual attitudes are placed horizontally on the outer zone in the figure, i.e. as both advantages and disadvantages of the general network approach.

The practical case approach has a 'two-by-one' vertical nature, with 'two' being the socio-individual value system and socio-personal pathology, and 'one' being human independence. The socio-individual value system is listed vertically on the right side of Figure 23.10 and is a social value system and individual attitude indicated by needs oriented, community participation, maximum use of resources, and co-ordination and integration which are the four major principles of the primary health care concept. Socio-personal pathology is placed vertically on the left side of the figure and personality is represented as egoist or an aggressive personality, and social environment is characterized by the bureaucratic organization and hierarchical structure which can be observed in some social settings.

The general networking model thus has a multi-dimensional nature which cannot easily be expressed on paper. Figure 23.10 should be regarded merely as an indicator of the principles of the approach.

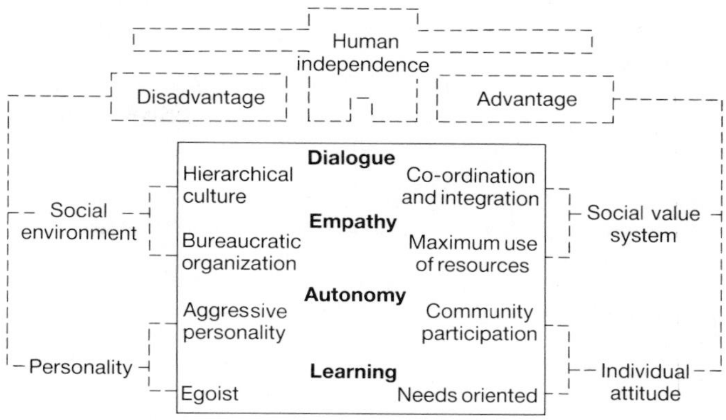

Fig. 23.10. Advantages and disadvantages of the general networking approach.

Those people whose interests lie to the right of Figure 23.10 have the appropriate attitudes to make use of the general network approach. They can use the approach usefully to concentrate essential information, as in the Asahi village. Those people whose interests lie primarily to the left side of the figure would experience communication problems in common problem-solving using the general network approach.

Bibliography

Foreign Press Centre (1982). *Facts and figures of Japan*, 1982 edition. Foreign Press Centre, Tokyo.

Hashimoto, M. (1982). National health administration in Japan. In *Seminar on national health administration*. International Medical Foundation of Japan, Ministry of Health and Welfare, and Japan International Co-operation Agency, Tokyo.

Hashimoto, M. and Tanaka, T. (1978). Recent trends in the regional health planning of Japan. In *Health planning and health information in southeast Asia* (ed. N. Maruchi and P. Tuchinda) p. 227. SEAMIC, Tokyo.

Kaihara, S., Fukushima, M., Maruchi, N., Sasaki, Y., and Uematsu, U. (1979). Some description on the community health activities in Japan with special emphasis on planning and evaluation. In *The development of operational, performance and impact indicators with special reference to community health*. Proceedings of the 6th SEAMIC Workshop, p. 147. SEAMIC, Tokyo.

Kiikuni, K. (ed.) (1977). *Health services in Japan*. Special documents distributed at the 20th International hospital congress, Tokyo.

Kokumin Eisei no Dohko (1981 and 1988). *Kosei no Shihyo*, **28(9)** and **35(9)** (special issue). Kosei Tokei Kyokai, Tokyo (in Japanese).

Maruchi, N. (1983). *Community health in Japan with special emphasis on the assessment and perspectives for the problems of community health in Asia in the 80s*, Vol. 1, p. 21. The Asian Health Institute, Nagoya.

Maruchi, N. (ed.) (1986). *Holistic approach for health network*, p. 129. International Nursing Foundation of Japan, Tokyo.

Maruchi, N. (ed.) (1987*a*). *Holistic approach for health network—experimental edition*, p. 85. Primary Health Care Study Group, Tokyo.

Maruchi, N. (ed.) (1987*b*). *Holistic approach for health network through general network approach*, p. 81. ASEAN Training Centre for Primary Health Care Development, Mahidol University, Bangkok.

Maruchi, N. (ed.) (1987*c*). *General networking (GN) in health and disease*, p. 64. School of Public Health, Seoul National University, Seoul.

Maruchi, N. (1987*d*). Theory and practice on GN theory. *Japanese Journal of Nursing Science* **12**, 186 (in Japanese).

Maruchi, N. (1987*e*). Prevention on communicable disease with special reference to AIDS control. *Japanese Journal of Nursing Science* **12**, 84 (in Japanese).

Maruchi, N. and Haruna, Y. (1987*a*). Manpower development for international health co-operation with reference to software development. *Iryo* **3(6)**, 16 (in Japanese).

Maruchi, N. and Haruna, Y. (1987*b*). Aetiology and epidemiology of chronic pancreatitis with special emphasis on prevention-oriented approach by means of the general networking approach. *Kan-tan-sui* **15**, 739 (in Japanese).

Maruchi, N., Haruna, Y., and Ishi, Y. (1987). An alternative approach on heuristic problem solving for human network. *Proceedings of 3rd Human Interface Symposium* **3**, 45.

Maruchi, N., Shimanouchi, S., and Matsuda, M. (ed.) (1986). *Total care through dialogue with subject matters*, p. 140. Igaku-shoin Ltd. Tokyo (in Japanese).

Ministry of Health and Welfare (1974). *Health services in Japan*. Ministry of Health and Welfare, Tokyo.

Ministry of Health and Welfare (1977*a* and 1988). *Health and welfare services in Japan*. Japan International Corporation of Welfare Services. Tokyo.

Ministry of Health and Welfare (1977*b* and 1981). *Guide to health and welfare services in Japan*. Ministry of Health and Welfare, Tokyo.

Ministry of Health and Welfare, Statistics and Information Department (1986). *Vital statistics for 1947 to 1985*, Statistics, No. 5. Ministry of Health and Welfare, Tokyo.

Ministry of Health and Welfare, Statistics and Information Department (1985). *Patient survey for 1953 to 1984*, Statistics, No. 66. Ministry of Health and Welfare, Tokyo.

Nishi, S. (1982). National health development plan in Japan and its problems. *Bulletin of the Institute of Public Health* **31**, 45.

Noh, T. and Gordon, D.H. (ed.) (1975). *Modern Japan: Land and man*. Teikokushoin, Tokyo.

Statistics Bureau (1982). *Statistical yearbook of Japan, 1982*. Statistics Bureau, Prime Minister's Office, Tokyo.

Statistics and Information Department (1979 and 1980). *Patient survey, 1979 and 1980*. Statistics and Information Department, Minister's Secretariat, Ministry of Health and Welfare, Tokyo.

Takai, T., Aiso, F., and Maruchi, N. (1983). National health system development in support of HFA/2000 with special emphasis on health manpower development in Japan. In *Health Manpower Development in SEA* (ed. P. Tuchinda, T. Tanaka, and N. Maruchi) p. 47. SEAMIC, Tokyo.

Tanaka, T. (1977). Medical care in Japan—yesterday, today and tomorrow. In *Health aspects of community development in southeast Asia—community health/medicine* (ed. H. Katsunuma, N. Maruchi, and M. Togo) p. 51. SEAMIC, Tokyo.

Village of Asahi (1987). *Guidebook of Asahi village*. Asahi Village, Nagano (in Japanese).

World Health Organization, Regional Office for Western Pacific Region (1975; revised 1979). *Country health information profile of Japan (WHO/WPR/HIN)*. World Health Organization, Manila.

24

Provision and financing of health care in Australia

JOHN MOSS and TIM MURRELL

Themes

Australian health services are organized in a mixed public–private system of ownership which is best known for a long-running and acrimonious conflict over hospital and medical finance. Although methods of payment are to the forefront of the debate, they are surface manifestations of underlying struggles over power and purpose, culminating in fundamental differences over the respective roles of Government and the private sector.

In this country, the arguments have eventually come to focus on payment for services. With a national health service along British lines apparently ruled out by the Constitution, Australia has become a social laboratory for different approaches to health insurance. It may not be so unusual for a nation to turn from voluntary to universal coverage, but Australia is unique in having done this twice over. A national health insurance mechanism has been created, dismantled, and restored.

These extraordinary events can shed light on the comparative cost-effectiveness and equity of the alternative policies. Even more importantly, the Australian upheavals clearly demonstrate how financial and administrative arrangements are not autonomous, but are determined by political, social, cultural, and economic forces (Table 24.1).

Important influences may also be found in the realm of ideas. The public health movement has resurfaced to challenge the conventional wisdom on the nature of the health service task. The question is whether health really is best advanced when biomedical scientific research enlightens restorative services provided in high technology hospitals by increasingly specialized medical practitioners and subordinate ancillary workers. No satisfactory answer can now ignore social factors, prevention, equity, team-work, and community participation. This chapter should therefore be read in conjunction with Chapter 30 in which recent developments in Australia's public health services are described.

Table 24.1. A multi-dimensional perspective of the Australian health system

Context
Environment: physical, chemical, microbiological
Society: networks of interaction, distribution of inequalities
Culture: popular beliefs about health
Politics: who has power? who gains and who loses?
Economics: inescapability of choice in resource allocation
Ideology: what is health? biomedical model or public health?
Mediating variables
Resources
Policy
Pressure groups
Technology
Information
Pattern of organization
Sickness services and public health
Institutions and agencies
Personnel
Access: who misses out?
Health outcomes
Physical, mental, and social health
Mean and distribution
Evaluation
Cost-effective?
Equitable?

Medicare: an end to instability?

On 1 February 1984, Medicare, described by the responsible Commonwealth Minister as a simple, universal, and fair system of national health insurance, commenced operations. After some 18 years of instability and the better part of a century of struggle, the Australian debate over health care finance might well have been regarded as settled. Nine years previously, in 1975, voluntary (though heavily taxation-subsidized) private health insurance had been supplanted by universal coverage under a taxation-funded health care payment mechanism entitled Medibank. Between 1976 and 1982, following a change in Government, Medibank had been progressively dismantled in a perplexing succession of seemingly *ad hoc* policy changes. Then, with the return to government of the Australian Labor Party (ALP), the spirit of Medibank coupled with better mechanisms of financial control had been restored under the revised name Medicare. Two essentially similar policy cycles had been completed.

Yet five years later, the system remains under stress and its overall affordability is questioned; power struggles seem inevitable so long as private medicine is publicly funded and the Commonwealth Parliamentary Opposition contemplates a return to private health insurance, although underpinned by public finance (O'Reilly 1988). At the time of writing in early 1989, the Royal Australian College of General Practitioners was about to implement a controversial new fee schedule for financially troubled general practice (Allender 1989). Meanwhile, public hospital budgets were under considerable strain and waiting lists were lengthening (Cooper 1988). Whether these are straws in the wind or precursors of a sustained momentum for further change remains to be seen.

Pluralist or merely disorganized?

At no stage has Australian health care been wholly private or wholly public. A mixed public–private system for both the provision and the financing of health care has survived throughout all the upheavals. This mixture might be regarded merely as disorganized or as reflecting a pragmatic pluralist compromise. Although short-term expediency undoubtedly has been to the fore, an alternative interpretation is that in a democracy multiple goals evoke multiple strategies, and lead to multiple foci of power over decision-making and the method of finance. Each nation, within its health system, must somehow reconcile several competing sets of objectives, whether or not these have been made explicit. Foremost amongst these are rival claims between:

(1) communal and individual rights and responsibilities;
(2) comprehensiveness and freedom of choice;
(3) responsiveness and accountability;
(4) professional interests and those of the general public;
(5) health promotion and relief of sickness;
(6) ambulatory and institutional care;
(7) personal care and technical prowess;
(8) value for money and social justice in the distribution of resources.

Hence, some conflict is inevitable in any health system. The task is to manage this for the best.

Moreover, this diversity of objectives ensures that, however persuasive the contending biomedical and public health models may be to their adherents, neither on its own can claim to be a sufficient blueprint for the design of a national health policy. Each describes an important facet of a wider reality that is incompletely grasped. Furthermore, each seeks to answer somewhat different questions.

Medicare in its political context

Australia has a mixed economy in which the nation's pursuit of efficiency has usually been moderated by concern as to who gains and who loses. Both sides of the Commonwealth Parliament have usually supported equality of opportunity as a broad social goal. This has required regulation of the market, income redistribution through progressive taxation, and, despite recent libertarian misgivings, public ownership of a few industries, including the major referral hospitals.

In health policy, this goal has translated into fair entitlement and access to sickness services. Thereby attention has been directed to financial and other barriers to use. In recent times, increasing support has been expressed for another kind of equality, that of outcome or health status. This in turn has focused a spotlight on to the substantive nature of the service provided, suggesting a need for more public health initiatives and for a social justice strategy.

It is essential to recognize that the set of programmes labelled Medicare is but one strand in an intricate web of financing packages designed to surmount the restrictions imposed by a Constitution whereby the Commonwealth is the major tax-gatherer and the States have most responsibility for service provision. Neither the Commonwealth nor the States can be completely dominant, medicine is largely private, and public hospitals have a considerable measure of autonomy.

The ALP, sponsor of universal health insurance, is affiliated with the trade unions and has a reformist democratic-socialist tradition. The proposals have been opposed by the coalition of a conservative but increasingly market-oriented Liberal Party and the rural-based National Party, both of whom prefer voluntary insurance. Public opinion polls have shown consistently high support for the ALP stance on this issue: see, for example, Hailstone (1988).

Moreover, financing cannot be considered in isolation from attempts by the ALP to counter the power of the organized medical profession. Sax (1984) documents the conservative stance of the Australian Medical Association (until 1961 the British Medical Association in Australia) in opposing what it has seen as socialist tendencies from the 1930s to the present day. The ideology of organized medicine has called for individuals to take responsibility for paying for their own health care, for the preservation of private fee-for-service medical practice, and for individual professional independence.

Currently, the battle for control of the system between a nominally socialist government and a conservative medical profession is complicated by deep splits within the profession itself. The medical establishment has been challenged within its own ranks by a libertarian ginger group. Further complications are the present dominance within the Australian Medical Association of procedural specialists over general practitioners; and break-away movements on both the right and the left politically. Although its support is recovering,

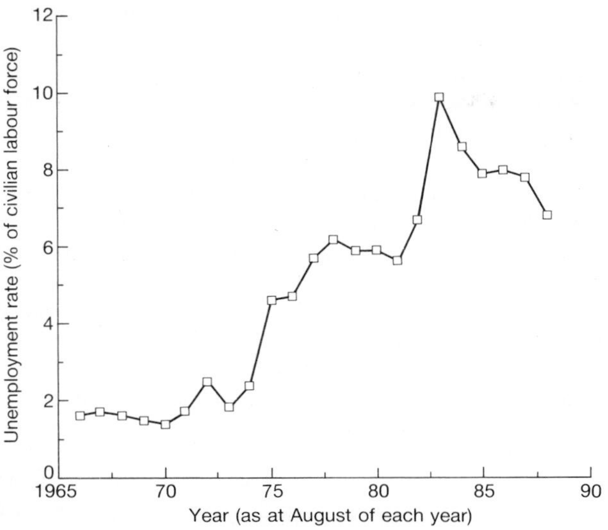

Fig. 24.1. Unemployment rate, Australia (percentage of the civilian labour force), as at August of each year. (Source: Australian Bureau of Statistics 1986, 1988*e*)

the Association (1987) for a while could speak for little more than 50 per cent of medical practitioners.

Meanwhile, other interests have become more vocal. Previously subordinate nurses, the most numerous occupational group, have embraced professionalism and industrial action. Community pressures for responsive management, and for consumer rights and social justice, are more keenly expressed.

Economic context

Sheer distance, both within the continent and to major overseas markets, has played a large part in shaping Australia's pattern of economic development. Dependent for its exports on agriculture and minerals, and hence a price-taker on world markets, the nation is quite vulnerable to international economic fluctuations. Balance of payments difficulties in recent years have led to heavy foreign debts and a decline in the value of the currency. A manufacturing sector developed behind a screen of tariff protection has been criticized for low productivity and is undergoing major structural adjustment. Nevertheless, most jobs are in secondary industry and the burgeoning services sector rather than on the land or in mining. Australia's uneasy prosperity is reflected in the rate of unemployment, which by August 1988 had recovered somewhat to 6.8 per cent, having exceeded 10 per cent in 1983 (Fig. 24.1).

From the end of the 1940s' wartime stringency until the impact of the international oil crisis in the early 1970s, economic growth usually allowed conflict to be managed by buying off competing interests, each in turn receiving a share of the increment of resources. With a strained economy over the past 15 years and a manifest unwillingness to bear higher taxes, this ploy has been less readily available. The various actors now struggle over relative shares from a bundle of real resources whose growth does not keep pace with demand. Where once it entailed identifying unmet needs, rationality in health service planning now equates to minimizing duplication and waste.

Priority health issues

Although a cursory map inspection reveals a huge land mass, sparse rainfall has limited the extent of intensive cultivation in Australia. Much of the continent is semi-desert or supports only open-range grazing. Despite its 'outback' image, the nation is highly urbanized, with 70 per cent of the population living in the 12 cities of 100 000 population or more in size.

Australia shares with its counterparts amongst the wealthier nations a high incidence of diseases attributable to the adverse effects of life in an industrialized urban environment (Hetzel and McMichael 1987). Ischaemic heart disease is the most common cause of death in males. Cancer mortality is high, especially breast cancer in women, lung cancer (where female rates unfortunately are catching up with their male counterparts), and colon cancer. Injury, largely due to motor vehicles crashes, is another major cause of lost years of life. Cerebrovascular disease, other circulatory disorders, and respiratory disease are other leading causes of death (Table 24.2).

Table 24.2. Major causes of death, Australia, 1987

Cause of death (ICD code)*	Rate per million of mean population	
	Males	Females
Infectious and parasitic diseases (001–139)	42	37
Malignant neoplasms (140–208)	1984	1497
Diabetes mellitus (250)	123	131
Diseases of the circulatory system		
Hypertensive disease (401–405)	51	79
Ischaemic heart disease (410–414)	2215	1732
Cerebrovascular disease (430–438)	625	920
All other diseases of the circulatory system (remainder 390–459)	567	658
Pneumonia (480–486)	89	104
Bronchitis, emphysema, and asthma (490–493)	180	110
Ulcer of stomach and duodenum (540–543)	45	58
Chronic liver disease and cirrhosis (571)	106	39
Nephritis, nephrotic syndrome, and nephrosis (580–589)	58	81
Congenital anomalies (740–759)	56	45
Certain conditions originating in the perinatal period (760–779)	58	42
All other diseases (remainder 001–799)	930	775
Motor vehicle traffic accidents (E810–E819)	246	97
Accidental falls (E880–E888)	52	69
Suicide (E950–E959)	218	57
All other accidents and external causes (remainder E800–E999)	189	64
All causes	7833	6596

Source: Australian Bureau of Statistics (1989).
* Classified according to the Ninth (1975) Revision of the International Classification of Diseases.

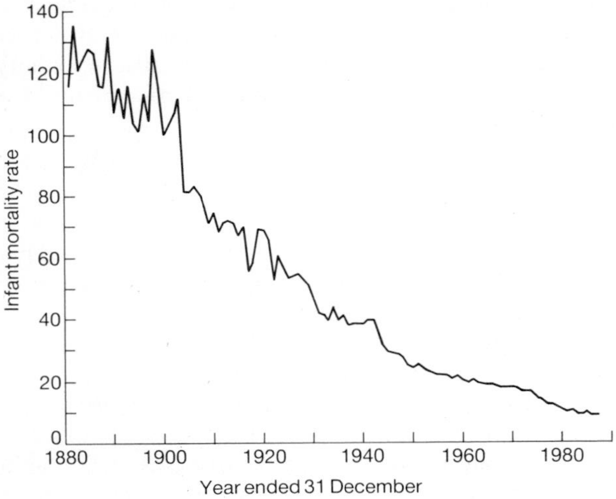

Fig. 24.2. Infant mortality rate per 1000 live births, Australia, by year. (Source: Australian Bureau of Statistics 1988*a*, and related publications in previous years)

Some notable successes have been achieved. Expectation of life at birth in 1987 had risen to 79.5 years for females and 73.0 years for males (Australian Bureau of Statistics 1988*d*). Coronary heart disease mortality has declined from its peak. This has been consequent upon both life-style change and coronary bypass surgery (Dobson *et al.* 1985). Perinatal mortality at 11.5 per 1000 live births (Australian Bureau of Statistics 1987) and infant mortality at 8.7 per 1000 live births (Australian Bureau of Statistics 1988*d*) are historically at their lowest (Fig. 24.2). Epidemics of infectious disease, prior to the advent of the acquired immune deficiency syndrome, had largely been controlled.

Nevertheless, serious health problems are readily identifiable. Alcohol and tobacco-related diseases are rife. Chronic disease and major disability are widespread. With increased longevity, the proportion of frail aged in the population is growing, although no more so than in comparably industrialized countries (Russell and Schofield 1986). Drug abuse and sexually transmitted diseases have become more prevalent in recent years. Suicide, domestic violence, and child sexual abuse are all more openly discussed.

Compared with the seven other industrialized countries in the Luxembourg Income Study, Australia has a relatively large percentage of children and of the elderly living in poverty (Smeeding *et al.* 1988). This same study suggests that 13.2 per cent of the overall population were living below a poverty standard in 1981. Furthermore, there can be little doubt that the social distribution of health is far from equal. Social class inequalities in adult male mortality and infant mortality have been well characterized (McMichael 1985; Taylor 1979) (Fig. 24.3). Sole parent families are increasingly common. Nearly 50 per cent of all single-parent families, mostly female headed, and consequently nearly 20 per cent of all Australian children, live below the poverty line. Despite this, few studies have been made of their health (Hicks *et al.* 1989).

Aboriginal habitation of the continent has so far been traced back at least 40 000 years. At the 1986 census, 1.46 per cent of the population was enumerated as Aboriginal. In the 1970s, the nation awoke to the scandalously adverse health status of its indigenous people, who were experiencing infant mortalities in some localities above 100 per 1000 live births, a life expectancy in the fifties, and grossly elevated prevalence in adulthood of chronic diseases such as diabetes and renal failure (Thomson 1984). A multi-form cultural, social, and economic deprivation is implicated. Public health initiatives have since made a modest inroad, but only a return of control over health services to the Aboriginal communities themselves can be expected to produce a substantial improvement.

Establishment of the colony of New South Wales in 1788 led to a predominantly Anglo–Celtic (English, Scottish, Welsh, and Irish) settlement, now constituting the ethnic majority. Since the end of the Second World War, large-scale Anglo–Celtic, continental European, and more recently some Asian immigration has augmented the low birth rate (15.0 per 1000 in 1987) so that by the end of 1987 the population was 16.4 million. Nearly 40 per cent of all present Australians were either themselves born overseas or have a parent who was. Migrants as a group can be physically more healthy than their counterparts in their home country (Powles *et al.* 1986), and experience greater longevity than native-born Australians of comparable socio-economic status (Dunt 1982). But, because they are congregated in lower socio-economic strata of society, they have, overall, lower health status and inferior access to services than the population in general. Moreover, there is concern that medical services be more culturally sensitive (Shoebridge 1980).

Allocation of responsibilities under federation

When Australia became a nation with the proclamation of the Constitution on 1 January 1901, specific powers were vested in the Commonwealth. An extensive and undefined residue was left with the States,* because of popular reluctance to cede powers centrally. This location of residual powers has constrained the role of the Commonwealth in health care.

Both the regulation of private sector health care and

* Australia is a parliamentary democracy with a federal system of Government. The six States (in order of population size) are New South Wales, Victoria, Queensland, Western Australia, South Australia, and Tasmania. The Northern Territory is self-governing. The Australian Capital Territory (surrounding Canberra) was governed by the Commonwealth until it became self-governing in May 1989. Local government, concerned with community-level issues within the municipalities and rural districts, is established under Acts of the State Parliaments. In this chapter, references to the States usually would apply to the Territories as well.

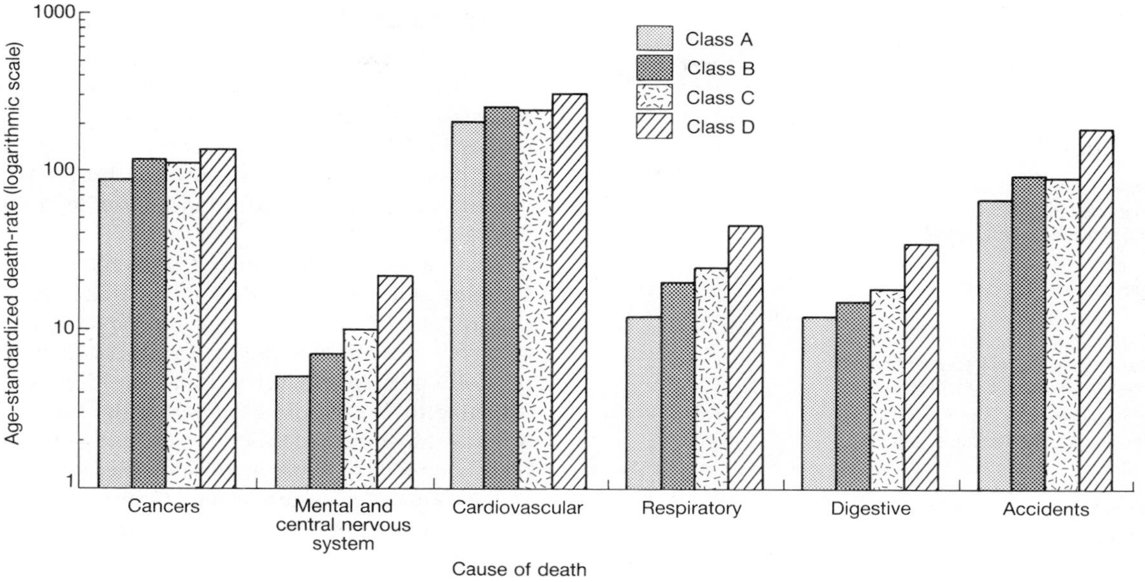

Fig. 24.3. Occupational class differences in male mortality. Age-standardized death-rates for selected causes of death according to the Congalton (1969) classification. Class A (high socio-economic status) to Class D (low socio-economic status). (Source: McMichael 1985)

Table 24.3. Commonwealth of Australia Constitution Act: an extract

Section 51 The Parliament shall, subject to this Constitution, have power to make laws for the peace, order and good government of the Commonwealth with respect to:

•

(ix) Quarantine:

•

(xxiiiA) [inserted by referendum 1946] The provision of maternity allowances, widows' pensions, child endowment, unemployment, pharmaceutical, sickness and hospital benefits, medical and dental services (but not so as to authorize any form of civil conscription), benefits to students and family allowances:

•

Section 96 . . . the Parliament may grant financial assistance to any State on such terms and conditions as the Parliament thinks fit.

Source: Sawer (1988).

governmental health care provision remained matters for the States. Indeed, the sole direct mention of health in the Constitution was for quarantine. Subsequently, Section 51 (xxiiiA), inserted by referendum in 1946, granted the Commonwealth substantial new powers for payment of hospital, pharmaceutical, and several social welfare benefits to persons, and provision of medical and dental services (Table 24.3).

Under the uniform taxation agreement dating back to the 1940s' wartime emergency, the Commonwealth Government is the main revenue raiser, relying disproportionately on personal income tax. The States in turn have the greater responsibility for health service provision. The Commonwealth provides general revenue assistance to the States by way of untied grants. The States also receive specific purpose payments under Section 96 of the Constitution for purposes designated by the Commonwealth. In addition, the States have their own sources of revenue. The Commonwealth also derives some power over domestic health issues from its international obligations, and from other special powers such as those with respect to immigrants and Aborigines.

Commonwealth–State relations

Each nation must reconcile the tensions between uniformity and diversity that characterize central–local relationships. However, under the Australian federal system, in which both the States and the Commonwealth are sovereign, special mechanisms have been necessary. Unfortunately, these do not as yet amount to a sufficiently integrated approach to health planning. On the other hand, there is also considerable duplication between the two tiers of bureaucracy (Table 24.4).

The Australian Commonwealth Grants Commission (1987) makes recommendations for the fair allocation between the States of Commonwealth monies including health expenditures. The National Health and Medical Research Council makes recommendations to the Commonwealth and the States on medical research and public health policy, and advises the Commonwealth Minister on disbursements from the Medical Research Endowment Fund, which is the nation's major source of biomedical research funding. The annual Australian Health Ministers' Conference is an avenue for consultation between the Commonwealth and the States to foster effective co-ordination of policy. Consequently its bureau, the Australian Health Ministers' Advisory Council, is taking an increasing role in health planning. The Australian Institute of Health has been established as a statutory authority under the Commonwealth to manage national health information collections.

Table 24.4. Australian health system: main elements

Commonwealth Government
Department of Community Services and Health
Health Insurance Commission
Department of Veterans' Affairs
Armed forces medical care
State Governments (and Territories)
Health Department (or Commission)*
Intersectoral activities in other departments
Local government
Health inspection, immunization
Commonwealth–State relations
Commonwealth Grants Commission
National Health and Medical Research Council
Australian Health Ministers' Conference
Institutions and agencies
Hospitals: public, private charitable, for-profit
Nursing homes
Community health centres
Home and community care services
Private practitioners
Medicine, dentistry, pharmacy, physiotherapy, optometry, clinical psychology, etc.
Specialist medical colleges and professional associations
Trade unions
Industry associations
Private sector health insurance
Workers' compensation and motor vehicle accident insurance
Special interest associations and patient support (self-help) groups
Social networks, family support

* Department of Health (or similar title) except for Commission in South Australia, Department of Community Services and Health in the Australian Capital Territory, and Department of Health and Community Services in Northern Territory.

Commonwealth responsibilities

In 1987, the Commonwealth Department of Community Services and Health was formed from the previously separate health and community services portfolios and the housing welfare function. The Department administers a wide range of national programmes, grants to the States, and benefits for individuals. The Health Insurance Commission is a Commonwealth Statutory Authority responsible for the administration of the medical benefits provisions of the national health insurance scheme Medicare. It also operates Medibank Private in competition with the other private health insurance funds. The Department of Veterans' Affairs, amongst its other functions, provides medical and hospital care for returned service personnel and their dependants. The Commonwealth has announced its intention to integrate repatriation hospitals with the State hospital systems by 1995. Through the Department of Defence, the Commonwealth also provides health services to the serving armed forces. Overall, more than 90 per cent of the Commonwealth's health function outlays are directed towards improving access to medical, public hospital, pharmaceutical, and nursing home services (Table 24.5). Its other health functions include standard setting, research funding, and promotion of better health.

State responsibilities

Most public sector health services are actually provided by the States and the Territories. This tier of government is responsible for the provision of public hospitals and community health services, for most public health protection services, for the licensing of private hospitals and nursing homes, and for the registration and/or regulation of health workers.

These activities are overseen by a minister accountable to Cabinet and the State Parliament. The minister, via the departmental head, is nominally in direct day-to-day control of policy and administration. In practice, the details of the portfolio are so extensive that considerable discretion is available to the middle echelons of the bureaucracy. Public hospitals, by virtue of their prestige and popularity as well as their complexity, retain a considerable degree of autonomy. In the 1970s, New South Wales, Victoria, and South Australia integrated their hospital and public health functions under a Commission structure. This allowed health units more autonomy by interposing a board between the minister and operational staff, and furnishing this board with a set of advisory committees. In the harsher economic conditions of recent years, with pressures on Ministers to take direct control over costs, only the South Australian Health Commission has survived in this format.

Service organization: different values and interests

Primary health care

Since the Australian population is subject to a disease pattern comparable to that of other urbanized and industrialized nations, the range of health services that has emerged would be readily recognizable in those countries.

Murrell (1985) has reviewed developments in primary health care in Australia. Traditionally from colonial times the general practitioner has been the back bone of ambulatory medical care in this country. General practitioners regard themselves as the personal doctor of the patient and the family. They attempt to provide continuing total-patient care, and aspire to be an integral part of the community in which they live and work (Jungfer 1974). Most families report considerable satisfaction in this continuing relationship (Australian Bureau of Statistics 1981).

The general practitioner is a self-employed professional registered under State government legislation and usually consulting on a fee-for-service basis. A greater number nowadays join with several partners in group practice. All major interest groups acknowledge that the general practitioner has a key role as the gatekeeper in referring patients to specialist medical practitioners or to public hospitals.

Reflecting the modern trend to specialization and hospital employment, the percentage of the medical profession working in general practice is now estimated to be about 40–45 per cent (Committee of Inquiry into Medical Education and

Table 24.5. Australian health expenditure, 1985–6: area of expenditure by source of funds ($ million)

Area of expenditure	Commonwealth Government	State/local government	Private	Total as percentage of recurrent expenditure
Institutional				
Recognized public hospitals	1116	4026	431	32.7
Repatriation hospitals	287	6		1.7
Public psychiatric hospitals	32	574	54	3.9
Private hospitals	168		810	5.7
Nursing homes	1086	72	306	8.6
Ambulance and other	77	114	105	1.7
Non-institutional				
Medical services	2686		405	18.1
Dental services	25	70	820	5.4
Other professional services	69		505	3.4
Community health services	120	351	1	2.8
Pharmaceuticals	702		737	8.4
Aids and appliances	43		246	1.7
Other	9	51		0.4
Health promotion and illness prevention	31	125		0.9
Research	167	24	42	1.4
Administration	249	73	233	3.3
TOTAL RECURRENT EXPENDITURE	6869	5485	4695	100.0
Capital consumption	35	381		
Capital expenditure*	49	498	452	

Source: Australian Institute of Health (1988*b*).
* Capital expenditure at $999 million was equal to a further 5.9 per cent on recurrent expenditure.

Medical Workforce 1988). General practitioners are at liberty to disperse widely throughout the community, but their concentration is lower in the poorer neighbourhoods of the major cities and in their rapidly expanding outer suburbs. Many rural districts have considerable difficulty in recruiting a local doctor because, in addition to the greater attractions that city life and work has long held for most of them, comparatively few medical practitioners now regard themselves as sufficiently trained to handle the range of procedural work that a country practice would demand of them. The famous Royal Flying Doctor Service covers the sparsely populated 'outback'.

However, although general practitioners are by far the most frequently consulted health professionals, they are by no means the only source of first-contact health care (see Table 24.6). The accident and emergency services of public hospitals are crowded with people whose problems are more in the nature of primary care than emergencies. Retail pharmacies are another major source of advice. Dentists, optometrists, physiotherapists, and clinical psychologists also provide first-contact health care and usually practise in the private sector. Community nurses do so from community health centres (see below). Many people consult complementary practitioners such as chiropractors.

The main alternative ideology resides in a primary health care model more redolent of the World Health Organization's (1978) global strategy. A Community Health Programme was begun by the Australian National Hospital and Health Services Commission (1973) as part of the Whitlam ALP Government's commitment to improve services to the outer suburbs of the major cities. Community health centres staffed by salaried professionals have been established in under-doctored localities.

Table 24.6. Health professionals consulted in a two-week period*

Health professional	Number of persons† per 1000 population having such a consultation
Hospitalization	8
Doctor, including specialist	174
Dentist	49
Chemist	26
Chiropractor	12
Physiotherapist	9
Optician	5
Community nurse	4
Other categories	each < 4

Source: Australian Bureau of Statistics (1984).
* During the two weeks prior to interview.
† Each respondent may have consulted more than one type of health professional.

This programme includes a network of women's community health centres. These have been created and maintained in the justifiable belief that the conventional health system lacks sufficient expertise and sympathy with the special biological, personal, and social factors in women's health.

In an attempt to redress their high prevalence of health problems, many of which are causally related to their marginal status in society, special community services are provided for Aborigines, both in cities and at outback settlements. These in no way entail racial segregation, but are intended to be culturally relevant and client directed.

Staffed by a range of health personnel, including community nurses and social workers but seldom medical practitioners, community health centres provide a broad range of primary health care. Services are available free or for a nominal charge at the point of delivery. These centres place more emphasis than do medical practitioners on health as a positive goal, and on prevention, equity, community participation, and multi-professional team-work. Tension between publicly funded community health centres and private general practitioners is keenly felt. The proximate cause appears to be differences over finance, leading to problems in referral mechanisms between the two. However, the deeper origins may well lie in a clash between conventional and dissenting ideas about sickness and health.

The dominance of the biomedical model and the belief that health is an individual responsibility are both challenged by activities under the Community Health Programme (Furler 1982). A widespread faith in the success of heroic, scientific medicine is being increasingly questioned in the mass media. So, also, is the emphasis placed on the financial aspects of the doctor–patient relationship for effective therapy. Ideologically squeezed between community health and high technology hospital medicine, general practitioners are struggling to reassert their role.

A note on prevention and public protection

Major elements of the administration and funding of the 'old' public health, especially sanitation, have traditionally resided outside the formal health system within State engineering departments. State health departments have a division for the investigation and regulation of environmental health hazards. Typical functions exercised there include occupational health, radiation safety, treatment and prevention of sexually transmitted disease, monitoring of drugs of dependence, and descriptive epidemiology. Australian local government has certain responsibilities for public and environmental health. These duties include garbage collection, the inspection of drains, commercial food preparation, and noxious wastes; and immunization. Today, local government's main preoccupation is cost-effective urban and regional planning, which includes provision of community services. Fears that funding from Commonwealth and State sources may be impermanent have tended to deter health initiatives, so that, for example, community health centres are largely outside the ambit of local government.

The 'new' public health, concerned with social determinants, is progressively developing its own infrastructure, very much in alliance with primary health care. Since first coming to power in 1983, the Hawke ALP Government, through the Health Minister Neal Blewett, has taken many substantial initiatives. In Chapter 31, Chapman and Leeder describe these contemporary developments in public health in Australia. Professing an alternative set of values, these programmes are already in competition for resources with established hospitals and clinical services.

Hospitals

Hicks (1981) has described how medical care and basic hospitalization were initially provided without charge by the colonial service. However, as emancipists and free settlers soon came to form the majority of the population, private practice began to thrive and those who could afford it were nursed at home. Reacting to observable squalor and poverty, prominent citizens subscribed towards the establishment of voluntary hospitals under independent incorporated boards. However, the burden of expense became such that they were forced to seek governmental subsidy. In contrast to the English poor law administration based on locality of domicile, it suited all parties to press the colonial administration for funds (Sax 1984). Although overcrowded and ill-equipped, these public hospitals provided acute nursing care free of charge to the indigent population. Medical care in the public hospitals was provided in an honorary capacity by private practitioners as a way of gaining recognition and hence private patients. Working people banded together in Friendly Societies for mutual support against the expense of sickness, contracting with medical practitioners on a capitation basis. Religious societies founded hospitals and nursing homes but their expense often forced them to seek financial assistance from the Government. Private for-profit establishments were also founded, often by medical practitioners. Mental asylums were kept separate from the general hospital system and were controlled by the Government.

In broad terms, the pattern of Australian hospitals established by the latter part of the nineteenth century has endured in its basic form to the present day. This is despite considerable change in the burden of illness, in technology, and in the internal dynamics of the organizations concerned. With the transmission of the seminal discoveries of anaesthesia and antiseptic surgery to the Australian colonies, developments in scientific management of illness ensured that public hospitals came to be regarded no longer as a refuge of last resort for the indigent sick, but as an essential service, attractive also to fee-paying patients. Nowadays, public hospitals are regarded as a universal entitlement as part of the social wage.

In June 1987, there were 720 recognized public hospitals (not counting psychiatric hospitals) and 333 private hospitals altogether providing 5.4 hospital beds per 1000 population. In the public hospitals, private patients accounted for 26 per cent of total bed–days (Australian Commonwealth Department of Health 1987*a*).

In each State capital city there are several teaching hospitals so designated because of their links with a university medical school. They are funded largely by the government of that State. Teaching hospitals are the major referral public hospitals and endeavour to provide state-of-the-art technology. They incorporate general medical and surgical units and super-specialty units such as neurosurgery, endocrinology, and radiotherapy. As well as general hospitals, there are special hospitals catering, amongst other needs, for confinement, for children, and for infectious diseases. Having been founded in the last century, most are situated close to the city centre. As the capital cities have expanded, residents in the outer suburbs have found themselves a considerable travelling time away from these services. Therefore, this has led to attempts to transfer teaching hospital beds closer to the bulk of the population, notably the building of Westmead Hospital in the western suburbs of Sydney. The honorary system of medical care prevailed until the 1970s when it was replaced by payment of visiting consultants. Appointment of full-time salaried specialists dates from much the same period.

Religious and charitable organizations have a long tradition of running private non-profit hospitals. More recently, chains of for-profit hospitals have been set up and local subsidiaries of American firms have made an appearance in the market. So far, the profitability of stock exchange listed hospitals has been equivocal (Jacques 1988).

Nursing home care for the frail aged has been established variously by religious organizations, by private individuals (a means whereby nurses could rise from employee to proprietor), and lately in chains run by private corporations. This form of care is heavily subsidized by the Commonwealth Government and heavily regulated. Hence, Australians may be over-serviced with 46.9 nursing home beds per 1000 persons aged 65 years and over (Australian Parliament 1982).

Health personnel and the professional identity

Members of different health-related professions acquire during their basic education different paradigms, in Kuhn's (1974) sense, on the nature of health, and hence on what the task of health care is. Their paradigm seems to attain a fundamental place in their conceptual armoury. Usually, it receives strong reinforcement from the work environment in the early years after graduation.

The explicit and implicit objectives of some occupations tend more towards a physical basis for health and disease, while others emphasize mental functioning or social processes. The point is not that practitioners are unaware of the full ramifications of a broad concept of health, such as in the well-known World Health Organization (1947) definition, but that they actually practise as though their beliefs were narrower. In particular, their professional associations tend to adopt a similar viewpoint. The strength of commitment is such that these paradigmatic differences may not be resolvable merely by improved communication (Moss 1984).

Huntington (1977) has identified substantial differences in social structure and culture between the Australian health professions, involving, for instance, gender and mode of employment. N.D. Hicks and J.E. Hiller (personal communication) have found that (in Adelaide) all members of the health team, including even social workers, have tended to be recruited as students in preponderance from the more affluent suburbs. Their social origins call into question the degree to which these professionals are able to identify with the needs of their less well-to-do clients.

In Australia, as elsewhere, since the latter half of the nineteenth century, the medical practitioner has risen from artisan to professional, enjoying a concomitant rise from menial to high social status. This has come about through advocacy of scientific practice, initially in antisepsis and anaesthesia, but also through asserting control over medical education and hospital practising rights against rival health practitioners such as the homoeopaths. Importantly, it also entailed crushing an alternative payment mechanism. Friendly Societies offered their working class members sickness benefits and contract medical care, known as 'lodge practice'. The latter was stoutly opposed by the Australian branches of the British Medical Association on the grounds that the constraints imposed on the doctor's freedom lowered the quality of care. After a prolonged dispute beginning before the Great War, the Societies ultimately lost most of their role in employing doctors by 1922 (Pensabene 1980).

Nevertheless, most Australians would seem likely to regard medical practitioner autonomy as desirable, despite having reservations about the monopolistic practices and condescension that sometimes accompany it. Countless anecdotes point to the general public's recognition of the importance of the doctor–patient relationship and their desire to be able to choose and hence to trust their doctor. Radical politicians may have misread public opinion in calling for stronger constraints on doctor autonomy. Besides which, the medical profession has been sufficiently powerful to thwart all attempts to do so.

In the wake of medicine, the other health-related professions are experiencing a growing trend to higher education and higher professional standards, with a higher income and status. Autonomy remains their treasured goal. An increasing proportion of nurses have become militant, their grievances being a subordinate occupational status and lower wages than other health professionals. Through the Royal Australian Nursing Federation, they have achieved better pay, college-based education, and the implementation of a nursing career structure. Similar trends can be seen in the other emergent allied health professions, and it is noteworthy that most are mainly female in composition. Social workers especially and also physiotherapists and occupational therapists are now adopting a management role. Most health professionals now require registration under State laws. Associates of the Australian College of Health Service Administrators as professional managers are replacing the lay secretary, although in turn are facing competition from the generic manager (i.e. one claiming skills in management applicable to diverse settings). The health industry currently amounts to 6.9 per cent of the total employed labour force (Table 24.7).

Accountability to the consumer

Central to the community health movement of the last two decades has been disquiet over the dependent, passive, and

Table 24.7. Persons employed in the health industry, Australia, 1986

	Persons employed per million of total population	
	Men	**Women**
Managers and administrators	282	367
General medical practitioners	1057	346
Specialist medical practitioners	450	86
Dental practitioners	329	49
Pharmacists	41	56
Occupational therapists	7	130
Optometrists	69	19
Physiotherapists	58	285
Chiropractors and osteopaths	72	12
Speech pathologists	1	54
Radiographers	93	165
Podiatrists	17	39
Registered nurses	608	7479
Other professionals and para-professionals	1075	1405
Enrolled nurses	122	1903
Dental nurses	2	514
Other personal service workers	65	357
Clerks	312	4051
Tradespersons	860	451
Plant operators, drivers, labourers and related workers	1418	3986
Inadequately described/not stated	87	183
Total health occupations*	7028	21 936

Source: Australian Institute of Health (1989).
* Comprising 6.9 per cent of the total employed labour force.

uninformed status of many patients in mainstream hospital and medical services. Lay perceptions of health appear to be changing with more awareness of prevention and more willingness to question the conventional wisdom of the medical practitioner. Certain health workers, notably from social work and community nursing, have come to regard themselves as patient advocates. Grass roots health care consumers' movements have nevertheless emerged only locally and sporadically. The beginnings of a concerted movement may well lie in the Consumers' Health Forum of Australia. Under this umbrella, many consumer and community groups with an interest in health can now contribute to Commonwealth policy-making through a close but independent relationship with the Commonwealth Department of Community Services and Health.

Calls for greater consumer participation indeed go beyond the inclusion of patients as equals in decision-making about their own health care. They extend to effective local community representation on boards of management of health services. In addition to ensuring more widespread representation on boards, governments have made concessions to the demand for participation in planning. The Victorian Health Department has instituted District Health Councils to foster new local health activities. South Australia has followed with Health and Social Welfare Councils. New South Wales has devolved some planning and administrative functions upon a system of Area Health Boards, having responsibility for hospital as well as community services. More recently, local government has started to make a direct contribution to health planning through the international Healthy Cities movement.

Concurrently, patient support or self-help groups have increased in numbers and influence. The current Directory of Social Welfare Resources (Community Information Support Service 1987) lists more than 80 such groups in South Australia alone. Many now receive some Government funding or receive professional advice. Their role is mutual social support, information dissemination, and consumer advocacy.

The multiplicity of health care arrangements in Australia does at least open alternative avenues for consultation. Under private health arrangements, consumers are sovereign to the extent that they select their source of primary health care and, except in isolated rural areas, if dissatisfied they may easily switch to another, provided always that they experience a sense of control. What is possible under pluralism might not have been so under a nationalized structure.

Health insurance under the Hawke ALP Government (from 1984)

Medicare cover

The most recent major structural change in the financing of health care occurred in 1984 with the introduction of Medicare by the Hawke Labor Government. This is a third party payment scheme, conventionally described as health insurance, but also incorporating redistributional elements. All Australian residents are eligible for medical and public hospital cover (Table 24.8).

Medicare covers the full cost of public hospital accommodation and treatment, when this is provided by hospital-appointed doctors. Free in-patient and out-patient public hospital care is thereby extended to all where previously it was means-tested in every State except Queensland. As always, patients are unfettered in their choice of referring general practitioner, but under this form of cover, they cannot choose their hospital specialist. Public psychiatric hospitals are outside the Medicare arrangements.

Medicare is now the sole form of basic cover for medical expenses (i.e. the fee of a registered medical practitioner, including consultations and procedures carried out in a hospital). Medical practitioners have firmly insisted on their right to bill patients as individuals and to set their own fees. Each year, the Australian Medical Association determines a list of fees for each item of service. The Commonwealth Government separately publishes the Medical Benefits Schedule Book (Australian Commonwealth Department of Health 1987*b*), which though not binding on doctors, forms the basis for health insurance calculations.

The 'gap' is the amount by which the Medicare benefit is less than the Schedule fee. It is intended as a financial disincentive to unnecessary use. The Medicare legislation initially allowed no insurance cover for this gap. This served to discourage application of the Australian Medical Association's own, more expensive, schedule. However, in the negotiations settling the New South Wales specialists' withdrawal from public hospitals (McKay 1986), the private funds were allowed to offer gap medical insurance for private hospital patients (for whom Medicare only covers 75 per cent of the fee).

Medical practitioners have the option to direct bill (formerly described as bulk bill) at 85 per cent of the Schedule fee by sending an account direct to the Health Insurance Commission, bypassing the patient (who pays nothing). Most doctors have agreed voluntarily to direct bill for pensioners and social welfare recipients, whose eligibility is recognizable by their holding concessional health cards. Doctors have the option to direct bill for any patient. The incentive for this is the consequent reduction in practice overheads. In 1986–7, 51.8 per cent of all Medicare services were direct billed (Australian Health Insurance Commission 1987).

Medicare is financed from Commonwealth general revenue, augmented by a 1.25 per cent levy on taxable income. A common misunderstanding suggests that the levy covers the total expenses of the scheme. In fact it attempts to raise sufficient revenue to cover the extra costs borne by the Commonwealth for Medicare over and above its previous commitments to publicly funded health care. The medical side of Medicare is administered by the Health Insurance Commission, whereas the hospital arrangements have been arrived at by Commonwealth–State negotiation.

Table 24.8. Health insurance cover: arrangements as at September 1988

Medicare (All Australian residents may enrol)

Medical benefits—cover the doctor's bill:

- Calculations are based on the Schedule of Medicare Benefits
- The 'gap' is the amount by which the Medicare benefit (i.e. refund) is less than the Schedule fee—the gap is the patient's own liability
- Patient of private doctor receives benefit equal to 85 per cent of Schedule fee, with maximum gap $20 for any one item of service
- To protect those needing multiple consultations, the maximum cumulative gap payment is $150 per person per year
- When a private doctor directs bills, the patient pays nothing (at the point of service)
- Medical services in a private hospital attract a benefit equal to 75 per cent of the Schedule fee, with no maximum gap
- Some oral surgery and optometrical benefits

Hospital benefits—cover 'hotel' expenses:

- In-patient and out-patient treatment at a public hospital, without choice of doctor, attracts no charge (at the point of service)
- Public hospital accommodation is free (at the point of service)
- Accommodation in a private bed or hospital is not covered by Medicare

Private health insurance (e.g. The Hospitals Contribution Fund of Australia Limited, Medibank Private)

- Contributions are paid from the contributor's private resources
- Benefits for private hospital accommodation and for ancillary services
- Gap medical insurance for private patients in hospital

Table 24.9. Pharmaceutical Benefits Scheme: most frequently prescribed drug groups, 1986–7

Drug group	Total benefit prescriptions (%)
Analgesics	13.8
Heart, drugs acting on	8.9
Diuretics	7.8
Bronchial spasms, preparations for	7.5
Blood vessels, drugs acting on	6.0
Penicillins	6.0
Anovulants	4.4
Tranquillizers	3.9
Antidepressants	3.2
Tetracyclines	3.0
Hypnotics and sedatives	2.9
Eye drops	2.9
Antacids	2.5
Sulphonamides	2.5
Water and electrolyte replacement	2.3
Skin sedative applications	2.2
Other drug groups	20.1
Total	100.0*

Source: Australian Commonwealth Department of Health (1987*a*).

* Representing 102.762 million benefit prescriptions.

Private cover

Private insurance is available through non-profit organizations and is subject to governmental regulation. Cover is available for private hospital accommodation, for gap medical insurance for private hospital patients, and for private treatment by various allied health professionals, including dentists, physiotherapists, optometrists, and chiropractors. Since private hospital accommodation is no longer covered by Medicare, it can require substantial additional out-of-pocket expenditure. This is why the price of a year's family private insurance cover at the top rate can exceed $1000 (equal to more than two weeks' earnings at the average wage before tax). Even so, while it is unclear whether subscribers are risk-averse or misinformed, loss of freedom of choice of medical practitioner and lengthy waiting lists for some procedures in public hospitals continue to be incentives for private insurance. It should not go unrecognized that workers' compensation and motor vehicle accident insurance are significant funders of private sector health care, nor that an income tax rebate is available for any out-of-pocket health care expenditure exceeding $1000 in a year.

Pharmaceutical, nursing home and community care benefits

The Pharmaceutical Benefits Scheme (PBS) covers a comprehensive list of drugs when they are medically prescribed (Table 24.9). A subsidy enables the price to be kept at no more than $11 per item for the general public, and at further reduced rates or even no cost for holders of concessional health cards and recipients of social security. Under the safety-net arrangements, a person or family group using more than 25 PBS prescriptions in a calendar year thereafter

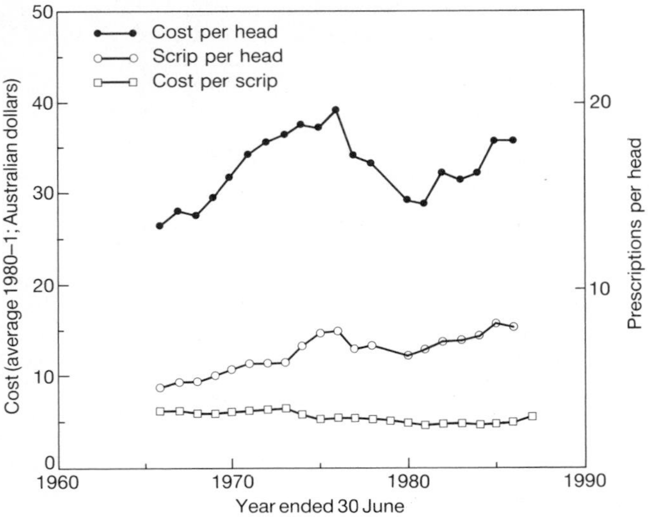

Fig. 24.4. Pharmaceutical Benefits Scheme prescriptions per head according to financial year (ended 30 June). (Private, repatriation, and public hospital prescriptions are excluded.) Average total cost per prescription and average cost per head of total population, both in average 1980–1 Australian dollars adjusted by the Consumer Price Index. (Source: Australian Commonwealth Department of Health 1987*a*, and earlier publications in that series, and Australian Bureau of Statistics 1988*c*)

qualifies for free pharmaceuticals for the remainder of that year. Patients pay the full cost of non-schedule drugs. In 1986–7, 102.8 million PBS prescriptions were filled at a total cost of $903.7 million (Fig. 24.4). This entails a substantial Government subsidy. However, since most drugs are imported, the nation as a whole has benefited from the Government's success as a monopoly buyer (monopsonist) in keeping down prices. Nevertheless, an acceleration in cost increases in the last couple of years, attributed to new drug developments, is leading to pressure for revision of the scheme.

The target of the Home and Community Care programme is both the frail aged and younger disabled people, including also their informal carers. Its aim is to consolidate and develop a comprehensive range of decentralized and co-ordinated services to enable them to continue to live in their own home, thereby avoiding premature or inappropriate admission to long-term institutional care. The programme is funded on a joint, or cost shared, basis between the Commonwealth and the States. The Commonwealth also pays personal, hostel, and respite care subsidies to eligible organizations providing hostel-type accommodation for aged or disabled people unable to remain in their home but whose level of dependency does not justify admission to a nursing home where they would receive continuing nursing care.

Since 1962 the Commonwealth Government has become the main source of funding for nursing-home care. Encouraged by the availability of a substantial subsidy, the number of beds has proliferated and the total amount of payment in real terms has become a major drain on the health budget. Concerned to achieve greater efficiency and equity, the present Government is moving to base its subsidy on standard fees rather than on the previous cost-reimbursement arrangements.

Development of health finance under the Commonwealth

In 1918, the Commonwealth acquired its first major stake in health service provision when the Repatriation Department was set up. With an emerging recognition, fostered by successes in preventing disease amongst the troops during the Great War, that national priorities should outweigh State differences over public health, the Commonwealth Department of Health soon followed in 1921 (Cumpston 1978). This development was partly in response to the havoc created by the influenza pandemic of 1919 (McQueen 1975), and at the urging of the US Rockefeller Foundation. Compared to its present workload, the main functions of quarantine, investigation, and prevention of disease, and health education were quite limited.

In the 1920s and again in the 1930s, efforts to establish national health care coverage were defeated. In 1926, the Commonwealth Royal Commission on Health recommended both public health services (many of which actually were set up with Commonwealth subsidy) and medical benefits. In 1928, the National Insurance Bill, including provision for sickness benefits, was introduced but did not proceed. By the mid-1930s, voluntary hospital contribution funds were operating in several States. Although passed in 1938, the National Health and Pensions Insurance Act, a contributing insurance scheme, was never implemented, being shelved indefinitely in 1939 in the face of opposition from the medical profession (Hunter 1984).

Hicks (1982) has suggested that the impetus for community and technical rationalization of ambulatory health services in Australia is not recent. Both the Commonwealth Royal Commission on Health (Australian Parliament 1926) and the Australian National Health and Medical Research Council's (1941) plan for a national health service proposed an integrated, national system of medical services based on primary medical care with ancillary support, backed up by specialist services in larger population centres.

Curtin–Chifley ALP platform

Between 1941 and 1945, the Labor Government under Curtin made plans for the introduction of a comprehensive national health service. However, the Pharmaceutical Benefits Act 1944, under which free medicines were to be dispensed provided the doctor's prescription was on a government form, was declared unconstitutional by the High Court. A referendum seeking 14 additional Commonwealth powers, including national health, allowed no choice between different measures and was decisively defeated. Nevertheless, in the same year the Queensland ALP Government achieved nationalization of that State's public hospitals to provide free in-patient and out-patient care without application of any means test (Bell 1968). This system was maintained until subsumed under Medicare. In the following year, the Commonwealth removed the charitable basis for public hospital care through the passage of the Hospital Benefits Act 1945, which provided a flat rate bed-day subsidy to ensure free public hospitalization without a means test, and reduced fees for public patients (Thame 1974). The Chifley Government was able also to introduce benefits for patients in mental institutions, a campaign against tuberculosis, and a Commonwealth Acoustic Service.

In 1946, the Chifley ALP Government promoted a successful referendum to change the Constitution. This was an uncommon achievement for, even to the present day, only eight referendum proposals have ever been agreed to, this one by the narrow margin of a 50.48 per cent affirmative vote. The Commonwealth was thereby granted power to make laws with respect to pharmaceutical, sickness, and hospital benefits, and medical and dental services. Insertion of the 'civil conscription' clause (see Table 24.3) was forced by the Opposition under Menzies. Another successful High Court challenge, this time by the British Medical Association in Australia (Federal Council of the British Medical Association v. Commonwealth 1949), held that the Pharmaceutical Benefits Act 1947–9 imposed a form of civil conscription (79 CLR 201). This ruling has remained a major impediment to the Australian Commonwealth Government introducing a counterpart to the English National Health Service (introduced in 1948). Despite the civil conscription clause, the States are believed to retain the power to impose price control on medical practitioners, but have been manifestly reluctant to do so.

This same referendum enabled the Commonwealth to take a much enhanced role in the development of social welfare. Unfortunately, since then, Commonwealth initiatives in social policy and health service finance have diverged with minimal integration.

Menzies–Page Acts, 1951–62

Ironically, Labor was unable before its defeat at the 1949 election to translate the powers granted by the 1946 referendum into a workable scheme that would pass High Court scrutiny. The incoming Menzies Government and Health Minister Page based their more enduring approach on the principle of individual responsibility, underwritten by subsidies to the private sector and to the States, rather than on direct Commonwealth control. This policy was consolidated under the National Health Act 1953 which allocated a subsidy to the voluntary hospital and medical insurance schemes. Contributions were now made tax deductible. Fee and benefit schedules came under Government regulation. Contributions were 'community rated' so that young and old, healthy and sick, rich and poor, each paid the same amount regardless of risk. The public hospital means test was reintroduced. The States received capital grants towards the costs of hospitals. The Pharmaceutical Benefits Scheme and a Pensioner Medical Service were also introduced. Nursing-home benefits followed in 1962.

It is arguable whether the Menzies–Page Acts were merely a conservative reaction to the Chifley platform or a positive expression of pluralism, attempting to work constructively with the medical profession. Whatever the case, in narrowing their attention to the method by which services are paid for, these Acts became the basis of the present Australian health system, subsequently to be augmented by Medibank/Medicare, the Whitlam ALP Government's use of Section 96 of the Constitution to provide more extensive tied grants to the States, and the recent Blewett public health reforms.

By the late 1960s, the shortcomings of the system of voluntary health insurance with governmental regulation were apparent. Many people could not afford cover. The poor and elderly were categorized, leading to fears of a second class service. There were low benefits for specialist services, limits on benefits to the chronically ill, and high cost. In 1968, the Commonwealth Committee of Enquiry into Health Insurance (Australian Parliament 1969) under Mr Justice Nimmo drew attention to these problems, and Scotton and Deeble (1968) published a plan for universal health insurance. The Gorton coalition Government introduced subsidized health insurance contributions subject to a means test, but many potential recipients seemed unaware of their entitlement. Benefits for specialist medical services were raised, leaving the patient to pay directly only a $5 maximum gap. In so doing the coalition conceded the notion of the 'most common fee' thereby raising costs and reducing price competition between doctors. This measure also ensured that services provided by a specialist received a higher rebate than those provided by a general practitioner, the source of much difficulty in the system today.

Medibank to Medicare

A major plank of the ALP platform at the 1972 election was a national health insurance scheme based on the Scotton and Deeble plan, and much influenced by the Canadian approach. Under the slogan 'It's Time', Labor returned to government after 23 years in opposition. Despite the Labor majority in the House of Representatives, a set of key Bills, including that dubbed 'Medibank', was rejected twice in the Senate. The ALP went to the polls again and was returned to government, but still without an upper house majority. The legislation was again rejected in the Senate, and then passed at the deadlock-breaking first ever, and to date only, Joint Sitting of the Australian Houses of Parliament in August 1974.

Medibank was a single universal national health insurance scheme for both medical and hospital expenses. It was funded from general taxation revenue. Most of its features need not be detailed here because they have re-emerged in Medicare. The scheme was implemented in stages barely four months before the defeat of the Whitlam administration, so there was not really sufficient time for a proper operational trial.

Under the Fraser coalition Government, Medibank was progressively dismantled in five separate steps between 1976 and 1982. Health sector objectives were subordinated to the imperative of controlling costs in line with macro-economic policy and the 'new federalism', meaning a return of some Commonwealth financial commitments to the States. The intention was to achieve these objectives by increased involvement of the private sector. Health care financing reverted to something akin to the prior system of subsidized voluntary health insurance with governmental regulation. Despite the Commonwealth's intentions, the private health insurance funds faced a continuing crisis of higher contribution fees and falling membership, thereby increasing the actuarial risk of those retaining membership, and hence exacerbating the increase in contribution rates. Although the move out of private insurance was eventually stemmed by a taxation rebate, the public perception was one of instability. By now, the major shortcomings of subsidized voluntary health insurance seen previously in the 1960s had again emerged.

In February 1984, following the election of the first Hawke ALP Government, universal national health insurance was reinstated under the revised name Medicare. Although the Commonwealth–State hospital cost-sharing arrangements are more tightly constructed and the private hospital accommodation subsidy has been removed, the implications for patients and private medical practitioners are similar.

Evaluation

Most authorities are agreed that the special qualities of health care as an economic good almost invariably require some form of governmental intervention into the interaction of supply and demand in the freely operating market. However, there are also problems in asserting that health care is a right and should be available to all in proportion to need, and free at the point of service. Proponents of the latter view ought to specify the incentives relating effectiveness to control of costs. Unfortunately, there is little objective evidence on the comparative efficiency of public and private health services in Australia.

With a national health service effectively blocked by the civil conscription clause, Australian sickness care policy has centred on finance. Until recently, in the absence of an alternative paradigm, decisions about the substantive nature of the service have been left largely in the hands of the medical profession. Official statistics in the health sector have dealt mainly with the amount of work done and its cost. With financial considerations determining the mix of services, few incentives to offer prevention have been generated. However, the Community Health Programme represents both a move to increased governmental concern for the style of service offered and a commitment to a positive view of health. Evaluating outcome in addition to process henceforth will require more of the statistical interest in health status and differences between groups that has emerged recently.

Because of the constitutional complexities, no one authority is empowered to conduct overall health planning. The Commonwealth and State Government departments each have planning units largely concerned with their respective public sector activities. Private medical specialists and general practitioners are autonomous. Several commentators on the health policy-making process, including Goode (1981), have described the dominant influence of medical ideas even in public sector policy. Apart from government medical officers, few senior health system bureaucrats are trained in health care, although many are now qualified in health administration. The generic managers now in vogue may well lack the background for planning where equity issues are entwined with efficiency. Hence they administer the system much as they find it, intent upon cost savings.

The Medibank/Medicare legislation acknowledges, some would say enshrines, the primacy of fee-for-service in medicine. Given the power and prestige of the established medical profession, any hopes that the ALP might have been able to implement a national health service staffed by salaried doctors would seem to have been misplaced. The scope and shape of medical care is largely mediated through the Medical Benefits Schedule. This rewards procedural and technological practice more highly than counselling. Hence pathologists, radiologists, and surgeons have prospered, while the incomes of physicians and general practitioners have not kept pace. Under fee-for-service, the rewards follow high patient turnover. By setting higher rebates for specialists than for general practitioners performing the same tasks, the Schedule reinforces the exclusion of general practitioners from new medical technologies. Specific prevention consultations have been removed from the Schedule. Moreover, the Schedule provides no incentives for doctors to call upon skills possessed not by them but by other health professionals.

The numerous changes in the Australian health system have not been evaluated in great depth. Indeed, the quality of Australian health statistics has been said to compare unfavourably with most other Organization for Economic Co-operation and Development (OECD) member nations. Under a federal system, the incentives to know would appear to be diminished, because this might sheet home responsibility. The paucity of well-found publicly available data on expenditure breakdowns, performance indicators, and outcome measures might well be said to have promoted inertia.

Medicare has a distinct advantage over its precursor voluntary schemes in generating a national database on service use. In addition, following representations by social scientists and epidemiologists, this deficiency was given prominence in the Kerr White Report (White 1986). Subsequently, the charter of the recently established Australian Institute of Health has been broadened to develop and maintain a range of national databases.

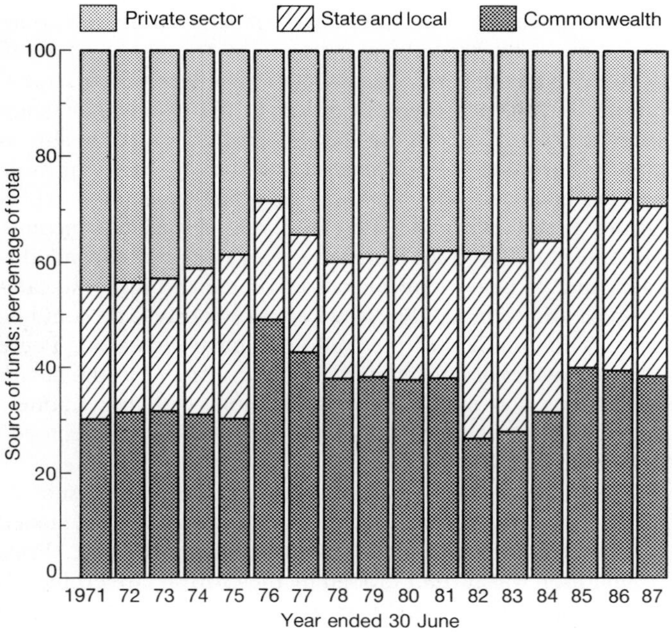

Fig. 24.5. Percentage of total recurrent health expenditure according to source of funds. (Source: Australian Institute of Health 1988*a*, 1988*b*)

The health dollar

Using comparable categories to those recommended for the World Health Organization by Abel-Smith (1969), Deeble (1978, 1982) was the first to construct a breakdown of total national Australian health expenditure across all areas. His methods have been followed by the Australian Institute of Health (1988*a*).

Most expenditure on health services is derived from Government revenue. In 1985–6, 40.3 per cent of all recurrent health expenditure came from the Commonwealth, 32.2 per cent from the States, and 27.5 per cent direct from private sources (i.e. out-of-pocket expenses, private health insurance premiums, and payments under workers' compensation and motor vehicle third party insurance) (Australian Institute of Health 1988*b*). These relative shares have fluctuated in line with changes to the system, particularly the method of public hospital financing (Fig. 24.5). Health accounted for 13.2 per cent of the Commonwealth's total budget outlays for 1988–9 (Australian Treasury 1988). Health expenditures also comprise about a quarter of each State's budget.

These latest figures show that hospitals and other institutions account for 54.3 per cent of total recurrent health expenditure; the bulk of this would be on salaries and related costs. The fees of medical practitioners in private practice represent 18.1 per cent. However, by their clinical decisions, these practitioners would influence expenditures in other categories, particularly the balance at the margin between hospital or nursing-home and community care. Dentists and other private practitioners account for 8.8 per cent, and pharmaceuticals 8.4 per cent. Despite complaints over the level of funding, the amount of expenditure on community health services can hardly be said to threaten a cost explosion. Expenditure on research, on health promotion by health workers, and on capital works in each instance seems to be lower than desirable (see Table 24.5).

In a comparison across 22 of the OECD nations, Australia ranked tenth highest for health care expenditure as a proportion of gross domestic product (GDP), and eighth highest for the share of public expenditure in total health care outlays (Poullier 1987).

The health care expenditures enumerated here represent only part of the nation's expression of concern for the well-being of its citizens. Income security payments are the largest in amount. Others are education, personal social welfare, public housing, and public transport, i.e. items now considered part of the social wage. Moreover, formal health expenditures are only part of the expenditures which advance health. Much of the old public health expenditure, especially provision of a clean water supply and hygienic disposal of sewerage, is the responsibility of civil engineering departments. Furthermore, many health problems arise in social processes against which the formal health services are but a feeble opponent. This calls into question the degree to which the limits of achievement of such services are known and understood; the appropriateness of existing services and the potential of an inter-sectoral approach; and the responsibility that should rest on families and communities.

Time trends in health expenditures

The costs of any system should always be considered in relation to effectiveness, and never in isolation. Regrettably in Australia as elsewhere, there are many gaps in what is known of the effectiveness of the health system. Life expectancy at birth has recently started improving again (Fig. 24.6), but it is difficult to disentangle the contributions of the health care industry from life-style change. Figures on morbidity trends are scanty. The numbers consulting a medical practitioner or admitted to hospital are known, but these are head counts not measures of outcome. Moreover, it is uncertain whether health inequalities have narrowed or widened (Moss and McMichael 1990). Since effectiveness is not measured in sufficient operational detail, the more practical policy objective (or the line of least resistance) has been to minimize the costs of the system, subject to constraints about the effectiveness of the care provided and its fair distribution. As in almost all countries, services are usually judged as efficient to the extent they contain costs.

Inflation in the general economy feeds into the overall costs of the health system and in turn is partly caused by health sector wage and commodity price rises. To allow for the effect of economy-wide inflation, the actual dollars spent can be expressed as a percentage of GDP. This is the usual form in which international comparisons are made. Economic growth rates as expressed in the GDP fluctuate from period to period with the trade cycle, and have been

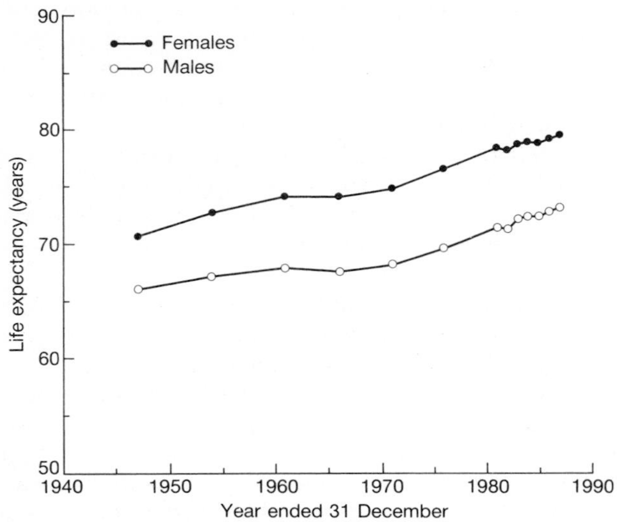

Fig. 24.6. Life expectancy at birth for males and females in Australia according to year of calculation (ended 31 December). (Source: Australian Bureau of Statistics 1988*a*,1988*d*)

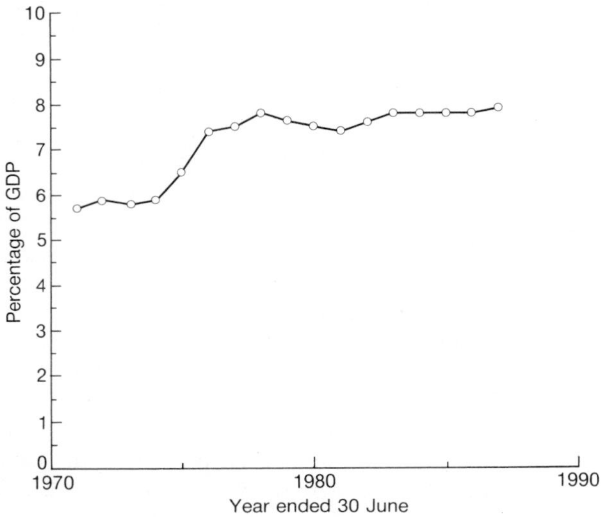

Fig. 24.7. Total health expenditure (current prices) in Australia from all sources as a percentage of gross domestic product (GDP) according to financial year (ended 30 June). The value for the most recent year is an estimate only. (Source: Australian Institute of Health 1988*a*, 1988*b*)

implicated in apparent rises in health expenditures, exemplified by those occurring in the mid-1970s.

Deeble's pioneering calculations (above) showed that total national Australian health expenditure in 1960–1 represented 5.0 per cent of GDP. This had risen by about one percentage point by the early 1970s. Thereafter, official Commonwealth figures show a substantial increase in the mid-1970s, peaking just below 8 per cent, followed by a sustained levelling off continuing to the present (Fig. 24.7).

Over much the same time span, most westernized countries have experienced an increase in the share of GDP devoted to health care. The association between these expenditures and the achievement of longer life expectancy is not straightforward, as would be anticipated since health is largely a product of environment and society. Many of these nations have felt the pressures of rising consumer expectations based partly on increased real disposable income. Most have suffered an increasing prevalence of chronic disease, and a heavy concentration of expenditure on catastrophic and terminal illness, where previously little could be done. The numbers of the frail aged have increased. These nations have experienced increased recruitment to the health professions and increased specialization, with practitioners dependent on increasingly complicated technology deployed in large hospitals.

The Menzies–Page legislation ushered in a large subsidy element to health care prices. Community rating removed incentives for people to improve their actuarial risk. Price competition was further eroded by the Gorton Government's conceding the notion of the 'most-common-fee', and this also became the price built into all subsequent negotiations. Private medical practitioner incomes benefited substantially (Taylor 1984). Where patients are still required to pay a moiety of their medical costs, this is usually a small percentage and not a strong disincentive to unnecessary use.

In Australia, the growth of GDP was strong in the early 1970s in the boom associated with the Vietnam war, and then restrained following the international oil crisis (Australian Bureau of Statistics 1988*b*). A belief in continuing growth reduced the incentives to economize and lasted beyond the point of actual downturn. As in other comparable countries, nominal wages rose from about 1968, although there was a reduction in real wage growth.

Being so labour intensive, health sector expenditures are affected by overall wages policy. Thus under the Whitlam ALP Government, the attainment of female wage equality, while undoubtedly both just and widely supported by the general public, inevitably led to rising health care costs since most major occupational categories in the health sector are mainly female. Moreover, visiting consultants ceased to have an honorary status, resident medical officers made substantial gains on their previously meagre pay, and the costs associated with staff ('on costs'), such as workers' compensation insurance, rose.

Under Medibank, Commonwealth–State hospital cost-sharing agreements were based on actual rather than budgeted expenditures, providing little incentive for the State-run hospitals to economize. Fulfilment of previously unmet need amongst those uninsured under the prior arrangements was a further upward pressure on total health care expenditures.

Thereafter, health financing was subordinated to macroeconomic policy. In the Fraser years, combating inflation became the priority, while the Hawke Government has been constrained by the balance of payments crisis. Both issues

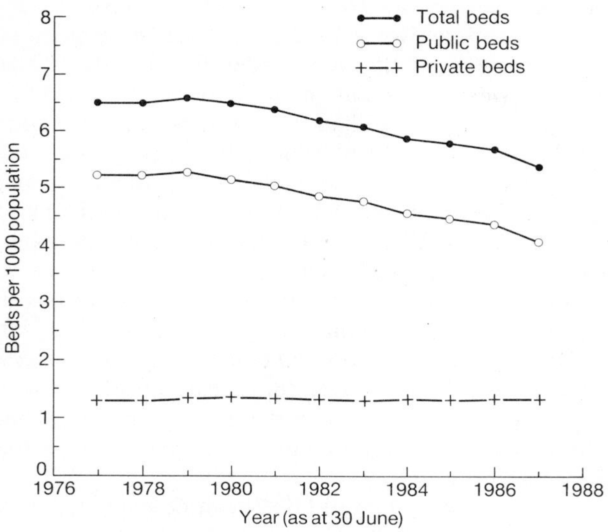

Fig. 24.8. Hospital beds in Australia per 1000 population as at 30 June of each year: beds in recognized public hospitals (including repatriation hospitals, but excluding public psychiatric hospitals), private hospitals, and total beds (sum of these two categories). (Source: Australian Commonwealth Department of Health 1987*a*, and earlier publications in that series)

have required restraint on health spending, although the mechanism adopted has been different. The Fraser Government under the policy of 'new federalism' sought to return the prime responsibility for health funding to the States, while the Hawke Government has been more centralist, relying on the Commonwealth's power of the purse.

The Medicare arrangements continue the Fraser policy that Commonwealth contributions to hospitals be based on budgets agreed in advance, with each State responsible for any extra cost incurred. Cost controls are mainly to be found on the supply side through the setting of fees, the establishment of hospital beds, and the imposition of quotas on medical schools. Arguably, the Accord between the Hawke ALP Government and the Australian Council of Trade Unions has moderated overall wage rises.

Public hospitals have been allocated reduced budgets in real terms. The total number of beds available per 1000 population has been substantially reduced (Fig. 24.8). In the face of an increased demand for their services, hospitals have increased their productivity. Average length of stay has been reduced remarkably. The decreased purchasing power of the Australian dollar has increased the local costs of drugs and medical supplies obtainable only overseas.

The changing health insurance schemes have caused marked swings in the relative magnitudes of private and public sector expenditures. In blaming Medicare for large cost blow-outs, the critics have overlooked the long-term trend for total health expenditures to remain stable as a percentage of GDP. In other words, health expenditures are taking no more than the previous relative share of total economic growth. Commonwealth Government outlays have risen substantially, but there has been an off setting reduction in private sector outlays transmitted through the health insurance funds.

The following points can be gleaned from a recent paper by Deeble (1988): services per person covered have been increasing at an average annual rate of 3.4 per cent over the last two decades; in its first year, Medicare covered some 2.3 million people previously without health insurance or pensioner entitlement; budgeted Medicare benefits were set with these increases (and general inflation) in mind; Medicare has strayed substantially (3.7 per cent) over budget in only one of its first four years of operation (1985–6); general practitioner attendances are not increasing appreciably (1.24 per cent in 1986), rather it is the total number of treatment procedures (largely surgery, 8.08 per cent in 1986) and of diagnostic procedures (10.7 per cent in 1986) that are increasing the most rapidly; the average cost per patient in general hospitals seems to have been kept well below the rate of inflation in 1985–6, by reducing average length of stay in line with an overall two-decade trend.

A considerable proportion of the population has relinquished private hospital insurance (Fig. 24.9). These people have increased the burden on the public hospital system. Waiting lists for standard ward beds have lengthened (Cooper 1988). It is arguable whether this reflects structural imbalances (e.g. shortages of surgeons and/or theatre nurses in public hospitals), or the interaction of increased demand (fewer people have private insurance) with decreased supply (closure of hospital beds). Except perhaps in New South Wales, private hospital capacity is underutilized. Having been deprived of their market niche in medical insurance, the

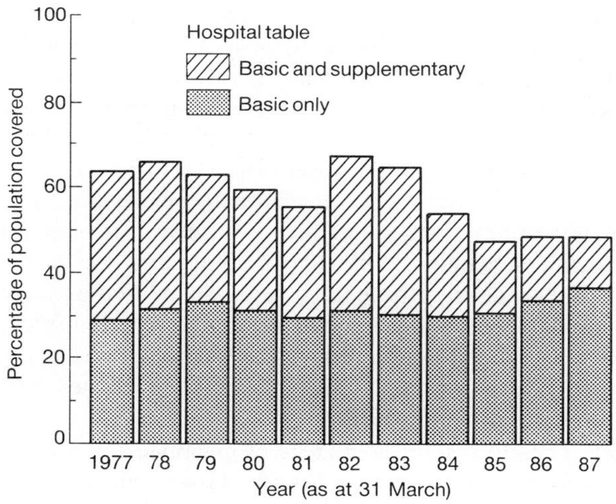

Fig. 24.9. Percentage of the Australian population (all persons) covered by voluntary health insurance as at 31 March of each year. Membership of a basic hospital table is a prerequisite for membership of a supplementary hospital table. Care should be taken when analysing trends because of the effects of major administrative changes such as the introduction of Medicare on 1 February 1984. (Source: Australian Commonwealth Department of Health 1986*a*, 1987*a*, and earlier publications in that series)

private funds are diversifying into health facility ownership (McHarg 1986) and general financial services.

New forms of practice organization are emerging in response to the particular pattern of incentives thrown up by the existing economic order in medicine. Entrepreneurial medical practitioners have established polyclinics. Since their profitability seems to derive from rapid patient turnover and a comprehensive internal referral system, questions of conflict of interest have been raised (Webster 1988). Moreover, ownership of private for-profit hospitals has swung from individuals to consortia, including health insurance funds, and chains owned by multinational corporations.

Perhaps an inevitable consequence of a fee-for-service system supported by insurance has been manipulation by some providers (Opit 1983). Payment by item of service has encouraged the deployment of more expensive technology, so that pathologists, radiologists, and some surgeons have prospered mightily. The few prevention items have been withdrawn from the Schedule because of alleged abuse. Medical practitioner entrepreneurs have set up chains of clinics, with mixed results for profitability. Surveillance has been instituted to curb unnecessary servicing and outright fraud, although many practitioners regard the measures as draconian. Meanwhile, other health professions in the private sector lobby Government in an attempt to attract the lucrative Medicare cover, especially that already available to medical practitioners performing similar functions in an era of role overlap.

Health services for the poor

Since the Harvester case before Mr Justice Higgins in 1907, Australians have been protected by the basic wage and its later variants. The Consitution provided for aged and invalid pensions and the necessary legislation was enacted in 1909–10. Antipoverty policy has henceforth concentrated on the availability of employment and, failing that, on means-tested income support; but the income minimum has not necessarily been above the level of poverty. The 1946 referendum enabled the Commonwealth to regularize its prior involvement in paying maternity allowances, unemployment and sickness benefits, child endowment, and widow's pensions. Subsequently, these have been augmented by the concept of the social wage, i.e. the worker's earned entitlement to publicly funded health, education, housing, and transport services. More recently (Graycar 1983) there has been a retreat from the welfare state.

In the health system, a mixture of means-tested social welfare benefits and service provision by Government has developed with little integrated planning. When voluntary health insurance prevailed, the contributions were heavily subsidized to encourage more use than would otherwise have occurred. Additionally, a largely Commonwealth-funded reinsurance pool provided benefits at a reduced rate to the chronically ill. Unfortunately, 14 per cent of the population may not have been covered by either private health insurance or a pensioner or repatriation entitlement (Adams *et al.* 1971). The subsidy was extended to cover most or even all the contributions of the poor under the subsidized health benefits scheme. This had a disappointingly low rate of acceptance. The Australian Government Commission of Inquiry into Poverty (1976) pointed to the economic, social, and geographical barriers to equal and easy access to primary health care services and hospitals. Under Medibank, the poor need pay nothing provided the doctor has bulk-billed. They could avail themselves like everyone else of free standard-ward public hospital treatment. During the dismantling of Medibank, a concessional health card system was introduced to identify pensioners and beneficiaries, war veterans, and the socially disadvantaged for special consideration. The concessional health card system has been retained under Medicare, and is administered by the Australian Department of Social Security (1987).

The analysis of Le Grand (1982) can be extended to suggest that access should be construed in terms of the total cost facing the potential consumer. Such an opportunity cost approach would include not only the payment for the actual service but also associated costs such as transport, loss of work time, search costs for information and interpreters, and child-care. Medicare entitles all Australian residents to an adequate level of medical and hospital care at low out-of-pocket cost at the point of service. But not all types of health-related service are covered. Unless the patient carries a concessional health card or the doctor direct-bills, some out-of-pocket expenditure will be incurred (up to the $150 maximum gap per person per year, and perhaps more if the doctor charges above the Schedule fee). This may loom large in a tight family budget.

Medicare has not entirely escaped the taint of residualism. The residual model is criticized for targeting of the poor and other powerless groups for separate servicing, thus making it difficult to avoid labelling them as second-class citizens and perhaps creating or reinforcing dependence. Under the alternative of a universalist model, services are provided for the community as a whole (Wilenski and Lebeaux 1958). Special needs are taken into account, but within a community framework. In community health centres, generalist workers handle the range of presenting problems, with back-up from team members having specialist expertise. When public hospitals and health services are open to all, there is a danger that those most in need will be crowded out because they are less articulate or because they lack contacts within the system. While Medicare is supposedly a one-class service, the poor are none the less categorized by the Health Card and/or direct billing.

Conventional illness-treatment services may have a small impact anyway. The Medicare objective is equity of access not equality of outcome. Social justice strategies will be more concerned with off setting the income and social differences between rich and poor. Hence, there is a need for policies on women's health, multi-culturalism, Aboriginal health, and on expansion of the Community Health Programme. The Better

Health Commission (1986) was an advocate for change in the substantive nature of the service, with a shift to prevention and genuine consumer participation. As part of the Bicentenary initiative on health, the Australian Health Ministers' Advisory Council (1988) has published a list of health targets which will provide valuable guidance in this endeavour.

Possible amendments

The longer the ALP Government retains power, the more entrenched universal health insurance would appear to be. Nevertheless, cost pressures could force a reversion to a means-tested or a needs-based scheme. The Opposition is considering a further reprivatization, which might entail dismantling Medicare and selling off Medibank Private. Yet any superior efficiency of the private health sector in Australia is a matter of faith rather than empirical demonstration. Moreover, equity might be compromised since some people would lack the ability to pay, resulting in a return to residualism. To be successful, reprivatization would have to ensure that there was an increase in competition rather than merely the transfer of a monopoly from public to private control. Furthermore, prior to Medicare, the private health insurers had higher administrative costs and no national data collection was feasible. Given the opportunities for cream-skimming contrasted with the burden of chronic illness, these proposals somehow smack of the time-honoured Australian tradition of privatizing the profits and socializing the losses.

The Opposition has also considered deregulation of private health insurance. As part of a less strict oversight, removal of the community rating principle has been proposed. Perhaps this would mean cheaper contributions for the young and the healthy, but at the expense of the elderly and of those at more risk. The mechanisms by which these arrangements might generate incentives for greater efficiency have yet to be spelled out.

Prepaid health plans such as Health Maintenance Organizations have been examined (Australian Commonwealth Department of Health 1986*b*; Douglas 1979). For this country, their advantages are believed to include the redirection of material incentives thereby encouraging cost-saving through reduced hospitalization, a private form of responsive organization, and perhaps even a more preventive approach. Perceived disadvantages include perpetuation of medical dominance yet the subordination of professional practice to management. It is doubtful whether a Health Maintenance Organization in a poor district would have the resource base to be able to provide the range and depth of services available in a well-to-do district, without some form of resource transfer. There are strong reasons for believing that these models may not translate well outside their country of origin to a different culture and set of institutions. As yet no project has materialized, largely because there is little incentive for either providers or consumers to join a Health Maintenance Organization under the present Medicare arrangements.

In the 1988–9 Budget, the Commonwealth Government set aside funds to prepare for the use of a Diagnosis-Related Groups admission taxonomy as the basis of public hospital payment. This will build upon their use especially in Victoria as a planning tool, where Diagnosis-Related Groups have provided a framework for costing and activity information for the evaluation of hospital performance and thence for budget allocation (Duckett 1986; Palmer and Reid 1986).

As with any form of insurance, Medicare is liable to the problem of moral hazard. This is the potential for over-use where the insured consumer faces a low cost at the point of service. Problems of this kind are exacerbated when insurance and social security are interwoven. The General Practitioners' Society in the recent past has proposed a form of insurance against major risks rather than first dollar coverage, thereby incorporating 'front-end deductibles' (see also McLeod 1987). The usual forms of care would be provided and paid for strictly through a private market mechanism. Insurance cover would be available should the costs to any patient exceed a predetermined and substantial maximum in any 12 month period. Subsidized or Government-provided insurance-cum-social security would protect the socially disadvantaged. While this scheme would impose the discipline of the private market, there would be a disincentive to seeking early treatment before problems became entrenched. The poor would be categorized, monopoly elements within the health system would face little price restraint, and questions would remain as to the fate of those who gamble and lose.

Meanwhile, policy-makers will face pressure to determine whether cost containment is better achieved by squeezing recurrent expenditure (such as by revamping the Medical Benefits Schedule, or by holding down nursing salaries), or by investment restriction (of the supply of hospital beds or of human capital). Pressure can be anticipated for adjustments to the Schedule to increase the benefit for general practitioner services. The absence of a mutually agreed arbitration process for reconciling fees and rebates has bedevilled the medical side of Medicare since the Australian Medical Association withdrew from the Medical Fees Inquiry in 1986.

Ultimately, the pressures of an increased demand run the risk of overwhelming a tightly held supply in the public hospital sector. Thus, it may make sense to rechannel demand by moving the routine care of chronic disease out of the hospitals back into general practice. This would augment the increased use of day surgery that is already occurring and also the greater use of rehabilitation facilities.

Conclusion

In 1988, Australians celebrated the Bicentenary of the settlement of the colony of New South Wales amidst growing dismay over national economic performance. Concern was also mounting over whether the costs and benefits of the readjustment process were being fairly distributed. On both counts—efficiency and equity—health service provision was again under scrutiny.

Williams (1988) makes the important point that so many

effective health care interventions are now available that no country can afford to provide all the potentially beneficial procedures to all people who might possibly benefit from them. Hence priority setting cannot be merely a matter of eliminating ineffective activities, but entails unavoidable sacrifice. In the Australian situation this analysis can be taken further. Public health and antipoverty measures must also come into this reckoning and are almost certainly more vulnerable to pressure from entrenched interests. To minimize unavoidable sacrifice, the overall economy must be maintained in as good a shape as is possible. In this light, the Hawke ALP government may have been justified in stimulating economic growth to create employment, and hence amongst other benefits reducing the burden placed on the state. Unfortunately this has been at the cost of huge overseas debts, which must in turn call into question present levels of health spending. In this way, economic performance and the fate of recent initiatives in universal health cover and public health, which have relied more on new money than on reallocation, are inextricably linked.

As in any advanced society, the Australian health system is notable for its complexity. To the usual range of competing sets of objectives that need to be reconciled must be added the opposing pressures for uniformity and diversity between the Commonwealth and the States. While policy-making is nominally rational, it is fed with indicators of illness and service provision rather than of health. Add to this the entrenched autonomy of the medical profession, and the increasing assertiveness of the nurses and other allied health professionals, and the system can be seen to express conflicts of values and interests between many parties.

Amongst the providers and policy-makers, none can be neutral. All stand to gain or lose career enhancement or income, be they in government or private service. Moreover, all the major actors are seeking to advance a legitimate sphere of influence, although some presently fare better than others. Inevitable resource constraints and different perspectives on priorities have brought them into conflict. No sweeping vision has triumphed, but no single ideology has become rampant. Perhaps in time they may all tolerate a pluralism based upon an enhanced understanding of the relationship between equity and efficiency in the health arena. To do so will, however, require much greater acceptance that the pursuit of legitimate interest can easily overreach into the entrenchment of power or monopoly.

Ironically, a medical profession that has fought governmental control for so long now also fears control by big business in the guise of entrepreneurs (Solomon 1985) and of Health Maintenance Organizations. Many general practitioners, while holding fast to private enterprise, seriously doubt the viability of their present role. Yet a considerable proportion of doctors already earn much of their income by way of a salary from a public hospital.

Since the realities of both voluntary and universal health insurance are now revealed, it can be cautiously predicted for the near future that muddling through will prevail over wholesale change. Enhanced cost-effectiveness probably will be sought by controlling investment in hospital bed stock, high technology and medical education, and by greater emphasis on same-day surgery and early discharge. Consequently, an adverse impact on nursing work load, and hence on morale, can be expected. Moreover, the measures will require sensitive handling if equity is not to be compromised.

Given the multiplicity of interests, it is impossible to please everybody all the time. Nevertheless, the office practice arrangements under Medicare appear to be an effective and workable compromise provided the Schedule is revised in favour of general practice and prevention. However, hospital insurance arrangements do not seem to have reached a settled form. Perhaps more open competition between public and private sectors can encourage both to be efficient. It can be anticipated that the shift in demand from private to public hospital care will continue to cause budgetary strain unless stemmed, perhaps by means testing, by even tighter budgetary controls based on Diagnosis-Related Groups, or perhaps by designing worthwhile incentives for the establishment of Health Maintenance Organizations. To remain viable, private hospitals will increasingly call for fair competition from an over-subsidized public sector. Although patients in private hospitals are covered by Medicare for 75 per cent of Schedule medical services, they receive no rebate for hospital accommodation. Prospective payment based on Diagnosis-Related Groups might remedy this without committing Government to an open-ended drain on funds.

The Australian system is sometimes compared unfavourably with the UK National Health Service because Government does not have direct responsibility for providing health services to all as a right, and with the US because market mechanisms are more frequently overridden. But, despite the pervading bitterness, it is arguable that the Australian system is a pragmatic pluralist model. From this perspective, the system has been assembled to meet the differing aspirations of the various actors within the context of politics, economics, environment, and society. However, an alternative interpretation is that conflict between the power blocs remains a barrier to a rational solution. From this opposing viewpoint, there is an uneasy stand-off between bastardized versions of libertarian and socialist programmes. As the dour struggles that have characterized the Australian health system continue to be worked out, it is clear that provision and finance cannot be understood in isolation from the differing sets of values they represent.

References

Abel-Smith, B. (1969). *An international study of health expenditures*. Public Health Papers no. 32. World Health Organization, Geneva.

Adams, A., Chancellor, A., and Kerr, C. (1971). Medical care in western Sydney: A report on the utilization of health services by a defined population. *Medical Journal of Australia* **1**, 507.

Allender, J. (1989). GPs give and take in new fees structure. *The Australian* 3 March 1989, p. 3.

Australian Bureau of Statistics (1981). *Australian Health Survey 1977–78: Doctor consultations* (cat. no. 4319). AGPS, Canberra.

Australian Bureau of Statistics (1984). *Australian health survey 1983* (cat. no. 4348.0). ABS, Canberra.

Australian Bureau of Statistics (1986). *The labour force, Australia: Historical summary, 1966 to 1984* (cat. no. 6204.0). ABS, Canberra.

Australian Bureau of Statistics (1987). *Perinatal deaths, Australia, 1986* (cat. no. 3304.0). ABS, Canberra.

Australian Bureau of Statistics (1988*a*). *Australian demographic statistics: March quarter 1988* (cat. no. 3101.0). ABS, Canberra.

Australian Bureau of Statistics (1988*b*). *Australian national accounts, national income and expenditure*. Special issue, March quarter 1988 (cat. no. 5206.0). ABS , Canberra.

Australian Bureau of Statistics (1988*c*). *Consumer price index—June quarter 1988* (cat. no. 6401.0). ABS, Canberra.

Australian Bureau of Statistics (1988*d*). *Deaths, Australia 1987* (cat. no. 3302.0). ABS, Canberra.

Australian Bureau of Statistics (1988*e*). *The Labour Force, Australia, August 1988* (cat. no. 6203.0). ABS, Canberra.

Australian Bureau of Statistics (1989). *Causes of death, Australia 1987*. (cat. no. 3303.0). ABS, Canberra.

Australian Commonwealth Department of Health (1986*a*). *Annual report 1985–86*. AGPS, Canberra.

Australian Commonwealth Department of Health (1986*b*). *Health maintenance organisations: A development program under Medicare*. AGPS, Canberra.

Australian Commonwealth Department of Health (1987*a*). *Health statistical supplement 1986–87*. AGPS, Canberra.

Australian Commonwealth Department of Health (1987*b*). *Medical benefits schedule book*. AGPS, Canberra.

Australian Commonwealth Grants Commission (1987). *Fifty-fourth report, 1987*. AGPS, Canberra.

Australian Department of Social Security (1987). *Annual report 1986–87*. AGPS, Canberra.

Australian Government Commission of Inquiry into Poverty (1976). *Social/medical aspects of poverty in Australia*. Third Main Report, G.S. Martin, Commissioner. AGPS, Canberra.

Australian Health Insurance Commission (1987). *Annual report 1986–87*. HIC, Canberra.

Australian Health Ministers' Advisory Council (1988). *Health for all Australians*. Report of the Health Targets and Implementation Committee. AGPS, Canberra.

Australian Institute of Health (1988*a*). *Australian Health Expenditure, 1970–71 to 1984–85*. AGPS, Canberra.

Australian Institute of Health (1988*b*). *Australian Health Expenditure, 1982–83 to 1985–86*. Information Bulletin no. 3. AIH, Canberra.

Australian Institute of Health (1989). *Medical workforce 1986* (Health Welfare Bulletin no. 14). AGPS, Canberra.

Australian Medical Association (1987). Report of the task force to review the structure, function and constitution of the Australian Medical Association. (R.C. Cotton, chairman). *Medical Journal of Australia* **146** (Special Supplement), 30.

Australian National Health and Medical Research Council (1941). *Report 12th session*. AGPS, Canberra.

Australian National Hospital and Health Services Commission Interim Committee (1973). *A community health program for Australia*. AGPS, Canberra.

Australian Parliament (1926). *Report of the Royal Commission on health*. Government Printer, Melbourne.

Australian Parliament (1969). *Report of the Commonwealth Committee of Enquiry into health insurance*. Parliamentary Paper no. 2. Commonwealth Government Printing Office, Canberra.

Australian Parliament (1982). *In a home or at home: Accommodation and home care for the aged*. Report from the House of Representatives Standing Committee on Expenditure. AGPS, Canberra.

Australian Treasury and Department of Finance (1988). *Budget statements 1988–89: Statement no. 3: Outlays*. AGPS, Canberra.

Bell, J. (1968). Queensland's public hospital system: Some aspects of finance and control. *Public Administration* (Sydney) **27**, 39.

Better Health Commission (1986). *Looking forward to better health*, Vol. 1, *Final Report*. AGPS, Canberra.

Committee of Inquiry into Medical Education and Medical Workforce (1988). *Australian medical education and workforce into the 21st century*. AGPS, Canberra.

Community Information Support Service (1987). *Directory of social welfare resources, South Australia* (12th edn). Community Information Support Service of South Australia, Adelaide.

Congalton, A.A. (1969). *Status and prestige in Australia*. Cheshire, Melbourne.

Cooper, J. (1988). Hospital waiting list grows to 40 000. *The Australian* 17 May 1988, p. 4.

Cumpston, J.H.L. (1978). *Health of the people: A study in federalism*. Roebuck Society, Canberra.

Deeble, J.S. (1978). *Health expenditure in Australia 1960–61 to 1975–76*. Research Report no. 1. Health Research Project, Australian National University, Canberra.

Deeble, J.S. (1982). Financing health care in a static economy. *Social Science and Medicine* **16**, 713.

Deeble J. (1988). Health care under universal insurance: The first three years of Medicare. In *Economics and health 1987: Proceedings of the ninth Australian conference of health economists* (ed. J.R.G. Butler and D.P. Doessel) p. 230. Australian Studies in Health Service Administration no. 63. University of New South Wales, Sydney. p. 230.

Dobson, A.J., Gibberd, R.W., Leeder, S.R., and O'Connell, D.L. (1985). Occupational differences in ischaemic heart disease mortality and risk factors in Australia. *American Journal of Epidemiology* **122**, 283.

Douglas, R.M. (1979). Prepaid health plans and control of health care costs. *Medical Journal of Australia* **2**, 478.

Duckett, S.J. (1986). Diagnosis related groups: Towards a constructive application for Victoria. *Australian Health Review* **9**, 107.

Dunt, D.R. (1982). Recent mortality trends in the adult Australian population and its principal ethnic groupings. *Community Health Studies* **6**, 217.

Federal Council of the British Medical Association in Australia v. Commonwealth (1949). 79 CLR 201.

Furler, E. (1982). Conflict and cooperation. In *Working papers in community health* (ed. J.D. Potter and A.M. Hodgson) p. 17. ANZSERCH/APHA, Adelaide.

Goode, J. (1981). The health policy process in Victoria. *Community Health Studies* **5**, 206.

Graycar, A. (ed.) (1983). *Retreat from the welfare state: Australian social policy in the 1980s*. Allen and Unwin, Sydney.

Hailstone, B. (1988). More support for Medicare. *The Advertiser* 27 August 1988, p. 15.

Hetzel, B. and McMichael, T. (1987). *The LS factor: Lifestyle and health*. Penguin, Melbourne.

Hicks, N. (1982). The 'community' in community health. In *Working papers in community health* (ed. J.D. Potter, and A.M. Hodgson) p. 12. ANZSERCH/APHA, Adelaide.

Hicks, N., Moss, J., and Turner, R. (1989). Child poverty and children's health. In *Child poverty* (ed. D. Edgar, D. Keane, and P. McDonald). Allen and Unwin, Sydney. pp. 92–103.

Hicks, R. (1981). *Rum regulation and riches: The evolution of the Australian health care system*. Australian Hospital Association, Sydney.

Hunter, T. (1984). The politics of National Health Insurance: 'Plus ça change'? In *Perspectives on health policy* (ed. P.M. Tatchell). Proceedings of a Public Affairs Conference, p. 28. Australian National University, Canberra.

Huntington, J. (1977). Social work and general medical practice: Towards a sociology of inter-professional relationships. *Community Health Studies* **1**, 17.

Jacques, B. (1988). Whatever happened to the hospital boom? *Australian Business* 7 September 1988, p. 48.

Jungfer, C. (1974). The role of the family doctor. *Australian Family Physician* **3**, 61.

Kuhn, T.S. (1974). *The structure of scientific revolutions* (2nd edn). University of Chicago Press, Chicago.

Le Grand, J. (1982). *The strategy of equality*. Allen and Unwin, London.

McHarg, M. (1986) Medicare: The funds strike back! *Health Action* McDonnell Douglas Information Systems, folio 11, p. 6.

McKay, B.V. (1986). A participant's account of 'the New South Wales doctor's dispute'. *Community Health Studies* **10**, 220.

McLeod, R.H. (1987). *Empty bed blues: Australian health-care policy* (policy paper number 9). Australian Institute for Public Policy, Perth.

McMichael, A.J. (1985). Social class (as estimated by occupational prestige) and mortality in Australian males in the 1970s. *Community Health Studies* **9**, 220.

McQueen, H. (1975). 'Spanish 'flu'—1919: Political, medical and social aspects. *Medical Journal of Australia* **1**, 565.

Moss, J. (1984). Analysing roles and interactions in the health system. *New Doctor* **13**, 16.

Moss, J.R. and McMichael, A.J. (1990). Social inequalities in health: The epidemiological evidence—and the gaps. In *Prospects for prevention in medicine* (ed. J. McNeil, R. King, G. Jennings, and J. Powles). Edward Arnold, Melbourne.

Murrell, T.G.C. (1985). Developments in primary health care in Australia. In *Oxford Textbook of Public Health*, Vol. 2, *Processes for public health promotion* (ed. W.W. Holland, R. Detels, and G. Knox) (1st edn.) p. 77. Oxford University Press.

Opit, L.J. (1983). Wheeling, healing and dealing: The political economy of health care in Australia. *Community Health Studies* **8**, 238.

O'Reilly, D. (1988). Medicare: A fitness report. *The Bulletin* 19, April 1988, p. 32.

Palmer, G.R. and Reid, B. (1986). Diagnosis-related groups, peer review and the evaluation of hospital activities. *Australian Clinical Review* **6**, 120.

Pensabene, T.S. (1980). *The rise of the medical practitioner in Victoria*. Australian National University Press, Canberra.

Poullier, J.-P. (1987). From risk aversion to risk rating: Trends in OECD health care systems. *International Journal of Health Planning and Management* **2**, (Special), 9.

Powles, J., Hage, B., and Ktenas, D. (1986). Who came? who stayed behind? selection and migration from the Greek island of Levkada (abstract). *Community Health Studies* **10**, 381.

Russell, C. and Schofield, T. (1986). *Where it hurts: An introduction to sociology for health workers*, p. 112. Allen and Unwin, Sydney.

Sawer, G. (1988). *The Australian Constitution* (2nd edn). AGPS, Canberra.

Sax, S. (1984). *A strife of interests: Politics and policies in Australian Health Services*. Allen and Unwin, Sydney.

Scotton, R.B. and Deeble, J.S. (1968). Compulsory health insurance for Australia. *Australian Economic Review* 4th Quarter, 9.

Shoebridge, J. (1980). The unspeakable in pursuit: The Australian health services and its migrants. *Community Health Studies* **4**, 236.

Smeeding, T., Torrey, B.B., and Rein, M. (1988). Patterns of income and poverty: The economic status of children and the elderly in eight countries. In *The vulnerable: America's children and elderly in an industrial world* (ed. J.L. Palmer, T. Smeeding, and B.S. Torrey) p. 89. Urban Institute Press, Washington, DC.

Solomon, S. (1985). Gatecrashers in medicine's cosy club. *Australian Society* **4**, 8.

Taylor, R. (1979). Health and class in Australia. *New Doctor* **13**, 22.

Taylor, R. (1984). Income from private medical practice in Australia 1966–1978: An analysis of taxation statistics. *Community Health Studies* **8**, 1.

Thame, C. (1974). Health and the state: The development of collective responsibility for health care in Australia in the first half of the twentieth century. Unpublished PhD thesis. Australian National University, Canberra.

Thomson, N. (1984). Aboriginal health: Current status. *Australian and New Zealand Journal of Medicine* **14**, 705.

Webster, R. (1988). Entrepreneurial dangers. *Medical Journal of Australia* **148**, 270.

White, K.L. (1986). *Australia's bicentennial health initiative: The independent review of research and educational requirements for public health and tropical health*. Report to the Hon. Neal Blewett Minister of Health. Commonwealth Department of Health, Canberra.

Wilenski, H.L. and Lebeaux, C. (1958). *Industrial society and social welfare*. Russell Sage Foundation, New York.

Williams, A. (1988). Priority setting in public and private health care: A guide through the ideological jungle. *Journal of Health Economics* **7**, 173.

World Health Organization (1947). The constitution of the World Health Organization. *WHO Chronicle* **1**, 29.

World Health Organization (1978). *Primary health care: Report of the international conference on primary health care, Alma-Ata, USSR, 6–12 September 1978*. WHO, Geneva.

25

Comparative analysis of approaches to the provision and financing of health care

JOHN FRY

Introduction

Any comparison of approaches to and systems of health care has to include an examination of some basic principles that apply to all systems in all countries. It is useful to begin by posing fundamental questions such as 'what is health?' and 'what is health care?'.

What is health?

The original World Health Organization's definition of health as 'a state of complete physical, mental and social well being and not merely an absence of disease' was a well-intended altruistic concept that concentrated minds, but which was soon realized to be a wishful and fleeting mirage, strivable but never attainable.

It has become clear that, in an imperfect world, present deficiencies have to be accepted and people have to be motivated to live optimally and on terms with their own personal environments. In Japan and Finland the importance of changes in personal behaviour and attitudes is demonstrated by a decline in mortality rates from cardiovascular disorders.

In addition to individual responsibilities, health, or non-health, is part of a much wider spectrum of life that involves many other sectors, such as agriculture and industry, employment, housing, and education. Consideration of the 'Health For All By The Year 2000' call by the World Health Organization shows that national health policies are part of broader national policies involving these other sectors.

What is health-care?

As already noted, health depends a great deal on personal and intimate behaviour and on attitudes in promoting well-being and preventing disease. Good health, therefore, is much more than provision and delivery of a health care system, but, at the same time, optimal health does require delivery of systems of health care through which health services, ancient as well as modern, can be made available and accessible to the people.

Health care has to be an integrated mixture of personal responsibilities to health maintenance and disease prevention; of public responsibilities to provide basic essentials for the public health; of allowing scope and choice for provision of private extra amenities; and of encouraging voluntary and charitable agencies to take community action.

National health systems are complexes of activities. They take note of these general matters, but also provide services that are seen to be sound, fair, and equitable, which is in keeping with principles of social justice, but which also must comply with national and local economic and social restraints, deficiencies, and difficulties.

There is no single 'best-buy' system of health care. There are almost as many national health systems as there are nations. Each system has tended to develop gradually and has been influenced by many characteristic factors. Among these are national history and culture; prevailing political ideologies and philosophies; strength of religious principles; local geography and climatic conditions; and, questionably of the greatest importance, national wealth, economic stability, available natural resources, such as oil and other raw materials, and industrial and agricultural developments.

What are the components of a health-care system?

However detailed the structure of a national health care system, each has to provide preventive services, therapeutic 'curing' services, and caring services for relief, comfort, and social support for 'non-curable' conditions and situations. Essential components of any health care system include resources, organization and administration, management, systems of financing, and health care delivery.

Resources, such as the health personnel available and required, and the appropriate education and training. The medical personnel requirements, both in terms of numbers and rates per population in the various grades and specialties, are difficult to estimate as there are many imponderables such as how should work be shared between various professionals, and how much work should be expected from workers in terms of time, quantity, and quality? How much time should be set aside for further education and training, committee work, and other para-professional activities? How much time and encouragement should be allowed for 'private work' outside the national health system?

Physicians are a most expensive resource in a health system, much more so than nurses. Countries where the cost of health care is high tend to have proportionately more doctors than nurses and vice versa. It may be that expensive doctors are carrying out duties that could well be done by less expensive nurses. An important question facing all countries is how many doctors and nurses and other paramedical workers are needed to be trained and employed?

In addition to personnel, attention has to be given to provision and availability of commodities, such as pharmaceuticals, and on whether there should be any controls or rationing.

Allocation of facilities has to try to achieve a balance between the various levels of care, i.e. primary, secondary, and tertiary medical care, and between community and institutional services.

Organization and administration involve making policies, planning, and setting goals for achievement of standards, and for striving to achieve value for money available and spent. Fair allocation of available resources, geographically and among social groups and classes, has to be attempted as well as sharing between the various medical personnel who are always striving to obtain perhaps more than their allocated rations.

Management, at every level and in every sphere, has the key role and the difficult tasks of striving to apply and to implement national and local policies and targets, and, at the same time, managing the professional and para-professional workers and keeping the public happy and satisfied.

A completely satisfactory and satisfying *system of financing* is a well-nigh impossible task. Wants and demands are infinite, but resources are truly finite. Whatever the source of the money, there will never be enough. The source may be government through direct or indirect taxes; compulsory or voluntary prepaid health insurance; private fees from patients; and charities or voluntary bodies. In no system is the total available sufficient to meet stated needs.

The forms of *delivery* of health care, although dependent on national and local customs, are basically similar and have to be related to the problems presented.

National health systems

National health systems, in details, may appear very different, but they have a common structure of essential levels of care each with similar roles and functions. Although there is such a recognizable common structure, there are differences related to national factors. Therefore, the similarities and the differences should be examined as they relate to structure, goals and financing, dilemmas, solutions, and immediate needs.

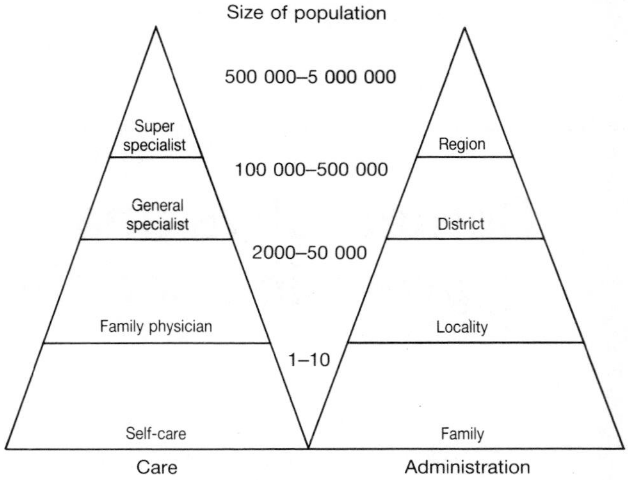

Fig. 25.1. The four levels of care found in all developed health systems and their correspondence with population size. (Source: Fry 1978.)

Common structure

In all health systems there are four essential and recognizable levels of care that tend to correspond with levels of population size (in developed countries) and administration (Fig. 25.1) (Fry 1985). These levels are:

1. Self-care in family units of one to ten persons.
2. Primary professional care in the community, involving physicians, paramedical workers such as nurses, social workers, and various other community workers. The population may range from a primary physician's practice of around 2000 persons to large groups providing services for 10 000–50 000 people. The administrative unit is a community neighbourhood or locality.
3. Secondary general specialist care is provided by district general hospitals which are staffed by specialists such as general surgeons, general medical physicians (internists), general paediatricians, general obstetricians/gynaecologists, trauma/orthopaedic specialists, general psychiatrists, and others. Alongside these clinical medical specialties will be generalist nurses, midwives, and other paramedical workers. The number of people served by a district general hospital in cities is about 250 000, and the administrative unit is the district.
4. Tertiary super (sub) specialist care is based on regions of 1–5 million people and includes units, such as cardiothoracic medicine and surgery, neurology and neurosurgery, organ transplantation, oncology, and others.

Each level has responsibilities for a given spectrum of common medical and social problems and situations. Thus individuals and their families can be expected to manage most minor ailments that do not require medical attention. They should also be expected to promote their own health and take steps to prevent disease. Such responsibilities cannot be carried out effectively without suitable encouragement, incentives, and public and personal health education.

Primary care is provided by physicians and other health workers who are the public's first professional contacts. Their workload includes minor, chronic, and major illnesses, and also a mass of social pathology (Fry 1985).

The primary physician acts as a gatekeeper to secondary and tertiary specialist services. Few of the patients who consult the primary physician need referral to hospital-based specialists.

There are important distinctions between secondary and tertiary specialist services. The district general hospital and its general specialists liaise closely with the primary physicians in its catchment area, and care for the common hospital disorders such as appendicitis, hernias, haemorrhoids, cancers, myocardial disorders, asthma, fibroids, and similar conditions, whereas at the tertiary level the 'bread and butter' disorders are relatively uncommon problems such as cerebral tumours, coronary artery bypass surgery, kidney dialysis transplant surgery, severe burns, etc.

Types of health care systems

Classification of national health care systems is not easy. They can be separated into political groupings, or in relation to national wealth, or by the extent of government involvement.

Politically the range is from the permissive entrepreneurial free enterprise system that exists in US to the centrally planned and administered systems of the USSR, Eastern Europe, Cuba, and China. Between these extremes are the selective welfare oriented systems of western Europe, Japan, and Australia, and the more comprehensive systems in the UK, Scandinavia, New Zealand, Saudi Arabia, and Kuwait.

National wealth as measured by gross domestic product, does not appear to be a major influence in determining the type of national health system. The US, Japan, the Federal Republic of Germany, Australia, the UK and the USSR are all wealthy countries with an annual GDP gross domestic product of more than US $5000 per person and yet their health systems span the political range, as is the case for the poorer developing countries with an annual gross domestic product of less than US $500, such as Nepal and Ghana with a free enterprise system, India with a welfare oriented system, Tanzania which seeks to provide a universal comprehensive system, and China which follows socialist principles.

Classification by the extent of government involvement through funding is probably the most useful as it may forecast the scope and extent of changes for the future. Table 25.1 shows the range of involvement, with the US at one end of the spectrum and the USSR at the other. It should be noted, however, that, even in US, almost one-half of all health care costs are met through various government agencies. The average for western developed countries is around 80 per cent (Organization for Economic Co-operation and Development 1987).

Table 25.1. Health care costs by direct government funding

Country	Health costs funded by government (%)
US	40–49
Japan, Australia, Federal Republic of Germany, France	50–74
UK, Norway, Sweden	75–89
USSR, Saudi Arabia	> 90

Source: Maxwell (1988).

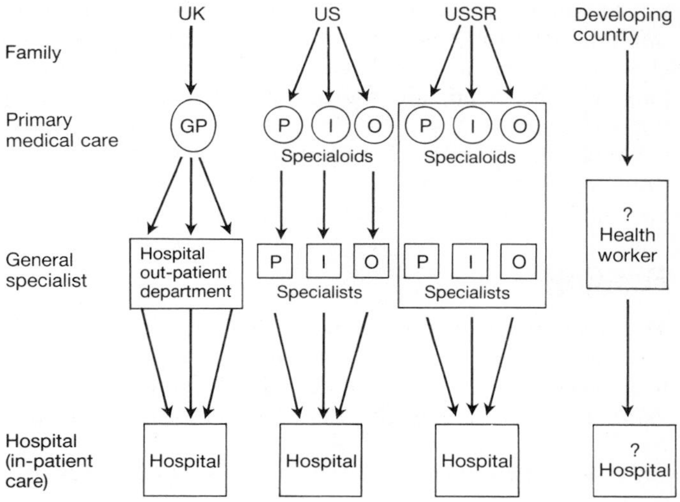

Fig. 25.2. The flow of care between the four definable levels. GP, general practitioner; P, paediatrics; I, internal medicine; O, obstetrics and gynaecology. (Source: Fry 1978.)

Ideally, there should be a public–private mix of health services to allow the consumers some free choice. The respective proportions of such a mix are influenced by particular national attitudes and beliefs. In no system can private health care be denied and every system has to make provisions for government-subsidized health services for those unable to pay for them.

Flow of care

Yet another way of comparing national health systems is to assess the flow of care between the four definable levels (Fry 1985). (Fig. 25.2.)

In the UK, every person can be registered with, and receive care from, a general medical practitioner (family doctor). General practitioners are the portals of entry into the health system, and are also the gatekeepers to the hospital services since patients can receive specialist care only after referral by their general practitioner, who will refer to a specialist of his or her choice.

Where there is no universally available and accessible primary generalist physician (family doctor), the family has to choose between a variety of primary 'specialoids', or has direct access to secondary or tertiary level specialists, as in the US, many western European countries, and Japan.

The flow of care in the USSR and allied systems is somewhat different. In the larger cities, primary care is sectorized and carried out from a 'policlinic' (health centre) where primary sector (uchastok) physicians, each with an allocated population, are based. Children, women, and men may have different uchastok physicians allocated to them. At the policlinic there are also ambulatory clinical specialists whom patients can attend, directly or by referral, from primary physicians.

Finally, in a poor developing country, primary care will be provided, if at all, by a non-medical health worker, or from a hospital emergency room that may be many miles away.

Funding

The costs of health care know no limits and the amounts spent depend on how much individuals can and are prepared to spend from their own resources and how much the other major funders, such as governments and insurers, are prepared to contribute.

The ultimate source of funds in most systems is 'the people' who pay taxes, insurance premiums, and direct fees. Although there are many variations in the patterns of national financing, one has to wonder whether the various schemes really affect the final outcomes?

Everywhere the expenditure on health services has increased and will continue to do so. All countries are facing the extra costs of having new medical technologies, drugs, and other therapies, as well as having to provide care for populations that are living longer with more old persons, and also for the disabled survivors of modern medical 'miracles' which may require lifelong treatment with expensive drugs and personal care. Inevitably more and better care leads to increasing costs.

Where do funds come from?

There are three main sources of funding: taxation, insurance, and direct fees from patients (Table 25.2). Taxation may be general, through central or local governments, or specified by hypothecated taxes. In the UK, in Eastern Europe, in other socialist countries such as Cuba, general taxation is the method, whereas in Canada and Sweden, local regional and county health taxes are levied.

Hypothecated taxation, or taxes pledged for a specific purpose, was tried in New Zealand for a while. It consisted of an extra tobacco tax which was intended to fund innovative community health services. A novel fiscal measure—it did not survive for very long.

Table 25.2. Types and sources of funding of health care

Funding source	Country
Taxation	
General	UK (central), USSR (central), Sweden (local), Canada (local)
Hypothecated	New Zealand
Insurance	
Compulsory	Western Europe, Japan, Australia
Voluntary	US, UK, France
Direct fees	All to a variable extent

Source: Roemer (1988).

Health insurance is compulsory or voluntary. In western Europe, Australia, and Japan compulsory insurance covers most of the population and is administered through government-approved sickness funds. Contributions are collected from employers and employees, together with some government input for certain services, such as the sick poor, the chronic sick, for mental disorders, and child and maternity services. Voluntary insurance exists in a free enterprise system such as US, but it is necessary, in addition, to have government-funded services to cover the 20 per cent of the population who are unable to purchase such insurance.

In the comprehensive UK National Health Service, 10 per cent of the population is covered additionally by voluntary health insurance and receives private care from chosen hospital specialists, in private hospitals and at convenient times (for both patients and doctors).

In all the countries under review voluntary health insurance exists in addition to compulsory health insurance. Voluntary insurance may be taken out to cover medical (doctors') fees as in France, or for special privileges such as private clinics in hospitals as in the UK, or because the compulsory system excludes much of the population, as in the US.

Direct payment of fees exists universally, even in socialist health systems, but can only meet the costs of occasional sickness events; few persons or families can be expected to pay the huge bills for modern clinical care using complex technological procedures.

How much is spent?

Comparisons of expenditures on health care in different countries are fraught with difficulties of definitions, measurements, and exchange rates. Nevertheless, data from the Organization for Economic Co-operation and Development (1987) can be taken as a guide-line and can be expressed as the total health expenditure per head (Table 25.3) or as percentages of gross domestic product (Table 25.4). Even between the five countries under review, there is a three fold difference in expenditure per head (see Table 25.3) and nearly a two fold difference in the proportion of gross domestic product.

Table 25.3. Total health expenditure per head in 1985

Country	Health expenditure	
	Annual expenditure per head (US $)	Ratios with UK as 1.0
US	1776	2.83
Australia	904	1.44
Western Europe	850	1.36
Japan	783	1.25
UK	627	1.00

Source: Organization for Economic Co-operation and Development (1987).

Table 25.4. Total health expenditure as a percentage of gross domestic product, 1985

Country	Health expenditure	
	GDP spent on health care (%)	Ratio with UK as 1.0
US	10.8	1.9
Australia	7.6	1.33
Western Europe	7.5	1.32
Japan	6.6	1.16
UK	5.7	1.00
Mean for all OECD	7.4	1.30

Source: Organization for Economic Co-operation and Development (1987).

Do such differences in health expenditure produce different degrees of health and disease in the community? Available health indices in these five countries are very similar and it does seem that greater health expenditure achieves better health overall.

How is it spent?

The UK National Health Service (NHS) provides the most detailed data on distribution of expenditure in a health system (Office of Health Economics 1987). The hospital service has always been the most expensive, accounting for about 60 per cent of total costs, with community health services consuming 6 per cent, general medical services (family doctors) 7.5 per cent, prescribing by family doctors 10.5 per cent, and other services, such as dental, ophthalmic, ambulance, public health laboratories, and blood transfusion, the remainder.

More than one-half (52 per cent) of NHS expenditure goes on wages and salaries. Nurses account for almost one-half, physicians for one-sixth, technicians and ancillaries for one-fifth, and administrators for one-tenth of total salary costs. Looked at another way, it is good to note that two-thirds of all NHS expenditure is on direct patient care and one-third on provisions, buildings, and administration.

A major influence on the way in which medical care is provided is the manner in which physicians are paid, for it is the physicians who are chiefly responsible for patient care. The various systems of payment include salaries, capitation fees, agreed fees for specified services, unrestricted private fees, and various others such as 'merit awards', and reimbursements for staff salaries and building costs.

Undoubtedly, different payment systems carry different incentives, motivations, and inducements, and these possibly explain different rates of surgical procedures, investigations, hospitalization, and attendance.

Who controls who is in charge?

Within such extensive and expensive national health systems there have to be social controls, delegation of responsibilities, and various degrees of market philosophy and intervention. At all levels, and in all systems, someone has to try to balance budgets and make 'profits'; someone has to ensure quality and quantity of care and its availability and accessibility; someone has to ensure professional standards and continually check outcomes, utilization, wastage, and value for money.

Although details may differ between systems, requirements for data collection and analysis, and for management skills within a complex system employing professional and non-professional workers are similar. Special managerial skills are required to deal with such disparate groups of workers and at the same time to provide satisfactory services to the public.

What outcomes and benefits?

Ultimately, everyone providing and using health services is interested in achieving maximum benefits at minimum costs. Such a philosophy must include efficiency, effectiveness, economics, and equity in providing fair shares for all in situations of finite resources.

The problems involving equity are immense. Although in theory it is possible to provide equal access to available health services, achieving 'health for all' appears to be a utopian mirage. In all countries, with all the variety of systems of health care, there are inequalities in morbidity and mortality rates between the various social classes, however, they may be defined. It is not 'social classification' itself that influences the differences but all the many other factors, such as life style, housing, diet, education, and even family and genetic traits. Although efforts to provide equal shares and opportunities for health care have to be a goal, we should not be disappointed if differences in health continue to be reported among social classes.

But how should various items of quantity and quality be measured, evaluated, and implemented? The answer involves the whole process of management of any health system and has to give attention to the system itself, the processes involved and carried out, the outcomes and the costs.

It has to be admitted that the methods and measures employed are still rather crude and controversial. Also, no single method or measure can provide all the necessary information. For example, mortality data are the final outcome of many factors besides the nature of the disease and the ways in which it is diagnosed and treated. There are exquisite and intricate personal, family, and social factors for which standardization is not possible. It is more meaningful to concentrate on premature and/or preventable deaths and to use their rates, albeit as rather insensitive indicators.

Morbidity data are beset by difficulties of agreement in nomenclature and definition, of diagnosis, and of precision in measuring outcomes. Sophisticated concepts, such as QALYs (quality adjusted life years) yield controversial results. Marker conditions, such as childhood infectious diseases, mental disorders, peptic ulcers, and disabling conditions, for which agreed positive diagnostic or therapeutic methods exist, can be used but require full co-operation from physicians and others.

Assessment of quality of care involves much more than measures of mortality and morbidity. It has to consider standards of professional competence and behaviour and how they are perceived by consumers as well as by providers. Measurements are largely subjective and depend on patient satisfactions or dissatisfactions leading to complaints and litigation, and on professional morale and satisfaction with their work, the facilities and resources available, and the rewards.

Measures of utilization of services can be related to individuals and groups. Thus utilization of primary care services by patients varies widely. As is well known by general practitioners in the UK, annual consultation rates per person and per family for minor, self-limiting disorders can vary by five or even tenfold. Likewise, prescribing rates and referral rates to hospitals also vary widely between general practitioners and are unrelated to any apparent measurable benefits to patients. In hospitals, specialists also vary widely in rates of investigations, prescribing costs, use of therapeutic procedures including surgical operations, and length of stay in hospital.

Obviously, high priority has to be given to collecting meaningful data on processes and outcomes in health care if they are to influence public and professional actions and habits for the better.

National systems

The structure of each national health system has to be related to national characteristics, including history, social philosophies, and wealth. The processes of care in the selected countries in this section are very similar and are based on accepted principles. The outcome of care, as measured by health indices, is also similar, suggesting high standards. One wonders, therefore, whether differences in structure, including provision and financing, are that important when countries with similar social and economic standards of living are compared. Nevertheless, it is useful to highlight some of the national differences, problems, and issues.

Australia

Australia is a vast country. Most of the interior is uninhabited and the majority of the population are urban dwellers in cities along the coast. The health care system has evolved along principles that originated with earlier British settlers, although it has shown changes since the Second World War.

Provision and financing of health care has become a 'political pendulum', with each of the two large political parties changing policies when in power. Power blocs have emerged, and differences and confrontations between the Government and the medical profession continue.

The national health insurance system is pragmatic and pluralistic, and appears complex and unwieldy. This has led to confusion over the public–private mix of involvement and muddles over competition and controls.

The medical profession is striving to maintain its independence and autonomy free from directives and controls. As well as fearing Government intrusion and supervision, it also fears the possibility of take-overs by private for-profit entrepreneurs, and the prospects of some form of American-style health maintenance organization.

Special features of the Australian health scene are the growing political militancy of health workers and power of nurses, and the growth of private nursing homes subsidized through the Government's health programme.

United States

The situation in the US appears to be that of a free enterprise system with the individual responsible for meeting most health costs (Fry 1969). However, a sizeable proportion of the population has no insurance cover to meet the increasingly heavy costs, despite the availability of systems such as Medicare, for the elderly, and Medicaid, for the poor.

Many schemes and ideas have been promoted and tried, based on private, employer/employee and medical insurance prepaid by the Government. None is universally possible. All have the ideologies of value for money, control of costs and expenditure, and increasing scrutiny of quality as related to quantity. Most have in-built profit motives and are organized and administered by corporations that apply modern business principles and methods.

Inevitably there has been a move away from the traditional independent practitioner and specialist towards large groups of physicians contracted by organizations to provide services.

The current trends are for developing systems such as health maintenance organizations (HMOs); management indemnity plans (MIPs), which include intensive monitoring; preferred provider organizations (PPOs), with arrangements between employers and insurers and groups of physicians, and with exclusive provider organizations (EPOs) as a variant.

In spite of the growth of such schemes, only one-quarter of

the population is covered by them, and by excluding many social groups and conditions, they cover a selective population. It is questionable whether the whole of the population can be covered for all its medical problems in this way.

The conclusion must be that the health care situation in the US is one of fragmentation with many mini-systems and no emergence of a 'best for all' system.

Japan

Japan is the world economic leader with the enviable problem of making too much profit and the embarrassment of huge annual trade surpluses. It has rather an old-fashioned system of health care, but has nevertheless achieved some of the best health indices, such as long life expectancy and low mortality rates for common causes of death such as heart disease and some cancers.

Historically, there have been three influences on the Japanese health system—ancient Chinese, late nineteenth century German, and American in the period after the Second World War.

The Japanese health system is a paradox of considerable central government planning and a national health insurance scheme and a sizeable free enterprise private system in hospitals and primary care. Local government, through the prefectures and municipalities, provides public health centres (1 per 100 000 population) for community health, disease prevention, and general health promotion and maintenance.

Of the hospitals, 70 per cent (45 per cent of hospital beds) are privately owned and run for profit. There is competition between private and public hospitals.

Primary medical care is not synonymous with family or general practice in either system. The Japanese primary physicians are 'specialoids' who provide, own and run their 'clinic', often from their own homes, where they also have beds for in-patient care. They have undergone training in a specialty but have not been able to follow a specialist hospital career. In their clinics, primary physicians continue with their specialty at a less intensive level and also undertake more traditional primary care.

The clinics tend to be well equipped with modern diagnostic technologies which duplicate those in hospital. The staff are employed by the physician. Eighty-five per cent of clinics are privately owned.

The national health insurance system that finances medical care has two sections: one is for employed persons (60 per cent) and the other is for the self-employed (40 per cent). Dependants are covered under both schemes for all medical and dental services, but some payment is required by the patients. The system is administered by many insurance groups. Hospitals and clinics are paid on a computed fee-for-service basis, which sometimes includes incentives to provide more services.

There is free choice of and free access to both clinics and hospitals, and this leads to competition and duplication of resources. There is no real need to wait for care, which is generally of good quality, although few checks on quality are made.

Europe

Western Europe comprises many small democracies, many with new socialist tendencies yet long-held customs and traditions.

Similarities between the health systems in Europe can be seen, but their differences are shaped by political, cultural, and economic factors. All are grappling with the same problems of attempting to control the high costs of modern medical care and social services, of deciding upon and controlling the appropriate levels of resources and medical personnel, of correcting inequalities of health and care, of creating national health policies, administration, and organization, and of checking and monitoring outcomes and value for the huge amounts of monies spent.

Eastern Europe includes the USSR, which is the largest country in the world, and many smaller satellite countries. The socialist health services are comprehensive and provide care for the entire population. They are administered in definite territorial hierarchies with national, regional, district, and neighbourhood levels. The emphasis has been on quantity with employment of large numbers of doctors and nurses, and attention to quality controls has lagged behind, although this is now changing (Fry 1969).

All European countries, including the Eastern bloc, have variable degrees of public–private mix of medical care, including government involvement, social insurance, and independent professional practitioners.

Scandinavian countries have some of the world's best health indicators and have strong governmental social insurance schemes. After the Second World War, Sweden put much emphasis on hospital building and development, and neglected primary health care, which it is now trying to correct with some success. Finland, on the other hand, actually passed legislation to make primary health care, with local involvement, the keystone of its national health system.

The processes and funding of health care in the rest of western Europe, i.e. the Netherlands, Belgium, France, and the Federal Republic of Germany, involve three groups: relatively independent insurance agencies who deal with administration; governments who decide national policies; and independent physicians who work on a fee-for-service basis in primary medical care and hospital doctors who are on good salaries. France, the Federal Republic of Germany, and Switzerland have some of the most expensive health systems (by percentage gross national product).

In Spain and Portugal the health systems are in transition as the countries, formerly dictatorships, adapt to democratic government, which endeavours to introduce more equitable services. Inequality of health and care are recognized, as is the over-supply of physicians and paucity of hospital beds and nurses.

A similar situation exists in Greece where too many

physicians in cities compete for private patients (40 per cent of medical costs are for private practice). The free market creates great inequalities which the socialist government is seeking to correct with proposals for a national health service.

Italy has a national health service but inequalities between the more affluent north and poorer south still exist. The excessive numbers of physicians also cause many difficulties for the new system.

The interface between primary and secondary (specialist) medical care varies between countries. Only in Denmark, the Netherlands, and the UK is there an accepted referral system which denies patients direct access to specialists. They have to be referred by their general practitioner or family doctor who is the gatekeeper, and this helps to prevent unnecessary use of more expensive health resources.

The UK National Health Service (NHS) was introduced in 1948. The NHS was not a new concept. It was a logical extension of the voluntary Sick Clubs and Friendly Societies of the late nineteenth century and the National Health Insurance Act 1911.

The NHS is now seen by most people in the UK as an integral part of the British culture. It provides 'free' health care for all. It now costs (1990) 6 per cent of gross national product and is very 'cheap' at this price. It is paid for by indirect (85 per cent) and direct (15 per cent) taxation.

Primary medical care is provided by general practitioners with whom patients are registered. The practitioners are paid by capitation fees and some extra fees for special preventive services, and they receive reimbursement for 70 per cent of their employees wages and also the cost of their premises.

Persons can be seen by specialists only if referred by their general practitioner. Hospitals are part of the NHS, and specialists receive salaries. Private hospitals provide services for those people (less than 10 per cent of the population) who wish to be so treated.

The NHS is sound in philosophy, but has some practical problems. Although relatively inexpensive compared with other national systems, the present Conservative Government believes the NHS can be more efficient, effective, and economic, and the government is trying to keep costs down by restricting the amount of money allocated to the NHS. This is causing particular problems in the hospital service, where hospital beds and wards are being shut down.

A feature of the NHS is the rationing of facilities, creating long waiting times for non-urgent hospital consultations and admissions.

The British system has managed to keep down costs through its general practitioners who carry out a gatekeeping role in controlling access to specialists in the hospital service; by cutting the number of hospital beds to a minimum (which many believe is too low); by having long waiting lists for surgical operations; and by exercising tight central financial controls on regional health authorities. A critical point may be reached when reductions in quantity of resources may lead to reductions in quality of care, depending on how 'quality' is assessed and measured, and by whom.

Common issues

Within the various national health systems there are many shared problems and dilemmas. Possible solutions also might be shared.

Common dilemmas

All health systems face the insoluble equation where 'wants' of the public and the professionals are always greater than 'needs' as perceived by politicians, planners, and managers, and which, in turn, are always greater than available 'resources'. The dilemma and challenge is to achieve a fair balance of service with finite resources when faced with infinite demands.

Another modern medical dilemma is how to provide personal patient care in an increasingly technological medical world with teams of specialists. Good care of the individual sick person has always, and will always, be an intensely personal affair. Personal doctoring and personal nursing should receive high priority and ideally each person should have a generalist personal doctor who provides long-term care and, when required, a specialist doctor who will treat the patient as a person rather than as a case.

Common problems

Many factors create more and more demands on health services which have to meet ever new and changing situations.

Demographic effects, resulting from social changes, and improved medical care have increased the numbers of elderly, and, with falling birth rates in most developed countries and with more women at work, care of the elderly is a problem.

Some age-old diseases, such as major infections, have been controlled, but into the vacuum come new ones, such as acquired immune deficiency syndrome (AIDS).

The explosion of new technological advances in diagnosis and treatment tends to make it difficult to assess and evaluate their usefulness before they are in demand.

Concepts, too, tend to be introduced and implemented before their true usefulness can be evaluated. For example 'community care' has become a bandwagon on to which are jumping more and more systems. The belief is that it is better and cheaper to discharge chronic mentally and chronic physically disabled persons from the long-stay hospitals into the community. It is now being realized that it may not be cheaper or beneficial for many such patients to be in the community.

'Prevention' and 'screening' are fine objectives when they achieve measurable benefits, but, unfortunately, many 'medical check-ups' are of questionable value for the patient, and the best forms of prevention are those that are well known to all and which should be the responsibility of the individual, namely, non-smoking, weight control, sensible diet, low alcohol and drug consumption, and regular exercise.

How 'free' should a health service be, and should consumer demands be restricted and controlled, and if so, how? It is by no means proven that a comprehensive health service, free at point of delivery, is open to consumer abuse. Certainly, the UK National Health Service is cheaper than other systems and there is no evidence of excessive use by the public.

Physicians are the most expensive health workers. How many do we need and how can their numbers be controlled? There appear to be no scientifically based policies on medical staffing. It is likely that too many doctors are being trained. It seems prudent to carry out services studies on medical staffing requirements, investigations on work content, and trials in which nurses and paramedical workers carry out some tasks that have traditionally been done by doctors.

The following questions have to be answered in every health care system. How can the public and the medical profession change set habits and adopt new and better ones? How can people be educated to accept more self-responsibility for their own health maintenance and prevention of disease? What incentives are most effective in motivating doctors? (Probably financial ones.)

Common solutions

There is no single 'best-buy' system of health care that is appropriate to all countries. Each has to evolve its own system. Nevertheless, each country can learn from the experiences of others, and adapt and adopt good ideas and methods.

References

Fry, J. (1969). *Medicine in three societies*. MTP, Lancaster.

Fry, J. (1978). *A new approach to medicine*. MTP, Lancaster.

Fry, J. (1985). *Common diseases* (4th edn). MTP, Lancaster.

Maxwell, R.J. (1988). Financing health care: Lessons from abroad. *British Medical Journal* **296**, 1423.

Office of Health Economics (1987). *Compendium of health statistics* (6th edn). Office of Health Economics, London.

Organization for Economic Co-operation and Development (1987). *Financing and delivering health care*. OECD, Paris.

Roemer, M.I. (1988). *National health systems as market interventions*. Rosenstadt lecture, University of Toronto.

E

Provision of Public Health Services

26

Public health services in the United States

ALAN R. HINMAN and WINDELL R. BRADFORD

C.-E. A. Winslow, a leading American theoretician in public health in the first half of the twentieth century, gave the following definition of public health:

> Public health is the science and the art of preventing disease, prolonging life, and promoting physical health and efficiency through organized community efforts for the sanitation of the environment, the control of community infections, the education of the individual in principles of personal hygiene, the organization of medical and nursing services for the early diagnosis and preventive treatment of disease, and the development of the social machinery which will ensure to every individual in the community a standard of living adequate for the maintenance of health (Winslow 1920).

Emerson described the essential programmes comprising the public health system in the US as including: vital statistics, communicable disease control, environmental sanitation, public health laboratories, maternal and child health, and public health education (Emerson 1948). In many parts of the US other activities have included provision of medical care to the indigent and provision of mental health services.

More recently, a study on the 'Future of Public Health' by the Institute of Medicine of the National Academy of Sciences defined the mission of public health as 'the fulfillment of society's interest in assuring the conditions in which people can be healthy'. The study further defined the substance of public health as 'organized community efforts aimed at the prevention of disease and the promotion of health' (Institute of Medicine 1988).

Public health services, emphasizing individual and community prevention, are typically provided through governmental agencies whereas personal medical care services are typically provided through the private sector. There are many exceptions to this principle, most notably in public provision of medical care services to the indigent or socioeconomically disadvantaged.

As described in Chapter 20, the States have primary responsibility for protecting and preserving the health of Americans. However, health services are typically delivered through local governmental agencies and the Federal Government plays an important role in policy development and financial support. This chapter describes the roles of various governmental and non-governmental organizations and the organizations, funding, staffing, and functioning of public health agencies in the US.

Because of the constitutional definition of a primary role for States, the role of the Federal Government might be viewed as secondary and supportive. However, the Federal Government plays a major role in defining health objectives for the nation (see Chapter 13), providing financial and technical support in achieving these objectives, and co-ordinating the efforts of the States. This is in part due to the importance of federal financial support for public health programmes (see below). The federal role involves both the executive and the legislative branches of government.

The States have the responsibility of assuring that necessary services are carried out in order to protect the health of their populace. The actual implementation of public health programmes usually falls on the shoulders of local government. Although public health services are commonly performed by governmental agencies, a wide range of non-governmental agencies plays an important role in assuring that the services are available and that the population is aware of them. Professional, voluntary, and community-based organizations are all important contributors and may also become involved in provision of services. Their activities range from public information and education to individual counselling of members of high-risk populations, and from professional education through lobbying for financial support.

Organization of public health services

Federal

At the federal level, the primary responsibility for public health services rests with the Department of Health and Human Services, which has five major operating components: the Office of Human Development Services, the Family Support Administration, the Public Health Service, the Health Care Financing Administration, and the Social Security Administration (Fig. 26.1). Although each of these plays a role in supporting public health services (primarily through financial support), the Public Health Service has the

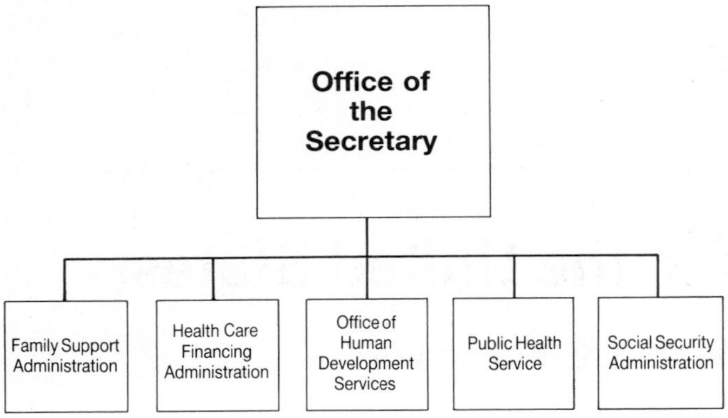

Fig. 26.1. Major operating components of the Department of Health and Human Services.

primary responsibility, under the direction of the Assistant Secretary for Health. The Public Health Service has seven agencies: Centers for Disease Control; Food and Drug Administration; National Institutes of Health; Alcohol, Drug Abuse, and Mental Health Administration; Health Resources and Services Administration; Indian Health Service; and Agency for Toxic Substances and Disease Registry (Fig. 26.2). In addition, there are several staff and programme offices in the Office of the Assistant Secretary for Health which have a direct bearing on public health programmes, notably including the Office of Disease Prevention and Health Promotion (see Chapter 31), the Office of Population Affairs, the National AIDS Programme Office, and the National Vaccine Programme Office.

Within the Public Health Service, the agencies primarily involved in provision of public health services are the Centers for Disease Control and the Health Resources and Services Administration. The Centers for Disease Control is the lead prevention agency of the Public Health Service and provides technical and financial support for a range of programmes, including surveillance and investigation of communicable diseases; immunization; sexually transmitted disease and human immunodeficiency virus (HIV) prevention and control; tuberculosis prevention and control; injury and violence prevention; health education and health promotion; and general support for prevention activities. In addition, the Centers for Disease Control provides reference laboratory support, environmental health programmes, occupational safety and health programmes, and training for public health workers. Finally, the Centers for Disease Control provides, through the National Center for Health Statistics, the central repository of vital statistics and other health information obtained through a series of surveys of health status, health knowledge, ambulatory care, hospital discharges, and a health examination survey.

The Health Resources and Service Administration provides general financial support for maternal and child health programmes as well as providing specific support for activities such as rural health, migrant health, and community health centres. In addition, the Office of Population Affairs provides financial support for family planning programmes. The National AIDS Programme Office co-ordinates all HIV/AIDS activities of the Public Health Service (from basic research to prevention and treatment programmes), and the National Vaccine Programme Office co-ordinates vaccine activities of both public and private sectors.

Other federal departments with major roles in public health include the Department of Agriculture, which plays a major role in nutrition with its Food Stamp Programme and its special supplemental nutrition programme for pregnant and lactating women, infants, and children, and the Environmental Protection Agency.

State

Each of the 50 States, the District of Columbia, and four US territories have established State health agencies with primary responsibility for public health (Public Health Foundation 1988). Of the 55 State health agencies, 32 are free-standing, independent agencies responsible directly to the Governor or State Board of Health and 23 are components of a super-agency which typically may include social and income maintenance services.

In fiscal year 1985, 42 State health agencies were designated as the State Crippled Children's Agency, 33 were the designated State Health Planning and Development Agency, 18 operated hospitals or other health care institutions, 14 acted as the lead environmental agency for the State, 13 were the designated State mental health authority, and nine were the designated Medicaid State agency.

The State health agency is headed by the State Health Officer, who is appointed by the Governor in approximately two-thirds of the States. Most States require that the State Health Officer has a medical degree. In 1986, salaries of State Health Officers ranged from a low of $35 957 (in Montana) to a high of $96 168 (in Alabama). Whereas State Health Officers used to be career professionals, there has been an increasing politicization of the position, and the average tenure of a State Health Officer now is only approximately two years.

Local

There are 3040 counties, 39 independent cities, 18 878 municipalities, and 25 city–county consolidations in the US. To serve these jurisdictions, in 1985 there were 2925 local health departments (Public Health Foundation 1987). Some of these serve a single municipality/county and others serve groups of counties. In approximately one-third of the States, the local health departments are district offices of the State health agency; in another third, they are autonomous; and in the remaining third, they are responsible to both local government and the State health agency. As is the case with State health agencies, many of the local health departments are separate agencies and others are components of local health and human service agencies. They carry out their activities

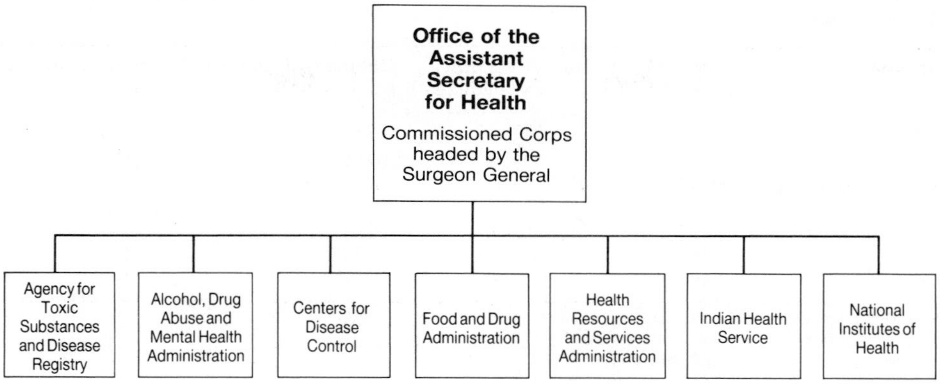

Fig. 26.2. The seven agencies of the Public Health Service.

Table 26.1. National health expenditures, by type of expenditure and source of funds, calendar year 1984 (dollars in billions)

	Total	Private	Public		
			Total	Federal	State and local
Total national health expenditures	387.4	227.1	160.3	111.9	48.3
Health services and supplies	371.6	220.3	151.2	105.4	45.9
Personal health care	341.8	206.5	135.4	101.1	34.3
Programme administration and net cost of health insurance	19.1	13.9	5.2	2.9	2.3
Government public health activities	10.7	—	10.7	1.4	9.3
Research and construction of medical facilities	15.8	6.7	9.0	6.6	2.5

Source: US Department of Health and Human Services (1985).

under authority delegated by the State or enacted by the local jurisdiction itself. Typically, local jurisdictions are empowered to enact requirements which are stricter than those imposed by the State but not to enact requirements less stringent than those of the State.

Directors of most local health departments are appointed by leaders of the local jurisdiction (mayor, county supervisor, council, etc.). Approximately two-thirds of local health department directors are physicians with nearly one-third having a master's degree in public health.

Funding

The vast majority of the funds for public health services in the US come from governmental sources. As an indication of the relative emphasis placed on public health services (which are typically preventive) compared with curative personal health services, in fiscal year 1984 the total national expenditures for health were in the range of $387 billion, approximately $1580 per person (Table 26.1). Of the total, approximately $112 billion came from the Federal Government; approximately $101 billion of this was for personal health care. Looking strictly at public health services (as reflected by expenditures through governmental agencies), approximately $11 billion were expended, approximately 3 per cent of national health expenditures and nearly 7 per cent of total governmental spending on health. This represents $45 per person. Of the public health total, approximately $1.4 billion came from the federal level and $9.3 billion from State and local sources.

As indicated above, most public health services are delivered by State and local governments. Federal funds in support of State and local health public health services come either in block grants, which are relatively undifferentiated, or in categorical grants, which are much more focused. In addition, federal funds can be expended as contracts or reimbursements. In recent years, there has been considerable discussion about the relative merits of block versus categorical grant funding. Block grants provide maximum flexibility for State and local jurisdictions, but are the least accountable. Categorical funding, while much more restrictive, provides for clear measurement of progress towards objectives and is much more accountable. In the early 1980s, there was considerable effort to consolidate all federal support into block grants. Four block grants were created: maternal and child health; preventive health and health services; primary care; and alcohol, drug abuse, and mental health. None the less, categorical grants remain popular, with the Public Health Service providing categorical support for immunizations, sexually transmitted disease control, prevention of HIV infection, tuberculosis control, family planning, diabetes prevention, community health centres, refugee health, and

Table 26.2. Expenditures of state health agencies and local health departments,* by source of funds, fiscal year 1986 (dollars in millions)

Programme area	Total SHA and LHD expenditures	Direct SHA expenditures	SHA inter-governmental grants to LHDs	Additional expenditures of LHDs
Total public health	9384.7	5765.4	1698.4	1920.9
State funds	4265.0	3212.5	926.7	125.8
Federal grant and contract funds	2859.0	2180.7	513.7	164.6
Local funds	1262.6	21.1	124.5	1117.0
Fees and reimbursements	619.9	236.6	118.5	264.8
Other sources	162.8	114.5	15.1	33.2
Source unknown	215.5	—	—	215.5

Source: Public Health Foundation (1987, 1988).
* Includes State funds and federal grant and contract funds.
SHA, State health agency; LHD, local health department.

Table 26.3. Expenditures of state health agencies and local health departments,* by programme area, fiscal year 1986 (dollars in millions)

Programme area	Total SHA and LHD expenditures	Direct SHA expenditures	SHA inter-governmental grants to LHDs	Additional expenditures of LHDs
Total public health	9384.7	5765.4	1698.4	1920.9
Personal health	6443.8	4381.5	1138.1	924.2
Environmental health	690.5	366.2	113.7	210.6
Health resources	757.3	390.7	260.2	106.4
Laboratory	284.4	213.9	24.0	46.5
General administration	561.1	413.1	51.9	96.1
Not allocated to programme areas	647.6	—	110.5	537.1

Source: Public Health Foundation (1987, 1988).
* Includes State funds and federal grant and contract funds.
SHA, State health agency; LHD, local health department.

migrant health, among others. In addition, only two States exercised the option to receive primary care block grants instead of categorical grants, and both of these have since switched to categorical funding. No such option was available to States for programmes consolidated into the other block grants.

Each year, the Public Health Foundation surveys all State health departments to obtain information on public health expenditures (Public Health Foundation 1988). In 1986, total State and local health department expenditures were approximately $9.4 billion, with federal funds accounting for approximately 30 per cent of the total. State appropriated funds represented approximately 45 per cent, with local funds representing approximately 13 per cent. Fees and reimbursements accounted for a little more than 8 per cent of total expenditures. Table 26.2 summarizes State and local health department expenditures in 1986 by source of funds. Table 26.3 summarizes these expenditures by programme area.

The largest proportion of State and local health expenditures is for personal health services. More information is available on expenditures at the State level than those at the local level. The largest single category of expenditure is for maternal and child health, representing 45 per cent of personal health expenditures and more than 25 per cent of all State health agency expenditures. The next largest category overall is for institutions operated, which accounts for 22 per cent of personal health expenditures and 13 per cent of all State health agency expenditures. In environmental health, the largest category is consumer protection and sanitation. Development represents the largest health resource expenditure, and general laboratory support is the largest category in laboratory programme expenditures.

Table 26.4. Expenditures of State health agencies (SHA) by programme area, fiscal years 1976 and 1986* (dollars in millions)

Programme area	1976		1986	
	($M)	%	($M)	%
Non-institutional personal health	1217	(47.9)	4285	(57.4)
SHA-run institutions	531	(20.9)	1235	(16.5)
Environmental health	199	(7.8)	480	(6.4)
Health resources	208	(8.2)	651	(8.7)
Laboratory	104	(4.1)	238	(3.2)
Other	281	(11.0)	576	(7.7)
Total	2540	(100.0)	7465	(100.0)

Source: Public Health Foundation (1987, 1988).
Values in parentheses are percentages of the total.
* Includes State funds and federal grant and contract funds.

Table 26.4 shows the changes in State health agency expenditures from 1976 to 1986. During this period, non-institutional personal health expenditures rose from 47.9 per cent of the total to 57.4 per cent of the total whereas expenditures for State health agency-operated institutions, environmental health, and 'other' (including expenditures for general administration and funds to local health departments which were not allocated to programme areas) represented a decreasing proportion of the total.

Although total expenditures nearly tripled during the period, when corrected for inflation the increase was only approximately 50 per cent.

Considering the source of funds for State health agency programmes, State funds accounted for 58.5 per cent of the total in 1976 and 55.4 per cent in 1985. Federal grant and contract funds rose from 31.4 per cent of the total to 36.1 per cent during the same period. Local funds declined from 3.8 to 2.0 per cent, and fees, reimbursements, and other funds increased slightly from 6.3 to 6.5 per cent.

Staffing

Overall, it is estimated that more than 500 000 persons spend at least part of their time in public health activities, with approximately one-half spending the majority of their time in public health (US Department of Health and Human Services 1988). This number accounts for less than 10 per cent of the entire health work-force. State health agencies employ more than 100 000 of these workers. Approximately 85 000 of those working in public health have graduate degrees in public health. Slightly more than one-half of these received their degrees from schools of public health.

Training in public health takes place both in formal educational institutions and on-the-job. There are currently 24 accredited graduate schools of public health in the US, with more than 10 000 students enrolled. These students are pursuing nine major subject areas: biostatistics, epidemiology, health services administration, public health practice and programme management, health education/behavioural sciences, environmental sciences, occupational safety and health, nutrition, and biomedical and laboratory sciences. The areas of specialization attracting more than 10 per cent of students are health services administration (27.0 per cent), epidemiology (15.2 per cent), environmental sciences (12.4 per cent), and health education/behavioural sciences (10.8 per cent).

In addition to the schools of public health, there are some 317 graduate programmes in public health located in other settings: 117 in health education, 52 in preventive dentistry (residency programmes), 49 in health service administration and planning, 37 in environmental and occupational health, 17 in health statistics, 15 in community health nutrition, 14 in community health and preventive medicine, 6 in epidemiology, and smaller numbers in other areas of specialization. In 1981–2, these programmes enrolled more than 14 000 students.

Between the dedicated schools of public health and the other public health programmes, there are approximately 25 000 students enrolled in graduate programmes in public health and more than 6500 graduates each year. As of 1985, it was estimated that approximately one-third of the total public health work-force had received some graduate training in public health.

The recent study of 'The Future of Public Health' by the Institute of Medicine commented that 'many observers feel that some schools [of public health] have become somewhat isolated from public health practice and therefore no longer place a sufficiently high value on the training of professionals to work in health agencies'. They recommended that 'Schools of public health establish firm practice links with state and/or local public health agencies . . . ' and that 'Schools of public health should provide students an opportunity to learn the entire scope of public health practice, including environmental, educational, and personal health approaches to the solution of public health problems; the basic epidemiological and biostatistical techniques for analysis of those problems; and the political and management skills needed for leadership in public health' (Institute of Medicine 1988).

Functions

By and large, and governmental public health agencies are providing the range of services defined by Emerson. Nearly all State health agencies provide services related to: maternal and child health; communicable disease control; chronic diseases; dental health; public health nursing; nutrition; health education; consumer protection and sanitation; water quality; health statistics; and diagnostic laboratory tests. As described above, some also provide direct medical care to the indigent, mental health, and environmental health services.

Local health departments are generally responsible for direct delivery of public health services to the population. More than 80 per cent of local health departments provide communicable disease control programmes (including immunization, tuberculosis control, and sexually transmitted disease control); environmental surveillance (including sanitation and inspection programmes); maternal and child health services; school health programmes; and chronic disease control programmes (Miller *et al.* 1977).

The recent Institute of Medicine study described three functions for public health agencies: assessment, policy development and leadership, and assurance (of access to environmental, educational, and personal health services). The traditional role of departments of public health has included assuring the safety of water and food and providing community and personal preventive services (e.g. prenatal care, immunization). To a variable extent it has also included provision of personal medical services to those who cannot afford them in the private sector. The core environmental and preventive services have become fairly well defined, have known efficacy, and have generally been highly successful.

In recent years, growing concerns about the health effects (both acute and long-term) of toxic substances in the environment have brought troublesome new issues including the difficulties of assessing the health impact of long-term low-level exposure, multi-factorial causation of many conditions, and practical problems of how to remove some of the pollutants from the environment and how to substitute other, less hazardous, substances if the current pollutants serve necessary functions.

At the same time, it has become increasingly clear that many of the current major killers are largely the result of individual behaviours such as smoking, drinking alcohol, over-eating, not exercising, etc. This was first laid out explicitly in a Canadian study which assessed the relative importance of four factors in bringing about disease or assuring health: genetics, life-style, environment, and medical services. Life-style was found to be the single largest contributor (Lalonde 1974).

More recently, a major conference assessed the gap between what is known about causation of the major killers and cripplers, what is known about prevention, and what is now being done. This conference was called 'Closing the gap: The burden of unnecessary illness' (Amler and Dull 1987). Fourteen health problems were addressed: alcohol dependence and abuse, arthritis and musculoskeletal disease, cancer, cardiovascular disease, dental disease, depression, diabetes mellitus, digestive diseases, drug dependence and abuse, infectious and parasitic diseases, respiratory diseases, unintended pregnancy and infant mortality/morbidity, unintentional injuries, and violence (homicide, assault, and suicide). In 1980, these conditions accounted for approximately 70 per cent of hospitalization days, 85 per cent of direct personal health care expenditures, 80 per cent of deaths, and 90 per cent of potential years of life lost before age 65. Roughly two-thirds of reported mortality (and potential years of life lost) was considered to be due to potentially preventable causes.

The principal risk factors associated with unnecessary deaths or potential years of life lost were: tobacco, alcohol, injury risks, lack of prevention services, high blood pressure, and improper nutrition. Unintended pregnancies were also considered major contributors to infant morbidity and mortality.

Single exposure, permanently effective, preventive services comparable to vaccines do not exist to reduce or remove these risk factors. Available approaches to prevention or modification of behaviour typically involve repetitive contacts (reinforcement), and many are of unproven or only partial efficacy. Additionally, behaviours are heavily affected by socio-cultural norms and community attitudes, and official agencies such as health departments may have only limited ties to the groups that help shape behaviour.

A result of these intertwined factors has been that health agencies have had to search for interventions that will be effective against these new challenges; implement interventions that may be only marginally effective; search for financial support for these repetitive contact interventions; and forge new alliances with community-based organizations, voluntary organizations, and minority groups. Their success in accomplishing these tasks has been quite variable.

As pointed out in Chapter 13, the Report of the Surgeon General on Health Promotion and Disease Prevention and the development of Prevention Objectives for the Nation have had a major impact in directing thought and effort in prevention during the past ten years. Another document, which has received less publicity, has also been important. Beginning in 1976, a collaborative effort was undertaken to develop model standards for community preventive health services. These were released in 1979 (Centers for Disease Control 1979). Major organizations involved in the effort were the Centers for Disease Control, the Association of State and Territorial Health Officials, the National Association of County Health Officials, the US Conference of Local Health Officers, and the American Public Health Association. The report was revised and updated in 1985 (Centers for Disease Control 1985). It provides goals, objectives, and indicators in 34 personal and environmental health programme areas in a way which permits each community to set its own target levels of disease incidence, programme activity, etc. One feature of the project was the enunciation of a governmental responsibility to assure the availability of preventive health services in every community, although not necessarily to be the direct provider of these services. This precept has acquired the acronym AGPALL (A Governmental Presence At the Local Level) and directly relates to the assurance function subsequently enunciated by the Institute of Medicine study. Another feature was the clear statement of the need for more and better data about health conditions and health programmes at the local level. This also was later reflected by the Institute of Medicine.

Since the original Model Standards were developed before the 1990 Prevention Objectives, they did not relate directly to them. The 1985 revision attempted to address this lack, and the process of developing Prevention Objectives for the Year 2000 is fully integrating the Model Standards. It is likely that the integrated Model Standards/Prevention Objectives will, in concert with the Institute of Medicine's report on the 'Future of Public Health', play a major role in the further evolution of public health services in the US. The Association of State and Territorial Health Officials has recently been drafting legislation which would establish a new grant programme to support State and local health department activities designed to achieve the Year 2000 Prevention Objectives (Association of State and Territorial Health Officials 1988).

The Institute of Medicine's report describes the current public health system in the US as a system 'in disarray' which has been unable to keep up with the changing demands and which does not presently have the capacity or the capability to carry out the actions necessary. This conclusion is very similar to that reached 11 years earlier by Miller *et al.*: 'The United States has in place an unevenly operative public infrastructure of community and personal health services—understaffed, underfunded, and widely ignored' (Miller *et al.* 1977). If the Institute of Medicine's report is used as a guide, substantial additional resources will be devoted to public health. In addition, there will be two major changes in the character of health departments: integration of mental health services into health departments and a much greater role in provision of medical care services to the indigent.

The report has stimulated considerable debate about the public health system in the US. Some feel the report is too negative and does not recognize the continuing effective work of health departments. Others are concerned about the potential dilution of the public health effort by incorporating mental health services and indigent medical care. It is too early to assess the full impact of the report but it seems clear that there will be more public and congressional interest in the public health system. This is likely to result in some increase in resources as well as to some (currently unpredictable) changes in structure and services.

References

Amler, R.W. and Dull, H.B. (1987). *Closing the gap: The burden of unnecessary illness*. Oxford University Press, New York.

Association of State and Territorial Health Officials (1988). *The National Health Objectives Act: A proposal by the Association of State and Territorial Health Officials*. ASTHO, Washington.

Centers for Disease Control (1979). *Model standards for community preventive health services*. Centers for Disease Control, Atlanta, GA.

Centers for Disease Control (1985). *Model standards: A guide for community preventive health services* (2nd edn). Centers for Disease Control, Atlanta, GA.

Emerson, H. (1948). The unfinished job of essential public health service. *American Journal of Public Health* **38**, 164.

Institute of Medicine Committee for the Study of the Future of Public Health (1988). *The future of public health*. National Academy Press, Washington, DC.

Lalonde, M. (1974). *A new perspective on the health of Canadians*. Health and Welfare Canada, Ottawa.

Miller, C.A., Brooks, E.F., DeFriese, G.H., Gilbert, B., Jain, S.C., and Kavaler, F. (1977). A survey of local public health departments and their directors. *American Journal of Public Health* **67**, 931.

Public Health Foundation (1987). *Public Health Agencies 1987: Expenditures and sources of funds*. Publication No. 103. Public Health Foundation, Washington, DC.

Public Health Foundation (1988). *Public Health Agencies 1988: An inventory of programs and block grant expenditures*. Publication No. 106. Public Heatlh Foundation, Washington, DC.

US Department of Health and Human Services (1985). *National Health Expenditures 1984. Health care financing review*. Health Care Financing Administration, Baltimore, MD.

US Department of Health and Human Services (1988). *Sixth report to the President and Congress on the status of health personnel in the United States*. DHHS Publication No. HRS-P-OD-88–1, Washington, DC.

Winslow, C.-E.A. (1920). The untilled fields of public health. *Science* **51**, 23. Quoted in Winslow, C.-E.A. (1923). The evolution and significance of the modern public health campaign. Yale University Press, New Haven. Reprinted by *Journal of Public Health Policy* 1984.

27

Provision of public health services in Europe

JOHANNES MOSBECH

Introduction

Europe has a population of 817 million inhabitants in 32 countries (including the USSR with 250 million). The political systems vary; so do standards of living; and there are considerable variations in the historical development of public health services in the different countries. Nine European countries are linked in the European Economic Community (EEC), within which an important process of harmonization has taken place in recent years; this includes the health area, and in the future this will undoubtedly lead to greater homogeneity.

It is beyond the scope of this chapter to give a detailed description of the public health care system in every European country. The intention is rather to give some examples, particularly from the Nordic countries, to illustrate trends and development and to exemplify how health problems are handled in different ways in the European area.

Very important changes in society have occurred in Europe in recent years: a falling birth-rate has resulted in small families where both parents work, and many children are cared for outside their home for most of the day. The divorce-rate is high, urbanization is increasing, and more and more people live in satellite towns with long travel times to their work. Further problems stem from the increasing proportion of older people in the population.

The changing disease and health care demand patterns, with increasing emphasis on the care of chronic illness, are reflected both in morbidity and mortality statistics. The balance between primary care and hospital care is everywhere under review, with increasing stress on the importance on the long-term care and a well-developed primary care system. Reliable statistical information is important for monitoring these changes as the need for planning and priority-making in public health grows.

Another essential issue is the balance between prevention and clinical care. The population increasingly demands prevention, but there are both practical and political limitations to its implementation, and detailed analyses of efficacy cost and benefit are needed before large-scale preventive action can be successfully organized. National responses vary greatly. In pre-school and school children as well as in the work-place there is, however, a well-established basis for preventive action in most of the countries of the region.

The financial implications of the operations of health organizations are enormous; painstaking planning, prior evaluation, and a detailed subsequent research are increasingly necessary. All recent experience shows how difficult it is to achieve a satisfactory balance between competing priorities in health care, between the demands of effectiveness and equity, and between competing attitudes of different health professions.

The World Health Organization programme 'Health for All by the Year 2000', has been adopted by the countries of the European region, and many of the 38 targets have already been built into the various national health programmes (Kleczkowski *et al.* 1984).

In the following sections the manner in which these various pressures are shaping the contemporary patterns of European public health services is discussed in more detail. The growth and growing emphasis upon preventive services and the pathways along which different countries are proceeding is examined in the section on preventive services. Finally, the problems of co-ordinating the work of different health agencies including both health services proper (including their separate components), social agencies, and the political and administrative infrastructures through which they work, are reviewed. Here the example of the Scandinavian health services is used as an illustration.

Responses to contemporary pressure

Demographic trends

Crude live births in most of Europe are about 13 per 1000 per year of the population, almost equal to mortality rates. As a consequence, the total population-size is essentially stable. In only a few countries is there a natural increase or an overall decline of the population (Herberger 1984). The population of Europe is, however, aging. The proportion of children in the age-group 0–14 is decreasing, and the high-age groups are increasing (Table 27.1) These demographic changes have

Table 27.1. Population by age in the Nordic countries (per cent) on 31 December 1985

	Finland	Sweden	Denmark[1]	Norway	Iceland
0–14	19.4	18.1	18.3	19.8	28.1
15–24	14.7	14.0	15.4	15.7	17.7
25–34	16.1	13.6	14.5	14.9	16.5
35–44	15.8	15.2	15.4	13.9	12.4
45–54	11.1	10.8	11.0	9.5	8.8
55–64	10.5	11.0	10.3	10.4	8.3
65–74	7.5	10.0	8.7	9.3	5.8
75–	5.1	7.4	8.5	8.6	4.4
	100	100	100	100	100
Total (1000)	4911	8358	5116	4159	242

[1] 1.1.1986.

Table 27.2. Legal abortion by age of mother in the Nordic countries in 1984[1]

	Finland	Sweden	Denmark	Norway	Iceland
15–19[2]	18.9	17.7	17.4	21.0	18.3
20–24	18.7	26.7	28.2	24.4	21.1
25–29	11.1	20.4	22.7	17.0	13.7
30–34	9.8	17.9	18.3	13.7	11.1
35–39	9.4	15.0	14.5	10.7	8.8
40–44	8.5	8.1	7.3	5.5	5.4
45–49	1.3	1.2	1.0	0.6	0.6
Per 1000 live births	209.7	327.8	400.4	279.9	180.4

[1] Per 1000 women in age group.
[2] Abortion to women under age of 20 per 1000 women aged 15–19.

important consequences for public health policy and planning. Low fertility will undoubtedly continue, and the number of families with few children will further increase. The number of large families will continue to be low, but they will tend to present health services with social, economic, and health problems (Klinger 1984; Lopez 1984; Teper and Backett 1984).

The modernization of family planning and the spread of more efficacious and less hazardous methods has contributed to a decrease in the number of unplanned pregnancies. The use of more dangerous methods such as abortions is being discouraged but it is still quite high in a number of European countries (Table 27.2). The youth group is declining in size but the problems facing young people are important for social and health policy. Accidents, drug abuse, smoking, unwanted pregnancy, and sexually transmitted disease are very important in youth groups as are the psychological and social effects of unemployment, family breakdown, loneliness, homelessness, and migration (Backett *et al.* 1984). The AIDS epidemic will almost certainly take its place among these major hazards.

The increase in the size of the older age groups also presents important specific health problems. These are due to higher chronic morbidity, the requirement for more visits by the physician and days in hospital, an increased use of drugs, and a heavier utilization of nurses, home-help, and nursing homes. These are all matters which will demand a high priority for resource allocations in the coming years (Skeet 1988).

Mean life expectancy at birth, in Europe, varies from 65 years (Portugal) to 74 years (Iceland) for men, and from 73 years (Portugal) to 80 years (Iceland) for women (Table 27.3). In all European countries women have a higher life-expectancy than men: on average 6.5 years more. The gap seems indeed to be widening; women are tending to live even longer, whereas the life-expectancy for men seems to be levelling off. The national differences in length of life are probably to some extent due to differences in the standard of public health services, but the contributions of economic variation and unhealthy life-styles are undoubtedly of much greater consequence. This is reflected, within different countries, in social class differences in mortality (Lynge 1984).

The main causes of death in the region in most age groups are diseases of the circulatory system, cancers, and accidents (Fig. 27.1). Suicides are important and so is mortality from traffic accidents. (Fig. 27.2: Table 27.4). The main causes of chronic disability are accidents, stroke and other vascular diseases, chronic lung diseases, mental illness and mental subnormality, senile dementia, arthritis, and the physical disabilities of extreme old age. The main determinants of health lie outside the traditional health sector. Health policy cannot remain a matter for health centres, hospitals, or other health-care services, alone. Yet there are still serious problems in mobilizing the expertise of health professionals and applying their findings and recommendations in health policy areas outside their traditional framework of employment. Meanwhile, the roles of national governments are chiefly restricted to controlling costs, guaranteeing equity in the distribution of resources, and developing local services. There is little evidence of engagement with true health objectives.

Table 27.3. Life expectation in the Nordic countries

	Finland	Sweden	Denmark	Norway	Iceland
Aged 0					
Males					
1961–65	65.4	71.6	70.3	71.0	70.8
1971–75	60.7	72.1	70.9	71.4	71.8
1984–85[1]	70.1	73.8	71.6	72.8	74.7
Females					
1961–65	72.8	75.7	74.5	76.0	76.2
1971–75	75.2	77.8	78.5	77.7	77.5
1984–85[1]	78.5	79.7	77.5	79.5	80.2
Aged 65					
Males					
1961–65	11.4	13.9	13.5	14.1	15.0
1971–75	11.8	14.0	13.7	14.0	15.0
1984–85[1]	12.9	14.7	13.9	14.4	15.8
Females					
1961–65	13.7	15.8	15.4	16.0	16.8
1971–75	15.1	17.2	17.0	17.0	17.8
1984–85[1]	17.2	18.5	17.8	18.5	18.6

[1] In Finland and in Sweden only for 1985.

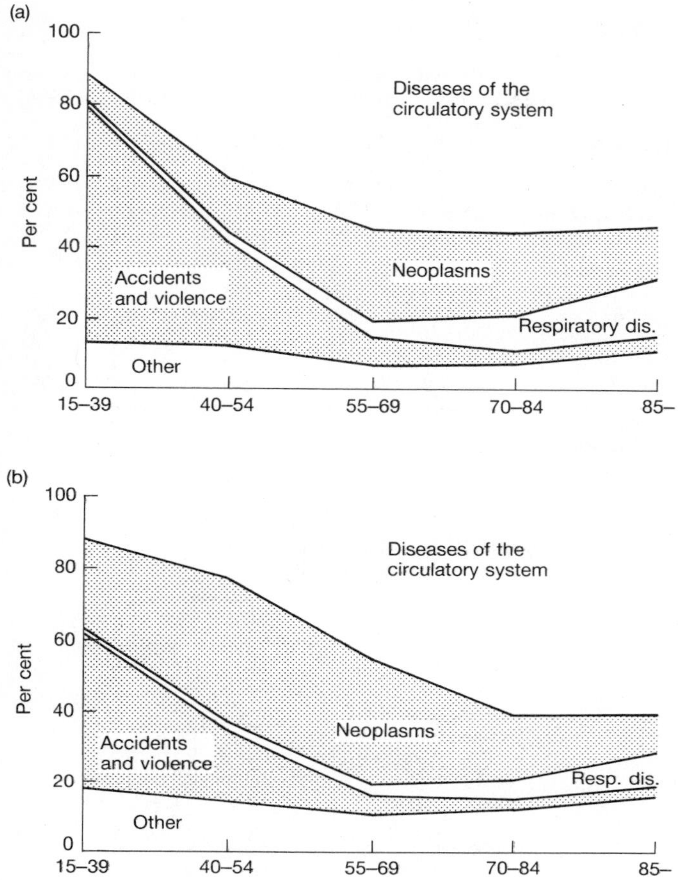

Fig. 27.1. Most frequent causes of death in 1986: (a) males and (b) females.

These deficiencies are serious, and acceptable solutions to these problems have not in general been found.

Organization of the health system

The health systems of the European countries are characterized by a relatively large supply of health manpower, particularly physicians and nurses; and a reasonable supply of adequate health facilities. In most European countries these resources function as part of a large organized system of health services under the central direction of a ministry of health. These services are then available to the whole population.

The pattern of health care delivery is, again, fairly uniform. For both in-patient and ambulatory patient care, most services are provided by salaried personnel employed in public facilities. In northern Europe, private practice contributes only a small part of the total service. Preventive services are generally integrated with treatment services, and financed through the same central administrative mechanisms. Economic support of the overall health system is derived almost entirely from general revenues, usually from the national government: as a result, health care has come to be regarded as a public service available without charge or for only a nominal payment. In recent years, however, there has been a substantial erosion of this 'principle' of public care 'free at the point of demand'. Experience has shown that introduction of even a moderate fee for service reduces the demand by 20–25 per cent, so there is a trend at present away from services which are completely free of charge. In some countries this movement has been re-enforced by conservative political ideologies.

Medical demography

There has been a large increase in the number of physicians in Europe in the last decade. Countries which have had insufficient control of the intake into their medical schools are now suffering from an over-production of physicians, which

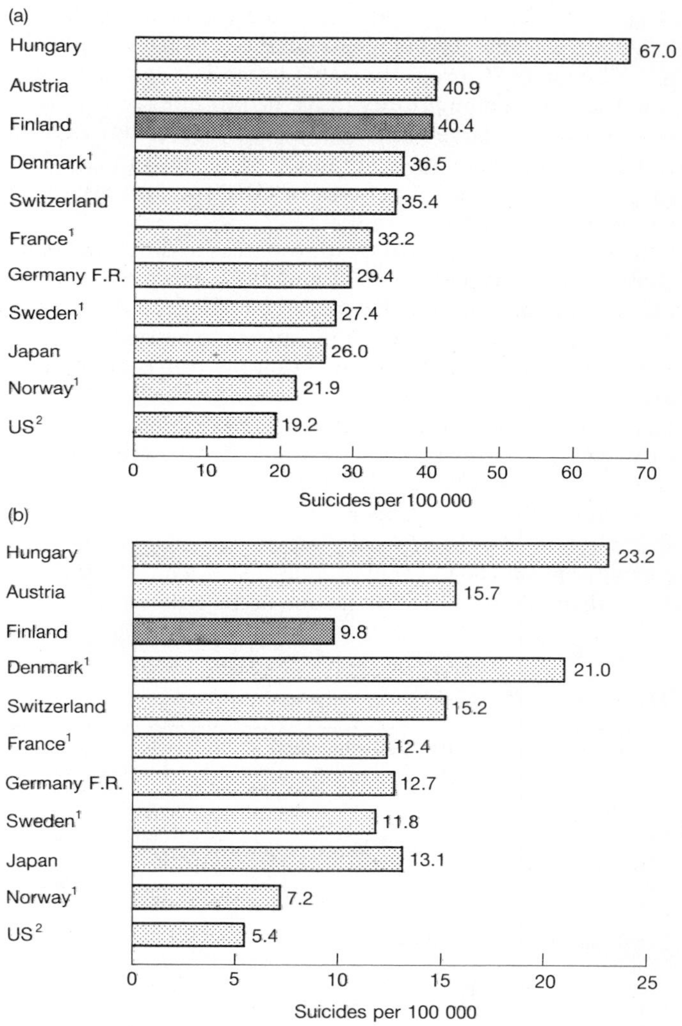

Fig. 27.2. Suicides in selected countries in 1985: (a) males, [1]1984, [2]1983; (b) females, [1]1984, [2]1983.

already has quite serious consequences (Table 27.5) (Viefhues 1988). The 'free' labour market of the EEC is now visiting these consequences upon neighbouring countries which have exercised control over the 'production' of doctors. There is now a movement within these 'receiver' countries towards restrictive regulation of physician-migration. Some unification of doctor-production policies will be necessary in the future if these conflicts are to be eased.

The cost of health care

The cost of health care is being given great attention in most European countries. Increasing costs are creating severe problems for many governments. The capacity of governments to finance total health care costs is limited and, given a harsher economic climate, the financial consequences on other fields of social endeavour are becoming quite serious (European Public Health Committee 1980).

Table 27.6 and Figure 27.3 give the total expenditure on health care as the percentage of GNP per country in Europe in 1972 and 1977. From this table it can be seen that many countries give a high priority to the health care sector. There is, however, a large variation among the countries with respect to the share of GNP allocated to health care. This ranged in 1977 from 2.7 per cent of GNP (in Greece), to 9.5 per cent in Sweden.

In Table 27.7 the allocation of resources to different health activities is presented for 1972 and 1977. From this table it can be seen that, in spite of widespread policies to strengthen the primary care sector, only Finland has to any extent raised the percentage of total costs spent on the provision of primary care.

The size of the hospital sector is a crucial determinant of total costs. The distribution of resources between hospital care and ambulatory care is a major policy question. When considering these problems it should be noted that most of the costs in the health care sector are manpower costs (between 55 and 80 per cent of total costs), which tend to rise faster than other production factors in the public sector.

Table 27.4. Traffic accidents in the Nordic countries

	Finland[1]	Sweden	Denmark	Norway	Iceland
Deaths and injured per 100 000 inhabitants					
1981	201	232	279	264	317
1982	201	241	275	283	328
1983	206	247	270	267	288
1984	200	257	281	278	329
1985	206	257	288	298	378
Deaths per 100 000 population					
1981	12	9	13	8	10
1982	12	9	13	10	10
1983	12	9	13	10	8
1984	11	10	13	10	11
1985	11	10	15	10	10

[1] Slightly injured are excluded.

Table 27.5. Health personnel in the Nordic countries

	Finland 1985[1]	Sweden 1985	Denmark 1984	Norway 1985	Iceland 1984
Physicians	10 193	21 000[2]	12 980	9 176	574
Dentists	3 916	9 338	5 100	3 702	193
Nurses	43 989	72 366	38 643	35 552	1 706
Practical nurses	28 366	33 179	49 456	36 896	1 119
Physiotherapists	4 557	5 889	3 111	3 701	119
Inhabitants per					
Physicians	481	398	394	452	419
Dentists	1 252	893	1 000	1 120	1 246
Nurses	111	115	132	117	141
Practical nurses	173	251	103	112	215
Physiotherapists	1 078	1 419	1 643	1 120	2 020

[1] All.
[2] Estimate.

Table 27.6. Total health care expenditure as percentage of GNP

Country	1972	1977
Austria	5.3	6.7
Belgium	—	7.1
Denmark	4.4[1]	5.4[1]
France	5.9	6.8
Fed. Rep. of Germany	7.3	9.4[2]
Greece	2.0	2.7
Ireland	4.5	6.1
Italy	—	6.6
Luxembourg	2.8	4.8
Netherlands	7.2	8.2
Norway	6.7	8.4
Portugal	2.8	—
Sweden	8.0	9.5
Switzerland	4.7	7.3
United Kingdom	4.8	5.5
Finland	6.5	7.5

[1] Expenditure on care of mentally retarded and on nursing departments in institutions for old people are not included.
[2] 1975.
Source: *The cost of health care in member states of the council of Europe and in Finland.* European public health committee, Strasbourg (1980).

Inequality in health care provision

All countries exhibit social and geographical variations in medical needs and in the quality and effectiveness of health care. Often, health care is weakest where the need is greatest. These inequalities and inadequacies of health care are expressed in terms of their accessibility, their acceptability to patients, and their relevance to special local needs (Hellberg 1984; Vogel *et al.* 1987; Köhler and Wahlberg 1984).

Equality in health care provision requires a proper balance between its different aspects: health promotion, prevention, cure, care, and rehabilitation. What are the right proportions for resource allocation in a given area? This is not an easy question to answer: indeed, there can never be an exact answer. A search for the best possible balance often involves a conflict between immediate demand and long-term need (Hjort 1984*a*).

The highly organized health care systems in the northern European countries have had almost all their costs absorbed by government: as a result health services are available to the entire population, but there is a need for improvement of quality and efficiency. Where resource demand and resource supply exert pressures upon each other, a third element of the equation is liable to suffer: this third element is the standard of the care provided. Greater shares of the national budget might be allocated to the health sector, in order to relieve these pressures, but this is always a sticky political problem—and political thinking is influenced more by questions of equity than questions of efficacy. To present excessive regional differences and an overall escalation of expenditure there may also be a need for national regulation of decentralized taxation.

Variations in utilization of health services

It is remarkable to find that there are great intra-national variations among county councils or other local administrative bodies in both the utilization and provision of health care facilities. For instance, there is a variation of provision, within Sweden, between limits of 14 and 20 beds per thousand inhabitants. A study of inequalities in short-stay 'somatic' hospitals in Norway has also shown that hospital provision varies widely in the 19 Norwegian counties, between 2.8 and 6.1 beds per 1000 population. There is no simple explanation in terms of varying needs for either of these phenomena. The primary factor that seems to affect the utilization of health care, is the mere existence of the facilities (Eklundh and Roos 1984). In economists' terms, demand is 'elastic', and it responds to the supply.

The geographic inequalities of hospitals care thus tend to be self-perpetuating. They depend ultimately upon the early

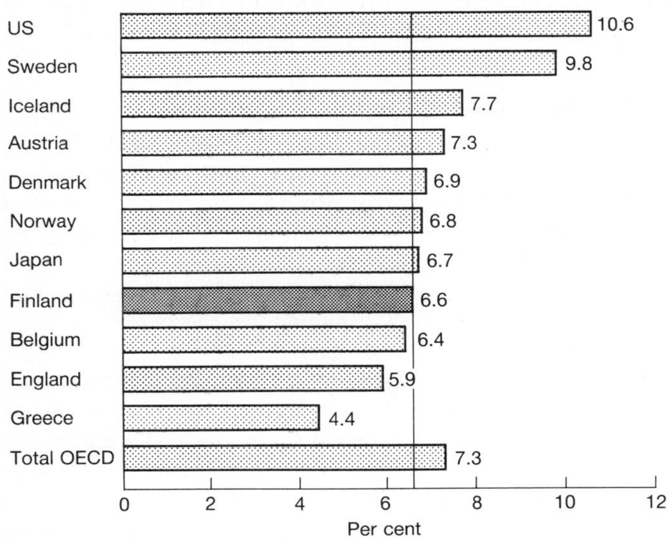

Fig. 27.3. Health expenditure in OECD countries as a percentage of GNP.

history of a district or a county. In general, the better off counties have and use more hospital resources than the poor ones, whether in terms of doctors, nurses, or out-patient clinics; because that is what they always did.

A good supply of beds tends to prolong the length of stay, but the main reasons for differences are traditions and habits. Hospitals, like doctors, have a traditional style and personality which tends to determine contemporary practice. The frequencies of operations vary widely between and within countries for reasons that appear to be entirely intrinsic. Norwegian surgeons, unlike surgeons in some other countries, are paid a fixed salary irrespective of how much they operate; yet one county performed 2.9 times more hysterectomies than another; for tonsillectomy the ratio was 4.6, and for cataract 2.0 (Hjort 1984*b*). The main explanations seem to lie in differences among surgeons and referring doctors in terms of their training, their experience, and local 'home-grown' attitudes and beliefs.

To summarize, there are major regional differences in resources, demands, utilization, and individual patterns of practice. The differences are large, with major medical and economic consequences. It is important to investigate them further and to identify and later eliminate unfair and unreasonable differences (Hjort 1984*b*; Kamper-Jörgensen 1984).

Reasons for differences in health care provision and uptake

Although a number of specific organizational and economic factors explaining some of the differences in health care can be identified, there are a number of general phenomena which underlie the trends. There are common demographic factors such as the ageing of the population which influence the demand for health care. As is evident from statistics on service utilization, people over 65 years of age have, on average, much greater rates of service utilization than people of lower age groups.

It has often been shown that the wealthier a country is, the more is spent on health care. More than 90 per cent of the variation in national health-care expenditures per capita can be 'explained' by the simple measure of GNP per capita (European Public Health Committee 1980).

Primary and secondary care

A major element of many declared policies in recent years, has been to reduce expenditure on secondary care (hospital beds) and to promote primary care. It has also been seen as important to strengthen the role and status of the primary care physician, who has a key role in the system (Kohn 1977, 1983). It is therefore essential to improve training, enhance research potential, and thus enhance his professional status.

It is also clear that in the future, primary care will increasingly be organized in group practices, with access to laboratory and X-ray facilities. Facilities for referral to specialists and hospitals will still be necessary if the primary care

Table 27.7. Expenditure in different health activities as percentage of total expenditure

Country[1]	1972				1977			
	Hospitals				Hospitals			
	Outpatient care	Inpatient care	Primary care	Other health care expenditure	Outpatient care	Inpatient care	Primary care	Other health care expenditure
Belgium	—	—	—	—	—	—	18.6	81.4
Denmark	—	—	27.0	73.0	—	—	28.0	72.0
Fed. Rep. of Germany	27.6[4]		35.5	36.9	28.3[5]		34.0[5]	37.7[5]
Greece	—	—	—	—	2.9	46.9	24.0	26.2
Ireland[2]	—	—	15.3	84.7	—	—	18.2	81.8
Luxembourg	—	12.1	—	87.9	—	20.1	—	79.9
Netherlands	38.3	25.9	16.1	19.7	39.3	22.0	16.1	22.6
Norway	47.4		33.9[6]	18.7	48.8		28.7	22.5
Sweden	5.7	46.2	8.9	39.2	7.5	52.3	9.9	30.3
United Kingdom								
England[2]	8.1	32.2	32.6	27.1	9.3	34.2	27.2	29.3
Northern Ireland[2]	6.3	54.2	25.9	13.6	6.9	57.5	20.9	14.7
Scotland[2]	5.3	32.0	22.2	40.5	5.5	33.1	20.4	41.0
Wales	—	—	29.0	71.0	7.7	33.5	24.0	34.8
Finland[3]	4.5	43.1	17.8	34.6	4.2	36.5	24.1	35.2

[1] United Kingdom split up into England, Northern Ireland, Scotland, and Wales.
[2] Public expenditure only.
[3] Long-term care included in primary and hospital care.
[4] Excluding prescribed drugs.
[5] 1975.
[6] Including prescribed drugs.
Source: *The cost of health care in member states of the council of Europe and in Finland.* European public health committee, Strasbourg (1980).

physician is to fulfil his duties. It is commonly supposed that this will of itself result in a need for fewer hospital beds and reduced hospital costs. This remains to be seen, however. It may not be so simple. At the very least, these new primary care arrangements may enforce different patterns of specialization within the hospital system.

Improved central planning of the geographical distribution of medical specialist facilities is already necessary in order to optimize the use of highly qualified manpower and to prevent inappropriate and inefficient acquisition and placing of costly technology; this is all too likely to occur if the system is left to the mercy of individual professional rivalries. With a change in the pattern of demand, arising from altered primary care services, the need for central planning is more necessary than ever.

All sectors of society are responsible for actively reducing health hazards, but the health care system holds a special responsibility, and has therefore to be equipped to assemble and present a range of important information about health and sickness and the reasons behind its health problem and to disseminate information in order to influence other sectors of society. The health care system has a specific obligation to provide information to those responsible in other areas for containing health hazards.

Preventive services

Preventive health care aims at timely prevention of illness along several different pathways. Its goal is the maintenance of the highest possible level of public health by means of general and specific preventive activities. They take the form of health information and education, the control of infectious diseases, technical hygienic care, and the quality-control of water, air, and foodstuffs. At a personal level these functions are executed through child health clinics, special clinics for the treatment of venereal diseases, school and university health care, school dental health care, and occupational health services. The activities include both primary and secondary (e.g. screening) preventive procedures. Preventive activities can be carried out and organized in a number of ways but, in Europe, services are in fact very similar. The pattern is illustrated with reference to Denmark (Office of the Director General 1984; Söndergaard and Kransnik 1984).

Maternity and child care

In Denmark, all pregnant women have the right to three medical examinations before giving birth, and two afterwards. Advice is also given on how to avoid future, unwanted pregnancies. As a rule, these examinations are carried out by general practitioners in close collaboration with hospital departments of obstetrics. There is further provision for at least five examinations by a midwife in the period of pregnancy and the first postnatal year.

Maternity care and delivery services are organized by county councils. All births are notified by the midwife to the health services authorities in the municipality where the mother lives. They, in turn, offer health care facilities to the mother in the form of home visits by a health worker. On average, each child is visited on between seven and nine occasions during its first year, but this can vary considerably depending on the needs of the individual child or family. All

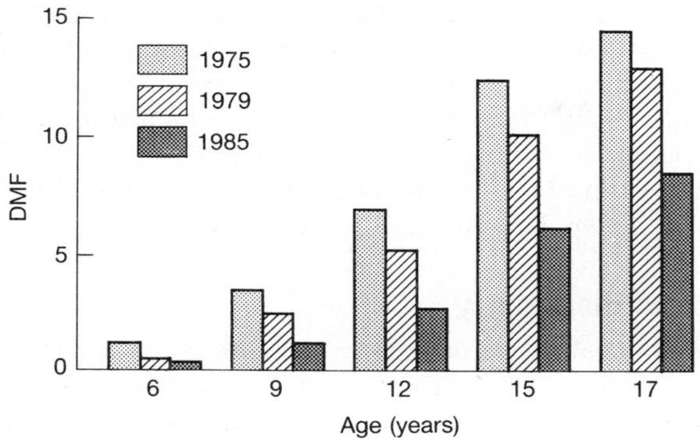

Fig. 27.4. Dental status among youth. DMF = number of decayed, filled. and extracted permanent teeth.

children are eligible for a programme which provides ten health examinations and all necessary vaccinations before school age. The municipalities are required by law to employ child medical officers and school health workers to undertake preventive medical examinations of all children of school and nursery school age.

Occupational health services

Legislation has been passed to give all industrial firms the opportunity to establish occupational health services. In factories with special safety problems, these services are mandatory (Nordic Statistical Secretariat 1988).

Dental care

Fluoridated water supplies have been introduced in a few European countries but topical application of fluoride through toothpaste and direct swabbing have been more generally preferred. School children and teenagers have been given priority in the development of dental health services. Preventive and treatment programmes for small children (0–6 years) have been introduced in most European countries in recent years, and free and systematic care, including topical application of fluorides, has been available to school children in public clinics in Scandinavia since the mid-1960s. Approximately 90 per cent of those aged 6 to 16 are now seen at least once a year.

There is substantial evidence for a decrease in caries in children during the 1970s (Fig. 27.4) (Heloe 1984). The participation in organized school dental programmes has undoubtedly had an important positive impact upon younger age groups, and is seemingly also reflected in the dental status of middle-aged people.

Smoking

Cigarette smoking is an established risk factor for lung cancer and other lung diseases as well as ischaemic heart disease; tobacco-related diseases are one of the most important challenges in health prevention. Smoking is targeted by means of a reduction or ban on advertisements, a prohibitive pricing policy, tar reduction, and widespread provision of information on risks of smoking (Peto 1985).

It is encouraging that the number of cigarette smokers has been reduced by about 20–30 per cent in Europe in recent years. The earlier increase in the smoking habit is levelling off among the young. It is no longer so 'fashionable' to smoke. But there is still a considerable number of 'habit' smokers and it continues to be vital to get the message across. It is especially important that all possible efforts are made to inhibit the youngsters from starting smoking. Unfortunately, the tobacco industry is still strong enough in some countries to avoid a total ban on advertising and thus largely to counteract such activities.

Screening

Screening has been practised in one form or another for many years. Antenatal care is a general example; screening for phenylketonuria is a specific one. The established guidelines for screening are: the condition should be numerically and medically important, an accepted treatment should be available, the test and the treatment should be free from serious risk, and facilities for post-screening diagnosis and treatment should exist.

Hypertension

A number of studies have been carried out and have shown that the complications of hypertension can be avoided through early detection and prolonged treatment; but at present the detection of severe hypertension is in most countries approached by opportunist case-finding rather than systematic screening; that is, by measuring blood pressure in those who present to their doctors for reasons other than raised blood pressure.

Breast cancer

The two main techniques for detecting breast cancer are physical breast examinations and mammography. Although there is evidence from randomized controlled trials that breast examinations and mammography delivered annually to women over 50 could reduce mortality from breast cancer (Eddy 1985), screening for breast cancer is not yet widely carried out in Europe. A recent Swedish study re-examined the efficiency of breast cancer screening. In a randomized study with 42 000 women who were followed for ten years, a significant difference in mortality was claimed and the cost-effectiveness of the procedure was considered to have been demonstrated. It is interesting therefore that while Sweden and Finland now carry out systematic screening for breast cancer, Denmark and Norway are awaiting further evaluation. The gains are thought to be marginal, and it may end up as a political issue.

Cancer of the cervix

The most effective means of screening for cancer of the cervix is the Pap smear, which is carried out widely in the European region. In areas with well-organized screening for cervical cancer, the mortality has been reduced. The cost-effectiveness of screening of average-risk women seems to be acceptable and, given the long pre-clinical history of the disease, a cost-effective method of screening is to give a three-yearly smear. This is generally the policy in the region.

The completeness of the screening services varies from one country to another and from one area to another, as does the manner of its organization, monitoring, and evaluation; and also the policy with respect to optimal ages, and with respect to the use of register-based invitations to attend. One of the important problems is that women with the highest risk are those less likely to present for screening of their own volition. In countries where these management and planning essentials are not well handled, the results have been disappointing.

Accidents

Accidents are an important health problem, with great potential for preventive activity (Table 27.5) (WHO European Region 1987). Developments in accident research have led to the development of some general concepts of the underlying processes which are responsible for an accident. It has become evident that accidents are determined by three interdependent components: human factors, the agent, and the environment. Up to now the importance of agent-related and environmental factors has been stressed, with little evidence of a concerted effort on human factors. However, the prevention of accidents and the reduction of their possible consequences is dependent on a well-balanced application of behaviour-modification, environmental-modification, and the use of safety devices.

Since accident research is multi-sectoral, it is of paramount importance that activities taking place in the various sectors are co-ordinated. This co-ordination results in the development of integrated national programmes which form the basis for more effective accident prevention and helps to enhance international cooperation (WHO European Region 1987).

There is much need for improved accident statistics, to elucidate causal factors, to detect high-risk groups, and establish preventive measures. The size of the problem clearly justifies the broad range of activities aimed at accident prevention apparent in most European countries (WHO European Region 1987; Kohler and Jackson 1987).

Co-ordinating health and social agencies

The co-ordination of health and social services is important, but in many places is not sufficiently developed; often the services do not jointly meet the people's needs, or they do so inefficiently. The situation reflects the growing complexity of modern society on the one hand, and a generally fragmented approach to dealing with newly identified problems, on the other. Development of medical and social sciences and of technology has resulted in a greater specialization and hence fragmentation of professional responsibilities. The expanding role of the state in various aspects of welfare provision also leads to a greater subdivision of services. The change from the extended family to the small modern family living in an urban setting has meant that many support services earlier provided by the family must now be provided collectively by the public: responsibility for action is being progressively shifted to the local community (Nordisk Medicinalstatistisk Komite 1988*a*; Brouwer and Schreuder 1987).

Care of the elderly

Care of the elderly is of concern to medicine because the likelihood of physical disability and mental illness is high at advanced ages. Prevention and the understanding of health risks are also important issues. This is very much a nursing concern, because one of the primary responsibilities of the nursing profession is to assist individuals and groups to make the most of their functions within varying states of health; nursing is involved in caring and in prevention, as well as in illness. The elderly are of concern to social workers because lack of knowledge and lack of contact between health and social personnel often results in inappropriate levels of care. The team approach is therefore important, where members of the primary health care team together have sufficient knowledge to make a total assessment either in the elderly person's home, and his or her health status and needs.

The most important source of informal care is still the family of an elderly person. Evidence from throughout Europe confirms that the greatest, most immediate and most continuous forms of support, come from relatives. Although these traditional community and family relationships still exist in many countries, modern urban living is characterized increasingly by mobility and anonymity. The future planning of public care must take such community changes into account (Skeet 1988).

AIDS

The spread of HIV infection in Denmark corresponds to that of other western countries. Thus, the majority of AIDS cases to date have occurred among homosexual/bisexual men. However, in 1987, a proportionately greater rate of increase among intravenous drug users was seen. As of 30 November 1989, the number of AIDS cases identified in Denmark was 494, with 277 dead. With a total population of 5.1 million, the number of cases per million inhabitants was 97: among the highest in Europe.

As an example of a co-ordinated activity concerning an important disease, we describe the practical strategy of AIDS control in a single European country, Denmark. The Ministry of Health established a National AIDS Committee with overall responsibility for formulating general policy and implementation of measures to reduce the number of AIDS cases. On 1 November 1986, an AIDS secretariat was established in the National Board of Health with the task of initiating and co-ordinating measures to combat AIDS in the widest sense, and to evaluate the initiatives in operation.

Attempts to control the HIV epidemic are based exclusively upon counselling and voluntary action. A strategy of voluntary participation, anonymity, and open direct and honest information, was confirmed by an almost unanimous Parliament in March 1987. Both before and after this time, information and education for all groups have been accorded highest priority. In 1983, the first brochure had been distributed to the gay community, and, in the following years, brochures were distributed to other people with high risk behaviour, to the rest of the population, and to health care and other personnel groups. This campaign is currently being evaluated through measuring people's knowledge and AIDS, and possible changes in sexual and other high risk behaviour before and after the campaign (Neergaard 1988).

Organization of health care services in the Nordic countries

The following examples illustrate the organization of public health services further.

Denmark

In Denmark there are three political and administrative levels involved in providing health services (Söndergaard and Kransnik 1984). The State is responsible for legislation and therefore bears overall responsibility for health care and its supervision. The counties are responsible for running and planning most health care services and other public services. The municipalities are responsible for the running and planning of certain local and public health care services, major parts of social care, and a number of other services. Individual health care, and most services traditionally classified under Public Health, are thus decentralized; they are administered and planned by counties and municipalities within the terms of central legislation, general guide-lines, and prescribed economic limitations. The hospitals are also run by the counties. There are few private hospital facilities in Denmark and they are run entirely within the public hospitals. Patients are referred by general practitioners, and the only hospital service a patient can turn to without referral is the emergency unit.

Professionals providing health care under National Health Insurance (general practitioners, practising specialists, dentists, physiotherapists) are paid by the counties, which also have a responsibility for running the establishments within which they work. The general practitioner is the key person in the Danish health care system. As a result the population is offered free access to what is considered the most accessible link of the service chain. About 90 per cent of patients are treated by general practitioners only. The remaining 10 per cent are referred by general practitioners to specialized treatment in hospitals or to practising specialists (Nordisk Medicinalstatistisk Komite 1988*b*).

Finland

Finland went through an interesting development in the period after the Second World War, when emphasis was placed on the development of the hospital system (Ministry of Social Affairs 1987; Central Statistical Office of Finland 1988). The 1950s saw the launch of an unparalleled hospital construction scheme in which 19 000 general hospital beds and 11 000 psychiatric beds were established. Weaknesses in the Finnish health care system were, however, soon evident. The primary health care services remained, by international comparison, inadequate and inconsistent. There was a high morbidity, and health indicators showed that progress was slow or non-existent and that regional differences in morbidity were widening. It was generally admitted that the inadequacy of primary health care was the principal structural defect of the health care system. Priorities were revised in the 1970s in order to redress the fact that 90 per cent of public health care resources were being used for specialized medical care in hospitals and only ten per cent for primary health care. Since then the proportion of public health care expenditure devoted to primary health care has risen from 10 per cent to 30 per cent.

Health care in Finland is mainly a function of the municipalities. There are now 215 health-centre districts in Finland for primary health care. A health centre is an administrative unit providing services at several locations. Health centres also provide more than 19 000 beds in wards that are used mainly for long-term care.

There is a hierarchical system of health care referral. Patients must first apply to their local centre for examination and treatment. If they cannot be treated there they are referred to the out-patient department of a hospital and, if necessary, then admitted to hospital. During convalescence the patient can visit the out-patient department of the hospital, but gradually the responsibility for care is transferred to general practitioners at the health centre.

The aim of this organization has been to transfer the focus on health policy to primary health care and the out-patient services by creating administrative and economic conditions for the speedy development of municipal primary health care systems. The objective is to build a nationwide network of health centres providing the initial contact for the whole population, who are then referred to specialized treatment if necessary. Health centres provide not only general medical care, but also dental care, ward treatment, rehabilitation, X-rays, laboratory work, and other services. A fee is charged for ward treatment and adult dental care; otherwise services are free of charge.

In 1985 a revised nationwide plan for the improvement of public health care provision was introduced. The priorities were defined in a more specific manner. Thus, the goal of the social welfare services was defined as 'to develop a network which offered alternative forms of service, taking into account both local conditions and the population's need'. The aim of environmental health protection was 'to prevent health hazards and unfavourable effects on health emerging in the environment'.

Within primary health care, the importance of healthy living habits and individual self-reliance in maintaining one's own health were stressed. Other important aspects were continuity of care, more humane and personal treatment of the increasing elderly population, prevention, care, and medical rehabilitation concerning the most prevalent illnesses, organization of levels of care and collaboration between the health centres, and reduction in the regional differences in primary health care services supplied. Efficiency and follow-up of results were also stressed, as well as co-operation with social services and the labour force authorities.

Within the hospital services, emphasis was laid on systematic preventive work directed against the most prevalent illnesses, the functional organization of specialized hospital care and co-operation between different hospitals and health centres, an increase in out-patient and rehabilitation services, especially in psychiatric care, and further development of health education and patient advice services.

Sweden

The responsibility for health services in Sweden rests upon the county councils (Rosen 1987; National Board of Health and Welfare 1985; Vogel *et al*. 1987; Andersson 1988). Private practice exists only on a limited scale. Only 8 per cent of doctors are mainly working as private practitioners). At county level, health care is divided into different levels: primary care, which covers all individual health care and all treatment that can be provided outside a hospital: and secondary care, which is concentrated in hospitals. The most specialized treatment is found at regional hospitals.

Primary health care responsibilities are divided into districts, mostly corresponding geographically with municipalities, covering medical treatment in a doctor's consulting room, in health centres, work by district nurses, a variety of services in the home, and most long-term care at nursing-homes. The county health service also includes work in general and specialist hospitals which account for more than half of the resources spent by the county councils. According to the Health Act, county councils are required to collaborate within six medical care regions to provide specialized care.

During the 1950s and 1960s concentrated efforts were made to improve and develop hospital facilities; while during the 1980s primary care and long-term care were given first priority. A considerable increase in primary care resources is now planned; a main aim is to shift emphasis gradually towards preventive measures. The service will deal not only with the procedural prevention of disease and the investigation and treatment of diseases and injuries, but also try to influence other factors in society to eliminate or reduce existing risks to health.

On an international scale, Sweden has a relatively low frequency of visits to physicians (only three per person per year) compared to hospital utilization. Medical demands and the increasing costs of the health service, however, have caused out-patient care to be given priority and the number of primary health care centres has been expanded considerably.

There has, however, long been and continues to exist a relatively strong private sector. Few private hospitals are in operation, but the number of part-time and full-time private practitioners is large. In fact, there are about 4 million physician visits to 'private' doctors per year (i.e. almost one per adult inhabitant, whilst the number of all physician visits in ambulatory care is estimated to have been 3.8 in 1989). Of private physician visits, about one-third are to general practitioners and two-thirds to specialists. In addition to doctors, there are private laboratories and physiotherapists and many other services. The majority of adult dental care is *not* provided by health centres but by private dentists. Part of the costs of visits to the private sector are covered (i.e. reimbursed to the patient) by sickness insurance. The level of reimbursement is supposed to be between 50 and 60 per cent; however, the true current level is around 35 per cent, though for dental care may be as high as 80 per cent. Occupational health services provided by employers are also an important provider of primary care services. Of these costs the Social Insurance Institution ('sickness insurance') refunds 55 per cent to employers.

In addition to the above care costs, sickness insurance also covers, for example, the costs of medicaments, and the costs of dental care in young adults.

General trends

In all the Nordic countries there is a clear trend towards channelling medical resources from large hospitals to primary care. There are many reasons for this re-allocation but among the most important are aspirations to reduce total costs for health and to combine this with more accessible and high quality services for patients.

Primary health care centres are being given responsibility not only for health care services but also for preventive medicine, and are being allocated a catchment area in which the characteristics of the population are known. Using this as a guide, primary health care can be provided with resources according to need.

In their efforts to fulfil this new role, primary health services would be assisted if they had access to information about the population's previous contact with the health service. This is an important point. With the patients' consent, information concering medical treatment given at referral hospitals and treatment obtained outside the local medical region could be collected and would enable the centres to serve the patients sufficiently. Thus patients' main medical records are to be kept at the primary health centre chosen by the patient, and other medical care units are regarded as referral institutions.

Health trends in Europe

The development of the health status of a country can be compared with its economic development. Both are the result of a nation's collective development policies and form the basis for the welfare of its citizens (Brouwer and Schreuder 1987).

All sectors of society are responsible for activity participating in reducing health hazards. A forceful health policy requires a health care system that works actively to prevent injuries and diseases. The health care system has an obligation to provide information to those responsible in other areas concerning existing health hazards. Developed public health care systems depend on an efficient information system.

We have seen an increased average life expectancy in Europe during the last decade. There are, however, important variations both in mortality and morbidity (Holland 1988). Men and women are affected differently; the varied social background is significant; there are differences between the different socio-economic groups; and there are geographical variations. That is, certain socially unprivileged groups run a substantially greater risk of falling ill and suffering a premature death or chronic disability.

Health is not solely a matter for the health care sector. A broader form of co-operation including all sectors of society is needed. The public health system has to develop and enhance its socially oriented preventive efforts particularly in those groups most susceptible to health hazards. The trend in Europe is towards equality of health, which requires equality in the supply of resources, utilization, and access to care.

Research in health care

Evaluation of the organizational and qualitative factors of the health care system is of major importance. Partly as a result of policy developments, health service research continues to have a prominent position. A central research issue concerns the way in which a health care system can be made to match the health problems and health care requirements of the population, bearing in mind organizational, staffing, accommodation, and financial aspects. Research into the relationship between the supply of and the demand for health services merits priority.

Another important research issue concerns the shift from secondary to primary health care which has occurred, and also the shifts within secondary care. Research into efficiency and effectiveness of health care, including an evaluation of care programmes, is needed (Ministry of Welfare, Health and Cultural Affairs 1988).

Conclusion

The general health situation in Europe has improved; the disease pattern has changed; infectious diseases have been reduced; and malnutrition has disappeared. As a consequence, the mean duration of life has increased from about 40 years in the middle of 1800s to 75–79 years at present in the Scandinavian countries. Better hygiene and a better general standard of living, improved housing facilities and workplace conditions, as well as advances in medicine, have played a major role in this improvement. Seen in a historical perspective, poverty and poor living conditions are the major reasons for high morbidity and early death.

Despite the improvements in social and medical services the demand for public health services has increased. The cost of improved standards is a public demand for high standards still. There is increasing demand for preventive measures in the broad sense: preventing physical and biological as well as chemical and environmental risk factors in the surroundings. This calls for a multi-sectoral approach in the public health service. Recent developments in public health policy reflect attempts to redress earlier proportional increases in hospital facilities, as a result of which prevention and primary care services were left behind.

References

Backett, E.M., Davies, A.M., and Petros-Barvazian, A. (1984). *The risk approach in health care, with special reference to maternal and child health, including family planning*. World Health Organization, Geneva.

Brouwer, J.J. and Schreuder, R.F. (ed.) (1987). *Scenarios and other methods to support long term health planning*. Steering Committee on Future Health Scenarios, Rijswijk, The Netherlands.

Central Statistical Office of Finland (1988). *Health in Finland*. Helsinki, Finland.

Eddy, D.M. (1985). Screening for cancer in adults. In *The value of preventive medicine*, Ciba Foundation Symposium 110. Pitman, London.

Eklundh, B. and Roos, K. (1984). A Swedish study of the utilization of health care facilities. *Scandinavain Journal of Social Medicine, Supplement* **34**, 29.

European Public Health Committee (1980). *Cost of health care in member states of the Council of Europe and in Finland*. 1979-Coordinated Medical Research Programme, Strasbourg.

Hellberg, H. (1984). General aspects of inequities in health and health services. *Scandinavian Journal of Social Medicine, Supplement* **34**, 7.

Heloe, L.A. (1984). *The development of dental health in Scandinavia in the 1970s*. Scandinavian Journal of Social Medicine, Supplement, **34**, 39.

Herberger, L. (1984). Trends and perspectives in family formation. In *Demographic trends in the European region, health and social implications*. WHO Regional Publications, European Series No. 17. Copenhagen.

Hjort, P.F. (1984*a*). Inequities in Medical Care: Consequences for health. *Scandinavian Journal of Social Medicine, Supplement* **34**, 75.

Hjort, P.F. (1984*b*). Inequities in short stay somatic hospital care in Norway. *Scandinavian Journal of Social Medicine, Supplement* **34**, 35.

Holland, W.W. (ed.) (1988). *The EC atlas of 'avoidable death'*. Oxford University Press.

Kamper-Jörgensen, F. (1984). Causes of differences in utilization of health services. *Scandinavian Journal of Social Medicine, Supplement* **34**, 57.

Kleczkowski, B.M., Roemer, M.I., and van der Verff (1984). *National health systems and their reorientation towards health for all. Guidance for policy making*. World Health Organization, Public Health Papers No. 77, Geneva.

Klinger, A. (1984). Trends and perspectives in fertility. In *Demographic trends in European region. Health and social implications*, WHO Regional Publications, European Series No. 17. Copenhagen.

Köhler, L. and Jackson, H. (1987). Traffic and children's health. The Nordic School of Public Health. The European Society for Social Paediatrics, Report 1987, 2.

Köhler, L. and Wahlberg, H. (1984). Inequities in preventive programmes for maternal and child services. *Scandinavian Journal of Social Medicine, Supplement* **34**, 43.

Kohn, R. (1977). *Co-ordination of health welfare services in four countries: Austria, Italy, Poland and Sweden*. Public Health in Europe 6. World Health Organization, Regional Office for Europe, Copenhagen.

Kohn, R. (1983). *The health centre concept in primary care*, Public Health in Europe 22. World Health Organization, Regional Office for Europe, Copenhagen.

Lopez, A.D. (1984). Demographic change in Europe and its health and social implications: An overview in demographic trends in the European region. Health and social implications. WHO Regional Publications, European Series No. 17. Copenhagen.

Lynge, E. (1984). Trends and perspectives in mortality. In: *Demographic trends in the European region. Health and social implications*, WHO Regional Publications, European Series No. 17. Copenhagen.

Ministry of Social Affairs and Health (1987). *Health for all by the year 2000*. The Finnish National Strategy, Helsinki.

Ministry of Welfare, Health and Cultural Affairs (1988). *Health research policy in the Netherlands*. Rijswijk.

National Board of Health and Welfare (1985). *The Swedish health services in 1990s*. Stockholm.

Neergaard, L.D. (1988). Preventive actions in relation to AIDS (personal communication).

Nordic Statistical Secretariat (1988). *Occupational mortality in the Nordic Countries 1974–1980.* Statistical Reports of the Nordic Countries, Vol. 49, Copenhagen.

Nordisk Medicinalstatistisk Komite (Nomensko) (1988*a*). *Health statistics in the Nordic Countries* (1986), Vol. 28. Copenhagen.

Nordisk Medicinalstatistisk Komite (Nomesko) (1988*b*). *Computerised information systems for primary health care in the Nordic countries*. Copenhagen.

Office of the Director General (1984). *Public administration and health care in Denmark*. National Board of Health of Denmark, Copenhagen.

Peto, R. (1985). Control of tobacco-related disease. In *The value of preventive medicine*, Ciba Foundation Symposium 110. Pitman, London.

Rosen, M. (1987). *Epidemiology in planning for health*. Spri, Stockholm.

Skeet, M. (1988). *Protecting the health of the elderly*. Review of WHO activities public health in Europe, Vol. 18. World Health Organization, Regional Office for Europe, Cophenagen.

Söndergaard, W. and Krasnik, A. (1984). Health services in Denmark. In *Comparative health systems. Descriptive analysis of fourteen national systems* (ed. M.W. Raffel). Pennsylvania State University Press, University Park and London.

Teper, S. and Backett, M. (1984). Implications of demographic change for the young population (0–19 years). In *Demographic trends in the European region. Health and social implications*, WHO Regional Publication, European Series No. 17. Copenhagen.

Viefhues, H. (ed.) (1988). *Medical manpower in the European Community*. Springer-Verlag, Berlin.

Vogel, J., Anderson, L.G., Davidson, U., and Hall, L. (1987). *Living conditions*, Report No. 51, *Inequality in Sweden 1975–85*. Statistics Sweden, Stockholm.

WHO European Region (1987). *Accident prevention and injury research*, Report of the survey within the WHO European Region. WHO, Copenhagen.

28

Public health services in the UK

R. GRIFFITHS and A. McGREGOR

Introduction

In this chapter we consider the public health services in the UK, drawing examples largely from England, which is indicative of the situation throughout the UK, as all the constituent countries are underpinned by much the same history, philosophy, and legislation.

It has to be said at the outset that the real world of public health in England, and no doubt everywhere, is extraordinarily untidy with doctors and a variety of other professionals and non-professionals working for the 'public health' within a seemingly random mix of authorities and laws. In order to achieve understanding the description that follows has therefore been simplified within a framework that is itself, of course, something of an abstraction from reality. The reason for the apparent 'untidiness' of public health activity stems from the nature of 'public health' itself.

In any society it is always possible to identify a state or condition of 'public health', and in the mind of the public there will always be a set of services or activities that are seen as important to the maintenance or improvement of the state of the 'public health'. At any given time these will be called the 'public health services' and they will change as society develops and the major threats to the public health change.

Public health has been defined (Secretary of State 1988) as the science and art of promoting health, 'preventing illness and prolonging life by the organized efforts of society'. Public health services thus cover a very broad range of activities, but, importantly, in any society there will never be just one agency that has responsibility for organizing them and yet practitioners of public health will often be assumed to have some responsibility for anything that appears to pertain to the public health.

Public health in the UK is underpinned by a substantial body of legislation. The legal framework permits ministers to spend public money on services, establishes a structure and system of accountability within which public bodies and public servants are required to carry out defined duties, and requires individual citizens and organizations to conform to certain standards and procedures.

The fundamental duty laid on the Minister of Health by law is to improve the health of the population through the provision of services. There is no similar general duty laid on local authorities because the body of legislation concerned is not consolidated within one act as health services are consolidated by the National Health Service Acts. In general terms the powers given to local authorities relate to the maintenance of public health through the enforcement of standards. These powers are frequently interpreted imaginatively so that efforts are made to prevent infringement of standards through education of potential offenders and other preventive activities. In general, health authorities are responsible for treatment and local authorities for enforcement. Both have responsibilities for surveillance with some overlap and both engage in health promotion and prevention. Health services tend to aim at preventing the need for later treatment, while local authorities aim at preventing infringement of regulations and standards. Health promotion crosses many boundaries and is underpinned by the most general of legislation. It is the area where authorities and their officers can find the easiest excuses for doing nothing or the most creative opportunities for working across many sectors of society.

Overlaid on all this are interplays between the rational and irrational in social processes and between the need to maintain the status quo and the need to change. All these factors are influenced by the day-to-day work of public health practitioners, and the practitioners themselves are influenced by the society in which they live and for which they work.

In this chapter we shall show that public health services and public health practitioners need both a legal basis for their actions and the power and mechanisms to spend public money and influence the sums that are voted to their cause. They need a management structure through which to employ and command relevant manpower, and they need a structure through which they are accountable for their actions. They must be able to acquire training and maintain their skills, and, in practice, they are underpinned by a common culture.

The actions possible with these basic tools will be modified by both professional and by perceptions and knowledge of the major threats to public health, and by the ways in which these perceptions and the problems themselves change. They will also be influenced by changes in society itself.

Just as breadth of vision is an essential part of the work of public health, so is the understanding and management of change and organizations. The work of practitioners of public health is constrained by the organizational position from which they work, but this need not be a limitation on their vision. By examining the science and art of public health from the point of view of the practitioners in the UK National Health Service (NHS) it is possible to extract many of the principles which will guide the work of any practitioner seeking to create public health services. A central part of public health medicine is the science of epidemiology which is used to analyse patterns of health and sickness and to determine strategy. The art in public health medicine is turning strategies into services and results, but the art and science must be intertwined and results must be re-evaluated and further change planned.

The public health services in England are a particularly useful example to study at the present time because in 1987 they were exhaustively reviewed by a government inquiry—the Acheson Inquiry—which not only led to the implementation of a number of major changes but also spelt out policy and practice in a way that is bound to guide development for many years to come. As this was only the second time in a hundred years that the public health function in England had been reviewed in this way it represented a major watershed. In this chapter the pressures that led to the inquiry will be examined and its major conclusions will be set out and discussed. In the context of existing services it is also possible to see how public health practice will work and develop and from this to deduce some general lessons.

These developments will take place as further substantial changes occur in the organization and management of the NHS stemming from the government's issue in early 1989 of a consultative document or White Paper entitled *Working for patients* (Secretaries of State 1989).

In the light of the comments made at the beginning of this introduction on the 'untidy' real world of public health, the chapter begins with a description of the issues addressed in the Acheson Inquiry and the White Paper *Working for patients*. This is followed by a description of the administrative structure of the public health services. The description of the practice of public health in England is supported by short sections on information and research. Finally the future of public health in the UK is considered and some conclusions are drawn.

The Acheson Inquiry into public health in England

The antecedents of the Inquiry

This inquiry was established by the Secretary of State for Health in January 1986 under the chairmanship of the Chief Medical Officer, Sir Donald Acheson, to undertake a 'broad and fundamental examination of the role of public health doctors' and to consider the 'future development of the public health function including the control of communicable diseases . . . '. It issued its report in January 1988 (Secretary of State 1988). The inquiry was part of the Government's response to a food poisoning outbreak at Stanley Royd Hospital in 1984—the other responses concerned the issue of administrative directions to tighten controls and similar actions. The outbreak resulted from poor kitchen practices, and 19 elderly patients died from salmonella infections. The report expressed anxiety about the state of medical manpower and training in public health. (Department of Health and Social Security 1986*a*).

This outbreak was followed by a further major infectious disease incident. Legionnaires' disease at Stafford General Hospital caused the deaths of 28 patients and visitors, and this also led to a public inquiry which expressed anxiety about the availability of medical expertise in the control of communicable disease (Badenoch 1986). Alongside these formal statements, the Secretary of State also received many complaints from within the NHS that medical efforts in public health were being undermined by manpower shortages and by some of the managerial changes that had recently been introduced.

Public health doctors seemed to vary greatly in their access to power within the NHS. It appeared that in consequence effectiveness often depended more upon the personal charisma of the individual practitioner than on some formal pathway. As a result, in different districts, doctors with apparently similar titles were doing very different jobs. This led to confusion about roles, particularly among health workers in other disciplines as they could not find a consistent model of public health medicine and hence were confused about what it was for and what it did. Some of this confusion spread back into the specialty itself, which also made recruitment and training more difficult. In turn this led to a real shortage of public health medical manpower in many areas.

The seeds of this confusion were probably sown in the NHS reorganization of 1974 which moved public health medicine from its longstanding location in elected local government authorities into the appointed authorities of the NHS, while leaving the legislative powers, enforcement, and scientific resources for all work outside hospitals with local authorities. That reorganization also saw the introduction of the term 'community medicine' which described a medical specialty which drew together traditional medical management skills with the sciences of epidemiology and sociology. This was an inspired theory but many former local authority medical officers were unable to make the transition. There were also huge differences in the new merged NHS in access to the basic tools of public health, ie, information, power, money, and accountability.

In local authorities, public health doctors frequently managed their own information services. Data came in through statutory notifications, particularly from the staff of public health departments, nurses, and environmental health officers. In the merged NHS community nurses reported to nurse managers who tended to be hospital based. Environmental health officers stayed in the local authorities, as did most of the information services. The information that did exist in the NHS concerned only the activities of hospitals—effectively the sole administrative component since 1974—and was related to sickness and its treatment. Public health doctors, even if they were skilled epidemiologists, had to invent entirely new methods of obtaining information.

Within the local authority the Medical Officer of Health was a powerful chief officer with a large department, but in the post-1974 NHS the new community physicians were members of a consensus team, most of whom were only interested in the affairs of hospitals. Access to power therefore required entirely new negotiation skills.

Money also turned out to be in short supply. There was no tradition within the NHS of spending on public health. All the financial systems were geared to the continuation of the status quo, the hospital service. The much vaunted planning system was frequently overwhelmed with demands from those who already had most of the resources. It is difficult to build up services from a very low base. Existing services advertise themselves and tend to build up waiting lists which appear to be measures of need. There is no waiting list for a non-existent service, and hence the argument for it seems less immediate.

The last weakness of community physicians was that they appeared to be accountable to no-one and hence had no political allies. This was also a major change from the situation in the local authority where the Medical Officer of Health (MOH) was a chief officer reporting to a committee with a major budget.

The community physician thus tended to lack access to information, power, and money, and his accountability was unclear. As this became clear to the Acheson Committee, it also became clear that the public health responsibilities of the NHS had never been made explicit. The various NHS Acts have always laid on the appropriate minister a general duty to improve the health of the population, but this had never been formally delegated to health authorities. Formal responsibilities in relation to treatment services had been delegated and had become the main preoccupation of the NHS, particularly as the dramatic advances of clinical medicine repeatedly captured attention.

These two sets of circumstances were enough to weaken the position of public health doctors inside the NHS seriously to the point where sooner or later a disaster such as the Stanley Royd outbreak was bound to occur.

The situation for public health in general was made worse by the changes taking place outside the NHS. Academic researchers and those in public health outside the NHS persistently pointed out that the major advances in public health came from changes outside the treatment services. Even among diseases in which treatment seemed important, it was often clear that diet and environmental factors were of greater consequence. These researchers tended to undermine further the credibility of public health doctors: why were they in the NHS, apparently a hospital organization, when the relevance of the hospitals to general advances in public health was questionable?

The Acheson Committee thus had to answer some fundamental questions. Did we need public health doctors at all? Should they work in the NHS? What was the role of the NHS in relation to public health? How did this role relate to other key actors and influences on public health?

The outcome of the Acheson Inquiry

The major conclusion of the Acheson Inquiry was that the public health role of health authorities should be made explicit. This conclusion was later encoded in legal form in Health Circular HC(88)64 (Department of Health and Social Security 1988) which stated that Regional and District Health Authorities have a responsibility to:

(1) review the health of the population for which they are responsible and to identify areas for improvement;
(2) within the planning and review framework, define policy aims and where necessary set quantified service objectives to deal with any problems in the light of national and regional guide-lines and available resources;
(3) relate decisions which they take about the distribution and investment of resources to their impact on the health of their population and the objectives so identified;
(4) monitor and evaluate progress towards their stated policy aims and objectives including the development of indicators of outcome;
(5) at a regional level make plans for dealing with major outbreaks of communicable disease and infection which span more than one district taking into account where necessary the need to liaise with the Public Health Laboratory service and the Communicable Disease Surveillance Centre, and ensure that each District Health Authority (DHA) in turn has made arrangements for the surveillance, prevention, treatment, and control of communicable disease within its boundaries.

The Committee concluded that it must be the duty of general managers to ensure that these functions of health authorities were carried out. The committee went on to examine whether general managers required any prescriptive advice about how they should ensure that the public health roles of authorities were carried out, but came to the conclusion that, as a general principle, it was better to lay down as clearly as possible the duties of the various public bodies and to avoid prescription as to the mechanisms. However, the Committee did conclude that there was a unique role for doctors trained in public health and went on to suggest the broad content of that training as well as the numbers of such doctors likely to be needed in the different parts of the Health Service.

In order to try and avoid some of the confusions of the past the committee suggested a common title—Director of Public Health—for the leaders of public health medicine at regional and district level. The Committee's suggestions as to the duties of the Director of Public Health, as the post was to be known, were also accepted by the government and incorporated into the implementation circular (Department of Health and Social Security 1988) which said that in order to discharge their public health responsibilities each authority should:

(1) ensure it has access to appropriate public health advice including that of a designated Director of Public Health (DPH);

(2) require the DPH to produce an annual report on the health of the population;

(3) review and make more specific arrangements for dealing with communicable disease and infection;

(4) maintain close collaboration with local authorities, Family Practitioner Committees (FPCs), the Health Education Authority (HEA), and other public and voluntary agencies in matters affecting the health of the public with particular reference to the prevention of disease and the promotion of health.

The role of the DPH is spelled out very clearly in the paragraph 3 of the report of the inquiry:

The central tasks of the DPH and his/her colleagues are as follows:

- To provide epidemiological advice to the DGM and the DHA on the settings of priorities, planning of services and evaluation of outcomes.
- To develop and evaluate policy on prevention, health promotion and health education involving all those working in this field. To undertake surveillance of non-communicable disease.
- To co-ordinate control of communicable disease.
- Generally to act as chief medical adviser to the authority.
- To prepare an annual report on the health of the population (or, to quote the former MOH duty 'To inform himself as far as practicable respecting all matters affecting or likely to affect the public health in the [district] and be prepared to advise the [health authority] on any such matter').
- To act as spokesperson for the DHA on appropriate public health matters.
- To provide public health medical advice to and link with local authorities, FPCs and other sectors in public health activities.

In summary the Acheson Inquiry both carried out a comprehensive review of the public health function and succeeded in making detailed proposals for action which were virtually all adopted by the Government and issued as formal circulars. The report of the Inquiry will be a source of reference and action for years to come.

The White Paper—Working for patients

Working for patients was a government consultative White Paper (Secretaries of State 1989) published in January 1989 which proposed a number of fundamental changes in the way that health services are organized and paid for in the UK. The most fundamental proposal is the introduction of an internal market. Under this system, health authorities, instead of being given an allocation by government to run their hospitals and other services, will be allocated a budget based on a weighted population calculation and will have to use this money to buy services from hospitals. It is presumed that in the long run most hospitals will be self-managing institutions run by public trusts, but initially most hospitals will be managed by health authorities (as they are now). In order to make sense of this market concept, authorities will have to achieve some separation of function within their own organization so that they can buy services from themselves as well as from units that are managed by other authorities.

The longer-term role of health authorities is defined in paragraph 2.11 of the White Paper in the following terms (Secretaries of State 1989):

> '. . . Like RHAs, DHAs can then concentrate on ensuring that the health needs of the population for which they are responsible are met; that there are effective services for prevention and control of diseases and the promotion of health; that their population has access to a comprehensive range of high quality, value for money services; and on setting targets for and monitoring the performance of those management units for which they continue to have responsibility. The Government will expect authorities to provide themselves with the medical and nursing advice they will need if they are to undertake these tasks effectively.'

This paragraph, together with the definition of public health functions in the circular HC(88)64—the implementation circular referred to above that took up the Acheson Inquiry recommendation—makes quite clear the vital role that public health advice is now presumed to play in the proposed redesigned organization. (Department of Health and Social Security 1988). To underline this, the government has funded an additional 40 training posts in public health medicine in order to rectify the manpower problems identified in the Acheson Report.

The intended benefits of the proposals fall into two main areas; on the one hand units that provide health services will need to become more business-like and, it is presumed, more efficient; on the other hand, the authorities which buy services—which may include some large general practices—will make better strategic judgements as to what the population really needs if they are freed from the day-to-day pressures of running particular services. The requirement to contract out all services could force examination of the way in which all money is spent. It should become possible to reappraise priorities and change the pattern of spending because decisions should be based on population benefit criteria rather than pressures from particular services with vested interest in particular developments.

At the time of writing it is not possible to predict whether the proposals will survive intact through the legislative process or whether they will achieve the intended benefits.

The implications of the White Paper for public health

The changes proposed are major and will require a new infrastructure, particularly for the considerable development of information and financial systems necessary to underpin the budgetary contracts. The change and unavoidable uncertainty is bound to put some existing services at risk, and the additional information and financial support needed may reduce actual spending on health care. These are important considerations, but it is more important, at least in this chapter, to examine the extent to which the changes will allow public health practitioners access to the essential tools, i.e. information, money, power, and accountability. It appears that the Government's intention is to increase the public health voice in the way that NHS money is used. It seems probable that information will flow with contracts and therefore will be at least as good as it is now.

The extent of access to power in the future is not certain as yet because the formal structure of the new health authorities does not contain a public health voice as of right (although it does in Wales at the time of writing). The proposals do not appear to change accountability: public health directors must still write a public annual report which is formally presented to a health authority whose members continue to be appointed rather than elected.

On balance then, the proposals seem to enhance the voice of public health medicine, but its impact on relationships with other practitioners of public health is less clear. Health promotion is regarded as an essential activity for the new health authorities and is unlikely to be contracted out. If this proves to be the case, then its status will probably be enhanced. The laboratory and enforcement sides of public health remain in the large service departments of the local authorities. In addition to infectious disease control many such departments are concerned with refuse collection and disposal, urban renewal, and some aspects of safety. If in the future the government insists that many of these services are provided on a contract basis, then the role of the key professionals who remain in public health in local authorities will become very like those in the new health authorities. Until that happens there is a danger that the local authority officers will be preoccupied with running services and may not easily identify with the new role of public health practitioners in health authorities. An optimistic view is that a common interest in health promotion might prove an important bridge.

The administrative structure of the public health services

In this section the administrative structure of public health services at national and governmental level and within the health authorities is described. Some of the other important organizations concerned with the public health are also described.

The governmental level

National policy is determined by ministers who are accountable to Parliament. The processes by which ministers come to suggest new policies are, of course, very diverse. Ministers are furnished with analyses of the effects of existing policies which are conducted by professional officers in their departments. They also receive political feedback through the views of other Members of Parliament and the pressure groups who lobby them. When parties are in opposition they have much less ready access to the official machinery, but the political processes are much the same. The 1980s have seen the emergence of 'think tanks' as the initiators of new policy ideas. These groups gain access to ministers by a variety of routes, and these ideas may then gain further momentum through the political process before they are subjected to any real scrutiny by professional officers. The eventual outcome of such political discussions is some sort of policy statement, either through a purely political route, such as a party manifesto or policy statement, or the more formal process of Green and White Papers, of which the latter are direct forerunners of legislation.

Professional experts in public health give their opinion on all aspects of proposed changes in policy either directly or through the continuing contact between the Chief Medical Officer and his senior colleagues and the ministerial team. As a result of another recommendation of the Acheson Committee, the English national Department of Health now has a small central unit directly involved with reviewing the state of public health and furnishing advice to the Chief Medical Officer on the state of public health in the country.

In addition to its role in advising ministers about policy development, the Department of Health has a major role in policy implementation. Ministers have a number of routes through which they can create change, the two most obvious being allocation of resources and legislation. In the public health arena legislation can be used to proscribe certain activities, such as driving without a seat belt, to lay down certain standards, such as the maximum permissible levels of toxins in the environment, or to allocate duties to public bodies and officials, such as the requirement that DPHs produce annual reports on the health of the population in their districts. Legislation is the primary route through which ministers influence the public health activities of local authorities, whereas allocation of resources to health authorities is a much more direct route which can constrain the activities of those who work in the NHS.

Table 28.1. Activity analysis, environmental health departments, and distribution of staff time (technical and professional only): Average for all local authorities in England and Wales, 1985–6

Function/activity	Proportion of total staff time (%)
Housing standards, fitness, etc.	24
Air pollution control	4.5
Noise control	6
Occupational health, safety and welfare, and Shops Act	10
Meat inspection	9
Food hygiene, inspection of foodstuffs, sampling	14
Port health	1
Infectious disease control	3
Control of other public health risks (including drainage, pest control, statutory nuisance, offensive accumulations)	22.5
Health education including home safety	2
Other—animal health, public entertainment licensing, etc.	4
Total	100

Source: Environmental Health Statistics. CIPFA Statistical Information Service. SIS Ref. No. 65:87.

In addition to resource allocation ministers also have available a formal review mechanism through which they can influence health authorities. Each Regional Health Authority is reviewed by the minister on an annual basis and the regions in turn review the districts. These reviews are an opportunity for ministers to establish targets for performance and to influence priorities.

The local authority

Local authorities in England are elected bodies whose members are usually drawn from the same political parties as the national Parliament. Inevitably they are influenced by local factors as well as national trends. They are constrained by national legislation and by the fact that as much as 40 per cent or more of their money comes to them from the national government through various grants.

Public health departments in local authorities are usually called 'environmental services departments'. Table 28.1 shows how staff time is deployed within such departments.

It can be seen that roughly a quarter of the work concerns housing, a quarter concerns environmental noise, pollution, and safety outside and in the workplace, a quarter concerns the traditional public health function of dealing with drains and pests, and the final quarter concerns food hygiene, meat inspection, and port health. Only 1 per cent concerns infectious disease control.

Most of this activity is determined by legislation rather than local political priorities, and much of the work is also reactive as it is determined by public requests for service. If the public are not satisfied with the service that they receive, then some of them will complain directly to local politicians and this in turn may influence the priorities and style of public health officers. Inevitably much of the time of senior officers is taken up with the interface with politicians and the interplay with other major departments in the local authority. New services or developments of existing services are funded from the small surpluses that can be generated around the margins of the overall budgets of local authorities. In harder times it is equally important for public health officers to compete effectively with other chief officers if they are to avoid their budges being cut.

Officers operating in environmental health departments thus have access to the prerequisites for public health action. They have power, money, a management structure, and accountability, but the mechanisms which furnish these things also constrain them and to be effective officers must possess management and political skills as well as professional ones.

Health authorities

The fundamental role of the DPH and his staff at regional or district level is to give advice. It is therefore much more difficult to produce a summary table which shows how these staff use their time. If they are effective they will range over the subject matter of the whole of the NHS. It is easier to describe the activity in the terms used in the Acheson Report (and detailed on pp. 402–4 above), and that description is now accepted as the standard model.

Advisory functions are time consuming because the advice has to be soundly based. It therefore requires data, analysis, and study; it must draw on local knowledge as well as national policies and established learning within the field; it must take account of professional views and it must be practical. This means that the DPH must spend a great deal of time meeting people, listening to relevant professional and managerial views, and directing his own staff to research particular topics. Much of this time-consuming activity is formal in that there are established channels and representative bodies through which professional opinion is brought together, and the DPH is expected to attend these bodies or read of their deliberations. The DPH at regional or district level is deluged with paper soliciting his support for particular causes, demanding his attendance at meetings, plying him with fact and opinion. Among all this he must manage his time so that he still has time to think and so that his own priorities are moved forward.

In addition to his role as a chief advisor, the DPH is usually responsible for the direct provision or co-ordination of some services. Most DPHs are responsible for health education departments although there may be a District Health Education Officer (DHEO) who manages the department on a day-to-day basis. The DHEO normally reports to the DPH. Many DPHs are heavily involved in the overall direction of screening programmes although they may have staff in their departments who take day-to-day responsibility for those programmes. At the regional level and in teaching districts medical manpower matters take up considerable time, especially when a case requires some disciplinary action.

Other important organizations

The Health Education Authority is a government-funded centrally based organization which exists to carry out health education campaigns. The Authority has done its best over the years to develop an integrated strategy to health promotion but this is often frustrated by the need of the government to have high profile campaigns on issues which temporarily capture the political limelight. Over the years the Authority and its predecessor, the Health Education Council, have sponsored training in health education through bursaries. The Authority funds and publishes research on health education topics, particularly evaluations of the campaigns it has led nationally such as those against AIDS and HIV infection. It has also campaigned over the years to reduce the prevalence of smoking, cut down salt in the national diet, sponsor exercise, and many other health education topics.

There are many voluntary organizations that see public health as part of their remit. These organizations may obtain their funds through public subscription, or they may obtain them from central and local government and from health authorities. Some may obtain funds from industry, with or without strings attached. These organizations range from major charities like the Cancer Research Campaign or the British Heart Foundation which raise millions of pounds to fund research, through to pressure groups like the Public Health Alliance or the London Food Commission which seek to create national pressure for new frontiers in various aspects of public health.

The practice of public health

In this section a brief discussion of training public health professionals is followed by a discussion of the skills seen to underly public health practice and some notes on the nature of the diagnostic process in public health. The science and art of public health are then illustrated by a discussion of three major public health activities—the promotion of health, the prevention of illness, and the prolongation of life. Of course, none of these can be carried out by a single isolated practitioner of public health—all require team action and all necessarily interact with the public and politicians as well as with health authorities.

Training in public health

Training in the UK is carried out in a variety of academic settings but is overseen and approved by professional bodies who set their own postgraduate examinations. The professional bodies also approve academic and service settings for training. Control over manpower is exercised in a different way in relation to the different professional groups although, at the end of the day, it is public money that pays for the training and provides employment for those who have been trained. It is therefore important in the long run that the number of training opportunities is in balance with the expected number of jobs when due allowance has been made for the various forms of wastage.

The current division of labour in public health in England sees the medical practitioners specializing in medicine and epidemiology, the local authority environmental health officers specializing in science and enforcement, and the NHS health educators specializing in education. All three groups use social sciences and management skills.

The Acheson Committee noted that there appeared to be some tensions between those who practised in health authorities and those who worked in local government. The committee believed that there was a need to build bridges between these two arms of the public health movement (and with those with similar interests in the non-statutory areas). The Committee proposed the establishment of Schools of Public Health where practitioners destined to practise in all the various areas could train together and maintain ties through continuing education and research. At the time of writing a number of such schools are being proposed but it is too early to know whether the concept will work or can be established within the existing educational and cultural pattern in the UK.

Public health skills

Public health is a multidisciplinary art and individual practitioners' personal training and experience leads them to make differing sorts of contribution. Nevertheless all recognize a common base. Public health is fundamentally a medical and scientific subject which uses epidemiology to relate diseases to populations. Prevention requires the practitioner to understand how laws are made and enforced. Educational techniques and the social sciences relate these to everyday life.

Finally it is necessary to understand management. According to Drucker (1974), managers have five functions: they set objectives, they organize, they motivate and communicate, they measure, and they develop people. Public health practitioners must understand these managerial functions in order to run their own departments, but they are also deeply involved in these activities in the larger organizations within which they practice, particularly in setting objectives and measuring. Clear statements about the health benefits of certain public actions do a great deal to motivate public employees, set a framework for much organization, and provide a setting within which people develop, thus facilitating better management.

The diagnostic process

A description of the various skills and their organizational setting is of little use without an understanding of the ways in which these skills are integrated in practice. It is easy to describe the executive actions that may be taken by the various practitioners, but this must be preceded by a diagnostic process. The diagnosis must be constantly reviewed as time passes, both in order to be sure that the proposed action is effective and in the light of changing knowledge. This process can most easily be illustrated by comparison with medical practice. There is a common core of skill which is a continuing feature of public health medical practice. The basic medical sciences and medical disciplines lie at the heart of the diagnostic process.

The way that these disciplines integrate can be illustrated through examples drawn from the three essential areas identified in Acheson's definition of public health—health promotion, prevention of ill health, and prolongation of life. It is vital to be able to understand and set in context current knowledge about the pathology, aetiology, natural history, and treatment of particular diseases because this, together with their epidemiology, determines the most appropriate strategy for public health intervention. For example, health promotion must be attempted wherever there is knowledge of factors such as diet or life-style that may affect health, preventive strategies must be determined where the cause of illness is known, and screening must be introduced when there is a proven pre-symptomatic phase during which treatment is known to be effective. Where none of these is possible, it is important that treatment is shown to be efficacious and that resources are used effectively and efficiently.

Similar diagnostic skills are needed in the other public health disciplines. Problems do not come ready packaged and frequently fall into the category of so-called 'wicked' problems, ie, problems that change their nature whatever solution is applied. Theoretical frameworks, whether derived from previous training or from organizational structures, usually give no more than an idea of where to start. The diagnostic process is a generalist skill which must be possessed by all public health practitioners and enables them to determine where to start tackling any new problem. In order to do this, they must have a good understanding of the abilities and capabilities of public health practitioners in other disciplines and the organizational settings within which they practise. It is for this reason more than any other that the curriculum for each of the public health disciplines must cover enough of the whole subject to guarantee that the different arms of the discipline can work effectively together.

The importance of time-scales

Objective setting is a key part of public health practice, but these objectives must relate to different time-scales at different organizational levels. In the middle of an outbreak of infectious disease it is important that the officer in executive charge is tailoring his or her actions and the objectives of others to what is known about the method of spread of the disease. He or she might have to define objectives that must be met in hours or days.

Objectives for a programme to control cancers or cardiovascular disease will have to be set on a time-scale of years as far as outcomes are concerned, but there may still be short-term objectives if the programme as a whole is to maintain momentum.

At the governmental level, the UK has adopted the objectives of the WHO 'Health for All' campaign which sets targets for the next decade (World Health Organization 1985). At the present time this has not been systematically broken down into intermediate tasks which can be delegated to individuals. Unless such objectives are set and applied to appropriate time-scales in the right part of each organization, further progress with 'Health for All' in the UK will depend solely upon local inspiration and luck.

Similar issues arise in other areas of public health practice. It may be perfectly clear that major urban renewal is the only long-term solution to particular housing problems, but it may still be necessary to catch the rats in the short term if public health is to be fostered. In the educational field the same concepts apply. International collaboration and policing may eventually remove the problems of drug misuse, but harm minimization programmes may be essential locally today.

After this preliminary discussion we turn to the major activity areas which illustrate the practical application of public health—the promotion of health, the prevention of disease, and the prolongation of life.

Promotion of health

Health education has been a prominent public health activity probably for as long as public health has existed as a specialty. A national health education body was set up under the auspices of the County Medical Officers Association before the NHS was formed, and the Health Education Council and its successor the Health Education Authority have existed throughout the life of the NHS. The more modern term, health promotion, derives from a deeper study of human behaviour and the more recent understanding that education alone is not enough to change behaviour. Individuals have to learn how to take power over their own lives if the knowledge gained from education is to be changed into different individual behaviour. This concept of 'empowerment', together with the appreciation that many of the most important illnesses of the latter half of the twentieth century are caused by or related to individual behaviour, has led to the emergence of the concept of health promotion rather than health education.

The concept of health promotion is thus much wider and it involves a different relationship between those who have obtained the relevant knowledge and those who might benefit from its application. In the commonly perceived 'education model' there are two clear roles, that of information provider and information receiver, and where the information provider has furnished information his job is seen to be done and any failure to change behaviour can thus be laid at the door of the information receiver who fails to act on it. At its most extreme, this model blames the victim for failing to prevent the disease. Health education unfortunately routinely uses this extreme educational model in that the health educator only needs some understanding of epidemiology and disease processes and a little educational science to be able to deliver the message. Responsibility then lies with the receiver.

In health promotion, however, an alliance is required between the person who has need of the information and the person who has the scientific skills to acquire it. A much greater scientific breadth is required, incorporating sociology, social sciences, and psychology. At the very minimum media and presentational skills are required, but at a wider level a drama is being acted out in which it is important that the information possessors are seen themselves to act upon the information. They have an exemplar role, and this needs to be acted out by both the individuals and the organizations within which they operate. A much wider understanding of society is required as well as far greater flexibility. If the 'victim' fails to act on the information provided, the blame must rest with the provider who has failed to understand the perspective of the victim, failed to mobilize an adequate range of skill and failed to promote health.

Health promotion deliberately seeks to set up its own jargon. The words 'promotion' and 'health' are used in order to give the whole subject a positive aspect. In the past health education had become endowed with a 'thou shalt not' image, and health educators were scorned as killjoys; health promoters, however, see themselves as 'selling' a positive good rather than warning against certain harm.

Health promotion services tend to be harder to define than health education services because they cover a much wider range of activities. Most health authorities now have health education departments, and many of these would see themselves as engaged in health promotion and may even call themselves health promotion departments. They would normally expect to employ at least one professionally trained health education or health promotion officer, but in addition they may employ teachers either permanently (or on secondment from the local authority), psychologists, dietitians, and people with a background in media skills as well as graphic artists, video technologists, and many others.

Local authorities rarely have specifically labelled health promotion departments, although they may have health education sections within their environmental health service which may sometimes be called health promotion sections. However, most environmental health departments would see themselves as having a major educative role and may frequently employ specialist staff or designate particular environmental health officers with a specialist brief in education. In addition, recreation and community service departments and education departments frequently have staff with a specific health bias who may have special training and work full time in this area. Health appears as a part of the school curriculum and teachers often receive additional in-service training.

Newspapers and other media have specialized health reporters and, although part of their time is spent reporting on the activities of health authorities and other health providers and on political debates about health, most of the mass media dedicate at least part of their output to genuine health promotion, partly because it is seen as making good copy and partly because of a perceived social obligation to the greater good of society.

Many other employers outside the media engage in some health promotion through occupational health departments or through sponsorship of particular programmes such as the Look After Your Heart Campaign (Chief Medical Officer 1987), and many industries promote health in a variety of ways partly because of perceived social obligations and partly because of a belief that healthy workers will get more work done. Many trade unions are very active in going beyond narrow safety at work considerations to the broad promotion of the health of their members.

A combination of medical knowledge, epidemiology and social, media, and publicity skills, is necessary to mobilize this range of resources. One major difference from much of medical practice is that this activity has to take place in the public gaze and is frequently opposed by those who feel threatened by the knowledge that is imparted either because business interests are threatened or because they find it hard to change their individual behaviour. In such circumstances it is not unusual to find demands for an absurd burden of proof which frequently goes well beyond that which would normally be demanded by a patient of a doctor. In their normal dealings with patients doctors are expected to make a judgement and fit the best available knowledge to the best understanding that they have of the patients' signs and symptoms. It is paradoxical that some of those who are most vociferous in demanding exact proof of health promotion measures are those same doctors who are quite content with approximate solutions to their everyday clinical problems. Public health practitioners must be aware of the social and psychological processes that lead to such paradoxes and be prepared to defend approximate solutions and promote action when they believe it to be necessary, whilst at the same time seeking more precise evaluation of the results of the advice that they offer.

Prevention of illness

Prevention requires that there is at least a working hypothesis as to the cause of the disease in question. It does not require that a detailed pathogenic mechanism is understood. The classic episode of the Broad Street Pump demonstrated how it was possible to develop a theory about the cause of disease based on epidemiology and to devise an intervention strategy despite the fact that the technology to recognize the relevant micro-organism did not exist. Subsequent analysis of the cause of that outbreak following the intervention measures added further support to Snow's hypothesis about the nature of the infective process (Snow 1855).

The more resources that are required for prevention the greater will be the burden of proof required that the particular preventive measure is effective. This will normally require a detailed knowledge of the causation and natural history of the disease in question. Like health promotion, prevention requires more than a simple technical knowledge of the relevant processes because it may still be necessary to secure public co-operation with the preventive activity.

In a very few cases public compliance with a preventive measure can be obtained through a legislative mechanism that does not then depend upon public acquiescence. Probably the best example is the fluoridation of water in order to prevent dental decay. Although it is well established that the water authorities have the power to put beneficial quantities of fluoride into the water supply, the relevant local action has still not been implemented throughout the country despite the fact that in those areas where this measure has been put into place very significant benefits have proved beyond any question. Some other legal measures, such as the Clean Air Act, are now accomplished with the minimum of enforcement procedures while others, such as the wearing of seat belts, are largely effective although no particularly aggressive policing is necessary. In general terms prevention is most easily accomplished where the necessary manoeuvre is the separation of the population from the hazard and least easily accomplished where the preventive measure requires the compliance of the individual with particular advice or treatment on one or more occasions.

As was pointed out earlier, in general terms in England legal powers tend to rest with local authorities or central government whilst duties to persuade and promote health tend to rest with health authorities. The emphasis with local authorities and central government is thus on enforcement procedures and education to prevent offences being committed, whereas the thrust of health education, promotion, and prevention by the health service tends to be directed towards individual benefit and individual behaviour. In many ways the most successful measures are those which most of the population have long since forgotten about, such as the creation of an effective sewerage system, the Clean Air Act, the provision of piped water and the reduction of overcrowding. Many other measures are only noticed by those who have a specific responsibility, such as those governing catering and the preparation of foods, road safety, fireproofing of children's night clothes, reduction in the explosive content of fireworks, reduction of lead in paint and petrol, and many hundreds of regulations governing health and safety at work. It is not possible to detail all of these and, whilst the regulations concerned differ vastly in technical content, the general provision of services to enforce them is similar in that there is a framework of legislation and the appointment of specifically trained officers employed in a given public service who have powers of enforcement and a duty to educate.

Preventive services provided as health services contrast very greatly with this. They include the mandatory examination of infants by health visitors, and the continuing screening of child developmental conditions by the school medical service, the provision of immunization against infectious disease, screening for eye and ear defects, cervical cancer, and breast cancer, examination for dental disease, and in some areas screening for other conditions such as hypertension and heart disease.

None of these examinations is enforceable in the UK and no formal duty is placed upon either the service to make sure of a high level of delivery or the customer to ensure uptake. The legislative framework presumes that, by conferring on the Secretary of State the power to spend public money on these services, the state has done all that is required in order to secure individual benefit. Although the National Health Service Act 1948 and all successive acts laid on the Secretary of State the duty to improve the health of the population, this was not made explicit for authorities acting for the Secretary of State until the implementation of the Acheson Report and the issue of Circular HC88(64) at the end of 1988. Even this circular only lays upon health authorities the explicit duty to improve the health of the population. It does not require compliance with a particular level of uptake or standard of performance. However, this circular does require the DPH of each district and region to produce an annual report on the health of the population, and it must be presumed that this will include, at least from time to time, a review of the effectiveness of preventive services which can be used as performance indicators. It is hoped that this will lead to central action to specify the need for improved levels of performance either generally or in particular areas.

Control of communicable disease

Infectious disease control requires more rapid action than other public health topics. Over the years Parliament has required certain diseases to be notified to statutory authorities at the local level (see Appendix). This assists surveillance and enables action to be taken in both individual cases, and outbreaks.

Although notification is a legal requirement bearing on any doctor treating a case of a notifiable disease, it remains a fact that in practice the majority of cases of notifiable diseases are not notified probably because of lack of interest and a derisory fee. In addition, notifications still go to the local authority although the responsible medical officer is now actually employed by the health authority. Implementation of the Acheson Report requires DHAs to specify a suitably trained doctor to be in executive charge of infectious disease control. In fact this means that this doctor will be responsible for making policy and co-ordinating the action of all the various agencies who may become involved in the control of a particular disease or outbreak. This person needs leadership skills because he or she will often have to lead people over whom he or she has no formal control. The way that the statutes are framed requires the various agencies to work together and therefore to respond to the leadership of the designated doctor. However, many of these staff will also have other duties and priorities and will, therefore, be in a position to frustrate plans if the leader does not carry their confidence. The doctor nominated in executive charge of infectious disease is responsible to the DPH who has to satisfy himself or herself that arrangements are effective. The regional DPH in turn has to ensure that district arrangements are working well and also make sure that there is a mechanism to call upon if an outbreak spills over several districts.

Prolongation of life

There are many diseases in which the role of health promotion or prevention is minimal or non-existent, in which there is no hypothesis as to cause but in which it is known that life can be prolonged by treatment. In all developed countries these treatment services consume the majority of health care resources and the greatest impact on public health is obtained if these resources are used to best effect.

Public health physicians in the UK contribute to this objective in two ways—by accepting direct managerial responsibility for hospital consultants' contracts and by improving the understanding of both consultants and management of the relationship between costs, activity, manpower, and results.

Despite far greater dependence of medical services on technological advancement it is still true that the most important determinant of effective hospital services is the appointment of properly trained senior medical staff who maintain their expertise and discipline throughout their professional life. Given the importance of this body of professional expertise to the prolongation of life and consequent improvement in public health it is appropriate that senior public health doctors are called upon to exercise the authority of the state in maintaining this particular professional resource. To do this effectively requires art and skill in understanding the professional culture of medicine as well as particular expectation of individual disciplines. The formal powers and procedures available to carry out this task are cumbersome and rarely lead to the successful resolution of conflict. It is therefore important that public health physicians who have to discharge these duties know how to do so by informal mechanisms and the use of effective management skills.

On a wider level it is becoming increasingly clear that treatment resources could be managed more effectively if there was a clearer understanding of the relationship between cost, activity, manpower, and results. To draw together these different threads requires input from medical information systems as well as from financial information systems, but this must be brought into a framework that allows consideration of the disease processes under treatment and builds in an understanding of the natural history of the relevant conditions so that expectations of outcome are appropriate and the relevant time-scale is considered. It is necessary for public health practitioners to have considerable skill in understanding different taxonomies of medical conditions. For instance, the international classification of disease is at root a biological classification whereas the more recently introduced diagnostic related groups (DRGs) are a treatment classification drawing together diseases which have similar characteristics in terms of their resource use in hospital. While convenient for management purposes, DRGs are liable to sudden changes when, for example, a new and costly diagnostic procedure is found for only one member of the group. The epidemiology of the disease changes in response to quite different rules. The modern public health practitioner must be able to adapt his science to cope with these different perspectives.

More traditional epidemiological methods can be used in controlled trials and clinical epidemiology generally as well as in medical audit, and in most of these cases the role of the public health practitioner may well be to facilitate the examination by the individual clinician of his own work rather than to act as an external referee. If treatment services are to be effective then the public health practitioner must understand the clinical epidemiology of the conditions in question and the possibilities for controlled trials of treatment and for clinical audit.

Information and research in public health

Information

Overwhelmingly, information about public health is collected as a side-effect of recording the delivery of a service, whether this is in local government or in the NHS. This has unfortunate consequences. It is obviously much easier to determine from the statistics available how many houses have been renovated than how many need renovating, or how many sick people have been admitted to hospital than how many need to go. The result is that the DPH, the government, and the health authorities know much more about sickness than about health. The government attempts to remedy this defect through the activities of the Officer of Population Censuses and Surveys (OPCS) which collates process information from many sources and also collects its own information. OPCS is responsible for the national census every 10 years and for nationally recording births, marriages, and deaths (although the data are collected locally). In addition, OPCS runs a number of national surveys on an annual basis as sample surveys. These include the General Household Survey and the Family Expenditure Survey which gather much useful information which is relevant to health. OPCS also commissions some *ad hoc* surveys when ministers require information on particular issues.

Information from local sources normally relates to some publicly funded activity. Where there is national interest the form of information collection is prescribed by the appropriate government department or OPCS. Nationally required returns may be enriched locally by additional data but these will only be available locally and cannot easily be compared with similar information from other parts of the country. Local public health practitioners may collect their own information in order to guide management or report to their authorities, but this information will almost always be invisible to other parts of the system.

Academic researchers are another important source of information as their studies appear in the scientific literature. The subject of research will most often be one which is attractive to research funding bodies whose priorities are not necessarily those of the government or public health practitioners.

The ability to gather information rapidly and efficiently from these various sources is vital to practitioners of public health wherever they are employed. Time and money can be wasted on collecting information just as easily as on services that are already proved to be a waste of time. To avoid both mistakes requires skill and good organization. Following from the decision to implement the Acheson Report, the government is now taking steps to make information from OPCS much more freely available to DPHs. It is likely that this will be routinely supplied in electronic form, thus facilitating some of the work required in order to produce an annual report.

Research

There is no single identified body whose remit is to fund research in public health, and therefore in practice some important public health topics are not researched at all. A large proportion of research is funded by medical research charities, which have specific interests such as cancer, leukaemia or heart disease, or through the Medical Research Council (MRC), which is the chief route for government research money. The MRC tends to give its money to scientifically excellent but rather narrowly based epidemiological studies rather than those that are most needed.

Large intervention studies are difficult to undertake because control groups are hard to find when the media publicize the research plan as soon as the work begins, but large studies are needed if the results are to be reliable. Small studies can be mounted from local health service or local authority funds but they tend not to produce reliable results.

Some research is directly funded by the Department of Health, but this tends to be through small grants although occasionally the Department will use the vehicle of a demonstration project as a means of taking a particular policy forward. These can supply large amounts of money, for instance the Worcester Development Project funded new forms of community care for the mentally ill in order to close down a large psychiatric hospital. Some of the findings of this project then became models for care in other parts of the country.

Such funding presents a dilemma. If it is too generous others will find it difficult to implement the results from normal funds. However, if no additional money is supplied, it may be difficult to do the research at all. When a change in health care is researched in this way the end-point produces a test of political will as well as a research conclusion. In the case of community care the government provided no additional funds nor enhanced management drive and so the process moved on very slowly after the research had shown superior models of care. In a more recent example, the government provided funds for six hospitals to undertake experiments in resource management which have needed large computers and additional staff (Department of Health and Social Security 1986*b*). Even before the final results of the experiment have become available the Government has embarked on funding 50 additional sites because improved resource management is essential for the implementation of the White Paper *Working For Patients*.

Politics and public health are inextricably bound together, and so it is inevitable that political considerations will influence both the research that is done and the way that the results are interpreted. However, there is some independent funding of research and many public health practitioners continue to pursue less popular fields of research which can find their way into policy, although it can take longer than it ideally should.

Public health in the future

The Acheson Committee was conscious that it had been over 100 years since the public health arrangements in England had been examined and in its report the Committee said that it hoped it would be as long before it was necessary to do so again. The fundamental requirements for public health practice are unlikely to change, namely the need to access information, power, money, and accountability, and public health practitioners will also need to maintain their skills. The legislative framework therefore needs to be underpinned with mechanisms to accredit training and maintain standards.

The definition of public health employed by the Acheson Committee embraces the whole of society, and it seems likely that society will become more complex.The fundamental approach of public health over the centuries has been to identify local problems through epidemiological studies and then to try to prevent the import of new problems.

The definition of 'local' becomes more difficult as more people live in cities and conurbations which are increasing in size. In 1989 there were several outbreaks of Legionnaires' disease in London when infected vapour from air conditioners fell onto streets below and caused cases of infection which appeared over a wide area (Westminster 1988). Such are the travel patterns in the UK that the case-finding enquiries covered the whole country.

It is clearly impossible to anticipate all the problems that will be encountered in the future, but some are already obvious. The Single European Market may well concentrate production of manufactured foods in a smaller number of larger factories. Hygiene failures will then exert their effects over a wider area. Safe transportation of foods will be more important. Freer movement of labour may make it harder to maintain professional standards. New hazardous chemicals will continue to come on the market, and the varying economies of Europe may encourage a trend to set standards of safety at the lowest level rather than the highest. It is inevitable that the levels set will depend as much or more on intergovernmental negotiations within the European Economic

Community as on strictly scientific evidence. It is inevitable that microbiological organisms will adapt to the changes in society and 'new' diseases or new variants of old diseases will emerge.

The fundamental skills needed to combat these changes are those of the epidemiologist. This science provides a method of combating disease even when there is no clue as to the cause. This must be underpinned by information systems and managerial and political capacities to bring about appropriate action.

It is vital that all the different agencies co-operate. The thalidomide problem was detected in the FRG but experienced all over the world. A recent food poisoning outbreak was detected in the UK but the factory producing the meat product was in the FRG (Cowden *et al.* 1989).

The challenges of the future are likely to spread across the full face of society and may require action on many fronts. Some commentators found it strange that the Acheson Committee was only empowered to examine public health within the remit of the Secretary of State for Health. The Committee did not receive evidence of any major problems outside these areas other than lack of communication between different sectors. If this lack of communication is not resolved then it will be unlikely that the current arrangements could last another hundred years because there will be pressure to bring all the aspects of public health into one unitary structure.

In fact, the argument was presented to the Acheson Committee that all the arms of public health should be brought together under one national public health service. The committee did not find the argument compelling for a number of reasons. It considered that formal integration would be difficult to implement because it would require primary legislation and parliamentary time would only be made available if there were strong political arguments in favour of the change or it could be demonstrated that public health was in great danger. Neither of these were currently true.

When public health problems are analysed they usually turn out to be multifactorial and to cross administrative boundaries. Even if the Prime Minister took personal charge of public health he or she would still need skills of persuasion in order to achieve the necessary action on most large problems.

In a democratic society it will never be possible to solve public health problems through the use of some absolute power and so it may well be better that the responsibilities for action are distributed across a number of individuals and organizations. At least if one fails there is a fair chance that another part of the system will respond sufficiently to hold the problem for a while. However, in this plural system it is important that the statutes should emphasize that the accountable persons in each of the different agencies must work together.

Similar considerations apply to the relationships inside different organizations. It is common for current management structures to be led by generalist managers. Such people are selected on the basis of their general management skills rather than their particular knowledge or profession. Public health specialists have to work within these frameworks and they have to adapt to changing circumstances. Drucker (1955) has pointed out that there are often people who are essential to the survival of an organization who are not line managers. Public health specialists frequently fall into this type of role. There is an obvious analogy with the relationship between the pilot and the navigator of an aeroplane or the conductor of an orchestra and the composer of the music. The generalist may be essential to driving the aeroplane or getting the orchestra to play, but there also has to be a course to fly or music to perform. Setting that course or writing that music is frequently the role of the public health specialist.

As the legislative changes to health services in England progress we may well see further development of this type of role. The trend seems to be towards public servants buying services from more independent agencies. This may happen in health services, but it also happens in other public-health-related fields such as refuse collection and waste disposal. The public health specialist of the future may be more involved in analysing public health needs and buying services to meet them rather than managing services. If similar trends occur in local authorities as well as health services it may become easier to bring these services together in the future.

Conclusion

The main question that arises is how public health practitioners can best promote the organized effort of society on which public health depends. The longest established tradition in public health practice is that of a report issued at regular intervals on the health of a defined population. If the public health practitioner makes it his business to inform himself as best he can about those things which influence the health of his population and makes that contribution visible to those responsible for organizing society, then it is possible to turn almost any aspect of society into a public health service and to discourage those aspects and activities that are public health hazards. It is important that there is a sound scientific basis at the root of such reporting. Data must be collected accurately and analysed correctly. It is also important that such reporting and reviewing draws together science and art. Clear communication is essential and careful judgement must be used to determine the appropriate way to publish reports and how to publicize them when published. Judgement must be exercised as to how strong or strident a particular report should be and how specific its praise or criticism.

Reporting alone does not create change but it may help to create a climate in which change is possible. Public health practitioners have to understand the management of change and have a bias for action if they are to convert their analysis into new results. The most simple analysis of the change process suggests that change only takes place when there is some pain or dissatisfaction with the present, where there is some clear vision of a different future, and where an affordable first step towards that future exists. The science of public health practice can show the pain in the present, and can help to suggest a different future and map the possible affordable steps towards that future. The art of public health medicine is to be able to bring those three elements together and lay them in front of those who organize activities that could be of service to the public health authorities so that they may change them into public health services. Practicing this art is difficult and change even for the better is frequently resisted. The practitioners of public health must also know themselves; they must know how to call on inner resources to sustain themselves when change is resisted even when they have demonstrated that it would be in the best interest of the public health. It is important therefore that practitioners of public health should have an understanding of the psychological needs of those who attempt to create change, should know how to set up support mechanisms for their own staff and themselves, and should try to develop clear insight which reveals what is possible.

While the arrangements for training public health practitioners in the UK no doubt still have weaknesses they clearly cover everything needed for practice today, and all concerned are well aware that continuous updating and improvement are essential. The Government's response to the Acheson Inquiry combined with the changes that the White Paper will bring offer a secure base for practice, although it will be practice in a sea of changes. Overall the public health services in the UK seem set for a bright future.

Appendix: notifiable diseases

The following diseases in England and Wales are at present subject to statutory provision requiring notification:

Acute encephalitis	Measles
Acute meningitis	Ophthalmia neonatorum
Acute poliomyelitis	Paratyphoid fever
Anthrax	Plague*
Cholera*	Rabies
Diphtheria	Relapsing fever*
Dysentery (amoebic and bacillary)	Scarlet fever
Food poisoning*	Smallpox
Infective jaundice	Tetanus
Leprosy	Tuberculosis
Leptospirosis	Typhoid fever
Lassa fever	Typhus*
Malaria	Viral haemorrhagic fever
Marburg disease	Whooping cough
	Yellow fever

Separate statutory provisions and regulations apply to Scotland and Northern Ireland.

* Notifiable under Sections 10 and 11 of the Public Health (Control of Disease) Act 1984.

The other diseases listed are required to be notified by virtue of the provisions of the Public Health (Infectious Diseases) Regulations 1968 as amended.

References

Badenoch, J. (1986). *First report of the committee of inquiry into the outbreak of Legionnaires' Disease in Stafford in April 1985*. HMSO, London, Cmnd 9772.

Chief Medical Officer, Department of Health and Social Security (1987). *Look after your heart. (A joint DHSS/HEA campaign)*. Circular CMO (87)6, Department of Health and Social Security, London.

Cowden, J.M., O'Mahoney, M., Bartlett, C.L.R., *et al.* (1989). A national outbreak of *Salmonella typhimurium* DT 124 caused by contaminated salami sticks. *Journal of Epidemiology and Infection* **103**, 219.

Department of Health and Social Security (1986*a*). *The report of the committee of inquiry into the outbreak of food poisoning at Stanley Royd Hospital*. HMSO, London, Cmnd 9716.

Department of Health and Social Security (1986*b*). *Health services development. Resource management*. Circular HN(86)34, Department of Health and Social Security, London.

Department of Health and Social Security (1988). *Health of the population: Responsibilities of health authorities*. Health Circular HC(88)64, Department of Health and Social Security, London.

Drucker, P.F. (1974). *The practice of management* Heinemann, London.

Secretary of State (1988). *Public Health in England. The report of the Committee of Inquiry into the future development of the public health function*. HMSO, London, Cmnd 289.

Secretaries of State (1989). *Working for patients*. HMSO, London, Cmnd 55.

Snow, J. (1855). *On the communication of cholera*. London, Churchill.

Westminster (1988). *Westminster Action Committee. Broadcasting House Legionnaires' Disease, London*. Westminster City Council, London.

World Health Organization (1985). *Targets in support of the European Regional Strategy for Health for All*. WHO Regional Office for Europe, Copenhagen.

29

Public health services in Japan

NOBUO ONODERA

Introduction

A description of the historical development and background of public health services in Japan is necessary for a good appreciation of public health services as they exist in the country today. There have been dramatic changes in public health services during the long history of Japan. A practical review of these changes, in the modern period will be useful in developing public health services and in ensuring the social equitability of such services in the twenty-first century.

Development of public health services in Japan

With the rapid expansion of international trade during the Meiji era (1868–1912) there were frequent outbreaks of infectious diseases of foreign origin, such as cholera and smallpox. Public health services were then under the control of the government, whereas medical care services were essentially in the hands of private practitioners. Japan underwent an industrial revolution during the latter half of the Meiji era; and with it environmental sanitation and occupational health became the important tasks of the government.

The Taisho era (1912–26) and the early years of the Showa era were periods of mounting social tension as a result of the economic chaos brought about by the world recession and the great Kanto Earthquake of 1923. During this period, various laws on public health and social services were enacted in order to cope with social needs and changes in disease patterns. In 1927 the first social health-insurance programme was initiated for industrial workers, and a number of health consultation offices were opened for these workers. This eventually led to the enactment of the Health Centre Law in 1937, which had the aim of creating a network of health centres throughout the nation.

In 1938 the government, with the aim of promoting health administration, social welfare, and labour services separated these activities from the Ministry of Home Affairs and created an independent Ministry of Health and Welfare. In the same year, the Institute of Public Health was established as an attached institute of the Ministry of Health and Welfare, with the purpose of providing advanced training and education to those engaged in public health, conducting research in public health, and participating in public health activities. In consideration of the distressing social and economic circumstances of Japan at the time the Rockefeller Foundation in the United States generously donated funds for the buildings and facilities of this Institute.

During the Second World War Japan's national policy was geared to promoting the health of its people and military forces in order to strengthen its war effort; but in 1945 the country was obliged to capitulate. The war brought with it an almost complete obliteration of the existing systems, the loss of almost two million precious lives, and the tragic destruction of Hiroshima and Nagasaki by the first nuclear weapons used in warfare.

Public health services during the early post-war period of confusion (1945–50) had to cope with the challenges of such epidemics as smallpox, typhus, cholera, and malaria, and with the problem of malnutrition. The first priorities of the public health services were tuberculosis control and the promotion of maternal and child health.

During the following period of rapid economic development (1960–73) a serious pollution problem arose in Japan as a result of rapid industrial development and growth, and therefore systematic and organizational measures were employed in public health services for the control of environmental pollution. Rapid social changes accompanying modernization and urbanization brought about various new problems and the need for developing new policies. As a response to the increased incidence of chronic degenerative diseases and problems arising from the ageing population, a need arose for comprehensive public health services, with their emphasis given to the promotion of health, prevention, the early detection and treatment of diseases, and rehabilitation.

And then during the period of steadier development following the oil shock of 1973 there was a demand for the development of more efficient and rational public health services which could be more reasonably supported by the different age groups and would be more equitable socially for the coming of the twenty-first century.

Figure 29.1 shows the developmental process of the

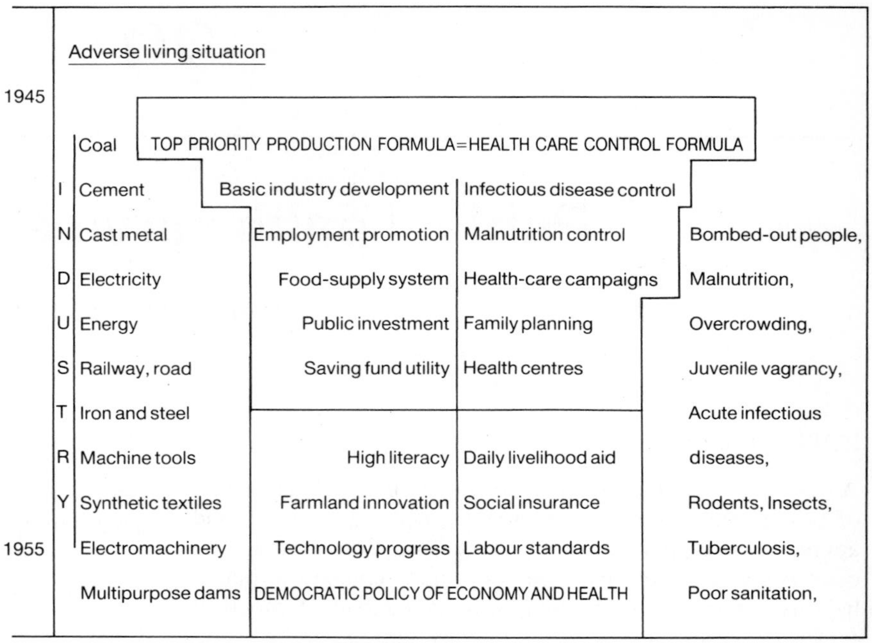

Fig. 29.1. Economic and health development process in Japan.

economy and of public health in Japan after the end of the Second World War, and suggests that there are items in common between the government measures aimed at fostering economic development and those aimed at improving public health. In order to cope with the extremely unfavourable post-war economic situation, the government measures aimed at fostering employment, securing energy sources, and producing basic construction materials were directed toward the development of basic industries engaged in the production of coal, cement, cast iron and similar primary materials, while the government measures for public health were focused on the control of malnutrition and the infectious diseases prevalent at the time. This common formula employed in Japan for economic development and for public health, which led to the dramatic socio-economic advancement of the country in the present century, can serve as an important key in analysing the process of Japan's remarkable socio-economic development to its present front-ranking position.

Present status of public health services in Japan

Present health status of the population

Accompanying the rapid socio-economic development of recent years, remarkable changes have been observed in the health status of the Japanese people. But though the health statistics of Japan have attained the highest level in the world, it remains a fact that many health hazards still exist, and many demands regarding health of the people remain unfulfilled.

Japan is one of the countries of the world with the most rapid changes in vital statistics. During a period of one decade after the war this country was transformed from one of a high birth-rate and a high death-rate to one of a low birth-rate and a low death-rate.

The birth-rate in Japan has decreased by half in recent years, with the rate dropping from 28.1 per 1000 population in 1950 to 11.1 per 1000 in 1987. The crude death-rate has dropped from 10.9 per 1000 population in 1950 to 6.2 per 1000 in 1987, while the adjusted death-rate showed a remarkable decline from 10.8 in 1950, through 6.9 in 1960, 5.2 in 1970, 3.6 in 1980, and 3.1 in 1985, to 2.9 in 1986. The infant mortality rate has steadily declined from 60.1 per 1000 live births in 1950 to 5.0 in 1987, with a neonatal mortality rate of 2.9 in 1987, while the stillbirth rate has decreased by half, from 84.9 per 1000 total births in 1950 to 45.3 in 1987. (However, the highest stillbirth rate, of 101.7 was observed in 1961, with the rate exceeding 100 from 1957 to 1961.) The perinatal mortality rate has also decreased fairly steadily from 46.6 per 1000 live births in 1950 to 6.9 per 1000 in 1987, with however a temporary reduction in the rate of decline discernible in the late 1950s and early 1960s, which parallels the higher figures for the stillbirth rate over the same period that has just been referred to.

The mean life-expectancy at birth has increased from 42.06 years in 1921–5 to 59.57 years in 1950–2 and 75.61 years in 1987 for males, and from 43.20 years in 1921–5 to 69.97 years in 1950–2 and to 81.39 years in 1987 for females.

In 1940 the principal cause of death was tuberculosis, with a death rate of 212.9 per 100 000, followed by pulmono-bronchial diseases, with a death-rate of 185.8 and cerebro-vascular diseases, with a death-rate of 177.7. In 1950 the principal cause of death continued to be tuberculosis, with a death-rate of 146.6, followed by cerebrovascular diseases, with a death-rate of 127.1, and pulmono-bronchial diseases, with a death-rate of 93.2. However, in 1987, the leading cause of death was malignant tumours, with a mortality rate of 164.2, followed by heart diseases, with a death-rate of 118.4, and cerebrovascular diseases, with a death-rate of 101.7. Deaths attributable to the three principal causes, the so-called adult diseases, accounted for more than 60 per cent of all deaths.

The disease prevalence rate ascertained by the annual National Health Survey has shown an increase from, 63.6 per 1000 population in 1965 to 287.8 (in-patients 7.8, out-patients 224.9, others 55.1) in 1986. When classified by disease, the disease category of the highest prevalence was cardiovascular diseases, followed by respiratory diseases. The disease prevalence rate by age showed a decrease to a minimum of 153.1 per 1000 in 1986 at the ages of 15–24 years, but increased with age thereafter, to 664.1 per 1000 at age 65 and over in 1986.

According to the National Patient Survey, during the last three decades there has been a remarkable increase in the rate of patient admissions (in-patients and out-patients in medical institutions) per 100 000 population. In 1948 the rate was 1306 per 100 000 population (175 in-patients and 1131 out-patients), which increased to 6403 (1118 in-patients and 5285 out-patients) in 1984, indicating that one out of every 15.6 persons was under medical care at a medical institution as of the date of the survey.

Organizational structure of public health services

Public health administrative structure at the central level is based on the National Administrative Organization Law, with the competent ministries being the Ministry of Health and Welfare, the Ministry of Education, the Ministry of Labour, and the Environment Agency. Public health administrative structure at the local level is based on the Local Autonomy Law, and the regulations for the establishment of such structures at the local level must be approved by the local assemblies.

Health centres have been established (851 of them by 1987 both in 47 prefectures directly under the Health Centre Law and in 32 ordinance-designated cities and 23 wards of the metropolice of Tokyo under the enforcement ordinance of the same law. Health centres are established as a rule at a rate of one per 100 000 population, with consideration given to transportation conditions, relation to other government offices, the condition of public health, and the distribution of population, for the purpose of improving and enhancing public health within their respective jurisdictions. Health centres are financed by local public expenditure and from subsidiary funds provided by national government, and are classified into five types: U (urban) type, UR (urban–rural) type, R (rural) type, and L (large area with small population) and S (small area with less than 30 000 population) type.

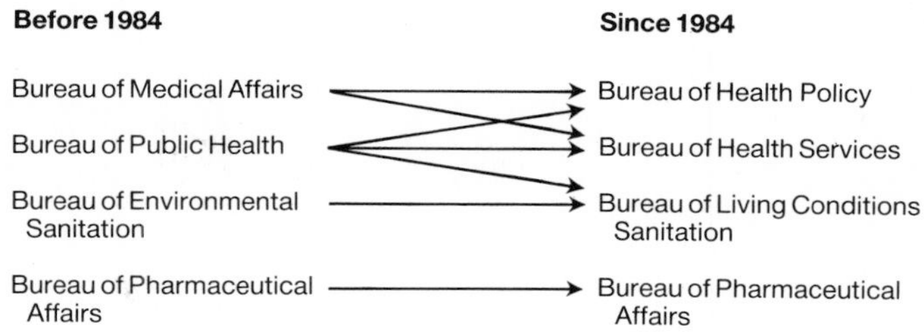

Fig. 29.2. The 1984 reorganization of public health services in Japan.

In 1984, the Ministry of Health and Welfare reorganized the structure for public health services at the central level for the enforcement of health policies and the effective integration of public health services. The nature of this reorganization is shown in Figure 29.2.

Health administration in Japan is basically divided into (1) general health administration; (2) school health administration; and (3) industrial health administration. Figure 29.3 shows the organizational structures for community health, school health, and industrial health from the central level to the local level. The area of responsibility at each level is spelled out to permit functional co-ordination in community activities.

Human development for public health services

In 1946, after the termination of the Second World War, educational training courses for public health services were resumed at the Institute of Public Health in order to cope with the needs arising from the poor post-war health conditions. Various refresher courses of short duration were made available for key public health officers and for public health personnel needed to staff the reorganized health centres in the free and democratic nation.

A course for public health nurses was commenced in April 1947, followed in 1948 by six special courses for doctors, pharmacists, veterinarians, sanitary engineers, health technicians, and others. After 1948 courses were initiated with a curriculum designed for public health statisticians and health educators assigned to health centres and for laboratory technicians of prefectural public health laboratories. In April 1949 a regular one-year course for graduates of medical schools was resumed, together with six-month courses for veterinarians, pharmacists, and dietitians.

The Institute of Public Health, relying on its wealth of experience over the years in educational and training programmes on public health and in the changing patterns of public health services, has trained to date more than 22 000 personnel engaged directly in public health or in public health research.

Development of information systems in public health

The advance of information and communication systems using computers is a remarkable development in every field

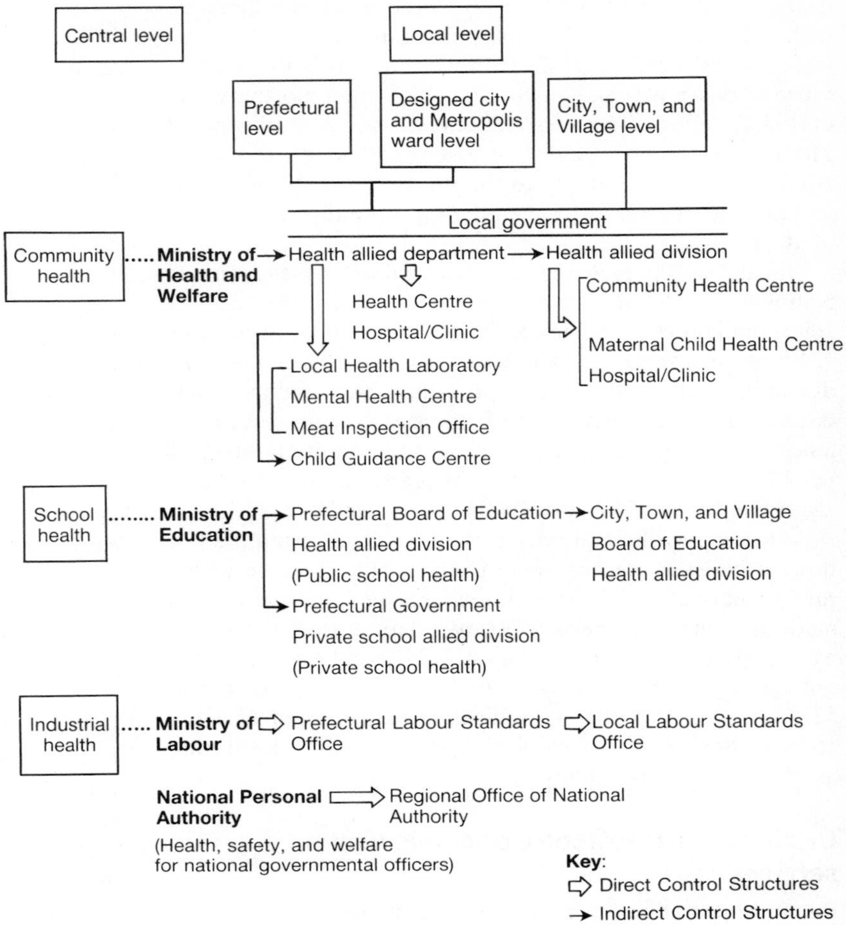

Fig. 29.3. Organization structures from the central to the local level.

of industry and service. In the field of public health services, the use of computers has had a great effect on information-processing in Japan.

The Ministry of Health and Welfare had originally established the Statistics and Information Department in the Ministry's Secretariat in 1949, since when this department has served as the national centre for health and welfare statistics and information.

In order to cope with the need for high technological development in the information systems required for the promotion and reinforcement of public health services the Ministry of Health and Welfare organized an expert committee for health and medical information systems in 1973, and set up a Medical Systems Development and Research Office in 1974.

To achieve governmental development goals for health and medical information systems on a basis of interdisciplinary approaches through the co-operation of public and private sectors, a legally independent foundation, Japan Medical Information Systems Development Centre (MEDIS-DC) was established with the joint approval of the Ministry of Health and Welfare and the Ministry of International Trade and Industry in 1974. The activities of MEDIS-DC consist mainly of the following elements:

1. Community Health and Medical Information Systems
   e.g. Health and Welfare I.S. (Information Systems), Health Care I.S., Health Institutions Linkage I.S., Emergency Medical I.S., and Regional Health and Medical Plan Support I.S.

2. Hospital Information Systems
   (a) Hospital Services Information Systems
       e.g. Pharmaceutical Substances I.S., Radiological Meter Calculation I.S., Meal Supply I.S., Clinical Records Management Systems
   (b) Hospital Management Information Systems
       e.g. Hospital Beds Management System, Bill Account System, Patient Registration I.S.

3. Medical Information Service Systems
   e.g. Medicines Information Service System, Resistance Bacillus Information Services System, Kidney Transplan-

tation Information Service System, Electro-cardiograph Telecommunication Analysis Service System, Medical Books and References Information Service System.

With regard to the application in public health services of network systems using computers and highly advanced telecommunication facilities, the surveillance of tuberculosis and infectious diseases through nation-wide channels of information collection and analysis sytems has been organized in practice by the Ministry of Health and Welfare, with the co-operation of every health centre and regional institute of health laboratory, each operating in local autonomy, and of approximately the 4000 designated medical facilities for infectious diseases reporting. This surveillance system has a very important role in the rapid collection and analysis of data concerning the rapid response necessary for a tuberculosis and infectious diseases control system, together with the additional fundamental and valuable approaches to public health services which it permits.

Public health clearly has a need for more detailed information-processing concerning the effects of its services, with a view to establishing a suitable health policy and strategy. During 1986 and 1987 scientific research in the form of a special research undertaking by the Ministry of Health and Welfare was conducted with an interdisciplinary approach on the subject of a 'Basic Study of a Network High Advanced Comprehensive Information-Processing System in Health and Medical Services'. Through this study, the surveillance system for tuberculosis and infectious diseases had a chance to attain effective implementation nation-wide, as well as to investigate the expected prospects, and a suitable system of integration of the channels of demand and supply, for information-processing systems on public health services.

Future prospects of public health services

Generally speaking, regional development of public health services in Japan has a long history of good relations between local participation in community activities and support from the government in such activities, though in a somewhat vertical structure. As for the health infrastructure of the country, it should be noted that grass-roots organizations at both the rural and urban levels were instrumental in meeting the needs of daily community life. These organizations, voluntary in nature, engaged in a variety of community activities, such as collecting town association fees on a household basis, carrying out various programmes such as recreation, community health festivals, and cleaning campaigns, and circulating notices issued by the local government authority. This active participation and involvement of the local residents in a broad spectrum of community public-health programmes was closely associated with the development of primary health care, such as village activities for health policy, activities of women's associations for the promotion of health, and women's campaigns for tuberculosis control and health education activities.

These programmes will be closely associated with the development and provision of primary health care in the twenty-first century. The fundamental policy for public health services should be established on the basis of the real health needs of the residents and of an action plan which takes into account these various levels of health needs. It is thus important to create effective organizations and functional structures for primary, secondary, and tertiary health-care systems in the community by the integration of social resources with existing infrastructures such as social insurance, welfare services, educational systems, labour standards and employment policies, communications and transportation, and local industrial development. Comprehensive health-care systems should promote a wide range of activities, such as promotion of health, prevention of diseases, medical care, and rehabilitation at all levels in the community, in schools, and in industry, and also the development of international health services.

For these organizations and structures to interact effectively, the three elements of a systematic approach, the scientific background, and the principles of human concepts and behaviours should be mutually linked for self-control systems and open systems. This is schematically presented in Fig. 29.4. The first package is the public-health and medical-care approach at the mass and individual levels; the second package is the socio-technological approach at the social and technological levels; and the third package is the contingency approach at the environmental and organizational levels. Although the principles of human concepts and behaviours for self-control systems and open systems involve a large variety of scientific backgrounds, the public-health and clinical approach, the socio-technological approach, and the contingency approach are all applicable to public health services in the future.

These approaches consist mainly of the following factors related to public health services:

Social level factors	**Technological level factors**
(a) Health planning	(a) Health information systems
(b) Health management	(b) Health engineering
(c) Health economy	(c) Health science research
(d) Health education	(d) Health examination systems
(e) Health legislation	(e) Health assessment and evaluation

The contingency approach of package 3 constitutes the valid and adjusted strategies between the environmental level and the organizational level. The environment around public health services can be classified into general, task, intersectional, and creative environments from the standpoint of development of comprehensive health-care systems. Better adjustments of the environment and organizations would more effectively and efficiently promote activities

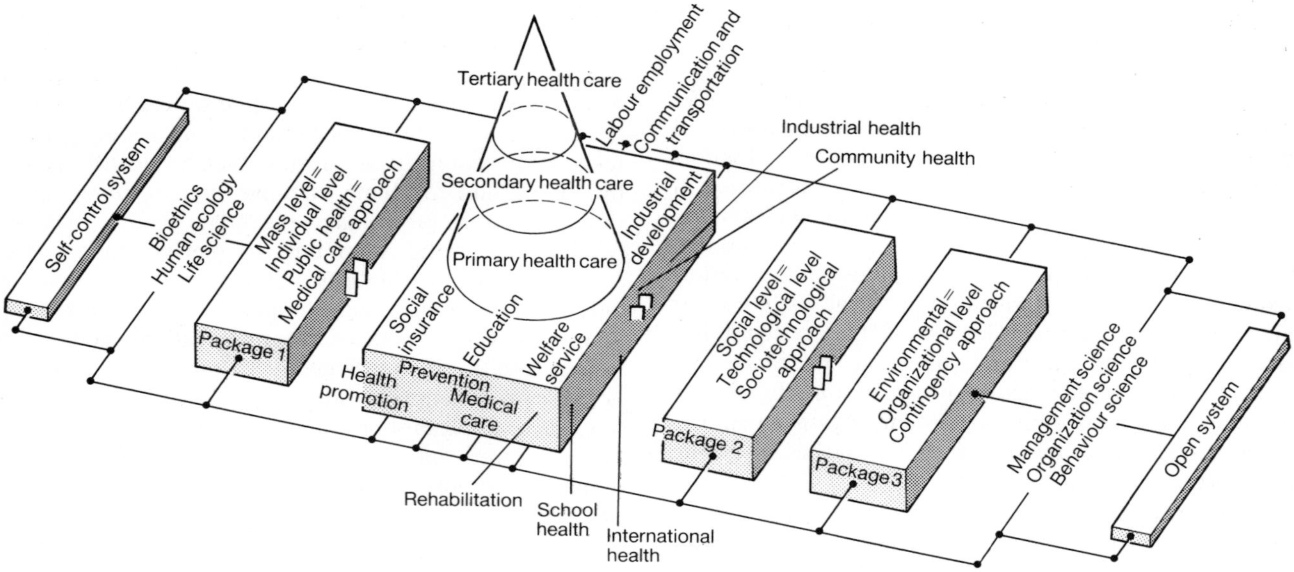

Fig. 29.4. Comprehensive health package system for primary health care.

utilizing the social resources in the community. The main focal points of the contingency approach for linking social resources with the consensus of the public can be considered as follows:

Contingency approach	**Consensus of the public**
(a) Perception of environment	(a) Promotion of health motivation
(b) Health decision-making	(b) Consciousness of integration
(c) Potential reconstruction	(c) Identification of high-risk groups and problem-points
(d) Continuity and stability	(d) Greater collaboration among health-service providers
(e) Integration and reasonable use of specialty	(e) Cultivation of teamwork and network systems
(f) Establishment of functional organization and structure	(f) Comparative analysis and information exchange

Public health services have made gigantic strides in the development of a comprehensive health care system during the past forty years since the end of the Second World War. During this post-war period the health pattern of Japan has shown drastic changes, beginning from the epidemics of acute infectious diseases such a smallpox and typhus and the problem of malnutrition. The people engaged in a long battle against tuberculosis, a disease of high prevalence and mortality rate, and continued to work towards improving the poor sanitary conditions that prevailed. Furthermore, public health services had to cope with episodes of serious pollution problems brought about by rapid economic and industrial growth and development, which eventually led to a dynamic control of environmental pollution. Also, in the face of the rapid changes in population structure and distribution, provisions had to be introduced to cope with degenerative diseases, mental health problems, medical-care problems developing in remote areas, and emergency care. Fortunately, public health services in Japan have been successful in overcoming these various difficult problems in health, as well as problems associated with the rapid economic and industrial development during the post-war period.

Needless to say, the most important problems in public health services in Japan can be said to be those associated with the rapid ageing of the population and related effects, changes in the disease pattern, increasing demand for medical care and welfare services, and limitation in social resources. These indicate the very important role that public health services must play, and the responsibility they have in comprehensive health-care systems.

In addition, Japan must devote greater effort to international co-operation and co-ordination in the field of public health. The focus of attention of international public health services is directed towards the development of primary health care in various countries in the world. The late Dr Haruo Katsunuma, Professor emiritus of the University of Tokyo, made the following remarks with regard to the position and nature of primary health care in a report entitled 'Primary Health Care in Japan':

'Today, primary health care is much at the focal point of

debate in WHO and throughout the world. Primary health care must be flexible so that it can be adjusted to the specific and varying needs in different countries.'

These many years of experience of public health services in national tuberculosis control, maternal and child health care, improvement in nutrition, infectious disease control, improvement in sanitation and pollution control, health-information systems and documentation, the activities of public-health nurses, and health care of the aged will be extremely valuable and useful in developing comprehensive public health services in terms of policies and strategies for public health systems and human development for the coming twenty-first century.

References

Hashimoto, M. (1977). A case study on health planning method in Japan—Comprehensive community health planning—Bull Institute. *Public Health (Tokyo)* **26**, (3, 4), 125–41.

Ministry of Health and Welfare (1985–7). *Health and Welfare Services in Japan.* Ministry of Health and Welfare, Tokyo.

World Health Organization (1978). Report of an International Conference on Primary Health Care, Alma-Ata, USSR, 6–12 September 1978. WHO, Geneva.

Katsunuma, H. (1981). Some topics from industrial health in Japan, American Industrial Hygiene Association, Journal Yant Memorial Lecture, 1981.

Onodera, N. (1985). *Future provision and present status of health services in Japan: The effectiveness of Nursing Education Programme for health for all*, IG 11–INFJ 85. The International Nursing Foundation of Japan.

Southeast Asian Medical Information Centre, Tokyo (1985, 1986). *SEMIC Health Statistics, 1985, 1986*. Japan International Medical Foundation, Tokyo.

30

Public health services in Australia

SIMON CHAPMAN and STEPHEN LEEDER

Australia, with a population in 1988 of some 16 million, is a federation of six states (New South Wales, Victoria, Queensland, Western Australia, South Australia, and Tasmania) and two Territories (the Northern Territory and the Australian Capital Territory), with a central (Commonwealth or Federal) government seated in Canberra. All nine jurisdictions—State, Territory and the Commonwealth—have departments or commissions of health. In the public health sphere, the Commonwealth Department of Community Services and Health's (CDCSH) principal functions are to control quarantine services, to maintain a communicable diseases intelligence service, and to fund—usually on a dollar-for-dollar basis with the States—numerous community health projects, including the Home and Community Care programme. It also serves as the secretariat to most national meetings on issues such as AIDS, drug and alcohol control. However, its main roles, via the Health Insurance Commission, are to administer the universal health insurance scheme, Medicare, and the linked Pharmaceutical Benefits Scheme, which provides rebates for a wide range of commonly prescribed drugs, and to distribute finance to the States and Territories to implement health services such as public hospitals and nursing homes.

States and Territories retain responsibility for all major implementation aspects of public health such as health and food inspection, the administration of radiation control and other environmental health standards, and the funding of State-wide and local area projects and campaigns. The latter vary from State to State, depending on differing perceptions of priorities, although nearly all States in recent years have run large-scale smoking control, immunization, and drug and alcohol abuse prevention campaigns. State and Territory health departments receive their budgets from their respective State or Territory treasuries, which in turn receive almost all of their funds from the Commonwealth government. There are few constraints or conditions applying to the manner in which State and Territory health departments disburse their budgetary allocations and, consequently, sometimes wide disparities occur among States in commitments to public health objectives and related programmes.

In this chapter, we describe and comment on the current and planned organization of public health policy determination, programming, research, and training. At the end of the chapter, we illustrate some of the more important themes underlying these areas through two case studies: smoking control and improved nutrition. Before proceeding, we will briefly describe some of the main tasks that face public health in Australia.

Health status in Australia

Two hundred years after white settlement, life expectancy of the Australian population as a whole compares well with that in countries of similar economic development, falling only behind Iceland and Japan. In 1986, life expectancy at birth for males was 72.77 years, and for females was 79.13 years (Australian Bureau of Statistics 1988). The infant mortality rate, 103.6 per 1000 live births at the turn of the century, has fallen to 8.8 per 1000 in 1986 (Commonwealth Bureau of Census and Statistics 1909; Australian Bureau of Statistics 1987a). With the exception of deaths from cancer and AIDS, rates for most leading causes of death are falling, often quite dramatically. Table 30.1 shows declines in mortality rates between 1972 and 1986.

Inequalities in health

However, these encouraging national statistics camouflage large differences in mortality, and presumably health, within the population; wide variations in the prevalence of some major disease risk factors and health-related preventive behaviours for which information is available; and, most importantly, the social and economic circumstances associated with these risk factors and behaviours (Health Targets and Implementation Committee to the Australian Health Ministers 1988, pp. 75–90). The quarterly poverty line calculations, supplied by the Institute of Applied Economic and Social Research, estimate that 2.7 million (many of them old) of 16.4 million Australians live below or just above the poverty line and that nearly one in five children lives in a

Table 30.1. Death rates for major and other causes by sex (per 100 000 population) 1972 and 1986*

	Males		Females		Persons	
	1972	1986	1972	1986	1972	1986
Ischaemic heart disease (410–4)	306	226	204	174	255	200
Other heart disease†	43	38	54	50	48	44
Lung cancer (162)	46	54	8	17	27	36
Breast cancer (female 174)	—	—	24	28	12	14
Cervical cancer (180)	—	—	6	4	3	2
Melanoma (172)	3	5	2	3	3	4
Other cancer (remainder 140–208)	113	137	89	100	101	118
Stroke (430–8)	101	63	141	93	121	78
Obstructive airways disease (490–6)	50	48	12	21	31	35
Motor vehicle accidents (E809–10)	38	27	15	11	27	19
Falls (E880–8)	8	5	11	6	10	6
Other accidents (remainder E800–949)	23	14	6	5	15	10
Suicide (E950–9)	17	19	8	6	13	12
Mental disorders (290–319)	6	10	6	10	6	10
Infectious and parasitic diseases (1–139)	7	4	6	3	6	4
Diabetes (250)	13	11	16	13	14	12
All other causes	162	118	145	113	153	115
All causes	938	778	752	658	845	718

* International Classification of Diseases codes in brackets.
† (393–8, 402, 404, 415, 416, 420–9).
Source: Australian Bureau of Statistics (1972, 1986).

family that depends on a pension or benefit as its sole source of income (Brotherhood of St Laurence 1987).

Narrowing health status differentials between population groups within Australia is fast becoming a central preoccupation for the public health fraternity, with a recent major Government report commissioned under the auspices of the World Health Organization's Health For All programme (see below) recommending 'that the reduction of inequalities should become a major goal of projects aimed at health promotion and disease prevention . . . [and that] actual differences in health status between groups in Australian society should be reduced by at least 25 per cent [by the year 2000]' (Health Targets and Implementation Committee 1988, p. 14). The 1987 annual conference of the Public Health Association chose social justice as its theme, reflecting a professional consensus that the reduction of inequalities in health status is perhaps the most outstanding goal now facing public health in Australia. Here, briefly, are some of the main dimensions of this inequality.

The health of Aborigines

The Aboriginal population (those describing themselves as having Aboriginal or Torres Straight Islander origins) was counted at 227 645 (1.46 per cent of the total population) in the June 1986 national census. There is no starker contrast between the extremes of health status to be found in Australia than that between Aborigines and other Australians. Recent data show life expectancy for Aborigines in Western Australia and the Northern Territory to be 15–20 years less than for non-Aboriginal Australians in the mid-1980s (Australian Institute of Health, unpublished data). Even their best levels of life expectation—61 years for males and 65 years for females living in the Kimberley district of Western Australia in 1983/4—are 11–14 years less than those experienced by other Australians (Lee *et al.* 1987). The lower expectations of life at birth for Aborigines largely reflect the much higher mortality experienced by young adults. The peak excess in deaths occurs in the 25–44-year age range. Circulatory and respiratory system diseases together account for 40 per cent of excess deaths.

Aboriginal infant mortality is also unacceptably high. Rates of 25–30 deaths per 1000 live births are about three times higher than that of non-Aboriginal Australians (Lee *et al.* 1987). Thomson (1984, 1985, 1986) has reviewed the litany of illness and deprivation experienced by Aborigines. Leading health problems include undernutrition in infants and obesity in adults, low birth-weight, communicable diseases including diarrhoea, trachoma, skin infestations, sexually transmitted diseases and hepatitis B, hypertension, alcohol abuse, petrol inhalation, and trauma from violence and motor vehicle injuries.

The poor health of Aborigines is associated with their low social and economic status, evidenced by indicators such as education, employment, income, and housing. In the 1981 census, 11 per cent of Aborigines reported never having attended school, compared with less than 1 per cent of non-Aboriginal Australians. Unemployment in the mid-1980s among Aborigines was about 50 per cent—six times higher than in the general population—and in places reached 90 per cent. In consequence, the median income of Aborigines is about 57 per cent that of other Australians. In 1985, it was estimated that 17 600 dwellings were required to overcome the perceived backlog in the provision of adequate housing for Aborigines (Thomson 1985).

Aboriginal health is widely considered to be a scandal of neglect although, because of its social and economic origins, it is a field that will not admit simplistic solutions. A process is now in operation for the development of a major new national plan for the improvement of Aboriginal health, and a National Aboriginal Health Strategy Working Party has been established by the Commonwealth Departments of Aboriginal Affairs and Community Services and Health and the State Departments of Health, to report in 1989.

Socio-economic status and health

Age-adjusted mortality data for Australia show that for every death from any cause in a professional or executive person, two deaths occur in unskilled or manual labouring workers (McMichael 1985). Smoking shows a similar pattern: lower blue-collar male smoking rates are 34 per cent higher, and in females are 44 per cent higher, than those in upper white-collar workers (Hill and Gray 1984). Although databases are often imperfect, available sources suggest that similar differentials apply to most of the leading diseases and accidents (Smith and Lee 1987), to nutrition (English and Bennett 1985), to hypertension (Dobson *et al.* 1985), and to immunization rates (Australian Bureau of Statistics 1985).

Sex differences in health status

Men in Australia die from nearly all non-sex-specific causes at much higher rates than do women (Table 30.2), whereas women have higher morbidity rates (measured by doctor consultation rates) for all causes except injuries (Table 30.3). Many of these differences suggest modifiable aetiological factors associated with sex role socialization and the division of labour, and with inequalities of opportunity. As such, the excess burdens of death and illness borne by men and women respectively represent obvious challenges to public health and, equally, illustrate how the policies and practices of other Government portfolios (e.g. education, media regulation, equal opportunity provisions across all areas) are of immense public health consequence.

Table 30.2. Age-specific death rates for leading causes of death in each age group, Australia 1986 (per 100 000 population)

Age group and cause of death*	Males	Females	F/M
1–14 years			
All causes	35	24	1:1.5
Accidents, poisonings, and violence (E800–999)	18	11	1:1.6
Motor vehicle accidents (E810–19)	9	6	1:1.5
Cancer (140–208)	5	4	1:1.3
15–24 years			
All causes	129	48	1:2.7
Motor vehicle accidents (E810–19)	58	19	1:3.1
Accidents, poisonings, and violence (E800–999)	101	30	1:3.4
Suicide (E950–9)	21	5	1:4.2
Cancer (140–208)	4	†	—
25–44 years			
All causes	155	76	1:2.0
Accidents, poisonings, and violence (E800–999)	74	21	1:3.5
Motor vehicle accidents (E810–19)	26	8	1:3.3
Diseases of the circulatory system (390–459)	27	10	1:2.7
Suicide (E950–9)	26	†	—
Cancer (140–208)	24	31	1:0.7
45–54 years			
All causes	479	281	1:1.7
Diseases of the circulatory system (390–459)	191	68	1:2.8
Cancer (140–208)	146	145	1:1.0
Accidents, poisonings, and violence (E800–999)	66	25	1:2.6
55–64 years			
All causes	1415	727	1:1.9
Diseases of the circulatory system (390–459)	635	248	1:2.6
Cancer (140–208)	488	327	1:1.5
65–74			
All causes	3583	1959	1:1.8
Diseases of the circulatory system (390–459)	1729	933	1:1.9
Cancer (140–208)	636	606	1:1.0
75+ years			
All causes	10008	7492	1:1.3
Diseases of the circulatory system (390–459)	5450	4784	1:1.1
Cancer (140–208)	2095	1075	1:1.9
Cerebrovascular disease (430–8)	1197	1386	1:0.8

* International Classification of Diseases codes are given in parentheses.
† Not in leading causes of death for this sex.
Source: Australian Bureau of Statistics (1988).

Specific public health challenges: selected preventable diseases

In addition to the challenges provided by health status inequalities, there remain many public health problems with large preventable components and unacceptably high prevalence and incidence rates which continue to be the focus of efforts by public health services. Several of these are briefly described below.

Coronary heart disease

Premature deaths (30–64 years) from cardiovascular disease, easily the leading cause of death in Australia, fell by 48 per cent in men and by 68 per cent in women between 1950 and 1986 (National Heart Foundation of Australia 1986, p. 8). However, compared with countries of similar industrialized status, Australia's cardiovascular disease mortality rates remain unacceptably high, ranking 13th in the world behind Japan (males) and France (females) (National Heart Foundation of Australia 1986, p. 14).

Hypertension

One in six men and one in eight women are hypertensive in Australia, with its prevalence reaching 60 per cent in older age groups (National Heart Foundation of Australia 1983). Some 16.5 per cent of those diagnosed as hypertensive fail to maintain adequately their antihypertensive regimens. In addition, 6 per cent of the 25–64-year-old population are thought to be hypertensive although not yet diagnosed as such (or 41 per cent of all estimated hypertensives in the population).

AIDS

By 1986, Australia had the third highest per capita incidence of AIDS in the developed world (Armstrong and Holman

Table 30.3. Reasons for most recent consultation with a doctor in two weeks prior to Australian Health Survey interview, Australia, 1983.

Illness conditions	Males	Females	M/F	Persons
Endocrine, nutritional, and metabolic diseases	31.7	41.2	1:1.3	72.9
Mental disorders	35.1	59.6	1:1.7	94.7
Nervous system and sense organs	106.1	138.4	1:1.3	244.5
Circulatory system	136.9	198.7	1:1.5	335.6
Respiratory system	278.6	313.0	1:1.1	591.6
Digestive system	77.2	83.2	1:1.1	160.3
Genito-urinary system	22.7	98.4	1:4.3	121.1
Skin	73.6	88.2	1:1.2	161.8
Musculoskeletal system	152.2	160.9	1:1.0	313.1
Injuries	127.2	76.5	1:0.6	203.7
Symptoms and ill-defined conditions	72.9	98.2	1:1.3	171.1
Check-up	59.5	85.9	1:1.4	145.4
Total*	1145.9	1525.7	1:1.3	2671.6

* Total includes conditions not in table.

1987). At 9 January 1989, 1168 cases had been reported, with 564 known deaths (National Health and Medical Research Council 1989). Of the 1168 cases, only 15 have been heterosexuals who seemed not to have contracted the disease via the classic high-risk behaviours of unprotected anal intercourse, needle-sharing, intravenous drug use, or who received a contaminated blood transfusion before May 1985, the date when all blood products in Australia began to be screened for human immunodeficiency virus (HIV) antibodies. Informed estimates put the likely number of HIV-positive people throughout the country at around 50 000. A national co-ordinating council, the Australian National Council on AIDS, has since 1988 taken responsibility for co-ordination of all Government policy on AIDS.

Breast cancer

Breast cancer is the leading cause of cancer deaths in Australian women with 2 085 deaths in 1984. One in fifteen Australian women will develop breast cancer, with the risk of dying from the disease being just under one in forty. Mammography for screening, rather than investigative purposes, is not rebatable under the universal health insurance scheme, Medicare, and as a result, only 18 per cent of women report having had mammography in a recent community survey (Cumming *et al.* 1988). Several mammographic screening trials are now under way examining cost, utility, and feasibility of widespread use including efficiencies in outreach into low socio-economic communities. Experience from these trials will help shape future policy and practices.

Motor vehicle deaths and injuries

Since 1980, more than 23 700 people have been killed in road crashes in Australia (Department of Transport and Communications 1987). However, there are encouraging signs of a significant decline in the road toll with fatalities per 10 000 vehicles between 1980 and 86 having fallen 28 per cent, and per 100 000 population by 19 per cent. The death toll in 1983 (2755) was the lowest figure recorded since 1963. The introduction of random breath-testing by most States (Arthurson 1985) is considered to be the principal factor responsible for this decline. Young men experience a greatly disproportionate burden of the road toll, and the continuing task for prevention to reduce their crash, fatality, and injury rates requires strategies that utilize the resources and powers of many different governmental sectors: roads, public transport, education, police, and the areas of Government responsible for liquor control.

These brief vignettes describe some of the leading disease and injury challenges facing public health in Australia. They are overlain by the ageing of the population, whereby an increasing number and proportion will be aged over 65 years (in 1976, 8.92 per cent of the population was over 65; in 1986, this had risen to 10.55 per cent. Between 1976 and 2021, the number of persons aged 65 years and over is predicted to double). If a significant number of older adults move into later life in a healthier state, the costs to the community for their health care will be less than if preventive efforts are neglected.

Recent initiatives in promoting health and preventing disease

The Health Targets and Implementation Committee Report

In April 1988, the Australian Health Ministers' Advisory Council's (AHMAC) Health Targets and Implementation Committee (HT & IC) published its report, *Health for all Australians* (Health Targets and Implementation Committee 1988) setting out Australia's progress in developing health goals and targets for the year 2000, and recommending several major organizational changes designed to facilitate the development of national health promotion policies and programmes. AHMAC is the penultimate forum for public health policy development at national level, beneath the Australian Health Ministers' Conference (AHMC) to which its recommendations are referred. The recommendations of the HT & IC report were given the assent of AHMC at their

March 1988 meeting, and implementation commenced in the later half of 1989.

Establishment of the HT & IC followed several significant developments both nationally and internationally in public policy for health promotion and disease prevention. At the forefront of these developments is the World Health Organization's Health For All By The Year 2000 programme which, at a bureaucratic level, has set an important agenda for the future development of preventive health services in Australia.

In keeping with the preoccupations of the Health for All programme, the HT & IC report was concerned not only with general health status improvement, but with themes and organizing principles considered vital to the process of attaining better health across all population groups. Five such themes were addressed in detail in the report: equity, health promotion; the need to develop primary health care, and especially to enhance preventive activity in primary care settings; the development of intersectoral approaches to improving health involving health departments, other governmental agencies, non-governmental organizations (NGOs) and the private sector; and the need to increase consumer participation in health planning.

Other recent initiatives

The HT & IC report followed the publication of the 1986 Report of the Better Health Commission, *Looking Forward to Better Health* (Better Health Commission 1986), a comprehensive analysis of the prospects and tasks for health promotion and public health in Australia, but a document considered by some in retrospect to have been published two years too early to have its recommendations funded and which required more extensive consultation with the States to win agreement to its recommendations.

Other significant developments in recent years relevant to improving Australia's health status include:

(1) the establishment over the past decade of health promotion or health advancement divisions within all Commonwealth, State, and Territory Health Departments;

(2) substantial increases in money made available to several (although not all) State and Commonwealth health promotion budgets. In Western Australia, the health promotion budget increased from A$100 000 in 1982 to about A$3 million in 1987. Legislation in Victoria and South Australia, in November 1987 and March 1988 respectively, banned outdoor and cinema advertising of cigarettes and hypothecated to newly established health promotion foundations all monies raised by 5 per cent and 3 per cent (respectively) increases in State tobacco taxes. This provision, unique in public health in the world, will raise some A$23 million (Victoria) and A$5.2 million (South Australia) to replace tobacco sporting and cultural sponsorships with health-related backing, and to fund health promotion projects. A$15 million was made available over 3 years through the CDCSH National Campaign Against Drug Abuse for preventive projects, and a significant part of the A$54 million Commonwealth/State cost-shared funds committed for 1985–89 for AIDS was allocated to health promotion projects;

(3) the publication of reports and goal-oriented planning documents on health promotion by State Governments and health departments (e.g. the Ministerial Review Health Education and Promotion in Victoria 1986; Our State of Health, Western Australian Department of Health 1986);

(4) the implementation of the recommendations of the Commonwealth Report on Public Health Training and Research Development (the Kerr White Report) on expanding public health training and research throughout Australia (see below).

Barriers to improved public health services in Australia

Before describing the principal recommendations arising from the HT & IC report, barriers to more efficient and effective public health services in Australia, which the report was commissioned to address, are identified.

Dominance of health care system by treatment and care services

As with most health care systems, the Australian health dollar and the systems it supports are deployed principally to treat illness and to care for the chronically ill, with most expense occurring at labour-intensive referral levels. By contrast, the WHO Health For All programme proposes reorientation of major sections of the health care system toward the lower cost prevention and health promotion areas, including primary care.

For over a decade there has been considerable political and senior bureaucratic posturing about the importance of prevention and health promotion in Australia, yet there remains a lack of high-level operational policy for improving health status (*cf* treating illness) and consequently for prevention and health promotion. Notwithstanding the expected implementation of the recommendations of the 1988 HT & IC report, to date there has been no powerful national co-ordinating concept or mechanism to plan or oversee progress in prevention nor to ensure its relative importance within the overall health care system. The substantial progress made in the past decade in health promotion, as exemplified by the developments listed above, has been a result of incremental policies and programming decisions, and not a consequence of any explicit major policy shift about the overall goals of the health system. Areas and periods of progress (and correspondingly, of inertia) have tended to reflect the influence of particular individuals in senior administrative and political positions, rather than being a consequence of any development of health system policy on prevention, certainly at any national level.

Following the example of the US Government's development of national health goals and targets (McGinnis 1985), the HT & IC report drew together many Australian health goals and targets already set by States, the CDCSH, non-governmental groups, and by the Better Health Commission. These had been developed for various population groups (e.g. women), for health problems (e.g. heart disease), and for selected risk factors (e.g. smoking). Their detail and quality reflected the large differences among health issues in the availability of population-based health data against which meaningful levels of improvement could be projected, and in the coherence of different professional groups agreeing on the realism of likely progress.

Attainment of the goals and targets set for health outcomes in the HT & IC report was seen to require a high-level, national mechanism that would declare their pursuit to be official Government policy, agreed to by all nine Governments in the Australian federation of States and Territories. The expectation is that all relevant Government health programmes would then address explicitly the goals in planning and resource allocations. As well it is hoped that, in keeping with the necessity to explore and solve health problems through intersectoral action, health departments would encourage all relevant non-governmental organizations and appropriate groups within the private sector to pursue these goals, to participate in intersectoral and interagency planning, and to report on their evaluations and monitoring of achievement. Most critically, with expenditure specifically allocated for prevention within total health budgets running at an estimated 0.5 per cent, it requires that Governments institute significant systems of funding at national and State levels to give incentive to Government departments, health service providers, community groups, and individual researchers to direct their efforts toward the goals. Finally, it proposes that a system of standardized monitoring of progress toward the goals must be established. The principal structural and finding reforms proposed by the HT & IC (described below) were attempts to satisfy these requirements.

Consequences of federalism for co-ordination and efficiency of effort in public health

National health goals and targets must eventually imply national, co-ordinated efforts directed at their attainment. However, such effort presents formidable problems in Australia's federal system of autonomous and often politically opposed Commonwealth, State, and Territory Governments—with their nine different Ministries and Departments of Health, a plethora of non-governmental health organizations dedicated to improving different aspects of community health, and a myriad of other governmental and non-governmental agencies whose policies and programmes have major implications for community 'health'.

This mosaic of responsibility and implementation can be both strong and creative, producing diversity, and weak because of inefficient multiplication of services, lack of co-ordination, and consequent fragmentation of effort. Both perspectives are valid, so the concern for planners becomes one of attempting to safeguard the strengths of diversity while lessening the wastage of unnecessary duplication. Implementation of preventive health programmes is generally considered inevitably to remain the responsibility of the States and Territories, yet there is much that can be done to improve national priority setting and communication between planners in developing national efforts in policy.

It is the reduction of inefficient multiplication of services wherein lies the incentive to States, Territories, and the Commonwealth to find agreement wherever possible. In addition, some things, such as universal health insurance, obviously are required to be organized nationally in order to facilitate the work of the States. The HT & IC sought to identify such issues in health promotion.

Attempts to respect and respond to locally variable 'needs' have often worked against the development of uniform community-wide responses to national priorities. Yet most major preventable health problems are common to all States and Territories: there are few important examples of health problems which are irrelevant in any particular State. While it is often important for effective programme implementation that local area planning occurs to provide community participation and 'ownership', this process is problematic if it produces a patchwork of different local priorities that, viewed from State or national perspectives, hides the common wood of national priorities in a proliferation of different trees.

The HT & IC recommendations

The HT & IC perceived the necessity to establish a permanent forum on health where the various major players could debate health goals and identify areas of common interest for common action. The forum, which includes governmental and non-governmental groups, is known as the Health For All Committee (HFAC). The principal function of the Health For All Committee, supported by satellite State and Territory level Health For All Committees, is to direct and co-ordinate an ongoing National Programme for Better Health. Under its auspices, Project Planning Teams have been established to develop strategic plans for each specific health issue nominated. Project Implementation Teams will then be selected from existing centres around the country to co-operate in the various facets of national programme development and implementation. From late 1989, these were the primary prevention of lung and skin cancer; the secondary prevention of breast and cervical cancer; the control of hypertension; improved community nutrition in accord with the Australian Dietary Guidelines; improved injury surveillance systems to enable more rational and planned approaches to injury prevention; and efforts at addressing the preventable components of the health problems of older adults.

Table 30.4. The number of people (thousands) consulting with different health care professionals in the two weeks prior to interview in the Australian Health Survey, 1983

	Males	Females	Total
Doctor	1145.9	1525.7	2671.9
Chemist	169.6	225.9	395.5
Optician/optometrist	31.6	40.0	71.6
Physiotherapist	73.7	68.0	141.7
Psychologist	6.0	7.7	13.6
Social worker/welfare officer	7.8	10.3	18.1
Chiropodist/podiatrist	8.6	21.4	29.9
Baby health nurse	8.5	9.6	18.1
School nurse	2.7	2.9	5.6
District, home or community nurse	24.5	38.9	63.4
Chiropractor	86.6	96.3	182.9
Osteopath	9.7	8.9	18.6
Naturopath	14.1	32.5	46.6
Herbalist	5.8	14.1	19.8
Acupuncturist	12.2	21.6	33.8

Source: Australian Bureau of Statistics (1986).

The strategies appropriate to the improvement of public health in each of these issues together encompass virtually the entire spectrum of strategies in the repertoire of public health interventions. Under Australia's federal system of government, the adoption of strategies within these five nominated priority areas, such as a comprehensive national ban on tobacco promotion, national call and recall systems for mammographic and Pap screening, uniform food labelling laws, and routine and standardized injury surveillance systems, presents formidable challenges to public health in the face of vested interest opposition and autonomy issues already discussed. By having in the national Health For All Committee a single point-of-reference for penultimate decision-making and planning on such issues, rather than the *ad hoc* fortuitous coalescence of views inherent in the previous arrangements, many are hoping for progress.

Enhancing prevention in primary health care

Two themes central to the Health For All strategy are the development of prevention via enhancing primary health care, and of intersectoral mechanisms for increasing the roles of other Government departments, non-governmental organizations, and the private sector in health promotion. Doctors are already engaged in considerable preventive activities, mostly at the secondary level. Most Pap smears are done by doctors in Australia, as is childhood immunization. Regular checks on blood pressure are carried out on most Australians when they visit a doctor. As shown in Table 30.4, doctors remain overwhelmingly the leading source of primary care professional help in Australia, with over two and a half times as many people consulting doctors as other health professionals combined. Chemists rank second to doctors in terms of frequency of consultation on health matters, but despite their proliferation, ease of access and no consultation fee, they are consulted nearly seven times less than doctors, according to the 1983 Australian Health Survey. Given the preventability of many problems presenting to doctors and that nearly 75 per cent of the population see a doctor annually, the role of doctors in prevention represents an unparalleled resource for addressing many priority issues in public health today. However, like many other primary health care workers (such as chemists, occupational health nurses, and home and community care nurses whose potential for involvement in prevention is also large) doctors are poorly trained in prevention and there is strong resistance from Medicare to remunerate preventive consultations under the present taxpayer-funded, fee-for-service system operating in Australia. In recent years, there have been numerous unsuccessful attempts by various public health interests to encourage the Commonwealth Government to extend its interpretation of legitimate rebate-eligible consultations into the preventive field. But on each occasion, the refusals have been explained against concerns about their potential to foster entrepreneurialism at huge public expense.

Assisting change in the role of primary health care workers, notably doctors, toward more prevention will be a formidable task necessitating a fundamental re-examination of the division of labour among various health professionals, medical training, and the remunerative relationships between health care providers, the public and Government. The monopolistic hold the medical profession has in some States of Australia over preventive issues, like the taking of Pap smears and the reading of mammograms, and the consequences for the health budget should they be allowed to expand their conduct of such services through recall arrangements, has prompted many to question such monopolistic control. The HT & IC has urged that such an enquiry be established.

Intersectoral efforts to improve public health

In discussions of 'public health', it is generally assumed that the policies, actions, and outcomes of importance are those originating from the public sector. It is the activities of health department bureaucracies and associated bodies, of publicly funded public health research and teaching institutions, and the laws and regulatory provisions generated by health ministers that are taken to be the obvious subject matter to consider when assessing the practice of public health in a country.

However, an emerging dialogue within public health circles is focusing on evidence that the health of the community and the fruits of the labours of those self-consciously engaged in explicit public health occupations are hardly co-extensive. An 'intersectoral' perspective on both analysis and action to improve the health status of populations is increasingly being recognized as fundamental to any consideration both of how the health status of populations does change, and of questions concerning efficiency in the roles and work of those public sector agencies that have traditionally addressed public health (World Health Organization 1986). The impacts, direct and indirect, on health resulting from the policies and actions or other (non-health) Government portfolios, such as employment, consumer affairs, education, housing, the environment, and agriculture; from non-governmental agencies such as pensioners' associations, leisure and sporting groups; and from the private sector (e.g. the food, pharmaceutical, sunscreen, and product safety design industries), are demonstrably of immense importance in variously promoting or retarding public health.

Where policies and actions of non-health agencies, institutions, and industries have a bearing on acute episodes of disease, illness, or injury (food and drug standards, product safety, water quality, and other aspects of environmental protection), statutory relationships between their activities and health departments have generally been long established in Australia. With the exception of burgeoning intersectoral developments in the injury prevention field (e.g. health, transport, and police on road safety), planned developments in such areas are not high on any political or health bureaucratic agenda in Australia, for it is chronic disease and quality of life (mental health) issues that dominate Australia's mortality and morbidity profiles (Tables 30.1, 30.2, and 30.3). These are largely seen as challenges in the provision of services, both hospital and community, rather than as opportunities for prevention.

By contrast to the environmental health response to acute threats and intoxications, the intersectoral response to the control of chronic diseases tends to be rather obtuse and speculative, with far fewer examples of significant institutionalized relationships between different Government departments on health issues in evidence. This reflects both the difficulty inherent in many of the preventive proposals (e.g. leaner meat production) and the comparative imprecision of chronic disease aetiology next to the urgent imperatives for action against acute illness and injury. In most cases of chronic disease, where complex interrelationships exist between host, agent, and environment—as, for example, with concerns for changing the diet of the population in the direction of the Australian Dietary Guidelines (Better Health Commission 1986, Volume II)—the rhetoric calling for intersectoral co-ordination is more obvious than tangible and important instances of the reality. None the less, the contemporary agenda for public health in Australia is showing encouraging signs of becoming increasingly less myopic and health bureaucracy-centred with themes such as intersectoral co-operation, community involvement, and the large-scale appropriation of health as a selling point by important sections of the private sector, such as the food industry, commanding the attention of planners and recent major public health reports.

A growing concern that a principal aim of public health efforts should be the facilitation of equity in opportunity to enjoy a healthy life, and for health status differentials to narrow between relevant population groups (see above), is also impelling consideration of intersectoral effort, as the futility and inappropriateness of interventions and policies driven solely through health departments is realized.

In Australia, the traditional seats of public health policy and practice (health bureaucracies, research institutes, university departments, and professional, provider, and consumer societies) remain the leading proponents, advocates, and monitoring groups of public health. For some issues, such as smoking control, efforts to prevent drug abuse, and AIDS, they are, and are likely to remain, the principal sources of preventive efforts. However, in a substantial and growing number of public health areas, these publicly funded bases could not be said to be the leading exponents of health-promoting practices or providers of information to the public. As will be shown, major sections of the food industry are contributing demonstrably far more, through advertising and labelling, and through ingredient changes, to the education of the public about reductions in fat, salt, and sugar, and increases in fibre and complex carbohydrate than the aggregated efforts of traditional public health agencies in Australia.

Such developments raise important questions about the future role of the state in advancing public health. The implications of a wider analysis of factors contributing to public health for the organization and functions of the so-called 'new public health' workforce is currently a central preoccupation for those now reforming the field in Australia.

Recent attempts to develop national public health programmes

The HT & IC report was essentially concerned with reforming the ways in which public health efforts are organiszed so that greater prominence might be given to agreed national programmes. In recent years, three important attempts at organizing truly national public health campaigns (as distinct from the aggregated efforts of different States) have been mounted. With some 30–40 per cent of 2–5-year-olds not immunized against measles (Australian Bureau of Statistics 1985), the bicentennial measles control campaign was organized to achieve its goal of eradicating indigenous measles by 1991. The 200th anniversary of white settlement in Australia (1988) served as the justification for many hundreds of civic, environmental, and in this case, public health bicentennial projects. There appears to be no rational reason why a major drive to eradicate measles was spotlighted through this process, rather than say, a major national campaign to reduce road deaths or smoking, both of which cause exceedingly more deaths and illness in Australia.

The selection of measles probably illustrates little other than that a proposal by communicable disease control interests about a sensible, achievable, and non-controversial issue was opportunistically raised in the right place at the right time. The same analysis could not be applied to Australia's two other recent instances of national public health efforts:

the National Campaign Against Drug Abuse (NACADA) and the Australian National Council on AIDS (ANCA).

The national drug offensive

NACADA, also known as the Drug Offensive, commenced in 1985 with a national drug summit of people involved in all aspects of drug control throughout Australia. The launch of the Drug Offensive followed an emotional television interview with the Australian Prime Minister, Bob Hawke, which related to his young adult daughter having a heroin problem. Some $100 million in addition to the routine expenditure already allocated to drug control was allocated to the campaign by the Commonwealth Department of Community Services and Health (CDCSH) on a cost-shared basis with the States—an unprecedented sum for any single field in the history of public health in Australia. Approximately $5 million is being spent on media awareness campaigns. While the campaign defines drugs broadly (including alcohol and tobacco), its main emphasis has been decidedly on narcotics abuse, with large-scale television advertising attempting to deglamorize heroin use in the young and a glossy booklet being distributed to every house in Australia.

Against all important criteria—availability, prevalence and incidence of use, physical and social morbidity and mortality—alcohol and tobacco are far larger public health problems in Australia than narcotics abuse (Australian Bureau of Statistics 1987*b*). While the Drug Offensive has included major components directed at these two licit drugs, their emphasis has been almost entirely on educational and persuasive efforts at building host resistance in potential users. While a major sub-campaign within the Drug Offensive entitled 'Stay in Control' attempts admirably and imaginatively to stigmatize teenage drunkenness, the campaign has never been empowered to address directly supply-side factors in the alcohol abuse equation—price, advertising and promotion, and availability. While the Drug Offensive offers television images of drunk teenagers vomiting in front of their embarrassed friends, Fosters Lager secured the main sponsorship rights to the Australian grand prix formula one motor race in Adelaide and publicized this association with a much vaster budget than the relevant segment of the Drug Offensive; the Bond Corporation launched airship advertising for Swan Lager and John Player Special cigarettes over major cities; 'Cooler' wine/fruit juice and 'Kix' spirits/fruit juice drinks were launched on the market; and the real prices of alcohol and cigarettes remained lower than they had been 20 years ago.

These two important shortcomings—the undue emphasis on narcotics and the limited, individual-directed focus—have evoked dramatic differences of opinion within the controlling committees of NACADA. With representation from each of the States, Territories and Commonwealth Governments, NACADA has proved to be a force-field of State/Commonwealth dissent, fuelling scepticism that, in a federal system that permits such differences, consistent and coherent national programmes will ever not be beset by disruptive hidden agenda, tempting States back to the relatively unproblematic autonomy of doing their own thing.

The Australian National Council on AIDS

ANCA, the Australian AIDS control programme, has had—like the disease it is designed to control—an explosive development as another major focus for public health action and research. ANCA has A$54 million (1985–89) allocated for all purposes, including immunological and other medical research. Its major public health commitment has been the funding of public awareness campaigns, the most celebrated of these being the so-called Grim Reaper television campaign broadcast for just one week in 1987 (Morlet *et al.* 1988). The advertisement, depicting a medieval grim reaper with scythe, bowling a ten-pin ball that skittled men, women, and small children in its path, was designed to raise concern and awareness throughout the community that AIDS can strike anyone. The advertisement was highly controversial, widely discussed, and generally agreed to have achieved its principal aim of raising community awareness. However, when the campaign moved into a further phase in 1988, focusing on the importance of condoms in prevention, again the problem of lack of unified agreement between the nine governments involved arose, with several States refusing to allow illustrations of condoms and colloquial slogans, both considered vital to their more effective promotion. This problem has apparently occurred elsewhere in the world (Baggaley 1988).

Both NACADA and ANCA illustrate important features of public health services in Australia. Their formations were precipitated not by any professional groundswell within the public health community or bureaucracy, but by two quite distinct factors that appear to contain few obvious lessons for the future development of other fields. AIDS is clearly a public health problem of epidemic and serious proportions, the ramifications of which, on worst-case projections, will affect most areas of the health care system. AIDS appears less to parallel aspects of chronic diseases than it does the more acute health problems addressed by public health, and therefore extrapolating lessons from its rise to prominence to the ongoing issues of chronic disease control in Australia is problematic. Partly as a consequence of these and other features, AIDS has attracted battle metaphors (*cf* Sontag 1979), resulting in appropriately extraordinary 'war cabinet' responses from governments and health bureaucracies.

The Drug Offensive shares many of these battle metaphors (Montague 1988). The enemy is identified and vilified, moral issues of right and wrong are inextricably entwined with the allegedly morally neutral concerns of public health, and, as a result, the neon issues for the Drug Offensive have been appropriated as much by the electoral sensitivities in Government for law and order issues as they have by their health portfolio administrative base. Here was an example of a truly intersectoral imperative for what has been legitimized as a public health benefit.

Australia's main preventable chronic disease problems (defined from within the professional field), such as the seven nominated by the HT & IC as the desired initial projects for the Health For All Committee, each have shopping lists of desired strategies. Many of these have existed for years, and have been the topics of earnest recommendations, motions, delegations, working parties, and committees, and of public debate. Most fields in Australian public health can describe a gradual and incremental history of the adoption of items on their respective strategic shopping lists (see, for example, the case study on smoking control below). But most also are well aware that they are unlikely to have their field elevated into the war cabinet luxury of urgent, well-resourced, and enduring commitment described above for AIDS and drug abuse.

Likely determinants of future progress

Progress in addressing Australia's main chronic disease problems is likely to be determined by three factors. The first might broadly be described as opportunism. Here, there are abundant examples of the predilections of a new health minister or senior bureaucrat for a particular issue, determining a flurry of activity (or, conversely, showing abject disinterest). A classic example is the wake of funding and interest in health promotion churned up by Mr. Bernie McKay. In the late 1970s, McKay was a regional health administrator of a coastal area in New South Wales. A personal interest in health promotion and news of the early progress in the Stanford Three-Town Heart Disease Prevention Program prompted McKay to fund a similar trial (Egger *et al.* 1983), the magnitude and likes of which had never been seen in Australia before. Over the next decade, McKay moved jobs to become head of two State health departments (South Australia and New South Wales) and, eventually, secretary of the (then) Commonwealth Department of Health. In each of these positions he funded major developments in health promotion, including the Better Health Commission mentioned above. Progress (or lack of it) in the public health aspects of many other fields can often be attributed to the presence of individuals like McKay.

A second determinant, and one that many would argue predicts the likelihood of the ascent of people like McKay, is the conduciveness of the social and political climate to preventive arguments being taken seriously. The case study below on smoking control in Australia illustrates the evolution of a rapid series of governmental control measures, all featuring the support and efforts of initiating individuals, but also characterized by strong public support. This support has not happened by serendipity, but through the skilful application of the principles and tactics of public interest campaigning (Wilson 1984).

A third determinant will be the extent to which deliberate and planned efforts now on track, following the acceptance of the HT & IC report by the Australian Health Ministers, will facilitate a significant, enduring, institutionalized level of response to prevention by governments, as opposed to the fate of preventive proposals falling into a policy vacuum determined by the day-to-day priorities of curative and care-based health systems. While supportive individuals and governments come and go, and the popularity of particular preventive efforts is subject to many vicissitudes, an institutionalized superstructure of policy commitment, funding, and reporting of progress toward goals and targets—as planned through the Health For All Committee—should provide a more supportive environment for public health than previously.

Public health research and training

Research and evaluation

Public health research and evaluation studies are conducted through a variety of institutional arrangements in Australia. Chief among these is funding provided through the auspices of the National Health and Medical Research Council (NH&MRC), the quasi-autonomous research arm of the Commonwealth Department of Community Services and Health (CDCSH). Systematic attempts to promote the research concerns of the public health sector within the NH&MRC began only in the early 1970s. The first public health scholarships and fellowships were awarded and special units were funded (Social Psychiatry Research Unit, Road Accident Research Unit, Health Economics Research Unit). In the early 1980s, a Special Purposes Committee was formed within the NH&MRC to select areas for priority research development funding. Areas currently recognized by the Committee as requiring special initiatives are the early stages of alcohol abuse, addictive behaviour, AIDS, ageing, behavioural medicine, environmental toxicology, Aboriginal health, health care evaluation, public and preventive health, and rehabilitation. Table 30.5 shows the distribution of NH&MRC-supported grants for disciplines relevant to public health between 1985 and 1987. The amounts funded can be contrasted with the total NH&MRC grants for 1985 for all aspects of health and medical research of A$32 814 000 (Commonwealth Department of Health 1985).

The Kerr White funding (see below) allowed for the establishment of the Public Health Research and Development Committee (PHR&DC) within the NH&MRC. The PHR&DC has been allocated A$2.85 million to distribute between 1986 and 1989 for research projects and fellowships.

Public health training: the Kerr White report

Following a January 1986 report to the Federal Minister for Health by Professor Kerr White, a major reorganization of public and tropical health teaching and research commenced in Australia. The Commonwealth Government is committed to providing A$26 million for the 3 years 1986–8, with extensions to approximately A$41.5 million over 7 years, for the development of these fields.

Eight universities and one research institute are being funded to develop programmes leading to Masters degrees in

Table 30.5. Research funded by the National Health and Medical Research Council, 1985–7

Reseach discipline	1985		1986		1987	
	No. per year	$(million)	No. per year	$(million)	No. per year	$(million)
Community medicine	0	—	4	0.16	11	0.45
Environmental health	7	0.23	11	0.49	20	0.87
Epidemiology	22	0.90	36	0.59	45	1.90
Health care evaluation	1	0.06	5	0.20	18	0.70
Health education and promotion	1	0.21	4	0.21	9	0.33
Occupational health	2	0.06	4	0.16	11	0.46
Primary health care	2	0.08	4	0.12	6	0.23
Public health (general)	15	0.60	25	1.11	34	1.52
Total	50	2.14	93	3.04	154	6.46

Source: Kalney (1987).

public health (MPH) and related areas. Under the new funding, the Australian National University in Canberra has established a National Centre of Epidemiology and Population Health; the Universities of Sydney and Adelaide and Monash University are expanding their existing MPH programmes; the University of Western Australia is establishing an MPH course; the University of Queensland has taken over the national tropical health teaching and research role, and in conjunction with the Queensland Institute of Medical Research is conducting Master and diploma-level courses in tropical health; the James Cook University of North Queensland has established a Tropical Health Surveillance Unit; and the University of Newcastle is expanding its post-graduate training in clinical epidemiology and biostatistics, previously open under Rockefeller Foundation sponsorship only to overseas scholars from the Asia and Pacific Region, to admit Australian students.

The Australian Institute of Health

The Kerr White funding also provided for the expansion of the Australian Institute of Health (AIH). The AIH began as a small bureau within the Commonwealth Department of Health in August 1984. In 1987, under the new funding, it became an independent statutory authority within the Community Services and Health portfolio of the Federal Government. Currently it has a staff of 54.

The AIH's functions include the collection and analysis of health-related statistics: the co-ordination of such data collected by other bodies and individuals; the study of cost and effectiveness of health services and health technologies; provision of information for researchers and health workers; and advice to the Minister on measures for improving public health. By these activities, the AIH will compile and improve the national health data base, with its proposed strategies for public health advancement having a firm basis in local information.

The AIH is developing a National Death Index, a National Cancer Statistics Clearing House, a National Nosology Reference Centre, and a compilation of national Aboriginal health statistics. Current research priorities include the provision and use of health and hospital services, storage/wastage of household medicines, health labour force supply and demand, use of discretionary surgery, and quality assurance. It has embarked on a major study of levels and trends in the health of the population, analysing differentials in health to identify both target groups for prevention programmes, and those socio-economic factors that appear more conducive to better health.

The AIH incorporates the National Committee on Health and Vital Statistics and provides the working Secretariat for the National Health Technology Advisory Panel which co-ordinates the evaluation of new health technologies. The AIH also funds the National Injury Surveillance and Prevention Project on a cost-sharing basis with the States, and supports the National Perinatal Statistics Unit at the University of Sydney and the Dental Statistics and Research Unit at the University of Adelaide.

Case study 1: smoking control in Australia

In 1984, tobacco smoking was estimated to have caused the deaths of 23 000 people in Australia, with 70 523 person-years of life lost between the ages of 15 and 69 years (Holman and Shean 1987). In 1945, cigarette smoking prevalence was estimated at 72 per cent (males) and 26 per cent (females). By 1962, cigarette smoking had declined in males to 58 per cent and risen slightly in females to 28 per cent (Woodward 1984). This male decline has continued during the 1970s and 1980s, to a present level of 31.9 per cent (males) with female rates remaining steady at 28.8 per cent (Table 30.6) (Hill *et al.* 1988). Table 30.7 shows changes in apparent annual per capita cigarette consumption, with a continuing decline after 1976, the year when tobacco advertising was banned from the electronic media.

Recent initiatives

During the 1980s, smoking control initiatives rose to a level of quite unparalled prominence in public health activity in Australia. The highlights of this activity have been:

Table 30.6. Age-standardized percentages of cigarette smokers, past smokers, and those who have never smoked, 1974–86

		1974	1976	1980	1983	1986
Cigarette smokers	M	42.2	40.5	40.6	37.9	31.9
	F	29.5	32.3	31.1	30.8	28.8
Past smokers	M	21.8	22.4	23.5	27.3	28.1
	F	9.8	11.6	14.1	16.6	16.8
Never smoked regularly	M	32.9	34.4	35.4	32.4	38.5
	F	60.6	55.9	54.7	52.5	54.3

Source: Hill *et al.* (1988).

Table 30.7. Changes in per capita consumption of cigarettes, Australia

Year	Cigarettes consumed
1962	2510
1973	3080
1976	3180
1977	3136
1978	3039
1979	2859
1980	2869
1981	2790
1982	2801
1983	2621
1984	2522
1985	2449

Source: Australian Bureau of Statistics (1987*a*).

(1) the mounting by most States of large-scale mass media public information and motivation campaigns with both cessation and preventive objectives (e.g. Egger *et al.* 1983; Pierce *et al.* 1986);

(2) following the banning of direct cigarette advertising on electronic media in 1976, cinema and billboard tobacco advertising was prohibited in Victoria and South Australia in 1988;

(3) a successful challenge to abuses of the self-regulatory code of cigarette advertising practice through the Trade Practices Tribunal, by a coalition of health groups led by the Australian Consumers' Association (Chapman 1988*b*);

(4) the incremental banning during the 1980s of smoking on nearly all forms of public transport and all domestic airlines (1988);

(5) a prohibition on smokeless tobacco products, first by the South Australian Government in 1986 (Chapman and Reynolds 1987) and then by the Commonwealth Government (1989);

(6) bans on the sales of packs of cigarettes containing less than 20 cigarettes in South Australia, followed by Victoria, Western Australia and New South Wales (Chapman 1988*a*);

(7) increases in State tobacco taxes in all States except Queensland, with a current range of 50 per cent of retail price (Tasmania) to 25 per cent (Australian Capital Territory); automatic indexation of federal tobacco tax to changes in the Consumer Price Index since 1983;

(8) the hypothecation of 5 per cent increases in State tobacco tax to Health Promotion Foundations in Victoria and South Australia (with others planned in other States) to enable the buy out of tobacco sponsorships and to fund health promotion efforts;

(9) the introduction of four explicit rotating health warnings on all tobacco product packs in 1987;

(10) a reduction in the sales-weighted tar content of cigarettes sold;

(11) a rapid growth in the number of public enclosed spaces where smoking is prohibited (including a total ban on smoking in all Commonwealth Public Service offices by employees and the public in 1988);

(12) a growing opposition to the acceptance of tobacco industry research money by scientists (Chapman *et al.* 1988);

(13) the development of a widespread civil disobedience movement dedicated to graffiting tobacco advertising and disrupting tobacco-sponsored events (Chesterfield-Evans 1987);

(14) changes in public opinion whereby large majorities support further restrictions on tobacco promotion and more vigorous efforts by governments to control smoking (Hill 1987).

Aggregated, these achievements in less than a decade represent remarkable progress in the face of opposition from the powerful tobacco industry and from associated industries such as advertising and publishing which have vested financial interests in the continuation of laissez faire, non-interventionist and de-regulatory policies generally in the ascendent in the Australian business regulatory environment. Professionally, the smoking control field in Australia has not been dominated by academically motivated researchers keen on examining the minutiae of cessation processes or other individual-directed interventions. Rather, smoking control efforts have been led and carried almost exclusively by public health workers and agencies in both governmental and non-govermental settings. The strategies utilized have been commensurate with a public health perspective on the size of the smoking population and the consequent need to address objectives using mass-reach strategies (Chapman 1985).

Since the formation in 1983 of a full-time lobbying office, Action on Smoking and Health (ASH), progress towards many legislative, regulatory, and fiscal policy objectives which had hitherto been piecemeal or dependent on the energy and opportunism of dedicated individuals has increased dramatically. ASH has played an important role in developing active, supportive networks of professional associations, trades unions, children's interest and consumer groups at both State and national levels. These networks have been invaluable in lobbying activities and in showing the flag in press and media debates where the tobacco industry has attempted to marginalize anti-smokers as narrowly based, unrepresentative minorities representing left-wing antibusiness, nanny State (Browning 1988) or neo-puritan ideologies. ASH has put great effort into educating political groups and editorial writers and into funding public opinion surveys at strategic times in campaign developments.

Educational and cessation efforts are largely the province of the States via mass mediated efforts. New South Wales was the first State to commence large-scale efforts in this regard, and was soon followed by most other States when it became apparent that there was widespread public and thus political support for such campaigns. This synergism between commitment by public health agencies to smoking control and the public support it both generates and subsequently thrives on suggests that public health objectives will continue to be met in years to come.

Case study 2: improving the nutrition of Australians

Progress towards healthier patterns of nutrition has been occurring in Australia for many years, although this has not happened uniformly across all social classes (Baghurst and Syrette 1988; Worsley and Crawford 1985). As with many western industrialized countries, the Australian diet remains aberrant with regard to its high average content of fat, salt, alcohol, and sugar and relatively low content of vegetable fibre.

Fifteen years ago in Australia, it was often difficult for the average suburban shopper to buy polyunsaturated cooking oils, low-fat diary products, whole-grain breads and cereals, saltless or salt-reduced groceries and no-added-sugar produce. This has changed dramatically, with such products being prominent on grocery shelves throughout the country.

Data derived from apparent consumption data on progress towards Australia's Dietary Goals (as determined by the Nutrition Task Force of the Better Health Commission (Better Health Commission 1986) between 1975/6 and 1983/4 suggest the following (Newell and Skurray 1987) (Better Health Commission targets are in brackets):

1. **FAT** The percentage contribution of fat to total energy has decreased from 39.1 to 38.13 per cent (target = 33 per cent by the year 2000).
2. **SUGAR** There has been an 11.1 per cent reduction in per capita sugar consumption from 54.1 kg to 48.1 kg per annum, with the contribution of sugar to total energy decreasing from 14.9 to 14.1 per cent (target = 12 per cent by the year 2000).
3. **ALCOHOL** Per capita alcohol consumption has decreased by 4.1 per cent from 9.7 to 9.3 litres of pure alcohol (target = 5 per cent reduction by the year 2000).
4. **COMPLEX CARBOHYDRATES** (breads, cereals, vegetables, fruit) increased by 6.7 per cent from 29.8 per cent contribution to total energy to 31.8 per cent (goal: 'to eat more'—consensus being to 48 per cent of total energy).

National data are not available on changes in fibre and salt consumption or on breast-feeding.

Thus, there are no national dietary goals that require a reversal of the current direction already in evidence in the community: where the goals are to reduce, reductions are occurring, and where goals are to increase, increases are evident. The tasks remaining for public health are therefore those of amplifying existing trends and, because of equity considerations, to ensure that these improvements extend fairly to all groups in the population.

There is a diverse range of interests that influence the diets of Australians, and whose roles are therefore relevant to concerns for intersectoral co-ordination. The principal interests involved are the many different sectors of the food and beverage industry; professional associations and agencies such as the Australian Nutrition Foundation, The Australian Dietitians Association, and the National Heart Foundation; and Government departments—health, agriculture, primary industry, and consumer affairs.

All these groups in different ways and for different motives seek to influence what Australians put in their mouths. This influence is extended to consumers in the form of advertising and marketing; via the provisions of laws and regulations about food standards, grocery labelling, and advertising; through price competition; and through various efforts at public education, whether this be in schools, in mass media, or through face-to-face work by nutritionists and dietitians. In Australia, the dialogue about improving community nutrition is dominated by professionals bureaucrats, and academics who have tended to imply by their preoccupation with their own initiatives, that advances in the community's nutritional status are largely dependent on the actions of what is clearly only one vehicle for change: professional and bureaucratic action.

Public and private sector influences on diet

Diets in non-subsistence economies such as Australia change through developments in the interdependent relationship of changing patterns of consumer demand with factors influencing the supply of food that consumers are able to choose. The role of the food industry, as the supplier of nearly all food consumed, is central to any analysis of who eats what, and why.

Table 30.8. Estimated advertising expenditure in selected media, Australia 1986 (millions of dollars)*

Category	All press	Television	Metropolitan radio	Total	Rank†
Food	21.874	190.037	12.985	224.896	1
Liquor	9.290	45.084	6.900	61.274	9
Confectionery	0.744	38.463	1.256	40.463	15
Soft drinks	0.568	15.840	2.967	19.375	22
				346.008	

* Does not include country radio, point-of-sale, sponsorship, or other promotional expenditure (figures unavailable).
† Rank order of all product-type advertising.
Source: Commercial Economic Advisory Service of Australia. *Foodweek* 15 September, 1987.

If people wish to eat a nutritionally balanced diet in Australia, and can afford to do so (a large proviso for very low income groups), there are few supply-side barriers to them altering their diet in the recommended ways. There is intense competition in the food industry for the dollar of the nutritionally concerned consumer. Recent features in the industry trade journals *Retail World* and *Foodweek* (Anonymous 1988) predicted that Australian supermarkets were showing signs of following the US where fresh fruit and vegetables are now the most popular and profitable sections in supermarkets, eroding the traditional position of frozen foods.

It is rather problems with the extent and strength of the wish to eat well, determined by demand-side factors such as taste preferences and dietary awareness and concern, that retards further change. There is no major publicly available research information in Australia on the determinants of contemporary dietary patterns. However, it is likely that for the majority of Australians whose diet is nutritionally poor, the main explanations will be found in the interplay of culturally determined taste preferences, lack of information about nutrition, lack of motivation about dietary change, and lack of skill in knowing what to do to change diet in the desired directions (e.g. not knowing what foods contain high levels of fat).

The issue of concern, then, becomes one of how the industry's aggregated willingness to meet demand and to obtain competitive edge over rival marketers can be made more compatible with the attainment of national dietary goals. Given that industry has virtual autonomy (within the bounds of food legislation) to produce and market whatever it judges to be profitable, and given its willingness in most cases to modify its products to meet changing consumer preferences for nutritionally improved foods, the challenging question becomes one of how the nutritional interests of health departments and professional non-governmental organizations might optimally facilitate or contribute towards the development of consumer demand in the desired directions. Education is critical in overcoming these problems and, below, the potential roles in nutrition education of governmental and non-governmental agencies and of the food industry are examined.

The role of the food industry in nutrition education

It is tunnel-visioned to perceive nutrition education solely in terms of the aggregated efforts of agencies that are self-consciously engaged in nutrition education efforts. There are abundant and increasing examples of the appropriation of (mostly quite legitimate) nutrition appeals as selling points for products, in addition to successful attempts to sell wholesome food using themes quite unrelated to nutritional themes. The budgets available to Government-run or sponsored nutrition projects in Australia are, compared with those available for advertising and promotion to industry, almost trivial. In 1986, total expenditure on nutrition education by Government health and agriculture departments throughout Australia approximated A$1 million. To this could be added several hundred thousand dollars from the aggregated educational efforts of non-governmental organizations.

By contrast, the food and beverage industry spends more than any other industry in Australia on advertising. Table 30.8 shows recent advertising expenditure by food and beverage manufacturers in Australia in 1986. The total expenditure A$345.8 million dollars, represents 21.6 per cent of total national advertising expenditure (A$1.6 billion).

So, clearly, while sections of the food industry and Government have the same goals, the aggregated labelling-and-advertising-as-education/persuasion efforts of marketers of salt-free, low-fat, no-added-sugar, no-additives, high-fibre, and calcium-fortified products, and the ability of the publishing industry to place daily nutritional information before the eyes of millions of consumers, make the comparative efforts of Government agencies rather paltry. Given this perspective, there are some fundamental questions worth asking of Government-run and sponsored nutrition education efforts. What, within a rational and co-ordinated intersectoral approach, should the contributions of non-governmental organizations and Government be to the education/information components of a national nutrition programme? What roles can Government and non-governmental organizations play that the food industry either cannot play, or can do more cost-effectively? These are fundamental questions which are increasingly dominating the planning agenda for nutritionists in the public sector in Australia. The range of possible roles for Government-sponsored nutrition policies and programmes includes:

(1) development and use of cost-efficient mass-reach strategies in nutrition education;

(2) administration of, and advocacy for, regulation of food standards, nutrient labelling, and advertising;

(3) formation of intersectoral mechanisms between Government departments, non-govermental organizations, and the private sector to promote nutritional concerns in policy-making (e.g. primary industry), to co-ordinate efforts/avoid duplication, and to advocate desired changes;

(4) subsidies for primary food industries to encourage product development consistent with dietary guidelines (e.g. leaner meat, improved fishing methods);

(5) development of policy and guide-lines for dietary practice in Government institutions serving and selling food (e.g. schools, hospitals, prisons, office canteens, interstate trains);

(6) 'honest brokerage' of information: opposing misinformation;

(7) development of and participation in a national research strategy in nutrition;

(8) training of health personnel in minimum standards of nutritional knowledge and skills.

References

Anonymous (1988). Fresh food 'predator' on frozens. *Foodweek* 5 April, 8.

Armstrong, B.K. and Holman, C.D. (1987). Acquired immunodeficiency syndrome in Australia. *Medical Journal of Australia* **146**, 61.

Arthurson, R.M. (1985). *Evaluation of random breath testing.* Research note 10/85. Traffic Accident Research Unit, Traffic Authority of New South Wales, Sydney.

Australian Bureau of Statistics (1985). *Children's immunization survey, November 1983.* Cat. No. 4352.0. Australian Bureau of Statistics, Canberra.

Australian Bureau of Statistics (1986). *Australian health survey, 1983.* Cat. No. 4311.0. Australian Bureau of Statistics, Canberra.

Australian Bureau of Statistics (1987*a*). *Deaths Australia.* Cat. No. 3302.0. Australian Bureau of Statistics, Canberra.

Australian Bureau of Statistics (1987*b*). *Statistics on drug abuse in Australia, 1987.* Commonwealth Department of Health, Australian Government Publishing Service, Canberra.

Australian Bureau of Statistics (1988). *Causes of death Australia, 1986.* Cat. No. 3303.0. Australian Bureau of Statistics, Canberra.

Baggaley, J.P. (1988). Perceived effectiveness of international AIDS campaigns. *Health Education Research* **3**(1), 7.

Baghurst, K.I. and Syrette, J.A. (1987). The influence of gender and social class on the nutrient contribution of meat and meat products to the diet of a group of urban Australians. *CSIRO Foods Research Quarterly* **47**:37.

Better Health Commission (1986). *Looking forward to better health.* Volumes I and II. Australian Government Publishing Service, Canberra.

Brotherhood of St. Laurence (1987). *Poverty update.* March 1987 figures. Fitzroy, Victoria.

Browning, B. (1988). *Politicising health issues—the hidden agenda.* Paper presented to 'The rise of the nanny state' seminar, Australian Chamber of Commerce, Adelaide, 6 April.

Chapman, S. (1985). Stop smoking clinics: a case for their abandonment. *Lancet* **i**, 918.

Chapman, S. (1988*a*). Smaller packs of cigarettes. *American Journal of Public Health* **78**, 92.

Chapman, S. (1988*b*). Case study in intersectoral coalition building: the successful appeal to the Australian Trade Practices Tribunal over abuses of the advertising self-regulation system. *Proceedings of 2nd International Conference on Health Promotion*, 5–10 April 1988, Adelaide, p. 71.

Chapman, S. and Reynolds, I. (1982). The mass media and smoking cessation: the ABC-TV Nationwide 'I Quit' program. *Community Health Studies* **6**, 247.

Chapman, S. and Reynolds, C. (1987). Regulating tobacco—the South Australian Tobacco Products Control Act 1986. *Community Health Studies Supplement* **11**, 9.

Chapman, S., Brown, R., Daube, M., McMichael, A.J., and Woodward, S. (1988). The Australian Tobacco Research Foundation. *Medical Journal of Australia* **149**, 46.

Chesterfield-Evans, A. (1987). *B.U.G.A. U. P.—the lessons learned four years on.* Paper presented to 6th World Conference on Smoking and Health, Tokyo.

Commonwealth Bureau of Census and Statistics (1909). *Commonwealth Demography Bulletin* No. 13.

Commonwealth Department of Health (1985). *Annual Report 1984–85.* Australian Government Publishing Service, Canberra.

Cumming, R.G., Barton, G.E., Fahey, P.P., Wilson, A., and Leeder, S.R. (1988). The Western Sydney Health Study: results of the shopping-centre survey. *Medical Journal of Australia* **148**, 277.

Department of Transport and Communications (1987). *Road crash statistics.* Department of Transport and Communications, Canberra.

Dobson, A.J., Gibberd, R.W., Leeder, S.R., and O'Connell, D.L. (1985). Occupational differences in ischemic heart disease mortality and risk factors in Australia. *American Journal of Epidemiology* **122**, 283.

Egger, G., Fitzgerald, W., Frape, G., *et al.* (1983). Results of a large scale media antismoking campaign in Australia: North Coast 'Quit For Life' program. *British Medical Journal* **287**, 1125.

English, R.M. and Bennett, S. (1985). Overweight and obesity in the Australian community. *Journal of Food and nutrition* **42**, 2.

Health Targets and Implementation (Health for All) Committee to the Australian Health Ministers (1988). *Health for all Australians.* Report to the Australian Health Ministers' Advisory Council and the Australian Health Ministers' Conference. Australian Government Publishing Service, Canberra.

Hill, D. (1987). *Public reaction to government initiatives to reduce tobacco consumption.* Centre for Behavioural Research in Cancer, Anti-Cancer Council of Victoria, July.

Hill, D. and Gray, N. (1984). Australian patterns of tobacco smoking and related health beliefs in 1983. *Community Health Studies* **8**, 307.

Hill, D.J., White, V.M., and Gray, N.J. (1988). Measures of tobacco smoking in Australia by means of standard method. *Medical Journal of Australia* **149**, 10.

Holman, C.D.J. and Shean, R.E. (1987). Tobacco-related hospital admissions. *Medical Journal of Australia* **146**, 117.

Kalucy, R.S. (1987). Public health funding within the National Health and Medical Research Council. In *Funding for Public Health Research in Australia.* Australian Epidemiological Association, Adelaide.

Lee, S.H., Smith, L., d'Espaignet, E., and Thomson, N. (1987). *Health differentials for working age Australians.* Australian Institute of Health, Canberra.

McGinnis, M. (1985). Setting nationwide objectives in disease prevention and health promotion: the United States experience. In *Oxford Textbook of Public Health*, Volume 3, p. 385. (ed. W.W. Holland, R. Detels, and G. Knox with assistance of E. Breeze). Oxford University Press, Oxford.

McMichael, A.J. (1985). Social class (as estimated by occupational prestige) and mortality in Australian males in the 1970s. *Community Health Studies* **9**, 220.

The Roy Morgan Research Centre Pty Ltd. (1987). *Survey on healthy foods. April 4/5, 1987. An Australia-wide probability sample of 1320 people.*

Montague, M. (1988). The metaphorical nature of drugs and drug taking. *Social Sciences and Medicine* **26**(4), 417.

Morlet, A., Guinan, J.J., Diefenthaler, I., and Gold, J. (1988). The impact of the 'Grim Reaper' national AIDS educational campaign on the Albion St (AIDS) Centre and the AIDS Hotline. *Medical Journal of Australia* **148**, 282.

National Health and Medical Research Council Special Unit in AIDS Epidemiology and Clinical Research (1989). Cumulative analysis of AIDS in Australia (roneo), 9 January.

National Heart Foundation of Australia (1983). *Risk factor prevalence study.* National Heart Foundation of Australia, Canberra.

National Heart Foundation of Australia (1986). *Heart Facts Report 1986.* National Heart Foundation of Australia, Canberra.

Newell, G.J. and Skurray, G.R. (1987). Changes in the apparent consumption of foods and nutrients in Australia 1975–6 to 1983–84. *Food Technology in Australia* **39**(5), 242; 253.

Pierce, J.P., Dwyer, T., Frape, G., Chapman, S., Chamerlain, A., and Burke, N. (1986). Evaluation of the Sydney 'Quit For Life' anti-smoking campaign. Part 1. Achievement of intermediate goals. *Medical Journal of Australia* **144**, 341.

Thomson, N. (1984). Aboriginal health and health care. *Social Science and medicine* **18**, 939.

Thomson, N. (1985). Review of available Aboriginal mortality data, 1980–1982. *Medical Journal of Australia* **143,** S46.

Thomson, N. (1986). Aboriginal health: current status and priorities. In *Proceedings of the second medical symposium, Royal Flying Doctor Service*, Sydney, p. 9.

Smith, L. and Lee, S.H. (1987). *Changing sex differentials in mortality: II. Recent changes in Australia.* Paper presented at the first Conference of the Public Health Association of Australia and New Zealand, Sydney, 24–26 August.

Sontag, S. (1979). *Illness as metaphor.* Vintage, New York.

Wilson, D. (1984). *The A–Z of campaigning.* Heinemann, London.

Woodward, S.D. (1984). Trends in cigarette consumption in Australia. *Australian and New Zealand Journal of Medicine* **14**, 405.

World Health Organization, Regional Office for Europe (1986). *Targets for health for all.* World Health Organization, Copenhagen.

Worsley, A. and Crawford, D. (1985). Awareness and compliance with the Australian Dietary Guidelines: a descriptive study of Melbourne residents. *Nutrition Research* **5**, 1291.

Young, S. (1987). Top national advertiser. *Broadcasting and Television Weekly*, 3 July, 25.

F

Co-ordination and Development of Strategies and Policy for Public Health

31

Co-ordination and development of strategies and policy for public health promotion in the United States

JULIUS B. RICHMOND and MILTON KOTELCHUCK

Introduction

In 1952 a conference was held in the US on 'Preventive medicine in medical schools' (Clark 1952). At that conference, the distinguished medical historian, Professor Richard Shryock observed:

> . . . it is still difficult for physicians today to conceive of care which transcends the traditional hospital type of diagnosis and cures. They were trained in the heyday of an almost exclusively somatic medicine based on a localized pathology and the related concept of disease specificity. Medicine of this type has achieved so much that some men quite naturally feel that it alone merits serious consideration. But in actuality, they are unable to view it critically because they are unfamiliar with any other type of medicine—that is, with the forms which preceded present medicine and which may, in part, supersede it in the future.
>
> The moment one examines historically the medicine which has been dominant in our own times, it becomes clear that this represents a late phase of medical development. Concepts now taken for granted hardly began to emerge before 1800, and did not command the thinking of most practitioners until about 1875. Prior to that time, most physicians were more concerned with the general condition of the patient in relation to his environment that they were with localized pathology or even specific forms of illness. They were interested in persons rather than in diseases and their medical education had implemented these outlooks through the apprenticeship, which had enabled them to see patients as persons in their normal surroundings.

The objectives of that conference were defined in terms that are relevant today:

> The goal of health now at mid-century calls for not only the cure or alleviation of disease. It calls for even more than the prevention of disease. Rather, it looks beyond, to strive for maximum physical, mental and social efficiency for the individual, for his family and for the community.
>
> To work towards this ideal, though, of course, never completely attainable end, requires the informed efforts of each individual person to improve his own health, requires provision of health resources by organized society, requires the skills and guidance of physicians and the assistance of various associated professional groups, and indeed of many other individuals, groups and agencies engaged in health activities in the community.
>
> Coordinated health services must be directed toward providing a wide spectrum of comprehensive health care, promoting health, preventing disease, detecting and treating disease at the earliest possible moment to prevent disabling sequelae, limiting disability to the greatest extent possible if disease becomes established, and restoring the individual to his most useful practicable functioning when permanent disability is inevitable.

These objectives remain to be fully achieved in the US, but much progress has taken place in the 30-plus years since that conference. This paper examines these developments historically, focusing principally on the role of the US Federal Government in organizing and promoting public health and disease prevention activities.

The first public health revolution: control of infectious disease

The improvement of the health of the people of the US over the past two centuries can be understood in the context of changes taking place in our society. The agricultural advances during this time were making it more possible to reduce hunger and concomitantly to improve nutrition. Industrialization and urbanization were associated with improved incomes and consequent improvement in housing and recreation, although, unfortunately, new occupational and environmental hazards were also introduced in the process (McKeown 1979).

The developments in the natural sciences gradually made it possible to gain a better understanding of the transmission of the infectious diseases. The awareness of the concept of communicability of disease fostered the development of public health practices which focused on improved sanitation and hygiene. The discovery of microbiological organisms in the latter part of the nineteenth century accelerated the understanding of transmissibility. A basic understanding of immunity led to the development of immunizing agents for some of the infectious diseases. Not too long afterward, the development of antimicrobial agents facilitated improved preventive and particularly therapeutic efforts to control further these diseases. The success of these efforts is often referred to as the first public health revolution.

The change in consciousness: towards a second public health revolution

These changes in nutrition, housing, and public health practices, characterizing the first public health revolution, were destined to have far reaching effects on morbidity and mortality patterns in the US. In turn these new morbidity patterns and the efforts to deal with them would provide the basis for the growth of a new consciousness about the need for a new and different public health practice in the US today.

Morbidity and mortality changes since 1900

The US Public Health Service's National Center for Health Statistics in recent decades reported that non-infectious diseases—particularly cardiovascular diseases and cancer—have become the leading causes of mortality (Fig. 31.1). In 1900, five of the ten leading causes of death, including the top three, were infectious (pneumonia and influenza; tuberculosis; diarrhoea, enteritis, and ulceration of the intestine); by 1980 only one of the ten leading causes of death was infectious (influenza and pneumonia ranked sixth). The major infectious diseases of childhood in the past—poliomyelitis, measles, rubella, mumps, diphtheria, pertussis, and tetanus—have been virtually eliminated; smallpox has been eradicated from the world; and there is now specific therapy for tuberculosis. This decline has brought about significant improvements in life expectancy and the quality of life in the US.

These improvements have also been associated with a change in the public's perception of its health needs. North Americans are no longer predominantly worried about infectious diseases; rather, the principle concerns have shifted to chronic diseases and the quality of life, especially for the elderly. Professor Archibald Cochrane has described this shift as one of moving from curing to caring (Cochrane 1972). For example, Susan Sontag, in her essay on 'Illness as metaphor' has noted that the central literary disease metaphor of the late 1800s was infectious tuberculosis or 'consumption' while the mid-twentieth century metaphor is cancer (Sontag 1978). The recognition of a problem and the ability to ameliorate it are, however, two distinct steps.

Growing public awareness that non-infectious diseases are amenable to public health intervention

During the early years of this century, while great strides were being made in reducing the incidence and severity of infectious diseases, little progress was being made in treating chronic, non-infectious diseases. Chronic illnesses were thought of as long term and incurable. In order to facilitate public health programmes for people with chronic illnesses, the public and professionals needed to be shown that the course of these diseases could be ameliorated.

Cardiovascular diseases, the leading causes of mortality since the mid-century, provided the critical example. As can be seen in Figures 31.2a and b heart and cardiovascular diseases have experienced sharp declines in mortality rates over the past 35 years (US Department of Health and Human Services 1987*a*). Figure 31.3 shows the dramatic declines in mortality from stroke (50 per cent) and coronary heart disease (35 per cent) in just the past decade and a half. This decline in cardiovascular diseases, probably resulting from a decreasing percentage of adult smokers, better hypertension detection and control, more attention to diet, exercise, and stress reduction, along with improvements in cardiac intensive care units, pharmacological agents, and surgical procedures, demonstrated that the incidence of non-infectious illnesses can be reduced and that therapies and rehabilitation can be effective.

The growing association of cigarette smoking with lung cancer also played a role in public perception about chronic illness. Epidemiological research of the 1940s (Doll and Hill 1952), and particularly the 1959 study of the American Cancer Society (Hammond 1966) which was based on a nation-wide survey of smokers and non-smokers, linked smoking and cancer in the minds of professionals. By 1957 the Surgeon General of the US Public Health Service, Dr Leroy Burney, issued a special statement on 'excessive cigarette smoking' based on the growing recognition of its hazard to health. The basic and simple recommendation—'do not smoke'—a straightforward individual health activity, could lead to a reduction in the risk of a major chronic illness. Public perception about the ameliorability of non-infectious illnesses was changing.

Financial pressures for health care cost reductions

The growing financial burdens of the present US health system—a system principally oriented towards acute medical care—has also provided a steady impetus to reorient public health activities in the US. Expenditures on health care as a percentage of gross national product (GNP) have been

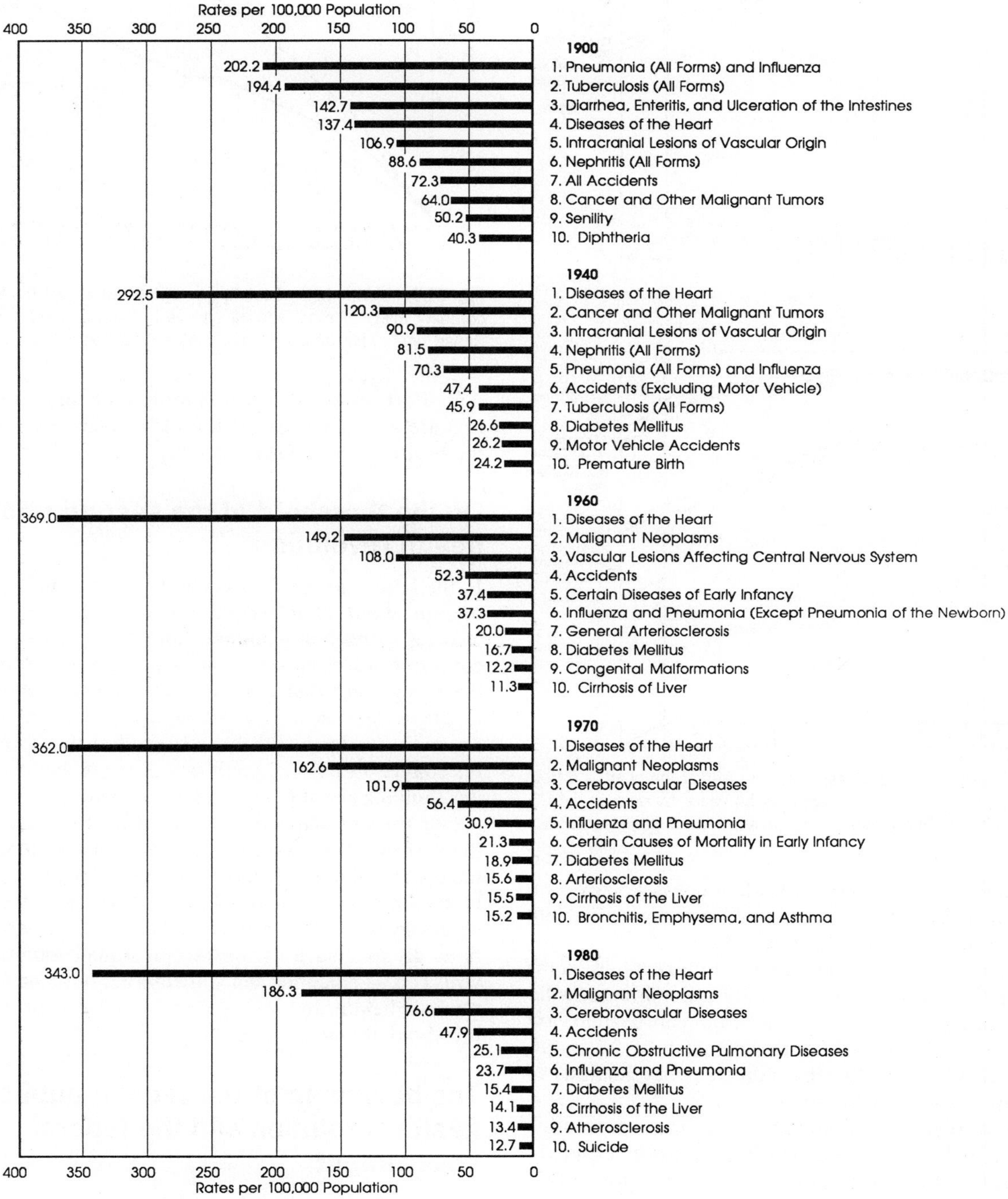

Fig. 31.1. The ten leading causes of death in the US, from 1900 to 1980. (Source: Levy and Moskowitz 1982.)

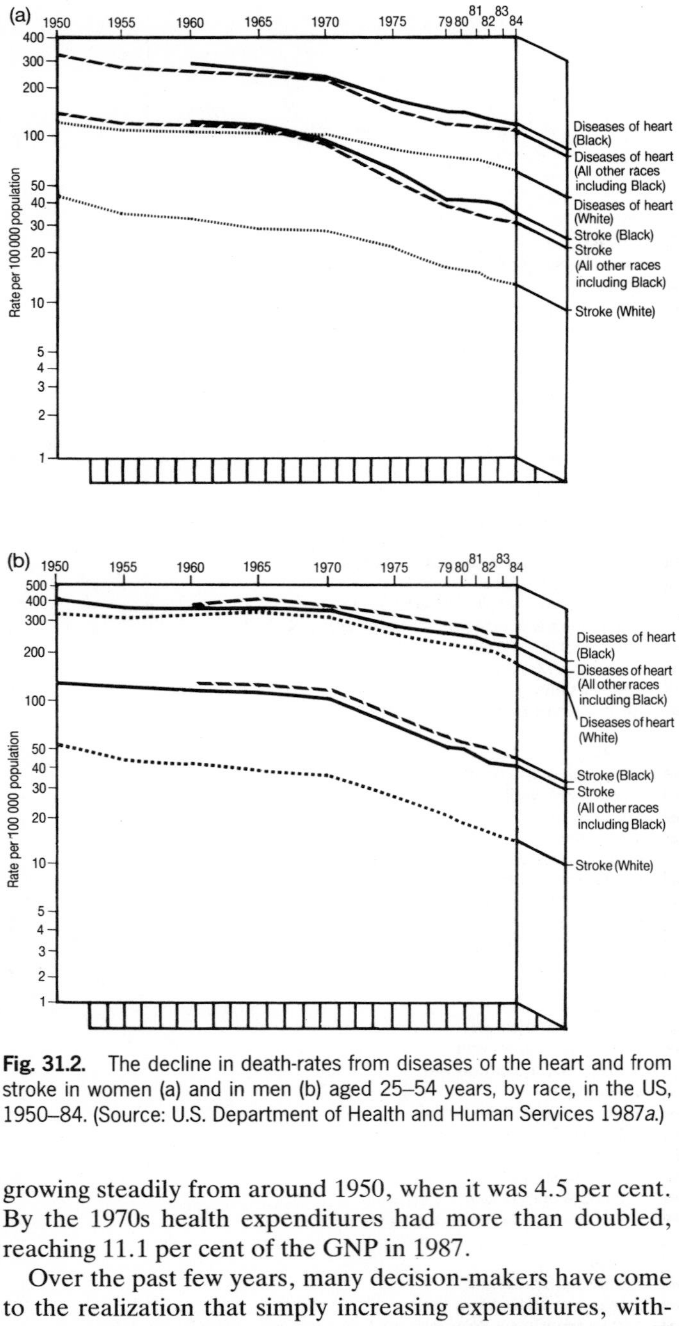

Fig. 31.2. The decline in death-rates from diseases of the heart and from stroke in women (a) and in men (b) aged 25–54 years, by race, in the US, 1950–84. (Source: U.S. Department of Health and Human Services 1987*a*.)

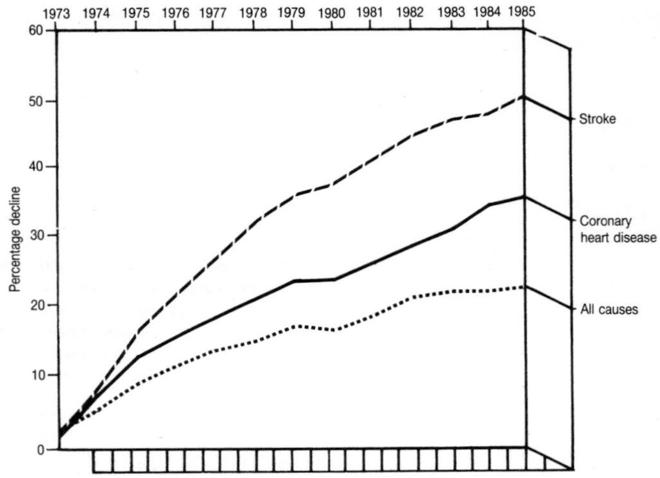

Fig. 31.3. Percentage reduction in age-adjusted death-rates for stroke, coronary heart disease, and all causes in the US, 1973–85. (Source: US Department of Health and Human Services 1987*a*.)

growing steadily from around 1950, when it was 4.5 per cent. By the 1970s health expenditures had more than doubled, reaching 11.1 per cent of the GNP in 1987.

Over the past few years, many decision-makers have come to the realization that simply increasing expenditures, without reorganizing the priorities of the health care system, will not necessarily lead to major improvements in health status. The appreciation of this fact has forced policy analysts to rethink where money is being spent and why.

For the past 20 years, national expenditures on prevention have represented only a very small proportion (approximately 4 per cent) of the total US expenditures for health. Yet it is perceived that these modest expenditures for health promotion and disease prevention have had a major influence on health status. This underlying fact has heightened the public, professional, and political consciousness that efforts directed at health promotion and disease prevention are a feasible and perhaps less costly way of improving the health of the nation.

On the threshold of the second public health revolution

In the latter part of the twentieth century, the US stands on the threshold of a second public health revolution. The changes in morbidity and mortality, the predominant shift of concern towards chronic illnesses rather than acute illnesses, the recognition that non-infectious illnesses were amenable to prevention as well as to intervention, and the growing expenditures for health have suggested that there is a need for change in public health practices. A dynamic change in our thinking about health has developed.

Our present acute care oriented health system stands on the verge of a conversion to a health care system focused on the control of non-infectious diseases, care for the chronically ill, health promotion, and disease prevention. Medical care is obviously still very much desired and needed, and new infectious diseases such as the recent human immunodeficiency virus (HIV) epidemic can still emerge, but a new orientation towards health and new public health practices are on the threshold of emerging.

The beginning of the second public health revolution and the federal government

The second public health revolution began slowly in the programme of the Federal Government, largely in a piecemeal fashion. There was no initial master plan and no dramatic call for action. Perhaps the seminal event of this new era was the publication of the 1964 Surgeon General's Report on Smoking and Health (US Department of Health, Education and Welfare 1964).

The first Surgeon General's Report on Smoking and Health, 1964

To provide a sound scientific analysis of all the data and to advise further the Public Health Service, the then Surgeon General, Dr Leroy Burney, appointed an Advisory Committee on Smoking and Health, which rendered its report *The Surgeon General's Report on Smoking and Health* on 11 January 1964, when Dr Luther L. Terry held the position. It concluded that 'Cigarette smoking is a health hazard of sufficient importance in the United States to warrant appropriate remedial action'.

This report had a remarkable impact on the public perception of what could be done to promote health and prevent disease. The following year Congress passed the Federal Cigarette Labeling and Advertising Act (Public Law 89–92) which required that all cigarette packages distributed in the US carry a Surgeon General's warning label concerning the hazard to health, and Congress not long after passed the Public Health Cigarette Smoking Act 1970 (Public Law 91–222) which banned cigarette advertising from radio and television. It is probably fair to say that this report had the effect of raising the level of consciousness concerning the health hazards associated with increasing affluence and industrialization as no other event had.

Several important implications derived from this report. The report educated the public about the importance of cigarette smoking as a cause of illness, and by implication suggested ways by which a personal health promotion activity—not smoking—could reduce the likelihood of illness. This report was a testament to the fact that non-infectious diseases, especially some forms of cancer, were amenable to public health intervention. It symbolized the first major federal initiative to develop new health promotion and disease prevention programmes.

Implementation of the Surgeon General's report

While this report was a historical moment for public health and the Federal Government, the limitations of its implementation must also be noted. The main effects of the report were educational in nature, summarizing a scientific consensus, restricting advertising on radio and television, and presenting a written warning to all users of cigarettes. No efforts were made to ban cigarettes, to stop federal subsidies to tobacco growers, or to increase the excise tax. This report did not crystallize a co-ordinated federal health promotion and disease prevention policy. Few resources were provided for these efforts. It seemed that issuing the report was all that could be achieved at that time, for the tobacco industry had formidable political influence.

Other federal activities in health promotion in the 1970s

Besides smoking prevention, several other programmes in health promotion and disease prevention began in the Federal Government during the 1970s. These programmes often began without fanfare and without a central theme guiding their development. They were more specific responses to specific health needs, rather than a unified national and comprehensive health promotional activity programme.

There was a growing awareness of environmental problems resulting from industrialization, urbanization, and the introduction of new chemicals and pesticides into the environment. As a consequence, during the decades of the 1950s and 1960s many narrow State and federal laws were passed dealing with efforts to establish standards for and control of air and water pollution and the use of chemicals such as insecticides, fungicides, and rodenticides. By the late 1960s it seemed clear that there was a need for a more orderly federal approach to environmental concerns. This need was met through the National Environmental Policy Act (Public Law 91–190) which established a Council on Environmental Quality to advise the President; it required an environmental impact statement to be submitted before any new major federal actions were undertaken. Soon after, the federal Environmental Protection Agency was formed to deal with the myriad of environmental problems that were surfacing.

Because of the increasing association of health hazards with the workplace, the US Public Health Service was directed by Congress to establish a National Institute of Occupational Safety and Health in 1970. It was charged with setting health standards for the workplace and conducting research on these issues. It provides data for the Occupational Safety and Health Administration of the Department of Labor, which has the responsibility for regulating health practices in work settings.

The growing interest in health promotion was further manifested by the establishment in 1972 of the National High Blood Pressure Education Program under the auspices of the National Heart, Lung and Blood Institute of the National Institutes of Health. Identification and referral of persons with high blood pressure for appropriate management was the programme's main emphasis. This programme further raised consciousness concerning prevention generally, while introducing specific programmes that have been associated with a decline in mortality for diseases related to hypertension–coronary heart disease and stroke. The rapid advances that were taking place in the management of hypertension enhanced the positive impact of this programme.

These several federal programmes of the 1960s and early 1970s—the Surgeon General's Report on Smoking and Health, the establishment of the National Institute of Occupational Safety and Health, the development of hypertension detection and control programmes—each entered the Federal Government further into the health promotion area.

Table 31.1. 1965 Congressional legislation

Medicare (Title XVIII)
Medicaid (Title XIX)
Regional medical programs
Comprehensive health planning assistance
Health professions education assistance
Maternal and infant care, children and youth projects (Title V)
Economic opportunity act
Neighborhood health centres
Head start

Yet it must be noted that each of these programmes was really not systematically co-ordinated with any of the other disease prevention programmes. No overall strategy for health promotion had yet developed in the Federal Government. Health promotion remained a series of somewhat disjointed activities.

The development of a national health policy

It is important to note that while health promotion and disease prevention activities of the Federal Government were not driven by a national health policy, neither were most of the other areas of health care activities. Indeed, prior to 1964, the Federal Government had only limited involvement in health care. Health was not seen as an area of major federal responsibility, but rather as a complex mix of private, state, local, professional, and agency responsibilities. As a result, there was not an explicit health policy of the US Government. Federal policy was simply the sum of its individual health programmes; conceptual themes were not enunciated. Publicly articulated policy in health is a relatively recent phenomenon (Richmond and Kotelchuck 1983).

The entrance of the Federal Government in a major way into the provision of financing of health care services came in 1965. In 1965 Congress passed more important pieces of health legislation than ever before. Table 31.1 indicates some of the important bills of this bumper crop of legislation (Richmond 1969). The enactment of these programmes reflected the social ferment which was taking place in the nation at that time in the form of the civil rights revolution and the national efforts to achieve greater equity in the distribution of human services, including health services. In particular, the political movements for a health insurance programme for the elderly (Medicare) and a publicly financed programme of health services for the poor (Medicaid) culminated in Congressional legislation in 1965. Moreover, Congressional perception of growing deficits in national health personnel, hospital beds, and health care facilities resulted in a legislative push to redress the perceived national health resources deficits through the authorization of programmes for Community Health Centers, Migrant Health Centers, a National Health Service Corps, and the Health Manpower Training Act.

These new programmes, moving toward equality in access to health care resources and funding of health services for the elderly and the poor, were associated with major improvements in the nation's health status (Aday *et al.* 1980; US Dept of Health and Human Services 1980*a*). The benefits of the American health care system were now potentially available to all members of our society. For the first time, starting in 1965, the US Government had become actively involved in providing direct financing of health care for major sections of the US population and in providing direct health services through specific programmes. Ultimately these programmes paid for approximately 40 per cent of medical care; private sources, especially voluntary health insurance, paid for the remaining 60 per cent.

The concept of a national health policy developed in this era. Equity of access to health care services through the provision of direct services and resources for the poor and elderly shaped the programmes of this initial era. Specific national programmes were mounted to overcome national deficits. Health promotion and disease prevention were not major issues during this initial period of federal health activities.

Federal interest in health promotion and disease prevention activities

Although there were a growing number of agencies in the Federal Government whose mission can be broadly interpreted as being in the health promotion and disease prevention area, it is important to note that there was also an equally pervasive lack of federal understanding and commitment to this area. This is perhaps best seen in the Medicare Program. The initial Medicare legislation specifically spelt out that preventive services were not to be reimbursed by this programme for older people; it focuses explicitly on curative services. Preventive medicine was not yet on the national agenda.

Moreover, there was no agency of the Federal Government in the 1970s which was mandated to be responsible for health promotion and disease prevention activities in any broad sense. While scattered programmes existed, no single agency guided them.

The growth of public awareness of the need for national action

During the 1970s the public perception grew that insufficient national or local attention was being paid to preventive efforts and health promotion. There were several reasons underlying this:

1. The continued growth of public awareness of the impact of cigarette smoking on health. As the fifteenth anniversary of the 1964 report of the Surgeon General's Advisory Committee on Smoking and Health approached, and a complete review of the literature since the publication of the 1964 Report was undertaken, it became apparent that the additional research overwhelmingly

ratified the original scientific indictment of cigarette smoking as a significant contributor to disease and premature death. The then Secretary of the Department of Health, Education and Welfare wrote in the forward to the Report:

> . . . Smoking accounts for an estimated $5 to $8 billion in health care expenses, not to mention the cost of lost productivity, wages, and absenteeism caused by smoking-related illness; an annual cost estimated at $12 to $18 billion. No person, given these staggering costs, can reasonably conclude that smoking is simply a private concern; it is demonstrably a public health problem also.
>
> *A second major problem is that our health care system over-emphasizes expensive medical technology and institutional care, while it largely neglects preventive medicine and health promotion.* [italics ours] Certainly, if the government is to shift its health strategy toward preventive rather than merely curative medicine, it cannot ignore smoking. (US Dept of Health Education and Welfare 1979*a*).

2. The striking decline in mortality from heart disease and stroke continued. As these trends persisted and became even more dramatic, it became more apparent to the political leadership in the executive and legislative branches of Government that health promotion efforts could be significant, even when multiple risk factors rather than single, specific agents (as with the infectious diseases) were responsible for causation. The importance of health education and behaviour was becoming more evident.
3. The growing public interest in physical fitness and improved health. Perhaps because of the growing awareness of the importance of exercise, diet, control of high blood pressure, smoking avoidance or cessation, and reduction of stress, a trend developed for the practical applications of this knowledge. Jogging almost replaced baseball as the national sport. Health promotion organizations and activities proliferated. These health practices were embraced initially by people with higher educational backgrounds, but have become more pervasive with time.
4. There was growing concern over environmental hazards. As epidemiological studies progressed indicating linkage between environmental issues and health, citizens showed more concern and commitment for action than in prior years. The proliferation of chemicals and other pollutants in the environment, and their potentialities for carcinogenesis, mutagenesis, sensory and psychological difficulties, as well as for aesthetic deterioration in communities, presented many complex problems for policy-makers. But public concern was significant, and increasingly sought greater protection, even though the knowledge base was often not entirely adequate for more effective action.
5. *A new perspective on the health of Canadians* was published (Lalonde 1974). In 1974, the Canadian Government under the leadership the then Minister of Health and Social Affairs, Mr M. Lelonde, called attention to the importance of a new approach to health problems. The recommendations centred around the following four categories: inadequacies in the existing health care system; behavioural factors or unhealthy life-styles; environmental hazards; and human biological factors. While this report has not been implemented as forcefully in Canada as many had hoped, it had a world-wide impact in stimulating a renewal of interest in health promotion and disease prevention.

The decision to have an explicit health promotion and disease prevention policy

By the latter half of the 1970s it was apparent that many changes had developed in the public perception about health, and that a redirection of health policy was indicated. With the change in federal administrations in 1977, a new administration committed itself to developing health policy in a more integrated way and to generating the political will to support the new directions. The federal administration formulated a health policy explicitly focused on improving the health of people and not predominantly on the financing of medical care (US Department of Health, Education and Welfare 1980).

The policy had three priorities which were designated to accomplish this:

1. Improving equity in the delivery of health services. Although there had been much progress since 1965, redistribution of resources for those in need still remained an important goal.
2. Increasing activities to promote health and prevent disease. For the first time, health promotion was to be a guiding principle of federal activities.
3. Increasing support for research in biomedical sciences, mental health, alcoholism, drug abuse, and health services. The federal commitment to expanding the health knowledge base continued.

While all three areas were thought of as equally important, there was a general perception that considerable prior federal programmatic attention had been given to the two priorities of extending health services and to biomedical and behavioural research, but that preventive efforts and health promotion had not received appropriate emphasis.

The Surgeon General's report on healthy people

Because of the clustering of these trends, federal health officials believed that the time had come to undertake 'consciousness raising' and 'strategic planning' efforts for health professionals and citizens broadly. The Surgeon General of the Public Health Service organized the efforts of the Service for the publication, in 1979, of a comprehensive report *Healthy people—The Surgeon General's report on health pro-*

Table 31.2. Health strategy recommendations from the Healthy people* report

Preventive health services for individuals
Family planning
Pregnancy and infant care
Immunizations
Sexually transmissible diseases services
High blood pressure control
Health protection for population groups
Toxic agent control
Occupational safety and health
Accidental injury control
Community water supply fluoridation
Infectious agent control
Health promotion for population groups
Smoking cessation
Alcohol and drug abuse reduction
Improved nutrition
Exercise and fitness
Stress control

* US Department of Health, Education and Welfare (1976*b*).

motion and disease prevention (US Department of Health, Education and Welfare 1979*b*).

The report represented the consensus of experts in the field of health promotion and disease prevention from both governmental and non-governmental organizations and, for the first time, established quantitative goals to be achieved in health over the course of the next decade. A national policy had emerged. These goals were specified by age group, and a brief condensation of these are presented:

(1) to continue to improve infant health, and, by 1990, to reduce infant mortality by at least 35 per cent, to fewer than nine deaths per 1000 live births;

(2) to improve child health, foster optimal childhood development, and, by 1990, to reduce deaths among children aged 1–14 years by at least 20 per cent, to fewer than 34 per 100 000;

(3) to improve the health and health habits of adolescents and young adults, and, by 1990, to reduce deaths among people aged 15–24 years by at least 20 per cent, to fewer than 93 per 100 000;

(4) to improve the health of adults, and, by 1990, to reduce deaths among people aged 25–64 years by at least 25 per cent, to fewer than 400 per 100 000;

(5) to improve the health and quality of life for older adults and, by 1990, to reduce the average annual number of days of restricted activity due to acute and chronic conditions by 20 per cent, to fewer than 30 days per year for people aged 65 years and older.

In addition to the goals defined by age groups the report systematically developed a health strategy with specifically defined targets (Table 31.2).

The publication of the report *Healthy people* was followed by the convening of a large group of experts in the nation to detail further the health objectives and strategy for the nation in quantitative terms. Their report entitled *Promoting health; preventing disease. Objectives for the nation* provided more detailed quantitative goals toward which the public and private agencies in the US should be striving (US Department of Health and Human Services 1980*b*). A total of 226 specific disease prevention and health promotion objectives were adopted. These objectives fell into four broad groupings: risk reduction; public and professional awareness; service improvement; and surveillance and evaluation. This report has served as a road map for health professionals, as well as for citizens generally, to determine whether or not we are fulfilling our greatest potentialities in attaining better health for the people of the US. It emphasizes the systematic collection of data to measure the effectiveness of the efforts.

Further complementing this effort was the preparation of two major federal reports, to help in the implementation of the many efforts in prevention which were ongoing through State and local health departments, as well as to focus them on to the national goals. The two reports were *Model standards for community preventive health services* (Centers for Disease Control 1979) and the later publication *Model standards: A guide for community preventive health services* (US Public Health Service 1985).

The value of the Surgeon General's report on healthy people

Healthy people may well be the symbolic initiation of the second public health revolution. For the first time a strategic document was prepared which:

1. Gives unity to the concept of health promotion and disease prevention. No longer was this an area of discrete programmes, but a framework exists for common philosophical and programmatic unity. Health promotion is not a discrete programme, but unifying philosophy.
2. Presents quantifiable goals. Health promotion is often unfairly characterized as being too general and virtuous and lacking in specific, practical recommendations. *Healthy people* explicitly set quantifiable goals which were deemed realistic and achievable by professionals. Programmes are defined so that evaluations of success and failure can be measured.
3. Focuses on specific age groups. Health promotion was examined developmentally. For every age group, goals for improvement of the health for that group were targeted. Benefits to all citizens are thereby highlighted for all stages of their life.
4. Represents professional consensus. The report derived from a process of professional consensus over feasibility. *Healthy people* was not simply a document reflecting the

aspirations of one set of administrators of public programmes.

Health promotion and disease prevention during the 1980s

Once a new national health policy is articulated and institutionalized, as in the late 1970s, it tends to have a momentum of its own, despite changes in federal leadership and ideology. Several major federal and local health promotion and disease prevention activities and trends initiated by *Healthy people* continued developing during the 1980s.

First, the 1990 Health Objectives for the Nation were actively monitored by the Federal Government. Numeric goals are powerful political and professional symbols. In 1986, a major analysis of the progress towards each of these goals was published by US Public Health Service in *The 1990 health objectives for the nation: A mid-course review* (US Department of Health and Human Services 1986*a*). The results, although mixed, were generally positive; nearly half of the 226 goals were on the way to being achieved, though nearly a quarter were not being achieved. For example, the goals of an infant mortality rate of 9.0 per 1000 births or the proportion of adults who smoke to be reduced to below 25 per cent appear to be achievable, while the goal of an infant mortality rate of 12.0 per cent 1000 births for all racial and ethnic groups appears not to be achievable. In general, risk reduction and public and professional awareness objectives were the least well achieved. Examination of the objectives not being achieved, and the reasons, is as important as the examination of the ones making progress. In areas where achievement languished, renewed efforts and changes in federal strategy were initiated. The existence of numeric goals has helped keep attention focused on health promotion and disease prevention efforts.

Second, the planning of numeric goals by decade for national health promotion and disease prevention achievements, has been firmly established. The US Department of Health and Human Services, with the advice of the Institute of Medicine of the National Academy of Sciences, is conducting public hearings and professional meetings to achieve a consensus on a new set of Year 2000 Health Objectives for the Nation. There is much competition. The value of being enumerated as a specific health objective has not been lost on specific interest groups. The implications for future federal priorities and funding are great. Quantifiable goals for guiding the Health Promotion and Disease Prevention programmes have now become institutionalized as an ongoing process of the Public Health Service.

Third, national data systems which can measure health promotion and disease preventive behaviours and attitudes have been developed throughout the 1980s. The National Center for Health Statistics of the Public Health Service, for example, regularly collects 'health promotion data' as part of its ongoing National Health Interview Survey (Thornberry *et al.* 1986). This kind of ongoing nationally representative population-based data collection allows for the monitoring of the national health objectives, for evaluating national programmatic strategies, and for educating consumers, professionals, and politicians about health promotion and disease prevention. The existence of routine national data collection efforts symbolizes the arrival and establishment of the health promotion and disease prevention field.

Fourth, while relatively few new federal funds were directed towards the health promotion and disease prevention field in the 1980s, several recent federal initiatives are worthy of note. All federal public health agencies now coordinate their efforts through the Office of Disease Prevention and Health Promotion, headed by a Deputy Assistant Secretary for Health. The US Centers for Disease Control has funded the establishment of five major national Centers for Health Promotion and Disease Prevention. These prevention centres are the loci for multi-disciplinary prevention research and for the training of health professionals. Health promotion and disease prevention are being institutionalized in our major university health centres. For the first time nationally, injury prevention is being viewed as possible and necessary; the US Centers for Disease Control are now funding eight Injury Prevention Research Centers nationally. One must also note, however, that during this same period many federal health promotion and disease prevention activities were seriously limited by restrictive budgets. Even without new funds, the Surgeon General's Office has become a 'bully' pulpit, especially for health promotion and disease prevention efforts, particularly in the areas of smoking, nutrition, and acquired immune deficiency syndrome (AIDS) (US Department of Health and Human Services 1986*b*, 1988*a*, 1988*b*).

Fifth, the 1980s have been characterized by a major diffusion of health promotion and disease prevention activities from the federal to the State and local levels. The growth in local initiatives has been dramatic. Many States are monitoring their own progress towards meeting the 1990 National Health Objectives (Dumbauld 1987; Surles and Blue 1988). Regional prevention efforts such as NECON (the New England Coalition for Health Promotion and Disease Prevention) have developed. States and even local communities are attempting to influence their citizens' health through local prevention policy and practice initiatives. Local smoking ordinances, seat-belt laws, funding for health promotion activities, air and water conservation measures, infant mortality reduction initiatives, health education, etc. are all increasingly common at the local level. Some States, such as North Carolina, have started new State Health Promotion Programs using State funds. Also several of the major private philanthropic foundations are now beginning to support health promotion and disease prevention; for example, the Kaiser Family Foundation's Health Promotion Program or the Health of Public Program, funded by the PEW and Rockefeller Foundations to train clinicians in primary prevention and population-oriented medicine. Even private industry, with its need for a healthy work-force, has entered this arena. A recent national survey of health promotion activities in work-sites with 50 or more employees reports that 65 per cent had at least one health promotion activity—with smoking control the most common activity (35 per cent) and risk assessment, back care, stress management, and exercise/fitness programmes not far behind (US Public Health Service 1987).

Sixth, and most importantly, consciousness about health promotion and disease prevention continues to grow. Knowledge about and participation in health promotion activities have continued to increase Antismoking initiatives have proliferated in the 1980s. Smoking has been increasingly restricted in public offices, restaurants, aeroplanes, and hospitals, to name but a few areas. National smoking rates have continued to decline slowly, but steadily, throughout this period. Consumer groups, such as MADD (Mothers Against Drunk Driving), have proliferated and fostered stricter State drunk driving laws—again reflecting the growing belief that injuries can be prevented. Nutritious foods are an increasingly large percentage of the US national food budget; with virtually all the major food companies now conscious of this growing market segment. Physical fitness activities have continued to grow during the 1980s. Conservation groups concerned with clean air and water and other environmental issues have flourished during the 1980s; they have had the ability to unite often disparate liberal and conservative political philosophies around common goals. Virtually every major newspaper has broadened its feature pages to focus, at least weekly, on 'health' which generally means prevention. The national policies of the 1970s and 1980s have fostered a continued growth in public consciousness about health promotion and disease prevention.

Progress in health promotion and disease prevention continued and expanded in the 1980s, but not all the trends of 1980s have been positive. Our medical care system is still not prevention oriented (Laurie *et al.* 1987). There has yet to be any major reorganization of health care service delivery institutions or, more importantly, of the financing of health care services to reflect the growing importance of health promotion and disease prevention activities. Medicare still does not routinely reimburse preventive health care visits. Even the apparent public emphasis of Health Maintenance Organizations on preventive out-patient care often masks a more fundamental strategy to save money by limiting hospitalization. Prevention remains a very low proportion of the total US health care costs.

There has been much more federal emphasis in the 1980s on life-style issues than on institutional and environmental regulatory changes to improve health promotion and disease prevention. Regulatory changes, such as air and water quality, workplace regulations, etc., have seen relatively little attention in this decade. Many critical environmental issues remain to be tackled.

The recent Institute of Medicine report *The future of public health* (Institute of Medicine 1988) noted a major erosion

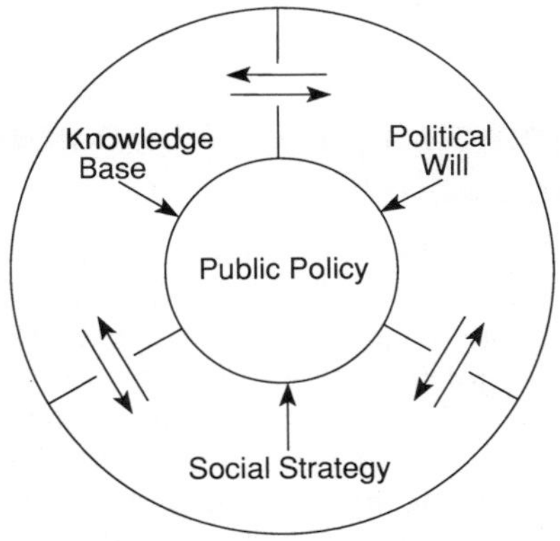

Fig. 31.4. The development of public policy: three-part health policy model. (Source: Richmond and Kotelchuck 1983.)

of the public health infrastructure in the US. The mission of State and local public health departments, as well as their funding, has not kept pace with changing health care needs in the country. The funding limitation of these agencies hinders the advancement of health promotion activities, since they are one of the main foci for these initiatives at the local level.

Healthy behaviours are not yet evenly distributed throughout the population. Changes in personal health prevention practices have followed the traditional route of diffusion of social practices in America. Many of the life-style changes initiated by more well-educated and financially able segments of the population have yet to be widely accepted or available to the poorer segments of the population.

New threats to the public health, such as AIDS, have emerged in the 1980s. The inadequacy and ambiguity of our knowledge concerning AIDS reminds us of the many public health challenges of past years. Yet, the absence of any known cures for AIDS re-emphasizes importance of the need for preventive efforts. We must act on the limited knowledge we have concerning prevention, while supporting research fully to inform us better of the fundamental nature of the disease.

The basis for further development of health promotion and disease prevention

Public policy emerges from the interactions of various forces in society. Health policy is no exception. If health promotion and disease prevention are to continue to move forward, three forces—the knowledge base, political will, and social strategy—need to be understood and brought into appropriate balance. This is shown visually in Figure 31.4.

Knowledge base

Recent biomedical research has improved our understanding of the multiple risk factors associated with cardiovascular diseases, strokes, cancer, and congenital anomalies; thus it has improved our understanding of how to approach problems of disease prevention and health promotion. Although progress in recent decades has been substantial, more can be accomplished. The knowledge base underlying health promotion and disease prevention is relatively weak.

Continuing biomedical research will undoubtedly further clarify the effects of the multiple risk factors for chronic illnesses. Three needed areas of research seem to hold much promise for furthering health promotion and disease prevention activities: toxicology, occupational and environmental health, and behavioural factors.

Research in the field of toxicology is not sufficiently well developed. There is a need to develop improved methodologies for determining, in an efficient and inexpensive way, the degree of human hazard associated with the introduction of new chemicals in the environment. We also need a better understanding of the toxic effects of two or more chemicals in combination. Recently the complexity of the numbers of chemicals located in toxic waste disposal areas has rendered it almost impossible to make any precise estimates on the hazards of these combinations.

The field of occupational and environmental health is undergoing a rapid transformation as it seeks to create safe and healthy working environments. The multiplicity of hazards in the work environment has been highlighted in recent years, ranging from reproductive hazards to respiratory difficulties and the development of occupationally caused cancer. The extent to which hazards in the workplace are controllable and the level of toxic agents which may be permissible remain to be more clearly determined.

Since so many issues related to health promotion and disease prevention are concerned with behavioural factors, it is apparent that we need to improve our competence in research in the behavioural sciences. We are only on the threshold of understanding effectively the issues related to habituation to cigarette smoking, alcohol consumption, drug dependence, eating patterns, and many other health-related behavioural issues. Clinical research programmes are developing in which the study of the modifications of behaviour can be observed systematically and quantitatively.

Awareness of the close relationship between health and behaviour influenced the Institute of Medicine of the National Academy of Sciences to undertake a study of the relationship of these issues. This report, published as a volume entitled *Health and behaviour* (Hamburg *et al.* 1982), establishes an agenda for research in this area and should serve to attract more behavioural scientists to research on the relationship between health and behaviour. It has become important to develop a deeper understanding of these relationships in order that citizens may use this knowledge to improve their health.

Political will

Political will stems from public understanding and support. The implementation of programmes depends upon a consensus which will provide the resources from public and private funding. In the US there is evidence in recent years that health promotion and disease prevention have broad public interest and support. It seems to be an idea whose time has arrived. What is the evidence for this?

First, there has been a growth in size and in number of voluntary health organizations. Organizations such as the American Cancer Society, the American Heart Association, and the American Lung Association have intensified their activities. Organizations concerned with antismoking efforts, improved nutrition, and other health promotional activities are proliferating and are playing an increasingly activist role in local communities. Non-smokers are becoming increasingly assertive about the right to be in a smoke-free environment. They are becoming more aggressive in efforts to enact ordinances prohibiting smoking in public places.

Second, many groups of employers are becoming aware of the importance of health promotion programmes for their industries. This is significant not only for maintaining productivity (e.g. happier and healthier workers), but it can, over time, potentially reduce industrial health care costs. Japanese industries have made excellent use of these highly popular employee programmes.

Third, the health insurance industry is coming to recognize the need to focus on health promotion. Some life insurance companies have reduced premiums for individuals who do not smoke and who otherwise reduce risk factors. Since such efforts which promote health are in the interest of life insurance firms, it is curious that they have not pursued these directions more aggressively in the past.

Fourth, the health professions are becoming more knowledgeable about health promotion. The developing knowledge base has largely dissipated the long-term concern among health professionals that we didn't know enough for such programmes to have some beneficial effects. There is a growing awareness that while we don't know everything we would like to know, it doesn't mean that we know nothing. Awareness of the limitations of professional knowledge is distinct from disinterest.

Fifth, legislators and other decision-makers are becoming more sophisticated concerning health promotion and disease prevention. The area of cost–benefit analysis will improve over time and become more helpful in the process of decision-making.

Finally, there is considerable individual interest broadly in personal health promotion efforts. This is manifested by the recent growth of activity in physical fitness, improved nutrition, and efforts at stress reduction. There has been an explosion of popular literature on health promotion activities.

It is appropriate to mention that there are also some strong sources of resistance to this growing movement. Some health professionals are concerned lest more allocation of resources to prevention will detract from resources for curative medicine. Also there has developed a large medical–industrial complex with large investments in technology, hospitals, nursing homes, and other health efforts which are largely curative in their orientation (Relman 1980). Certainly, the tobacco industry, the alcoholic beverage industry, and the automobile industry, concerned with the costs of safety appliances, and others, view these augmented interests with some alarm.

Social strategy

A sound knowledge base and political will are prerequisites for formulating public policy, but a sound social strategy is necessary for effective implementation. *Healthy people* (US Department of Health, Education and Welfare 1979*b*), provides a model social strategy for implementing the second public health revolution of health promotion and disease prevention. *Healthy people* suggested three strategic thrusts to health promotion and disease prevention activities:

(1) key preventive services that can be delivered to individuals by health professionals;

(2) measures that can be used by governmental and other agencies, as well as by industry, to protect people from harm;

(3) activities that individuals and communities can use to promote healthy life-styles.

Health promotion and disease prevention are not the sole responsibility of a single programme or a single sector of the health care system. Health care professionals, federal and State health regulatory agencies, and individuals themselves must co-ordinate their activities to bring the second public health revolution to fruition. No sector alone is sufficient.

Health professionals have a major and continuing role in the provision of specific health promotion and disease prevention activities. Although people ordinarily perceive no need for preventive health services when they feel well, there is evidence now that certain key services can do much to preserve health—and that people can be attracted to using them, with striking benefits, when they are offered. Preventive services tailored to pertinent screening, detection, diagnosis, and treatment of specific risks for individuals at specific ages is the most effective. *Healthy people* highlighted five critical priority preventive services: family planning, pregnancy and infant care, immunizations, sexually transmissible diseases services, and high blood pressure control. These are typical of the range, the settings for delivery, and the blend of public and private, individual and organized efforts necessary for a well-rounded prevention strategy.

Health protection activities are the second critical component of a health promotion and disease prevention social strategy. The American environment today contains health hazards, such as nuclear wastes, with the potential to kill, injure, and disable individuals and substantially affect the health of entire communities. It is estimated that 20 per cent of all premature deaths—and a vast amount of disease and disability—could be eliminated by protecting our people from environmental hazards (US Department of Health, Education, and Welfare 1980).

We have seen past improvements in protection—through better sanitation, better housing, better water—contribute greatly to the increased life expectancy of the last 80 years. But during this same period, rapid industrial and technological developments have increased the complexity of maintaining a healthy and safe physical environment. Measures are available to communities to provide better health protection. *Healthy people* highlighted five areas: toxic agent control, occupational safety and health, accidental injury control, fluoridation of community water supplies, and infectious agent control. Federal and State regulatory agencies are the key to the development of active health protection efforts.

Individual life-style activities are the third key element in a social strategy for health promotion and disease prevention. Health promotion begins with people who are basically healthy and seeks the development of community and individual measures which can help them to develop life-styles that can maintain and enhance their state of well-being.

From early childhood and throughout life, each of us makes decisions affecting our health. They are made, for the most part, without regard to or contact with, the health care system. Yet their cumulative impact has a greater effect on the length and quality of life than all the efforts of medical care combined. Many factors increasing risk of premature death can be reduced without medication. *Healthy people* highlighted five types of behaviours which affect health and are the targets for health promotion programmes: smoking; alcohol and drug abuse; nutrition; exercise and fitness; and the management of stress.

All three strategic areas—health profession-delivered preventive services, regulatory health protection activities, and individual healthy life-styles—are necessary to promote health and to prevent disease. All programmatic efforts should be seen as fitting into one of these three distinct categories.

It is also critical that certain other sectors of the society be drawn into these health promotion and disease prevention activities to maximize their potential influences. This three-part social strategy for health promotion and disease prevention must not be simply an analytical description discussed in only a few offices in Washington, but a conscious national public consensus. The first health revolution gained such a pubic presence, the second must also. In particular, the following non-health sectors must become involved.

The educational institutions can foster a positive attitude toward health promotion and disease prevention, and should begin early. The schools have a major role to play. This should have long-term positive effects. Textbooks and other teaching aids merit more attention than they have received in the past.

The mass media reflect the growing public interest in health promotion. It is being utilized in increasingly effective ways. Research efforts such as the Stanford Community Study (Macoby *et al.* 1977) and the Karelia Study (Puska *et al.* 1980) in Finland, for the prevention of cardiovascular disease, are evidence of the fact that it is possible to learn to use the media with greater effectiveness. There has been a remarkable proliferation of health programmes recently on radio and television. The mass media can be used to foster health promotion and disease prevention activities.

Direct citizen involvement in protecting health is perhaps of greatest potential importance. Grass roots organizations along with existing voluntary organizations are providing the impetus for local institutions and individuals to reorient their activities to foster health promotion and disease prevention through educational efforts and consciousness-raising.

Federal activity complements and facilitates a national strategy of health promotion and disease prevention. The leadership role of federal efforts should not be underestimated. The publication in 1979 of *Healthy people* and related publications, along with complementary new federal programmes, pointed the way for the rest of the nation to move in the direction of improving the health of its people. Federal financial resources are critical to a sustained prevention effort.

Public policy in health grows out of the three components just discussed: political will, knowledge base, and social strategy. All three are necessary to adopt new public policies. All three must interact constructively for the adoption of the new consensus on health promotion and disease prevention.

Conclusion

Recent trends in the US indicate a remarkable interest in health promotion and disease prevention among citizens broadly and health professionals. This has been stimulated by the remarkable advances in the control of the infectious diseases and a more recent trend toward a significant reduction in mortality from non-infectious causes, particularly of cardiovascular disorders.

The first federal efforts were disease-specific and uncoordinated. In 1979, the Surgeon General's Report, *Healthy people*, proclaimed a new national health policy: Health Promotion and Disease Prevention. For the first time, quantitative concrete health goals by age groups were established, and a three-pronged strategy for preventive health services for individuals and health promotion for population groups was enunciated. The importance of health promotion and disease prevention in shaping policy is becoming increasingly evident.

The shaping of health policy is dependent upon improving our knowledge, generating the political will for appropriate allocation of resources, and the development of social strategies to bring our political will into line with our knowledge base. This cannot be exclusively a publicly sponsored effort. The social strategy requires the combined efforts of the many private and public agencies acting to bring better health information and programmes to all the people.

The US has embarked on the second public health revolution. Our present health institutions and our present individual health attitudes, which developed in the era of the first public health revolution and focused on acute illnesses, are now beginning to change in order to deal with the new morbidities. The second pubic health revolution with its emphasis on health promotion and disease prevention foretells a healthier future.

References

Aday, L., Anderson, R., and Fleming, G. (1980). *Health care in the United States, equitable for whom?* Sage Publications, Beverly Hills, California.

Centers for Disease Control (1979). *Model standards for community preventive health services*. CDC, Atlanta, Georgia.

Clark, K. (1952). Preventive medicine in medical schools. Report of the Colorado Springs Conference, November 1952. *Journal of Medical Education* **28** (10), Part 2.

Cochrane, A.L. (1972). *Effectiveness and efficiency; random reflections on health services*. Nuffield Provincial Hospitals Trust, London.

Doll, R. and Hill, A.B. (1952). Study of the aetiology of carcinoma of the lung. *British Medical Journal* **ii**, 2171.

Dumbauld, S. (1987). *California's progress in meeting the national health status objectives for 1990*. Department of Health, California.

Hamburg, D.A., Elliot, G.R., and Parron, D.L. (1982), *Health and behaviour, frontiers of research in the biobehavioural sciences.* Institute of Medicine, National Academy Press, Washington, DC.

Hammond, E.C. (1966). *Smoking in relation to death rates of one million men and women.* National Cancer Institute Monograph 19, p. 127. National Cancer Institute, Bethesda, Maryland.

Institute of Medicine (1988). *The future of public health.* Publication No. IOM 88–02. National Academy Press, Washington, DC.

Lalonde, M. (1974). *A new perspective on the health of Canadians. A working document*. Government of Canada, Ottawa.

Laurie, N., Manning, W.G., Peterson, C., *et al*. (1987). Preventive care: Do we practice what we preach? *American Journal of Public Health* **77**, 801.

Levy, R.I. and Moskowitz, J. (1982). Cardiovascular research: decades of progress, a decade of promise. *Science* **217**, 121.

McKeown, T. (1979). *The role of medicine: Dream, mirage or nemesis?* Princeton University Press, New Jersey.

Macoby, N., Farquhar, J.W., Wood, P.D., and Alexander, J. (1977). Reducing the effects of cardiovascular disease. Effects of a community-based campaign on knowledge and behaviour. *Journal of Community Health*, **3**, 100.

Puska, P., Tuomilehto, J., Nissinen, A., Salonen, J., Maki, J., and Pallonen U. (1980). Changing the cardiovascular risk in an entire community. The North Karelia Project. In *Childhood prevention of atherosclerosis and hypertension* (ed. R. Lavar and R. Sherelle). Raven Press, New York.

Relman, A.S. (1980). The medical–industrial complex. *New England Journal of Medicine* **303**, 963.

Richmond, J.B. (1969). *Currents in American medicine*. Harvard University Press, Cambridge, Massachusetts.

Richmond, J.B. and Kotelchuck, M. (1983). The effects of political process on the delivery of health services. In *Handbook of health professions education* (ed. C. McGuire, R. Foley, D. Gorr, and R. Richards) p. 386. Jossey-Bass, San Francisco, California.

Sontag, S. (1978). *Illness as metaphor*. Farrar, Strauss and Giroux, New York.

Surles, K.B. and Blue, K.P. (1988). *The 1990 health objectives for the nation: The North Carolina course*, Special Report Series

No. 44. North Carolina State Center for Health Statistics, Raleigh, North Carolina.

Thornberry, O.T., Wilson, R.W., and Golden, P.M. (1986). *Health promotion data for the 1990 objectives. Advanced data from vital and health statistics*. National Center for Health Statistics. No. 126. DHHS Pub. No. (PHS)86–1250. Public Health Service, Hyattsville, Maryland.

US Department of Health, Education and Welfare (1964). *Smoking and health. Report of the Advisory Committee to the Surgeon General of the Public Health Service*. PHS Pub. No. 1103. US Public Health Service, Washington, DC.

US Department of Health, Education and Welfare (1979*a*). *Smoking and health—A report of the Surgeon General*. DHEW Pub. No. 79–50066. US Public Health Service, Washington, DC.

US Department of Health, Education and Welfare (1979*b*). *Healthy people. The Surgeon General's report on health promotion and disease prevention*. DHEW Pub. No. 79–55071. US Public Health Service, Washington, DC.

US Department of Health, Education and Welfare (1980). *Improving health in America—US Public Health Service highlights of 1977–80*. Office of Assistant Secretary for Health. HE 20.2: H 34/4. US Department of Health, Education and Welfare, Washington, DC.

US Department of Health and Human Services (1980*a*). *Health United States 1980*. National Center for Health Statistics. US Public Health Service Pub. No. 81–1232. US Department of Health and Human Services, Washington, DC.

US Department of Health and Human Services (1980*b*). *Promoting health; preventing disease. Objectives for the nation*. US Government Printing Office, Washington, DC.

US Department of Health and Human Services (1986*a*). *The 1990 health objectives for the nation: A mid-course review*. US Public Health Service, Washington, DC.

US Department of Health and Human Services (1986*b*). *Surgeon General's report on acquired immune deficiency syndrome*. US Department of Health and Human Services, Pubic Health Service, Washington, DC.

US Department of Health and Human Services (1987*a*). *Prevention 86–87. Federal programs and progress*. G.P.O. No. 1987–184–847. US Department of Health and Human Services, Public Health Service, Washington, DC.

US Department of Health and Human Services (1987*b*). *Health United States 1987*. Public Health Service Publication No. 88–1232. US Department of Health and Human Services, Hyattsville, Maryland.

US Department of Health and Human Services (1988*a*). *The health consequences of smoking: Nicotine addiction. A report of the surgeon general*. Public Health Service Pub. No. 88–84006. US Public health Service, Washington, DC.

US Department of Health and Human Services (1988*b*). *Surgeon General's report on nutrition and health*. Public Health Service Pub. No. 88–50210. US Public Health Service, Washington, DC.

US Public Health Service (1985). *Model standards: A guide for community preventive health services*. US Public Health Service, Washington, DC.

US Public Health Service (1987). *National survey of worksite health promotion activities: A summary*. Office of Disease Prevention and Health Promotion Monograph Series. HE. 20.34: W 89/snm. US Public Health Service, Washington, DC.

32

Public health policy in The Netherlands

A.J.P. SCHRIJVERS and C. SPREEUWENBERG

Introduction

Since the 1960s many European countries have been confronted with new problems, such as increasing multipathological needs among the elderly, the availability of more medical technologies for diagnosis and therapy, an increasing health-consciousness in the population, and growing health-care costs. For these reasons many countries have changed the organization of their health-care systems (see Volume 1, Chapter 21). A good example is the changing organization of health-care in The Netherlands since the end of the 1960s. At that time most acute hospital buildings had been renewed, a social insurance system with broad coverage existed, and the first nursing homes had been opened. Health-care costs had increased from 4.9 billion guilders in 1968 to 6.6 billion in 1970, an increase of about 35 per cent (CBS 1974*a*).

There have been three distinct periods of organizational change in Holland since 1970. During the first period (1970–82) new organizational structures for the health services were prepared, and short-term measures for cost-containment were introduced. This phase ended with the decision of Parliament in 1981 to pass a Health Services Bill which to introduce health regions and to give more power to local government. During the second period (1981–7) attempts were made to implement that law. In 1987 these efforts were abandoned, and new legislation was prepared to introduce an organizational structure based on competition both among health-insurers and among health-care providers.

Volume 1, Chapter 21 by Spreeuwenberg and Schrijvers presents a history of the development of the Dutch health-care system (see also Table 32.1).

The end of the 1960s

In 1966 the Dutch government published the Public Health Report (*Volksgezondheidsnota*, 1966). Several years earlier the Minister of Health had introduced new laws on health-care insurance, by which the whole population was covered for very expensive health-care and 70 per cent of the population was compulsorily insured for all health-care costs. The Minister of Health at that time, Mr Veldkamp, did not succeed in getting a proposal to introduce a National Health Insurance scheme accepted by Parliament. However, in 1966, he formulated new principles for heath-care policy in the Public Health Report (of *c.* 500 pages). The key concept of this was 'shared responsibility' between the government and the private/(non-profit) sector with respect to the functioning of health services. The Public Health Report described the existing system, urged the introduction of *regionalization*, and emphasized the need to introduce nursing units on a community level and the necessity of mutual supplementation of community and hospital care (*Volksgezondheidsnota* 1966, p. 160). After the publication of the Public Health Report many advisory agencies and authors supported this governmental point of view. (See, for a description of these opinions, Schrijvers 1980). Nevertheless, consensus about the ideal financing system did not emerge. A request by Minister Veldkamp in 1968 to the Social and Economic Council (the most important advisory agency to the government, consisting of employees, employers, and some independent, Crown-appointed members) to give an opinion as to the most desirable system for financing health-care system, was answered *five years* (!) later with four proposals supported by four different minorities.

The 1960s can be characterized as a period of prosperity for the Dutch health services. The number of doctors increased from 7909 in 1958 (4320 general practitioners and 3589 medical specialists) to 10 173 in 1970 (4470 GPs and 5703 medical specialists). The capacity of acute hospitals grew from 54 748 beds in 1958 to 72 359 beds in 1970 (CBS 1974). Annual health-care costs exploded from 1.4 billion guilders in 1958 to 6.4 billion guilders in 1970. Most of this explosive growth was attributable to inflation. At 1970 prices, health-care expenditure grew from 4.5 billion guilders in 1963 to 6.4 billion guilders in 1970 (CBS 1974*a*).

The preparation of new organizational structures: 1970–1982

During the years following the publication of the Public Health Report, an Act creating new hospitals was prepared.

Table 32.1. A timetable of health policy in Holland: 1965–1992

1965–6	HEALTH INSURANCES ACT and the ACT ON EXCEPTIONAL MEDICAL EXPENSES (Algemene Wet Bijzondere Ziekte Kosten 1966) pass Parliament. A proposal for a NATIONAL INSURANCE FOR HEALTH is rejected.
1966	PUBLIC HEALTH REPORT is published.
1968	The Minister of Health asks advice of the Social and Economic Council about a National Insurance for Health.
1971	The HOSPITAL PROVISIONS ACT passes Parliament.
1971–9	Amendments to the HOSPITAL PROVISIONS ACT, by which the scope of this law is broadened.
1972–3	Advice from commercial consultancy firms to the Minister of Health on limitation of the growth of health-care expenditure: they advise the introduction of health regions.
1973	The Social and Economic Council gives four different forms of advice about a system of National Insurance for Health. Each view is supported by a minority.
1973	The Feston Committee advises the introduction of 4 acute hospital beds per 1000 inhabitants of a health region. Parliament adopts this guideline.
1974	STRUCTURE REPORT on the desirable structure for the Dutch health services.
1974–81	Preparation of the HEALTH SERVICES ACT, the FRAMEWORK ACT FOR SPECIFIC WELFARE, and the HEALTH-CARE CHARGES ACT.
1975	A ministerial concept-proposal for a NATIONAL HEALTH INSURANCE ACT creates much uproar inside and outside Parliament and disappears from the political agenda.
1980	The HEALTH-CARE CHARGES ACT (Wet Tarieven Gezondheidszorg) passes Parliament.
1982	The HEALTH SERVICES ACT (Wet Voorzieningen Gezondheidszorg) and the FRAMEWORK ACT SPECIFIC WELFARE pass Parliament.
1982–7	Sectoral and integral implementation of the HEALTH SERVICES ACT.
1983	The Government withdraws the FRAMEWORK ACT ON SPECIFIC WELFARE.
1983–6	Preparation of the HEALTH CARE AND SOCIAL SERVICES ACT.
1985	REGIONAL HEALTH COMMITTEES proposed by the Department of Welfare, Health, and Cultural Affairs.
1986	A General Practitioner licensing system is introduced.
1986	Withdrawal by the Government of the proposals contained in the HEALTH CARE AND SOCIAL SERVICES ACT.
1986	REPORT-2000 (the Dutch implementation of the WHO-strategy HEALTH FOR ALL BY THE YEAR 2000).
1987	DEKKER REPORT, which advises *competition instead of planning, no local authorities in health services,* and a *Basic Insurance for Health.*
1987	Implementation of the HEALTH SERVICES ACT is stopped.
1989	The Government publishes CHANGES ASSURED in which a BASIC INSURANCE FOR HEALTH is worked out.
1989	Parliamentary election. New committees, reports, and structures?

The fear of overcapacity expressed in the Public Health Report, was translated into the Hospital Provisions Act, which passed through Parliament in 1971. This still-functioning law gave the provincial authorities (at a level between national government and the municipalities) the responsibility of presenting a plan to the Minister of Health describing the hospitals and beds needed within each province. The provincial role was regarded as a fundamental part of this law, because through it local hospital needs could be determined (*Wet Ziekenhuisvoorzieningen* 1971). The provincial plans had to be approved by the Minister of Health. As part of such approval the Minister could license the building or alteration of a hospital; but operating any hospital which was not described in a provincial plan was forbidden. During the 1970s this Act was amended in 1972, 1978, and 1979. Its scope was broadened to include all hospitals and the distribution of high technology. The Minister of Health was also given the legal power to close a hospital.

In the early 1970s cost-control became more and more a political issue. Health-care expenditures kept growing at yearly rates of 20 per cent and more. In 1972 the then Minister of Health, Mr Stuyt, asked three major commercial bureaux of management consultants for recommendations on how to limit this growth. They suggested the creation of a regional structure for the health services, although they foresaw problems arising around the issues of the autonomy of health-care institutions, the definition of the desirable size of a region, and the lack of co-ordination between the regional health-insurance agencies (sick funds) and the provincial and local authorities (Berenschot Bureau 1973). After a change of government in 1973 the new State Secretary, Mr Hendricks, appointed a special independent committee (the Festen Committee, named after its chairman) with the aim of advising him in the short term about measures for cost-control. This commission recommended the concept of a maximum of 4.0 acute hospital beds per thousand inhabitants. This meant a considerable reduction in bed density: previously, the number of beds had far exceeded 5 per thousand inhabitants. Hendricks adopted this norm, and since 1974 the hospital sector has been governed by this type of capacity norm. Table 32.2 shows some of these bed norms in 1988, which are generally even lower than those of 1974.

In 1974 State Secretary Hendricks wrote his Structure Report, which described the long-term structure of Dutch health services (*Structuurnota Gezondheidszorg* 1974). Until the Dekker Report in 1987 (see below), this was the most important health-policy document for many years. The Structure Report identified many bottle-necks in health-care: a lack of cost-controlling instruments; inadequate co-ordination between different types of agencies concerned with health-care provision; a poorly organized financing system; too much emphasis on the hospital sector and too little on primary health-care; an unbalanced geographical distribution of health-care provision, and a lack of involvement of the population. Because of these problems, State Secretary Hendricks argued first for a four-tier arrangement of health-care facilities, and second for a division of the system into health regions. We shall discuss both organizational principles below.

Services in the *basic tier* were provided by local public health services and non-government organizations, such as Cross Unions (home-nursing organizations), and included preventive health services for the whole community, including vaccination, school medical services, baby and child-welfare clinics, school dental services, and mass radiography.

Table 32.2. Maximum number of hospital beds per 1000 inhabitants

Type of health-care facility	Maximum no. of beds per 1000 population
Acute hospitals	3.7
Psychiatric hospitals	1.1
Nursing-homes for somatic patients	1.2 (per 100 inhabitants over 65 years) plus 0.35 per 1000 of the whole population
Nursing-homes for psychogeriatric patients	1.25 per 100 inhabitants over 65 years
Institutions for mentally retarded people	1.8 places per 1000 inhabitants

Source: Wet Ziekenhuisvoorzieningen 1971, artikel 3.

The Cross Unions' main function was to provide a home-nursing service, which was carried out by district nurses. Their task was to visit patients' homes, to perform light nursing duties, to give bed baths, and to do some preventive work. Midwifery services were provided at home or in the hospital. General practitioners rendered obstetric assistance if no midwife was available. There were also maternity services which provided help in the home for a short period after childbirth.

General practitioners occupied a central position in the *first-tier*, or primary health-care. Drugs were supplied to patients by doctors with their own dispensaries or by dispensing chemists.

The *second-tier* included both in-patient and out-patient services provided by all categories of medical specialists. Out-patient clinics in The Netherlands have always been accommodated in hospitals. The second tier was also held to include the services of mental-health workers, including psychotherapy, the treatment of drug addiction, and other services which lie between health and social work. In-patient treatment in general hospitals, psychiatric hospitals, residential homes for the mentally deficient and the blind, and nursing homes also formed part of the second tier.

The *third tier* included highly specialized facilities in academic and specialized hospitals, which, in principle, extended beyond regional boundaries, for example neurosurgery, kidney transplantation, and special cancer treatments.

The Structure Report proposed regionalization (the second organizational principle) as one of the structuring instruments necessary for an effective and efficient health-care system (*Structuurnota Gezondheidszorg* 1974, p. 10). Responsibility for the planning of facilities in the basic and primary tiers was given to the Municipalities. The second and third tier were to be planned by the Provinces and the State, which had final responsibility for the health of the population and for the health-care system. This was a policy shift from the Public Health Report of 1966, with its emphasis on shared responsibility (see above).

In the years after 1974 many agencies, institutes, and organizations were discussing the Structure Report (Schrijvers 1980). All parties endorsed the two organizational principles it embodied, but consensus about the autonomy of facilities, the size of the regions, and a satisfactory relation between planning and financing could not be achieved, the three consultancy firms had earlier predicted it might not be. Exploratory studies in some regions illustrated the problems (Tweede Kamer 1976 and 1978).

In 1974 Parliament began the preparation of the Health Services Act, in which the principles of the Structure Report were given a statutory foundation. The law was seen as an extension of the earlier-mentioned Hospital Provisions Act. The key provisions in the Health Services Act were as follows. Municipalities and Provinces had to approve four-year plans for health-care facilities. All these services were to be financed by insurance companies, and to be provided by autonomous health-care organizations. The plans set out what facilities were required and described the system which would obtain during the four years of the plan's operation.

Health-care facilities dealt with in the local authority plans comprised the first-tier facilities provided by general practitioners or by institutes. The planning of similarly provided facilities classified under the second and third tiers were carried out by the provincial authorities, with the exception of those which were the responsibility of central government. For those facilities plans were established at the national level. The Minister provided guide-lines for the drafting and adoption of provincial and local plans. The guide-lines dealt with a number of matters, including the need for and distribution of services. The guide-lines also stated what claims the services could make annually on staff and material and financial resources during the period covered by the plan.

During the preparation of the Health Services Act a new policy topic arose: the integration of health and social services. In the 'Bottle-necks Report' (*Knelpuntennota* 1974) many overlaps were observed between, for instance, district nursing and home help, which is a social service. Another example was the overlap between nursing homes and old people's homes. The Government and Parliament began the preparation of a new bill, the Framework Act For Specific Welfare (Kaderwet Specifiek Welzijn 1982). During the preparation period, the planning system of the Health Services Act was adjusted to take into account this Framework Act, which was intended in the long run to replace the former. According to the recommendation of the Structure Report the Government in 1974 also started the preparation of a National Health Insurance Act, which was supposed to replace the existing Health Insurance Act, the Exceptional Medical Expenses Act, and the private insurance schemes. In 1975 a draft of the National Health Insurance Act was leaked to the press (*Wet Gezondheidsverzekering* 1984). After an outcry inside and outside Parliament the concept disappeared from the political agenda. No new efforts to introduce such legislation were made until 1987.

From 1974, health-care costs were better controlled than in the years before, as Figure 32.1 shows. From 1963 to 1975 these costs increased from 4.6 per cent to 7.7 per cent of GNP. From 1975 the growth was slower, reaching 8.2 per

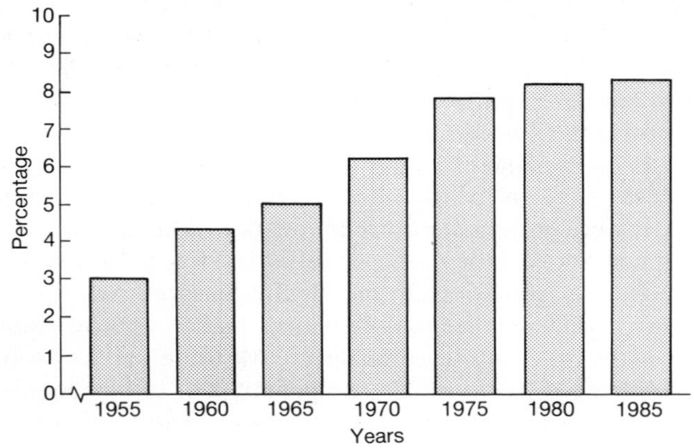

Fig. 32.1. Health-care expenditure in The Netherlands as a percentage of Gross National Product at market prices. Source: CBS, Kosten en financiering van de gozondheidszorg, 1970–1986, Den Haag, Staatsuitgeverij, 1974–1988.

cent of GNP in 1980 and 8.3 per cent in 1986. The most important factor in this was the application of more restrictive guidelines for the calculation of the number of required posts in each individual hospital's budget, as part of a more restrictive financial policy. Below we shall discuss the financial regulation of the Dutch health services between 1970 and 1982.

Health-care charges (including independent practitioners' fees) were agreed upon in consultations between health-care organizations and organizations of independent practitioners on the one hand, and the health-insurance funds and companies on the other. In the period 1965–82 the rates that could be charged for health-fund insurance were subject to approval by the Health Insurance Funds Council. The rates for privately insured persons, however, required the approval of the Minister for Economic Affairs. During this era, hospital charges were under a different system, and were fixed by a separate body: the Central Council on Hospital Charges.

Each of the aforementioned bodies had their own criteria for assessing charges, and the existence of separate bodies made it very difficult to co-ordinate charges properly, as the system of fees and charges affects the functioning of institutions and workers in the health services. As a result the Government began preparing a new law. The Health-Care Charges Act, which was to replace the existing regulations. The objectives of this Act were twofold: to promote a balanced system of health-care charges, and to control rising costs. The new Act provided for the setting-up of a Central Council For Health-Care Charges that either approved or rejected all charges, fees, estimates, etc. Initially representatives of the health-insurance funds, health-care organizations, and health-service staff convened in 'Chambers' to discuss guidelines on charges, and then they were passed to the Central Council to be adopted. On a regional level, health-insurance funds and private insurance companies had to negotiate with health-care institutes about the charges. The results of these negotiations were subject to approval by the aforementioned Central Council.

In 1980, the Health-Care Charges Act was the subject of discussion in Parliament. Disagreement existed between the main political parties about the relationship between this Act and the Health Services Act. In November 1980 the Charges Act passed through Parliament. In the second half of 1982 (after many debates, comments in professional magazines, and much advice from statutory bodies both in and outside the health-service system) the Health Services Act and the Framework Act on Specific Welfare were enacted. A period characterized by preparation of a great deal of new legislation was closed with the passing of these laws.

The implementation of new organizational structures: 1982–1987

The Health Services Act (1982) did not provide rules for decision-making at the regional level. The municipalities and cities were empowered to plan health services. The health-insurance agencies kept their responsibilities in the budgeting and financing process. The mostly private health institutions retained their responsibilities for the provision of care and for structuring their own organizations. To find out how these different specialties could be co-ordinated the government decided to undertake a sector-by-sector and an across-the-board implementation of the Health Services Act. The former was nationwide, but covered only the basic and first-tier health-care provisions. The latter was limited to eight experimental regions (see Fig. 32.2), and covered all health-care facilities.

In the years 1982–7 a licensing system was prepared as a part of the sectoral implementation of the Act. This system prohibited all independent health-care providers from practising without a licence from the Municipality where they worked. In 1986 the same system was introduced for general practitioners.

At the same time the payment of 'goodwill sums' (payments by a doctor taking over the established practice of another to his or her predecessor on the latter's retirement) was abolished. For other professional workers comparable regulations were under discussion until the publication of the Dekker Report (see below). The sectoral approach can be seen as a part of an overall policy since 1974 to promote and restructure primary health-care. Facilities within the basic and the first tiers and ambulatory psychiatric care were allowed to grow faster than new hospitals and new facilities were built or established. Reorganizations took place to create regional working organizations for public health, district nursing, and ambulatory psychiatric care.

The integral implementation of the Health Services Act in the eight experimental regions aimed at the realization of a regional planning system. Five conditions were seen as necessary for such a system. They were: (1) a regional health

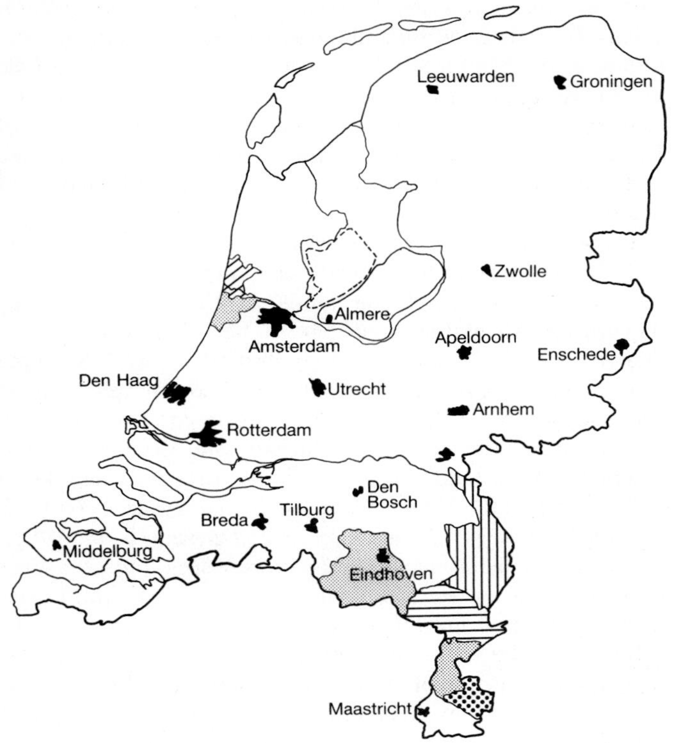

Fig. 32.2. The eight experimental regions of the internal implementation of the Health Services Act.

policy; (2) well-defined health regions; (3) block grants; (4) planning procedures; and (5) well-defined links with the financing system.

Condition 1. A regional health policy

The collection of regional data on patient flows, costs, building initiatives, and institutional activities was the first step towards a regional health policy. Later steps were the designing of a regional health profile, which described the health status of the regional population and established an inventory of the policy goals of the different regional participants. This work began in 1985/6. One of the first problems encountered was the variety of definitions of statistical data in use. Another problem was that not all participants were eager to distribute their data about their 'market share' and 'financial position' to other institutes, as this could weaken their position.

Condition 2. Well-defined regions

Regions were chosen with approximately 200 000 to 300 000 inhabitants. This scale is big enough for the evaluation of the facilities of the basic and first tiers, but too small for general hospitals and more specialized institutions. Because of the promotion of primary care, the Department of Welfare, Health, and Cultural Affairs concentrated on regions with this number of inhabitants. Many problems arose with hospi-

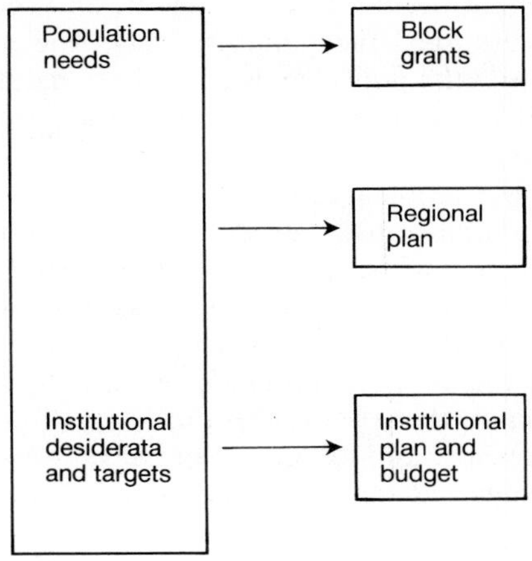

Fig. 32.3 The predominant pattern of regional planning procedures.

tal boards. Another bottle-neck was caused by a lack of co-operation on a regional level between municipalities. Sometimes, too, there was friction between villages and towns, because the former were afraid of annexation by the latter. This was not a health-services problem, but one of public administration; but it still held up the intended across-the-board implementation.

Condition 3. Block grants

A new venture was the creation of block grants for the experimental regions. The Dutch health-care system is financed by three sources: social insurance, private insurance, and taxation. The aim was to allow regions the possibility of substituting social insurance payments and taxation for each other. A national mechanism had been developed, whereby a lower or higher social-insurance rate was compensated by a reciprocally higher or lower taxation rate. With this system, the eight regions would know the size of their block grant. For the allocation of this grant, a formula was developed based on the number of inhabitants and their age-distribution. The implementation of the formula was nevertheless tempered, because this formula would have led to too many geographical reallocations. The regional costs over the previous year, and whether they had decreased or increased by a certain percentage, were important for the calculation of the block grant. Another problem encountered was the opposition of the health-insurance organizations (and the institutions financed by them) to the possibility that their insurance money might be used for state-financed facilities. The insurers feared a loss of identity.

Condition 4. Planning procedure

Figure 32.3 shows the predominant pattern of the planning procedure, which starts with a decision of the Ministry of Health on the size of the block grant. A regional plan is designed within the limits of this grant, and this forms the basis of institutional plans and budgets. The needs of the population and the policy goals of institutional management play a role in the decisions concerning block grants, planning, and budgeting.

Some of the problems encountered in the period 1982–7 were of an administrative nature: for instance, how does such planning function within the period of one year? Or how can one organize enough forums for health-providers, health-insurance organizations, and patient groups to influence the decision-making process of municipalities and provinces? On the other hand, there were also problems of differences in culture, language, knowledge, and trust between the different participants who met each other in the planning process.

Condition 5. Well-defined links with the financing system

The most difficult implementation problem was to define the relationship between the planning and the financing systems. The dilemmas were manifold. The financing system had been developed outside public administration, and had its own national agencies, regulations, and local executive organizations (in the form of sick funds). Introducing regional planning meant that some of this freedom of decision-making had to be handed over to the provinces and municipalities, or at least, that the ambitions of health-insurance companies to achieve more freedom in their decision-making had to be tempered. Although the links were well defined on paper, with the financial needs of the regional plan based on estimates from health-insurance organizations and institutions, the struggle between the leading actors (sick funds, health-care providers, and local authorities) intensified in the years 1984 and 1985. The history of the Dutch health service and its funding institutions involves both private institutions and public agencies such as municipalities and provinces. The differing aims and backgrounds of the two groups created much distrust and many misunderstandings in the collaboration between the regional actors in the implementation process.

In order to communicate on these points and learn from each other, the Department of Health, Welfare, and Cultural Affairs proposed a regional forum in which all the 'actors' participated. This *Regional Health Forum* (RHF), or in Dutch *Regionaal Overleg*, was made obligatory by Article 41 of the Health Services Act. Representatives of patient groups, care-providers, funding agencies, municipalities, and provinces have to meet each other regularly in the Regional Health Forum. The forming of the forum itself was a delicate process: since 1986 it had only functioned in three of the eight experimental regions. Then in 1986 a new government came into office. The new State Secretary for Public Health, Mr Dees, was a member of the Conservative Party, while his predecessor Mr Van der Reyden had been a Christian Democrat. A new philosophy was expressed in a government programme, subscribed to by the two coalition partners (Christian Democrats and Conservatives). The keywords were *Competition instead of planning* and *No local authorities*. Both the sectoral and the across-the-board implementations of the Health Service Act were stopped. In June 1986 the Government installed The Dekker State Commission (named after its chairman, ex-President of the Board of Directors of the multinational firm 'Philips'). This committee had the task of 'advising on the structure and finance of health-care'.

There were two other policy implementations in the period 1982–7: *the integration of health and social services* and the introduction in The Netherlands of the WHO strategy 'Health for All by the Year 2000'. The Framework Act on Specific Welfare passed through Parliament in 1982. One year later, the government (newly installed after elections in 1982) withdrew this law. The proposed integration of health and social services was too ambitious, and the law was too complicated. Instead, the Government prepared new legislation, entitled the Health-Care and Social Services Act (*Wet Gezondheidszorg en Maatschappelijke Dienstverlening* 1985). This law covered not only the health services, but also some parts of social services: home helps, social work, and home-care for different groups in the population, for instance the elderly and the mentally handicapped. This Act could be seen as an expansion of the Health Services Act; but the local authorities were now given more planning power. The publication of advance drafts of the Health-Care and Social Services Act created much uproar in the health sector, and delayed both the sector-by-sector and the across-the-board forms of implementation of the Health Services Act. In 1986 the proposed new law was removed from the political agenda and the implementation of the existing (1982) Act was halted.

Health Policy rather than *Health-Care Policy* was a subject of greater political emphasis during the period 1982–7 than in the years before. The interest of political parties, advisory agencies, and professional organizations switched from structures to targets, such as life-styles, environmental health risks, and the health of the population in the long run. Ethical questions were also on the agenda. Many professional journals, newspapers, and television programmes discussed euthanasia, genetic manipulation, and other new technologies, and also the criteria used to ration scare resources for health services. Among other policy documents the publications of the Steering Committee on Future Health Scenarios (Schreuder 1988) were important. The Government itself, inspired by the WHO initiative, 'Health for All by the Year 2000', in 1986, published its 'Report-2000', (Nota-2000 1986; cf. also *Health as a focal point* 1987), containing many ideas on the achievement of healthier life-styles and, more generally, a multisectoral health policy. The previously mentioned eight regions selected for across-the-board implementation under the Health Services Act were also selected as experimental areas for the implementation of this policy. There was some reluctance on the part of the eight areas themselves to participate in yet another experimental scheme.

Towards competition? 1987–1992

In March 1987, the Dekker Commission published its report (*Bereidheid tot verandening* 1987). The three central premises of it were as follows:

- the integration of provisions in the field of health-care and social care;
- improvements in the efficiency and flexibility of health-care; and
- a shift of emphasis from the government regulation to market regulation and self-regulation.

In the Committee's own view, the essence of its proposals lay in the restructuring of the finance and insurance system. The Committee proposed the amalgamation of most health and social-care facilities into a single basic insurance scheme, covering about 85 per cent of the total cost of health and social-care facilities. In addition to the basic insurance, the Committee advocated an optional supplementary insurance for example, to cover physiotherapy, dental care for people over eighteen years of age, and medication).

The introduction of a basic and supplementary insurance would mean the end of the current division of provisions into those funded by private health insurance, by the Health Insurance Act, by the Exceptional Medical Expenses Act, and by the central government budget. The basic insurance would mainly be paid for by an income-related premium.

The Committee envisaged insurances being provided by competing insurers. The present distinction between health-insurance funds and private insurers would no longer apply. Every citizen would be free to choose his or her insurer.

Another important proposal was the shift from government-imposed rules and regulations to regulation by the market. The idea is that the parties involved will have an incentive to become cost-conscious and to act in an efficient manner.

This shift of emphasis in the regulation of health-care is reflected in reduced planning by the government, a lower level of political involvement in determining charges and prices, and the opening up of wider options for the insurers.

The preparatory measures for the new system could be taken under the present legislative and regulatory framework. Thus, the Dekker Committee proposed the elimination of 4000 hospital beds between 1987 and 1991. The Committee's proposals mean that the number of beds will fall to approximately 49 000, or 3.2 beds per 1000 members of the population. The Committee suggested that the new system should be introduced within a period of three to five years.

The publication of the Dekker Report provoked much discussion. There was general agreement with the central premises, but doubts existed about many individual aspects, and about the feasibility of implementing the proposals in three to five years. The Government responded in 1988 (after two preliminary reactions) with the document *Change Assured* (*Verandering Verzekerd* 1988). The new paradigm is the distinction of three markets within the health services: between insurers and patients; between doctors and patients; and, thirdly, between doctors and insurers. All the proposals in the document aimed at strengthening the functioning of these markets. The Government followed the financing system with some amendments, as suggested in the Dekker Report: a basic and a supplementary insurance system will exist in 1992.

Table 32.3. Future insurance system

1. Basic insurance
   - compulsory
   - health care and social care (85%)
   - mainly income-related premiums and fixed premium
2. Supplementary insurance
   - optional
   - all provisions not covered by the basis insurance (15%)
   - fixed premium

Table 32.3 summarizes the governmental proposals. There will no longer be any distinction between sick-funds and private insurers: only one type will exist. The insured will be free to choose his or her insurer, who will be obliged to accept all applicants.

To strengthen the market forces between health-care providers and insurers and between the former and the insured patients, the government is arguing for a functional rather than an institutional approach. All types of services will be provided within generally defined quality standards, by all types of institutions: for instance, home-care may be delivered by primary health-care organizations as well as by hospitals. The government also wants to stimulate the interaction between health and social services. For this reason home help and care for the elderly and for the mentally handicapped will be financed by the proposed basic insurance. *Change Assured* also promotes primary health-care. General practitioners will be financed partly on a capitation fee, and partly on a fee-for-service basis paid by the basic insurance. This will mean a change for many; up to the present, privately insured persons, have mostly paid their family doctor out-of-pocket.

The Government has opted for a step-by-step approach for the implementation of the ideas in *Change Assured*. The plans will be implemented in 1992.

Will The Netherlands experience competition between the health services during the 1990s? We are not sure. In 1989 there were new elections, and there is now a new State Secretary for Health, who is again proposing new legislation, a Commission, and a Report. The situation is that current ideas are partly out-of-date following the elections. On the other hand, there is broad public support for a basic health insurance, which did not exist in the past. So such an insurance will probably be a reality in 1992.

The way back towards regional planning and towards a significant exercise of influence by local authorities is blocked: there were too many traumatic experiences in reorganization of the health sector in the 1970s and 1980s. Maybe the time for structural policy (which was used for instance for the reorganization of the British National Health Service in 1974, and which was the basis of many bills introduced in the Dutch Parliament) is over. At the end of the 1980s in the Netherlands we can see many small-scale home-care innovations, based on health promotion and on co-operation between primary health-care and hospitals. An innovative, comprehensive health system is planned for Almere, a new town in one of the recently drained Dutch polders for which more than 100 000 inhabitants are anticipated by the 1990s. The government and other regulatory bodies are following Almere and other projects to try to adapt the system as much as necessary, and to support this expansion of home-care. From this point of view health policy will be an end, rather than a starting point, for developments in the health services: structures follow strategy.

Acknowledgement

With many thanks to Professor J. van Londen, Director-general for Health in the Ministry of Welfare, Health and Cultural Affairs in Rijswijk, for his comments on earlier drafts of this article.

References

Algemene Wet Bijzondere Ziektekosten (1967). 14 December 1967, Stb 1967, 617.

Bereidheid tot verandering (1987). Advies van de Commissie Structuur en Financiering; voorzitter W. Dekker. Distributiecenttrum Overheidspublikaties (DOP), The Hague.

Berenshcot Bureau (1973). *Onderzoek Beheersbaarheid Gezondheidszorg*: rapport uitgebracht aan de Minister van Volksgezondheid en Milieuhygiene door Bureau Berenschot, GITP (Gemeenschappelijk Instituut voor Toegepaste Psychologie). Bosboom en Hegener, Bureau Berenschot, Utrecht.

CBS (Centraal Bureau voor de Statistiek) (1974*a*). *Kosten en financiering van de gezondheidszorg in Nederland, 1970.* Staatsuitgeverij, The Hague.

CBS (Centraal Bureau voor de Statistiek) (1974*b*). *Compendium Health Statistics of the Netherlands, 1974.* Staatsuitgeverij, The Hague.

CBS (Centraal Bureau voor de Statistiek) (1979). *Compendium Health Statistics of the Netherlands, 1979.* Staatsuitgeverij, The Hague.

Health as a focal point (1987). An abridged version of the Memorandum Health-2000. Ministry of Welfare, Health, and Cultural Affairs. Without publisher, The Hague.

Kaderwet Specifiek Welzijn (1979–84). Part III (looseleaf publication). VUGA, The Hague.

Knelpuntennota (1974). Knelpunten harmonisatie welzijnsbeleid en welzijnswetgeving, rapport van de beraadsgroep, Tweede Kamer der Staten-Generaal, Vergaderjaar 1973–1974, 12968, nr. 2, 's-Gravenhage. zitting 1973–1974. nr. 12.968 nr. 2. Staatsuitgeverij, The Hague.

Nota-2000 (1986). Over de ontwikkeling van gezondheidsbeleid: feiten, beschouwingen en beleidsvoornemens. Tweede Kamer der Staten-Generaal. Vergaderjaar 1985–1986, 19500, nrs. 1–2. The Hague.

Schreuder, R.F. (1988). Scenarios for health planning and management: The Dutch experience, *International Journal of Health Planning and Management*, **3**, 74–87. (1980).

Schrijvers, G. (1980). *Regionalisatie en financiering van de Engelse, Zweedse en Nederlandse gezondheidszorg.* De Tijdstroom, Lochem.

Structuurnota Gezondheidszorg (1974). Ministerie van Volksgezondheid en Milieuhygiëne, Staatsuitgeverij. The Hague.

Tweede Kamer der Staten-Generaal (1976 and 1978).Vergaderjaar 1975–1978, 13.600, hoofdstuk 17, nr. 32; zitting 1977–1987, stuk nr. 14.800, hoofdstuk 17, nr. 13. Staatsuitgeverij, The Hague.

Verandering Verzekerd (1988) Stapsgewijs op weg naar een nieuw stelsel van zorg. Tweede Kamer der Staten-Generaal. Vergaderjaar 1987–1988, 19945. nrs. 27 en 28, S.1, S.1., SDU, 1988.

Volksgezondheidsnota 1966 (1966). Tweede Kamer der Staten-Generaal. Vergaderjaar 1965–1966; 8462, nr. 1. Staatsuitgeverij, The Hague.

Wet Gezondheidszorg en Maatschappelijke Dienstverlening (1985).

Wet Gezondheidsverzekering (1984). Memorie van Toelichting. In *Medisch Contact*, **30** (1984), p. 1164–83, and in *Unie*, **1975**, (nos. 7/8).

Wet Tarieven Gezondheidszorg (1980), regelen met betrekking tot de tarieven van organen voor gezondheidszorg, 27 januari 1982, Stb. 1982, 24. 17 november 1986, Stat. 1987, 26.

Wet Voorzieningen Gezondheidszorg (1982), regelen ter bevordering van een doelmattig stelsel van voorzieningen van gezondheidszorg, Tweede Kamer der Staten-Generaal, 9 september 1982, Stb. 563.

Wet Ziekenhuisvoorzieningen (1971), 25 maart 1971, Stb. 268, Gewijzigd op 21 december 1972, Stb. 773.

Ziekenfondswet, 15 oktober 1964, Stb. 392. 16 juni 1988, Stb. 1988, nr. 305.

33

Co-ordination and development of strategies and policy—the UK example

L.B. HUNT* and J.H. JAMES*

Introduction and summary

Earlier chapters in this volume have indicated the wide range of activities that contribute to the promotion and maintenance of a good state of health of a population. These include measures to maintain a healthy environment, provision of a system of microbiological surveillance to prevent and control outbreaks of communicable disease, and provision of medical services for the treatment of illness and injury. This chapter describes those processes, mainly of an organizational nature, that are needed to develop and implement policies for the effective performance of certain of these functions and to co-ordinate the work of those engaged in the diverse tasks that contribute to them. Most of what follows relates to the provision of curative and preventive services, but policies and strategies for the promotion of health in a broader sense are also touched on.

The UK is an example of a country which has adopted a co-ordinated form of service provision on a national scale funded mainly out of taxation (although charges play a significant part in certain sectors: dentistry, provision of spectacles, and prescription medicines). Central government determines the national strategy for the maintenance of health and the provision of services and distributes Exchequer monies to a network of operational authorities. In England the central department from 1968 to 1988[1] was the Department of Health and Social Security (DHSS)[2]. The several elements in this integrated system will be described in succeeding sections.

The following two sections, which give an account of this organization and the steps by which it took shape, are included, first, to give the reader a picture of the administrative machine within which the policy-making processes to be described later take place and, second, because the way in which successive changes in this administrative structure have been brought about serve to illustrate some important facets of the policy-making process itself. These two sections lead naturally into a systematic account of the way in which the central government health department operates. This is followed by further examples of the way policies are developed, with particular emphasis on the processes of consultation in widely different fields.

After summarizing the policy process, the concluding sections deal with the problems of co-ordinating the range of activities required at the local level to give effect to national strategies.

Background to the development of health policies

The constitutional setting

The National Health Service (NHS) is repeatedly referred to throughout this chapter. Although expanding in recent years, the private sector remains small and the NHS is the main vehicle through which medical services are provided in the UK; hospital and related community health services are provided by District Health Authorities (Area Boards in Scotland), and general medical services are provided by general practitioners operating under the aegis of local Family Practitioner Committees. Some health services, because they are more specialized or are better organized on a larger scale, may serve larger areas or, in a limited number of cases, all or most of the country. These, however, are the exception. In the main, services are local and policies for their development and co-ordination reflect this. None the less, in each of the countries of the UK there is a strong central role in the development and co-ordination of strategies and policies. In

---

* Any views expressed in this chapter are those of the authors alone and not of the Department of Health.

[1] A number of administrative differences distinguish the way health and social services are provided in the constituent countries of the UK; details given in the remainder of this chapter relate to England unless otherwise stated.

[2] In July 1988, the Prime Minister announced that the DHSS would be split into two separate Departments of State, respectively Health and Social Security. The examples in this chapter predate this announcement, and are therefore described against the background of the DHSS rather than the Department of Health.

England, that role centres on the DHSS and the Secretary of State for Social Services.

The Secretary of State's role is enshrined in the legislation under which the NHS is provided. It is his or her duty, one which has remained unchanged since the NHS was founded in 1948, 'to continue the promotion . . . of a comprehensive health service designed to secure improvement . . . in the physical and mental health of the people . . . and . . . in the prevention, diagnosis and treatment of illness, and for that purpose to provide or secure the effective provision of services . . . ' (NHS Act 1977, S.1 (unchanged from the NHS Act 1946)). Two consequences arise from this statutory position. First, health authorities may provide services only because the Secretary of State delegates his or her powers and responsibilities to them. While they enjoy considerable discretion in practice in the exercise of these delegated functions, health authorities are none the less accountable for them to the Secretary of State. Second, the NHS is conceived within a national framework which centres on the Secretary of State's office. There are therefore strong central pressures making for conformity to national policies, although, as we shall see later, these are counterbalanced by elaborate arrangements for local consultation; successive administrations have laid emphasis on local autonomy in decision-making.

The machinery

The DHSS was a large central government department. Its range of responsibilities covered the social security system as well as the provision of health and personal social services, administration of the Medicines Acts and oversight of some public health matters including environmental toxicology and control of communicable disease. Although this book is not concerned with social security matters, it is worth noting that of the DHSS's total staff of approximately 96 000, the great majority were employed in administering, through a network of some 550 local offices through the UK, the payment of some 34 different benefits which were being paid at any one time to more than 22 million persons and dependants. Less than 3500 members of the DHSS Headquarters staff were concerned with the co-ordination and development of health policies and services.

For the provision of health services, England is divided into regions, each composed of districts providing hospital and related services for populations mostly between 150 000 and 500 000. Each district may contain one or more district general hospitals with a fairly comprehensive range of facilities. Thus hospital and community health services are provided in England by 14 Regional and 192 District Health Authorities together with eight Special Health Authorities or Boards of Governors for postgraduate teaching hospitals. Ninety Family Practitioner Committees have responsibility for the oversight of general practitioner services as well as dental, pharmaceutical dispensing, and ophthalmic services. None of those bodies is part of the Department but all are wholly or largely financed by it and operate within a nationally approved framework. Health services in Scotland, Wales, and Northern Ireland are organized on a broadly similar basis but are the responsibility of the Scottish, Welsh, and Northern Ireland Offices respectively.

The Secretary of State for Health is now the government Minister responsible for the activities of the Department. The Office of Population Censuses and Surveys, headed by the Registrar General, is separate from the DH but reports to the Secretary of State for Health. The Secretary of State is always a member of the Cabinet and is assisted by three (on occasions reduced to two) junior Ministers. The Secretary of State will in practice delegate the major day-to-day responsibilities for health services to the Minister for Health, although the extent and nature of delegation have varied according to circumstances and to the individuals holding the office.

Both Parliament and the Secretary of State may be the focus upon which a wide range of what may be loosely termed 'pressure groups' concerned with all aspects of health promotion come to bear. These range from the national bodies representing the various categories of health service practitioners (Eckstein 1960; Jones 1981) through interest groups covering dozens of particular conditions, groups or services, to the various outside agencies or industries which have an involvement with the NHS. Local pressure groups may also choose to press a particular case at national level, e.g. when seeking to have local decisions overturned nationally.

The tendency for pressures to converge on the centre is reinforced by the way in which NHS services are financed. Over 85 per cent of costs are financed by taxation levied by central government and about 10 per cent through the health element in the National Insurance contribution. Most of the remainder comes from charges to users (mainly of the Family Practitioner Services) which are collected locally but at levels set in the main by the Government. The whole of the cost of the general medical practitioner service (other than income derived from patients who pay for private treatment) is borne by national taxation. The whole of NHS expenditure is classified as public expenditure, and thus is subject, firstly, to whatever limitations Government may decide to place upon its expenditure programmes as a whole and subsequently to annual Parliamentary approval of expenditure and the general processes by which the spending of public monies have to be accounted for. Over the years the DHSS developed a number of systems to enable both itself and the NHS to ensure that the latter can account satisfactorily for the services provided and for expenditure. These systems are discussed in detail later in this chapter, but first we turn aside briefly to explain how this complex and highly integrated system came to take its present form.

How the system evolved—a historical note

Ministry of Health to DHSS

In the UK social diversity and a wealth of independent initiatives in all fields of social welfare have, for long, been combined with a tradition of looking to central authority to promote the national interest. Chave (1984) described how the gross deficiencies in social provision of all kinds, characteristic of the eighteenth and nineteenth centuries at a time of expanding population, led to numerous attempts on the part of private philanthropists, religious, and other bodies to meet the nation's needs for social support and medical care in diverse and often original ways. The voluntary hospitals, numerous institutions for the care of deprived children, and many attempts to provide help for the handicapped and the mentally ill began in this way, and the loyalty and devotion which have sustained them reflects their origin in voluntary philanthropic efforts. In many cases, the need which such institutions were set up to meet turned out to be so great that some more formal, established type of organization to sustain them was quickly seen to be required. This process explains the origin from entirely separate roots of organizations providing, for example, hospital treatment, domiciliary nursing, the child care. At the same time it became necessary for the public authorities to step in to plug gaps left by voluntary efforts, particularly as concerned the care of the elderly and the infirm and the sick who are poor. As a result the overlapping jurisdiction of competing agencies was a perennial sources of confusion and waste. In order to rationalize the delivery of services at the local level, a central co-ordinating focus proved to be essential.

Until 1919 no such central focus existed in the UK. Several attempts were made during the nineteenth century to create a governmental body capable of co-ordinating the national effort towards sanitary reform (see Chave 1984). First, the General Board of Health, then the Medical Department of the Privy Council, and later still the Local Government Board were successively entrusted with this task. Meanwhile, personal health services were being developed from many different roots, among them the Poor Law, the National Health Insurance scheme covering general practitioner services, and the school health provisions of the Education Acts. The more or less random allocation to older, pre-existing bodies, of central responsibility for these newer services, or the creation, as in the case of National Health Insurance, of totally new, virtually autonomous, bodies for the purpose, led to an administrative patchwork which was not only confusing to the ordinary person but seriously defective in terms of its impact on the major health problems of the day.

The establishment in 1919 of a Ministry of Health is a major landmark in the development in the UK of policy-making machinery in the health field. Full accounts of the public and political campaign which led up to it are available (Gilbert 1973; Wilding 1967), and Honigsbaum (1970) provides a vivid short account of the interplay of forces and personalities involved. These publications should be referred to for an account of the administrative structure of the new Ministry at its inception; they also provide a valuable insight into the process of policy-making itself.

By contrast, the reformist and idealistic motives inspiring many of those who conceived and worked for the new Ministry are more prominent in the personal account left by Sir George Newman, its first Chief Medical Officer. Amongst many other interesting and important matters, Newman drew attention to the significance of the enhanced status now accorded to the Chief Medical Officer, carrying with it the right of direct access to the Minister (Newman 1939). This change ensured that, in future, full account would be taken of medical and scientific factors in determining policy.

Thus, with the consolidation of many (although still not all) governmental health responsibilities in a single Ministry we begin to see for the first time the outline of a modern central department for co-ordinating policies dealing with the nation's health as a whole. It became possible to co-ordinate the work of the several divisions dealing, for instance, with public health matters, the control of infectious diseases, and personal health services. Information flowed in from many sources including the national and international statistical agencies, the office of the Registrar General, and the Medical Officers of Health. Advice could be obtained from expert advisory committees and administrative action taken or legislation framed, as appropriate.

Establishment of the National Health Service

Few events give a clearer insight into the process of formulating and implementing health policy in the UK than the establishment in 1948 of a national health service. Now that the confidential papers relating to the period have become available under the 30-year rule, it is possible to follow in some detail the development of policy in the years preceding 1948. Those wishing to study the matter in greater depth should consult the book by Pater (1981) based, among other sources, on a careful study of documents in the Public Record Office.

Quite commonly the 'broad strategic' approach to a health problem is formulated at a relatively early stage in the form, perhaps, of a 'blueprint' developed by some authoritative expert group or body independent of government. The strategy which emerges may attract a broad measure of agreement but, over the years that follow, repeated shifts and changes of policy may be necessary to put it into effect. The effort to introduce a comprehensive medical service in the UK followed this pattern.

The strategic blueprint in this case was the Dawson Report (Ministry of Health 1920), produced in 1920 by a consultative council of medical and non-medical experts set up by the Minister of Health. 'What they had produced was nothing less than the outline of a national health service, and in doing so they laid down the main principles and raised the main issues which governed the pattern of discussion for nearly thirty years' (Pater 1981, p. 7).

Throughout the inter-war years there was widespread dissatisfaction with the available medical services (Eckstein 1959), most notably because of the restricted coverage of general practitioner services. The future of the hospitals gave cause for concern too, since many were in severe financial difficulties, even insolvent. The resistance of the voluntary hospital interests to submerging their identity in a unified hospital service was sufficient to thwart all efforts towards greater co-operation with the public hospitals and in the end was overcome only by the overriding needs of wartime.

Even before the end of the Second World War it had become abundantly clear that quite radical changes both in hospital management and in the structure of medical practice were called for. But the more urgent it became to bring in measures of reform, the more hotly was the precise form of those measures disputed. Major policy issues were the terms of service of general practitioners, the role of health centres in the new service, the structure of local administration, and the place of the voluntary hospitals. The principal interest groups involved were the local authorities, the voluntary hospital managers, and the medical profession. Each in turn presented proposals to the Minister and each in turn consulted its own constituency regarding its response to the counter-proposals put to it.

A process of negotiation of this kind has inevitably to be conducted within a political context. The concessions that Ministers are able to make in order to carry through their government's main strategic aims are crucially dependent on the support they are able to muster among cabinet colleagues. In this respect the cabinet papers, now so thoroughly analysed by Pater, reveal very clearly the sharp difference in this respect before and after the election of 1945.

During the closing months of the war and the period of the caretaker Government a rather prominent role was taken by the Ministry itself and its senior officials. They assessed, by means of informal contacts with the interested parties, the obduracy with which stated positions on particular points were held and prepared a series of discussion papers setting out feasible schemes for further discussions. The range and variety of the possible options considered at this stage was remarkable and it seems clear that, far from there being an official or departmental 'policy', official backing would have been given to whatever approach seemed likely to gain sufficient support to enable the very broad aims of Government's strategy to be realized.

After the sweeping Labour victory in the election of 1945 the whole climate of negotiations changed. The new Minister, Aneurin Bevan, set about the implementation of his Government's policy in a very different style. 'The main complaint . . . was that the Minister refused to negotiate: he simply presented his proposals and then listened to the comments without argument. In this, Bevan's stance was clear. His view was that Parliament must be the first to know his proposals and only after that would negotiation be proper; and this view he maintained throughout.' (Pater 1981, p. 117).

The radical solution to the problem of the hospitals: wholesale nationalization with all hospitals, voluntary and municipal, placed under the jurisdiction of regional boards, was enthusiastically adopted by Bevan on the basis of suggestions put to him by Sir John Hawton, a deputy secretary in the Ministry (Pater 1981). We now know that Bevan had more difficulty in carrying this particular proposal in Cabinet than in persuading the medical profession of its merits (Pater 1981). However, it is Bevan's conflict with the doctors regarding their terms of service, both before and more particularly after the passage of the Act, which has attracted most attention. Full accounts of these events exist (Foot 1973) and most commentators concur in according to Bevan himself a large measure of the credit for finally bringing to fruition a policy which had been under discussion since at least 1920.

It is to a large extent through these two linked organizations, the NHS and the DH, that national policies for the promotion of health are formulated and carried into effect. Our brief description of how they came into being has itself illustrated many of the features of the policy-making process, in particular the tendency for broad agreement at an early stage on principles and aims, to be followed by the emergence of widely divergent views on their means of achievement; the prolonged consultations needed to find a practical way forward; the skilful use by interested parties of the bureaucratic mechanisms of the governmental machine; and the crucial importance of the key political figures when the final decision point is reached.

The next section will set out a more detailed account of the mechanism of the DHSS and how it operates today, as the Department of Health, while the following section illustrates the system at work.

How the Department of Health operates

Organization

The creation of the DHSS, which we have briefly described, brought together two very different Ministries—Social Security with a relatively small policy-making headquarters and a large local and regional office network, and Health with virtually no local elements, but a considerable central policy-making role. Following the separation of the two Departments, the Department of Health has adopted a broadly tripartite structure. The NHS Management Executive under its Chief Executive is responsible for the operational management of the NHS. Policies relating to health and the delivery of services are formulated within multi-disciplinary policy divisions reporting to the Permanent Secretary; while medical matters of a scientific or professional nature having an impact on the nation's health are the responsibility of the Chief Medical Officer and his staff. Liaison and co-ordination takes place at every level while final decisions on implementation of specific policies are taken by the Secretary of State with advice from his NHS Policy Board.

Development of policies

The way in which policies towards health promotion were developed in the DHSS and now in the DH is in part typical of Government generally, and in part special to the health field. As in all governmental departments, major policy developments, although prepared by officials, are usually initiated by and ultimately decided by Ministers, in the light of their political philosophy, overall Government objectives, the availability of resources, and their assessment of the situation. Major announcements will be made by Ministers, usually in the first instance in a statement to the House of Commons or in a Parliamentary Answer, and communicated afterwards to the NHS. Communications with the NHS, reflecting the fact that its constituent bodies are not part of the Department, are usually in circulars of guidance, which less frequently may direct a particular course of action. Nearly always there is prior consultation with NHS interests before guidance is issued and this is often the case before major policy developments are announced. This particular relationship is one of the features of DH working that is only loosely paralleled in other governmental departments. The other is that the DH has within its staff members of all the main professions in the NHS, and policies are developed throughout the health and personal social services field on a multi-professional basis. We shall consider these two features in turn.

Consultation

The particular relationship of the Secretary of State to the NHS, which we described earlier, results in a close interdependence. The NHS is dependent on the Secretary of State for resources and in the main for its powers to provide services. The Secretary of State also appoints the Chairmen and Members of Regional Health Authorities (RHAs) and the Chairmen of District Health Authorities (DHAs). On the other hand, though the Secretary of State has powers of direction, he or she ultimately relies on health authorities and the health professions for the provision of services and for the promotion of health.

Views as to the best way to work together have varied over time. There have been periods when the emphasis was on devolution, others when central planning or accountability have been stressed. At all times great attention has been given to trying to secure mutual understanding—that the Secretary of State of the day and the Department understand what is going on in the NHS, and that the NHS understands Ministerial policies and priorities. Regular meetings, of Ministers with RHA Chairmen, and of DH officials with regional officers, in each of the main disciplines, are the main fora for discussions about national policies and priorities. More recently, the creation within the DH of the NHS Management Board (see below) and contemporaneously the introduction of general management at each level within the NHS (see Volume 1, Chapter 15) gave a sharper focus to this process, between the Department and the regional tier and between RHAs and DHAs. Annual Review Meetings were introduced in 1982 between each RHA and the Department at which the Chairman and a Minister considered the Region's plans and progress in the light, among other things, of national policies and priorities. From 1986 the Management Board introduced an additional element of performance appraisal stretching across the Region's responsibilities. Comparable developments have occurred between Regions and Districts (see Volume 1, Chapter 15).

Over the years formal arrangements have evolved to enable the major professional bodies to be consulted on matters which concern them, and their views are sought before policies are finally settled. The Chief Medical Officer, or a deputy, meets regularly with the Joint Consultants Committee representing hospital-based specialists to discuss all matters of NHS policy which might affect their work. Terms and conditions of service for the many groups of people working in the NHS are negotiated nationally within Whitley Councils composed of management and staff representatives from the NHS, but all agreements require the approval of the Secretary of State.

The Standing Medical Advisory Committee (SMAC) is a statutory body which advises the Secretaries of State for Health and for Wales on clinical medical matters. Its Chairman is a leading medical consultant or general practitioner. Twenty-two of its 32 members are appointed in a personal capacity by the Secretary of State; the remaining ten are *ex officio* members and include the presidents of the medical Royal Colleges and deans of faculties. SMAC is an important source of advice to Ministers on clinical matters. It offers an authoritative medical view based on a balance of experience between individual hospital specialties and general practice, between teaching and non-teaching hospitals, and between regions. In recent years SMAC has provided advice on, among other topics, heart transplant services and the diagnosis of child sexual abuse. A second Standing Advisory Committee deals with nursing and midwifery.

On matters of wider interest, such as proposals for changes in health policy or organization, extensive written consultation may be undertaken (see below for examples). Perhaps the greatest volume of comments—some 5000—arose from Patients First, the consultation document which preceded the 1982 reorganization. Another major avenue for establishing a common NHS/DH/Ministerial view on subjects is through the establishment of working parties charged, usually, to report to the Secretary of State. Although the variations are wide, such committees often have joint NHS/Departmental membership and secretariat. Their reports are ultimately published, for information or for action, usually, if action is required, with a Departmental circular.

The views of the health authorities, and of doctors and other professional workers who provide the direct clinical and treatment services to patients, naturally exert powerful influences on Health Ministers. But there are many other pressure groups and individuals whose views Ministers may need to weigh on particular issues. Examples of these (an exhaustive list would be impossibly long) are Members of Parliament (MPs), the Trade Union Congress (TUC) Social Services Committee (see later in this chapter), and bodies such as the National Association for Mental Health (MIND) and the National Society for Mentally Handicapped Children and Adults (MENCAP) concerned with particular groups of patients.

Multi-professional working

In any modern, technologically complex organization, the role of the specialist or technical expert is crucial. While the British health service is not managed exclusively by 'experts', the system enables appropriate specialized knowledge to be drawn upon freely, continuously, and at every level. The First Permanent Secretary, who heads the Department's administrative hierarchy, is a senior civil servant of wide experience. The Chief Medical Officer has the responsibility of providing expert, objective, medical assessments and advice, direct to Ministers if need be, not only on matters relating to overall management of the NHS but also in relation to the many factors affecting the nation's health.

Within the Department of Health, multi-professional working is the key feature of the policy-making side of the organization. The Chief Medical Officer, the Chief Nursing Officer, and other senior professional officers are each the head of a group of professional colleagues who include amongst their number specialists in all the relevant disciplines. Small multi-disciplinary teams form the basic building blocks of policy-making in the health field. A typical team consists of an administrator, a doctor, a nurse, and a social service inspector. Depending on the weight of work involved, an individual may work entirely in one team or be a member of several. Multi-disciplinary teams cover the major client groups for whom health and social services are provided (children, the elderly, the mentally ill, the mentally handicapped, and the physically handicapped), the major service sectors such as primary care and acute hospital services, and policy areas such as health promotion and disease prevention. It is axiomatic that working groups of this kind function as teams of equals.

Management arrangements

The management arrangements of the Department are intended to provide a framework within which the work of the various multi-disciplinary groups can be co-ordinated and the overall consistency of policy development ensured. Senior officials, both administrative and professional, then have the collective responsibility of ensuring that policies are practicable and worthwhile from a wider viewpoint before submitting them to Ministers.

Following the report of an inquiry team led by Mr (now Sir Roy) Griffiths (Griffiths 1983), an NHS Management Board was set up at the beginning of 1985 to be responsible for advising and supporting Ministers on all management matters affecting the NHS. The Board is chaired by the Minister for Health and its Chief Executive is a businessman formerly with a large industrial corporation. The membership includes civil servants and individuals with NHS or private sector experience. Those parts of the Department that had previously been involved in various aspects of NHS management were regrouped into Directorates of the Management Board.

A Health Service Supervisory Board was also established at that time, but following separation into two Departments this was replaced by the NHS Policy Board, the main function of which is to advise the Secretary of State on broad strategic and policy issues. At the same time, the NHS Management Board was replaced by a smaller NHS Management Executive under the chairmanship of the Chief Executive, which, as its name implies is responsible for the day-to-day management of the NHS including implementation of Ministers' strategic and policy objectives.

Policy-making in action—some examples

In the preceding section the machinery for establishing health policies was outlined. In this section some specific examples are briefly described as illustrations of the process at work. We have chosen to illustrate the process principally by examples from the field of preventive health care, because this field shows in microcosm all the key features of the process. It should, however, be borne in mind that a vast range of policy issues also arises in relation to the provision of treatment and curative services, and each of these may make substantial demands on the policy-making capacity of the Department.

Control of infectious diseases

A number of developments over recent decades, including higher living standards, the widespread use of antibiotics, and the availability of more effective methods of immunization, have combined to reduce the significance of infectious diseases as causes of serious morbidity and premature death in industrialized countries. This has inevitably reduced the opportunity for health workers to obtain first-hand experience in managing infectious cases and controlling outbreaks, and may have induced a false sense of security about the likelihood of future threats to the public health from this cause in the minds of many people. In this situation the emergence of 'new' infections such as legionnaire's disease and particularly acquired immune deficiency syndrome (AIDS) has been a cause of concern and has predictably been given extensive media coverage.

Legionnaire's disease has attracted widespread attention since it was first identified among delegates at an American Legion convention in Philadelphia in 1976. The organism has been implicated in numerous outbreaks since then in several countries, and in the UK a particularly large outbreak, causing about 100 cases with 28 deaths, occurred in the town of

Stafford in 1985. But more familiar microbes also continue to cause trouble. Outbreaks of *Salmonella* infection, often with some deaths, have occurred in long-stay institutions for the elderly and handicapped in several countries of Europe and North America. In the UK, an outbreak in Stanley Royd Psychiatric Hospital in 1984 resulted in 19 deaths. Such events inevitably arouse public alarm which receives expression both politically and through the media, and such expressions of public concern, in their turn, act as powerful agents for the development of policies in the health field.

The British health system makes provision for the Secretary of State to set up formal inquiries into events and episodes which, in his or her view, warrant detailed public scrutiny (e.g. S. 84 of the NHS Act), and these powers were used in both the Stafford and the Stanley Royd outbreaks. The detailed investigation of these outbreaks recorded in the Reports of the official inquiries (DHSS 1986*a*, *b*) not only provides a wealth of information about the difficulties encountered in investigating and controlling these two particular outbreaks of infectious disease but also throws significant light on the difficulties of establishing and implementing effective policies even in a field assumed to be so well understood as the control of infectious disease.

Both inquiries were held in public; both took evidence from the participants; and both explored in detail the pattern of the respective outbreaks as they unfolded day by day, together with the responses of those involved. It has been common knowledge for many years that if rapidly developing outbreaks like these are to be brought under control, many individuals from a wide range of disciplines must collaborate in a systematic way and according to an agreed protocol. It had been assumed that the Public Health Acts and related legislation provided just such a framework. In the event, however, the reports of these two inquiries, carried out quite separately and independently, found deficiencies in the way the two outbreaks had been managed—deficiencies, moreover, of a remarkably similar kind in each of the two episodes. Lack of a previously agreed plan for handling contingencies of this kind resulted in lack of clear guidance, failures of communication between the responsible agencies, confusion about the division of responsibilities, and delay in summoning expert help from the appropriate specialized agencies.

Each inquiry report made detailed recommendations relating to the specific place and situation it had been concerned with, but taken together they threw up issues of a more general nature which could be tackled only by action at national level. These had to do with resolving the confusion about the allocation of public health responsibilities which appeared to have entered the system concurrently with the series of reorganizations and management changes which had taken place since 1974, and with repairing the loss of specialized medical expertise in this field since the abolition of the office of Medical Officer of Health—a loss which had been only partially compensated by the creation in 1979 of a new national agency, the Communicable Disease Surveillance Centre, an arm of the well-established Public Health Laboratory Service.

The findings of two public inquiries, so close together in time, with implications of a similar kind for the proper management of an area of the public health so unambiguously identified as a governmental responsibility, is something no health administration could ignore. Even so it proved difficult for the Secretary of State in his capacity as the responsible executive authority in such matters to take appropriate remedial action without the benefit of further expert advice of a specialized kind, targeted at the particular problems confronting central Government.

To correct the deficiencies in the supply and deployment of appropriately trained personnel and to create a suitable management context in the NHS within which they could operate effectively, it therefore proved necessary to institute yet a further inquiry, this time by an independent committee chaired by the Government's Chief Medical Officer, to review these problems in the context of the public health function as a whole. We shall be looking in greater detail below at the wider implications of the Report of this committee, published in January 1988 under the title 'Public Health in England' (Committee of Inquiry into the Future Development of the Public Health Function, 1988). Suffice it to say at this point that its appearance at that moment of time helped to create a climate of opinion from which a wider understanding of the need for change could emerge. The committee recommended, for example, that clearer guidance should be issued emphasizing the public health responsibilities of health authorities and that, in line with the findings of the outbreak inquiries mentioned above, each health authority should appoint a named person, a District Control of Infection Officer, with the appropriate qualifications and experience, to be held accountable for ensuring that proper plans were made and the necessary measures were taken to ensure the effective collaboration of the various agencies concerned with the control of infectious diseases, and to take charge of the management of emergencies.

Acquired immune deficiency syndrome (AIDS)

During 1984 it became clear that the UK would face a growing problem with the spread of the human immunodeficiency virus, as was already the case in the US and, although this was less well understood at the time, in Africa. During 1985 the Department created a wholly new mechanism—an AIDS Unit—to enable it to plan action on all fronts: public education about the disease and methods of prevention, research, provision of testing and treatment services, care and support for the dying, and sponsorship of voluntary bodies. In developing this programme it was supported by a multi-disciplinary Expert Advisory Group drawing upon experts from outside the Department with experience in a wide variety of fields. In addition, and because AIDS raises issues well beyond health—housing and insurance, for example—an Inter-Departmental Committee was established under the leadership of a senior (non-DHSS, initially) Cabinet Minister to co-ordinate action across Government.

Immunization

A striking feature of the policy-making process is the emphasis placed on consultation. This may take a diversity of forms, ranging from negotiations with staff representatives about working conditions to consultation with expert advisory bodies on scientific and technical matters. The more technical the topic the greater the reliance placed on the advice of expert committees. In the preventive field one of the most important aspects of the Department's work is to establish a national policy for immunization, and for this purpose it takes advice from the Joint Committee on Vaccination and Immunization (JCVI) on which sit the country's leading experts on the subject. Departmental observers are able to contribute authoritatively regarding the cost and practicability of proposed programmes of immunization, but the advice on policy is given by the independent scientific members of the committee and, provided a consensus exists amongst them, it is difficult to visualize a situation in which their views would be seriously challenged. Examples of this process can be seen in the establishment of the national policy for vaccination against rubella (DHSS 1983) and in the introduction of combined measles, mumps, and rubella (MMR) vaccine (DHSS 1988*a*)

Where available evidence is insufficient to enable an agreed view to be arrived at on an important subject, the committee itself can take steps to commission further enquiries. Public anxiety about the possibility of brain damage following the use of whooping cough vaccine led to the setting up in 1976, on the initiative of the JCVI, of the National Childhood Encephalopathy Study. The results, along with much other survey material and the conclusions which the committee advised could be drawn from them, were published in an authoritative report in 1981 (DHSS 1981). The study revealed a not inconsiderable amount of serious neurological disease occurring in early childhood, a lot of which occurred independently of whooping cough vaccination while some coincided with vaccination. In a subsequent court case it was stated that it was impossible to determine whether encephalopathy leading to permanent brain damage was attributable to, or coincidental with, vaccination. This conclusion is quoted in the 1988 edition of the official JCVI handbook 'Immunisation against Infectious Disease' (DHSS 1988*b*) which states (p. 16, para. 3.4.1 (c)): 'Neurological events, including convulsions and encephalopathy, may rarely occur after whooping cough vaccine. The best estimate of risk to an apparently normal infant of suffering a neurological reaction such as a prolonged febrile convulsion after whooping cough vaccine is about 1 in 100 000 infections, almost all of those occurring without permanent consequence. Neurological complications after whooping cough disease are considerably more common than after immunisation. However in both immunised and unimmunised children, encephalopathy resulting in permanent brain damage or death may occur in the first year of life; no wholly reliable estimate of such complications after whooping cough vaccine can therefore be made. There is no specific test to identify those cases of encephalopathy which may be due to whooping cough vaccine'.

Expert committees of a comparable nature deal with other technical and scientific aspects of health policy. Examples are the Committees on Medical Aspects of Food Policy (COMA) and on Medical Aspects of Radiation in the Environment (COMARE).

Reorganization of the NHS

We have seen that machinery exists for, first, formulating policies by seeking advice from the relevant experts and then for implementing them through legislative or administrative measures after consultation with the parties likely to be most closely affected. This machinery is well adapted to dealing with situations where the major considerations are technical in nature and the impact of the policy is circumscribed. But what of developments of a more strategic character whose influence may be far-reaching, and where factors of a more general or political kind come into play? Such an issue is the structural form to be taken by the NHS as a whole.

Quite early in the history of the NHS its divided, tripartite structure became a focus of criticism. The hospitals under their management boards, the general practitioner service organized by executive councils, and the community health and social services provided by the local authorities were not only separately administered in a day-to-day sense, they were separately planned and financed too. Many deficiencies were attributed to these divisions. The 'Porritt Committee', set up jointly by the Royal Colleges and some other medical bodies in 1962, came out strongly for the view that all medical and social services in each area of the country should be brought together under the control of a single area health board. 'In the earlier sections of this report', the Committee wrote, 'we have drawn attention to one major and recurring criticism of the National Health Service—its division into three watertight compartments. Nowhere has this defect been more strikingly apparent that in the maternity services, where the patient is often in need of help from at least two, if not all three, branches of the Service. In this sphere cooperation between the family doctor, hospital consultant and the health workers of the local authority, is quite essential if the maternity service is to be effective. Our plan for integrating the Health Service locally under one overall administration should do much to prevent the confusion and frustration which has existed in the past'. (Medical Services Review Committee 1982, para. 296).

This firm advocacy from the Porritt Committee lent weight to the view, which grew from the early 1960s onwards, that administrative integration would bring benefits in both the planning and the delivery of services: at the planning level because it would encourage more rational allocation of resources between services and at the local level because it would permit more effective co-ordination of the separate elements which together make up effective patient care. In spite of this high degree of consensus, a decade elapsed before the NHS Reorganization Act reached the statute book and it is a matter of some interest to enquire why it took so long to put together a policy to achieve an aim on which such a broad measure of agreement existed. The initial response to the wave of opinion favouring integration came from within the central health department and took the form of the publication of a Green Paper.[3] But the subsequent process of consultation provides an illustration of the difficulties to be overcome in introducing a management innovation into a publicly provided service.

The local authorities, who would be losing their community health services to the new, all-embracing health authorities, were concerned, firstly, about where the dividing line was to be drawn between the health services they were to lose and the non-medical social services they were determined to retain. Secondly, they were concerned about how to retain as influential a position as possible in the health field by establishing their rights to membership of the new controlling bodies. They therefore sought to formulate a consistent view on these matters and to make this known to the Secretary of State. To this end they worked within the political parties at constituency level, lobbied MPs, and used the national bodies representing the local authorities to promote their collective views.

An additional relevant factor was the concurrent reorganization of social work and social services following the report of the Seebohm Committee (1968) (Committee on Local Authority and Allied Personal Social Services 1968). Seebohm had recommended that social services for all client groups should be pulled together and organized within a single department of the local authority. The committee had itself been the focus of a major constellation of policy-making interests, both in government and in the country, and the completion of its work in this authoritatively written report, with its clear and unambiguous recommendations, proved to be a significant turning point in the policy-making process. As well as welfare services and social services for children, the consolidated departments of social services which it proposed were to embrace all medical social work including that with the elderly, the handicapped, and the mentally ill. The implementation of this recommendation in the Local Authority Social Services Act 1970 thus brought about a clear administrative separation between health services and social services provided for the same groups of people and lent an added urgency to the need for a health service reorganization that would include proper links between the two separate structures.

At the same time, medical interests were taking steps to ensure representation on the new controlling bodies whilst seeking to safeguard their position in the future structure. This confusing array of cross-currents and contending forces came to a focus within the DHSS and particularly on the person of the Secretary of State himself. One fascinating, albeit intensely personal, insight into the dynamics of the situation can be gained from a number of passages in the diary of the late Richard Crossman (1977), the Secretary of State from 1968 to 1970, in which he describes his personal reactions to the conflicts of opinion and interest exposed during the consultation process. He summarized his view of the reorganization problem as a whole in an entry dated 4 August 1969: 'I have never had a more intractable problem, because everybody believes in the integration of the hospital and GP service with the local government health services but both sides assume they can integrate wholly their way. Wherever we draw the line will be arbitrary and detested by both sides and anyway we shall be leaving unresolved the major problem that on one side of the line the services are financed by the ratepayer and on the other side by the taxpayer'.

Several different proposals were drawn up, only to be revised later in the light of further consultation (DHSS 1970, 1971; Ministry of Health 1968) before the Government was ready to commit itself to a firm policy set out in a White Paper (DHSS 1972) and subsequently embodied in legislation. The interested reader will find a useful short account of these and other Parliamentary proceedings relating to health matters in a publication by Ingle and Tether (1981). The main feature of the House of Commons' debate on the Reorganization Bill was the opportunity it offered for general issues of principle to be aired. Many of the speeches revolved around broad themes such as 'local democracy' and the role of 'managerialism'. In Committee and in the Lords, minor modifications were introduced but these authors conclude from their analysis that the impact of Parliamentary debate on the practical details of reorganization policy was small compared with that of the consultation and negotiations which had taken place during the pre-legislative phase.

Although the 1974 reorganization, like every other policy issue, had its own unique features, we have used it here to illustrate the more 'political' aspects of the policy-making process which inevitably become prominent in any situation which raises important general issues affecting the public interest. This is in contrast to the more technical matters mentioned earlier where policy is more likely to be determined by the advice of specialized experts.

Policy development in the area of health promotion—the process summarized

The examples given above illustrate many of the features of policy-making in the UK, including the emphasis placed on

[3] A Green Paper is a document presented to Parliament as a basis for consultation and comment. Although produced by a governmental department, the Government itself is uncommitted to any policy proposals it might contain. It is to be distinguished from a White Paper which sets out the Government's proposed policies, often as a prelude to legislation.

wide consultation, the crucial part played by expert advisory bodies, and the multiplicity of pressures to which the Secretary of State may be subjected. We are now in a position to sum up the main features of the process and to mention briefly the additional problems that arise when wider health issues are involved.

There are many fields in which the technical aspects of health policy possess a momentum of their own—where 'policy' is a process of responding to technological change. In the discussion above on the establishment of a national policy for immunization, we described a little of the framework of specialist advisory bodies within which this process takes place. In the case of subjects of wider public interest the ideas which develop into health-related policies originate and develop like any other aspects of social policy as part of the political process, taking shape within political parties, interest or pressure groups (Yarrow 1986).

To transform ideas into effective policies two additional ingredients are essential: information and expert advice. Sometimes policies are advocated in their initial stages by a tripartite alliance of journalists, self-help groups of patients, and interested specialists or research workers. In recent years the press and television have come to play an increasingly influential role in policy formation by disseminating ideas and in generating public support for them. In many other cases the Department, with a small core of professional opinion in the NHS, has itself acted to educate, persuade, and stimulate a greater general awareness of a need for change. When the DH takes up a new concept, it subjects it to a number of tests for its political implications in relation to other activities within the health or social services, for its costs, and for other aspects of feasibility such as acceptability within the existing structure of services and framework of staff activities. Perhaps the most crucial test is its acceptability or otherwise to the established sources of advice, which may include individual advisers in the field in question as well as the appropriate specialist advisory committee. Once a new policy idea has attracted the attention of Ministers, the focus of attention of the interested groups concerned to promote it tends quite naturally to switch from the public to the narrower political arena. There follows an increase in Parliamentary activity: Parliamentary Questions will be asked, the subject may be raised in debates, and deputations may attend upon the Minister. If Ministers feel that the evidence and advice are insufficient for a decision, some kind of formal enquiry such as an expert committee or working group may be set up.

The distinction that we have suggested above, between policies of a strictly technical nature and those affecting the wider public interest is, of course, far from clear-cut, and the difficulties that may arise when a matter arousing widespread public concern is the subject of conflicting expert opinion can be particularly difficult to resolve. The question of the harmful consequences of lead additives in petrol is an example of this. Not only is the technical problem of establishing the nature and degree of the toxic hazard involved particularly difficult but the policy implications in this case affect not only the health field but the fields of energy and industry too. Thus we begin to see that difficulties can arise when different but equally valid priorities conflict, particularly when more than one sector of governmental policy is involved. Inter-sectoral policy co-ordination in order not only to promote health but to protect the environment is a live issue in many technically advanced countries today. The next section looks briefly at some of the measures already under way in the UK to improve policy co-ordination, before going on to consider the wider aspects of implementation, with which it is closely linked.

Policy co-ordination at the central level

There are many other examples of contemporary health problems which cannot be dealt with by action limited to the field of health services. Efforts to influence the use and abuse of tobacco and alcohol by fiscal means and by regulation of advertising and sale raise issues affecting trade and industry and the sponsorship of sport, as well as implying legislative changes which require that the views of other governmental departments including the Treasury, the Departments of Trade and Industry, the Environment, and the Home Office should be sought. Similarly, measures to reduce road deaths by making the wearing of seat belts compulsory fall within the jurisdiction of the Department of Transport, not the Health Department. All the departments concerned have strategic aims which they are pursuing over the longer term, and specific policies may have incompatible consequences in different fields, e.g. health and employment. In all these areas a strategy to promote health can be implemented only by enacting policies across a wide span of governmental activities; the initiative may well come from DH but effective action calls for close inter-departmental and inter-ministerial consultation and collaboration.

The 44th Report of the Committee of Public Accounts (1986) was devoted to prevention, and an appendix to that report consists of a Supplementary Memorandum submitted by DHSS in which are listed all activities in connection with the promotion of health and prevention of disease being undertaken by Government. Some 230 individual projects are listed, involving at some point virtually every governmental department. The prevention of accidents, for instance, involves the Department of Trade and Industry, the Home Office, and the Department of the Environment, while the health of schoolchildren, including the issue of guidance to schools on subjects such as solvent abuse and AIDS, requires collaboration between the health and education sectors. In the UK, little difficulty exists in bringing about administrative collaboration over the implementation of agreed policies. This is achieved by routine contacts between officials in different departments. At the earlier stage where policies are determined, although little in the way of formal standing machinery exists to ensure effective inter-departmental co-ordination, *ad hoc* arrangements are brought into play at a more senior ministerial level whenever the occasion requires it. For instance the level of excise duty to be levied on tobacco products and the relevance of this to the adverse effects of smoking on health form the subject of representations from time to time, from health Ministers to their Treasury colleagues. Policy considerations arising from medical aspects of diet and pollution have also formed the subject of inter-ministerial contacts.

The issue of inter-departmental liaison was taken up by the independent committee appointed in 1986 to look into various aspects of public health in England, the Report of which is referred to in more detail below. It recommended that a small unit should be established within DHSS, bringing together relevant disciplines and skills to monitor the health of the public (Committee of Inquiry into the Future Development of the Public Health Function 1988). The main purpose lying behind this recommendation was to ensure that changes in patterns of morbidity and mortality were promptly noted and their implications for health policies assessed but the report went on to say that: 'More sharply focussed health monitoring at DHSS will also be helpful to the work of other Government departments. To this end, and reflecting the underlying public health responsibilities of the Secretary of State, the unit should have (echoing the approach in the Ministry of Health Act 1919) a co-ordinating brief in respect of other Government departments. In particular, it will help maintain consistency of public health policy across Whitehall, for example, when other Government departments are considering decisions (eg on food and agricultural policy or on tobacco and alcohol) which might impinge upon health policy'. ('Public Health in England', para. 4.8).

Implementing health policies

Earlier sections in this chapter have described how policies are formulated. Now we turn to the development of strategies for putting those policies into effect.

The consultation which has been a part of the policy-forming process will have ensured the policy is known to those who will have to implement it, and that it is feasible and acceptable. But that alone does not ensure that action will follow. Some mechanism to co-ordinate the process is required and also some form of monitoring to provide information about progress and to permit modification or corrective action to be taken in the light of experience. We shall now consider in turn how each of these necessary procedures is provided for.

Co-ordination at operational level

Before the introduction of the NHS, attention in the field of health policy was directed towards two main strategic areas: (i) the improvement of the environment and the control of infectious disease; and (ii) the development of personal medical services for specific vulnerable groups such as mothers and young children, the tuberculous, etc. In each case the policies to be pursued were enacted in legislation and the responsibility for their implementation was laid, in the majority of cases, upon the local health authorities, i.e. the County, Borough, or District Councils, according to the particular measure. It was to the local authorities that central Government looked at that time to co-ordinate the national strategy for health, while during the early decades of the Ministry of Health the principal channel for ensuring close consultation with local authorities on health matters was the Medical Officer of Health (MOH). However, the succession of administrative changes which have taken place in the years following the setting up of the NHS, of which the separation of health and social services and the abolition of the office of MOH in 1974 were probably the most significant, have had far-reaching effects on the ability of central Government to co-ordinate the local implementation of its health strategy in this way.

The necessity for collaboration between health and social services was evident even before the NHS reorganization of 1974 came into effect, and a working party set up to resolve the difficult problems it entailed (DHSS 1973) produced a formidable set of reports. The National Health Service Act itself places a statutory obligation on both sets of authorities to provide for collaboration by setting up Joint Consultative Committees (JCCs) to advise them on the planning and operation of services of common concern. Their membership and duties are prescribed in a Statutory Instrument (SI, 1974, No. 190). Joint Care Planning Teams (JCPTs) of officers advise their Committees on the development of strategic plans and guide-lines covering priority services identified by JCCs as requiring a joint approach to planning (DHSS 1977). The difficulties, however, remain intractable.

The structure of local government is such that not all the functions in respect of which a health authority might seek collaboration are performed by the same local authority. In many parts of the country County Councils are responsible, amongst other services, for education and social services, and thus have the primary responsibility for the care and welfare of children, whilst the lower tier of District Councils is responsible for environmental health matters including housing, pollution, and the control of communicable disease. Nor do the boundaries of the different authorities always coincide. Some counties, for instance, include the territories of several DHAs and may themselves have delegated certain social services responsibilities to smaller areas or divisions which, in turn, may overlap a number of DHAs. Under such circumstances the difficulties in establishing an effective administrative basis for joint working can be formidable. An article by a local government official (Smith 1983) describes how the county of Surrey attempted to cope with a 'situation in which social services operates through three divisions, education through four . . . housing through eleven district councils and health through seven DHAs'. In addition, the two sets of authorities have different constitutions, a different form of financing, and a different corporate culture. The resulting difficulties have been discussed and analysed by a number of commentators, e.g. Norton and Rogers (1981), Lee and Mills (1982), Sainsbury (1982). In yet a further attempt to resolve the problems a working group representing health and local authorities and the DHSS came together in 1984 'to review the working of the present arrangements . . . and to consider what steps could be taken to improve joint planning . . . '. Their report made no suggestions for radical change but gave useful practical advice on how to make the joint planning procedures operate more smoothly and to better effect. That this was going to prove an inadequate solution to the problem of collaboration, particularly as it applied to the provision of community care, soon became apparent and at the end of 1986 Sir Roy Griffiths was commissioned by the Secretary of State to look into the possibility of more specific management action to deal with the problem. He reported in 1988 (Griffiths 1988), and the Government is currently considering its response to his recommendations.

Monitoring performance

A largely monopolistic system of health care delivery requires constant monitoring in order to ensure that policies are being put into effect and to enable them to be modified in the light of experience. Within the UK health care system, monitoring takes place at both the political and the operational levels.

Parliamentary scrutiny of the DHSS

The Select Committee on Social Services, established in 1979, is a committee of members of the House of Commons charged with responsibility for examination of the work of the whole of the DHSS.[4] It takes anything from three to six topics a year, invites written evidence from the Department and outside agencies, examines Ministers and officials, and produces a written report of its conclusions. Conventionally, although there are no statutory requirements, the Government publishes a written response. Health topics examined have included perinatal and infant mortality, community care, and AIDS. Each year the Department's expenditure plans are scrutinized and additionally there was in 1988 an inquiry into resources for the NHS. Substantial continuity of membership has led to a considerable build-up in expertise and knowledge of the Department's work. The Committee is able to supplement this by employing outside advisers, although it has nothing like the scale of assistance given, for example, to a Senate Committee in Washington.

The Committee of Public Accounts, which was established in the nineteenth century, is again a committee of members of the House of Commons. It is charged with examination of the accounts of all Government Departments. It is assisted in this work by an official known as the Comptroller and Auditor General, formally a servant of the House of Commons. This official and his or her staff formally audit the accounts of all Government Departments and of health authorities collectively. They also carry out examination of specific topics, for example where a programme or project over-ran cost-estimates, increasingly of a value-for-money nature. Recent enquiries in the health field have included topics such as prevention, use of operating theatres, or the cost and operation of a scheme for premature retirements. In each case the Comptroller and Auditor General produces a written report of his findings agreed as to fact with the responsible accounting officer in the Department (normally at Permanent Secretary level). The accounting officer is examined on that report by the Committee which publishes a report of its findings. The Government is required to respond in writing, but it does not have to agree to any of the Committee's recommendations.

Monitoring the delivery of health care

By management

The accountability and review system was set up in 1982 as a means of ensuring that greater freedom for local management was matched by arrangements to call health authorities to account for their performance. The process is described in detail in Chapter 22 of this volume. In brief, it consists of two parts: annual management meetings take place between the NHS Management Board led by the Chief Executive and a team of Senior Officers from each Region led by their general manager, to discuss major issues of performance in relation to strategic and short-term plans. Selected issues arising from those meetings are then taken forward to Annual Ministerial Review meetings where they are further discussed, together with other issues of concern to Ministers. The outcome of each regional review is an Action Plan, a copy of which is placed in the library of the House of Commons and thus forms a link between operational management of the NHS and the political process. The process is continued by each Region carrying out a comparable set of reviews with each of its component Districts and producing in each case a District Action Plan, which is also made public.

By inspectorates

The establishment of inspectorates as a public health measure is traditionally associated with the control of hazardous substances in the environment (see Chapter 9). From time to time the question is raised whether the method might also be applicable to the delivery of personal services. Against the suggestion is the desirability on managerial grounds of decentralizing the locus of responsibility and encouraging local autonomy in the delivery of what are essentially local services. On the other hand bureaucratic agencies, lacking the immediate awareness of customer perceptions which the discipline of the market brings, are notoriously weak in developing a culture of accountability to the public they serve.

Against this background the experience of the Health Advisory Service (HAS) should be noted. It was originally set up in 1969 as the Hospital Advisory Service to keep under review the services provided for mentally ill and elderly

[4] At the time of writing the question of appropriate Select Committee arrangements following the separation into two Departments had not been settled.

people in long-stay institutions. In 1976, with the change in title, it expanded its remit to include both health and social services provided in the community for the same client groups.

It operates by setting up multi-disciplinary teams to visit and report on the services in a particular area with a view to providing an objective assessment of those services, coupled with advice on how standards might be improved. HAS is independent of the DHSS and reports directly to the Secretary of State and also to the authorities visited. It does not investigate individual complaints. Over the years the HAS has become well accepted and is looked to as a source of guidance on good practice in the fields with which it has concerned itself. Much of its success probably has to do with its method of working whereby visits are undertaken less as an inspectorial exercise and more as a form of peer review.

A key feature of the work of the HAS has been its attempt to promote collaboration at the operational level between health and social services. The latter are provided by the elected local authorities, not directly by the Secretary of State through specially constituted agencies as is the case with health services. However, the Local Authority Social Service Act 1970 requires them to do so 'under the general guidance of the Secretary of State'. In 1985 the professional social work service of the DHSS which advised the Secretary of State in this field was reconstituted, with the agreement of the local authorities themselves, as the Social Services Inspectorate (SSI).

The SSI, by virtue of its constitution, cannot enjoy the same degree of independence as the HAS; nevertheless, in practical terms it has drawn upon the experience of HAS which has demonstrated the benefits to be obtained from fostering the greatest possible degree of collaboration between inspectorate and the service inspected. The need for a body of this kind has been recognized for many years and derives from the fact that the social services often assume responsibility for the most handicapped and dependent members of society and that special measures are needed to protect their interests. Such persons might include the severely mentally handicapped, the demented elderly, and delinquent juveniles under detention. On appropriate occasions the inspectors can employ statutory powers of entrance, inspection, and examination on behalf of the Secretary of State. Much more often, though, they carry out supportive and advisory inspections, at the request of and in agreement with a local authority in order to assist with a problem identified by that authority.

Conclusion: the future of health promotion

What the UK example demonstrates is that an integrated, centralized health care system can operate effectively and efficiently, perhaps better in a relatively small and homogeneous country than elsewhere, but at a cost. The continuous effort needed to develop, promulgate, and implement policy seems to demand a complex bureaucracy, providing extensive opportunities for consultation, the setting up from time to time of expert advisory working parties and committees of inquiry carrying high prestige in order to carry policy forward, and an effective communications network linking the centre and the periphery to enable progress to be monitored and appropriate action to be pushed ahead. At the same time, such a system is liable, as are all bureaucracies, to lack sensitivity to consumer needs, and so countervailing mechanisms, which may include inspectorates, need to be built in. Only in these ways can the cycle of policy development, information flow, and implementation be maintained.

Attention has already been drawn to the key role in this cycle played by the Medical Officer of Health prior to 1974. Although the administrative responsibilities of the local authorities in the health field were greatly reduced in 1948 when all hospitals were transferred to the Regional Hospital Boards, the co-ordinating role of MOsH continued until the NHS reorganization of 1974. Their continued responsibility for measures of both control and prevention in the communicable disease field, and for management of the community health services, enabled those services to be related to the health needs of the local population. The position of MOsH within local government enabled them to take an overview both of the health status of the local population and of the particular environmental situation as it affected health. They were in a strong position to engage in health-promoting activities, including health education and the use of publicity via the local media. While the integration of hospital and community health services within a single authority in 1974 gave the opportunity for more comprehensive planning of services, the abolition of the office of MOH left a gap in the local co-ordination of community services and in the implementation of community-wide policies for the promotion of health.

It has been claimed by some that wider social issues as they affect health, particularly the necessity to modify contemporary lifestyles if the diseases that now pose the greatest threats to health are to be prevented, are less amenable to locally based action. As a response to this there has been a marked expansion in recent years in the work of national bodies such as the Health Education Authority (formerly the Health Education Council). The Authority's current programme includes a national publicity campaign against smoking, a public information campaign about AIDS, the distribution of films on baby care, contributing to the training of teachers in health matters, and working with the trade unions on the special problems of health at work. The DHSS, too, took active steps to heighten public awareness of the extent to which personal behaviour could affect health by publishing a series of booklets on the subject under the general title 'Prevention and Health'.

However, serious problems of co-ordination remain. Health promotion and preventive activities, whether undertaken by educational or other means, cannot always be effectively carried out in total isolation from the treatment services being provided by the health authority: some measure of integration is called for. Early detection of disease by screening, for example, is increasingly being considered as a means of promoting health. The development of policies in this field calls for careful assessment of research findings as they become available so that decisions can be taken on the most cost-effective way to incorporate a screening programme into the existing local pattern of services. This was part of the background leading up to the establishment in 1985 of a committee of inquiry 'To consider the future development of the public health function, including the control of communicable diseases and the specialty of community medicine, following the introduction of general management into the Hospital and Community Health Services'.

The inquiry was wide ranging and highlighted a number of fundamental problems, including:

(1) a lack of co-ordinated information about the health status of the population both nationally and locally;

(2) a lack of emphasis on promotion of health and prevention of disease;

(3) weakness in the capacity of health authorities to evaluate their activities;

(4) widespread confusion about the role and responsibilities of public health doctors;

(5) confusion about responsibility for the control of communicable disease.

The committee published its report in January 1988 under the title 'Public Health in England' (Committee of Inquiry into the Future Development of the Public Health Function 1988) and in announcing its publication the Secretary of State said: 'as required by its terms of reference, the report concentrates on the control of communicable disease and the role of public health doctors within the framework of general management in the NHS and examines the wider issues of the public health function. It also suggest ways in which surveillance of the health of the population might be improved, a greater emphasis on the prevention of illness and premature death achieved, and the effectiveness of health services better evaluated'. We have noted in an earlier section the report's recommendation that a central health monitoring unit should be set up in DHSS. Amongst other proposals, it also recommended that:

(1) guidance should be issued to all health authorities updating them on their public health responsibilities;

(2) a public health doctor (normally the existing District Medical Officer) should be appointed as Director of Public Health; and that he/she should be required to produce an annual report on the state of health of the District or Region.

The report devotes close attention to the future of the specialty of community medicine, which, in view of widespread misunderstanding, it suggests should be known in future as the specialty of public health medicine. 'The role of the community physician is considered both in respect of prevention of illness and promotion of health, and in relation to the planning and evaluation of health services and the need to improve their balance, effectiveness and efficiency. At a time of growing . . . demand for health services, techniques for evaluating outcomes are assuming increasing importance'. (para. 1.7). The report finds that 'epidemiological skills are relevant to monitoring the health of the population, analysing the pattern of illness in relation to its causes and evaluating services—all of which are helpful in seeking to make best use of finite resources.' (para. 3.12). The overriding aim of the recommendations is to ensure that the state of the nation's health is constantly monitored and that services provided, both for prevention and treatment, are evaluated in terms of their likely impact on that state of health.

This important report serves to demonstrate a widespread recognition of the importance of some of the themes considered in this chapter. The main principles which it embodies were accepted by the Government in July 1988, and later the same year a circular to health authorities began the process of giving effect to its main recommendations (Department of Health 1988*c*).

Tailpiece

At the time of writing the outcome of a fundamental review of the NHS, set up by the Prime Minister in January 1988 and conducted under her personal leadership, is not known. Interested parties have been invited to submit evidence, and a substantial number of organizations and groups both from within the NHS and outside it have done so. A number of broadly based working parties have contributed to this process. The evidence has been characterized by a continued commitment to the basic principles of the NHS, and notably the provision of comprehensive health care regardless of ability to pay, and by a willingness to examine and in many cases to endorse radical changes in the means by which those principles might be secured. It is idle to speculate whether such changes will be deemed either workable or beneficial, but it is an interesting comment on the dynamism of policy formulation within the British health care scene that change is so readily and freely considered.

References

Chave, S.P. (1984). The origins and development of public health. In *Oxford textbook of public health* (ed. W.W. Holland, R. Detels, and G. Knox) Vol. 1, p. 3. Oxford University Press, Oxford.

Committee of Inquiry into the Future Development of the Public Health Function (1988). '*Public health in England*' Cmnd 289. HMSO, London.

Committee on Local Authority and Allied Personal Social Services (1968). *Report (The 'Seebohm Report')*. Cmnd 3703. HMSO, London.

Committee of Public Accounts (1986). 44th Report: *Preventive medicine*, Appendix: '*Health promotion and prevention of ill-health—Note on Government activity*'. HMSO, London.

Crossman, R. (1977). *The diaries of a cabinet minister, Vol. 3. Secretary of State for Social Services 1968–70*. Hamish Hamilton, London.

Department of Health and Social Security (1970). *National Health Service. The future structure of the National Health Service.* HMSO, London.

Department of Health and Social Security (1971). *National Health Service reorganisation consultative document*. DHSS, London.

Department of Health and Social Security (1972). *National Health Service reorganisation: England*. Cmnd. 5055. HMSO, London.

Department of Health and Social Security (1973). A report from the Working Party on Collaboration between the NHS and Local Government on its activities to the end of 1972. HMSO, London. (Also 2nd report, 1973; and 3rd report, 1974.)

Department of Health and Social Security (1977). *Joint care planning: Health and local authorities*. Health circular, HC(87)17.

Department of Health and Social Security (1981). *Whooping cough. Reports from the Committee on Vaccination and Immunisation*. HMSO, London.

Department of Health and Social Security (1983). *Rubella immunisation*. Health notice, HN(83)83.

Department of Health and Social Security (1986*a*). *The Report of the Committee of Inquiry into an outbreak of food poisoning at Stanley Royd Hospital*. Cmnd. 9716. HMSO, London.

Department of Health and Social Security (1986*b*) *First report of the Committee of Inquiry into the outbreak of legionnaire's disease in Stafford in April 1985*. Cmnd. 9772. HMSO, London.

Department of Health and Social Security (1988*a*). *Immunisation*. Executive letter. EL(88)P/125.

Department of Health and Social Security, Welsh Office, Scottish Home and Health Department (1988*b*) *Immunisation against infectious disease*. HMSO, London.

Department of Health (1988*c*). *Health of the population: Responsibilities of health authorities*. Health circular, HC(88)64.

Eckstein, H. (1959). *The English health service: Its origins, structure and achievements*. Oxford University Press, London.

Eckstein, H. (1960). *Pressure group politics: The case of the British Medical Association*. Allen and Unwin, London.

Foot, M. (1973). *Aneurin Bevan: A biography, Vol. 2. 1945–1960*. David-Poynter, London.

Gilbert, B.B. (1973). *British social policy, 1914–1939*. Batsford, London.

Griffiths, R. (1983). *NHS management inquiry: Recommendations for action*. DHSS, London.

Griffiths, Sir Roy (1988). '*Community care: Agenda for action*' a report to the Secretary of State for Social Services. HMSO, London.

Honigsbaum, F. (1970). *The struggle for the Ministry of Health, 1914–1919*. (Occasional papers on social administration No. 37). Bell, London.

Ingle, S. and Tether, P. (1981). *Parliament and health policy: The role of MPs 1970–75*. Gower, England.

Jones, P.R. (1981). *Doctors and the BMA – A case study in collective action*. Gower, England.

Lee, K. and Mills, A. (1982). *Policy-making and planning in the health sector*. Croom Helm, London.

Medical Services Review Committee (1962). *A review of the medical services in Great Britain. Report of Medical Services Review Committee*. (The 'Porritt' Committee.) Social Assay, London.

Ministry of Health, Consultative Council on Medical and Allied Services (1920). *Interim report on the future provision of medical and allied services*. Cmnd 693. HMSO, London.

Ministry of Health (1968). *National Health Service. The administrative structure of medical and related services*. HMSO, London.

Newman, G. (1939). *The Building of a Nation's Health*. Macmillan, London.

Norton, A. and Rogers, S. (1981). The health service and local government services. In *Matters of moment: Problems and progress in medical care, 13th series* (ed. G. McLachlan) p. 107. Published for the Nuffield Provincial Hospitals Trust by Oxford University Press, Oxford.

Pater, J.E. (1981). *The making of the National Health Service*. King Edward's Hospital Fund, London.

Sainsbury, E. (1982). United Kingdom. In '*Linking health care and social services: International perspectives*' (ed. R.A. Ritvo and M.L. Hokenstad) p. 189. Sage Publications, Beverly Hills.

Smith, J. (1983). Joint Planning: Surrey with the forum on top. *Health and Social Services Journal* . **93**, 114.

Wilding, P.R. (1967). The genesis of the Ministry of Health. *Public Administration* **45**(2), 149.

Yarrow, A. (1986). *Politics, Society and Preventive Medicine*. Occasional papers No. 6. Nuffield Provincial Hospitals Trust, London.

34

Coordination and development of strategies and policy: the Japanese example

ATSUAKI GUNJI

Introduction

The Japanese constitution states that the government must make efforts to promote public health. Preventive services in Japan are provided by the government as a public service. However, public health as defined by the constitution also includes medical care, which, in Japan, is largely supplied by hospitals and clinics in the private sector. Therefore public health tends to be regarded in terms of research and activities pertaining to disease prevention and health promotion, rather than provision of diagnosis and treatment.

Nowadays, however, with the problems of an ageing population, the government is placing the highest priority on policies to adapt the social system to this demographic change. One of their most urgent tasks is to reorganize various community services and establish a comprehensive care system, of which medical care is an indispensable element. Therefore, in discussing strategy and policy for public health in Japan, prevention and health promotion are of central concern, but medical care is not excluded.

As the government inevitably has the greatest role in establishing public health strategy and policy, the nature of the Japanese administration is described in detail below.

Determination of policy requires a higher level of decision-making and organization than the development of strategy, since it involves the political process. Political decisions can be defined as those which involve disputes regarding the desirability of the objectives in question. Selection of the strategy to achieve a policy objective does not require the same level of political discussion. Political decisions ('policy') often require changes in the legislation or new legislation whilst strategic decisions do not.

Although the concepts of co-ordination and development are essentially different, it is very difficult to differentiate between them in the actual process of policy and strategy formulation. Therefore they are not described separately in this chapter.

The structure and characteristics of the Japanese Government

Central government and local government

Japanese government is structured in a three-tiered hierarchy: central government, prefectures, and municipalities. There are 47 prefectures, most of which have a population of over a million. Municipalities are defined on the basis of their population and include villages, towns, and cities.

The Japanese government has separate systems for legislation, administration, and the judiciary. Each level of government has a legislative and administrative body, but local government does not have the right to enact any regulation which contravenes the laws of national government.

The function of the administration

The central government is administered through the Prime Minister's Office, 12 ministries, 34 agencies, and some standing committees. Even though the structure of local government differs in each prefecture and municipality, they have similar hierarchical structures divided into many separate sections, each of which corresponds to a section of the upper tier of the administration.

The size of the administration has consistently been increased. This trend has become more marked and rapid since the Second World War as a result of increasing administrative demands, not only in the financial management of the country in a competitive world but also in improving public health and welfare services. Along with this trend, the strengths and weaknesses of the administration have become increasingly important features in policy development and the co-ordination process of this country.

The process of decision-making and co-ordination

As in most countries in the world, the structure of the Japanese administration is very hierarchical. This kind of organization is notorious for its many shortcomings, such as slow decision-making, widespread sectionalism and lack of co-ordination. These deficiencies are rife in the administration of Japan. However, the Japanese administration actually has more power than those of the UK, US, and some other western countries for example. Most of the bills proposed for legislation to the Diet are actually prepared by the administration; only a very small number of bills are prepared by the members of the Diet itself.

The administration works actively to ensure the smooth passage of bills through the Diet. This means that the administration often has to work in a co-ordinating capacity with other administrations from different sectors of society. The interests of different sectors are often contradictory. This hierarchical structure hinders co-ordination, often causing the policy-making process to be slow and stagnant.

The ministers and heads of agencies are usually nominated by the Prime Minister from among the members of the Diet. This system works as the co-ordinating machinery for ministries and different interest groups.

The Japanese administration has a characteristic system of decision-making known as *Ringi*. This system was originally used to avoid calling meetings for unimportant matters. However, it is now used for fairly important decisions and has a co-ordinating role in the administration. When an administrative decision is made, a *Ringi-sho* has to be written. This is a document with a standard format. An official in the lowest tier of administration from the section directly concerned with the matter writes the first draft. After outlining the decision or sanction to be made and the reasons for it, it is officially stamped and then transferred to a superior. When the document is agreeable to the superior, it receives another stamp. This process continues until the document finally reaches its destination (also prescribed by the first writer), although it may sometimes be altered by superiors according to the level of importance of the matter.

This is the formal system of decision-making in the bureaucracy, but there is also an informal system. If the decision is on a fairly important matter or needs some explanation, the subordinate has to explain the matter to the superior before the document arrives. This process is known as *nemawashi*. This system works very effectively for co-ordinating the Japanese bureaucracy. When there is disagreement among sections relating to a decision, the person or section responsible has to talk to the other sections until all reach agreement.

Therefore decision-making in the Japanese administration is based on consensus. This may be a good system from the viewpoint of co-ordination, but there are also shortcomings. The system is slow and inefficient. Responsibility is shared among the people committed to the decision, but shared responsibility can often result in irresponsibility. The incidence of error may decrease but decisions tend to be conservative.

Organization of public health

The role of the government in public health

The constitution states that health is one of the basic human rights and that the government must make every effort to promote social security, public health and welfare. In Japan the preventive services are generally provided as public services directly or indirectly by the government, but medical services are provided by clinics and hospitals with various kinds of ownership. The majority of them belong to the private sector.

The central government owns about 450 hospitals throughout the country. They are former army hospitals which were transferred to the Ministry of Health and Welfare after the war. They do not differ from hospitals in the private sector in terms of content of services or prices, but some of them are exclusively for the care of patients with chronic and intractable diseases and others provide special services such as schooling for dystrophic children.

Local government also owns hospitals which are not significantly different from those of the private sector. It is a characteristic of the Japanese medical care system that hospitals and clinics with varying forms of ownership are in competition with each other, especially in city areas. Patients are free to visit any medical facility and doctors of any speciality; there are no official rules governing referral of patients.

Thus the Japanese government does not take a central role in providing hospital services.

Ministries of central government

Almost all preventive services are provided directly or indirectly by the government. There are three ministries and one agency which have responsibilities for this function: the Ministry of Health and Welfare, the Ministry of Labour, the Ministry of Education, and the Environmental Agency.

The Ministry of Health and Welfare has prime responsibility for health administration and for the actual services provided for people living in a community. The Ministry of Labour is responsible for the health and safety of workers. The Ministry of Education is responsible for school health. The Environmental Agency is responsible for the control of water, rivers and lakes, the atmosphere and the conservation of nature.

Local government organizations

Each ministry except the Ministry of Labour has a corresponding section in the prefectural government. The prefecture has some health centres as 'front–line' health facilities. Municipalities are below the prefecture in the administrative organization but some large cities are given responsibility for their own health administration.

Health centres

The prefectural government and some large cities have health centres which are the front-line machinery of health

Table 34.1. The activities of health centres

Average number of staff	
Medical doctors, dentists	1.5
Public health nurses and midwives	10
Pharmacists, veterinarians, nutritionists, radiologists, other paramedics	7
Medical social workers and other clerical workers	12
Total number of staff	30.5

Source: Ministry of Health and Welfare (1987).

administration. Thirty-two large cities and 23 special districts in Tokyo have been given substantial responsibilities for health administration, including provision of health centres. There are 851 health centres at present and among them, 637 belong to prefectures, 161 to large cities, and 53 to special districts of Tokyo. The health centre is a fairly large organization. Table 34.1 shows the types of personnel who were working in health centres in 1989. The number is the average number of personnel per health centre. On average about 40 professional and clerical staff work in each centre. In principle, each health centre serves a population of about 100 000, but in fact the size of the population differs between rural and urban areas. Government regulation states that the head of the centre must be a medical doctor. The range of activities of the health centre is very wide but does not include medical services.

Health centre activities

Health centre activities include the following:

(1) control of infectious diseases;
(2) food and environmental hygiene;
(3) activities of public health nurses;
(4) mental health;
(5) dental health;
(6) health education;
(7) improvement of nutrition;
(8) collection of vital statistics;
(9) administrative jobs;
(10) other public health activities.

Municipal health centres

Decentralization is a basic principle underlying national policies. The central government encourages municipalities to have their own 'municipal health centres' by subsidizing a third of the standard construction budget. By 1988, 926 municipal health centres had been built and already outnumbered the prefectural health centres. There were about 10 000 public health nurses working in municipal government and about 8400 public health nurses working in prefectures.

This policy left some strategic and co-ordinating problems to be solved between prefectures and municipalities. In 1982, The Act for Health of the Elderly was passed which placed the responsibility for implementation not on the prefectures but on the municipalities. It was a bold decision to omit prefectures. Compared with prefectures, municipalities vary widely, and range from large cities, such as Osaka and Kyoto, to small villages on remote islands with populations of only a few thousand. Such villages often have no public health nurse or midwife. In these cases, therefore, the prefectural health centre has to take over the responsibility for implementing the Act.

Non-governmental groups promoting public health

Non-governmental groups are very important in both the implementation and the development of strategy and policy. These include many professional associations, non-governmental organizations (NGOs), and volunteer groups.

The Japanese Medical Association is the 'central government' of local medical associations whose constituencies are almost coterminous with those of governmental administration. Local government must depend upon the medical associations to carry out public health activities such as immunization of children, mass screening, and health checks on the elderly and children. Therefore co-ordination with the medical associations is important from the developmental stage of policy and strategy planning through to its implementation.

There are similar professional associations for pharmacists, dentists, nurses, and nutritionists. There are many NGOs such as the Association of Tuberculosis Control, the Cancer Prevention Society, many organizations for promoting maternal and child health, the Benevolent Society, the Red Cross, and so on. Many volunteers work actively in communities throughout Japan. There are about 170 000 who promote improvement of nutrition, for instance, and comparable numbers for tuberculosis control and maternal and child health activities. They provide health in their communities and co-operate with professional health workers. Not only are these public health groups important for the development of public health strategy and policy, but co-ordination with them is vital for implementation of public health programmes.

Strategy and policy formulation

From the bottom up

New ideas often come up from the bottom of an organization where people are actually working and encountering problems in implementation of policies and strategies. Therefore the bottom-up system is important in formulating strategy and policy. In carrying out day-to-day activities, staff of health centres communicate with the corresponding sections of local government. A medical doctor must be in charge of the health centre by regulation. Around the time of the Second World War, when many infectious diseases were endemic (for example tuberculosis), many doctors devoted themselves to preventive activities against these diseases. Nowadays, many of these disease have been controlled, and it may be true that the health centre has become a less attractive place for the medical professions than before. In many areas local government has difficulty in recruiting medical doctors to run health centres, despite the fact that this is a key position in the management and co-ordination of various public health activities in communities. They are given very wide-ranging functions, in both the promotion and regulatory fields. They are involved in almost all the health programmes in the community, and are in charge of the inspection of medical facilities, food industries, restaurants, public baths, and other facilities which directly influence public health. They are given authority to order the improvement of facilities and/or management, and to close businesses if necessary.

Usually, the governing body of a health centre includes the Chair of the Standing Co-ordinating Committee for Community Health, and the members include not only delegates from the professional associations but also delegates representing residents and many sectors of the community which have some association with public health activities. The health centre issues an annual report containing a record of the committee's activities, opinions, and proposals. It is submitted to local government and finally the data presented reach the Ministry of Health and Welfare.

In addition to these routine and formal channels of communication, there are many informal communications which are often more important than formal meetings or documents. It is quite common for active communication among people to occur even before a subject is put on the agenda of a formal meeting.

Public health problems identified by this system are finally divided into two categories: those which can be solved only by the local government and others which require some measures from central government.

In the former case, new measures are produced in the health section of the prefectural government. When new budgeting for the measure is required, the health section has to negotiate with the financial section and the budget has to be approved by the local government assembly. The matters which can be dealt with by local government are mostly of a strategic nature.

The cases requiring the commitment of central government are treated as follows. Requests from local government are submitted to the central government at regular meetings. In addition to this process, the directors of the health sections of local government bodies meet voluntarily to discuss the strategies and policies they should adopt. They compile a document of requests and submit it to the Minister of Health and Welfare before the annual budgeting activities begin.

When the matters are of great importance, the governors of the prefectures and/or the mayors of the municipalities become involved. As they are elected political figures, they are more powerful than the administrators in the bottom-up process, not only in the development of a new policy but also in challenging central government.

From the top down

In fact, most new policies and strategies are formulated by central government and their implementation is delegated down to local government. This is particularly true of administration. The ministries take the lead in this process for the following reasons: First, bills are usually formulated by ministries and then proposed to the Diet, where they are debated by the members and the minister or specially appointed ministry staff; second, administrative power is concentrated in central government; third, the information system is concentrated in central government.

The central government is also influential because it has access to greater information resources and so can identify problems and solutions more easily than local government. The census is carried out by the Prime Minister's Office every five years. Other statistics such as vital statistics, morbidity, health resources, nutrition, and surveillance of infectious diseases are also collected, tabulated, and published by the Statistics and Information Department of the Ministry of Health and Welfare. In the past there has been criticism that data gathered by central government are not returned to local government for their own usage, but this has now been rectified. The data on the surveillance of infectious diseases are sent back to local government through on-line information systems and reach hospitals and clinics through various channels.

In addition to these standard statistics, each department and division of the central government has its own channels and methods for gathering the necessary data for planning activities. Therefore, the Japanese central government has far more data of importance for the development of policy and strategy than is possessed by the local government.

Relationship between central and local administration

The local administration has two functions: their own discrete functions and those delegated by central government. The latter are classified into two categories: those in which the local administration is required to carry out the orders of the central administration, and those in which the decision to implement the activity is left to the Mayor or governor. In the former case there is no requirement on the part of the local government to report these activities to the assembly of the local government, and indeed there could be some form of sanction if governors or mayors refused to carry them out.

In Japan, the majority of these activities are of the former category. Public health services must be equally available to every citizen. Problems could arise if any governor were to refuse to immunize children, for example. Therefore, local governments are commanded by the central government to implement most public health tasks. In the recent trend toward decentralization, the Japanese government has made efforts to give local government more autonomy in these areas, but this centralized structure is still the salient feature of the administration of Japan.

The process of developing new policy and strategy in each ministry in central government begins about June following the end of the fiscal year in March. Plans for new policy and strategy are contemplated in each division using information and statistics from the various sources mentioned above. Then they are submitted to the higher level of the ministry (bureau level), co-ordinated, and forged into bureau policy. Each bureau submits its plan to the accounts division where it is co-ordinated at a ministry level within the conditions prescribed by the Ministry of Finance. By the end of August, every ministry and agency has submitted its plans to the Ministry of Finance. Negotiations between the Ministry of Finance and each of the other Ministries continue until the end of the year when the Ministry of Finance and the Cabinet finalize the budgetary plan for the next fiscal year. Then the plan is submitted to the Diet. After debate in the Diet, the plan is usually carried so as to be in place for the new fiscal year. Throughout this process, it is often necessary for the various divisions, bureaus, ministries, and political parties to co-ordinate with each other. After submitting the plan to the Diet, the administration makes strenuous efforts to achieve an agreement from all sides, but the final decision is made by the Diet, if necessary by vote.

The development of the plan for the next fiscal year in local government begins a little later and takes into account the decisions of the central government, because most local governments draw financial support from the central government. However, the process of compiling the budgetary plan for local government is very similar to that for central government.

Legislative process

An important policy formulation usually requires new legislation or amendment of one or more existing acts. In Japan, it is the administration's responsibility to propose a bill to the Diet. The legislative process is much more complex than that of budgeting because legislation usually presupposes the existence of groups of people with different interests. After clearing the discussions through the ministry, the bill has to be submitted to other ministries if it infringes on their jurisdiction, and it is finally examined by the Attorney General's Office before it is proposed to the Diet.

Administrators work actively to solve any problems both before and after submitting the bill to the Diet. That means that administrators often have to work in a co-ordinating capacity with other ministries which administer different sectors of society. The interests of different sectors are often in conflict. Such a hierarchical organization is not ideally suited to co-ordination. Therefore the policy-formulating process is often delayed and decisions cannot be reached. However, ministers and heads of agencies are usually nominated from the members of the Diet by the Prime Minister. This system promotes a co-ordinated approach by ministries and different interest groups.

Scrutiny of bills proposed to the Diet for legislation begins after the budget is determined, usually in April each year. After a bill is carried in the Diet, the Cabinet and the ministry prepare more detailed regulations for its implementation. Then local governments are notified when they are responsible for its implementation.

Commissions and committees

Japanese ministries often use various commissions and committees other than the statutory standing committees in order to transform the opinions of specialists and experts into actual policies and strategies. The members are usually appointed by the ministry. Some committees are specifically designed to co-ordinate and negotiate the views of different interest groups. In these cases, the delegates of the group would be appointed by the group.

Project teams and reorganization of administration

When infectious diseases were endemic, medical technology was a major element in public health activities. Immunization could save many lives at a comparatively low cost. In other words, at that time a categorical approach was possible and effective. In this era of chronic disease, the public health approach needs to be more comprehensive. That means that the range of co-ordination in developing public health policies and strategies must inevitably be wider than before. However, the structure of the administration is a typical hierarchy. Its bottom tier consists of divisions, each with a fairly narrow range of responsibility. Therefore, the current decision-making process is not very compatible with the current need for wider co-ordination.

Nowadays it has become common for the vice-minister to organize a cross-sectional project team to facilitate the development and co-ordination of new policies and strategies. An example of this is shown in measures to curb soaring medical expenditure, which will be carried out by this kind of organization in the Ministry of Health and Welfare.

With an ever-increasing proportion of elderly people in the population, it is essential to promote amalgamation of health and welfare services in Japan. To facilitate this process, the divisions responsible for health and welfare services for the elderly in the Ministry of Health and Welfare have been combined with the divisions responsible for health services for the elderly and put under the control of the Minister's Secretariat. The new Department of Health and Welfare for the Elderly was formed recently and its structure is not significantly changed. Reorganization in local government has not yet followed.

The need for further co-ordination and development of public health policy

Along with increased life expectancy in many industrialized countries, the spectrum of diseases has also changed. Nowadays, the major risk factors to health in such countries are those associated with life-style. The Ottawa Charter for health promotion was correct in its view that the health problems in such countries, cannot be solved by the health sectors alone. This is certainly the case in Japan.

Mass screening has been the major form of preventive activity in Japan, for many years. They major fatal diseases in Japan are malignancies, of which gastric cancer has been the most prevalent. Mass screening for gastric cancer is one of the major programmes supported financially by the Act for the Health of the Elderly. However, the incidence of death from this disease is decreasing and in the near future is going to be overtaken by lung cancer. This means that this secondary preventive approach is becoming less and less efficient. However, the incidence of lung cancer is rising and the epidemiological causal relationship between lung cancer and cigarette smoking has been clearly proven. Thus the most urgent and important public health priority in Japan is control of the current epidemic of smoking. It is apparent that if responsibility for antismoking measures is given to the Ministry of Health and Welfare, their actions will be ineffective. Cigarette commercials on television are still allowed in Japan. This is the responsibility of the Ministry of Telecommunications. The enforcement of non-smoking in trains is the responsibility of Ministry of Transportation. The Ministry of Agriculture and Forestry administers tobacco farms, and the Ministry of Finance and the Ministry of the Local Government collect taxes from tobacco industries, which is an important revenue source. Not only the Ministry of Health and Welfare but also the Ministry of Labour and the Ministry of Education are concerned about the control of smoking in the work-place and by young people respectively. Therefore, in order to solve one of the major health problems in Japan, lung cancer, it is quite obvious that wider co-ordination than before beyond the boundaries of individual ministries is crucially important.

Summary

Government administered public health in Japan is primarily concerned with disease prevention and health promotion, whereas disease diagnosis and treatment are primarily provided by the private sector. However, concern with the increasing age of the Japanese has resulted in the government taking greater responsibility for diagnosis and treatment, particularly of the elderly.

Government in Japan is very hierarchical. Bills, including those dealing with public health issues, often originate in the administration, but are approved by the legislative body, the Diet. The process of policy formulation incorporates the concept of consensus. Policy formulation is usually the result of negotiation and co-ordination between all the parties concerned with the issue. In the implementation of public health policy, however, there has been an increasing trend towards decentralization to the prefectural and municipal level. Further, the administrative structure encourages recommendations from those implementing public health programmes at the lowest levels.

The shift of public health concerns from infectious diseases which responded to advancement in technology to chronic diseases which are often the result of life-style characteristics will require those responsible for public health in Japan to coordinate their efforts with an even greater number of ministries and government agencies.

35

Approaches to the co-ordination and development of strategies and policy for public health: a comparative analysis

KENNETH LEE, NAOKI IKEGAMI, and CALUM PATON

Introduction

Earlier chapters in this volume have considered a number of themes common to many countries: the origins and development of public health; specific public health services; public health policies and strategies; and, in respect of the four chapters preceding this, co-ordination and development of strategies and policy in four selected countries. This chapter seeks to provide a synoptic review of these developments across the countries surveyed and beyond.

What is immediately obvious is that different countries have tackled the co-ordination and development of strategies and policy for public health in different ways. Undoubtedly, this reflects different political traditions, different economic situations, and different social priorities, which all have an impact upon health policy. None the less, insights and warnings from one country to another are important: countries can learn lessons from the different methodologies employed to improve public health and from the different forms of organization and agency used to address public health needs.

This chapter therefore addresses a number of themes by which the development of policy can be analysed. First, it considers the structure of public health policy—in other words, the characteristics of health systems and the organization of health services. This analysis considers the degree of centralization as opposed to devolution in policy-making and implementation; the degree of public financing as opposed to private financing; the degree of public provision as opposed to private provision; and the overall nature of the system. Is it, for instance, a National Health Service? Is it a system of National Health Insurance or comprehensive social insurance? Is it a comprehensive or fragmented system? It is also important to consider how co-ordination and development of strategies and policy embrace acute and curative services, community services, services for specific client groups such as the elderly, and priorities such as prevention and promotion.

Second, methodologies are considered for identifying the need for health care—on a public and population basis—and resulting systems of resource allocation. This review is complemented by consideration of different approaches to allocating resources, and identifying priorities which are born of different political and social features in different countries. For example, in some countries, such as the UK, policies for prevention and promotion are rarely treated separately, but are implicitly linked into overall systems of resource-allocation; in other countries, such as the US and Japan, they are identified and treated separately as major priorities. Likewise, strategies for public health in some countries incorporate the use of information on factors such as demography, social class, and inequality, and their effect upon the health status of the public; other countries, however, do not attempt to use formulae in such an ambitious manner.

Third, an assessment is made of the degree to which overall health-systems planning is seen as a way of co-ordinating and developing policy for public health. Over the last twenty years, a number of countries (such as the US and The Netherlands) have seen a rise and then a decline in the apparent attractions of a planning-based approach; while Japan has just started implementing regional health-planning. Different types of planning are identified, and their goals for public health are assessed. Complementary to this discussion is a consideration of the increasing use made of competitive mechanisms and of incentives to improve the efficiency and the effectiveness of health care. This debate is given a contemporary context by reference to a number of policies emerging in different countries, and lessons are drawn.

Fourth, the chapter identifies the need for inter-departmental and inter-agency co-operation within countries, in order both to set public health objectives and to achieve them. A comparative analysis is made of alternative approaches to the development of policy for public health. No simple answers are obtainable as to 'the right system'.

There are however some lessons to be learned, both from conceptual analysis and from experience in different countries. In pursuit of the co-ordination and development of policy—as opposed to merely the maintenance of existing public health services, policies, and structures—the challenge for all countries is to reconcile efficiency and equity in health care.

The structure of public health policy

The debate as between centralism and devolution of responsibility for policy in health-care systems is an important theme addressed in this section. For, on one argument, it is necessary to centralize responsibility for policy within the national government, to generate national objectives and to monitor their achievement. On another view, however, the decentralization and devolution of responsibility for policy is argued to provide a more effective link between policy and implementation, and a greater sense of 'ownership' of policy by those directly responsible for service-delivery.

In the UK, for example, a National Health Service which is publicly financed and publicly provided has resulted in a large amount of central responsibility for expenditure and the allocation of resources, given the derivation of those resources mostly from general taxation. In contrast, in the US a pluralistic and fragmented health-care system, both of financing and of provision, linked to a fragmented political system, has ensured a correspondingly lesser role for central government in determining global criteria for the allocation of resources.

Between these poles, one finds the 'National Health Insurance' systems of Western Europe, which combine a significant role for public *financing* and a significant role for private *provision* of services. Here, the setting of public-health objectives by central government may be significant, yet their achievement is generally sought by regulation.

The recent introduction and development of general management in the British National Health Service illustrates the complexity of the debate concerning centralism and devolution. Some have interpreted general management, following the recommendation of the Griffiths Inquiry (Griffiths 1983), as a move to greater central direction of health-care objectives and their implementation. Others have seen it primarily as a devolution of responsibility to Unit General Managers (for example, of hospital services and community health care), who, although responsible to District and Regional General Managers respectively, are, nevertheless, in greater day-to-day control than in the pre-general-management period. Indeed, although responsibility for the National Health Service would inevitably lie ultimately with politicians as before, a separation of 'management' from 'politics' was an important part of the Griffiths agenda.

In theory, the NHS Management Board would be responsible for the setting of overall objectives for the health service. These would be communicated to health authorities and general managers down the line of control, and be implemented by the latter. Thus a mix of central control and devolved responsibility for implementation was suggested. The main implication for public health policy was that consensus management, especially involving the diverse viewpoints of the various health-care professions, was to be replaced by general management.

Existing political lines of control, for example, from the Secretary of State to a Health Authority Chairman, were not however to be replaced. Neither were existing professional lines of control: for example, from the Department of Health's Chief Medical Officer to Regional and District Medical Officers. Rather, a new line of accountability—based on the principles of corporate management—was to be added to the existing disparate lines of responsibility.

In practice, the NHS Management Board was incorporated into the Department of Health, and was a less radical initiative than early commentators had perhaps expected. Following the issue of the White Paper 'Working For Patients' (Department of Health 1989), the Management Board has been replaced by an NHS Management Executive responsible to an NHS Policy Board. This is intended to clear up some of the ambiguity as to whether the Griffiths reforms were about 'removing politics from the health service' or not. The UK will now witness the setting of health-service goals by the Policy Board, chaired by the Secretary of State for Health, and the responsibility for their achievement by the NHS Management Executive.

The implications for other countries' public-health systems of instituting or reforming a system of general management for health services may now be considered. What are the implications for the setting of clinical priorities? What are the objectives general management is intended to meet, and how are these to be monitored? How are efficiency, effectiveness, and equity to be defined? What is the arena for general managers', as opposed to health professionals', decision-making?

In the US, the complex mosaic of government institutions is responsible for the nature of health policy, and especially for its implementation. The system of federalism ensures that the federal government shares power with state governments and with local governments in making policy. Separation of powers at federal and other levels ensures that policy is often assembled according to the principle of the aggregation of interests, rather than of what might be termed 'top-down' central objectives. In consequence, there is not a publicly financed, publicly provided service, as there is in Britain and in many countries in Eastern Europe.

In the US, the main federal responsibility for public health goals lies with the Public Health Service (PHS), located within the Department of Health and Human Services. It is the PHS which has been responsible for 'comprehensive health care', and for the goal of preventing disease, as opposed to planning a hospital-based or curative health-care delivery system. The main role of the US federal health-care bureaucracy has been to conduct and publicize research, and also to institute significant prevention and promotion programmes, rather than to seek, through central resource-allocation and provision of services, the direct attainment of public-health objectives. Traditional curative medicine in the US still dominates the private, fee-for-service system, although there are significant trends to investment, both in the growing for-profit sector and in the non-profit sector, in prevention and promotion.

In the case of Japan, resource-allocation for curative medicine is centrally, but indirectly, controlled by a nationally based tariff, determined by the Central Social Medical Care Council. That is to say, although each provider is free to allocate resources in the way it chooses, the tariff ensures that a balance is maintained between the actual cost of the services and the revenue received. Thus the tariff has acted as an instrument for total cost-capping, and for providing economic incentives to services which are given priority (Ikegami 1988). In preventive medicine, the allocation is largely a 'top-down' process from the central government. However, the fact that it is centralized does not necessarily mean that it is well co-ordinated: the Ministries of Health and Welfare, Labour, and Education and the Environmental Agency may well have differing objectives and channels of command.

Allied to the centre-versus-periphery debate about objectives, and about the ownership of the means of production of health care, is the issue about the public financing for health care. Arguably, where publicly financed health care dominates the health-care system, there is the prospect of overt political or managerial choice between broad strategic priorities. That is, the percentage of the budget going to curative or acute medicine can be traded off against investments in public-health programmes, such as preventive and promotive programmes, or programmes of care for chronic conditions. There is, however, no guarantee that this will be done in practice, or will be conducted through a 'rational' methodology. The central determination of objectives can mask who determines objectives, and why and how objectives are determined. Domination of the policy-making machinery by traditional providers can still be envisaged, as well as its domination by political viewpoints.

In decentralized political and health-care systems, where public financing is either less evident or less centralized, it is likewise an open question as to who dominates the decision-making machinery, although the means of so doing will be different (Paton 1990). The mere fact that a system is decentralized does not mean that it is less public or more dominated by private interests; but, in practice, there is the possibility of a high correlation between such characteristics.

It is important to consider the degree of public provision of health care. At one end of the spectrum is the health system, characterized by largely publicly-owned hospitals and community services. But the other end of the spectrum is the system characterized by for-profit private providers. In between one finds systems dominated by voluntary, or non-profit but private, providers (as in much of Europe); varying degrees of regulation and control of private providers; and varying obligations of private providers to combine publicly stipulated objectives with their own self-determined institutional objectives.

Across the globe, one can think in terms of a spectrum, with the publicly owned and publicly financed National Health Service at one end, and with the wholly privately financed, wholly privately providing system at the other end. Between these polarities all countries can be said to be situated, and can be analysed in terms of how they combine public or private financing with the provision of services; how they combine central and local determination of objectives; how they combine central and local mobilization and allocation of resources; and the degree to which they are characterized by interventionist criteria for the planning or regulation of services.

Apart from what might be termed the issues of political economy, and the structural aspects of countries' health care systems, one can also analyse them in terms of their global allocations of resources as between acute and curative medicine, and between caring and chronic services (including community care) and preventive and promotive health services. In a public system, these choices are overt and global; whereas, in a system characterized by large-scale private financing and consumption of health-care, such choices will be determined by the priorities of individuals, groups, businesses, and trade unions rather than by politicians and managers. This does not mean that one can generalize as to which type of system finds it more easy to move, for example, from a 'traditional' hospital-based curative system to what might more fully be termed a 'public health' system.

Allocation of resources for public health

General issues

Although, in the developed world, reductions in the levels of certain infectious diseases are changing the nature of public health, new infections—most notably AIDS—remind one that infection is not altogether yesterday's problem. At the same time, the main challenges for public health are seen as chronic diseases, cancers, cardiovascular and cerebrovascular diseases (the so-called diseases of affluence), and the major challenges of 'ageing'. Compounding these difficulties are the pressures from increased expectations, increased capacity on the supply side due to advances in medical technology and other forms of treatment, and cost- and price-inflation in the health sector which can be seen to be larger than that affecting the economy as a whole.

In considering how best to allocate resources to address the needs of public health a number of complex factors must be taken into account. First, there are many determinants of health and disease, as earlier chapters of this volume have made clear: personal characteristics are only the tip of an iceberg, which includes the physical environment, the social and economic environment, and various aspects of behavioural culture and education. Determining which of the wide range of social factors should be addressed by the wide range of social policies available makes the articulation of a broad-based strategy for health very difficult.

Second, the fact that 'basic public health' measures have been tackled—though not necessarily successfully or appropriately—leaves the remaining challenges divided—in an often politically contentious manner—into the personal and social, without the mediating variable of the physical environment. Thus, paradoxically, debates have been revived about the responsibility of the individual for his own health versus the responsibility of society and the state for the health of populations (Green 1988).

Major causes of ill health (whether caused by smoking or by poverty or by both, to take typical challenges that provoke intense controversy) may require both more subtle and more politically ambitious reforms if they are to be met successfully. Debates about prevention and promotion point to individual responsibility, but also to social, cultural, and economic barriers to individual action. Access to healthy lifestyles may be difficult for the poor, and unhealthy behaviour may be 'rational' for them given the range of constraints they face. Both material deprivation and social deprivation can increase, and reinforce, health disadvantage.

Third, a decline in the large-scale killers and sickeners of yester-year means that huge geographical, demographic, and international variations in mortality and morbidity have declined, at least in the developed world. A consequence of this is that national systems of resource-allocation that rely on measures of morbidity or mortality may be more difficult to institute—if only for the reason that smaller variations between regions result in more ambiguous statistical variations in, for example, death-rates.

Fourth, choosing between different measures of 'need' to underpin the allocation of resources for public health may often be a contentious task: as, for instance, in the choice between mortality figures; in direct measurements of morbidity as a surrogate for mortality; and in the use of measures of social factors involving material or social deprivation, such as income, housing, amenities, and environment. Furthermore, acquiring robust information on such factors, and then tracing their link to health status, is difficult. Distinguishing between the 'use' of health services and the 'need' for health services—let alone measuring the effectiveness and efficiency of such services in achieving favourable outcomes—is a barrier faced by any reforms geared to improving resource-allocation formulae (NHS Management Board 1988).

This however presupposes that need—defined on a social or population basis—is to be the criterion for allocating resources. For a health-care system which does not allocate public resources on a basis which attempts to be equitable may not be susceptible to an expert's definition of need. In other words, demands by private individuals in the market-place, whether or not supplemented by public and private insurance programmes, may lead to certain 'needs' being ignored, if the purchasing power is not there to transform them into demand. Furthermore, it is not just a question of purchasing power. Different attitudes to health, health status, and health care by different social classes mean that an expert's attribution of need may not be shared by the individual or group under consideration. What may be called 'false consciousness', or difference in opinions, applies equally in health as in other social spheres.

For instance, in the UK, the Resource Allocation Working Party's Report (Department of Health and Social Security 1976) led to the adoption of a formula for allocating resources from the Department of Health to England's health regions (similar formulae being applied in Wales, Scotland, and Northern Ireland) grounded on the principle of equality of opportunity of access to health care for those at equal risk. Despite some essentially contestable features of this definition, the social or ethical principle was clear, and a practical version of equity—in allocating financial resources—was established.

In the US, on the other hand, there is no central system of allocation, for public-health services or for health services more generally. Maybe occupying the middle ground are European countries such as The Netherlands, where public health-insurance for those earning below a fixed amount is intended to ensure that all in society have access to health care when needed. Here, there is neither a 'scientific' formula for allocating resources in lieu of the market-place, nor a reliance on private means augmented by fragmented specific government programmes, as in the US.

In Japan, the social insurance scheme is even less discriminatory, because payment to the provider is made according to the nationally uniform tariff, and the provision of additional benefits through private insurance is prohibited except for room charges and a very restricted range of specialized services. That is why there is a fundamental difference between countries such as The Netherlands and Japan, which have compulsory health-insurance schemes, and the US, which has a pluralistic and partial health-insurance system.

Measures of need

If there is to be an attempt at an objective system of resource-allocation, based on a formula instituted by a country's Ministry of Health, a decision has to be taken as to how need is to be measured. There would be wide consensus that the overall objective of public-health policy is to maximize the health status of a population without prejudice to group or class: that is, to improve the health status of the population as a whole, and that of all its individuals and groups. In ethical terms, this is more than a utilitarian strategy, as it incorporates a theory of justice involving equity.

Whether, however, it is possible to maximize health status in practice seems to be debatable. For the moment, this issue tends to be eclipsed by the more immediate technical question of defining, identifying, and measuring health status or its absence, and also of differentiating between different geographical areas, demographic groups, or social classes within the population. It is widely acknowledged that the direct measurement of morbidity is the best way forward; yet, the difficulty, cost, and complexity faced tends to lead to surrogate measures for morbidity.

These may involve specific measures of social deprivation and of morbidity's impact on health-care utilization. The difficulty with this approach is that the uptake of health services depends on the existing supply and provision of health services; in consequence, a geographical variation in uptake may reflect a geographical variation in supply, for example, rather than a geographical variation in need. Thus, complex statistical adjustments may be necessary to control for supply factors. Gravity models are commonly advocated, but alternatives exist (Mayhew 1986).

In terms of mortality measurement the Standardized Mortality Ratio (SMR) is popular. This takes crude death-rates—either in aggregate, or with specific reference to medical conditions—and adjusts them for the age- and sex-distribution of the populations under consideration. However, it is sometimes objected that even these adjustments still leave a bias towards 'elderly diseases', in that the SMR is likely to bias the measurement of need to elderly populations in general. Thus one finds that the measure known as Years of Potential Life Lost (YPLL) is becoming increasingly popular (World Health Organization 1989). YPLL is considered, in some quarters, to be a *policy-relevant* indicator by comparison with crude or age-standardized death-rates, because it can produce a different ranking of the major causes of death. For example, on SMRs motor-vehicle accidents may be ranked low as a cause of death in a country; on the basis of YPLL, motor-vehicle accidents may become much more significant.

Another popular (morbidity) indicator of need is 'avoidable deaths'. An avoidable death can be defined as a death from a condition amenable to medical intervention. The *European Community atlas of 'avoidable death'* (Holland 1988) has recently received prominence, given international recognition of the need to develop measures of the outcome of health-care services, as well as indices of need for health services. If an avoidable death is not avoided, there are a number of potential causes: for instance, resources may not be available; or, on the other hand, the organization of health services may be inefficient or inappropriate.

In countries such as The Netherlands and Japan, where a national health service does not exist, yet where the goal of universal access to health services is accepted, such indices of need may be useful. Although a national system of resource-allocation may not exist, the health-planning system may still be in a position to use such measures to reorientate and redistribute health-care facilities to the areas and populations most in need. Even in countries such as the US, where neither methods of resource-allocation nor methods of planning exist, such indicators may be useful in identifying the scope for regulatory action or exhortation.

In the UK, a review of the then existing formulae for resource-allocation from the Department of Health and Social Security to Regional Health Authorities (known as the RAWP formula) was completed in 1988, and its final report advocated that, instead of relying solely on SMRs to quantify the need for health services, a combination of SMRs and a weighting for social deprivation (the so-called Jarman Score) should be used (Jarman 1985). The 1989 White Paper on the future of the NHS tacitly accepted one part of this recommendation, though not in the form recommended, and the debate is destined to continue (*Lancet* 1988). In Italy and New Zealand, too, RAWP-type formulae have been developed and partially implemented; and there is increasing interest in such approaches in countries as diverse as Saudi Arabia and Thailand.

Yet the basic philosophical question remains: how do we measure 'need' for health services, in order to meet it, by reference to 'objective' criteria? How can data on morbidity and mortality best be used and (for planning purposes) compared with measures of usage? How can measures of usage, even if one adjusts for 'supply', be disentangled from 'demand' as opposed to 'need'?

Planning as a means of co-ordinating strategies for public health

Models of planning and resource allocation

This section of the present chapter reviews a number of strategies for health-service planning which have been adopted in developed countries in recent decades. It does not set out basic conceptual approaches to planning (see, for instance, Bevan and Spencer 1984), but refers to them where appropriate. At the outset, none the less, it is important to offer the definition of planning adopted in this chapter: at the broadest level, it is the determination of the pattern and mix of services (both spatial/geographical and qualitative) to serve acknowledged principles of distribution and redistribution of services, and the corresponding determination of strategy to achieve that pattern of services.

There are, of course, different philosophies of planning. 'Rational' planning has frequently been contrasted with 'incrementalist' planning, with the 'mixed' scanning approach seen as the middle ground between these two extremes. Different approaches to planning involve different actors: some planning processes, whether deliberately or not, are 'politically' dominated; others are intended to be dominated by the 'neutral scientists'.

Some systems of planning are, in effect, subservient to systems of resource-allocation, in that resources may be allocated by a formula or methodology such as those described above. Alternatively, one can have a process of resource-allocation which is subservient to planning. That is, decisions can first be taken about a pattern of services, including hospitals, community services, primary care, and preventive services. Capital budgets and revenue budgets can then be determined subsequently.

Thinking along these lines is perhaps only practical if a central body, whether a Ministry of Health or a country's health region, has a responsibility for global planning and/or resource-allocation. Conversely, in a system dominated by private health-insurance, private purchase of health care, and, for example, private subscription to organizations paid in advance, global planning will not, and cannot, exist. Planning, in this context, will tend to take the form of indicative or technical planning, with the providing organization gearing its services to market demand.

In mixed systems, such as that of The Netherlands, the public-health insurance system and private health-insurance schemes are the main sources of income for the providing institutions and for hospital facilities. However, even The Netherlands recently, in common with the Federal Republic of Germany and other European countries, has instituted national planning structures which attempt to influence the pattern and distribution of services over and above what would be determined purely by a purchasing market. Thus, in such mixed systems, one can distinguish between the insurance system, the overall financing system for providers, and the planning system as it affects providers.

In essence, these systems combine elements of a global resource-allocation system with a market system. For example, following recent changes, the new budgeting model for Dutch hospitals will include:

(1) a fixed component related to the number of inhabitants in the hospital's catchment area (25 per cent of the budget will be allocated on this basis);
(2) a 'semi-fixed component' based on the number of specialties and beds of the hospital (35 per cent of the budget is allocated by this factor); and
(3) a variable component—based on negotiations between the hospital management and the local insurance companies and sick-funds dealing with the hospital—related to the number of patients to be treated in the following year.

In some ways, this model resembles the system of resource-allocation employed in Britain from the late 1960s until 1976, when health regions were reimbursed on the basis of population, facilities, and activity.

It is interesting to note that, in The Netherlands, despite the existence of a public–private mix in financing, there is (still) a strong element of public planning. That is, a hospital cannot merely maximize its budget by selling itself in the market-place, whether to payers, insurance companies, or sick-funds. Furthermore, payment can be prospective rather than retrospective (that is, merely reimbursing costs); this reflects the increasing trend towards seeing planning very much in terms of cost-control.

This trend has been, arguably, at the expense of the school of thought that advocates comprehensive health-service planning which is charged with measuring the need for different types of services and redistributing resources across countries and across regions where necessary. However, in practice it is not surprising that such lofty goals will often not be met. In the United States, for example, there was a move up to the late 1970s to incorporate a form of planning into what was largely a private system. Hospital-development programmes after the Second World War such as the Hill–Burton Act of 1946 were followed by the Regional Medical Programs Act 1965, the Comprehensive Health Planning Act of 1966, and, allegedly more significant, the National Health Planning and Resources Development Act of 1974 (Paton 1990). However, planning has not been global: it has been capital-based planning; it has been voluntaristic rather than mandatory, increasingly so as the 1980s wore on; and it has been diverted from 'public health' goals by provider coalitions at the level of the planning agency (the Health Systems Agency).

Both proponents and opponents of planning have been disappointed with results: planning has been discredited, and a renewed faith in competition has emerged instead. This has raised a lively debate about whether profit-seeking health-care institutions are cheaper and more cost-effective, or the reverse, and whether access to health care is easier or not. What has happened, both in the US and Netherlands, is that planning has developed into a restrictive activity, with resources-development and redistribution very much played down in recent years (Kirkman-Liff *et al.* 1988).

In the case of Japan, which passed the legislation for regional health-planning in 1985, the articles on comprehensive health-planning are discretional; in contrast, that is, to the mandatory requirement to set the regional limit on the number of hospital beds. Although the prefectural governments have been given the responsibility for drawing up and executing the plans, no special funds from the central government have been set aside for their implementation.

Competition or planning?

One can identify 'competition in the market' and 'bureaucratic planning' as opposites in health-care systems. In The Netherlands recently the Dekker Commission has advocated increased competition between insurance companies and sick-funds on the financing side, and decreased planning and increasing reliance on competition on the provider side. However, this is still expected to coexist with the basic hospital planning law (the WZV law) instituted back in 1971.

In the UK, the recent White Paper (Department of Health 1989) has advocated a number of competitive measures on the supply side of health care: primarily, a system of trading for clinical, medical, community, and even preventive services between Districts within the National Health Service; optional budgets for general practices above a certain size to pay for the hospital care of their patients; and provision for hospitals and indeed community units to 'opt out' of health-authority control and act as independent providers, albeit they still remain NHS institutions. Arguably, facilities in the 'self-governing' category would be subject to only minimal planning or none (excepting an obligation to provide 'core services' to their local catchment populations), and would act as economically free institutions allowed to raise money on capital markets and determine their own budgets and basis 'for profit'. Thus, the language of the business plan and the concept of marketing are rapidly being introduced in the British National Health Service.

However, this does not imply that planning will no longer be relevant. District Health Authorities will still receive budgets based broadly on their population's needs. It is then up to Districts to guide potential providers to appropriate services and appropriate location of services. The fact that there may be competition in providing these services does not alter this fundamental fact. Furthermore Regions, above the level of the District, may be required to give guidance as to the Region-wide distribution and pattern of services, to prevent unnecessary duplication or wasteful competition—that is, required to indulge in indicative planning at the very least, and comprehensive planning at the most. Thus to see a simplistic distinction between competition and planning may be naïve.

Indeed, in looking at the respective merits of either competition or planning, it is important to avoid over-complexity or over-bureaucratic means of achieving one's aims. For example, if the catchment populations served by health facilities do not closely mirror the health authorities which receive resources, there will be problems about reconciling reimbursement for patients living outside the authorities' areas with reimbursement for patients within the authorities' boundaries. Bureaucratic systems of so doing, as with some elements of subregional resource-allocation in Britain, may develop perverse incentives. Similarly, one must exhibit caution in implementing competitive solutions which may, at the end of the day, be more bureaucratic to police than the bureaucratic systems they are intended to replace.

Techniques of planning

The technical basis of planning has changed considerably over the years. There is a discernible movement away from norms-based approaches, which are now seen as simplistic and inappropriate, as they deal with inputs rather than outputs or even outcomes from health-care systems.

Thus, planning for bed-norms, a variant of capital planning, has been replaced in the British National Health Service with variants of service-planning and planning for client groups. Broadly speaking, service-planning accompanied the 1974 National Health Service reorganization, and attempted to unify hospital and community health services (many of which had previously been the responsibility of local government). However, regional-based service-planning was often comprehensive in theory, but weak in practice on how objectives were to be met. In 1979, the emphasis moved to a devolved management-based planning approach within the District, in response to the view that planning had hitherto been a bureaucratic and remote activity.

Since then, UK health planning has been ambitious. For example, sophisticated systems of planning have been developed which attempt: to use local and national sources of data on morbidity and social deprivation; to model the willingness to travel normatively-defined distances to health care in different specialties; and to model that distribution of facilities which will combine need and equity with economy. The umbrella of a publicly financed, publicly provided health-care system make these technical approaches possible.

Yet, at the broadest level, no technical solution can exist to determine the strategic choices between, for example, acute and curative medicine and preventive and promotive health-care. Analysis can help, and specific research can, for example, outline tangible health-benefits of preventive health programmes; for example, anti-smoking programmes, which are allegedly very effective when employing the cost per Quality Adjusted Life Year approach. Yet, at the end of the day, social and political choices are bound to inform planning choices and health priorities. The most promising way forward, achieving the greatest consensus, appears to be two-fold: first, to stress changes in health status (whether for populations or for individuals) as the unit of achievement; and, second, to reconcile planning with more sensible economic incentives for providers, as the means to achieving improvements in health status.

The purpose of reform in public health policy

A number of countries—by no means only Western countries such as the UK, the US, and The Netherlands—have recently instituted reforms both in the provision and financing of health care. These reforms reflect features common to a number of health-care systems.

Whatever the source and method of financing, there are inevitable restrictions on the overall percentage of the GNP or GDP which can go to health care. Naturally this varies considerably: for example, the percentage in the UK is approximately 6 per cent, whereas the percentage in the US is approximately 12 per cent (of a GDP which is one-and-a-half times higher per caput than the UK's GDP). Most European percentages are between these poles, with France and Sweden at the higher end (as a result, respectively, of cost-inflation and of a deliberate political decision to spend more). Canada has limited cost-inflation by combining decentralization of responsibility to its provinces with efficient systems of financing which combine elements of National Health Insurance and the payroll tax. Japan has about the same ratio as the UK, despite the fact it has a fee-for-service system, although differences in calculation and its still relatively young population must be taken into consideration.

There is great scope for both different political priorities and different economic capacities to vary the amounts spent on health care. However despite this, it is difficult to envisage a society in which 'hard choices' about priorities do not have to be made. Different societies ration resources for health care in different ways.

For example, in the UK, rationing is conducted primarily through decision-making as to priorities by clinicians (often informally and tacitly, rather than by overt criteria, such as economists might wish to develop through tools such as the QALY (the Quality Adjusted Life Year)). In the US rationing occurs primarily on the basis of ability to pay or of eligibility for categorical government programmes of one sort or another. In Japan, the setting of each procedure's reimbursement price in the tariff acts as an indirect form of rationing, since the provider will face economic constraints in performing those which are not listed, or, even if listed, have a price lower than their cost.

A tendency which is common to many so-called mature health-care systems is that of bureaucratic ossification. Established structures may lead to standard operating procedures, to a lack of flexibility, and to a resulting diminished capacity to respond to changing health-care challenges—not the least of which is co-ordination with other social agencies which have a bearing on the health care of the population.

This is one of the reasons why so-called pro-competitive reforms have been instituted in a number of countries recently. One can point to the White Paper of 1979 in the UK; to various strategies for health-care reform in the US, such as Alain Enthoven's plan for Health Maintenance Organizations to compete in delivering health care for enrolled populations (Enthoven 1989); and to recommendations of the Dekker Commission in The Netherlands (Dekker 1987).

Further afield, however, one can point to significant reforms in countries such as the USSR and Hungary in the Eastern bloc, and in developing countries also. Recent pilot projects in the Soviet Union have attempted to increase efficiency and competition on the 'supply side' of health care, by giving budgets to the managers of primary-care institutions (polyclinics, in the USSR, which combine primary care and out-patient services), who then purchase care from competing hospitals and other health-care facilities. Such reforms are at an early stage (Paton 1989).

In Hungary, dissatisfaction with rigid planning norms is leading to an investigation of both more decentralized responsibility for providers of health-care, and an attempt to stimulate competition among health-care providers—not necessarily on the basis of price (as Diagnosis Related Groups are advocated as a means of national reimbursement, as in the West), but on the basis of efficiency and quality.

In many developing countries of varying political persuasions, greater flexibility in the provision of health care, including greater investigation of possible public, private, and voluntary links, is either advocated or implemented. This is not to argue that planning and national criteria ought to be supplanted; but rather that assumptions as to most forms of provision ought periodically to be challenged.

Health manpower is another arena which is witnessing demands for increased flexibility. Demarcation between areas of work done by specialized doctors, general doctors, nurses and other cadres of manpower are being questioned by the criteria of both medical effectiveness and economic effectiveness. The issue concerns not only loss due to the inappropriate division of labour, but also the difficulty of integrating the health-care organization in the face of the exclusive nature of professional identity (Ikegami 1985). In particular, in many developing countries, financial constraints upon the level of resources going to the health-care sector are leading to investigations of new mixes of manpower which are both more affordable and either more effective than, or equally as effective as, existing patterns.

Such changes are often perceived as radical, and may of course challenge the institutionalized interests of both provider groups and financing and providing bureaucracies. That is why a period of international cost-containment in health care is also a period of increasing general management of health-care priorities.

Prevention, promotion and primary care

The need to reconcile objectives such as the World Health Organization's 'Health for All by the Year 2000' with the realities of fiscal constraint and international political economy leads to a stress on both primary care and 'prevention and promotion' as alternatives to expensive high-technology medicine. It is naturally important in such areas to go beyond the level of generalization, piety, and cliché. Just as community care may not always be cheaper than institutionalized care, prevention may not always be cheaper than cure. And of course cheapness is not the only criterion by which one judges a policy. Desired outcomes must be designed (probably in terms of a mix of effectiveness, equity, efficiency, and quality). The best strategy to meet these objectives within given budget constraints must then be developed, and gradually translated into specific health-care plans.

For example, in the area of prevention, screening programmes have to be investigated as to their intended objectives and as to both their cost and effectiveness. There is little point in undertaking mass screening if the consequences of discovering potential morbidity cannot be acted upon because of lack of medical capacity. Similarly the cost-effectiveness of screening as opposed to alternative remedies must be investigated. Furthermore different types of screening (as in the varying options in undertaking screening for breast cancer) should be investigated by a number of criteria (Department of Health and Social Security 1987). These aspects have not been given adequate consideration in Japan, where the mass-screening methodology first developed for tuberculosis has come to be utilized for a wide spectrum of chronic diseases (Ikegami 1988).

It may be possible to build incentives into the organization of primary care and the reimbursement of primary-care physicians to achieve objectives in the promotion of preventive health schemes. The recent White Paper in the UK has stimulated a debate in this direction. As well as seeking to increase the importance of capitation in paying general practitioners, the White Paper recommends that screening services be rewarded not on a fee-for-service basis per individual, but by a reward to the practice if they reach norms of population screened in particular areas. In favour of this proposal it can be argued from a public-health viewpoint that it replaces a notion of medicine for the individual with medicine for populations. Against the proposal it can be argued that practices serving a catchment population with special problems may find it more difficult to meet targets for screening. Therefore flexibility is required.

Perhaps the areas most in the public eye internationally concerning prevention and promotion are smoking; various addictions, including alcohol; diet; and—most fundamentally—social conditions such as the degree of inequality in society and its effect upon health status. Disease patterns in so-called advanced societies often lead to certain diseases being categorized as the diseases of affluence, in that they are associated with the habits of richer societies (such as rich diets and high alcohol consumption).

However, slogans such as these are often misleading. For example poor diet, high levels of smoking and drinking, and generally unhealthy patterns of living may also be associated with relative poverty within certain societies. Recent research has clarified this debate to a considerable extent (Whitehead 1987). Often preventive and promotive campaigns will not be effective if conducted on a single-issue basis. A complex pattern of need and dependency may be created by social position, affecting life-style and health status.

What is more, research has shown, for example, that even if one controls for factors such as smoking, diet, and drinking, inequalities in health status between social classes may remain almost as significant. Thus other causes must be investigated—such as poor housing, low income, poor environment, and stress.

This tends to argue for well-rounded preventive and promotive health programmes which are integrated with the overall objectives of the health-care system, which itself must be integrated with broader social objectives.

If primary care is to be stressed, co-operation between hospitals and primary health-care services can adopt the philosophy that the hospital is a community resource. Often 'the hospital' has been seen as the antithesis of community work—perhaps a false antithesis (Black 1984). This view has reinforced the perception of 'priority services' as in conflict with acute medicine.

In practice, co-operation can take the following forms, which are merely illustrative of the possibilities:

- increased use of day-care and short-stay facilities in the hospital, co-operation with general practitioners by hospital doctors, and liaison with community services;
- open access to hospitals for pathology, radiology, and other services;
- buildings and equipment shared by hospital doctors and community services;
- wards in hospitals for the use of primary-health-care doctors (general practitioners);
- domiciliary visits by hospital doctors;
- the development of databases to link information on usage of hospitals and health services, social data, social services usage data, and profiles of general-practice populations; and
- facilitation of links between preventive programmes in general practices and specialist doctors.

Development of co-ordinated public-health policy

A major challenge in many countries is to secure adequate co-operation between different ministries and government departments in both the development and the implementation of a consistent and coherent public-health strategy. A strategy on diet advocated by a Department of Health may, for example, be undermined by the priorities of the Department of Agriculture. Policies on industrial health may be undermined by other industrial objectives which, for example, result in pollution and unhealthy environments as by-products. Objectives of equalizing as much as possible the health status of different social classes as they are currently organized may be undermined by policies which maintain or worsen social divisions, with an effect on relative health status. Again it is a question of overall social priorities.

However, the hugeness of such public-health challenges does not prevent lessons being learned from practical policies in different countries. What is more, just as there is a huge challenge in co-ordinating policies across ministries, there is an almost equally significant challenge in co-ordinating policies within ministries and within the health sector.

In the US, for example, the historical insignificance of primary care and the 'general practitioner' by comparison with, for example, the UK, has recently been challenged somewhat by the development of the type of Health Maintenance Organization (HMO) which is dominated by the 'primary physician gatekeeper'. The HMO, which takes individuals on to its books for a stipulated amount and then provides care, thereby unifying the provider and the insurer, who thus employs a primary physician as a key decision-maker, is effectively seen to limit access to unnecessary specialist care. This type of HMO has a good record of cost-control. Worries however about the denial of necessary care have been raised; and the RAND corporation has conducted a ten-year research project into the effect of pre-paid health care upon health status (Paton 1988).

Co-ordination between primary and secondary care is an important challenge. International lessons may be important here: for example, the Soviet system of polyclinics, which brings together under one roof the general practitioner and the provision of certain out-patient services, may limit the more expensive usage of hospitals' facilities, and could well be worthy of investigation. Achieving the World Health Organization's international health objectives through an increasing stress on primary care may lead to investigations as to how hospitals can best co-operate with agencies of primary care in co-ordinated strategies.

In the fields of mental health and care of the elderly, moves to deinstitutionalization and the development of community care create the necessity of inter-agency co-operation on a number of fronts, as, for example, between Departments of Health, Departments of Social Services, and local authorities. In the UK, the two recent reports by the Audit Commission (Audit Commission 1986) and Sir Roy Griffiths (Griffiths 1988), respectively entitled 'Making a Reality of Community Care' and 'Community Care: Agenda for Action', identify the importance of co-ordinating finance and decision-making in order to provide, or to contract for, care on rational criteria.

In this field, as in others, an increasing stress on competition in provider markets to supply care to a public purchaser, may represent a theme around which international consensus is developing. Naturally, directions from which this consensus is developing vary considerably.

Provision and financing reviewed

In the UK, the separating of financing and provision of health care is a new development, with both opportunities and dangers. In the US, competition has been developing on the 'supply side' for some time, albeit unevenly in different parts of the country; but there has always been an absence of the public purchaser who identifies society's needs, and particularly the needs of the under-served with little access to health care. In The Netherlands there is an attempt to combine the benefits of planning with the benefits of competition, instilling competition both on the demand side through competing insurers and also on the supply side. In Japan, hesitant steps in this direction have been taken, beginning with the relaxation of regulations for contracting out to for-profit organizations in the fringe areas of 'hotel' and social services.

The existence of provider markets must be distinguished from the relative role of private *finance* in health care. What is more, provider markets may not even entail an increasing role for private provision. In the UK, for example, the development of the 'internal market' is intended to lead to competition substantially within the public sector of provision.

In the US, the moves to state-wide health insurance in Massachusetts promoted by former presidential candidate Dukakis have to some extent reopened the dormant debate about the need for national health insurance in that country.

Disillusionment with bureaucratic systems for capital planning in the health sector has led to a stress on greater decentralization in decision-making as to capital investment and greater access to capital markets *whether* by public or private health agencies. The growth of the for-profit sector in US health care has led to a much greater use of investment capital. In the UK the 1989 White Paper has advocated that hospitals that free themselves from health-authority control should have virtually unlimited access to capital markets, and hospitals and units remaining within health-authority control should no longer treat capital as a 'free good', but should account for it, using interest rates and depreciation rates when making decisions.

The making of policy

Naturally international generalization as to development of policy is not possible. It is, however, worth pointing out that, as well as formal policy-making institutions, the influence of 'think-tanks' and policy units, interest groups, and other less formed sources of policy should be gauged. For example, the recent Prime Minister's Review of the health service in the UK did not follow hitherto familiar paths of consultation by the Department of Health, but instead established its objectives and sought blueprints for reform from think-tanks sympathetic to the government's objectives. This naturally raises the question of whether partisan leadership or widespread consensus is the best means of assembling policy, to which there is no simple answer.

In the US, health policy has, interestingly, rarely developed from presidential priorities. The recent DRG policy was sponsored by Congress, following research at Yale University, before being taken up by the Reagan Administration; the Health Maintenance Organization policy, despite early expressions of interest by President Nixon, developed in the voluntary sector rather than as a political initiative; and various regulatory and planning policies, which were forerunners to the 1974 Planning Act, generally developed at the level of the states. The most recent example of *significant* Presidential activism in fact lies in the Johnson Administration's sponsorship of Medicare and Medicaid in 1965. The political structure in the US, with its emphasis on decentralization and individualist policy-making in Congress, leads to specific policy-makers being identified with specific policies. For example, Senator Kennedy has been identified as the main access reformer, with his variants of national health-insurance proposals over the years.

In the case of Japan, radical policy formulations are rarely made, least of all in the health sector. This is because decisions are generally made by a widespread process of consensus-building within the ministerial bureaucracy, and a closed, well-established network of the concerned parties. Since each negotiates on the basis of pragmatic collectivism rather than their personal convictions, this has led to an extremely conservative approach, with power politics as the ruling principle. The fact that there has been no change in the ruling party for over forty years, together with the highly centralized form of government, has also contributed to the maintenance of the status quo. In the case of the health sector, these tendencies have been enhanced by the control exercised by the professional societies and the lack of international competition. However, the present health structure has become increasingly ill-suited to meet the demands of the rapidly ageing and affluent society (Steslicke 1987).

Who is the sovereign decision-maker in health?

Perhaps the most basic question, in an age when public-health objectives have to be reconciled with moves to increasing involvement of the 'consumer' in health care, is who is actually sovereign as to health-care decision-making.

Is it the patient? Is it the general practitioner or physician? Or is it the manager or bureaucrat? The answer to this question will depend upon how health care is both financed and provided. If the patient pays for his or her own health care, or receives a voucher for it, the patient will have greater scope for sovereignty than in other systems. If budgets are held by general practitioners (as advocated recently both in the UK and the USSR), then the primary-care physician will be a surrogate decision-maker on behalf of the patient. If budgets are held by district health authorities or management teams, then managers will have greater control. Finally if services and access to them are stipulated nationally bureaucrats will naturally have a greater say as to priorities and decision-making.

One of the greatest challenges in the provision of health care is to reward the provider (whether public or private) so that workload is adequately reimbursed and 'the money follows the patient', without creating an incentive to over-supply or to distort the pattern of care based on financial rather than medical criteria. It is with this objective in mind that recent reforms in the UK will be assessed. In the US, judgements as to the suitability of fee-for-service medicine on the one hand and the Health Maintenance Organization on the other hand are increasingly being made on the basis of the type, quality, quantity, and cost of care associated with the alternative means of provision. And, as regards the broadest public-health objectives, the need to reconcile incentives to providers with overall social objectives—such as the prevention of infection and the promotion of healthy populations—represents the greatest challenge for health-care systems.

Key questions in co-ordinating policy

Are the dictates of strategic planning adequately combined with incentives to providers? Are markets in provision adequately regulated to ensure the achievement of planning objectives? Is the management of public-health functions adequately clear at health-authority level (a debate raised recently in the UK in connection with the Acheson Committee recommendations)? (Acheson Committee 1988). Is there adequate co-ordination between the Department of Health and other government agencies both at the centre and throughout the country? Is there adequate co-ordination in the financing of health care to achieve objectives (as in the promotion of community care for priority groups)?

Are the boundaries between 'politics' and 'management' adequately clear (a debate raised in Britain following the Griffiths Inquiry's advocacy of general management, and its subsequent implementation, after 1983)?

Are the dictates of medical education, teaching, and research adequately reconciled with the provision of services to local populations? Are the special problems for health created by social deprivation, for example, in inner cities or sparsely populated rural areas, adequately recognized in resource-allocation? Is there adequate research into the relative benefits of cure, care, and prevention in specialties? What capacity exists for technology assessment in health care? On what criteria are priorities determined?

These questions are both general and selective. However, the challenge of the 1990s in health care in a variety of countries, both 'developed' and 'developing', lies in a two-edged strategy of both mobilizing resources for health care and making choices within constrained resources as to priorities. In consequence global resource-allocation and planning are likely to be of increasing importance in a variety of different health-care systems. What may be most needed is the establishment of an open forum concerning these key issues in a language which can be made comprehensive to a broad audience of policy-makers.

Conclusion

The main challenges confronting public-health policy internationally are the reconciliation of overall cost-control and equity, at the broadest level; the eradication of perverse incentives, at the 'micro' level, such as a lack of correlation between workload and financial reward for hospitals and other providers; the reconciliation of planned objectives and efficient means of provision; and an avoidance of fragmentation in the financing of health care.

Different ideologies, different forms of political structure, and different political characteristics (such as the nature of parties, interests, and pressure groups) all play a part in determining health policy as part of the arena of public policy.

Wider social and environmental factors, such as changing social and economic structures, also play their role, as do demographic factors which change both the demand for, and the supply of, health care: increases in the number of elderly affect demand, and changes in the labour market affect supply, for example, as does technology development. 'The technology push' which is constantly advancing the frontiers of alleged high-technology medicine (Jennett 1985) makes a framework for securing effective choice between curative, caring, and preventive strategies, and between priorities as to client groups, all the more vital.

If equitable access and reasonable cost-control are to be combined through effective providing and financing systems, resource-allocation and planning of objectives must be reconciled, as must basic incentives to providers in health-care systems, whether or not these are to be addressed through competitive strategies.

References

Acheson Committee (Committee of Inquiry into the Future Development of The Public Health Function) (1988). *Public Health in England*. HMSO, London.

Audit Commission (1986). *Making a reality of community care*. HMSO, London.

Bevan, R.G. and Spencer, A.H. (1984). Models of resource policy of health authorities. In London Papers in Regional Studies, 13,

Planning and analysis in health care systems, (ed. M. Clarke), Pion, London.
Black, D. (1984). *An anthology of false antitheses*. Nuffield Provincial Hospitals Trust, London.
Dekker, W. (Commissie Structuur en Financiering) (1987). *Bereidheid tot verandering*. Staatsdrukkerij, The Hague.
Department of Health (1989). *Working for patients*. HMSO, London.
Department of Health and Social Security (DHSS) (1976). *Sharing resources for health in England* (the RAWP report). HMSO, London.
Department of Health and Social Security (DHSS) (1987). *Objectives of the breast cancer screening programme*. Regional Representatives Meeting (July), London.
Enthoven, A.C. (1989). A consumer choice health plan for the 1990s, Universal Health Insurance . . . , Parts 1 and 2. *New England Journal of Medicine*, **320**, (1 and 2), 29–37, 94–101.
Green, D.G. (ed.) (1988). *Acceptable inequalities? Essays on the Pursuit of equality in health care*. Institute of Economic Affairs Health Unit, London.
Griffiths, R. (1983). *Letter to the Secretary of State* (6 October). The Griffiths Report. Department of Health, London.
Griffiths, R. (1988). *Community care: Agenda for action*. HMSO, London.
Holland, W. (ed.) (1988). *European Community atlas of 'avoidable death'*. Oxford University Press.
Ikegami, N. (1985). The concept of Gemeinschaft in health planning and management: A Japanese perspective. *International Journal of Health Planning and Management*, **1**, 27.
Ikegami, N. (1988). Health technology development in Japan. *International Journal of Technology Assessment in Health Care*, **4** (2), 239.
Jarman, B. (1985). Underprivileged areas. In *The Medical Annual* (ed. D.J. Pereira Gray), pp. 224. Wright, Bristol.
Jennett, B. (1985). *High technology medicine*. Nuffield Provincial Hospitals Trust, London.
Kirkman-Liff, B.L., Lapre, R., and Kirkman-Liff, T.L. (1988). The metamorphosis of health planning in the Netherlands and the U.S.A. *International Journal of Health Planning and Management*, **3** (2), 89.
Lancet (1988). Another shock to the system for the NHS (unsigned editorial). *Lancet*, **ii**, 260.
Mayhew, L. (1986). *Urban hospital location*. Allen & Unwin, London.
NHS Management Board (1988). *Review of Resource Allocation Working Party final report*. HMSO, London.
Paton, C.R. (1988). The trouble with Health Maintenance Organisations. *British Medical Journal*, **297**, 934.
Paton, C.R. (1989). Perestroika in the Soviet Union's health system. *British Medical Journal*, **299**, 45.
Paton, C.R. (1990). *U.S. health politics*. Chapter 1. Gower, Aldershot.
Steslicke, W.E. (1987). The Japanese state of health: A political-economic perspective. In *Health, illness and medical care in Japan* (ed. E. Norbeck and H. Locke), p. 24. University of Hawaii Press, Honolulu.
Whitehead, M. (1987). *The growing health divide*. Health Education Authority, London.
World Health Organization (WHO) (1989). Preventable mortality. *World Health Statistics Quarterly*, **42** (1), 4. WHO, Geneva.

G

The Ethics of Public Health

36

Public health and the law

RUTH ROEMER

Public health law, like public health, concerns the health of populations as contrasted with the health of individuals. Thus, public health law concerns the legal aspects of providing preventive, curative, and rehabilitative services to populations, although it also has an important impact on health protection and health care for individuals.

As the role of public health has expanded from its early function of preventing the spread of communicable diseases to encompass the development of resources, the organization and financing of the delivery of care, surveillance of the health care system, and overall protection of community health, so public health law and legislation have expanded to provide authorization, direction, and regulation of many fields of environmental and personal health services. As Grad has written:

> The reach of public health law is as broad as the reach of public health itself. Public health and public health law expand to meet the needs of our society. (Grad 1986.)

In this chapter, we begin by setting forth the functions of law and legislation in protecting the public health. Then we examine several fields of public health law, presenting for each: (1) the general scope of law and legislation; (2) methods of implementing law and policy; (3) selected examples of legislation; and (4) current legal issues.

In each field the review of the law is necessarily brief, and not all fields of public health law are covered. For further information, the reader is directed to the primary texts of the health laws of the world, published by the World Health Organization in its quarterly journal, the *International Digest of Health Legislation*, and to the many legal treatises, texts, journals, and reports of court decisions available in law libraries.

Perhaps a caveat or two are in order here. First, it should be noted that law and legislation can play a negative or positive role with respect to public health. Some laws may be adverse to public health by imposing restrictions on health services based on the knowledge, social conditions, or fiscal constraints obtaining at the time that legislation was adopted. Such negative laws are illustrated by the criminal laws outlawing abortion that are being replaced by legalized abortion in most countries of the world (Cook and Dickens 1988). Fortunately, laws also play a constructive role supportive of public health by authorizing measures to protect health, to increase access to health services, and to assure the quantity and quality of health care that a society needs. This chapter is replete with such examples.

Second, one should bear in mind that while law performs a technical function of expressing health policy and setting forth procedures for implementing it, the content of the law is determined by the nature and orientation of the political power in the country at any time and place. The element of political will is crucial to the enactment of health legislation. Legislation can serve as an instrument of change to improve health services and health protection, but only if policymakers have the necessary political will.

Functions of health laws

Health laws perform various essential functions in protecting the health of populations.

1. *Laws and legislation prohibit conduct that is injurious to the health of individuals and communities.* Examples of this function are environmental health laws that prohibit the dumping of toxic chemicals in the environment; traditional public health legislation to prevent the spread of disease; laws to control drug abuse (Porter *et al.* 1986); laws regulating smoking in public places; and laws to regulate the quality of health care.
2. *Health legislation authorizes programmes and services that promote the health of individuals and communities.* Many diverse, categorical programmes authorized by law provide health services for specific persons (mothers and children, the military, veterans, the mentally ill, the handicapped, the elderly); for specific diseases and conditions (heart disease, cancer, stroke, sexually transmitted diseases, mental illness and retardation, alcoholism and drug abuse, AIDS); and for specific services in various fields (environmental and occupational health, nutrition, mental health, dental health, and ambulatory, hospital, and long-term care).

3. *Legislation regulates the production of resources for health care.* Laws authorize or provide financing for the construction of hospitals, health centres, and other health facilities. Legislation provides support for education of the health professions and occupations. Regulation of the production and importation and export of drugs, medical devices, and medical equipment and supplies is carried out under the authority of law. Financial support of research—the production of knowledge—may require legal action.
4. *Legislation provides for the social financing of health care.* This function is carried out by laws establishing systems of national health insurance or a national health service. It is also expressed in government grants for specific health programmes, in the imposition of special taxes for health purposes, and in tax exemption for non-profit health facilities.
5. *Legislation authorizes surveillance over the quality of health care.* Examples of this function are licensure laws establishing minimum standards for health personnel and facilities, legislation providing for peer review of the quality of care, and financing programmes that regulate the quantity and quality of care. The judicial system, in handling malpractice suits, also carries out this function.

In the process of performing these various functions, health law faces the challenge inherent in all law—that of balancing the interests of the individual and the interests of society. This overriding issue faced by legislators, administrators, and judges is both legal and ethical. To what extent may individual rights be curtailed in order to promote the general welfare? The answer to this question is 'it depends'. It depends on the degree of risk to the community and the degree of intrusion on individual rights. It depends on the scientific and epidemiological evidence pertaining to the issue being legislated or litigated. It depends on the nature of the legal system and its protection of individual rights.

Environmental health and the law

Environmental health was one of the earliest concerns of public health because of the basic need for a safe water supply and waste disposal in all societies. Still a marvel today are the remains of the *cloaca maxima* from ancient Rome, the antecedents of contemporary engineering projects for waste disposal. With the growth of industrialization, modern law to assure a healthful living and working environment has expanded to include control of air and water quality, regulation of domestic waste and industrial and agricultural effluents, management of solid waste disposal, control of marine pollution, regulation of radiation emissions, control of toxic substances in industry and the community, regulation of the use of pesticides in agriculture, and noise abatement. Each of these branches of environmental law is based on the need to protect the public health. In addition, other branches of environmental law, while not directed solely to protecting health, have an important impact on health and the quality of life. These include conservation of natural and environmental resources, land use control and regulations governing housing, and measures to meet population growth and power needs (Grad 1985).

Public health personnel are involved to varying degrees in each of these problems in environmental health. They may be called on to set standards for air and water quality, to treat water to make it potable, to add fluorides to a public water supply to prevent dental caries, to inspect factories for toxic chemicals, to enforce sanitation regulations in markets and restaurants, to develop solid waste disposal systems, and for other tasks. As they undertake their varied functions, they encounter legislation and regulations designed to control contaminants and prevent harm to the health of the community.

A number of mechanisms are used for environmental control. The most important of these is not a legal mechanism but rather economic considerations that promote compliance with environmental standards. For example, an industry may find it cheaper to clean up its wastes than to pay the penalties for pollution. Or a government may find it advantageous to subsidize practices that will improve environmental quality.

Among the legal remedies used to implement environmental legislation are inspections and citations for violations of established standards of environmental quality; civil penalties for pollution; effluent charges; licensing of businesses and withdrawal of the licence in the event of violation of standards; criminal prosecutions to punish violators; injunctions to prevent future harm; and seizure and forfeiture of property in cases of egregious pollution (Grad *et al.* 1971, pp. 29–38).

Examples of environmental legislation

Environmental health legislation may be either categorical or comprehensive. Categorical legislation deals with one type of problem, such as air or water quality or solid waste disposal. Comprehensive legislation is designed to provide integrated control of the many and often interrelated insults to the environment having an impact on health.

Choosing examples of categorical environmental legislation at random, we may cite Norwegian legislation on control of oil slicks (Norway 1980*a*), on levels of lead compounds and benzene in motor fuel (Norway 1980*b*), and on the dumping and incineration of substances and objects at sea (Norway 1980*c*), as laws designed to protect health and safety in specific fields. Virtually all countries have categorical legislation dealing with various specific aspects of environmental control to promote health.

Both industrialized and developing countries have also enacted comprehensive environmental legislation addressed to multiple aspects of the environment. In the US, the National Environmental Policy Act 1969 was designed to create a means for integrating and co-ordinating the many programmes affecting environmental protection (US 1969).

The most important provision of the legislation requires the filing of environmental impact statements before major federal projects with significant impact on the environment can be undertaken. The Act requires agencies of the Federal Government to use an interdisciplinary approach in actions to achieve national environmental goals and requires these agencies to consider the environmental consequences of agency action (Grad *et al.* 1971, p. 1035).

Another example of comprehensive environmental legislation is the 1980 legislation of Sri Lanka which establishes a Central Environmental Authority for protection and management of the environment (Sri Lanka 1980). Among the powers and duties of the Authority are the duty to specify standards, norms, and criteria for protection of the quality of the environment, to conduct and promote research to prevent environmental degradation, to undertake investigations and inspections to ensure compliance with the Act, to promote and co-ordinate long-range planning in environmental protection and management, and to recommend national environmental policy and criteria for environmental protection.

Issues in environmental legislation

Many issues face policy-makers and public health administrators in the field of environmental control. Each of these issues merits lengthy analysis, which is not possible here. But even brief mention of the issues shows the magnitude and complexity of the problems in this field.

A priority for all countries, both industrialized and developing, is to balance the interest in a healthful environment and the need for employment and industrial development. This conflict is expressed by Algeria in its law on environmental protection which provides:

> National development implies the necessary equilibrium between the imperatives of economic growth and those on environmental protection and the preservation of the living conditions of the population (Algeria 1983).

The tension between the need for economic growth and the need to protect the quality of the living environment underlies all regulation of environmental pollution.

Management of environmental problems requires a high degree of scientific knowledge and technical sophistication in various specialized fields. At the same time, the interrelations among the various ambient elements, e.g. the impact of water pollution on land use, requires an inter-sectoral approach involving both health and non-health agencies. These environmental health interfaces and interactions have implications for the geographic jurisdiction of environmental agencies, for the responsibilities of various levels of government, for the functions of environmental health personnel, and for the role of public health personnel in large environmental management agencies (Roemer *et al.* 1971; Goldsmith 1970).

In the operation of any environmental management system, agencies responsible for regulating substances harmful to health face the difficult question of what limits exist on the agency's regulatory power in the light of scientific uncertainty. What are the powers of the agency if there are conflicting scientific opinions or if the evidence is based solely on epidemiological data? A case study of the regulation by the US Environmental Protection Agency restricting the amount of lead additives in gasoline provides important insights on the scope of judicial review and the role of the courts in cases of great technological complexity (Silver 1980).

Finally, an issue that is assuming increasing importance in the field of occupational and environmental health is worker and community involvement in assuring a healthful working and living environment. In the US, worker and community right-to-know laws impose on employers and manufacturers the duty to disclose hazards in the workplace or activities involving toxic exposures. These laws are not a substitute for enforcement of environmental protection laws, but they are an important aid to better regulation of the environment. The principle is embodied in an international convention, the Convention concerning Occupational Safety and Health and the Working Environment, adopted by the International Labour Organization in 1981 (Ashford and Caldart 1985). The ILO Convention requires employers and workers and their representatives to co-operate in protecting occupational safety and health. Measures taken by the employer to protect occupational safety and health must be disclosed to workers' representatives, and the workers have a right to know all aspects of occupational safety and health associated with their work. The Convention also requires training of workers and their representatives in occupational safety and health (International Labour Organization 1981).

Regulation of food and drugs

Laws to prevent the adulteration of food and medicines originated centuries ago (Christoffel 1982, p. 213). Today in industrialized countries, and to an increasing extent in developing countries, people are dependent on commercially produced food and manufactured drugs that they are unable to evaluate themselves. They must rely on governmental regulation of these goods.

Legislation related to nutritional quality and food safety regulates the hygienic standards for production and marketing of foods; control of equipment, utensils, and containers; hygiene and health of food handlers, storage and vending places; methods of testing and inspection; requirements for labelling of contents and shelf-life; and advertising of foods. More specialized legislation deals with such matters as food additives, including what additives are allowed, maximum permissible levels, and requirements for package labelling. These regulations include the requirement for iodization of salt to prevent goitre and in some countries labelling of foods for salt content to promote uniformity in definitions of low salt content.

Comprehensive drug control legislation, which exists generally in all industrialized countries and in many developing countries, provides authority to control the importation and production of drugs; the licensing of manufacturers, wholesalers, and distributors; drug registration; and the distribution, sale, labelling, advertising and promotion of drugs. A national drug control programme regulates the quality, safety, and efficacy of prescription drugs and over-the-counter drugs, and also shares in responsibility for control of narcotic drugs (Chapman 1976).

Enforcement of food and drug laws is carried out through inspections of the manufacturing process, recall or seizure of defective products, civil and criminal penalties for violations of established standards, and injunctions to prevent marketing of food and drugs found unsafe or unsanitary. Enforcement relies heavily on rule-making by the food and drug agency and on administrative hearings on violations of standards. Use of administrative law in this field so critical to health hastens the disposition of cases, provides expertise on the technical issues involved, and introduces flexibility in the process of adjudicating cases and designing sanctions.

Examples of food and drug legislation

In 1976, Norway became the first country in western Europe to establish a national nutrition and food policy (Norway 1981–2). Its objectives are to encourage healthful dietary habits, develop nutrition and food policy in accord with the recommendations of the World Food Conference, increase production and consumption of Norwegian food products, and improve self-sufficiency in food products. Numerous laws are implemented by various governmental agencies to these ends. The Food Control Co-ordinating Act 1978 established a Food Control Board with representation from the various ministries and interests. The Inter-Ministerial Co-ordinating Committee on Nutrition, composed of leading civil servants in the Ministries of Fisheries, Consumer Affairs and Government Administration, Trade, Industry, Church and Education, Agriculture, the Environment, Health and Social Affairs, and Foreign Affairs, is charged with defining tasks, preparing long-term plans, and implementing policies. The National Nutrition Council is composed of members who represent research and teaching in the fields of nutrition, diet, dietetics, food hygiene and technology, food production, and the food industry. The mandate of the Council is to advise the authorities, industrial organizations, large households, and food producers on nutrition and to disseminate information on diet. Implementation of the national nutrition policy thus depends on permanent inter-ministerial bodies with well-staffed secretariats.

In addition, the national nutrition policy of Norway is elaborated through a health policy on nutrition which includes preventive work and training of personnel; agricultural, fisheries, and price policies that affect production and subsidies of foods; and a consumer and school policy to present a more appropriate range of foods and develop sound attitudes towards diet and nutrition. The aim is to help the population consume more cereals, skimmed milk, fish, fruits, and vegetables, less fat, sugar, and meat, and to substitute fish for some meat, starch for some sugar, and skimmed milk for most whole milk. The goals of the Norwegian nutrition policy are to motivate individuals to adopt a healthful diet and to create a situation in which the individual is able to act favourably (Willumsen 1983).

Illustrative of comprehensive drug control programmes is the legislation of Australia, which regulates the manufacture, importation, sale and distribution of drugs and also establishes a Pharmaceutical Benefits Scheme providing publicly financed prescribed drugs to the entire population (Roemer and Roemer 1976, p. 7–8). About 90 per cent of all drugs prescribed in Australia are available on the approved list, and the patient generally pays only a fixed, small co-payment, regardless of the cost of the prescription.

The 1978 legislation of Tanzania provides for regulation of the manufacture and importation of pharmaceuticals, labelling and advertising, registration of pharmacists, and control of poisons (United Republic of Tanzania 1978).

Issues in food and drug control

Control of such essential consumer products as food and drugs requires constant vigilance to monitor the safety and nutritional values of foods and the safety, efficacy, and quality of prescription and over-the-counter drugs. Public health agencies are concerned with surveillance of the production and marketing of both food and drugs, with labelling and advertising, and with provision of information to the people. Mandatory labelling of the sodium content of foods is a form of consumer education. Warning labels and package inserts in pharmaceuticals are an important part of patient education.

Drug regulation begins with establishing and implementing protocols for research on and testing of new drugs and proceeds to evaluation of drugs for safety and efficacy. The process of evaluating animal and clinical data and of determining health risks from drug trials may be quite drawn out, so that regulatory agencies may be accused of unreasonable delays in approval of new drugs and of 'drug lag'.

Associated with the drug approval process is the determination of national policy governing the import and export of therapeutic drugs. The multiple national standards on acceptability of different drugs affect the availability of drugs in international trade—an increasingly critical problem in a shrinking world (Cook 1987). Fortunately, in 1975 the World Health Organization established a Certification Scheme on the Quality of Pharmaceutical Products Moving in International Commerce (a revised version of the Scheme was adopted in 1988) and regularly disseminates information on drugs and drug quality throughout the world, notably through the journal, *WHO Drug Information*. Cook suggests that use of international standards may alleviate the problems related to variance in standards among countries.

A major problem on which there is great variation in the laws of different jurisdictions is that relating to product liability and compensation for adverse effects of drugs. The tensions that exist between the interests of the pharmaceutical industry in marketing new drugs, the interests of the consumer in compensation for damages suffered, and the interests of society in promoting development of pharmaceutical products and assuring their availability have led to varying solutions. These have included, to take the example of the US, decisions holding the manufacturers strictly liable, with damages assessed according to their share of the market where the supplier of the drug could not be identified (*Sindell* v. *Abbott Laboratories* 1980); decisions holding a manufacturer liable only for failure to follow state-of-the-art manufacturing practices (*Brown* v. *Superior Court* 1988); and, in the case of vaccines, development of a no-fault, federally funded compensation system for untoward outcomes of childhood immunizations (US 1986).

Finally, an increasingly important issue relates to regulation of drug prices. The high cost of drugs has led to the use of generic drugs and to the repeal of laws banning substitution of generic equivalents for brand-name drugs prescribed by the physician (DeMarco 1975, p. 65). Another strategy for controlling drug costs in health care programmes is development of a drug formulary or a list of essential drugs for which reimbursement is provided .

Licensure of health personnel

Licensing or registration laws for physicians were designed originally to protect the public against quacks, charlatans, and incompetent practitioners. Over the years the function of licensing laws for a wide variety of health professionals has expanded to specify minimum qualifications for practice, to regulate educational programmes, to define the scope of practice for each profession, and to set forth requirements for continued competence. Some licensing laws are mandatory, such as medical and dental practice acts, requiring all who practise the profession to be licensed and making it illegal to practise without a licence. Voluntary or permissive licensing laws protect only the title and prohibit unlicensed personnel from holding themselves out as licensed personnel but do not require licensure in order to practise the profession. As professions grow in strength, voluntary licensure tends to be replaced by mandatory licensure.

While licensing or registration laws have been criticized as protecting the professions rather than the public, and as creating monopolies of the licensed professions, all countries have a governmental system for regulating the qualifications of medical, dental, and a varying number of other health personnel. In some countries licensure may be granted without further examination on completion of an approved educational programme. In others a separate examination may be required after completion of the approved educational programme. In recent years, recognition of the capacity of licensing laws to do more than specify minimum qualifications for practice—to influence the geographical location of physicians, dentists, and others, to affect the proportions of generalists and specialists, to influence the pattern of practice, and to promote the continued competence of practitioners—has given a new importance to licensing laws for the health professions.

Implementation of licensing laws is carried out by licensing boards composed originally largely of members of the profession to be licensed. In response to the demand for greater public accountability, members of other professions, consumers, and representatives of governmental agencies have been added to the boards in many countries.

The example of nursing licensing laws

To present some insight on the role of licensing laws in regulating the qualifications and functions of health personnel we may take the example of nursing licensure. Nursing practice Acts generally provide for personal and educational qualifications of nurses, prescribe the content of nursing curricula, including practical experience, define the scope of nursing practice, specify grounds and procedures for disciplinary action, and provide for renewal of licences.

As nursing education has been expanded and technology in health care improved, it became clear in the US and in other countries that nurses were being underutilized. Although they were equipped by enriched training for new nursing roles, the licensing laws generally barred nurses from undertaking 'diagnosis and treatment', which were defined as medical functions. Beginning in 1971, the American states adopted various legislative strategies to authorize an expanded role for nurses (Bullough 1975, p. 153). These included authorization by the medical and nursing licensing boards of expanded functions, amended definitions of professional nursing to include autonomous functions, adopting standardized procedures and protocols to authorize expanded nursing functions, and allowing individual physicians to delegate the right to diagnose and treat. The legislative changes made in nursing practice Acts in State after State expanded the contribution of nurses to patient care and made the profession of nursing more interesting and rewarding.

Legal barriers to extension of the nurse's role in developing countries are particularly grave because the nurse is often the only health professional available in rural areas to provide primary health care. Yet often the medical and pharmacy Acts bar the nurse from diagnosing and treating and from prescribing medication. The nursing practice Acts restrict not only the scope of nursing practice but also the training that nurses receive (World Health Organization Study Group 1986, p. 32). In order to alter this negative impact of the law, countries have enacted new statutes to reorient nursing education and to authorize functions for nurses that were formerly the exclusive province of the physician. For example, in 1977 Senegal issued a decree adapting its nursing education to the needs of the country so that nurses will be prepared to serve the rural population (Senegal 1977). Also in 1977 the Ministry of Health of Israel authorized nurses to carry out certain kinds of medical activity, such as drawing blood for tests, immunizations, and suturing wounds (World Health Organization Study Group 1986, p. 45). In 1982 Dominica authorized its family nurse practitioners to prescribe drugs from the Dominica Nurse Practitioner Formulary (Dominica 1982).

Legal issues in personnel licensure

The issue of scope of practice of health personnel recurs periodically as new types of health workers are introduced in a country, such as pharmacy technicians or acupuncturists, and as strengthened preparation of existing categories of personnel warrants expanded functions, as in the case of professional nurses discussed earlier. In fact, an overriding issue in this field is how allied and auxiliary personnel should be credentialed, whether by a governmental mechanism, such as licensure or some form of officially required registration, or by a voluntary mechanism, such as certification by a professional association.

Another concern is to assure equitable and rational geographical and specialty distribution of health professionals. Some countries, such as Norway and Mexico, require physicians to contribute a period of service in an underserved area as a condition of licensure. Other countries, such as the US, use governmental funding for medical education as leverage for specialty distribution and as an economic incentive for settlement in rural areas.

The problem of assuring continuing competence has challenged many countries. Various strategies have been adopted, including voluntary and mandatory educational programmes, periodic re-examinations, and further clinical training. But no consensus has been reached on the best way to achieve the objective of updated knowledge and skills.

Not only licensing laws but various other regulatory mechanisms affect the qualifications and functions of health personnel. These include the educational system, the policies of professional associations, the regulation of work settings, the requirements of payment programmes, and judicial decisions in malpractice suits and other legal cases. While the licensing laws impinge directly on the qualifications and functions of health personnel, indirect influences through the methods of providing and paying for care can also shape the health staffing component of a national health system.

Regulation of health care facilities

Various mechanisms regulate the quality of the care provided by hospitals, health centres, and long-term care facilities—hospital licensing laws, requirements of the financing system, court decisions, actions of voluntary accrediting bodies, standards of professional associations for specialty training, and rules of the facilities themselves. Legislation sets minimum standards that facilities must meet in order to operate. Legislation also governs the planning, construction, and distribution of facilities.

The corner-stone of this multi-faceted regulatory system is hospital licensure. Hospital licensing laws are important because such a large proportion of care—and care for serious illness—is provided in hospitals and because the costs of hospital and long-term care represent such a large proportion of health care expenditures. As Anne Somers has written: 'it is difficult to exaggerate the importance of the hospital in contemporary society' (Somers 1969).

Originally, hospital licensing laws were concerned solely with the physical conditions in the hospital—safety, sanitation, and space. But over the years, health facility legislation has expanded to cover many types of health facility and to prescribe requirements to assure not only the safety of the patient but the quality of her care (Lander 1980). Government sets standards for both public and private institutions. The public purpose of health facilities and the public interest in their use are the basis for public regulation of private institutions. For non-governmental agencies and the private market, legislation may regulate performance to protect public health, may provide support, and may define interrelationships of private institutions with Government.

Implementation of hospital licensing laws is carried out through rule-making by governmental agencies, inspections of facilities, consultations to remedy deficiencies, administrative hearings, injunctions, denial of reimbursement, licence suspension and, if all else fails, through closure of the facility.

Examples of health care facility regulation

In the US, hospital licensing laws are fairly recent, having been enacted in 1946 after the Second World War in response to the Hill-Burton Hospital Survey and Construction Act which required States to specify minimum standards for facilities receiving federal subsidies.

Since 1968, a number of States have amended their State hospital licensing laws to enact what were termed comprehensive laws encompassing both physical and patient care standards. For example, the modernized facility licensing law of New York State, enacted in 1969 (New York 1985), contains detailed provisions governing construction, financial reporting, and patient care. The law specifies, among other matters, requirements for ambulatory care, calls for a comprehensive evaluation of patients on a periodic basis, requires continuity of care when patients are referred outside the hospital, mandates full-time medical staffing in emergency rooms, specifies rules for surgical consultation, and requires general hospitals to admit patients in need of immediate hospitalization without advance inquiry as to their ability to pay. Thus, the provisions of modernized facility licensing laws have moved far beyond bricks and mortar.

In France, an order promulgated in 1982 established a Commission on Standards for the Equipment and Operation of Public and Private Hospital Establishments. The Commission was made responsible for developing minimum technical operating standards and minimum equipment standards to be fulfilled by public and private hospital establishments in order to guarantee patients the quality of care and accommodation required by their conditions and to ensure that staff work under proper circumstances (France 1982).

Issues in health care facility regulation

Implementation of requirements for facility licensure is the principal issue facing governmental agencies responsible for standards in health care facilities. Frequent inspections and time-consuming consultations require trained staff, often in short supply. Although licensing statutes provide legal mechanisms for enforcement, less onerous strategies are generally preferred. Particularly in the field of long-term care, where often the need is to upgrade the quality of care, tying standards of the facility to the financing mechanism is increasingly preferred as an effective sanction.

Control of communicable diseases

Prevention of the spread of communicable diseases was one of the earliest functions of public health. In this effort, two types of law have been employed: laws to assure a sanitary environment (discussed above) and laws to regulate human conduct to control the spread of disease.

Turning off the Broad Street pump through which cholera was spread was an ideal public health measure because it cut off the source of the disease and benefited the whole population served by that water supply. Such a solution is not always available, however. Therefore, other measures to prevent epidemics have been adopted. Such laws authorize public health officials to ascertain the incidence of communicable disease, to regulate the conduct of those who are infected, and to require measures to prevent its spread. Because these actions involve some restriction of the rights of individuals, the law in this field seeks to balance the need of society for protection against disease and the rights of the individual to privacy and liberty.

Traditional methods of preventing and controlling communicable disease have evoked statutory responses. To assist epidemiological investigation of the incidence of communicable diseases, laws mandating reporting to public health officials of specified communicable diseases by physicians, school authorities, and laboratories have been passed. To prevent and control communicable disease, the law may provide, as in the US, for compulsory examination of individuals who are in a position to spread disease (e.g. food handlers) and of individuals to whom communicable disease presents special hazards (e.g. schoolchildren, applicants for a marriage licence, pregnant women). A health officer also generally has power—although it is rarely exercised—to order a person suspected of being infected with a contagious disease to submit to a physical examination (Grad 1988). Similarly, rarely used today is the health officer's power to isolate and quarantine an infected individual, although the power continues on the statute books. The most important legal measure for control of communicable diseases is certainly compulsory immunization.

Statutes authorizing these measures for controlling communicable diseases generally provide a civil or criminal penalty for violators, but much preferred are other strategies for implementation of the laws, such as exclusion from school or work. Compliance with a specified immunization schedule may be required for school attendance, as in Ontario, Canada and the German Democratic Republic. (Ontario, Canada 1982; German Democratic Republic 1982). Compulsory examinations may be required for a marriage licence or to obtain a certain job, as in the US (Grad 1990).

The tragic epidemic of Acquired Immune Deficiency Syndrome (AIDS) presents the classical public health problem of a conflict between the welfare of the community and the rights of the individual, but with significant differences. Like other sexually transmitted and communicable diseases, AIDS calls for measures to prevent its spread and also for protection of the privacy and other civil rights of persons afflicted with the disease. But AIDS presents grave and different problems because as yet there is no cure and no vaccine for prevention, and because the incidence of AIDS is particularly high in the US and some other countries in certain groups engaging in high-risk behaviour—male homosexuals and intravenous drug users—who are particularly vulnerable to discrimination.

Legislation has been a significant component in response to the AIDS epidemic (World Health Organization 1988; Gostin and Curran 1987). An early response to the epidemic was to require testing of all blood and blood products provided by blood donors and confidential reporting of results. A most effective measure to protect the blood supply was the public health strategy, often adopted without the necessity of legislation, of establishing alternate testing sites (i.e. other than blood collection centres) enabling persons seeking information on their antibody status to have confidential or entirely anonymous testing without endangering the blood supply.

Most jurisdictions having legislation on AIDS require reporting of cases of AIDS to a health agency. Some statutes classify AIDS as a sexually transmitted disease, as in the State of Idaho (US), Iceland, Sweden, Chile, and Guatemala. Such an approach permits the testing of prostitutes and tracing of contacts to advise testing and provide counselling. A few jurisdictions require reporting on AIDS related complex (ARC) or of seropositive test results. In order to protect the confidentiality of test results, reporting may be anonymous or by code, number, or initials. Much of the AIDS legislation contains specific protection of requirements to safeguard the confidentiality of information. And stringent systems have been put in place to prevent any breach of confidentiality.

Protection of the confidentiality of test results has conflicted with the need to protect other members of society. In order to cope with this problem, some laws provide for very limited disclosure of identifying information, for example to a health care professional engaged in care of a patient with AIDS and to a medical facility that will receive blood, organs, semen, or breast milk from an infected individual.

Opinion is growing that a physician should be authorized to disclose positive human immuodeficiency virus (HIV) antibody status of an individual to a spouse or sexual partner when the physician has reason to believe that the individual will not inform the spouse or sexual partner.

A controversial legal issue concerns the recalcitrant patient or seropositive person who does not respond to education, counselling, medical direction, and community pressures to stop infecting others but knowingly exposes others to HIV infection. Legal remedies may exist in the general civil and criminal law. Regulations in the UK authorize an order by a Justice of the Peace to detain the patient if the JP is satisfied that the patient will not take proper precautions to prevent the spread of the disease (UK 1985). Hopefully, such cases will be few, and indeed only one case has been reported of use of these powers in England.

Examples of legislation to control communicable disease

Finland enacted in 1986 a comprehensive ordinance concerned with prevention of communicable diseases through vaccination, distribution of antibody preparations and medicines, the provision of measures related to individuals and their environment that are intended to prevent the development or spread of communicable disease, early diagnosis, screening, and treatment, and medical rehabilitation (Finland 1986). The duties of the National Board of Health, the county councils, the communes, the hospitals, the pathology laboratories, and the Institute of Public Health are set forth in this contemporary regulatory system to control communicable diseases.

Rather than select examples from the many statutes on AIDS, compiled on an ongoing basis by the World Health Organization (1987), we refer the reader here to several documents illustrating guide-lines for action at different levels of Government. First, on the international level, are various authoritative policy statements of the World Health Organization, for example on the safety of blood and blood products. (Petricciani *et al.* 1987) and on international travel and HIV (International Digest of Health Legislation 1987).

On a national level, one may turn to the US Public Health Service Plan for Prevention and Control of AIDS and the AIDS Virus (Institute of Medicine 1988; Report of the Presidential Commission 1988; US Public Health Service 1986). The Plan calls for a strategy to control the disease based on voluntary counselling and testing, conducted with confidentiality and encompassing the following five spheres:

(1) an information base to determine the size of the population at greatest risk, particularly the numbers of homosexual men, bisexual men, intravenous drug abusers, and heterosexuals who have multiple partners:

(2) national information and education campaigns on AIDS targeted to the currently uninfected population;

(3) prevention of intravenous drug abuse transmission;

(4) prevention of sexual transmission through voluntary serological testing and self-referral of sexual and drug abuse contacts, notifications and counselling of contracts by health authorities, and targeted education programmes;

(5) prevention of transmissibility by blood and blood products.

At the local level of government, one may examine the comprehensive New York City Strategic Plan for AIDS released in May 1988 (New York City Interagency Task Force on AIDS 1988). The plan covers the dimensions of the problem faced in New York City and the principles on which the plan is based, notably inter-governmental co-operation, co-ordination of the public and private sectors, the crucial role of community-based organizations in providing services, and the need for significantly increased drug treatment resources for intravenous drug users as a way of preventing the spread of AIDS. The plan outlines in detail new initiatives in the spheres of prevention, clinical and social services (both acute care and long-term care), AIDS care and services in the community, surveillance and research, and, most importantly, protection of human rights to prevent HIV-related discrimination.

Issues in communicable disease control

Immunization and other measures to control communicable diseases are so well accepted today that the principal problems in this field are spin-offs from effective immunization: the high cost of vaccines and the question of how to compensate those few patients who suffer an untoward outcome of immunization. These problems are handled in different fashions by the health and legal systems of each country.

One of the most controversial issues in the law governing AIDS is the extent to which screening for AIDS—the systematic application of the ELISA test and confirmatory, supplemental tests to specific populations—should be undertaken. While some countries have mandated tests for particular groups, such as prisoners, prostitutes, or immigrants, or even, in some jurisdictions, intravenous drug users or applicants for a marriage licence, the weight of authority favours voluntary testing for several reasons: (1) it encourages the co-operation of high-risk groups in case-finding, testing, and counselling, and notification of sexual contacts and persons with whom needles have been shared; (2) it facilitates protection of privacy; and (3) in a low-prevalence population, a test with a high degree of sensitivity and specificity, as the serological tests for HIV antibodies are, will produce a large proportion of false positive responses, causing great anxiety, providing misleading information, and requiring confirmatory tests. Also, a certain number of false negatives will occur. Therefore, the WHO Global AIDS Strategy strongly favours voluntary testing, counselling, and protection of confidentiality. The highest ethical and legal imperative is to prohibit and penalize discrimination in employment, housing, health insurance, public accommodations, governmental services, and schooling solely because the person has AIDS or is believed to be seropositive . If effective treatment for AIDS or a vaccine to prevent it becomes available, then the law and policy on screening, record-keeping, and contact tracing will undoubtedly change.

Legislation on mental illness

If one were to select a single sector of health services in which to see the field of health law in microcosm, one should examine health services for the mentally ill. This sector illustrates with particular sharpness the conflict between health needs and legal rights, between protection of the patient and protection of society, and ways to resolve these conflicts.

The scope of legislation affecting the mentally ill is very broad. It includes laws governing admissions to mental hospitals, standards for mental health facilities and care of patients, organization of community mental health programmes, legal protection of the person and property of the mentally ill, the doctrines of the right to treatment in the least restrictive alternative setting, legal aspects of deinstitutionalization, and mental illness and the criminal defendant (Curran and Harding 1978). Clearly, we cannot discuss the many legal aspects of mental illness. Here we shall restrict the discussion to mental hospital admission laws.

With the advent of the tranquillizing drugs, the development of the concept of the open hospital and the therapeutic community, and with increased awareness of the civil rights of patients, many jurisdictions have amended their centuries-old statutes governing criteria for hospitalization of the mentally ill. Definitions of who is mentally ill have moved away from vague standards, such as 'in need of care and treatment', to more precise standards, such as 'dangerous to others', 'dangerous to self', and 'gravely disabled'. In addition to changes in the grounds for admission, the procedures for admission have been modified to assure prompt and non-traumatic admission to mental hospitals when needed, to require periodic review of the need for continued hospitalization, and to assure prompt discharge as soon as the patient is ready.

Many of the old commitment laws, as they were called, resemble criminal proceedings. They required a petition to the court, notice of hearing, representation by counsel, a hearing before a judge, often with a jury trial, testimony by witnesses, and even sometimes a written opinion by the judge as to the necessity for hospitalization. These laws, it was found, provided only the illusion of due process and actually were adverse to the health needs of patients in many cases for prompt and non-traumatic hospitalization (Association of the Bar of the City of New York 1962). Modern statutes have replaced this legalistic procedure with new administrative mechanisms for protecting both the health needs and the legal rights of mental patients. At the same time, since an individual's liberty is at stake in an involuntary admission to a mental hospital, the role of the courts in overseeing the propriety of retaining an individual in a hospital has been strengthened.

Examples of mental hospital admission laws

The first country to enact a modernized mental health law was the UK, which adopted legislation along lines recommended by the Royal Commission on the Law Relating to Mental Illness and Mental Deficiency in 1957. In the past, in England and Wales involuntary patients were admitted to mental hospitals on an order from a Justice of the Peace based on one medical certificate from a medical practitioner. This procedure was viewed as providing inadequate safeguards because the magistrate could not form any sound, independent opinion on the patient's mental condition and because the judicial order associated mental hospitalization with the courts and with punishment of crime (Maclay 1960).

The Mental Health Act of 1959, applicable to England and Wales, abolished the judicial order and made compulsory admission, when necessary, a medical matter, requiring two medical opinions, including one from a doctor with special experience (UK 1959). The doctors recommending compulsory detention must specify the grounds for their opinions and state whether alternative methods of dealing with the patient are available and, if so, why they are not appropriate and hospitalization is necessary. The hospital must confirm the need for hospitalization, and on the basis of these three certifications the hospital is authorized to retain the patient for specified time limits. Most importantly, an administrative agency to which patients and their families have access—the Mental Health Review Tribunal—is established in each hospital region, with power to review the appropriateness of hospitalization and to discharge the patient.

Twenty-five years after the Mental Health Act was adopted, in 1983, amendments to the law were adopted to strengthen protection of the civil rights of mental patients (UK 1983). These amendments require, among other things, consent to treatment, assurance of patient rights, such as the right to visitors, to pocket money, etc., and establishment of a Board of Visitors to provide surveillance of the quality of care in mental hospitals.

In the climate of opinion created by new methods of treatment of the mentally ill and new public attitudes towards mental illness, New York State revised its mental hospital admission law after an extensive study which found great variations, inequities, and injustices in the involuntary admission of patients to mental hospitals. The measures intended to protect the rights of patients had become a rubber stamp by the judges of the decision of doctors (Association of the Bar of the City of New York 1962).

Accordingly, in 1964 the New York State Legislature unanimously passed a new Mental Hygiene Law (New York 1978). It abolished civil judicial certification of an involuntary patient to a mental hospital and provided that the initial admission of an involuntary patient to a mental hospital is a medical matter, on the application of a near relative or other interested person and on the recommendations of two physicians, with the concurrence of the admitting hospital. Immediate and periodic legal reviews of the propriety of hospitalization are required. The rights of the patient are protected by an arm of the court, the Mental Health Information Service, which faces towards the patient and his or her family to inform them of the patient's rights and alternatives and towards the court to inform it of the patient's condition and alternative treatment resources.

Issues in mental hospital admission laws

Modernized mental hospital admission laws reflect the revolution that has occurred in the care of the mentally ill. But not all problems have been resolved. What standard of proof should be required for involuntary hospitalization? Do patients have a right to treatment? Do they have a right to refuse a particular kind of treatment? What safeguards are afforded for minors deemed in need of mental hospitalization? (*Parham* v. *J.R. et al.* 1979). Probably in no field of health law is the conflict between the rights of the individual to liberty and confidential treatment and the right of society to protection from harm so sharp as in the field of mental illness. Resolution of this conflict in various contexts will depend on further advances in psychiatric diagnosis and treatment and on imaginative legal strategies to protect the individual and society.

Legal problems in human reproduction

A priority for public health through the years has been protection of the health of mothers and children. In all countries emphasis has been placed on prenatal and maternity care and on breast-feeding, immunization, and well-baby care. A new dimension was added to maternal and child health efforts with the recognition of the importance of birth control and abortion to prevent unwanted pregnancy and assure proper child spacing. Deaths from illegal abortion were the largest single cause of maternal mortality in many countries and still are in some. Yet laws making abortion a crime except when performed to save the life of the woman barred prevention of such tragedies.

To tackle the enormous toll in preventable maternal deaths from dangerous illegal abortion, countries turned to legal action to shift abortion from the illegal to the legal sector of medical practice (Cook and Dickens 1988). To promote family planning programmes, many countries enacted laws removing barriers to access to birth control and providing educational and financial support for contraceptive services, for example France, the Federal Republic of Germany, Italy, Morocco, and Spain (Isaacs 1981, pp. 199–257; Paxman 1980; Mason *et al.* 1987). Laws also authorized voluntary sterilization, as in Singapore, Panama, Japan, and the Scandinavian countries (Isaacs 1981, pp. 147–98; Stepan 1981). Statutes mandating sex education in schools were adopted (Roemer and Paxman 1985). Also, as another aspect of the woman's choice in reproductive matters, the law has been called on to authorize means to reduce infertility and has addressed alternative or assisted means of reproduction (Andrews 1984; Annas 1984; Annas and Elias 1983; Swiss Institute of Comparative Law 1986; Warnock Committee 1984).

These legal changes were not achieved without opposition. A minority of the population in a number of countries opposed legalized abortion and even attempted to restrict contraceptive and sex education programmes. Despite their efforts, advances in the technology of contraception and changed social attitudes concerning sexual behaviour and the rights of women have impelled modernization of the laws governing human reproduction and reproductive choice.

Implementation of these laws has been undertaken in various ways. The right to family planning has been embodied in the constitutions of some countries (Cook and Dickens 1988). Legislative enactments have authorized, outlined, and provided financial support for family planning programmes. Barriers to voluntary sterilization have been removed, and voluntary sterlization has been specifically authorized. The statutes governing abortion have been amended to set forth a wide range of criteria. Abortion may be legalized at the request of the woman as in virtually all the most populous countries of the world—China, the USSR, the US, and Japan—or legalized in the first three months of pregnancy, as in France and Italy. Abortion may be authorized on social or socio-medical grounds as in India, the UK, the Scandinavian countries, and most of the socialist countries of Eastern Europe. Abortion may be allowed on medical grounds, as in Algeria, Israel, and Switzerland. Abortion may be allowed only to save the life of the woman, as in Honduras and Turkey. Abortion is still prohibited in Ireland and the Philippines (David 1984; Mason *et al.* 1985).

Examples of legislation on human reproduction

The 1978 law of Luxembourg relating to sexual information, prevention of illegal abortion, and regulation of pregnancy termination is a comprehensive approach to family planning services (Luxembourg 1978). The legislation provides for consultation and information on contraception, sterilization, and abortion. Informational and consultative services are provided without charge to everyone, and for minors all services are provided without charge to everyone, and for minors all services and medications also are free.

The Italian law of 1978 illustrates the modern trend of laws legalizing abortion at the request of the woman in the first three months of pregnancy, with abortion permitted on specific grounds later in pregnancy (Italy 1978). The law requires local and regional health authorities to promote services and other measures to reduce the demand for abortion. The enactment of a national health service covering the total population of Italy may eliminate financial barriers to contraceptive services and abortion, but the reluctance of some physicians to perform abortions because of conscientious objection or threatened sanctions by the Catholic Church impedes access to abortion.

Issues in laws affecting human reproduction

The turnabout in the law governing birth control and abortion that has occurred in the last two decades all over the world has brought significant public health benefits to women and their families. Deaths from illegal abortion of desperate women faced with unwanted pregnancies have been prevented. Infant mortality and morbidity have been combatted by improved spacing of pregnancies (Maine and McNamara 1985). Many adolescents have been able to defer childbearing to a time more appropriate for parenting (Paxman and Zuckerman 1987; Roemer 1985).

Still, many problems remain. Geographical access to family planning and abortion services is uneven. Financial access is a serious problem where universal financing of health services is not available. Required third party authorization by a spouse or a parent blocks or delays services. Further work is needed to prevent teenage pregnancy by improved sex education in the schools and improved access to contraceptive services and abortion. Finally, an unknown element in this field is the impact that the tragic epidemic of AIDS will have on the use of condoms and on the option of abortion for seropositive women.

Legislative approaches to health promotion

The historical role of public health of preventing disease and disability received a new impetus in 1974 with the publication in Canada of the Lalonde report *A New Perspective on the Health of Canadians*. This report launched a world-wide effort for health promotion, examining and improving the judgments that

> must be made by individuals in respect of their own living habits, by society in respect of the values it holds, and by governments in respect of both the funds they allocate to the preservation of health and the restrictions they impose on the population of whose well-being they are responsible. (Lalonde 1974, p. 9.)

How does health promotion differ from the time-honoured function of public health, the function of health education? Horowitz defines health education as consisting of any combination of learning experiences designed to facilitate voluntary adaptations of behaviour conducive to health. By contrast, health promotion is the process of advocating health to enhance the probability that personal, private, and public support of positive health practices will become a societal norm (Horowitz 1981).

In this effort to establish societal norms that contribute to healthful life-styles, legislation has proved to be an essential component. This does not denigrate the role of education, which is important to engender the motivation to change behaviour. That motivation, however, can be encouraged by societal norms and expectations which, in turn, are promoted by governmental policy expressed in legislation.

Prevention of motor vehicle accidents

The fields in which legislation can be directed to promoting health are many and varied. A high priority has been given to laws to prevent the enormous toll in death and disability from motor vehicle accidents. These include laws setting standards for fitness to drive, mandating the use of child passenger restraint systems, seat belts, and motor cycle crash-helmets, and requiring manufacturers of automobiles to install air bags. With the technological innovation of blood alcohol measurement, stringent laws make the presence of a given level of alcohol in the blood a conclusive presumption of drunk driving entailing mandatory fines, jail sentences, licence restriction, and rehabilitation programmes (California Vehicle Code 1985).

Control of alcohol abuse

Drunk driving laws are but one type of law to control alcohol abuse. There are two major types of legislation: (1) control of the availability of alcoholic beverages; and (2) influence on drinking practices.

In the category of controlling the availability of alcohol are laws to control places of sale, hours of sale, and sales to minors, and to provide controls through taxation and prices (Addiction Research Foundation 1981; Institute of Medicine 1982; Moser 1980). While the evidence on the effectiveness of some of these measures is equivocal, it is generally agreed that decreasing the availability of alcohol generally is important. The experience of the Province of Ontario, Canada showed that lowering the drinking age increased alcohol-related traffic accidents and admissions of teenagers to alcoholism and rehabilitation facilities (Addiction Research Foundation Volume 2, p. 151; Single 1984). Increase in price of alcohol relative to income has been associated with a decline in consumption (Moser 1980, p. 109).

In the category of regulating drinking practices are control of advertising, punishing public drunkenness, control of drinking and driving, education and information about alcohol, and counselling, treatment, and rehabilitation. Again, although the evidence on the effectiveness of these measures is conflicting, there are indications that drinking practices can be affected by legislation, combined with education, provided the laws are enforced.

Control of the smoking epidemic

Smoking is the largest, single, avoidable cause of ill-health and premature mortality in the world, accounting for an estimated two and a half million deaths a year from smoking-related diseases. In response to the world-wide smoking epidemic, Governments have intensified their efforts to control smoking by legislation, combined with education and smoking cessation programmes.

Two broad categories of laws have been enacted: (1) laws

to bring about changes in the production, manufacture, promotion, and sale of tobacco—laws to control the supply (or 'production') side; and (2) laws to achieve changes in practices among smokers—laws to control the 'demand' (or 'consumption') side of tobacco use (Roemer 1982, 1986).

In the category of bringing about changes in the production, manufacture, promotion, and sale of tobacco are:

(1) control of advertising and sales promotion of tobacco products;

(2) requirements for health warnings and statement of tar and nicotine contents on cigarette packages and other tobacco products;

(3) control of tar and nicotine contents;

(4) restrictions on sales to adults;

(5) taxation and price policy;

(6) economic strategies relating to subsidies and crop substitution.

While all these measures contribute to decline in tobacco consumption, the most important are control of advertising and tax and fiscal policies. The tobacco industry spends more than $2 billion a year globally to lure consumers to its products, not counting the cost of indirect advertising through sponsorship of sports and cultural events by tobacco companies. Through advertising the industry conveys the message, especially to young people, that smoking is associated with success, pleasure, relaxation, sports, freedom, beauty in nature, sophistication, virility, and sexuality. Moreover, the substantial revenues received by newspapers and magazines from advertising of tobacco products have a chilling effect on their editorial policies and deter publication of articles on tobacco and health.

By 1985, at least 55 countries in the world had enacted controls on advertising of tobacco. Twenty countries had banned all advertising of cigarettes; 15 had enacted strong partial bans limiting the contents of advertisements to facts, barring the depiction of adolescents or children associated with sports, or restricting the amount of space devoted to advertising; and 20 countries had enacted moderate partial bans prohibiting use of the electronic media for advertising of tobacco products. Countries with total bans on advertising and strong antismoking policies, for example Norway and Sweden, have succeeded in lowering smoking rates markedly, particularly among young people (World Health Organization 1987).

Experience in the UK, US, and Hong Kong has shown that raising taxes on tobacco decreases sales. An important report from Finland confirms the need for substantial and repeated increases in taxes and prices of tobacco products if consumption is to be significantly reduced (National Board of Health of Finland 1985).

In the category of legislation to change smoking behaviour are:

(1) restricting smoking in public places;

(2) restricting smoking in the workplace;

(3) preventing young people from smoking (by prohibiting sales to minors and restricting sales of cigarettes from vending machines);

(4) mandating health education.

Because of the mounting evidence on the dangers of passive or involuntary smoking, the most important measures are restrictions on smoking in public places and the workplace. Such legislation has been enacted at the national level of Government, as in Belgium, Iceland, Ireland, and Norway, for example, and it has also been successfully implemented when enacted by States and cities, notably in the US (Roemer 1988). These laws contribute to the creation of a non-smoking environment and the view that smoking is socially unacceptable. It is generally agreed that social acceptability is the ground on which the battle against the smoking epidemic will be decided.

Other health promotion legislation

To promote healthful nutrition, legislation may regulate the quantitative declaration of calories, fats, cholesterol, sodium, sugars, and other nutrients to promote safe and wholesome foods, as discussed earlier. Legislation may also establish supplemental feeding programmes, such as school lunches and nutritional supplements for low-income pregnant women and infants.

Terris has suggested the following legislative measures to prevent coronary heart disease and cerebrovascular disease:

(1) requiring use of unsaturated fats in commercially produced foods;

(2) subsidizing the production of foods rich in unsaturated fats in order to lower the price;

(3) assisting farmers to change feed in order to produce beef low in saturated fats;

(4) providing scientific, technical, and other assistance to agriculture and the food industry to enable them to shift as painlessly as possible to production of foods compatible with the 'prudent diet' (Terris 1983).

These are ideal public health measures because they benefit whole populations and do not require individual behavioural change.

Another such measure is fluoridation of public water supplies to prevent dental decay. Community water fluoridation improves the oral health of an entire population regardless of socio-economic level, education, individual motivation, or the availability of dental personnel (Murray 1986).

Many of the problems discussed in connection with health promotion are manifestations of stress. Certainly, smoking and alcoholism are largely responses to stress. Stress is also produced by adverse environmental and life situations, such as poor working conditions, poor housing, and unemployment. Stress has so many causes that one needs to address specific problems to ascertain what legislation can do to alleviate them.

The list of health problems associated with stress is long and daunting. It includes drug abuse, depression, suicide, mental illness generally, child abuse, and family violence. Societal responses to these difficult problems lie in social institutions and programmes that can be assisted by legislation.

Examples of health promotion legislation

One could select many examples of legislation to promote health, such as legislation to assure product safety to prevent accidents in the home, laws to prevent road accidents, or laws to reduce alcohol consumption. The Norwegian national nutrition and food policy, discussed earlier, expresses an important policy to promote health. Here we choose two examples of laws to control the smoking epidemic.

A multi-faceted law to control smoking was enacted in Iceland in 1984 (Iceland 1984). On the 'supply' side, the Icelandic law bans all advertising of tobacco products, requires eight rotating warnings on packages of cigarettes with the innovation of effective depictions of the different parts of the human body affected by smoking, and provides that 2 per cent of the revenue from sales of tobacco is to be spent on antismoking activities.

On the 'demand' side, the Icelandic law bans smoking in a wide variety of public places and in specified work sites, and other places are subject to further regulation. It prohibits sales to persons under 16 years of age, all sales of tobacco from vending machines, and all sales in schools or institutions for children and young people. The law mandates health education on the harmful effects of tobacco in schools, with particular emphasis on such instruction in primary schools and schools preparing personnel for the fields of education, teaching, and health care. The law of Iceland is a model of a comprehensive law to control smoking.

A single-purpose statute with far-reaching impact is the 1987 Belgian Crown Decree banning smoking in all indoor places to which the public is admitted, including post offices, hospitals, schools, universities, homes for the elderly, theatres, concert halls, and sport arenas (Belgium 1987). In premises where the service to the public consists of furnishing food or drink, smoking may be allowed in a clearly defined part of the premises, but such permission may be denied if the premises are in buildings where students are taught, lodged, or cared for.

Issues in health promotion legislation

A number of general issues or recurring questions plague the field of legislation for health promotion. Here are some of them:

1. How should the interest in social control and individual liberty be reconciled?
2. What types of legislative control are acceptable and effective?
3. How do various legislative approaches affect different socio-economic groups, and are they consistent with our notions of equity?
4. What legislative measures are effective in motivating individuals to change their behaviour? And how do we evaluate them?
5. Who should bear the costs of risk-taking behaviour?
6. Are environmental measures available that lessen or eliminate the need for behavioural change?
7. What strategies for implementation of legislation are useful?
8. What is the responsibility of Government for the people's health *vis-à-vis* powerful commercial interests?
9. How should one legislate in the face of scientific uncertainty?

Ethical issues in public health law

Let us begin this brief discussion of the large and profound topic of ethical issues in public health law with some definitions. Ethics is a set of philosophical beliefs and practices concerned with questions of justice, fairness, equity, rights, allocation of resources, and costs. Law is a system of principles and rules devised by organized society for the purpose of controlling human conduct. Law may be described as a vital process or group of processes by which people living together in a society meet their common problems and solve them to promote the common good (Christoffel 1982, p. 3; Wing 1985, p. 1). Law and ethics converge because both are concerned with rights and duties.

Public health is concerned with ethics because, as Beauchamp points out: (1) public health is concerned not only with explaining the occurrence of disease but also with ameliorating it; and (2) public health is concerned with integrative goals expressing the commitment of the whole people to face the threat of death and disease (Beauchamp 1985).

A well-known ethicist in the field of health in the US, Daniel Callahan, has pointed out that the most significant development in the field of bioethics in the decade from 1970 to 1980 was the development of a closer interface between ethics and regulation (Callahan 1980). Formerly, ethics was discussed in terms of individual choice and personal morality. But we have come to examine ethical issues in terms of their legislative, regulatory, and judicial implications.

For example, Peter Barton Hutt, an eminent US expert in the field of food and drug regulation, has stated that all regulation of food and drugs and of the environment as a whole has emerged from our collective sense of societal ethic. He specifies the following five moral imperatives of government for its regulatory process (Hutt 1980):

(1) to protect the public from harm;

(2) to preserve maximum individual choice;

(3) to guarantee meaningful public participation in the decision-making process;

(4) to promote consistent and dependable rules applicable to everyone;

(5) to provide prompt decisions on all issues that arise in a regulatory context.

While ethical issues pervade much of governmental regulation of health services, there are a few fields in which ethical issues predominate. These may be described as:

(1) clinical decision-making on life-and-death issues involving providing medical treatment to severely disabled neonates and continuing life-support measures for terminally ill and comatose patients;

(2) regulation of clinical experimentation;

(3) equitable allocation of all resources for health care, including scarce resources, such as equipment for kidney dialysis and organ transplants;

(4) control of non-coital methods of human reproduction, such as *in vitro* fertilization and surrogate motherhood.

A voluminous literature is available on each of these topics and on many others involving ethical issues (Bankowski and Bryant 1985; Bankowski *et al.* 1989; Capron 1983, 1984; Ladimer and Newman 1963; Reich 1978; Walters 1985). It is not possible here to do justice to the complexities involved in any one of these issues. Sensitive and difficult questions are involved, such as the best interests of the child and the family, the length of life and the quality of life, assuring individual integrity and autonomy, protecting confidentiality, equitable distribution of resources, competition for scarce resources, and societal values in individual choice and collective conduct.

The development of law governing ethical issues has provided guide-lines to physicians and has narrowed their sphere for making judgments (Grad 1978). In fact, one of the most significant contributions of the law to this field is its capacity to create procedures and processes to resolve ethical questions in medicine and public health. Ethics committees of hospitals, ombudsmen to resolve disputes, patient advocates, review tribunals to monitor involuntary admissions to mental hospitals, protocols to govern care of the terminally ill, and manuals of procedures developed by the medical and nursing professions are all feasible mechanisms promoted and assisted by the law for resolving ethical issues in health care.

For the 'Number One' public health problem in the world—the threat of nuclear war—international law is all-important. No better illustration of law in solving ethical problems can be found than our efforts to achieve international agreements to stop the production and deployment of nuclear armaments and eventually to reduce conventional weapons of war. If we can secure the peace of the world through international law, we can surely solve the other ethical problems in public health.

Conclusion

Laws can be inspired by a sense of justice and right. We hope that this is generally the case. Law provides the rules that all in society must play by. Health law provides the regulatory system for patients, providers, and governments in the sphere of health protection and health services.

While the law may, in some instances, stand as a barrier to innovations in health service, in the main it has proved responsive to new scientific discoveries and to changed social and economic conditions. Where existing policies are inequitable or outmoded, the law has the capacity to serve as an instrument for change. If the outlines for change are uncertain, the law can facilitate the development of mechanisms for defining and achieving solutions.

Working in tandem with public health experts, the health lawyer can assist and clarify the definition of sound health policy and can promote its implementation for the benefit of individuals and society.

Acknowledgements

The author is indebted to Rebecca J. Cook and Bernard M. Dickens of the Faculty of Law, University of Toronto, Canada for review of the issues selected for discussion in this chapter. Their encouragement and support at an early stage of this work are much appreciated. For information and insights generously provided over the years on health legislation of countries throughout the world and for incisive review of this chapter, the author is deeply indebted to Mr S.S. Fluss, Chief, Health Legislation, World Health Organization, Geneva.

References

Addiction Research Foundation (1981). *Alcohol, society and the state, a comparative study of alcohol control* Volumes 1 and 2. Addiction Research Foundation, Toronto.

Algeria (1983). Law No. 83–03 of 5 February 1983 on environmental protection, Chapter I(3). *International Digest of Health Legislation* (1984) **35** (1), 176.

Andrews, L.B. (1984). *New Conceptions*. St. Martin's Press, New York.

Annas, G.J. (1984). Making babies without sex: the law and the profits. *American Journal of Public Health* **74** (12), 1415.

Annas, G.J. and Elias, S. (1983). Medico-legal aspects of a new technique to create a family. *Family Law Quarterly* **XVII** (2), 199.

Ashford, N.A. and Caldart, C.C. (1985). The 'right to know': toxics information transfer in the workplace. *Annual Review of Public Health* **6**, 383.

Association of the Bar of the City of New York in co-operation with the Cornell Law School (1962). *Mental illness and due process*. Report and Recommendations on Admission to Mental Hospitals under New York Law, Bertram F. Willcox, Director of Study. Cornell University Press, Ithaca.

Bankowski, Z. and Byrant, J.H. (ed.) (1985). *Health policy, ethics and human values. An international dialogue*, XVIIIth CIOMS

Round Table Conference. Council for International Organizations of Medical Sciences, Geneva.

Bankowski, Z., Barzelatto, J., and Capron, A.M. (ed.) (1989). *Ethics and human values in family planning*. Council for International Organizations of Medical Sciences (CIOMS), Geneva.

Beauchamp, D.E. (1985). Community: the neglected tradition of public health. *The Hastings Center Report* December 1985, pp. 28–36.

Belgium (1987). Crown decree on the banning of smoking in certain public places, 31 March 1987. *International Digest of Health Legislation* **38** (3), 542.

Bullough, B. (1975). The third phase in nursing licensure: the current nurse practice Acts. In *The Law and the Expanding Nursing Role* pp. 153–170. Appleton-Century-Crofts, New York.

Brown v. Superior Court of the City and County of San Francisco (1988). *Daily Journal* Daily Appellate Report 4211, 31 March 1988.

California Vehicle Code (1985). Sections 23152–3.

Callahan, D. (1980). Ethics and regulation. *The Hastings Center Report* February 1980, p. 25.

Capron, A.M. (1983). *Report of the President's Commission for the Study of Ethical Problems in Medicine and Biomedical and Behavioral Research*. US Government Printing Office, Washington, DC.

Capron, A.M. (1984). The new reproductive possibilities: Seeking a moral basis for concerted action in a pluralistic society. *Law, Medicine and Health Care* **12** (5), 192.

Chapman, R.A. (1976). *Development of a national drug control agency: Need, legislative authority, responsibilities, organization, and operation*. World Health Organization, PDT 76.1.

Christoffel, T. (1982). *Health and the law, a handbook for health professionals*. The Free Press, New York.

Cook, R.J. (1987). The US export of 'pipeline' therapeutic drugs. *Columbia Journal of Environmental Law* **12** (1), 39.

Cook, R.J. and Dickens, B.M. (1988). International developments in abortion laws: 1977–1988. *American Journal of Public Health* **78**, 1305–1311.

Curran, W.J. and Harding, T.W. (1978). *The law and mental health: Harmonizing objectives* World Health Organization, Geneva.

David, H.P. (ed.) (1984). *Abortion research notes* Volume 13, Nos. 1–2, pp. 1–2. Transnational Family Research Institute, 8307 Whitman Drive, Bethesda, Maryland 20817, USA.

DeMarco, C.T. (1975). *Pharmacy and the law*. Aspen Systems Corporation, Germantown, Maryland.

Dominica (1982). An Act (No. 34 of 1982) further to amend the Medical Ordinance, Cap. 149. Dated 21 October 1982 (The Medical Amendment) Act 1982. *International Digest of Health Legislation* (1983) **34** (2), 241.

Finland (1986). Ordinance No. 786 of 31 October 1986 on communicable diseases. *International Digest of Health Legislation* (1987) **38** (2), 244.

France (1982). Law No. 82–916 of 28 October 1982 amending Article L. 680 of the Public Health Code relating to activities in the private sector in public hospital establishments. *International Digest of Health Legislation* (1983) **34** (1), 65.

German Democratic Republic (1982). Order of 29 July 1980 on vaccination of children and adolescents. *International Digest of Health Legislation* (1982) **33** (3), 470–473.

Goldsmith, J.R. (1970). Changing concepts of environmental health. *The education and training of engineers for environmental health*. World Health Organization, Geneva.

Gostin, L. and Curran, W.J. (ed.) (1987). AIDS, law and policy. *Law, Medicine and Health Care*. **15** (1–2), 3–89.

Grad, F.P. (1978). Medical ethics and the law. *The Annals of the American Academy of Political and Social Science* **437**, 19.

Grad, F.P. (1985). *Environmental law* (3rd ed). Matthew Bender, New York.

Grad, F.P. (1986). Public health law. In *Maxcy-Rosenau preventive medicine and public health* (ed. J.M. Last), (12th ed). Appleton-Century-Crofts, New York.

Grad, F.P. (1990). Restrictions of the person. In *Public health law manual* (2nd edn). American Public Health Association, Washington, DC.

Grad, F.P., Rathjens, G.W., and Rosenthal, A.J. (1971). *Environmental control: Priorities, policies, and the law*, pp. 29–38. Columbia University Press, New York.

Horowitz, H.S. (1981). Promotion of oral health and prevention of dental caries. *Journal of the American Dental Association* **103**, 141.

Hutt, P.B. (1980). Five moral imperatives of Government regulation. *The Hastings Center Report* February 1980, pp. 29–31.

Iceland (1984). Law No. 74 of 28 May 1984 on the prevention of the use of tobacco. *International Digest of Health Legislation* (1984) **35**, 772.

Institute of Medicine (1982). *Legislative approaches to prevention of alcohol-related problems*. National Academy Press, Washington, DC.

Institute of Medicine (1988). *Confronting AIDS—Update 1988*. National Academy Press, Washington, DC.

International Digest of Health Legislation (1987). **38** (2), 409.

International Labour Organization (1981). Convention concerning occupational safety and health and the working environment. Convention No. 155 adopted by the International Labour Conference on 22 June 1981. *International Digest of Health Legislation* (1981) **32** (3), 549.

Isaacs, S.L. (1981). *Population Law and Policy, Source Materials and Issues*. Human Sciences Press, New York.

Italy (1978). Law No. 194 of 22 May 1978 on the social protection of motherhood and the voluntary termination of pregnancy. *International Digest of Health Legislation* (1978) **29** (3), 589.

Ladimer, I. and Newman, R.W. (ed.) (1963). *Clinical investigation in medicine: Legal, ethical, and moral aspects, an anthology and bibliography*. Law-Medicine Research Institute, Boston University, Boston.

Lalonde, M. (1974). *A new perspective on the health of Canadians, a working document*. Government of Canada, Ottawa.

Lander, L. (1980). Licensing of health care facilities. In *Legal aspects of health policy: Issues and trends*, pp. 129–72. (ed. R. Roemer and G. McKray). Greenwood Press, Westport, Connecticut.

Luxembourg (1978). Law of 15 November 1978 concerning information on sexual matters, criminal abortion, and regulation of pregnancy termination. *International Digest of Health Legislation* (1979) **30** (2), 253.

Maclay, W.S. (1960). The new Mental Health Act in England and Wales. *The American Journal of Psychiatry* **116** (9), 777.

Maine, D. and McNamara, R. (1985). *Birth spacing and child survival*. Center for Population and Family Health, Columbia University, New York.

Mason, P.E., Boland, R., and Stepan, J. (1987). *Annual review of population law, 1984* Volume 11 (and earlier volumes), United Nations Fund for Population Activities and Harvard Law School Library, Cambridge, Massachusetts.

Mason, P.E., Yackle, J.F., and Stepan, J. (1985). *Annual review of population law, 1983*, Volume 10, p. 74, United Nations Fund for Population Activities and Harvard Law School Library, Cambridge, Massachusetts.

Moser, J. (1980). *Prevention of alcohol-related problems, an international review of preventive measures, policies, and programmes*. World Health Organization and Addiction Research Foundation, Ontario.

Murray, J.J. (1986). *Appropriate use of fluorides for human health*, pp. 38–73. World Health Organization, Geneva.

National Board of Health of Finland, Advisory Committee on Health Education (1985). *An evaluation of the effects of an increase in the price of tobacco and a proposal for the tobacco price policy in Finland in 1985–87* Helsinki.

New York (1972). In 1972 the State of New York revised its definition of professional nursing as follows:

> The practice of the profession of nursing as a registered professional nurse is defined as diagnosing and treating human responses to actual or potential health problems through such services as case-finding, health teaching, health counseling, and provision of care supportive to or restorative of life and well-being and executing medical regimens prescribed by a licensed or otherwise legally authorized physician or dentist. A nursing regimen shall be consistent with and shall not vary any existing medical regimen.

New York Education Law title 8, article 139, sec. 6902. In Sermchief v. Gonzales, 660 S.W. 2d 683 (Mo. banc 1983), the Missouri Supreme Court upheld the authority of nurses to perform diagnostic and treatment functions in family planning services under a modernized nursing practice act prescribing a broad spectrum of nursing functions qualified by the phrase 'including but not limited to'.

New York (1978). N.Y. Mental Hygiene L., sec. 9.01 et seq. McKinney's.

New York (1985). N.Y. Public Health L., sec. 2805 et seq. McKinney's.

New York City Interagency Task Force on AIDS (1988). *New York City strategic plan for AIDS*. New York.

Norway (1980*a*). Regulations of 2 February 1980 on the composition and use of dispersants for the control of oil slicks. *International Digest of Health Legislation* (1981) **32** (1), 156.

Norway (1980*b*). Regulations of 16 May 1980 on levels of lead compounds and benzene in motor fuel. *International Digest of Health Legislation* (1981) **32** (1), 156.

Norway (1980*c*). Regulations of 8 July 1980 on the dumping and incineration of substances and objects at sea. *International Digest of Health Legislation* (1981) **32** (1), 156.

Norway (1981–2). Report No. 11 to the Storting (1981–2) on the Follow-Up on Norwegian Nutrition Policy, Recommendation of 10 July 1981 from the Ministry of Health and Social Affairs, approved by the Council of State on the same day.

Ontario, Canada (1984). An Act (Chapter 41) to promote the health of pupils in schools. Dated 7 July 1982 (The Immunization of School Pupils Act, 1982), *International Digest of Health Legislation* (1984) **35**(1), 57–59; Canada (Ontario) Regulation made under the Immunization of School Pupils Act, 1982. Ontario Regulation 23/83. Dated 13 Jan. 1983. *International Digest of Health Legislation* (1984) **35**(1), 59–60.

Parham v. J.R. *et al*. (1979). 442 U.S. 584 (1979); Secretary of Public Welfare v. Institutionalized Juveniles, 442 U.S. 640 (1979).

Paxman, J.M. (1980). *Law and planned parenthood*. International Planned Parenthood Federation, London.

Paxman, J.M. and Zuckerman, R.J. (1987). *Laws and policies affecting adolescent health*. World Health Organization, Geneva.

Petricciani, J.C., Gust, I.D., Hoppe, P.A., and Krijnen, W.H. (1987). *AIDS: the safety of blood and blood products* published on behalf of WHO by John Wiley & Sons, Chichester. Discussed in *International Digest of Health Legislation* (1987) **38** (3), 654.

Porter, L., Arif, A.E., and Curran, W.J. (1986). *The law and treatment of drug—and alcohol—dependent persons, a comparative study of existing legislation*. World Health Organization, Geneva.

Reich, W.T. (1978). *Encyclopedia of bioethics*. The Free Press, a division of Macmillan Publishing Company, New York.

Report of the Presidential Commission on the Human Immunodeficiency Virus Epidemic (1988). J.D. Watkins, Chairman, Submitted to the President of the United States, 24 June 1988, US Government Printing Office, Washington, DC.

Roemer, R. (1982). *Legislative strategies to control the world smoking epidemic*. World Health Organization, Geneva.

Roemer, R. (1985). Legislation on contraception and abortion for adolescents. *Studies in Family Planning* **16** (5), 241.

Roemer, R. (1986). *Recent developments in legislation to combat the world smoking epidemic*. World Health Organization, Geneva.

Roemer, R. (1988). *Legislative strategies for a smoke-free Europe*, pp. 26–30. World Health Organization Regional Office for Europe, Copenhagen.

Roemer, R. and Paxman. J.M. (1985). Sex education laws and policies. *Studies in Family Planning* **16** (4), 219.

Roemer, R. and Roemer. M.I. (1976). *Health manpower in the changing Australian health services scene*. DHEW Publication (HRA) 76–58, pp. 7–8. US Department of Health, Education, and Welfare, Washington, DC.

Roemer. R., Frink, J.E., and Kramer, C. (1971). Environmental health services: Multiplicity of jurisdictions and comprehensive environmental management. *The Milbank Memorial Fund Quarterley* **XLIX** (4), 419.

Senegal (1977). Decree No. 77–017 of 7 January 1977, discussed in *Regulatory Mechanisms for Nursing Training and Practice*, WHO Study Group, 1986.

Silver, L. (1980). An agency dilemma: Regulating to protect the public health in light of scientific uncertainty. In *Legal aspects of health policy: Issues and trends*, pp. 61–93. (ed. R. Roemer and G. McKray). Greenwood Press, Westport, Connecticut.

Sindell v. Abbott Laboratories (1980). 26 Cal. 3d 588 (1980).

Single, E. (1984). International perspectives on alcohol as public health issue. *Journal of Public Health Policy* **5** (2), 238.

Somers, A.R. (1969). *Hospital regulation: the dilemma of public policy* p. ix. Industrial Relations Section, Department of Economics, Princeton University.

Sri Lanka (1980). An Act (No. 47 of 1980) to establish a Central Environmental Authority (and) to make provision with respect to the powers, functions, and duties of that Authority; and to make provision for the protection and management of the environment and for matters connected therewith or incidental thereto. Dated 29 October 1980. *International Digest of Health Legislation* (1981) **32** (3), 565.

Stepan, J., Kellogg, E.H., and Piotrow, P.T. (1981). Legal trends and issues in voluntary sterilization. *Population Reports*, Vol. IX, Number 2, pp. E75–E102. Population Information Program, The Johns Hopkins University.

Swiss Institute of Comparative Law (1986). *Artificial procreation, genetics and the law*. Lausanne Colloquium of 29–30 November 1985. Schulthess Polygraphischer Verlag, Zürich.

Terris, M. (1983). The complex tasks of the second epidemiologic revolution. *Journal of Public Health Policy* **4** (1), 8.

UK (1959). Mental Health Act (England and Wales), 7 & 8 Eliz. 2, Chapter 72, Section 5 (1959).

UK (1983). An Act (Chapter 51) to amend the Mental Health Act

1959 and for connected purposes, 28 October 1982, superseded by the Mental Health Act 1983. *International Digest of Health Legislation* (1983) **34** (3), 524.

UK (1985). *International Digest of Health Legislation* (1985) 36(4), 970.

United Republic of Tanzania (1978). An Act (No. 9 of 1978) to repeal the Pharmacy and Poisons Ordinance, and the provisions of the Food and Drugs Ordinance relating to drugs; to provide for the control of the profession of pharmacy, and of matters relating to dealings in pharmaceuticals and poisons. *International Digest of Health Legislation* (1981) **32** (3), 503.

US (1969). The National Environmental Policy Act 1969, P.L. No. 91–190, 83 Stat. 852, 42 U.S.C.A. secs. 4321 et seq.

US (1986). National Childhood Vaccine Injury Act 1986, 42 U.S.C. 300aa–10 et seq., 100 Stat. 3755–3784 (P.L. 99–660), 14 November 1986.

US Public Health Service (1986). The US Public Health Service Plan for prevention and control of AIDS and the AIDS virus. Published in full in *Public Health Reports* (1986) **101** (4), 341. Also published in *Confronting AIDS: Directions for public health, health care, and research*, pp. 326–333. Institute of Medicine, National Academy of Sciences, Washington, DC.

Walters, L. (1975). *Bibliography of bioethics*. Center for Bioethics, Kennedy Institute, Georgetown University, Gale Research Company, Detroit, Michigan.

Warnock Committee (1984). *Report of the Committee of Inquiry into Human Fertilisation and Embryology* (Dame Mary Warnock, Chairperson). UK Department of Health and Social Security, Her Majesty's Stationery Office, London.

Willumsen, E. (1983). The Norwegian Nutrition and Food Policy, Address to the Conference on Primary Health Care in Industrialized Countries, Bordeaux, 14–18 November 1983.

Wing, K.R. (1985). *The law and the public's health* pp. 1–5. Health Administration Press, Ann Arbor, Michigan.

World Health Organization (1987). *Success against smoking, the story of four countries*. Tobacco or Health Programme, World Health Organization, Geneva.

World Health Organization (1988). *Tabular information on legal instruments dealing with AIDS and HIV infection*. Geneva, June 1988, updated periodically.

World Health Organization Study Group (1986). *Regulatory mechanisms for nursing training and practice: Meeting primary health care needs*. World Health Organization Technical Report Series 738, p. 32. World Health Organization, Geneva.

Index

NOTE: Since the major subject of this book is public health, entries under this keyword have been kept to a minimum, and readers are advised to seek more specific index entries. Likewise, with United Kingdom and United States of America references. This is an index to all three volumes of the *Oxford Textbook of Public Health*. The first digit in any reference gives the **volume number** (1–3); the digits after the decimal point give the page number.

Abbreviations used in subentries:

CHD	Coronary heart disease	MHC	Major histocompatibility complex
EPA	Environmental Protection Agency	NHS	National Health Service
GDP	Gross domestic product	UK	United Kingdom
GNP	Gross national product	US	United States of America